Power Electronics

Issa Batarseh • Ahmad Harb

Power Electronics

Circuit Analysis and Design

Second Edition

Springer

Issa Batarseh
University of Central Florida
Orlando, FL, USA

Ahmad Harb
German Jordanian University
Amman, Jordan

ISBN 978-3-319-68365-2 ISBN 978-3-319-68366-9 (eBook)
https://doi.org/10.1007/978-3-319-68366-9

Library of Congress Control Number: 2017955223

Printed on acid-free paper

This Springer imprint is published by Springer Nature
The registered company is Springer International Publishing AG
The registered company address is: Gewerbestrasse 11, 6330 Cham, Switzerland

This book is dedicated to our:
late former advisers, Professor C.Q. Lee,
Professor Ali Nayfeh, and
to our families

Preface

In the past three decades, the field of power electronics has witnessed unprecedented growth in research and teaching worldwide, and has emerged as a specialization in electrical engineering. This growth is due to expanding market demand for integrated and networked power electronics-based circuits and systems in all kinds of energy processing and conversion applications. Moreover, the need for power electronics engineers and researchers equipped with knowledge of new energy conversion technologies has never been greater.

Power Electronics is intended as a textbook to teach the subject of modern power electronics to senior undergraduate and first-year graduate electrical engineering students. Because of the breadth of the field of power electronics, teaching this subject to undergraduate students is a challenge. This textbook is designed to introduce the basic concepts of power electronics to students and professionals interested in updating their knowledge of the subject. The objective of this textbook is to provide students with the ability to analyze and design power electronic switching circuits used in common industrial applications.

The prerequisites for this text are a first course in circuit analysis techniques and a basic background in electronic circuits. Chapter 3 gives an overview of diode switching circuits and basic analysis techniques that students will find useful in the remaining chapters.

Material Presentation

Unlike many existing texts in power electronics, *Power Electronics* targets mainly senior undergraduate students majoring in electrical engineering. Since the text is intended to be used in a three-credit-hour course in power electronics, topics such as power semiconductor devices, machine drives, and utility applications are not included. Because of limited lecture times, one course at the undergraduate level cannot adequately cover such topics and still present all power electronic circuits used in energy conversion. This text contains sufficient material for a single-semester introductory power electronics course while giving the instructor flexibility in topic treatment and course design.

The text is written in such a way as to equip students with the necessary background material in such topics as devices, switching circuit analysis techniques, converter types, and methods of conversion in the first three chapters. The presentation of the material is new and has been recommended by many power electronics faculty. The discussion begins by introducing high-frequency, non-isolated dc-to-dc converters in Chap. 4, followed by isolated dc-to-dc converters in Chap. 5. Resonant soft-switching converters are treated early on in Chap. 6. The traditional diode and SCR converters and dc-ac inverters are presented in the second part of the text, in Chaps. 7, 8, and 9, respectively.

Examples, Exercises, and Problems

Unlike many existing texts, this text provides students with a large number of examples, exercises, and problems, with detailed discussions on resonant and soft-switching dc-to-dc converters.

Examples are used to help students understand the material presented in a given chapter. To drill students in applying the basic concepts and equations and to help them understand basic circuit operations, several exercises are given within each chapter. The text has more than 250 problems at different levels of complexity and difficulty. These problems are intended not only to strengthen students' understanding of the materials presented but also to introduce many new concepts and circuits. To help meet recent Accreditation Board for Engineering and Technology (ABET) requirements for design in the engineering curriculum, special emphasis is made on providing students with opportunities to apply design techniques. Such problems are designated with the letter "D" next to the problem number, such as D5.32, which is Design Problem # 32 in Chap. 5. Students should be aware that such problems are open-ended without unique solutions.

A bibliography is included at the end of the text, and a list of textbooks is given separately.

Web-Based Course Material

Ancillaries to this text are available on a dedicated website, www.fpec.ucf.edu, where both faculty and students can access additional material such as a complete set of lecture notes, additional sample quizzes and problems, up-to-date text corrections, and the opportunity to submit new ones.

Orlando, FL, USA
Amman, Jordan

Issa Batarseh
Ahmad Harb

Acknowledgments

Over the years, many of our power electronics students have had the opportunity to read our lectures notes and solve many problems. Their feedback and comments were instrumental in presenting, to the best of our knowledge, error-free examples, exercises, and problems. We are grateful for reviewers' constructive suggestions for the students' sake. We are highly indebted to Khalid Rustom, Chris Iannello, and Osama Abdel-Ruhman for spending many hours reviewing the examples, exercises, and problem solutions. Our special thanks to our current students Anirudh Pise, Seyed-milad Tayebi, Xi Chen, Ala'a Amarine, and Rand Mdanat who spent so much time helping us revise this edition of the book. Without their tireless work, this text would suffer from numerous errors and omissions! We are very fortunate to have such dedicated people work with us. We were also fortunate to have Dr. Kazi Khairul Islam visit our power electronics laboratory at UCF. He helped greatly in providing insightful suggestions on the inverters chapter, as well as in developing the appendixes. His suggestions and comments are greatly appreciated.

Special thanks go to the young, bright students from Princess Sumaya University for Technology (PSUT) in Amman, Jordan, who helped in typing and solving text problems. Their names are Reem Khair, Ruba Amarin, Amjad Awwad, Bashar Nuber, Abeer Sharawi, Rami Beshawi, and Mohhamad Abu-Sumra. Specifically, we wish to acknowledge the help of two of Dr. Batarseh's students, Khalid Rustom and Abdelhalim Alsharqawi, in solving problems and providing helpful insights throughout their undergraduate and graduate study at UCF. We would also like to thank our former graduate students Guangyong Zhu, Huai Wei, Joy Mazumdar, Songquan Deng, and Jia Luo for their help and suggestions. Many thanks are due to Shadi Harb, Maen Jauhary, Samanthi Nadeeka, and William Said for their help in typing the solutions manual and developing the text website. Finally, thanks are due to Banah Harb, from Jordan University of Science and Technology (JUST), for her help in typing some examples and redrawing some figures. Indeed, it was a pleasure working with all our students who made writing this text a gratifying and enjoyable experience.

Also, we would like to express our sincere thanks to the editors for her understanding and for keeping me on schedule.

Finally, Dr. Batarseh will never forget his late advisor, Professor C. Q. Lee, for introducing him to the field of power electronics and for being his advisor, teacher, mentor, and friend. He taught me the art of academic mentoring. On behalf of all his students, Dr. Batarseh is dedicating this text to him. Dr. Ahmad Harb also dedicates this book to his late former advisor Professor Ali Nayfeh who has reshaped his life not only technically but personally and spiritually. Finally, of course, without our parents' and families' love, patience, and support, this text would have never been written.

Orlando, FL, USA
Amman, Jordan

Issa Batarseh
Ahmad Harb

Contents

About the Authors

Issa Batarseh received his Ph.D. and M.S. in electrical engineering and a B.S. in computer engineering and science from the University of Illinois at Chicago in 1990, 1985, and 1983, respectively. In a career spanning three decades in education and research, Prof. Batarseh has served in numerous academic and administrative positions at the University of Central Florida (UCF) where he is currently a professor of electrical engineering and computer science. In 1998, he established the UCF's Florida Power Electronics Center and has been serving as its director. Dr. Batarseh has pursued power electronics research that focuses on the development of advanced systems for solar energy conversion to reduce cost and improve power density, efficiency, and performance. He has published over 300 journal and conference papers and a textbook entitled *Power Electronic Circuits* in 2003 with John Wiley. He is a fellow member in National Academy of Inventors (NAI), AAAS, IEEE, and IEE and has served as an associate editor for *IEEE Transactions on Aerospace and Electronic Systems* and *Transactions on Circuits and Systems*. He also served as a reviewer for the National Science Foundation and several IEEE Transaction journals and was a member of the program committees of many international conferences. Through his academic career, Dr. Batarseh has supervised 34 Ph.D. dissertations and 43 M.S. In 2017, Dr. Batarseh was inducted in Florida Inventors Hall of Fame.

Ahmad Harb received his Ph.D. degree from Virginia Tech., Virginia, USA, in 1996. Currently, he is a professor at German Jordanian University (GJU), School of Natural Resources Engineering. Dr. Harb is an IEEE senior member. Dr. Harb is the editor in chief for two international journals, *IJMNTA* and *IJPRES*. Dr. Harb served as the dean of the School of Natural Resources Engineering at GJU between 2011 and 2013. Dr. Harb has published more than 70 journal articles and conference proceedings. His research interests include power system, renewable energy and power electronics, modern nonlinear theory (bifurcation and chaos), and nonlinear control.

Chapter 1
Introduction

1.1 Introduction

No doubt that power electronics is now considered one of the most vital enabling technologies in electrical engineering. In fact, large part of all electrically powered devices, circuits, or systems has close connection with the field of power electronics. Its scope is broad and covers very wide spectrum, with the paramount among them is its ever-increasing role in integrating renewable energy sources and electric storage to the grid. Power electronics is the "glue" that makes the ushering of a new kind of smart energy technology revolution possible. It is because of the engineering field of power electronics that we are able to encompass the efficient and cost-effective use of electronic components, circuit and control theory, modern analytical tools, and design techniques to make this smart energy revolution possible. This revolution will modernize our electric grid, give birth to massive electric transportation, allow for large solar energy penetration, help solve climate change, and enable the deployment of the highest possible energy efficiency systems. In short, power electronics has emerged as the enabling technology that transformed the field of energy and power engineering from a high-tech frontier to smart-tech frontier. Arriving at today's remarkable important role of power electronics took more than 100 years of innovation and hard work by many scientists and engineers coupled with strong partnerships between the private sector, professional societies, and governments.

This chapter is intended to give the reader a broad introductory overview about the field of power electronics and its applications. Basic block diagrams for a power electronic system and its major functions will be given. We also present different types of power electronic circuits used to achieve power conversion that will be studied throughout the text.

I. Batarseh, A. Harb, *Power Electronics*,
https://doi.org/10.1007/978-3-319-68366-9_1

1.2 What Is Power Electronics?

To date, there is not a widely accepted statement that can clearly and specifically define the field of power electronics. In fact, many experts in the academic and industrial communities feel that the name itself does not do justice to the field that is application oriented and multidisciplinary in nature that encompasses many sub-areas in electrical engineering.[1] Because of this multidisciplinary nature of the field of power electronics, experts must have commanding knowledge in several electrical engineering fields such as electronic devices, electronic circuits, signal processing, magnetism, electrical machines, control, and power. In a very broad sense, power electronic circuits have the task to process one form of energy supplied by a source to a different form required at the load side. Hence, power electronics can be closely identified with the following subdiscipline areas in electrical engineering: electronics, power, and control. Here, *electronics* deal with the semiconductor devices and circuit topologies for signal processing in order to implement the control functions, and *power* deals with both static and rotating equipment that uses electric power, whereas *control* deals with the steady-state stability of the closed loop system during power conversion process. Hence, the subject of power electronics deals specifically with the application of power semiconductor devices and circuits for conversion and regulation of electric power. In summary, power electronics is an enabling technology that brings together three fundamental technologies: power semiconductor devices technology, power conversion technology, and power control technology, as illustrated in Fig. 1.1.

A final observation to make is that in power electronic circuits, there exist two types of switching devices: one type exists in the power processing stage which handles high power up to hundreds of gigawatts which represents the muscle of the system and the other type located in the feedback control circuit which handles low-power signal processing up to hundreds of milliwatts, representing the brain or the intelligence of the system. Hence, today's power electronic circuits are essentially digital electronic circuits whose switching elements manipulate pulsed power from the milliwatts to gigawatts range. As a result, one may conclude that the task of power electronics is to convert and control power using low-power switching devices to process power at much higher power levels of these devices (hundred times or more).

[1]Many universities today offer power electronic discipline either under the "power" or "electronics" area, with limited number of universities have it separately.

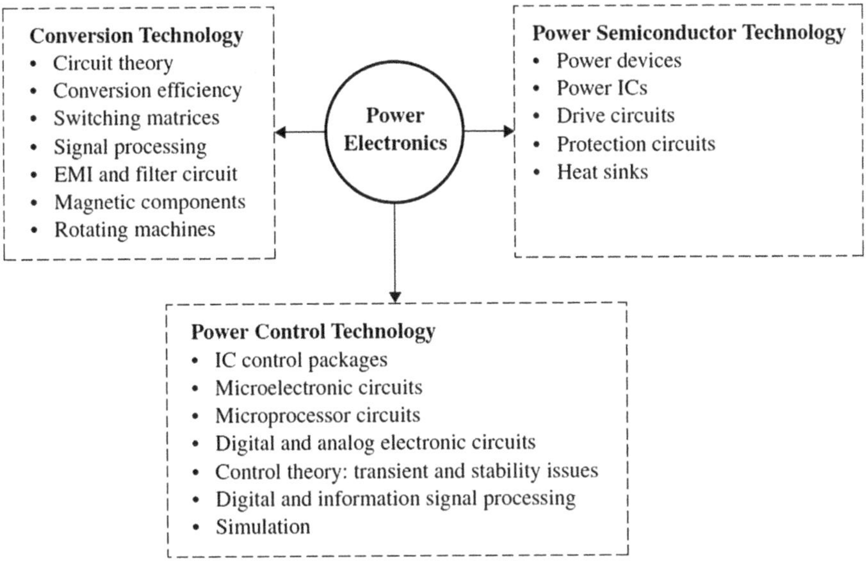

Fig. 1.1 Power electronics is a systems solution that encompasses three technologies: conversion, power semiconductor, and power control technologies

1.2.1 Recent Growth in Power Electronics

The field of power electronics has recently experienced unprecedented growth not only in terms of research and educational activities, but in diverse applications. Its application has been steadily and rapidly expanded to cover many sectors of our society. This growth is due to several factors, paramount among them is the growing markets in renewable energy applications, coupled with technological advancement made by the semiconductor device industry which led to the introduction of very fast, high-power capabilities and highly integrated power semiconductor devices. Other factors include (1) the revolutionary advances made in the microelectronic field which led to the development of very efficient and highly integrated circuits (ICs) used for generation of digital control signals for processing and control purposes; (2) the ever-increasing demand for smaller size and lighter weight power electronic systems; (3) the expand market demand for new power electronic applications in wind and solar energy conversion and other applications that require the use of variable-speed motor drives, regulated power supplies, robotics, and uninterruptible power supplies; and (4) a result of this increasing reliance on power electronic systems made it mandatory that all such systems have radiated and conducted electromagnetic interference (EMI) be limited within regulated ranges. The industry's interest in developing power systems with low harmonic contents with improved power factor and reduced cost will continue to place the field of power electronics on the top of the research and development priority list.

1.3 The History of Power Electronics

Before presenting the history of power electronics in this century, it might be useful to the reader to know the history of the development of what is called the alternating current (*ac*) and the direct current (*dc*) electricity in the last two decades of the nineteenth century. This is because the inventions of the 1880s resulted in the present worldwide *ac* electric power system, providing the energy form that must be processed for any power electronic applications.

1.3.1 The History of dc *and* ac *Electricity in the Late Nineteenth Century*

It was decided in the late nineteenth century that the electrical form of energy is the most practical and economic way to produce energy for human use. This is because electricity is an excellent form of energy when it comes to generation, transmission, and distribution. However, this realization not before a heated debate was underway among scientists and engineers whether the future of transmitting and distributing electricity to industries and homes would be based on alternating type of current flow known as (*ac*) or the direct type of current flow known as (*dc*). It was George Westinghouse and Nikola Tesla (1856–1943), representing the *ac* camp, and Thomas Edison (1847–1931), representing the *dc* camp. After more than 15 years of intellectual debate, supported by new inventions and developmental and experimental studies, the *ac* advocates won; consequently, the entire world today is using an *ac*-based power distribution system.[2]

Thomas Edison was a self-educated inventor who was awarded 1033 patents over 50-year period. He is best known for the invention of the phonograph and incandescent lamp, which was invented in 1879 after many years of repeated experiments. In 1878, he formulated the concept of a centrally located power station from which power can be distributed to surrounding areas. In September 4, 1882, using *dc* generators (at that time called dynamos) driven by steam engines, Edison opened Pearl Street Station in New York City to supply electricity to 59 customers in a one-square-mile area. It was the first *dc*-based power station in the world with a total power load of 30 kW only. In fact, it was the beginning of the electric utility industry that grew at a remarkable rate. In 1884, Frank Sprague produced a practical *dc* motor for Edison's *dc* systems. This invention, coupled with the development of three-wire 220 VDC, Edison succeeded to distribute *dc*

[2]Tesla and Edison worked together for a short time, and soon both developed hatred for one another, resulting in Tesla opening his own business, believing in *ac* transmission systems. Several rumors in the press stated that both were nominated for Nobel Prize in physics, and because of the feud between them, the prize was given to a third party. These rumors were all false since no one is asked whether to accept or decline a Nobel Prize!

electrical power to cover larger areas and supply heavier loads and consequently more customers. By doing so, Edison prompted the *dc*-based power distribution systems. As transmission distances and load demands start to increase, Edison's *dc* systems ran into troubles. The *dc* distribution lines suffered from very high-power losses because of the high voltage and current that existed simultaneously. This severely limited the transmission distance and resulted in highly inefficient systems. So in order to sustain power level, Edison had to build *dc* power station every 20 km! This was costly and very impractical. However, he did not give up the *dc* transmission idea and insisted that these problems can be overcome.

George Westinghouse and Nikola Tesla did not waste time to develop *ac*-based power distribution systems, despite Edison's plans to continue to construct *dc* transmission systems in New York. In 1885, a major step was taken by Westinghouse to develop *ac* systems when he bought the American patents of L. Gaulard and J.D. Gibbs of Paris for the design of transformers. Westinghouse, backed by Tesla's patents and the new transformers designs, challenged the *dc* transmission system and went ahead in developing it.

A major step in supporting *ac* systems was in 1885, when William Stanley, an early associate of George Westinghouse, developed a commercially practical transformer, allowing the possibility of distribution of *ac*-based electricity. This was the first challenge to Edison's idea of *dc* power systems. Using transformers, it was possible to transmit high-level voltages at a very low-level current, resulting in a very low voltage drop (low-power dissipation) in the transmission line. In winter of 1886, Stanley installed the first experimental *ac* distributed system in Great Barrington, Massachusetts, supplying 150 lamps in the covered area. In 1889, the first single-phase distributed power system was operational in the United States between Oregon City and Portland, covering a 21 km distance with 4 kV power rating.

Second major step that boosted the potential of using *ac* systems took place on May 16, 1888, when Tesla presented a paper at the annual meeting of the American Institute of Electrical Engineers, discussing two-phase induction and synchronous motors. Basically, he had shown that it is more practical and more efficient to use polyphase systems to distribute power. The first three-phase *ac* transmission power system was installed in Germany in 1891 rated at 12kv and transmitted over a distance of 179 km. Two years later (1893), the first three-phase power transmission system in the United States was installed in California, rated at 2.3 kV and a distance of 12 km. Moreover, a two-phase distributed system was demonstrated at the Colombian Exposition in Chicago in 1893. At this time, the apparent advantages of *ac*, especially the three-phase systems, over the *dc* system lead to the gradual replacement of *dc* by *ac* systems. Presently, the transmission of electricity is done almost entirely by means of *ac*. However, *dc* transmission of electric power is used in some locations in Europe and is rarely used in the USA. Since the late nineteenth century, economic studies have shown that *ac* transmission is much economical, hence receiving worldwide acceptance.

1.3.2 *The History of* dc *and* ac *Electricity in the Late Twentieth Century*

Over the last two decades, the ever increase of deployment of renewable energy sources to the power grid, coupled with technological advancement made by the semiconductor device industry and the revolutionary advances made in the micro-electronic communication and sensing technologies, has led to renewed interest in using *dc* transmission systems which is renewed. This time, many experts believe that because of the new technological advances, it is possible to develop *dc* transmission electric power systems economically and efficiently. Today's conversion systems from *ac* to *dc* and back to *ac* can be done using very fast, high-power rated, and highly integrated power semiconductor devices. What we can achieve using today's technology was not imaginable only 10 years ago. This is why many power electronic researchers believe that the old debate between *dc* and *ac* camps is coming back under a new set of technological rules.

Today, because of power electronics more than 100 years later, the argument whether *dc* or *ac* should be the way to go in future home and industry is resurfacing. In other words, in the turn of the twentieth century, *ac* was declared a winner over *dc*, and in the turn of the twenty-first century, the *dc* promoters, mostly power electronic experts, have had another shot at *ac*! Sounds familiar... history repeats itself! A century later, the *dc* advocates might win, and the twenty-first century might very well be friendlier to *dc* transmission system advocates! Who will win the new century? Only time will tell!

1.3.3 *History of Modern Power Electronics*

Many agree that the history of power electronics began in 1900 when the glass bulb mercury-arc rectifiers were introduced for the first time, signaling the beginning of the age of vacuum tube electronics or what was also called glass tube-based industrial electronics. This period remained until 1947 when the germanium transistor was invented at Bell Telephone Laboratory by the three physicists Bardeen, Brattain and Shockley, signaling the end of the age of vacuum tubes and the beginning of the age of transistor electronics. Between the 1930s and 1940s, several new power electronic circuits (then known as industrial electronics) were introduced including the metal-tank rectifier, grid-controlled vacuum-tube rectifier, thyratron motor, and gas/vapor tubes switching devices such as hot cathode thyratrons, ignatrons, and phanotrons. In the 1940s and early 1950s, solid-state magnetic amplifiers, using saturable reactors, were introduced.

The age of modern era of power electronics began in 1958 when General Electric Company introduced a commercial thyristor, 2 years after it was invented by Bell Telephone Laboratory. Soon, all the industrial applications that were based on mercury-arc rectifiers and power magnetic amplifiers were replaced by SCRs. In less than 20 years after commercial SCR was first introduced, significant improvements in semiconductor fabrication technology and physical operation were made, and many different types of power semiconductor devices were introduced. The growth in power electronics is made possible with the microelectronic revolution of the 1970s and 1980s by which the low-power IC control chips provided the brain and the intelligence to control the high-power semiconductor devices. Moreover, the introduction of microprocessors made it possible to apply modern control theory into power electronics. In the 25 years, the growth in power electronic applications was noticeable because of the introduction of very fast, high-temperature, and high-power switching devices, coupled with the utilization of advanced digital control algorithms. Today, power electronics is a mature technology. The future direction of the new era of power electronics is hard to predict, but one is certain that as long as humans seek to improve the quality of life and cleaner environment and implement energy-saving measures, the growing demand for clean energy will continue. This in turn implies that power electronics must be used to address the tremendous changes in the way we generate, transmit, and distribute electricity as we cross the bridge into the new century. For more detailed discussion of the modern history of the power electronics, see the paper by D. Wyke, and see this web site that was originally written by Prof. Bimal k. Bose http://ethw.org/Power_electronics.

1.4 The Need for Power Conversion

Since the invention of a practical transformer by Stanley in 1885 and polyphase *ac* systems by Tesla in 1891, the advantages of low-frequency *ac* over *dc* were compelling to power engineers at that time. The basis of utility power system generation, transmission, and distribution since the beginning of this century has been *ac* at fixed frequency either 50 or 60 Hz. The most outstanding advantage of *ac* over *dc* was the high-voltage over long transmission lines and the simplicity of designing distribution networks. However, the nature form of electricity being distributed is totally different than the nature of energy required by the electrical load.

At the consumer end, many applications may need *dc* power or *ac* at line, higher, lower, or variable frequencies. Therefore, it became necessary to convert the available *ac* systems into *dc* and must be controlled with precision. Furthermore, in some cases, the generated power is from *dc* sources such as batteries, fuel cells, or photovoltage, or in other cases, the available power is generated as variable frequency *ac* from sources such as wind and gas turbine. This power conversion, *ac-dc*, became more acute with the invention of *vacuum* tubes, transistors, ICs, computers, servers, smart appliances, and data centers. Moreover, modern electric conversion goes beyond *ac-dc* conversion, as we shall shortly address.

In the late 1880s, power conversion from *ac-dc* was done by using *ac* motors along with *dc* generators in series (motor-generator set). Such motor-generator arrangement was still operational and was used in *dc* and 50/60 Hz motors and generators. The difficulties of using the electromechanical conversion system include large weight and size, noisy operation, servicing and maintenance problems, short lifetime, low efficiency, limited range of conversion, and slow recovery time. To avoid problems of the electromechanical conversion systems, industrial engineers turned into linear electronics in the late 1960s, where power semiconductor devices were operated in their linear (active) region. To obtain electrical isolation, input line-frequency transformers were used, resulting in bulky, heavy, and large size power converters systems. Furthermore, since power devices are operating in the linear region, the overall efficiency of the system was low. Unlike electromechanical systems and linear electronic systems, power electronics has many advantages including the following: (1) high energy conversion efficiency, (2) results in highly integrated power electronic systems, (3) reduced EMI and electronic pollution, (4) higher reliability, (5) utilizes environmentally clean voltage sources such as photovoltaic and fuel cells to generate electric power, (6) allows for the integration of electrical and mechanical systems, and (7) allows for maximum adaptability and controllability.

In short, all forms of electrical power conversion will always be needed as long as the consumers keep living in homes and use light, heat, electronic devices, equipment, and interface with industry.

1.5 Power Electronic Systems

Most of the power electronic systems consist of two major modules: (1) the power stage (forward circuit) and (2) the control circuit (feedback circuit). The power stage handles the power transfer from the input to the output, whereas the feedback circuit controls the amount of power transferred to the output.

Typical generalized block diagram of a power electronic system is given in Fig. 1.2 below.

where.

$x_1, x_2, \ldots x_n$: Inputs signals (voltage, current, or angular frequency)
$y_1, y_2, \ldots y_n$: Output signals (voltages, currents, or angular frequency)
$p_{\text{in}}(t)$: Instantaneous input power in Watts
$p_{\text{out}}(t)$: Instantaneous output power in Watts

$f_1, f_2, \ldots f_n$ are feedback signals: voltages or currents in electrical system or angular speed or angular position in mechanical systems.

Efficiency, η, is defined as follows:

$$\eta = \frac{p_{\text{out}}}{p_{\text{in}}} \times 100\%$$

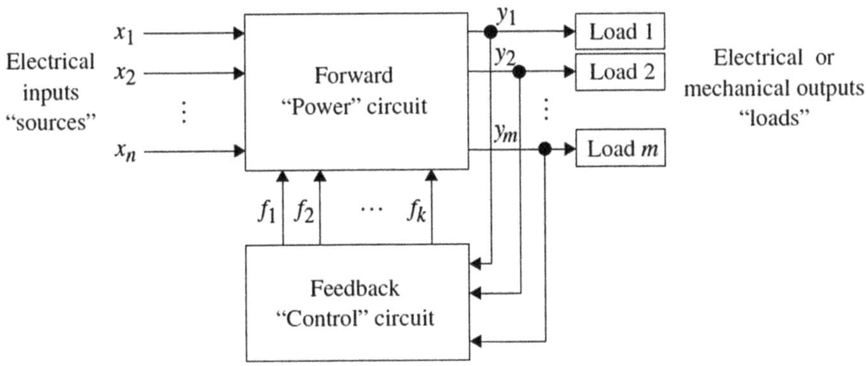

Fig. 1.2 Simplified block diagram for a power electronic system

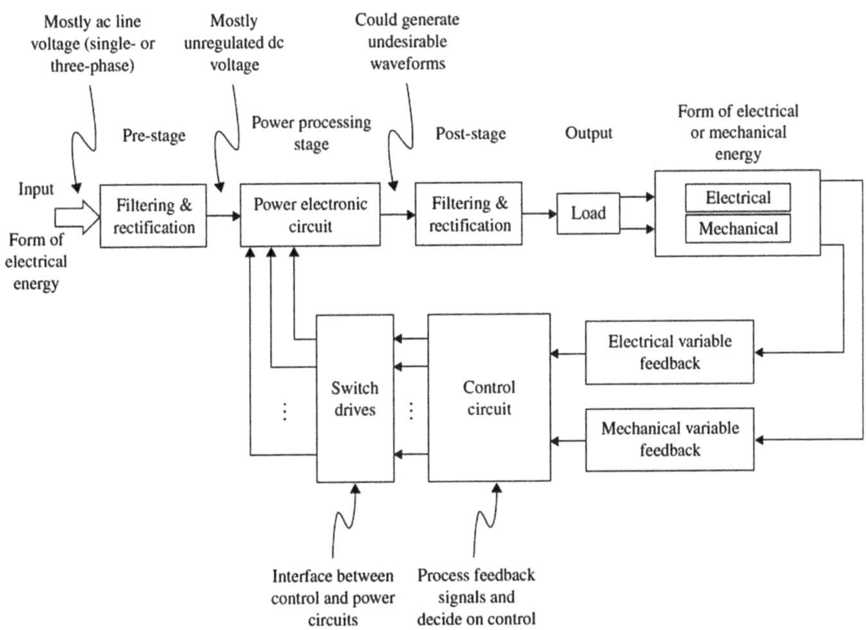

Fig. 1.3 Detailed block diagram of a power electronic system

Figure 1.3 shows a more detailed description of a block diagram for a power electronic input system with electrical and mechanical output loads. The main objection of the power electronic circuits is to process energy from a given source to a required load. In many applications we illustrated earlier, the conversion process concludes with mechanical motion.

1.5.1 Classification of Power Converter Circuits

The function of the *power converter stage* is to perform the actual power conversion and processing of the energy from the input to the output by incorporating a matrix of power switching devices. The control of the output power is carried out through control signals applied to these switching devices. Broadly speaking, power conversion refers to the power electronic circuit that changes one of the following: voltage form (*ac* or *dc*), voltage level (magnitude), voltage frequency (line or otherwise), voltage waveshape (sinusoidal or non-sinusoidal such as square, triangle, sawtooth, etc.), and the voltage phase (single or three phase).

Broadly speaking, there are four possible *conversion circuits* that are used in the majority of today's power electronic circuits:

(a) *ac-ac*
(b) *ac-dc*
(c) *dc-ac*
(d) *dc-dc*

In terms of the functional description, modern power electronic systems perform one or more of the following conversion functions:

1. Rectification (*ac-dc*)
2. Inversion (*dc-ac*)
3. Cycloconversion (*ac-ac* different frequencies) or *ac* controllers (*ac-ac* same frequencies)
4. Conversion (*dc-dc*)

1. Rectification (*ac-dc*)

   The term "rectification" refers to the power circuit whose function is to alter the *ac* characteristic of the line electric power to a "rectified" *ac* power at the load site that contains *dc* value. Figure 1.4a, b shows the block diagram representation of an *ac-dc* converter and its typical input and output waveforms, respectively. To smooth out the output voltage by removing the unwanted *ac* component, additional "filtering" circuit is added at the output side. Depending on the switch implementations, these converters are further divided into two types: *diode converter circuits (uncontrolled)* and *thyristor converters (phase controlled)*; each type can have either single-phase or three-phase input voltages. Both types are extensively used in various offline applications as shown in Table 1.1. Rectification circuits will be discussed in Chaps. 5 and 6. The topologies that perform the rectification function include half-wave, full-wave (full-bridge), semi-bridge, and transformer-coupled center-tapped. From the beginning of the industrial electronics area, the *ac-dc* line commutation converter class, utilizing thyristors, has been the largest among power electronic converters because of their simplicity in design, efficiency, and higher current and voltage ratings.

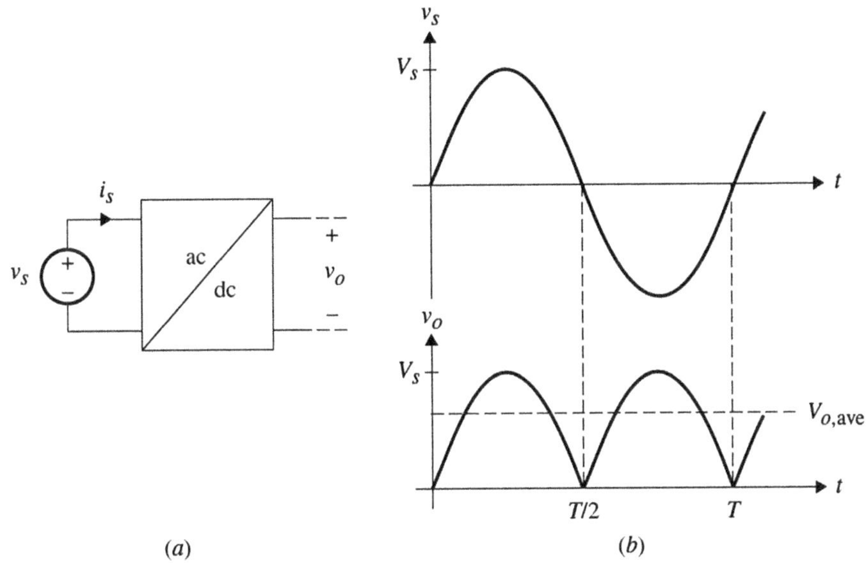

Fig. 1.4 *ac-dc* rectification

Table 1.1 Applications of power electronics by conversion functions

Conversion function	Application
• Uncontrolled *ac-dc* converters (diode circuits)	– Front-end offline regulated dc-ac power supplies and dc-ac inverters – Battery chargers – Welding – dc motor drives
• Phase-controlled converters (thyristor circuits)	– Regulated dc power supplies – ac and dc variable-speed motor control – Battery chargers – Flexible ac transmission system (FACTS) – Utility interface of photovoltaic systems – Regulated ac inverters – Solid-state circuit breakers – dc motor drives – Induction heating – Electromechanical processing (electroplating, anodizing, metal refining) – HVDC systems – Light dimmers – Active power line conditioning (APLC) (VAR compensator, harmonic filters) – Induction heating

(continued)

Table 1.1 (continued)

Conversion function	Application
• *dc-dc* converters	– High-frequency regulated dc power supplies using both isolated and non-isolated switch-mode and soft-switching resonant topologies – Digital and analog electronics – Solar energy conversion – High-frequency quasi-resonant converters – Electric vehicles and trams – dc-fed forklifts – Fuel cell conversion – dc traction drives – Distributed power systems – Power factor correction – Solid-state relays – Bidirectional dc-dc converters
• Linear-mode *dc-dc* converters	– Low-power linear dc regulators – Audio amplifiers – RF amplifiers
• Cycloconverters and *ac* controllers (*ac-ac*)	– ac motor drives – Rolling mill drives – Static Scherbius drives – Aircrafts – Frequency changers – Solid-state power line conditioners – Variable-speed constant frequency (VSCF) systems – Fluorescent lighting – Light dimmers – Induction heating
• *dc-ac* inverters	– Grid-tied inverters and microinverters – Aircraft and space power supply systems – ac variable-speed motor drives (lifts) – Uninterruptible power supplies(UPS) – Power factor correction – Light dimmers – Electric railroad systems – Magnetically levitated (MAGLEV) high-speed transportation systems – Electric vehicles
• Static switching	– ac and dc circuit breakers – Circuit protection – Solid-state relays
• Power ICs	– Home and office automation – Automobiles – Telecommunications – ac and dc drives – dc power supplies

2. Inversion (*dc-ac*)

The term *inversion* is used in power electronic circuits for the function that alters the *dc* source (like a battery) with no *ac* components into an "inverted" *ac* power at the load that has no *dc* components, as shown in Fig. 1.5a. Typical input and output waveforms are shown in Fig. 1.5b. The *ac* output can have an adjustable magnitude and frequency. Additional filter is normally used to extract the fundamental component of $v_o(t)$ *at* $\omega=2\pi/T$. Generally speaking, *dc-ac* inverters are classified as voltage-fed and current-fed inverter types. Also, resonant link technology that has been successful in the design of PWM power supplies is applied to the design of *dc-ac* converters, producing *ac* outputs at variable voltages and variable frequencies. Resonant link *dc-ac* inverters are a two-stage conversion circuit, one that takes the *dc* voltage and changes it to a high *ac* resonant voltage, which in turn is changed to a variable low-frequency output as shown in Fig. 1.5c. Generally speaking, since cascaded systems involve a two-stage conversion, power processing passes through more than one switching device and, therefore, increases conduction losses. However, in some cases, by using intermediate stages, it is possible to insert an electrical isolation transformer, and in other cases, cascaded stages produce high-

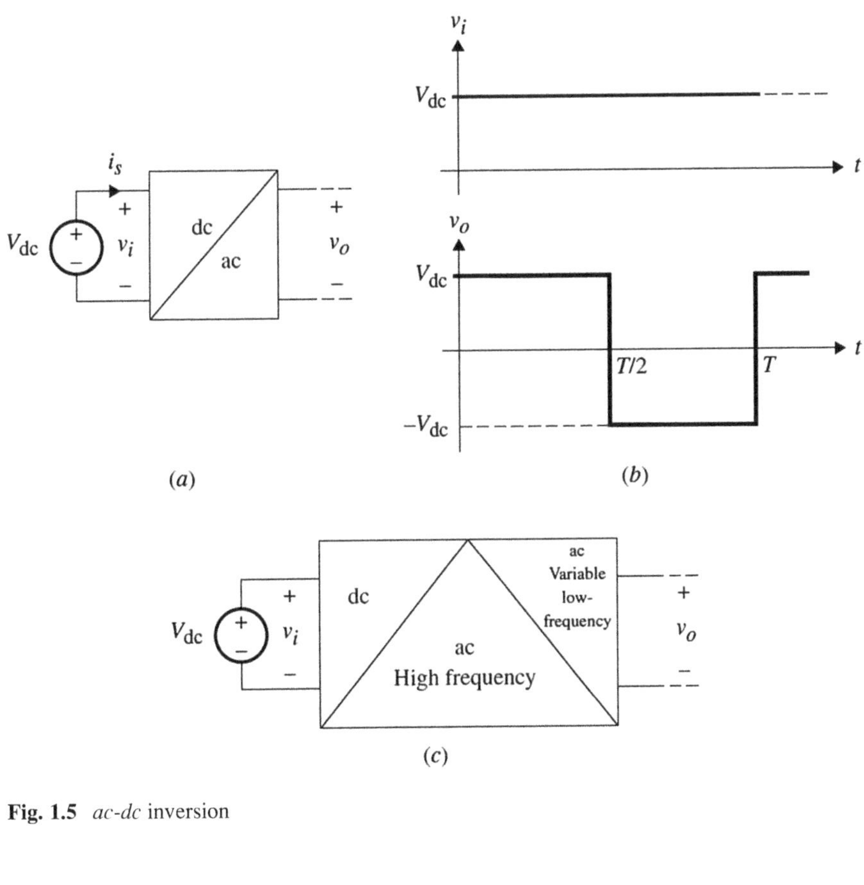

Fig. 1.5 *ac-dc* inversion

frequency resonant waveforms that could result in the soft switching of the power devices that in turn could reduce the overall switching losses of the cascaded system. Both *ac-dc* rectification and *dc-ac* inversion circuits represent the broadest functions of power electronic circuits. Finally, it is important to note that interest in *dc-ac* inversion has been highly intensified over the last 10 years due to the ever-increasing penetration of photovoltaic (PV) systems. Special interest has been in the design and development of a new class of inverters known as microinverters. For more details, please see the web site www.fpec.ucf.edu.

3. Cycloconversion or Voltage Controllers (*ac-ac*)

   The *cycloconversion* term is used for power electronic circuits that convert the *ac* input power at one frequency to an output power at a different frequency using a one-stage conversion as shown in Fig. 1.6. However, a two-stage conversion is also possible as shown in Fig. 1.6c, i.e., *ac-dc* and then *dc-ac*,

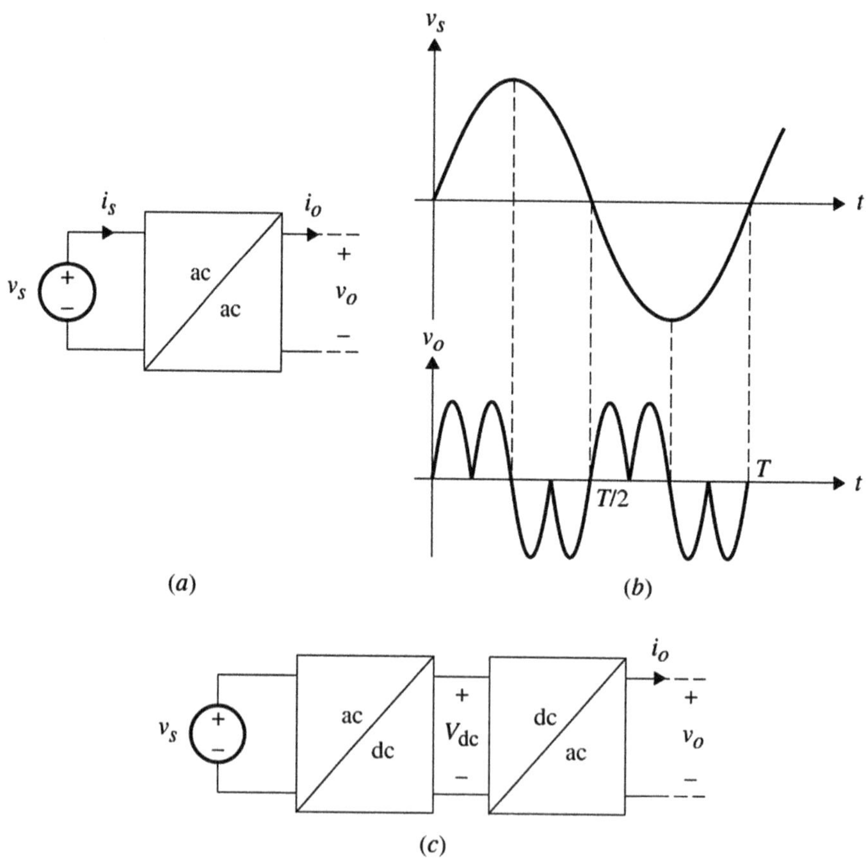

Fig. 1.6 (**a**) One-stage *ac-ac* cycloconversion, (**b**) typical waveforms, and (**c**) two-stage *ac-ac* cycloconversion

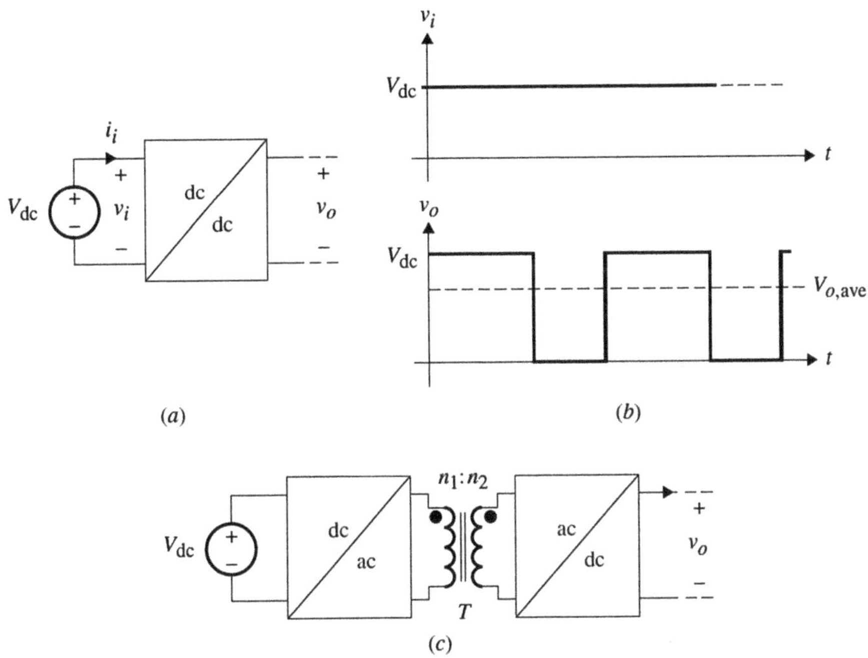

Fig. 1.7 (**a**) One-stage *dc-dc* conversion, (**b**) typical waveforms, and (**c**) two-stage *dc-dc* conversion

resulting in what is called *dc*-link converters, whereas an *ac* controller refers to a power electronic circuit that alters the *rms ac* input at the same frequency.

4. Conversion (*dc-dc*)

   dc-dc converters are used in power electronic circuits to convert an unregulated *dc* voltage to a regulated or variable *dc* output voltage as shown by the block diagram and its waveform of Fig. 1.7a, b, respectively. These circuits dominate the power supply industry such as the switch-mode power supplies (SMPS). In high-power *dc* traction drive applications, *dc-dc* converters are known as *choppers*. The high-frequency pulse width modulation (PWM) converters with or without output electrical isolation will be discussed in Chaps. 7 and 8, respectively. The high-frequency resonant-type *dc-dc* converters, which have a two-stage conversion, will be discussed in Chap. 9, where transformers can be used to provide electrical isolation and step-up/step-down features. Figure 1.7c shows such implementation.

To achieve this, we need to design converters in the hundreds of kilohertz and up to a few megahertz. In resonant converters, the switching devices are used in such a way that the turn-on and/or turn-off losses can be reduced or eliminated, depending on the converter operation. Such converters are known as "soft switching" converters. These converters are used in the design of high-power density *dc* power

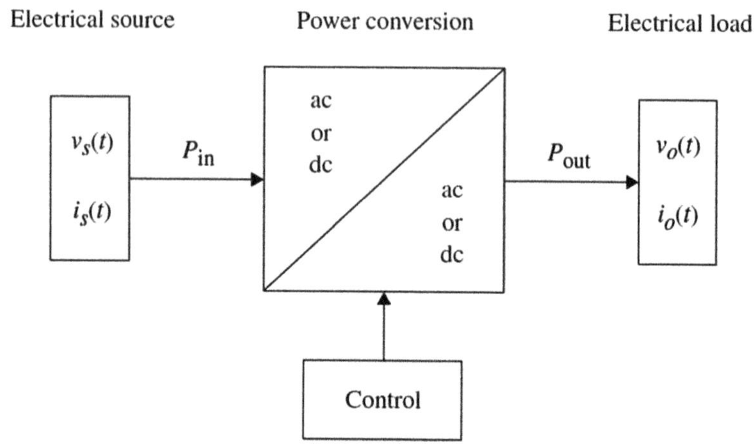

Fig. 1.8 Simplified block diagram representation of power electronic conversion function

supplies for laptop computers, adapters and notebook computers, aerospace, and communication instrumentation.

Figure 1.8 shows a simplified block diagram of a power electronic conversion system representing four possible conversion functions. The open literature is rich with all kinds of power electronic circuit topologies for various applications. When selecting a topology for a given application, one has to consider several factors including identifying the basic conversion function required and understanding the available switching devices and their switching characteristics, driving circuits, control and protection, and their maximum switching losses, and finally, the cost, size, and weight must also be considered.

Depending on the topologies used and the types of loads, power electronic circuits are capable of transferring power in only one direction, i.e., unidirectional power flow from the source to the load, and others are capable of transferring power in both directions, i.e., bidirectional power flow from the load to the source. The latter is known to operate in the regenerative mode. Since the polarities of the load current, i_o, and voltage, v_o, shown in Fig. 1.9a can be either positive or negative, there exist four modes of operation, as shown in Fig. 1.9b.

The first quadrant (Mode 1) suggests that the output voltage and current are always positive and the power flows unidirectionally to the load. Examples of such converters are the switch mode *dc-dc* buck and boost converters. Similarly, quadrant III means that the converter allows unidirectional power flow with both the output voltage and output current which are negative such as in the *dc-dc* buck-boost converter. Another example is the power electronic circuit with a *dc* motor, which can operate in quadrants I and II since the motor is capable of supporting the forward and reverse directions. Quadrants II and IV indicate that the power is being transferred from the load to the source such as in the case of a *dc* generator. Converters that can operate in quadrants I and II or III and IV can support bidirectional power flow. Examples of converters operating in the four quadrants will be discussed in the following chapters.

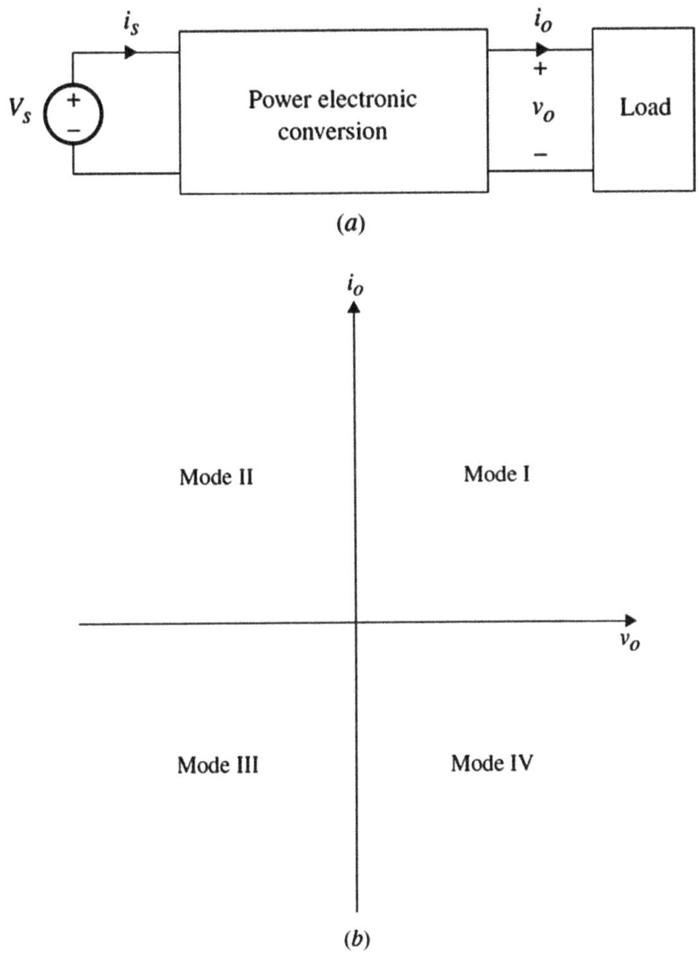

Fig. 1.9 (**a**) Simple power electronic circuit with single voltage source and single load, (**b**) possible converter modes of operation

1.5.2 Power Semiconductor Devices

In order to achieve high conversion efficiency in power electronic circuits, semiconductor switching devices are considered the heart of power electronic circuits, as we will see throughout the textbook. The control of power flow from the input to the output is done through a power processing switching network made of switching devices and energy storage elements. Detailed discussion on the switching network, or what is known as the switching matrix, will be given in the next chapter.

The need for new power electronic circuits to address the growing market demand for new power electronic applications resulted in intensive research activities in the semiconductor industry to make available all types of semiconductor devices with a wide range of power-handling capabilities and switching speeds.

The history of power electronics has been, and will continue for many years to come, very much tied to the advancements made in the power semiconductor technology. Due to the development of high-power, fast-switching, and high-efficient thyristor-based, unipolar and bipolar devices in the late 1950s and 1960s, the field of power electronics has reemerged as a separate subarea in electrical engineering. Today, semiconductor switching devices have achieved unprecedented power-handling capabilities and switching speeds.

In the next chapter, we will study the *i-v* characteristics for five basic power devices widely used as switching elements in power electronic circuits. These devices include (1) power diode, (2) bipolar junction transistors (BJT), (3) metal oxide semiconductor FET (MOSFET), (4) thyristors or silicon controlled rectifiers (SCR), and (5) insulated gate bipolar transistor (IGBT). Many other available devices such as TRIACs, DIACs, gate turn-off (GTO) thyristors, static induction transistor (SIT), static induction thyristor (SITH), and MOS-controlled thyristor (MCT) are, in one way or another, from the same family of the five basic devices listed above. Ratings of many of the above devices are in the 100 s of kWatts! Even higher current and voltage rating values can be achieved by connecting devices in parallel and series, respectively.

1.5.3 Converter Modeling and Control

In any power electronic circuit, the way to control and stabilize power flow from the source to the load through the system is extremely important, since the output must be regulated, whether it is voltage, current, or frequency. Over the last 30 years, tremendous efforts went into developing converter modeling methods that can be understood using well-known linear circuit analysis techniques. Moreover, the control of power electronic circuits has been greatly simplified by using microcomputer and commercial ICs that improve reliability, reduce weight and size of hardware, and provide flexibility to the designer by changing the software algorithms. Also, the analysis and design of control circuits in power electronics have been made even easier by the availability of several simulation packages such as PSPICE, MATLAB, Saber, and many others. A wealth of knowledge is available in the open literature in this subject. In this text, various converter modeling and analysis techniques as well as control design will be presented for a class of converters known as *dc-dc* pulse width modulation (PWM) converters. Still, the work focuses on an individual converter with little emphasis on developing converter modeling and control at the system level.

1.6 Applications of Power Electronics

Power electronics covers a diverse and wide range of applications in residential, commercial, industrial applications, computers, transportation, aircraft/aerospace, information processing, telecommunication, and power utilities. Broadly speaking, these applications may be classified into three categories:

1. *Electrical Applications*
   Power electronics can be used to design *ac*- and *dc*-regulated power supplies for various electronic equipment including consumer electronics, instrumentation devices, computers, aerospace, and uninterruptible power supplies (UPS) applications. Also power electronics is used in the design of distributed power systems, electric heating and lighting control, power factor correction, and static VAR compensation.
2. *Electromechanical Applications*
   The electromechanical conversion systems are widely used in industrial residential and commercial applications. These applications include *ac* and *dc* machine tools, robotic drives, pumps, textile and paper mills, peripheral drives, rolling mill drives, and induction heating.
3. *Electrochemical Applications*
   The electrochemical applications cover chemical processing, electroplating, welding, metal refining, production of chemical gases, and fluorescent lamp ballasts.

Table 1.1 gives a list of some possible power electronic application categories according to their conversion functions. We should mention that the examples given in Table 1.1 are not inclusive of all power electronic products and applications available today but rather an illustration of how wide the application spectrum of the field is. Finally, we note that the energy spectrum of the application of power electronics extends from a few watts such as a switching regulator to few megawatts such as in high-voltage *dc* (HVDC) systems.

1.7 Future Trends

As stated earlier, it is hard to predict the direction of the future research of the field of power electronics or any field for that matter. However, based on today's research and teaching activities in power electronics, which are driven by market demands, energy conservation, and cost reduction, it is possible to foresee possible short-term future-research activities in power electronics that may include:

1. Continuity in the technological improvement of high-power and high-frequency semiconductor devices
2. Continuity in the development of power electronic converter topologies to attain further size and weight reduction with increased efficiency and performance

3. Improvement in the design of driver circuits for newly introduced switching devices
4. Improvement in digital control techniques including optimal and adaptive control
5. Integration of power and control circuitry on "smart power" IC and further development of application specific modules
6. The integration of renewable energy stems to the power grid and its transformation into a smart grid
7. Power factor correction techniques and EMI reduction
8. More applications of power electronics in flexible *ac* transmission systems (FACTS)
9. Interacting electric storage to the grid
10. Enabling the electric vehicle adoption and integration into the grid

As power electronics becomes cheaper, it will penetrate into various new industrial, residential, aerospace, and telecommunications applications. Based on the growth of power electronics in recent years, future growth is projected to be even greater.

As long as we continue to seek to improve our standards of living, our quest for cheap and environmentally clean energy will continue. Power electronics will be used widely to address energy conservation and conversion efficiency. It was reported that more than 20% energy savings could be achieved with the help of power electronics. The role of power electronics will be greater as it becomes cheaper and more devices and systems become available. The challenge for power electronic circuit engineers is to keep developing new and optimum topologies to match market applications, and the challenge for power device engineers is to come up with new devices that can be used in these new topologies!

1.8 About the Text and Its Nomenclatures

1.8.1 About the Text

This text is written in a way that undergraduate students in electrical engineering can benefit the most. It is designed to be used in two-semester power electronic courses, the first part of the text is intended to be covered in the first undergraduate introductory course in power electronics, and the second part is to be used for first-year graduate power electronics course. Because of the interdisciplinary nature of power electronics, an expanded introduction (Chaps. 1, 2 and 3) has been presented to cover a brief introduction about the field, a review of power semiconductor devices, and general basic concepts that represent the cornerstone of power electronics.

1.8.2 Nomenclature

To avoid confusion and repetition, we establish some definitions, terminology, and notations to be used throughout the text:

1. *Time-dependent variables (pure ac)*

   All instant time-dependent variables including current (i), voltage (v), and power (p) will be presented as lowercase letters with lowercase subscripts, i.e.:

   $$i_a(t), v_b(t) \text{ and } p_i(t)$$

   The "(t)" indicates "function of time." For simplicity, the time notation is dropped, and the variables are given by i_a, v_b, and p_i.
2. *Average and constant variables*

   All average or constant variables will be given in uppercase letters and lowercase subscripts. Such as V_s, I_o, V_o, rms, and $I_{1,\min}$.

   To distinguish between *peak*, *dc*, *rms*, and constants, additional subscripts will be added when necessary.
3. *The ac voltage source and the dc voltage source will be given as follows:*

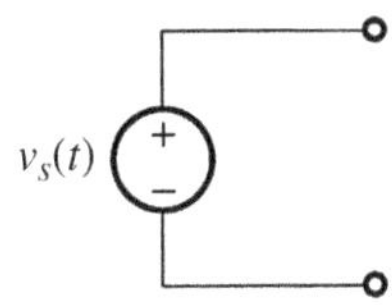

$$v_s(t) = V_s \sin \omega t \qquad \omega = 2\pi f = 2\pi/T$$

   where

   V_s: peak voltage
   ω: angular frequency in rad/sec
   f: frequency in Hz
   T: period in seconds

   Rectified *dc* source:

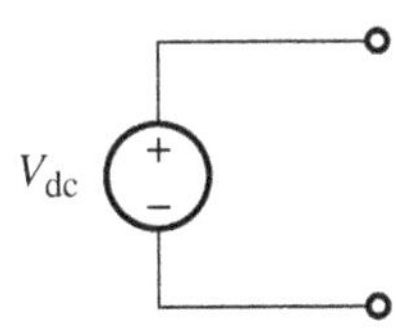

Constant *dc* source (battery):

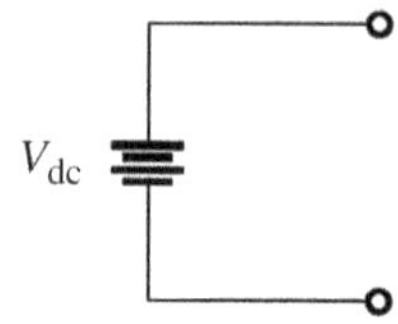

4. *Time-dependent variables (ac and dc)*

   Like the conventional notation in electronics, a lowercase variable and an uppercase subscript indicate that the time variable has both *ac* and *dc* components such as

$$i_L(t) = I_L + \hat{i}_L(t)$$

   where

   I_L: *dc* component
   $\hat{i}_L(t)$: *ac* component

   The "^" notation indicates that the source of the *ac* components is perturbation around the *dc* value I_L. These notations will be used when we discuss the small-signal analysis of *dc-dc* converters.

5. *The current and voltage sinusoidal harmonics of a periodical signal will be as follows:*

$$i_s(t) = I_{dc} + I_{s,1}\sin\omega t + I_{s,2}\sin 2\omega t + \ldots + I_{s,n}\sin n\omega t$$

   where

   I_{dc}: the average value of the signal
   $I_{s,n}$: the peak value of the *nth* harmonic component of the signal, where $n = 1, 2, \ldots, \infty$

6. *We will interchange the integration variables between time "t" and angular frequency ωt. For example:*

$$\frac{1}{T}\int_0^{T/2} V_s \sin\omega t dt = \frac{1}{2\pi}\int_0^{\pi} V_s \sin\omega t d\omega t$$

7. *Three-phase circuit representation*

   Balanced three-phase voltage sources consist of three equal sinusoidal voltages each shifted by 120° from the other in such a way that their phase sum is zero.

Two well-known three-phase configurations will be used in this text:

-Wye (Y) connected three-phase source as shown in Fig. 1.10a

$$\text{Phase voltages:} \quad \begin{cases} v_a = V_s \sin \omega t \\ v_b = V_s \sin\left(\omega t - 120^\circ\right) \\ v_c = V_s \sin\left(\omega t - 240^\circ\right) \end{cases}$$

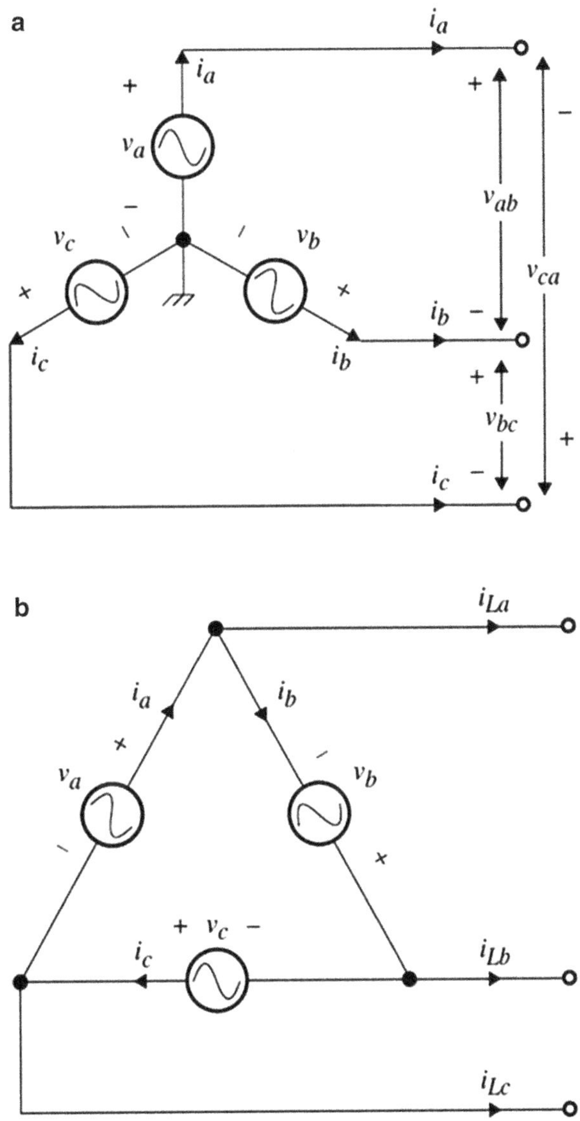

Fig. 1.10 (**a**) Wye (*Y*) connected three-phase voltage source, (**b**) Delta (*Δ*) connected three-phase voltage source

$$\text{Line voltages:} \quad \begin{cases} v_{ab} = \sqrt{3}V_s \sin\left(\omega t + 30^\circ\right) \\ v_{bc} = \sqrt{3}V_s \sin\left(\omega t - 90^\circ\right) \\ v_{ca} = \sqrt{3}V_s \sin\left(\omega t - 210^\circ\right) \end{cases}$$

Line (phase) currents: i_a, i_b, i_c

-Delta (Δ) connected three-phase source as shown in Fig. 1.10b

$$\text{Phase (line) voltages:} \quad \begin{cases} v_a = V_s \sin \omega t \\ v_b = V_s \sin\left(\omega t - 120^\circ\right) \\ v_c = V_s \sin\left(\omega t - 240^\circ\right) \end{cases}$$

$$\text{Phase currents:} \quad \begin{cases} i_a = I_s \sin \omega t \\ i_b = I_s \sin\left(\omega t - 120^\circ\right) \\ i_c = I_s \sin\left(\omega t - 240^\circ\right) \end{cases}$$

$$\text{Line currents:} \quad \begin{cases} i_{La} = \sqrt{3}I_s \sin\left(\omega t + 30^\circ\right) \\ i_{Lb} = \sqrt{3}I_s \sin\left(\omega t - 90^\circ\right) \\ i_{Lc} = \sqrt{3}I_s \sin\left(\omega t - 210^\circ\right) \end{cases}$$

Chapter 2
Review of Switching Concepts and Power Semiconductor Devices

2.1 Introduction

In this chapter, an overview of power semiconductor switching devices will be given. Only devices that are available in the market and are currently used in power electronics applications will be considered. These devices include unipolar and bipolar devices such as the power diode, bipolar junction transistor (BJT), metal oxide semiconductor field-effect transistors (MOSFETs), and the insulated gate bipolar transistor (IGBT) and thyristor-based devices such as silicon-controlled rectifier (SCR), gate turn-off (GTO) thyristor, TRIAC, static induction transistor and thyristors, and MOS-controlled thyristor (MCT). Detailed discussion of the physical structure, fabrication, and physical behavior of these devices and packaging are beyond the scope of this text. The emphasis here will be on the terminal *i-v* switching characteristics of the available devices and their current, voltage, and switching limits. Even though most of today's available semiconductor power devices are made of silicon or germanium materials, other materials such as gallium arsenide, diamond, and silicon carbide are currently being tested.

As stated in Chap. 1, one of the main contributions to the growth of the power electronics field has been the unprecedented advancement in semiconductor technology, especially with respect to switching speed and power-handling capabilities. The area of power electronics started with the introduction of the SCR in 1958. Since then, the field has grown in parallel with the growth of the power semiconductor device technology. In fact, the history of power electronics is very much connected to the development of switching devices, and it emerged as a separate discipline when high-power BJT and MOSFET devices were introduced in the 1960s and 1970s. Since then, the introduction of new devices has been accompanied with dramatic improvements in power rating and switching performance.

In the 1980s, the development of power semiconductor devices took an important turn when new processing technology was developed that allowed the integration of MOS and BJT technologies on the same chip. Thus far, two devices using

I. Batarseh, A. Harb, *Power Electronics*,
https://doi.org/10.1007/978-3-319-68366-9_2

this new technology have been introduced: IGBT and MCT. Many of the IC processing methods and equipments have been adopted for the development of power devices. However, unlike microelectronic ICs, which process information, power devices, ICs process power; hence, their packaging and processing techniques are quite different.

Since the development of thyristors in the late 1950s, the power semiconductor device technology has been going through dynamic evolution, always following the evolution of microelectronic technology and, in the process, introducing many different kinds of devices with wide ranges of power ratings and frequency. Because of their functional importance, drive complexity, fragility, and cost, a power electronic design engineer must be equipped with the thorough understanding of the device operation, limitation, drawbacks, and related reliability and efficiency issues.

Power semiconductor devices represent the "heart" of modern power electronics, with two major desirable characteristics guiding their development:

1. Switching speed (turn-on and turn-off times)
2. Power-handling capabilities (voltage blocking and current carrying capabilities)

Improvement in semiconductor processing technology as well as in manufacturing and packaging techniques has allowed the development of power semiconductors for high voltage and high current ratings and fast turn-*on* and turn-*off* characteristics. The availability of different devices with different switching speeds, power-handling capabilities, sizes, costs, and other factors makes it possible to cover many power electronic applications, so that trade-offs must be made when it comes to selecting power devices.

2.2 The Need for Switching in Power Electronic Circuits

Do we have to use switches to perform electrical power conversion from the source to the load? The answer, of course, is no; there are many circuits that can perform energy conversion without switches, such as linear regulators and power amplifiers. However, the need for semiconductor devices to perform conversion functions is very much related to the converter efficiency. In power electronic circuits, the semiconductor devices are generally operated as switches – either in the on-state or off-state. Typically, the interface between the power devices and control circuit is done by a circuit known as the "drive" circuit. Such circuits are designed for minimum cost and high efficiency. This is unlike the case in power amplifiers and linear regulators, where semiconductor devices operate in the linear mode. As a result, a very large amount of energy is lost within the power circuit before the processed energy reaches the output. This is why special packaging, protection, and heat sinks are found in power electronics that are not present in information processing circuits. The need to use semiconductor switching devices in power electronic circuits is based on their ability to control and manipulate very large amounts of power from the input to the output with relatively very low-power dissipation in the switching device, resulting in a very high-efficiency power electronic system.

Efficiency is an important figure of merit and has significant implications on the overall performance of the system. A low-efficiency power system means that large amounts of power are being dissipated in the form of heat, with one or more of the following implications:

1. The cost of energy increases due to increased consumption.
2. Additional design complications might be imposed, especially regarding the design of device heat sinks.
3. Additional components such as heat sinks increase the cost, size, and weight of the system, resulting in low-power density.
4. High-power dissipation forces the switch to operate at low switching frequencies, resulting in limited bandwidth and slow response, and most importantly, the size and weight of magnetic components (inductors and transformers) and capacitors remain large. Therefore, it is always desirable to operate switches at very high frequencies. But we will show later that as the switching frequency increases, the average switching power dissipation increases. Hence, a trade-off must be made between reduced size, weight, and cost of components and reduced switching power dissipation, which means inexpensive low switching frequency devices.
5. Component and device reliability is reduced.

For more than half a century years, it has been shown that switching (mechanical or electrical) is the best possible way to achieve high efficiency in energy conversion. However, electronic switches are superior to mechanical switches because of their speed and power-handling capabilities as well as their reliability.

We should note that the advantages of using switches come at a cost. Because of the nature of switch currents and voltages (square waveforms), high-order harmonics are normally generated in the system. To reduce these harmonics, additional input and output filters are usually added to the system. Moreover, depending on the device type and power electronic circuit topology used, driver circuit control and circuit protection can significantly increase the complexity of the system and its cost.

Example 2.1

The purpose of this example is to investigate the efficiency of four different power electronic circuits whose function is to take power from a 24 V *dc* source and deliver a 12 V *dc* output to a 6 Ω resistive load. In other words, the task of these circuits is to serve as *dc transformers* with a ratio of 2 : 1. The four circuits are shown in Fig. 2.1a–d representing a voltage divider circuit, Zener regulator, transistor linear regulator, and switching circuit, respectively. The objective is to calculate the efficiency of those four power electronic circuits.

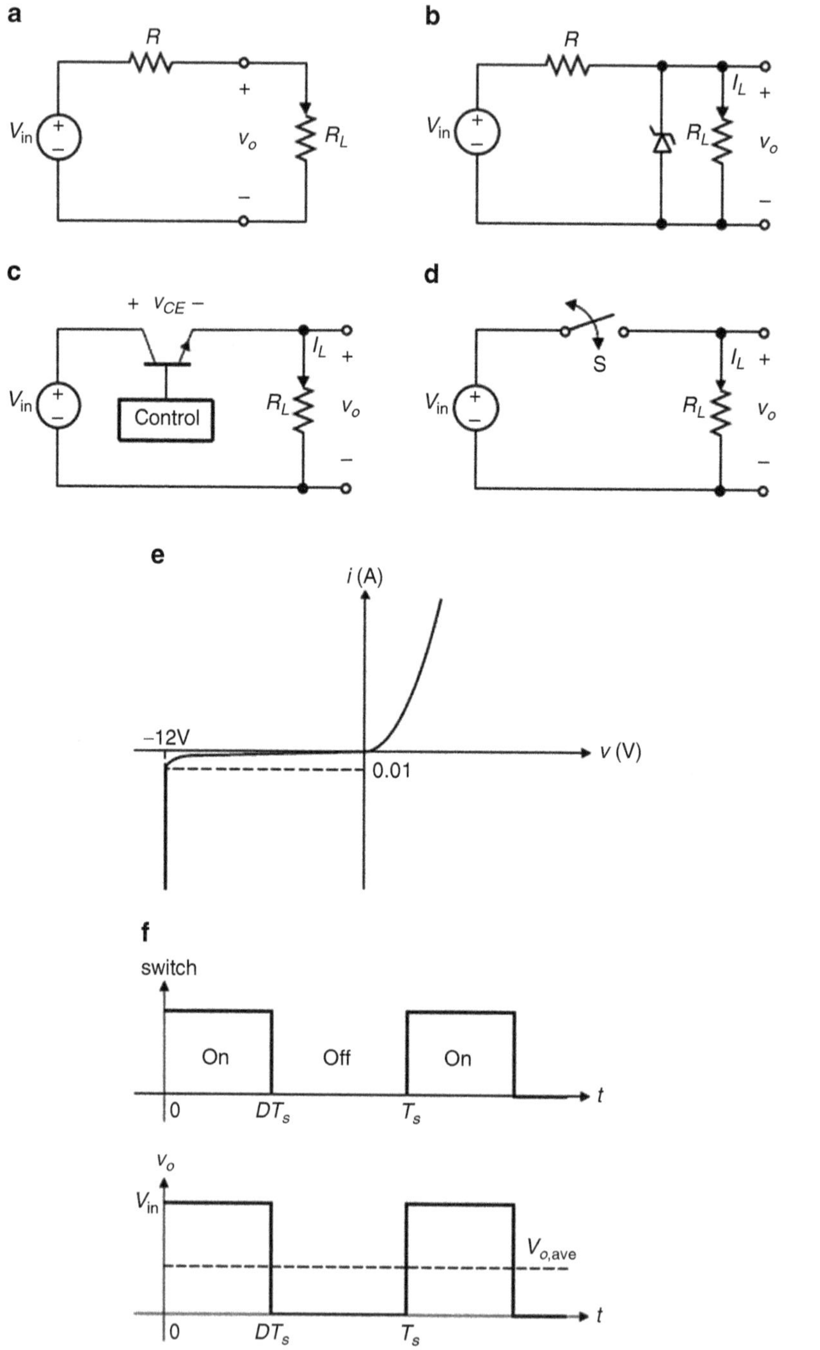

Fig. 2.1 (**a**) Voltage divider, (**b**) Zener regulator, (**c**) transistor regulator, (**d**) switching circuit, (**e**) diode i-v switching characteristics, (**f**) switching waveforms for (**d**)

Solution

(a) **Voltage Divider *dc* Regulator**

The first circuit is the simplest, forming a voltage divider with $R = R_L = 6\ \Omega$ and $V_o = 12$ V. The efficiency, defined as the ratio of the average load power, P_L, to the average input power, P_{in}, is:

$$\eta = \frac{P_L}{P_{in}}\%.$$
$$= \frac{R_L}{R_L + R}\% = 50\%$$

In fact, the efficiency is simply $V_o/V_{in}\%$. As the output voltage becomes smaller, the efficiency decreases proportionally.

(b) **Zener *dc* Regulator**

Since the desired output is 12 V, we select a Zener diode with Zener breakdown $V_Z = 12$ V. Assume the Zener diode has the *i-v* characteristic shown in Fig. 2.1e. Since $R_L = 6\Omega$, the load current, I_L, is 2 A. If we calculate R for $I_Z = 0.2$ A (10% of the load current), this results in $R = 5.45\ \Omega$. Since the input power is $P_{in} = 2.2\ \text{A} \times 24\ \text{V} = 52.8$ W and the output power is $P_{out} = 24$ W, the efficiency of the circuit is given by:

$$\eta = \frac{24\ W}{52.8\ W}\% = 45.5\%$$

(c) **Transistor *dc* Regulator**

It is clear from Fig. 2.1c that for $V_o = 12$ V, the collector-emitter voltage must be around 12 V. Hence, the control circuit must provide a base current, I_B, to put the transistor in the active mode with $V_{CE} \approx 12$ V. Since the load current is 2A, then the collector current is approximately 2 A (assume small I_B). The total power dissipated in the transistor can be approximated by the following equation:

$$P_{diss} = V_{CE} I_C + V_{BE} I_B$$
$$\approx V_{CE} I_C \approx 12 \times 2 = 24\ \text{W}$$

Therefore, the efficiency of the circuit is 50%.

(d) **Switching *dc* Regulator**

Let us consider the switching circuit of Fig. 2.1d by assuming the switch is ideal and periodically turns *on* and *off* as shown in Fig. 2.1f. The output voltage waveform is shown in Fig. 2.1f. Even though the output voltage is not constant or pure *dc*, its average value is given by:

$$V_{o,ave} = \frac{1}{T}\int_{o}^{T_sD} V_{in}dt = V_{in}D$$

where D is the duty ratio which equals the ratio of the on-time to the switching period, T_s. For $V_{o,\,ave} = 12$ V, we set $D = 0.5$, i.e., the switch has a duty cycle of 0.5 or 50%. In this case, the average output power is 48 W, and the average input power is also 48 W, resulting in 100% efficiency! This is, of course, because we assumed the switch is ideal. However, let us assume a BJT switch is used in the above circuit with $V_{CE,\,sat} = 1$ V and I_B is small, then the average power loss in the switch is approximately 2 W, resulting in an overall efficiency of 96%. Of course the switching circuit given in this example is oversimplified, since the switch requires additional driving circuitry that was not shown that also dissipates some power. But still, the example illustrates that the high efficiency can be acquired by a switching power electronic circuit when compared to the efficiency of a linear power electronic circuit. Also, the difference between the linear circuit in Fig. 2.1b, c and the switched circuit of Fig. 2.1d is that the power delivered to the load in the later case is pulsating between 0 and 96 W. If the application calls for constant power delivery with little output voltage ripple, then an LC filter must be added to smooth out the output voltage. This class of *dc-dc* converters will be studied in Chaps. 4 and 5.

A final observation is regarding what is known as load regulation and line regulation. The line regulation is defined as the ratio between the change in the output voltage, ΔV_o, and the change in the input voltage ΔV_{in}. This is a very important parameter in power electronics since the *dc* input voltage is obtained from a rectified line voltage that normally changes by$\pm$20%. Therefore, any off-line power electronic circuit must have a limited or specified range of line regulation. If we assume the input voltage in Fig. 2.1a, b is changed by 2 V (i.e., $\Delta V_{in} = 2$ V), with R_L unchanged, the corresponding change in the output voltage ΔV_o is 1 V and 0.55 V, respectively. This is considered a very poor line regulation. Figure 2.1c, d has much better line and load regulations since the closed-loop control compensates for the line and load variations.

Exercise 2.1

Consider the Zener regulator shown in Fig. 2.1b in Example 2.1 by assuming that minimum Zener current required to keep it operating in the breakdown region is 0.1 A as shown in Fig. 2.1e. Determine the range of R_L so that the output voltage will be regulated at 12 V.

Answer: $R_L \gg 5.7182$

2.3 Switching Characteristics

2.3.1 *The Ideal Switch*

It is always desired to have the power switches to perform as close as possible to the ideal case. For a semiconductor device to operate as an ideal switch, it must possess the following features:

1. No limit on the amount of current (known as forward or reverse current) that the device can carry when in the conduction state (*on*-state)
2. No limit on the amount of the device voltage (known as forward or reverse blocking voltage) when the device is in the nonconduction state (*off*-state)
3. Zero *on*-state voltage drop when in the conduction state
4. Infinite *off*-state resistance, i.e., zero leakage current when in the nonconduction state
5. No limit on the operating speed of the device when it changes state, i.e., zero rise and fall times

Typical switching waveforms for an ideal switch are shown in Fig. 2.2, where i_{sw} and v_{sw} are the current through and the voltage across the switch, respectively, and DT_s is the *on* time[1]. During the switching and conduction periods, the power loss is zero, resulting in a 100% efficiency. With no switching delays, an infinite operating frequency can be achieved. In short, an ideal switch has infinite speed, unlimited power-handling capabilities, and 100% efficiency. It must be noted that it is not surprising to find semiconductor switching devices that can almost, for all practical purposes, perform as ideal switches for a number of applications.

2.3.2 *The Practical Switch*

The practical switch has the following switching and conduction characteristics:

1. Limited power-handling capabilities, i.e., limited conduction current when the switch is in the *on*-state and limited blocking voltage when the switch is in the *off*-state.
2. Limited switching speed that is caused by the finite turn-*on* and turn-*off* times. This limits the maximum operating frequency of the device.
3. Finite *on*-state and *off*-state resistances, i.e., there exists forward voltage drop when in the *on*-state and reverse current flow (leakage) when in the *off*-state.

[1]Only the on time is shown since the emphasis is on the on and off times. The off time, $(1 - D)T_S$, has only conduction loss

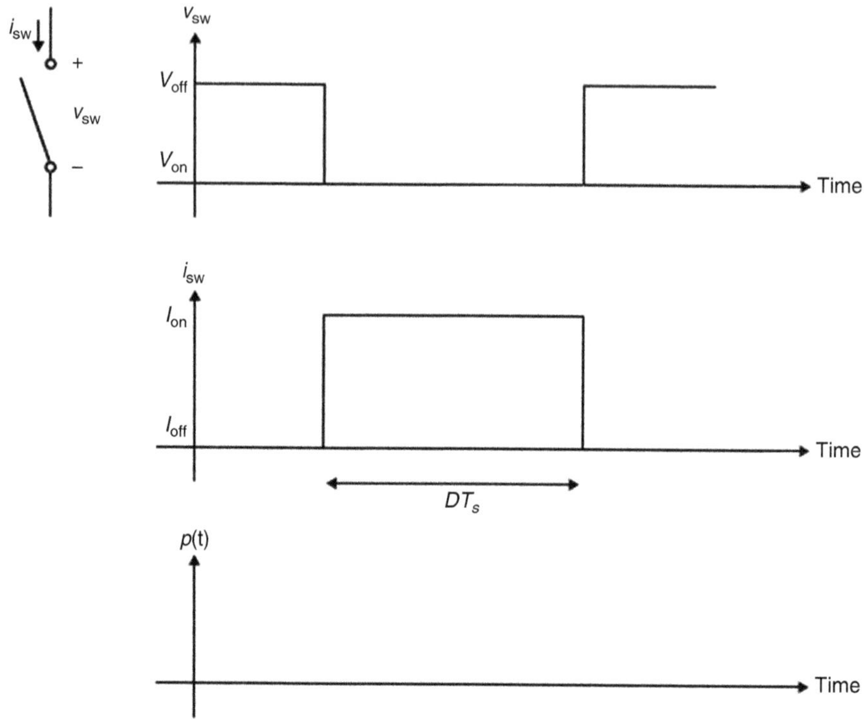

Fig. 2.2 Ideal switching voltage, current, and power waveforms

4. Because of characteristics 2 and 3 above, the practical switch experiences power losses in the *on*- and the *off*-states (known as conduction loss) and during switching transitions (known as switching loss).

Typical switching waveforms of a practical switch are shown in Fig. 2.3a. The average switching power and conduction power losses can be evaluated from these waveforms. The exact switching waveforms vary from one device to another, but Fig. 2.3a is a reasonably good representation. For simplicity, DT_s is shown to include the *off*-time and the *on*-time period of the switch. Moreover, other issues such as temperature dependence, power gain, surge capacity, and over voltage capacity must be considered for specific devices in specific applications. A useful plot that illustrates how switching takes place from *on* to *off* and vice versa is called a *switching trajectory*, which is simply a plot of i_{sw} *vs.* v_{sw}. Figure 2.3b shows several switching trajectories for the ideal and practical cases under resistive loads.

The average power dissipation, P_{ave}, over one switching cycle is given by:

$$P_{ave} = \frac{1}{T_s}\int_0^{T_s} i_{sw} v_{sw}\, dt = P_{ave,swit} + P_{ave,cond}$$

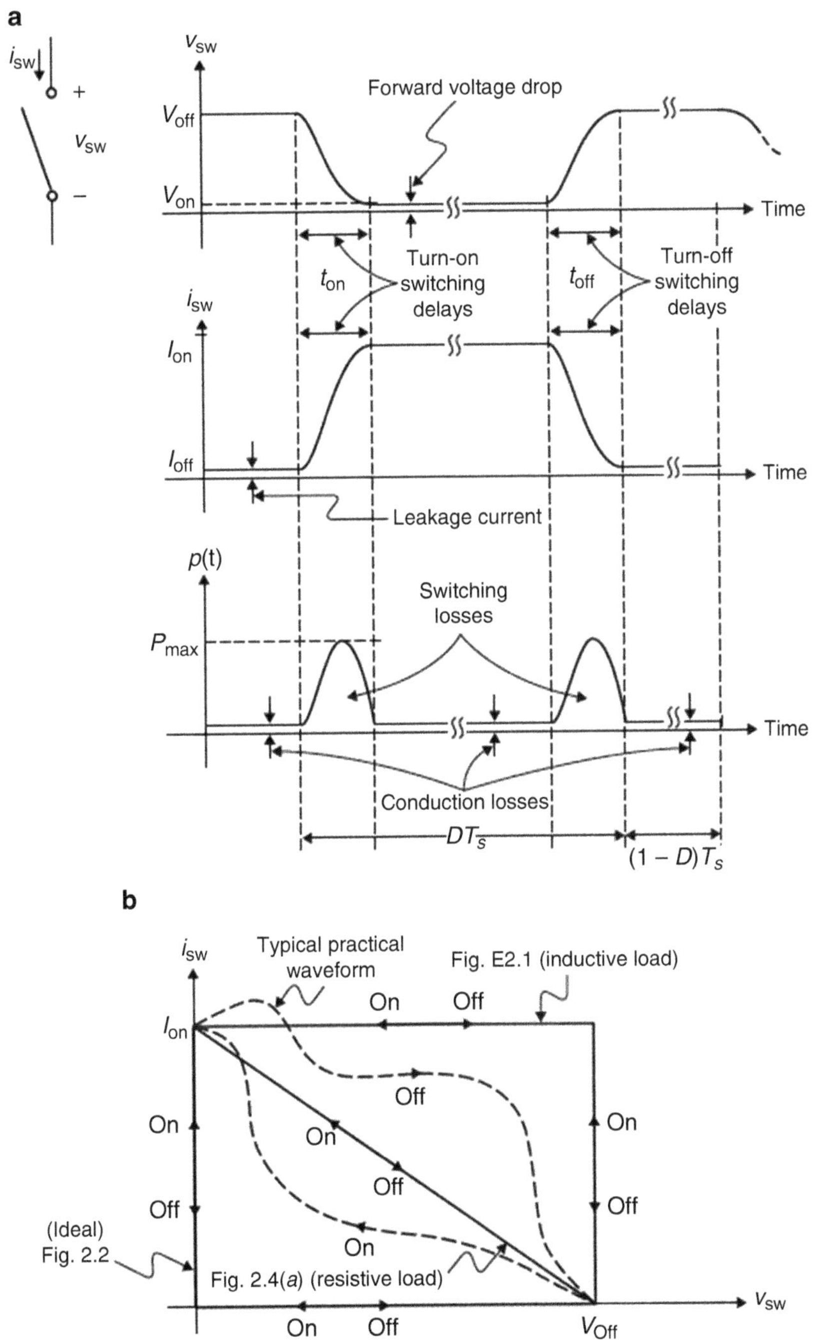

Fig. 2.3 (**a**) Practical switching current, voltage, and power waveforms. (**b**) Switching trajectories under different load conditions

where $P_{\text{ave, swit}}$ is the average switching losses and $P_{\text{ave, cond}}$ is the average conduction losses, given by:

$$P_{\text{ave, swit}} = \frac{1}{T_S}\left[\underbrace{\int_0^{t_{on}} i_{SW}v_{SW}dt}_{\text{on switching losses}} + \underbrace{\int_{DT_s-t_{off}}^{DT_s} i_{SW}v_{SW}dt}_{\text{off switching losses}}\right]$$

$$\begin{aligned} P_{\text{ave, cond}} &= \frac{1}{T_S}\left[\int_{t_{on}}^{DT_s-t_{off}} I_{on}V_{on}dt + \int_{DT_s}^{T_s} I_{off}V_{off}dt\right] \\ &= I_{on}V_{on}\left(D - \frac{(t_{on}+t_{off})}{T_s}\right) + I_{off}V_{off}(1-D) \end{aligned}$$

If we assume the *on* and *off* times are small compared to T_s, then we have

$$P_{\text{ave,cond}} \approx \underbrace{I_{on}V_{on}D}_{\text{on conduction loss}} + \underbrace{I_{off}V_{off}(1-D)}_{\text{off conduction loss}}$$

Example 2.2
Consider a linear approximation of Fig. 2.3a as shown in Fig. 2.4a with $D = 1.0$ (this assumes that T_s is the *on*-time).

(a) Give a possible circuit implementation using a power switch whose switching waveforms are shown in Fig. 2.4a.
(b) Derive the expressions for the instantaneous switching and conduction power losses and sketch them.
(c) Determine the total average power dissipated in the circuit during one switching period.
(d) Find the maximum power.

Solution

(a) First let us assume that the turn-on time, t_{on}, and turn-off time, t_{off}, the conduction voltage, V_{on}, and the leakage current, I_{off}, are part of the switching characteristics of the switching device and have nothing to do with the circuit topology.

When the switch is *off*, the blocking voltage across the switch is V_{off}, which can be represented as a DC voltage source of value V_{off} reflected somehow across the switch during the *off*-state. When the switch is *on*, the current through the switch equals I_{on}; hence, a DC current is needed in series with the switch when it is in the *on*-state. This suggests that when the switch turns *off* again, the current in series with the switch must be diverted somewhere else (this process is known as *commutation* and will be discussed later). As a result, a second switch is needed to carry the main current from the switch being investigated

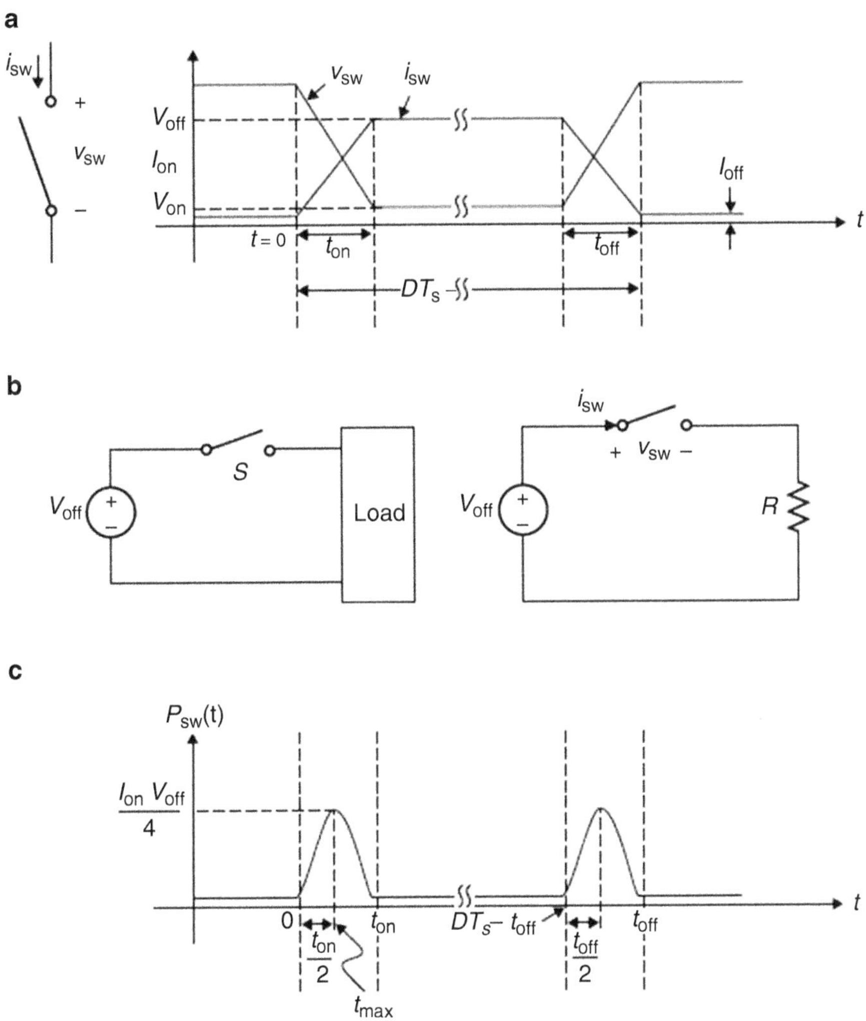

Fig. 2.4 (**a**) Linear approximation of typical current and voltage switching waveforms. (**b**) Circuit implementation. (**c**) Instantaneous power waveform

when it is switched *off*. However, since i_{sw} and v_{sw} are linearly related as shown in Fig. 2.4a, a resistor will do the trick and a second switch is not needed. Figure 2.4b shows a one-switch implementation where *S* is the switch and *R* represents the switched load.

(b) The instantaneous current and voltage waveforms during the transition and conduction times are given as follows:

$$i_{sw}(t)=\begin{cases}\frac{t}{t_{on}}(I_{on}-I_{off})+I_{off} & 0\le t\le t_{on}\\ I_{on} & t_{on}\le t\le T_s-t_{off}\\ -\frac{t-T_S}{t_{off}}(I_{on}-I_{off})+I_{off} & T_s-t_{off}\le t\le T_s\end{cases}$$

$$v_{sw}(t)=\begin{cases}-\frac{V_{off}-V_{on}}{t_{on}}(t-t_{on})+V_{on} & 0\le t\le t_{on}\\ V_{on} & t_{on}\le t\le T_S-t_{off}\\ \frac{V_{off}-V_{on}}{t_{off}}(t-(T_s-t_{off}))+V_{on} & T_S-t_{off}\le t\le T_s\end{cases}$$

It can be shown that if we assume $I_{on}\gg I_{off}$ and $V_{off}\gg V_{on}$, then the instantaneous power, $p(t)=i_{sw}v_{sw}$, can be given as follows:

$$p(t)=\begin{cases}-\frac{V_{off}I_{on}}{t_{on}^2}(t-t_{on})t & 0\le t\le t_{on}\\ V_{on}I_{on} & t_{on}\le t\le T_S-t_{off}\\ -\frac{V_{off}I_{on}}{t_{off}^2}(t-(T_s-t_{off}))(t-T_s) & T_S-t_{off}\le t\le T_s\end{cases}$$

Figure 2.4c shows a plot of the instantaneous power where the maximum power during turn-on and turn-off is $V_{off}I_{on}/4$.

(c) The total average dissipated power is given by:

$$P_{ave}=\frac{1}{T_s}\int_0^{T_s}p(t)dt=\frac{1}{T_s}\left[\int_0^{t_{on}}-\frac{V_{off}I_{on}}{t_{on}^2}(t-t_{on})t\,dt+\int_{t_{on}}^{T_s-t_{off}}V_{on}I_{on}\,dt\right.$$
$$\left.+\int_{T_s-t_{off}}^{T_s}-\frac{V_{off}I_{on}}{t_{off}^2}(t-(T_s-t_{off}))(t-T_s)dt\right]$$

The evaluation of the above integral gives:

$$P_{ave}=\frac{V_{off}I_{on}}{T_s}\left(\frac{t_{on}+t_{off}}{6}\right)+\frac{V_{on}I_{on}}{T_s}(T_s-t_{off}-t_{on})$$

The first expression represents the total switching loss, whereas the second expression represents the total conduction loss over one switching cycle. We notice that as the frequency increases, the average power increases linearly. Also, the power dissipation increases with the increase in the forward conduction current and the reverse blocking voltage.

(d) The maximum power occurs at the time when the first derivative of $p(t)$ during switching is set to zero, i.e.:

$$\left.\frac{dp(t)}{dt}\right|_{t=t_{max}}=0$$

Solving the above equation for t_{max}, we obtain the following values at turn-on and turn-off, respectively:

$$t_{max} = \frac{t_{on}}{2}$$

$$t_{max} = T_s - \frac{t_{off}}{2}$$

Solving for the maximum power, we obtain:

$$P_{max} = \frac{V_{off} I_{on}}{4}$$

Exercise 2.2
For Example 2.2, assume the parameters are given as follows: $V_{on} = 0$ V, $I_{off} = 0$A, $I_{on} = 100$A, $V_{off} = 80$ V, $T_S = 10$ μs, and $t_d = 100$ ns. Determine the average switching and conduction losses, the maximum instantaneous power, and the switch voltage and current at 50 ns.

Answer: 26.67 W, 0 W, 2000 W, 40 V, 50 A

Exercise 2.3
Repeat Example 2.2 for the switching waveforms shown in Fig. E2.3.

Answer:

(c) $P_{ave} = \frac{V_{off} I_{on}}{2T_s} (2\, t_d + t_{fall} + t_{rise})$
(d) $P_{max} = V_{on} I_{on}$

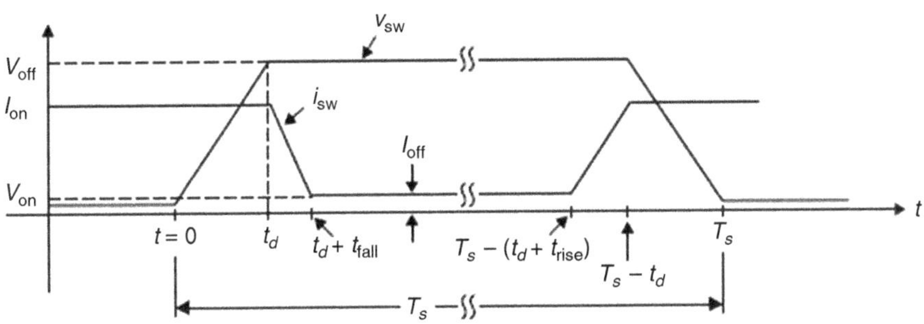

Fig. E2.3 Waveforms for Exercise 2.1

Exercise 2.4
Find the efficiency of the circuit in Fig. 2.1d assuming the switching characteristics for S are shown in Fig. 2.4a with $t_{on} = 100$ ns, $t_{off} = 150$ ns, $T_s = 1$ μs, $I_{off} = 0$, $V_{on} = 0$, and $D = 1$.

Answer: 80%

2.4 Switching Functions and Matrix Representation

Since switches perform the duties of conversion, rectification, inversion, regulation, and so on, it is possible to use the block diagram of Fig. 2.5 as a useful representation of quite many power electronic circuits. This system has n inputs and m outputs that can be either voltages or currents. There are $n \times m$ switches, where each of the n input lines could be connected to the m outputs, resulting in what is known as a *switching matrix*, with the control of switches described by a *switching function*. For illustration purposes, Fig. 2.6a, b shows the switching matrix representation for the single-phase full-bridge and the three-phase full-bridge, respectively. The switching function is a mathematical model for the switching matrix, describing the operation of the switches in the matrix. The literature is full of different techniques and technologies for the generation of switching functions. The switching function approach provides a compact matrix representation for the power converter and serves as a convenient tool for modeling all kinds of power conversion circuits. Due to the fact that no resistors are included in the structure,

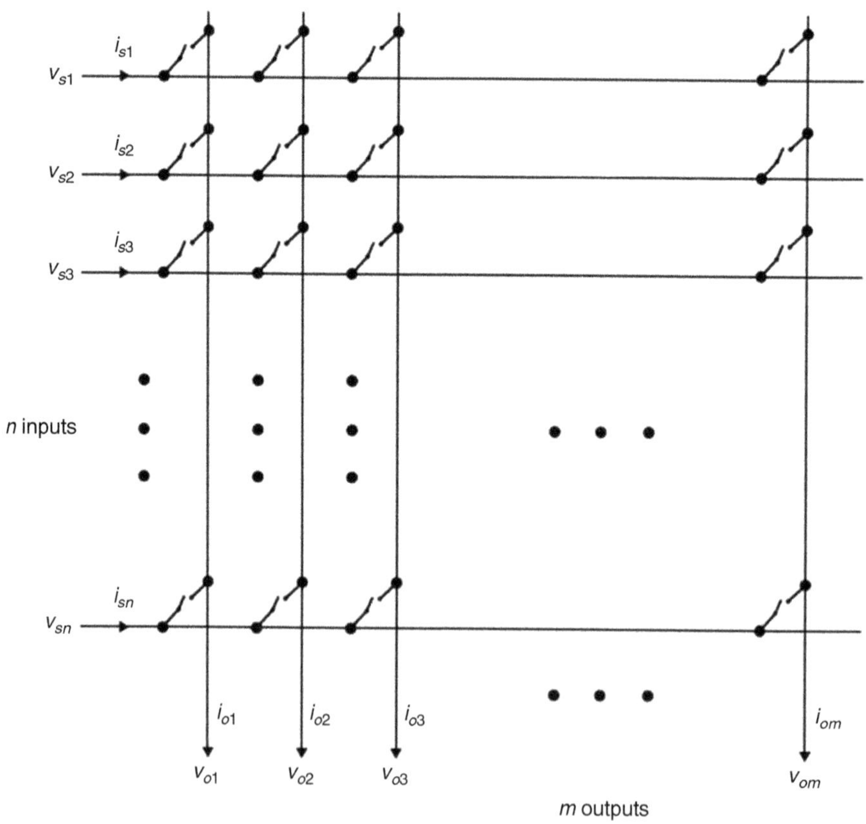

Fig. 2.5 Switching matrix representation

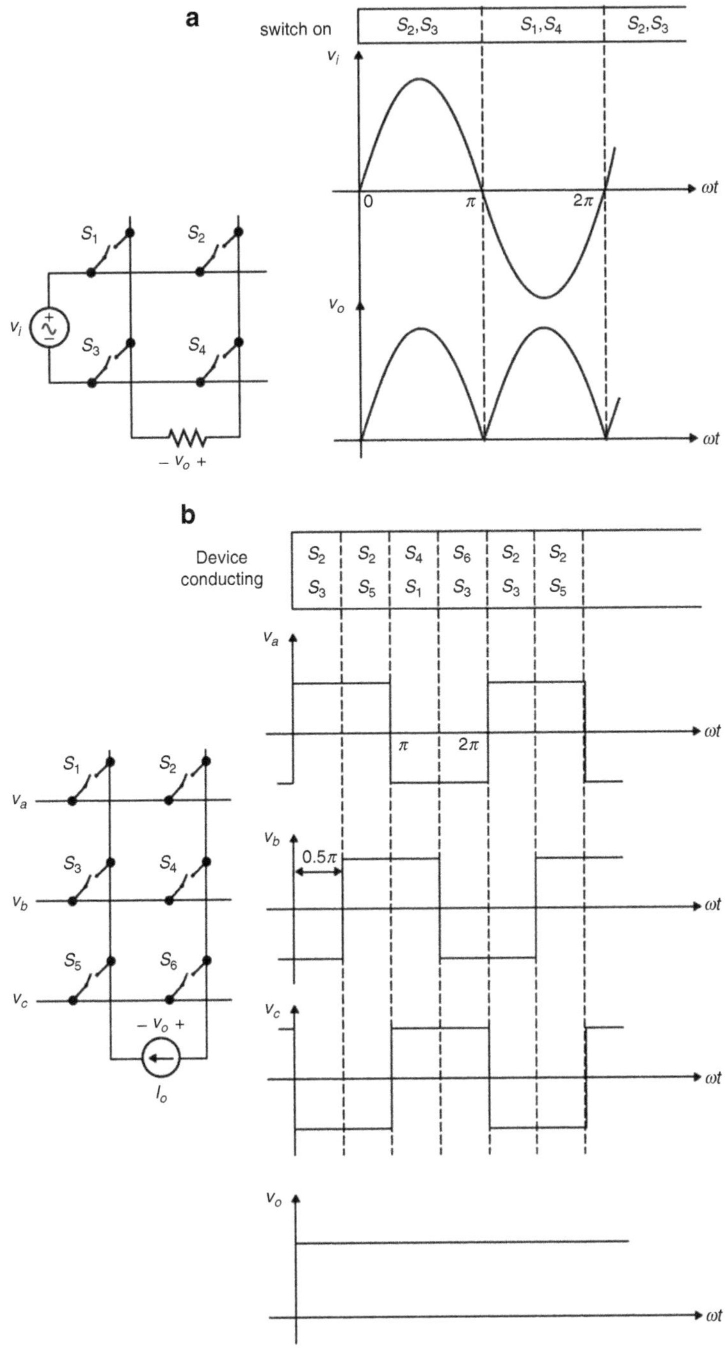

Fig. 2.6 Examples of power electronic circuits. (**a**) Single phase Examples of power electronic circuits. (**b**)Three-phase

there is no power dissipation. In practical systems, additional energy storage elements are present to facilitate energy transfer; however, losses are part of these components and the parts of the switching devices. If we assume ideal storage elements and ideal switches, then it is conceivable to achieve 100% efficiency in the switch matrix arrangement shown in Fig. 2.5. In order to process power bidirectionally, switches must be able to block voltages of either polarities and conduct currents in either direction. One of the most challenging problems in designing and analyzing the system is the design and implementation of the switching network within the storage elements.

Theoretically, since we assume ideal switches and because there are no energy storage elements, the instantaneous input power must be equal to the instantaneous output power. Also there are no restrictions on the form and frequency of the sources. However, only one variable at the source terminal can be fixed (either voltage or current), and the other corresponding terminal variable is determined by the switching function. For example, if $v_{s1}, v_{s2}, \ldots, v_{sn}$ represent fixed voltage sources, their corresponding currents $i_{s1}, i_{s2}, \ldots, i_{sn}$ are determined by the switching as will be illustrated in Example 2.3. Similarly, if the output variables are represented by fixed output current sources $i_{o1}, i_{o2}, \ldots, i_{om}$, their corresponding terminal voltages are also determined by the switching functions. The reverse is also true for both the input and output terminals. Even though the switching matrix and its function determine the type of power conversion in a given power processing circuit, the detailed implementation and terminal characteristics of the source and load sides are also a major part of the power conversion circuit. Normally the energy source is represented by an ideal voltage source that supplies a constant voltage over a wide range of currents. Similarly, the ideal current source can provide a constant current over a wide range of voltages. We will be using both types of energy sources throughout the book. As for the load side, the power conversion circuit must be designed to provide a stable and fixed output that can be represented by either a current source or by a voltage source. If the output is desired to be a current source, the load is connected in series with an inductor, and for a voltage source output, a capacitor is used.

Two important design issues need to be addressed when designing a power electronic switching circuit: (1) the "hardware" or physical implementation of the semiconductor switching matrix and (2) the "software" or logical implementation that guarantees the operation of switching matrix. The hardware implementation of the switching matrix is restricted by Kirchhoff's voltage law (KVL) and Kirchhoff's current law (KCL). Both of these circuit laws must be preserved at all times. KVL states that the algebraic sum of voltage drops around a closed-loop must be zero. Hence, there should not be a switching sequence that will allow two unequal voltage sources to be connected in parallel nor allow a short circuit across a voltage source (please see the reference by Philip Krein). For example, to avoid establishing a short circuit across v_i in Fig. 2.6a, S_1 and S_3 or S_2 and S_4 are not allowed to close simultaneously. Similarly, S_1 and S_3 of Fig. 2.6b cannot be closed simultaneously so that two unequal voltage, v_a and v_b, sources are not connected in parallel.

Because KCL guarantees that the algebraic sum of currents entering a node is zero, no switching sequence should allow two unequal current sources to be connected in series. For example, to avoid establishing an open circuit in series with a current source in Fig. 2.6b, KCL dictates that neither S_1, S_3, and S_5 nor S_2, S_4 and S_6 are allowed to be opened simultaneously. For more discussion about switching matrices and their associated switching functions, readers are encouraged to see the reference by Wood.

Example 2.3

Consider the single-switch, single input power processing circuit given in Fig. 2.7. Assume the source voltage, $v_s(t)$, is a triangular waveform with a peak voltage, V_p, and frequency, $f = 1/T$, as shown in Fig. 2.7b. Assume the switch is ideal and initially was off and its control works in such a way that it toggles every time $v_s(t)$ crosses zero. Use $V_p = 12$ V, $R = 10\ \Omega$, and $T = 1$ ms.

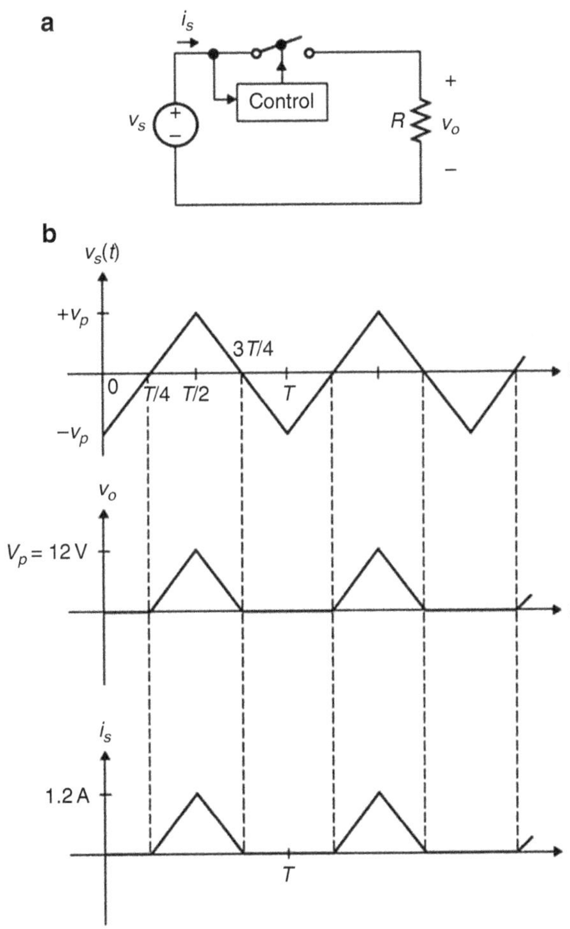

Fig. 2.7 Figures for Example 2.3

(a) Sketch the waveforms for i_s and v_o.
(b) Calculate the average and *rms* values for the output voltage.
(c) Calculate the average input power, average output power, and efficiency.
(d) Repeat parts (a)–(c) by assuming $T = 1$ μs.
(e) Repeat parts (a)–(d) by assuming the switch has 1 volt voltage drop when closed. (a) The output voltage and the source current waveform are shown in Fig. 2.7b.

Solution

(a) The output voltage and the source current waveform are shown in Fig. 2.7b.
(b) The average output voltage is given by:

$$V_o = \frac{1}{T}\int_0^T v_o(t)dt = \frac{1}{T}\left(\frac{1T}{22}V_P\right) = \frac{V_P}{4} = 3 \text{ V}$$

The *rms* is given by:

$$V_{\text{o,rms}} = \sqrt{\frac{1}{T}\int_0^T v_o^2(t)dt} = \sqrt{\frac{1}{T}\left(\int_{T/4}^{T/2}\left(\frac{4V_P}{T}t - V_p\right)^2 dt + \int_{T/2}^{3T/4}\left(-\frac{4V_P}{T}t + 3V_p\right)^2 dt\right)}$$
$$= \frac{V_p}{\sqrt{6}} \approx 4.9 \text{ V}$$

(c) The average input power is calculated from:

$$P_{\text{in}} = \frac{1}{T}\int_0^T i_s(t)v_s(t)dt = \frac{1}{T}\left(\frac{1}{R}\int_{T/4}^{T/2}\left(\frac{4V_P}{T}t - V_p\right)^2 dt + \int_{T/2}^{3T/4}\frac{1}{R}\left(-\frac{4V_P}{T}t + 3V_p\right)^2 dt\right)$$
$$= \frac{1}{4RV_P}\left(\frac{2V_P^3}{3}\right) = \frac{V_P^2}{6R} = 2.4 \text{ W}$$

and the average output power:

$$P_{out} = \frac{1}{T}\int_0^T i_o(t)v_o(t)dt = \frac{V_P^2}{6R} = 2.4 \text{ W}$$

and the efficiency is:

$$\eta = P_{out}/P_{in} = 100\%$$

(d) Same as above (because the results are independent of T).
(e) The average output voltage can be approximated by the following integration:

$$V_{\text{o,ave}} = \frac{1}{T}\int_0^T v_o(t)dt = \frac{1}{T}\left(\int_{T/4}^{T/2}\left(\frac{4V_P}{T}t - V_p - 1\right)dt + \int_{T/2}^{3T/4}\left(-\frac{4V_P}{T}t + 3V_p - 1\right)dt\right)$$
$$= \frac{1}{4}(v_p - 2) = 25\ V$$

And the *rms* voltage is:

$$V_{\text{o,rms}} = \sqrt{\frac{1}{T}\int_0^T v_o^2(t)dt} = \sqrt{\frac{1}{T}\left(\int_{T/4}^{T/2}\left(\frac{4V_P}{T}t - V_p - 1\right)^2 dt + \int_{T/2}^{3T/4}\left(-\frac{4V_P}{T}t + 3V_p - 1\right)^2 dt\right)}$$
$$= \sqrt{\frac{1}{6V_P}\left((V_P - 1)^3 + 1\right)} \approx 4.3\ \text{V}$$

It can be shown that the average input power is:

$$\begin{aligned} P_{\text{in}} &= \frac{1}{T}\int_0^T i_s(t)v_s(t)dt \\ &= \frac{1}{T}\left[\int_{T/4}^{T/2}\left(\frac{4V_P}{T}t - V_p - 1\right)\frac{1}{R}\left(\frac{4V_P}{T}t - V_p - 1\right)dt \right. \\ &\quad \left. + \int_{T/2}^{3T/4}\left(-\frac{4V_P}{T}t + 3V_p - 1\right)\frac{1}{R}\left(-\frac{4V_P}{T}t + 3V_p - 1\right)dt\right] \\ &= \frac{V_P(2V_P - 3)}{12R} = 2.1\ \text{W} \end{aligned}$$

and the average output power:

$$\begin{aligned} P_{\text{out}} &= \frac{1}{T}\int_0^T i_o(t)v_o(t)dt \\ &= \frac{1}{T}\left(\frac{1}{R}\int_{T/4}^{T/2}\left(\frac{4V_P}{T}t - V_p - 1\right)^2 dt + \int_{T/2}^{3T/4}\frac{1}{R}\left(-\frac{4V_P}{T}t + 3V_p - 1\right)^2 dt\right) \\ &= \frac{1}{6RV_P}\left((V_P - 1)^3 + 1\right) \approx 1.85\ \text{W} \end{aligned}$$

Resulting in efficiency of $\eta = P_{\text{out}}/P_{\text{in}} \times 100\,\% = 1.85/2.1 \times 100\,\% \approx 88.2\%$.

Exercise 2.5
Repeat Example 2.3 for (a) $v_s = 12 \sin \omega t$, and (b) v_s is a square wave with peak-to-peak voltage equals ± 12 V.

Answer:

(a) −3.82 V, 6 V, 3.6 W, 3.6, 100%; same; −3.32 V, 5.37 V, 3.22 W, 2.89 W, 89.7%
(b) −6 V, 8.5 V, 7.2 W, 100%; same;−5.5 V, 7.8 V, 6.6 W, 6.05 W, 91.7%

Example 2.4
One of the growing applications of the power electronics is light dimming, where the average power applied to the light is varied to change its brightness. Figure 2.8a shows a simplified circuit for conventional incandescent lamp with dimming functionality. Figure 2.8b shows the typical input and output voltage waveforms for such dimming circuit of Fig. 2.8a. By varying the phase angle α, it can be shown that average power delivered to the lamp may be varied from 0% to 100% of the available rated power. Hence dimming is achieved theoretically for the entire range. Assume that the incandescent lamp acts like constant pure resistance at it maximum rated power, with $v_s(t) = V_p \sin \omega t = 120\sqrt{2} \sin(120\pi t)$ V, and the maximum rated lamp power, $P_{\mathrm{L,max}} = 100$ W.

(a) Discuss the type of switch is required to input this circuit functionality.
(b) Determine the equivalent rated lamp resistance.
(c) Derive the expression for the rms value for the output voltage, ($V_o(t)$), as a function of the control angle α. What are the rms values for $\alpha = \frac{\pi}{3}, \frac{\pi}{2}, \frac{3\pi}{2}$ and π.
(d) Derive the expression for the average output power, $P_{\mathrm{o,ave}}$, as a function of α. What are the values at for $\alpha = \frac{\pi}{3}, \frac{\pi}{2}, \frac{3\pi}{2}$ and π.
(e) The percentages of the power delivered to the lamp with respect to its maximum rated value for $\alpha = \frac{\pi}{3}, \frac{\pi}{2}, \frac{3\pi}{2}$ and π.

Solution

(a) Since the load (incandescent lamp) is resistive, then the load current flows in both directions given the input source is sinusoidal. This means the switch should allow bidirectional current flow. Moreover, when the switch is not conducting, it should be able to block positive and negative voltages. Fortunately, there is one device in the market that can support bidirectional current flow and block the voltages bidirectionally as well. This device is known as the Triac and is discussed in the next section.
(b) Since the maximum rated power is given and load assumed resistive, the equivalent load resistance, R_{L}, is obtained from:

$$P_{\mathrm{o,max}} = \frac{V^2_{\mathrm{o,rms,max}}}{R_L}$$

where $V_{\mathrm{o,rms,max}}$ is the maximum available rms output voltage across the lamp.
So the load resistance is given by:

$$R_{\mathrm{L}} = \frac{V^2_{\mathrm{o,rms,max}}}{P_{\mathrm{o,max}}} = \frac{V^2_{s,\mathrm{rms}}}{P_{\mathrm{o,max}}} = \frac{120^2}{100} = 144\ \Omega$$

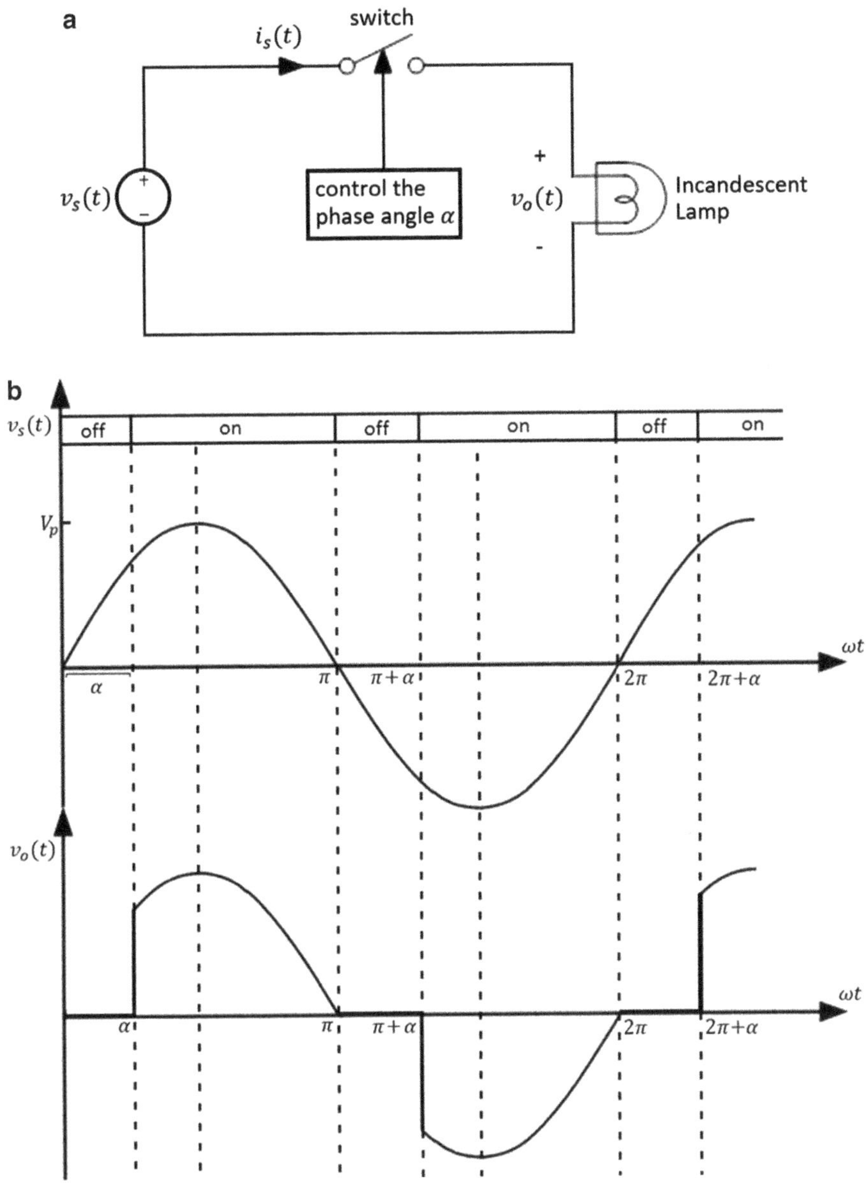

Fig. 2.8 (**a**) Simple control circuit used in conventional light dimming. (**b**) Typical switch waveforms

(c) The output voltage $v_s(t)$ is obtained by inspection, as follows:

$$V_o(t) = \begin{cases} 0 & 0 < \omega t \leq \alpha \quad \text{and} \quad \pi < \omega t \leq \pi + \alpha \\ v_s(t) & \alpha < \omega t \leq \pi \quad \text{and} \quad \pi + \alpha < \omega t \leq 2\pi \end{cases}$$

Because of the symmetry of $V_o(t)$, it can be shown that its rms is given by the following equation:

$$V_{o,\mathrm{rms}} = \sqrt{\frac{1}{\pi}\int_{\alpha}^{\pi} V_s^2(t)d\omega t} = \sqrt{\frac{1}{\pi}\int_{\alpha}^{\pi} V_p^2 \sin^2\omega t d\omega t}$$

Students are invited to verify that $V_{o,\mathrm{rms}}$ is given by:

$$V_{o,\mathrm{rms}} = V_p\sqrt{\frac{1}{2}\left(1 - \frac{\alpha}{\pi} + \frac{\sin 2\alpha}{2\pi}\right)}$$
$$= V_{s,\mathrm{rms}}\sqrt{1 - \frac{\alpha}{\pi} + \frac{\sin 2\alpha}{2\pi}}$$

For $\alpha = \frac{\pi}{3}, \frac{\pi}{2}, \frac{3\pi}{2}$ and π, we have:

$V_{o,\mathrm{rms}}\left(\frac{\pi}{3}\right) = 107.632\ \mathrm{V}$	$V_{o,\mathrm{rms}}\left(\frac{\pi}{2}\right) = 84.85\ \mathrm{V}$
$V_{o,\mathrm{rms}}\left(\frac{3\pi}{2}\right) = 84.85j\ \mathrm{V}$	$V_{o,\mathrm{rms}}(\pi) = 0\ \mathrm{V}$

(d) The average output power, $P_{o,\mathrm{ave}}$, is given by:

$$P_{o,\mathrm{ave}} = \frac{V_{o,\mathrm{rms}}^2}{R_L}$$
$$= \frac{V_{s,\mathrm{rms}}^2}{R_L}\left(1 - \frac{\alpha}{\pi} + \frac{\sin 2\alpha}{2\pi}\right)$$

Notice when $\alpha = 0$, no dimming, the output power is at its maximum ratio. For $\alpha = \frac{\pi}{3}, \frac{\pi}{2}, \frac{3\pi}{2}$ and π, the average output power is given by:

$P_{o,\mathrm{ave}}\left(\frac{\pi}{3}\right) = 80.45\ \mathrm{W}$	$P_{o,\mathrm{ave}}\left(\frac{\pi}{2}\right) = 50\ \mathrm{W}$
$P_{o,\mathrm{ave}}\left(\frac{3\pi}{2}\right) = -50\ \mathrm{W}$	$P_{o,\mathrm{ave}}(\pi) = 0\ \mathrm{W}$

(e) The percentage of the power delivered of the lamp with respect to the maximum rated value is given by:

$$\%P = \frac{P_{o,ave}}{P_{o,\max}} \times 100\% = \left(1 - \frac{\alpha}{\pi} + \frac{\sin 2\alpha}{2\pi}\right) \times 100\%$$

For $\alpha = \frac{\pi}{3}, \frac{\pi}{2}, \frac{3\pi}{2}$ and π, the average output power is given by:

$\%P\left(\frac{\pi}{3}\right) = 80.45\%$	$\%P\left(\frac{\pi}{2}\right) = 50\%$
$\%P\left(\frac{3\pi}{2}\right) = -50\%$	$\%P(\pi) = 0\%$

2.5 Types of Switches

To implement a given switching function, the ideal switches in the switching matrix must be realized by practical power devices. Functionally speaking, any switch must have the ability to conduct current and/or the ability to block voltage by means of control signals. When in the *on*-state, the current conduction state is the task under consideration, and when in the *off*-state, the voltage blocking state is what we are considering. Practical switches have limitation in their conduction current and limitation in their voltage blocking. Since the switch current can flow in the forward, reverse, or both directions and the voltage can be blocked in the forward, reverse, and both directions, there are nine different combinations of current carrying and voltage blocking directions. Four of these combinations are duplicates of the other four, i.e., forward current carrying and reverse voltage blocking is the same as reverse current carrying and forward voltage blocking. As a result, switches are classified into five general types of restricted switches as shown in Table 2.1 (for more details, please see the reference by Philip Krein).

Table 2.1 Types of semiconductor switches, their controllability features, and their possible switch implementation

Type	Current flow	Voltage blocking	Switch implementation
1	Forward	Reverse	 Diode
2	Forward	Forward	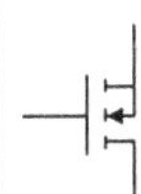 Transistor
3	Forward	Bidirectional	 SCR
4	Bidirectional	Forward	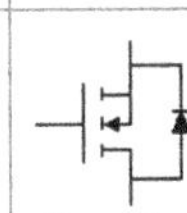 Transistor With flyback Diode
5	Bidirectional	Bidirectional	 or 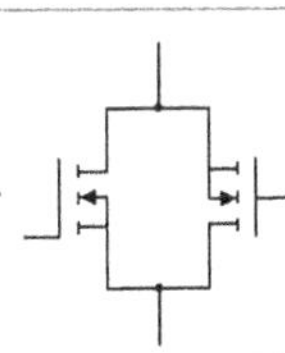 Triac

1. *Forward current carrying and reverse voltage blocking*. This is an uncontrolled device with unidirectional current flow. Uncontrolled device means the turn-*on* and turn-*off* are not controlled by an external control signal but rather by the power circuit itself. An example of this type of switch is the diode since it carries the current only in the forward direction when the anode-cathode voltage is positive. The diode is an uncontrolled device since no external control signal can be applied to initiate *carrying* the forward current or to initiate *blocking* the reverse voltage. The controlling voltage is derived from either the source or the load or both.
2. *Forward current carrying and forward voltage blocking*. This is a controlled device with unidirectional current flow. Such a switch should be able to carry the current in the forward direction and block voltage in the forward direction. Of course the diode is unable to block a forward voltage across it. Another type of switch is needed that has an external control signal that allows the device to decide whether the forward current to flow or not even when a forward voltage is applied. This is the same as implying the switch performs a current conduction delay function. An example of such a switch is the transistor which is able to block voltage in the forward direction when the base (gate) current is absent.
3. *Forward current carrying and bidirectional voltage blocking*. This switch can block current flow in both directions (i.e., supports forward and reverse voltage blocking) but carries the current only in the forward direction. An example is the silicon-controlled rectifier (SCR), to be studied later in this chapter.
4. *Bidirectional current carrying and forward voltage blocking*. This switch is similar to type 2, except the current can flow bidirectionally. Hence, the implementation is a transistor with a diode connected as shown in the figure in Table 2.1 to allow reverse current flow. The diode is known as flyback or body diode since it picks the current in the reverse direction. This switch can carry the current in the forward direction through the transistor and in the reverse direction through the flyback diode. The base (gate) signal is used to allow the switch whether to carry the current in the forward direction or to be in the voltage blocking state, because of the presence of the diode the switch is unable to block reverse voltage.
5. *Bidirectional current carrying and bidirectional voltage blocking*. This switch is the most general power electronic switch and is similar to type 3 with additional characteristics that it supports current flow in both directions. Unlike type 4 where a flyback diode is added to allow reverse current direction, here the forward and reverse current flow must be controlled. An example of this type is known as two MOSEFT on a single device known as the TRIAC (TRIod AC), which is simply two SCRs connected in parallel and in the opposite direction as shown in Table 2.1.

2.6 Available Semiconductor Switching Devices

In this section, the emphasis will be on the *i-v* switching characteristics of these devices and their corresponding power ratings and possible applications. Selecting the most appropriate device for a given application is not an easy task, requiring knowledge about the device characteristics, its unique features, innovation, and engineering design experience. Unlike low-power (signal) devices, power devices are more complicated in structure, driver design, and understanding of their operational *i-v* characteristics. This knowledge is very important for power electronics engineers to design circuits that will make these devices close to ideal. In this section, we will discuss briefly two broad families of power devices:

i. ***Bipolar and unipolar devices:***

   1. Power diodes
   2. Bipolar junction transistors
   3. Insulated gate bipolar transistors (IGBTs)
   4. Metal oxide semiconductor field-effect transistors (MOSFETs)

ii. ***Thyristor-based devices:***

   1. Silicon-controlled rectifiers (thyristors)
   2. Gate turn-off (GTO) thyristors
   3. Triode AC switch (TRAIC)
   4. Static induction transistors (SITs) and static induction thyristors (SITHs)
   5. MOS-controlled thyristors (MCT)

2.6.1 Bipolar and Unipolar Devices

2.6.1.1 The Power Diode

The power diode is a two-terminal device composed of a *p-n* junction and whose turn-*on* state cannot be controlled (uncontrolled switch). The diode turn-*on* and turn-*off* is decided by the external circuitry: a positive voltage imposed across will turn it *on* and a negative current imposed through it turns it *off*.

The symbol and the practical and ideal *i-v* characteristic curves of the power diode are shown in Fig. 2.9a–c, respectively. When in the conduction state, the forward voltage drop, V_F, is typically 1 V or less. The diode current increases exponentially with the voltage across it, i.e., a small increase in V_F produces a large increase in I_F (see Problem 2.5). When in the reverse bias region, the device is in the *off*-state, and only a reverse saturation current, I_s, exists in the diode (also known as leakage current). The breakdown voltage, V_{BR}, is the maximum inverse voltage the diode is capable to block. V_{BR} is a diode rated parameter with values up to a few kilovolts, and in normal operation, the reverse voltage should not reach V_{BR}. Zenor diodes are special diodes in which the breakdown voltage is approximately 6 – 12 V, controlled by the doping process.

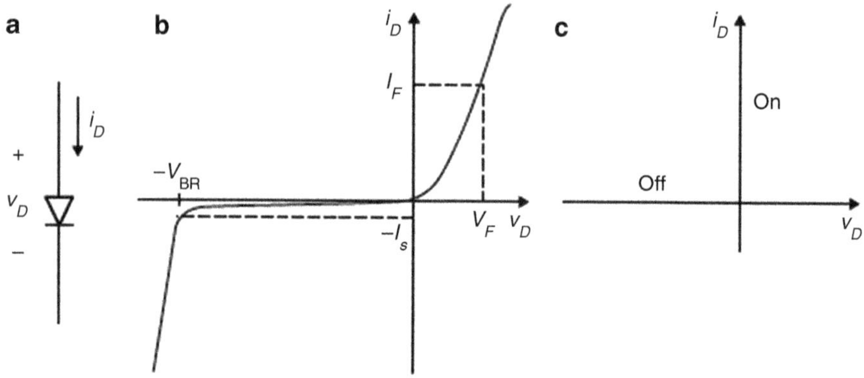

Fig. 2.9 Power diode: (**a**) Circuit symbol, (**b**) practical, and (**c**) ideal switching characteristics

In power circuits, power diodes have two important features:

1. Power-handling capabilities including the forward current carrying and the reverse voltage blocking
2. Reverse recovery time (t_{rr}) at turn-*off*

The parameter t_{rr} is very significant because the speed of turning off the diode could be large enough to affect the operation of the circuit. At turn-*on*, the delay time is normally insignificant compared to the transient time in power electronic circuits.

Broadly speaking, two types of power diodes are available:

1. The bipolar diode, which is based on the p-n semiconductor junction. Depending on the applications, the bipolar diode can be either the standard line frequency type or the fast recovery high frequency type with its t_{rr}varying between 50 ns and 50 μs, respectively. Typical voltage drop is 0.7 – 1.3 V and reverse voltage blocking of 3 kV and forward current of 3.5 kA.
2. The Schottky diode, which is based on the metal-semiconductor junction. It has a relatively lower forward drop voltage than the bipolar (about 0.5 V or less). Unlike the bipolar diode whose current conduction depends on the minority and majority carriers, the Schottky diode current depends mainly on the majority carriers. Hence, Schottky diodes have no reverse recovery time delay. Finally, the *i-v* characteristics of the Schottky diode are similar to the i-v characteristics of the bipolar diode. Unlike bipolar diodes, Schottky diodes generate less excess minority carriers than the majority carriers; hence its current is primarily generated due to the drift of majority carriers. Due to its large leakage circuit, it is normally used at the output stage in low-voltage, high-current *dc* power supplies.

To study the reverse recovering characteristics of the diode, we consider the circuit of Fig. 2.10a, which has a typical diode current waveform during turn-*off* as shown in Fig. 2.10b. For simplicity we assume the switch is ideally switched on and off for the duty cycle d and $(1 - d)$, respectively, with T_S as the switching period.

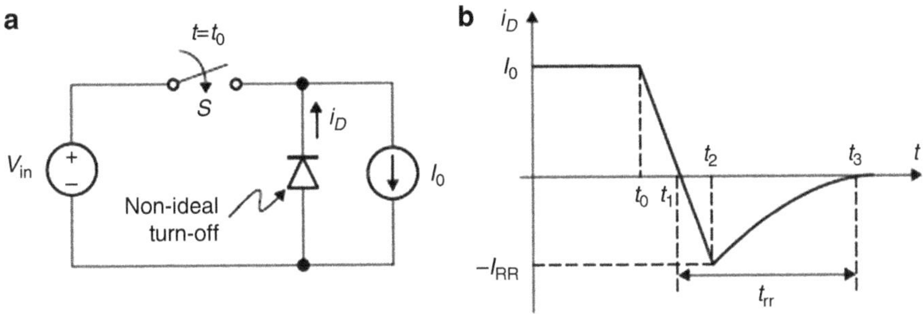

Fig. 2.10 Typical diode switching characteristics (**a**) Switching circuit with *S* closed at $t = t_0$ (**b**) diode current

Such a circuit arrangement is normally encountered in switch-mode *dc*-to-*dc* converters with the switch replaced by either a BJT or a MOSFET. Initially we assume the diode was conducting with forward current I_0. At $t = t_0$, the switch is turned *on*, forcing the diode to turn off due to the *dc* input voltage V_{in}. The turn-*on* characteristics of the diode are simpler to deal with since turn-on only involves charging the diode depletion capacitor. The diode's forward conduction begins when its depletion capacitor has been charged. At turn-*on*, the diode voltage drop is larger than the normal forward drop during conduction. This transient voltage exists due to the large value of diode resistance at turn-*on*. This is why during the diode's turn-*on* time the power dissipation is much larger than the case when it is in the steady-state conduction state. The diode turn-*off* characteristics are more complex since significant stored charges exist in the body of the p-n junction and at the junction.

As shown in Fig. 2.10b, during turn-*off*, the diode current linearly decreases from its forward value, I_0, at $t = t_0$, to zero at $t = t_1$ and then continues to go negative until it reaches a negative peak value at $t = t_2$ known as the *reverse recovery current*, I_{RR}, at which the current starts to rise exponentially to zero at $t = t_3$. The time intervals can be broken down as follows: between t_0 and t_1, the diode current is positive and the diode forward voltage is small. Hence we assume the rate of change of diode current is constant and is determined by the total circuit inductance in series with the diode. In our example, since the switch is ideal, di_D/dt and I_{RR} are limited by the diode and lead to parasitic inductances. At $t=t_1$, the current becomes zero and the diode should begin turning off by supporting reverse voltage. But because of the excess minority carriers in the p-n junction that need to be removed before the diode's reverse voltage begins to rise, the diode remains in the conduction state for longer time, i.e., until $t = t_2$.

The delay between t_1 and t_2 is due to the minority carriers in the depletion region, whereas the delay between t_2 and t_3 is caused by the charge stored in the bulk of the semiconductor material. At $t = t_3$, all charge carriers are removed, causing the device to be fully switched *off*. The time it takes from the moment the diode current becomes negative until it becomes zero is known as the *reverse recovery time*, t_{rr}, as shown in Fig. 2.10b.

Between t_2 *and* t_3, the junction behaves like a capacitor whose voltage goes from zero to the reverse voltage via a charging current in this interval. The total charge carriers that cause a negative diode current flow when it is turned *off* constitute the *reverse recovery charge*, Q_{rr}, and can be expressed in terms of I_{RR} and t_{rr}.

The time between t_2 and t_3 could be very short when compared to t_{rr}, resulting in high *di*/*dt*. The ratio between $(t_3 - t_2)$ and t_{rr} is a parameter that defines what is known as *diode snappiness*. The smaller this ratio, the quicker the diode recovers its reverse blocking voltage, resulting in what is known as *fast recovery* or *hard recovery* diode. Meanwhile, a diode with a high ratio of $(t_3 - t_2)$ to t_{rr} takes a relatively long time to bring its forward current to zero from its negative peak value. These diodes are known as *soft-recovery* diodes. The fast-recovery diodes have high *di*/*dt* and normally experience oscillation at turn-off. Standard or general-purpose diodes have soft recovery time and are used in low-speed applications where the frequency is less than a few kHz.

In general, an attempt to reduce either t_{rr} or I_{RR} will result in the increase of the other. The forward recovery voltage limits the efficiency because of device stresses and higher switching losses, and t_{rr} limits the frequency of operation. Power electronics engineers should keep in mind that the transient voltage at turn-on and the transient current at turn-off might affect the external circuitry and cause unwanted stresses. External snubber circuits are added to suppress these transient values.

Example 2.5

Consider the switching circuit shown in Fig. 2.11 by modeling the circuit parasitic inductance as a lumped discrete value, L_s. Assume the switch was opened for a long time before being turned-*on* at $t = t_0$. Assume the same diode switching characteristics of Fig. 2.10b, except that it is a fast-recovery diode with $t_3 - t_2 \approx 0$. Derive the expressions for I_{RR} and the peak switch current in terms of the diode reverse recovery time.

Solution

While the diode is in the conduction state, its forward current is I_0. When the switch is closed at $t = t_0$, the diode voltage remains zero, and its current is given by:

$$i_D = I_0 - i_s$$

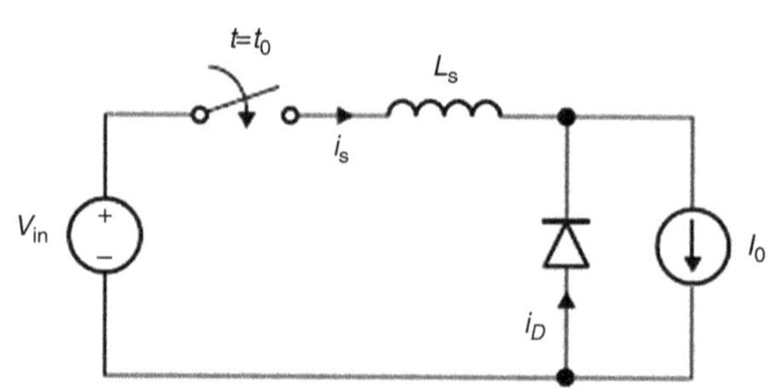

Fig. 2.11 Diode switching circuit with parasitic inductor

and i_s for $t \geq t_0$ is given by:

$$i_s(t) = \frac{V_{in}}{L_s}(t - t_0) \qquad t_0 \leq t \leq t_2$$

At $t = t_1$ the diode current becomes zero and i_s becomes I_0. Hence, the interval $t_1 - t_0$ is given by:

$$t_1 - t_0 = \frac{I_0 L_s}{V_{in}}$$

Since $t_3 - t_2 \approx 0$, then $t_2 - t_1 \approx t_{rr}$ and I_{RR} is given by:

$$I_{RR} = \frac{V_{in}}{L_s} t_{rr}$$

The peak switch current occurs at $t = t_2$ when $i_D = -I_{RR}$ and is given by:

$$I_{s,\ peak} = \frac{V_{in}}{L_s} t_{rr} + I_0$$

At this point the diode is turned *off* and the peak inductor current is higher than the load current I_0. Since the load is highly inductive, its value cannot increase suddenly by the amount $V_{in}t_{rr}/L_s$ without creating high reverse voltage across the diode. As a result, a snubber circuit must be added across the diode to dissipate excess stored energy in the inductor.

Another important point to make is that when L_s becomes very small, a very large reverse recovery current occurs that could damage the diode and cause large switching losses.

The peak value of the reverse current,$-I_{RR}$, is a very important parameter, and it can be less than, equal to, or larger than the forward current I_0, depending on the external circuitry connected to the diode and the diode parasitic inductance. The fast-recovery diodes have low recovery time, normally less than 50 ns, and are used in applications such as high-frequency *dc-dc* converters where the speed of recovery is critical.

Exercise 2.6

Consider the dynamic switching characteristics for a diode during turn-off to be approximated as shown in Fig. E2.6; assume $V_m = 3.5$ V, $V_F = 1$ V, $t_f = 90$ ns, $I_F = 10$ A, and $t_m = 45$ ns.

(a) Determine and sketch the instantaneous power $p(t)$ across the diode for $0 < t \leq t_f$.

(b) What is the maximum power dissipated in the diode during turn-on interval.

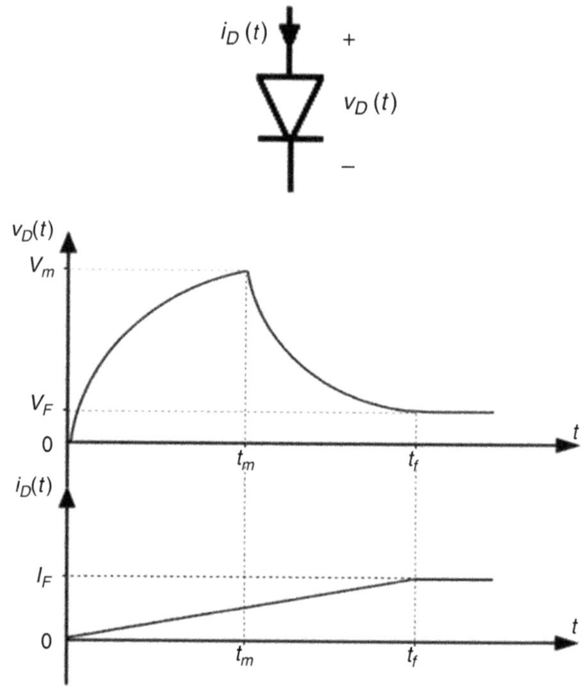

Fig. E2.6 Dynamic switching waveform for Exercise 2.6

2.6.1.2 The Bipolar Junction Transistors (BJTs)

The schematic symbol and the i-v characteristics for the bipolar junction transistors (BJTs) are shown in Fig. 2.12a–c, respectively. It is a two-junction, three-terminal device with the minority carriers being the main conducting charges. The switching speed of the BJT is much faster than thyristor-type devices. A major drawback is its *second breakdown* problem.[2]

Unlike the SCR, the BJT is turned on by constantly applying a base signal. Power BJTs have two different properties when compared to the low-power BJTs and logic transistor: large blocking voltage in the *off*-state and high forward current carrying capabilities in the *on*-state. BJT power ratings reach up to 1200 V and 500 A. These high rating values suggest that the power BJT's driving circuits are more complicated.

Because the BJT is a current-driven device, the larger the base current, the smaller β_{forced}[3] and the deeper the transistor is driven into saturation. In saturation, the collector-emitter voltage is almost constant and the collector current is

[2]Normally, the first breakdown voltage refers to the avalanche breakdown caused by the increase in the reverse bias voltage, which can be nondestructive. The second breakdown voltage is a destructive phenomenon caused by localized overheating spots in the device.

[3]$\beta forced$ is defined as the ratio I_C/I_B when the transistor is operating in the saturation mode.

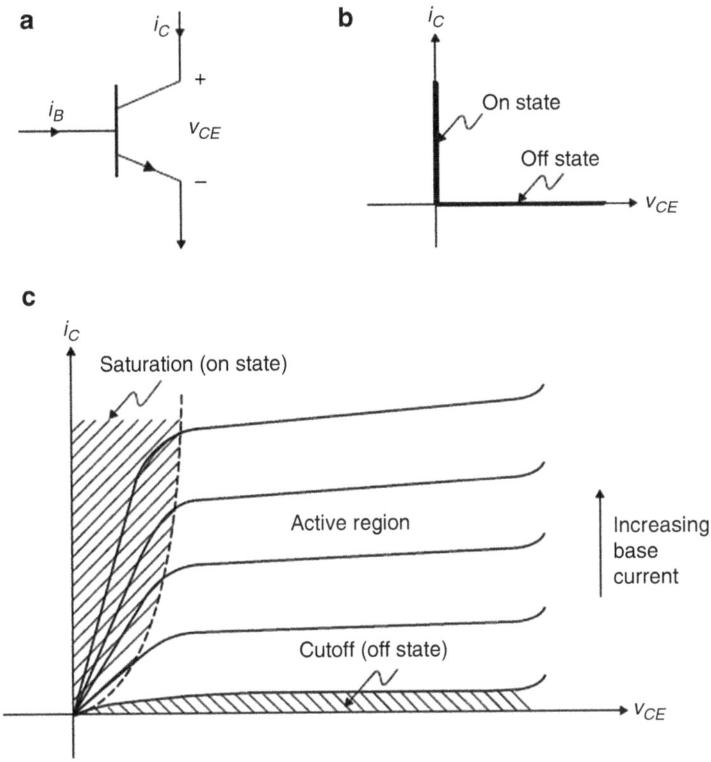

Fig. 2.12 BJT switching characteristics: (**a**) npn transistor, (**b**) ideal i-v characteristics, (**c**) practical i-v characteristics

determined largely by the external circuit to the switch. It is sometimes useful to define what is known as an *overdrive factor*, which gives a measure to how deep in saturation the transistor is. For example, if the transistor is at the edge of saturation with given base current I_B, then with an overdrive factor of 10, the base current becomes $10I_B$, and the transistor becomes deeper in saturation.

Since the base thickness is inversely proportional to the current gain β, Darlington-connected BJT pairs have been developed in which the collector of two devices is joined and the base of the first is connected to the emitter of the second, as shown in Fig. 2.13. This arrangement results in an overall gain that equals approximately the product of the individual βs of the two transistors. Transistor Q_1 serves as an auxiliary transistor, which provides the base current to turn on Q_2. Because there is a high current gain, a smaller base current to Q_2 is needed to drive the power Darlington pair. Darlington power transistors are widely used in UPSs and various AC and DC motor drives up to hundreds of kilowatts and tens of kilohertz. A modern Darlington pair has ratings up to 1.2 kV with current reaching up to 800 A and operating frequency up to several kilohertz.

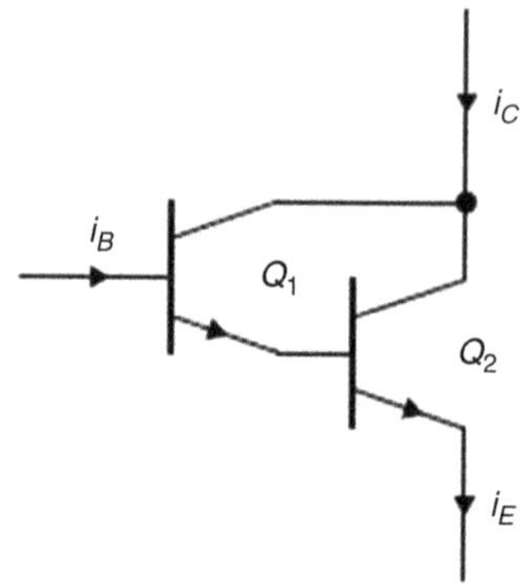

Fig. 2.13 Darlington-connected BJT

Triple Darlingtons are available, in which the current gain becomes proportional to the product of the three individual current gains of the transistors. To turn *off* the Darlington switch, all base currents must become zero, resulting in slower switching speed compared to a single transistor. Also the overall collector-emitter saturation voltage $V_{CE,\,sat}$ is higher than for a single transistor, as will be illustrated in Exercise 2.5.

There are three regions of operation: saturation, active, and cutoff. As a power switch, the BJT must operate either in the saturation region (*on*-state) or in the cutoff region (*off*-state). The third state is when the transistor is in the linear region and is used as an analog amplifier.

To investigate the turn-on and turn-off processes, we consider a simple inverter circuit shown in Fig. 2.14a with its switching waveforms shown in Fig. 2.14b. The voltage v_I is the base driving voltage with positive polarity, V_1, to push positive current into the base, $I_{B1} = (V_1 - V_{BE})/R_B$, and a negative polarity, V_2, to quickly discharge the base current, $I_{B2} = -(V_2 + V_{BE})/R_B$. At time $t = t_0$, V_1 is applied with positive *dc* voltage,+V_1, because it takes time to charge the internal depletion capacitor to turn the junction *on* at $V_{BE} = 0.7$ V; a delay time, t_d, elapses before the collector current starts flowing. After the junction is turned *on*, the collector starts flowing exponentially through R_B and the emitter-base junction capacitor. During this period, the minority carriers are being stored in the transistor base region. The collector current increases until it reaches its maximum saturated value, I_{on}, determined by:

$$I_{on} = \frac{V_{in} - V_{CE,\,sat}}{R}$$

The time it takes for the collector current to rise from 10% to 90% of its maximum value, I_{on}, is called the rise time. For simplicity, Fig. 2.14b shows the rise time from $I_C = 0$ to I_{on}. The total switching *on*-time is given by $t_{on} = t_d + t_r$. To turn off the transistor, a negative (or zero) base voltage is normally applied, resulting in a base current I_{B2} being *pulled* out of the base as shown in Fig. 2.14b. The collector current does not start decreasing until sometime later after the stored saturation charge in the base has been removed. This time is called the storage

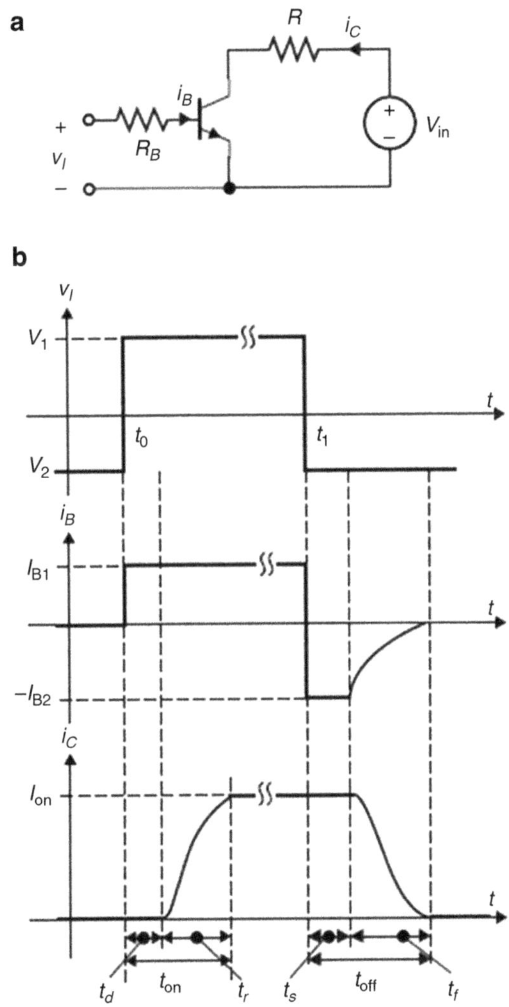

Fig. 2.14 Switching characteristics for the BJT. (**a**) Circuit. (**b**) Switching waveforms

time, t_s; and it is normally longer than the delay time, t_d, and usually determines the limiting range of the switching speed. If the base voltage is not negative (i.e., in the absence of I_{B2}), the entire base current must be removed through the process of recombination.

To turn on the BJT, a large current must be pushed to the base. This base current must be large enough to saturate the transistor. In saturation region, both base-emitter and base-collector junctions are forward biased. This is why the BJT is known as a current-driven device. When operated in the saturation region, $I_B > I_C/\beta$ where β is the *dc* current gain. In this region a new *dc* current gain β is defined to indicate the depth of the transistor saturation. The saturation collector-emitter voltage is given as $V_{CE,\,sat}$, and β_{forced} is defined as

$$\beta_{\text{forced}} = \frac{I_c}{I_B}$$

where I_c and I_B are the collector and base currents in saturation, respectively, and $\beta_{\text{forced}} < \beta$. The smaller β_{forced}, the deeper the transistor is driven into saturation. Typically, β_{forced} can be as low as 1. Ideally, $V_{CE,\,sat} = 0$, but in practice, this value varies between 0.1 and 0.6 V, depending on how deep in saturation the device is driven. The new ratio of collector to emitter currents is much smaller than the case when the transistor is operated in the active mode. At the edge of saturation, $\beta_{\text{forced}} = \beta$.

The total power dissipation in the transistor is obtained by adding the input power supplied by the collector current and the input power supplied by the base current; hence, the total power dissipation is defined as follows:

$$P_{\text{diss}} = V_{CE} I_C + V_{BE} I_B$$

Exercise 2.7

Consider the transistor circuit shown in Fig. E2.7. Assume the transistor is operating in the saturation region with $V_{CE,\,sat} = 0.5$ V, $V_{BE} = 0.75$ V, $V_D = 0.7$ V, and $D = 0.5$. Sketch i_B, v_{CE}, and i_D. Determine the overall efficiency of the circuit.

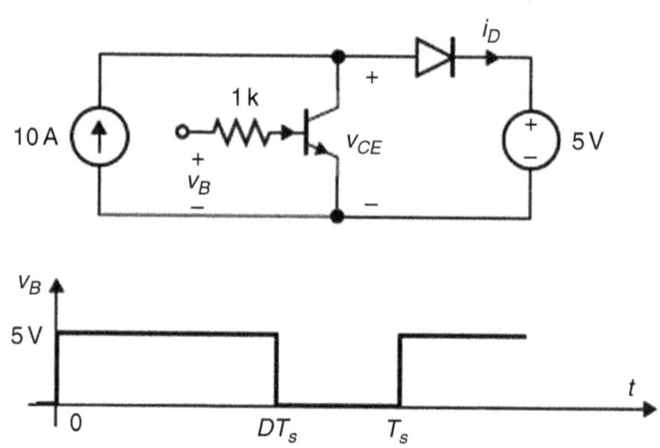

Fig. E2.7 Switching circuit and its waveforms for Exercise 2.7

Answer: $\eta = 80.6\%$

To determine the voltage, current, and power operational limits, normally a plot of i-v characteristics is given, as shown in Fig. 2.15. It gives the region in which the transistor can operate within its limits; the region is known as the safe operation area (SOA). It represents the permissible range of current, voltage, and power of the device when in operation. The switching locus of switch voltage verses switch current during turning *on* and *off* must lie within the SOA.

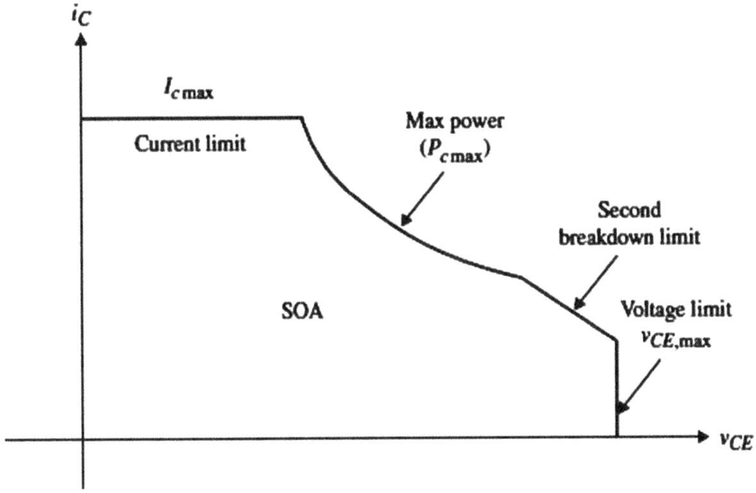

Fig. 2.15 Safe operation area (SOA) for a BJT

Exercise 2.8

(a) Show that the current gain of the triple Darlington transistors shown in Fig. E2.8 is given by:

$$\frac{i_{\mathrm{C}}}{i_{\mathrm{B}}} \approx \beta_1 \beta_2 \beta_3$$

(b) Assume transistor Q_1 has collector-emitter saturation voltage $V_{\mathrm{CE1,\,sat}}$. Show that $V_{\mathrm{CE,\,sat}}$for transistor Q_3 is given by:

$$V_{\mathrm{CE3,\,sat}} = V_{\mathrm{CE1,\,sat}} + 2V_{\mathrm{BE}}$$

Assume identical V_{BE} for the three transistors.

Fig. E2.8 Triple Darlington transistors

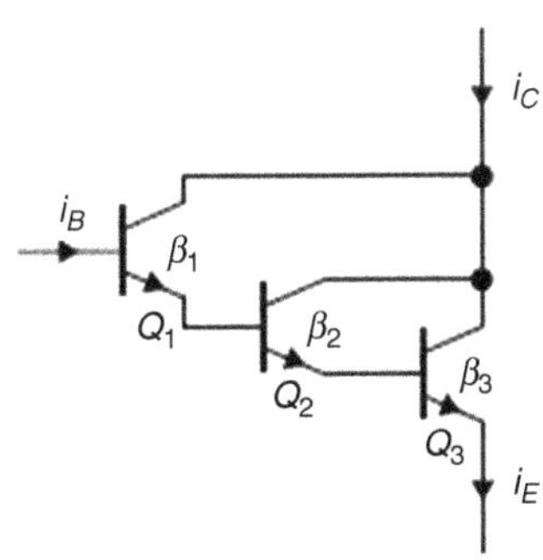

Exercise 2.9

Consider a simple BJT switch shown in Fig. E2.9. Determine β_{forced} for $R_B = 1$ kΩ, $R_B = 10$ kΩ, and $R_B = 20$ kΩ. Use $V_{BE} = 0.7$ V and $V_{CE1,sat} = 0.3$ V and assume ideal diode.

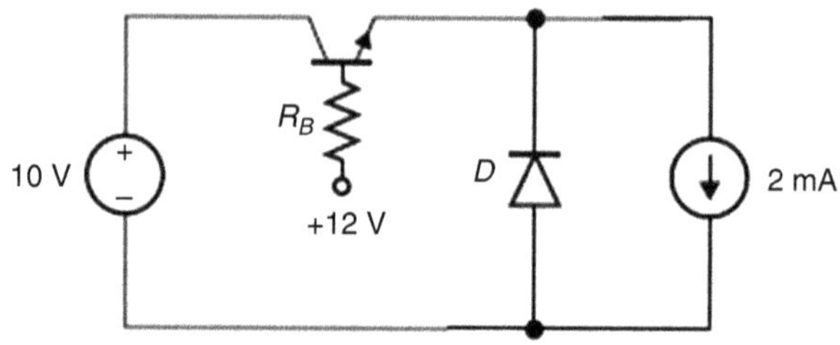

Fig. E2.9 Switching circuit for Exercise 2.9

Answer: 1.25, 12.5, 25

Initially, the BJT was developed to be used in linear audio output amplifiers. Soon BJT devices were used in switch-mode and high-frequency converters for aerospace applications to reduce the size and weight of magnetic components and filter capacitors. In applications where self-turn-*off* devices are needed such as in *dc* choppers and inverters, BJTs quickly replaced thyristors.

2.6.1.3 The Power MOSFET

In this section, an overview of power MOSFET semiconductor switching devices will be given. A detailed discussion of the physical structure, fabrication, and physical behavior of the device and its packaging is beyond the scope of this chapter. The emphasis here will be on the device's regions of operation and its terminal *i-v* switching characteristics.

Unlike the bipolar junction transistor, the metal oxide semiconductor field-effect transistor MOSFET device belongs to the *unipolar device family*, since it uses only the majority carriers in conduction. The development of the metal oxide semiconductor technology for microelectronic circuits opened the way for developing the power MOSFET device in 1975. Selecting the most appropriate device for a given application is not an easy task, requiring knowledge about the device characteristics and unique features as well as innovation and engineering design experience. Unlike low-power (signal) devices, power devices are more complicated in structure, driver design, and operational *i-v* characteristics. This knowledge is very important in enabling a power electronics engineer to design circuits that will make these devices close to ideal. The device symbols for a p- and n-channel enhancement and depletion types are shown in Fig. 2.16. Figure 2.17 shows the *i-v* characteristics for the n-channel enhancement-type MOSFET. It is the fastest power switching device, with switching frequency more than 1 MHz, with voltage power ratings up to 600 V and current rating as high as 40 A. MOSFET regions of operations will be studied shortly.

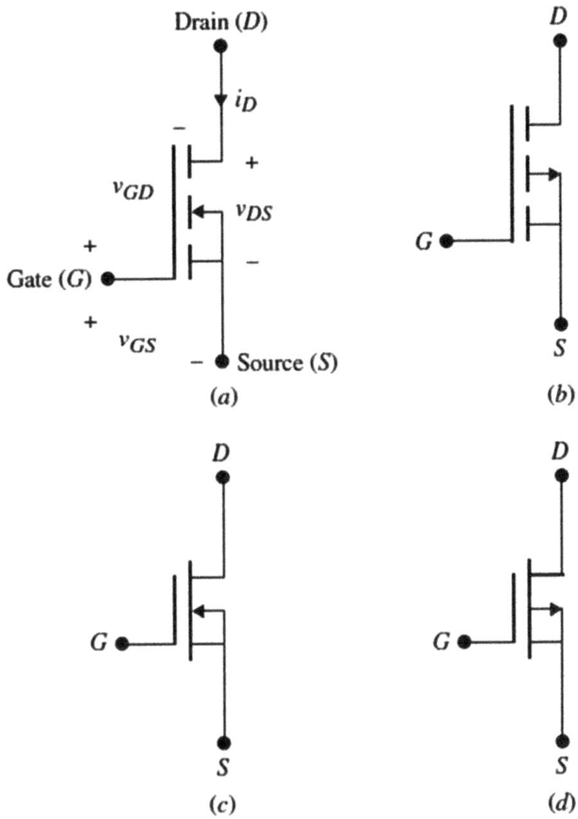

Fig. 2.16 Device symbols: (**a**) n-channel enhancement mode, (**b**) p-channel enhancement mode, (**c**) n-channel depletion mode, (**d**) p-channel depletion mode

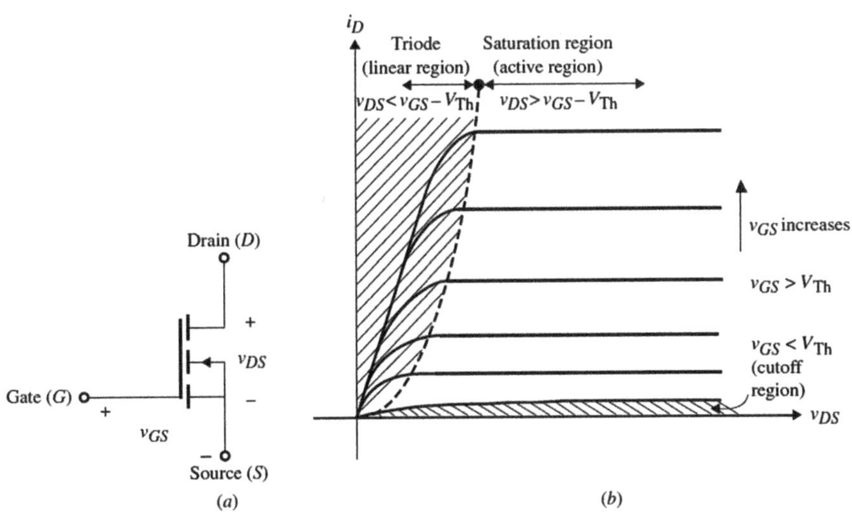

Fig. 2.17 (**a**) n-channel enhancement-mode MOSFET and (**b**) its i_D vs. v_{DS} characteristics

2.6.1.4 MOSFET Structure

Unlike the lateral channel MOSET devices used in many IC technologies in which the gate, source, and drain terminals are located in the same surface of the silicon wafer, power MOSFETs use vertical channel structure to increase the device's power rating. In the vertical channel structure, the source and drain are on opposite sides of the silicon wafer. There are several discrete types of the vertical-structure power MOSFET available commercially today, such as V-MOSFET, U-MOSFET, D-MOSFET, and S-MOSFET.

2.6.1.5 On-State Resistance

When the MOSFET is in the on-state (triode region), the channel of the device behaves like a constant resistance, $R_{DS(on)}$, that is, linearly proportional to the change between v_{DS} and i_D, as given by the following relation:

$$R_{DS(on)} = \left.\frac{\partial v_{DS}}{\partial i_D}\right|_{V_{GS}=\text{constant}}$$

The total conduction (on-state) power loss for a given MOSFET with forward current I_D and on-resistance $R_{DS(on)}$ is given by:

$$P_{on,\,diss} = I_D^2 R_{DS(ON)}$$

The value of $R_{DS(on)}$ can be significant and varies between tens of milliohms and a few ohms for low-voltage and high-voltage MOSFETS, respectively. The *on*-state resistance is an important data sheet parameter, since it determines the forward voltage drop across the device and its total power losses.

Unlike the current-controlled bipolar device, which requires base current to allow the current to flow in the collector, the power MOSFET is a voltage-controlled unipolar device and requires only a small amount of input (gate) current. As a result, it requires less drive power than the BJT. However, it is a non-latching current like that of the BJT, i.e., a gate-source voltage must be maintained. Moreover, since only majority carriers contribute to the current flow, MOSFETs surpass all other devices in switching speed, with speeds exceeding a few megahertz. Comparing the BJT and the MOSFET, the BJT has higher power-handling capabilities and lower switching speed, while the MOSFET device has less power-handling capabilities and relatively fast switching speed. The MOSFET device has a higher *on*-state resistance than the bipolar transistor. Another difference is that the BJT parameters are more sensitive to junction temperature when compared to the MOSFET parameters. Unlike the BJT, MOSFET devices do not suffer from second breakdown voltages, and sharing current in parallel devices is possible.

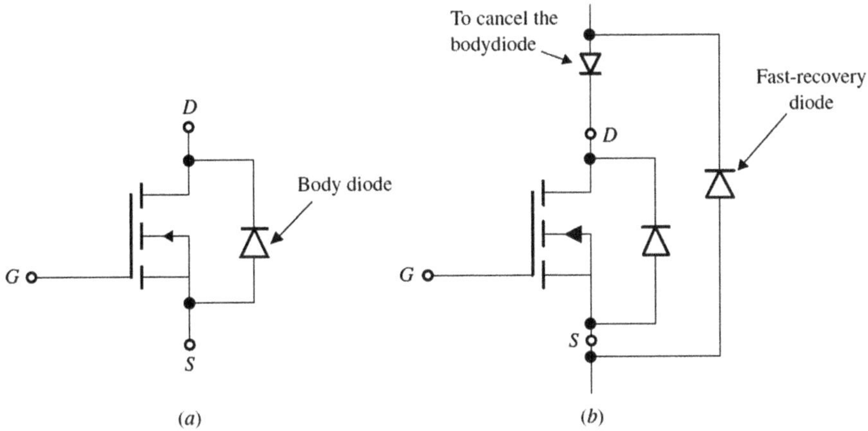

Fig. 2.18 (**a**) MOSFET internal body diode. (**b**) Implementation of a fast body diode

2.6.1.6 Internal Body Diode

The modern power MOSFET has an internal diode called a body diode connected between the source and the drain, as shown in Fig. 2.18a. This diode provides a reverse direction for the drain current, allowing a bidirectional switch implementation. Even though the MOSFET's body diode has adequate current and switching speed ratings, in some power electronic applications that require the use of ultrafast diodes, an external fast recovery diode is added in an antiparallel fashion, with the body diode by a slow recovery diode, as shown in Fig. 2.18b.

2.6.1.7 Internal Capacitors

Another important parameter that affects the MOSFETS's switching behavior is the parasitic capacitances between the device's three terminals, namely, gate-to-source (C_{gs}), gate-to-drain (C_{gd}), and drain-to-source (C_{ds}) capacitance as shown in Fig. 2.19. The values of these capacitances are nonlinear and a function of the device's structure, geometry, and bias voltages. During turn-on, capacitors C_{gd} and C_{gs} must be charged through the gate; hence, the design of the gate control circuit must take into consideration the variation in this capacitance. The largest variation occurs in the gate-to-drain capacitance as the drain-to-gate voltage varies. The MOSFET parasitic capacitances are given in terms of the device's data sheet parameters C_{iss}, C_{oss}, and C_{rss} as follows:

$$C_{gd} = C_{rss}$$
$$C_{gs} = C_{iss} - C_{rss}$$
$$C_{ds} = C_{oss} - C_{rss}$$

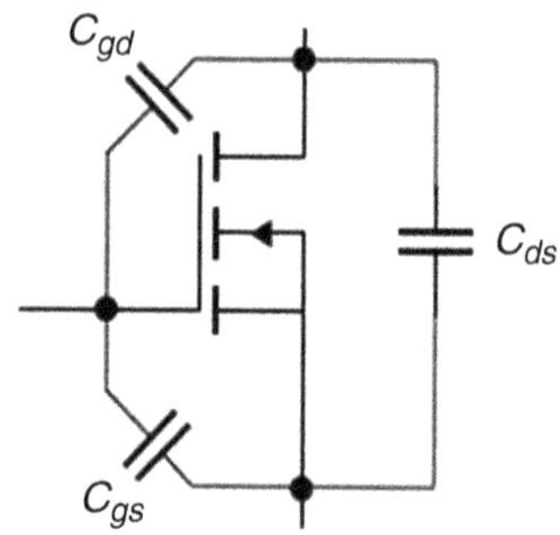

Fig. 2.19 Equivalent MOSFET representation including junction capacitances

where

C_{rss}= small-signal reverse transfer capacitance.

C_{iss}= small-signal input capacitance with the drain and source terminals shorted.

C_{oss}= small-signal output capacitance with the gate and source terminals shorted.

The MOSFET capacitances C_{gs}, C_{gd}, and C_{ds} are nonlinear and are a function of the dc bias voltage. The variations in C_{oss} and C_{iss} are significant as the drain-to-source and gate-to-source voltages cross zero, respectively. The objective of the drive circuit is to charge and discharge the gate-to-source and gate-to-drain parasitic capacitances to turn *on* and *off*, respectively.

In power electronics, the aim is to use power switching devices to operate at higher and higher frequencies. Hence, size and weight associated with the output transformer, inductors, and filter capacitors will decrease. As a result, MOSFETs are used extensively in power supply designs that require high switching frequencies, including switching and resonant-mode power supplies and brushless *dc* motor drives. Because of the device's large conduction losses, its power rating is limited to a few kilowatts. Because of its many advantages over the BJT devices, modern MOSFET devices have received high market acceptance.

2.6.1.8 Regions of Operation

Most of the MOSFET devices used in power electronic applications are of the n-channel, enhancement type like that shown in Fig. 2.16a. For the MOSFET to carry drain current, a channel between the drain and the source must be created. This occurs when the gate-to-source voltage exceeds the device threshold voltage, V_{Th}. For $v_{GS} > V_{Th}$, the device can be either in the triode region, which is also called "constant resistance" region, or in the saturation region, depending on the value of v_{DS}. For a given v_{GS}, with a small v_{DS} ($v_{DS} < v_{GS} - V_{Th}$), the device operates in the triode region (saturation region in the BJT), and with a large v_{DS} ($v_{DS} > v_{GS} - V_{Th}$), the device enters the saturation region (active region in the BJT). For $v_{GS} < V_{Th}$, the device turns off, with the drain current almost equals zero. Under both regions of operation, the gate current is almost zero. This is why the MOSFET is known as a voltage-driven device and, therefore, requires a simple gate control circuit.

The characteristic curves in Fig. 2.17b show that there are three distinct regions of operation labeled as triode region, saturation region, and cutoff region. When used as a switching device, only triode and cutoff regions are used; when it is used as an amplifier, the MOSFET must operate in the saturation region, which corresponds to the active region in the BJT.

The device operates in the cutoff region (off-state) when $v_{GS} < v_{Th}$, resulting in no induced channel. In order to operate the MOSFET in either the triode or saturation region, a channel must first be induced. This can be accomplished by applying a gate-to-source voltage that exceeds v_{Th}, i.e.,

$$v_{GS} > V_{Th}$$

Once the channel is induced, the MOSFET can either operate in the triode region (when the channel is continuous with no pinch-off, resulting in the drain current being proportional to the channel resistance) or in the saturation region (the channel pinches off, resulting in constant I_D). The gate-to-drain bias voltage (v_{GD}) determines whether the induced channel undergoes the pinch-off or not. This is subject to the following restrictions.

For the triode mode of operation, we have

$$v_{GD} > V_{Th}$$

And for the saturation region of operation, we have

$$v_{GD} < V_{Th}$$

Pinch-off occurs when $v_{GD} = V_{Th}$.

In terms of v_{DS}, the preceding inequalities may be expressed as follows:

1. For the triode region of operation:

$$v_{DS} < v_{GS} - V_{Th} \qquad v_{GS} > V_{Th}$$

2. For the saturation region of operation:

$$v_{DS} > v_{GS} - V_{Th} \qquad v_{GS} > V_{Th}$$

For the cutoff region of operation:

$$v_{GS} < V_{Th}$$

2.6.1.9 Input Capacitances

Because the MOSFET is a majority carrier transport device, it is inherently capable of a high-frequency operation. Still, the MOSFET has two limitations:

1. High input gate capacitances
2. Transient/delay due to the carrier transport through the drift region

As stated earlier, the input capacitance consists of two components: the gate-to-source and gate-to-drain capacitances.

The frequency response of the MOSFET circuit is limited by the charging and discharging times of C_{in}. Miller affect is inherent in any feedback transistor circuit with resistive load that exhibits a feedback capacitance from the input and output. The objective is to reduce the feedback gate-to-drain resistance. The output capacitance between the drain and source, C_{ds}, does not affect the turn-on and turn-off MOSFET switching characteristics. Figure 2.20 shows how C_{gd} and C_{gs} vary under increased drain source, v_{DS}.

In power electronic applications, the power MOSFETs are operated at high frequencies in order to reduce the size of the magnetic components. In order to reduce the switching losses, power MOSFETs are maintained in either the on-state (conduction state) or the off-state (forward blocking state).

2.6.1.10 Safe Operation Area

The safe operation area (SOA) of a device provides the current and voltage limits the device must handle to avoid destructive failure. Typical SOA for a MOSFET device is shown in Fig. 2.21. The maximum current limit while the device is on is determined by the maximum power dissipation.

$$P_{diss,on} = I_{DS(on)} R_{DS(on)}$$

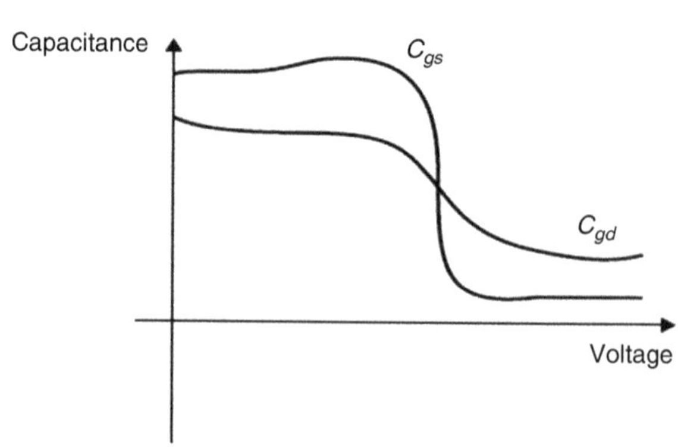

Fig. 2.20 Variation of C_{gd} and C_{gs} as a function of v_{DS}

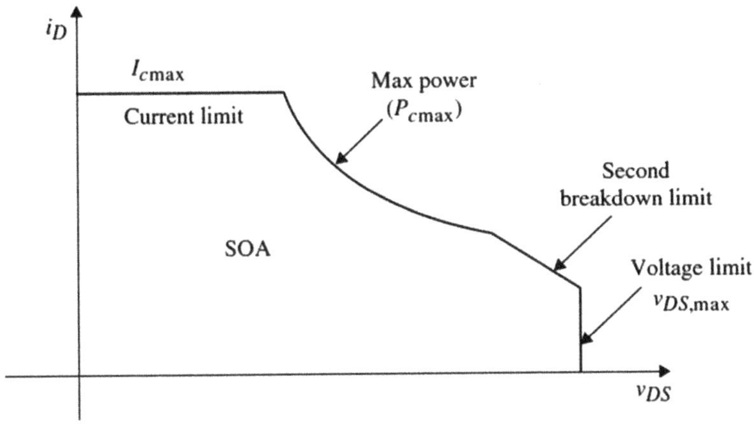

Fig. 2.21 Safe operation area for a MOSFET

Fig. 2.22 MOSFET equivalent circuit including the parasitic BJT

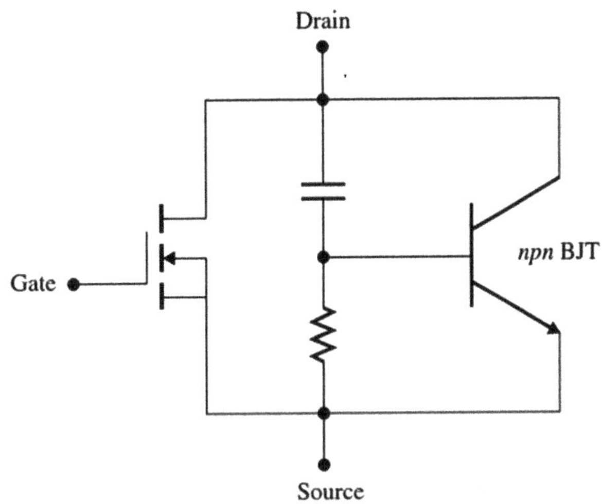

As the drain-source voltage starts increasing, the device starts leaving the on-state and enters the saturation (linear) region. During the transition time, the device exhibits large voltage and current simultaneously. At higher drain-source voltage values that approach the avalanche breakdown, it is observed that a power MOSFET suffers from a second breakdown phenomenon. The second breakdown occurs when the MOSFET is in the blocking state (off), and a further increase in v_{DS} will cause a sudden drop in the blocking voltage. The source of this phenomenon in MOSFETs is caused by the presence of a parasitic n-type bipolar transistor as shown in Fig. 2.22. The inherent presence of the body diode in the MOSFET structure makes the device attractive to applications in which bidirectional current flow is needed in the power switches.

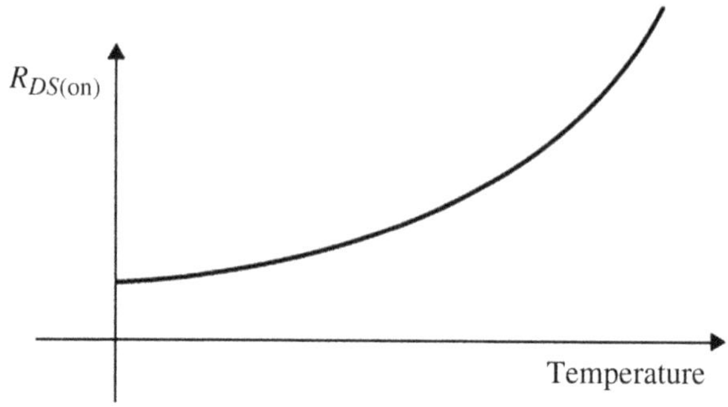

Fig. 2.23 The on-state resistance as a fraction of temperature

2.6.1.11 Temperature Effect

Today's commercial MOSFET devices have excellent response for high operating temperatures. The effect of temperature is more prominent on the on-state resistance as shown in Fig. 2.23. As the on-state resistance increases, the conduction losses also increase. This large $v_{\mathrm{DS(on)}}$ limits the use of the MOSFET in high-voltage applications. The use of silicon carbide instead of silicon has reduced $v_{\mathrm{DS(on)}}$ by manifold.

As the device technology keeps improving in terms of improving switch speeds and power-handling capabilities, it is expected that the MOSFET will continue to replace the BJTs in all types of power electronic systems.

2.6.1.12 The Insulated Gate Bipolar Transistor (IGBT)

The detailed equivalent circuit model, the simplified two-transistor circuit model, and the schematic symbol of the insulated gate bipolar transistors (IGBTs) are shown in Fig. 2.24a–c, respectively. Its *i-v* characteristic is similar to the MOSFET device and is not shown here. Since the IGBT architecture consists of a MOSFET and a BJT as shown in Fig. 2.24b, it is clear that the IGBT has high input impedance of the MOSFET along with the high current gain and the small on-state conduction voltage of the BJT. The device was commercially introduced in 1983, and it combines the advantages of MOSFETs, BJTs, and thyristor devices: the high current density allowed in BJT devices and low-power gate drive needed in MOSFET devices.

The device is turned off by zero gate voltage, which removes the conducting channel. However, a negative base cannot turn off the *pnp* transistor current. As a result, the turn-*off* time is higher in IGBT than the bipolar transistor. However, like the GTO (to be discussed shortly), the IGBT has a tail current at turn-*off* due to the recombination of carriers from the base region. At turn-*on*, a positive gate voltage is applied with respect to the emitter of the *npn* transistor, creating an n-channel in the MOS device, that causes the *pnp* transistor to start conducting.

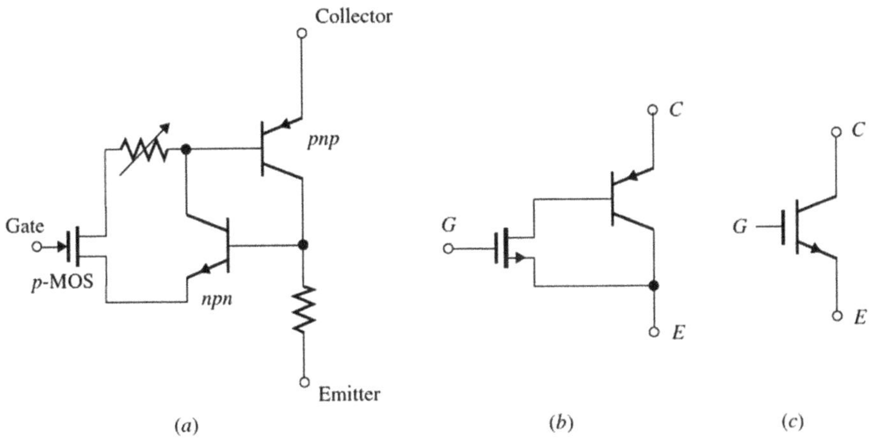

Fig. 2.24 (**a**) IGBT equivalent circuit, (**b**) simplified equivalent circuit, and (**c**) symbol

Its input capacitance is significantly smaller than the input capacitance of the MOSFET device, and the device does not exhibit second breakdown phenomena. It is faster than the BJT and can operate up to 20 kHz in medium-power applications. Currently, it is available at ratings as high as 1.2 kV and 400 A. The improvement in its fabrication is promising, and it is expected that it will replace the BJTs in the majority of power electronic applications.

2.6.2 *Thyristor-Based Devices*

The generic term *thyristor* refers to the family of power semiconductor devices made of three p-n junctions (four layers of *pnpn*) that can be latched into the *on*-state through an external gate signal that causes a regeneration mechanism in the device. In this section, we will discuss six main members of the thyristor family that are currently used in power electronic circuits: the silicon-controlled rectifier (SCR), the gate turn-*off* thyristor (GTO), the triode AC switch (TRIAC), the static induction transistor (SIT), the static induction thyristor (SITH), and the MOS-controlled thyristor (MCT).

2.6.2.1 The Silicon-Controlled Rectifiers

The silicon-controlled rectifier (SCR) is the oldest power controllable device utilized in power electronic circuits, introduced in 1958. Unlike the diode, the SCR can block voltages bidirectionally and carry the current unidirectionally. Until the 1970s, when power transistors were presented, the conventional thyristor had been used extensively in various industrial applications. The SCR is a three-

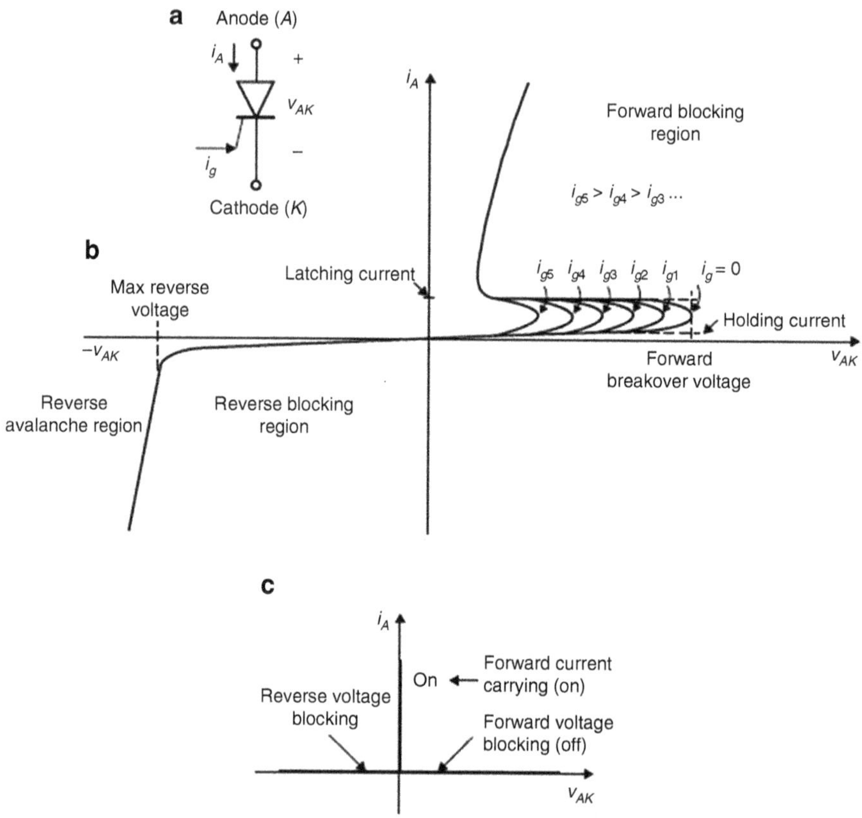

Fig. 2.25 SCR switching characteristics: (**a**) symbol, (**b**) i-v characteristics, (**c**) ideal switching character

terminal device composed of a four-semiconductor *p-n* junction. Unlike the diode, the SCR has a third terminal called the "gate" used for control purposes.

The symbol and *i-v* characteristics for the SCR are shown in Fig. 2.25a, b, respectively. The ideal switching characteristic curves are shown in Fig. 2.25c, where v_{AK} and i_A are the voltages across the anode-cathode terminals and the current through the anode, respectively.

The *latching current* is always less than the minimum trigger current specified in the device's data sheet. The holding current is the minimum forward current the SCR can carry in the absence of a gate drive. The *forward-breakover* voltage, V_{BO}, is the voltage across the anode-cathode terminal that causes the SCR to turn on without the application of a gate current. *Reverse avalanche* (breakdown) occurs when v_{AK} is negatively large.

The normal operation of the SCR is when its gate is used to control the turn-on process by injecting a gate current i_G to allow the forward current to flow; v_{AK} is positive and can be turned off by applying a negative v_{AK} across it.

It must be noted that once the SCR is turned on, the gate signal can be removed. For this reason, this device is also known as a *latch device*. The gate current must be applied for a very short time and normally it can go up *to* 100 mA. Once the SCR is turned on, it has a 0.5 – 2 V forward voltage.

When the thyristor was invented, all schemes for force-commutating mercury-arc rectifiers of the 1930s soon became thyristor-based circuits with expanded applications to include *ac* drives and UPS. However, because of their cost and low efficiency, thyristor circuits did not penetrate the adjustable-speed drive application area. Today's applications range from single-phase-controlled rectifier circuits to static VAR compensation in utility systems. Because of its limited frequency of operation, the application of thyristors has reached saturation.

2.6.2.2 The Gate Turn-off Thyristor (GTO)

The schematic symbol and the practical and ideal switching *i-v* characteristics for the gate turn-off thyristor (GTO) are shown in Fig. 2.26a–c, respectively.

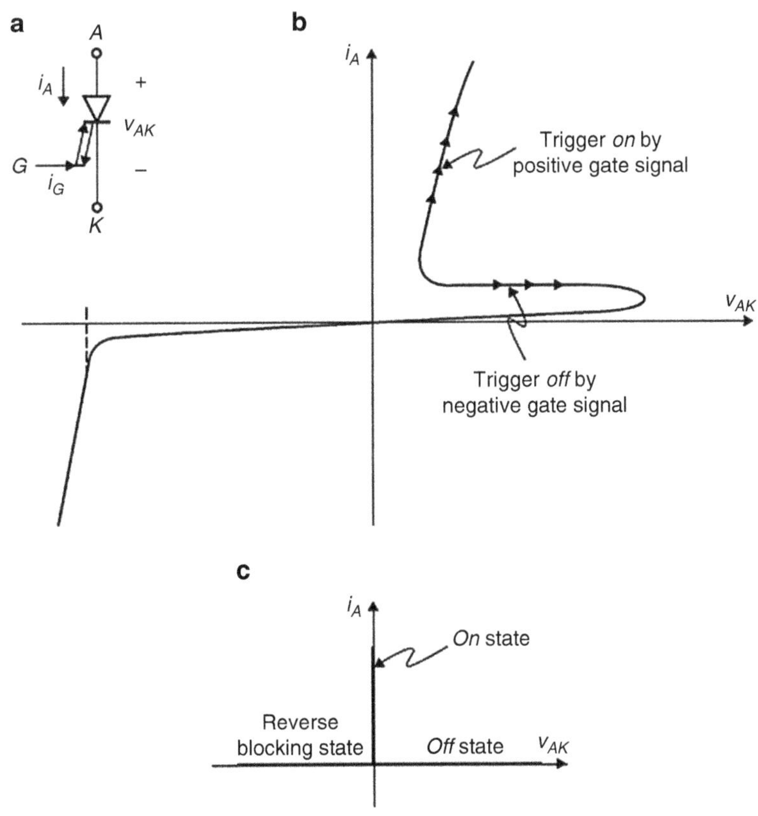

Fig. 2.26 GTO switching characteristics: (**a**) symbol (**b**) i-v characteristics, (**c**) ideal switching characteristics

The device is as old as the SCR and was introduced commercially in 1962. Like the SCR, it can be turned *on* with a positive gate signal, but unlike the SCR, applying a negative gate signal as shown in Fig. 2.26b can turn off the GTO. Once the GTO is turned *on* or *off*, the gate signal can be removed. The device has a higher *on*-state voltage than the *on*-state voltage of the SCR at comparable currents. The GTO is normally an *off* device and has a very poor turn-*off* current gain, and it exhibits a second breakdown problem at turn-*off*.

Because of its high switching power dissipation, the GTO's frequency of operation is limited to less than 1 kHz, and modern GTO devices are rated at 4.5 kV and at currents as high as 3 kA. The GTO is used in high-current and voltage applications, such as voltage-fed inverters and induction heating resonant converters.

2.6.2.3 The Triode AC Switch (TRIAC)

Like the GTO, the TRIod AC (TRIAC) switching device was introduced immediately after the SCR. In fact, the TRIAC is nothing but a pair of SCRs connected in reverse parallel on one integrated chip, as shown in Fig. 2.27. It is also known as a bidirectional SCR. The TRIAC's equivalent circuit and the circuit schematic symbol are shown in Fig. 2.27a, b, respectively. The device can be triggered in the positive and negative half-cycle of the AC voltage source by applying either a positive or a negative gate signal, respectively. Today's TRIAC ratings are up to 800 V at 40 A. The i-v characteristics and the ideal switching characteristics are shown in Fig. 2.27c, d. The use of the TRIAC is considerably limited due to low rates of rise of voltage and current. Applications include light dimming, heating control, and various home appliances.

2.6.2.4 The DIAC

Finally, we should mention another power device known as the DIAC which is essentially a gateless TRAIC constructed to break down at low forward and reverse voltages. The DIAC is mainly used as a triggering device for the TRAIC.

2.6.2.5 The Static Induction Transistors and Thyristors

In 1987, a new device known as the static induction transistor (SIT) was introduced. One year later, the static induction thyristor (SITH) was introduced. The symbols for the SIT and SITH are shown in Fig. 2.28a, b, respectively.

The SIT is a high-power and high-frequency device. The device is almost identical to the JFET, but with its special gate construction, it has a lower channel resistance when compared to the JFET.

Fig. 2.27 TRIAC switching characteristics: (**a**) equivalent representation using two SCRs, (**b**) symbol, (**c**) i-v characteristics, (**d**) ideal switching characteristics

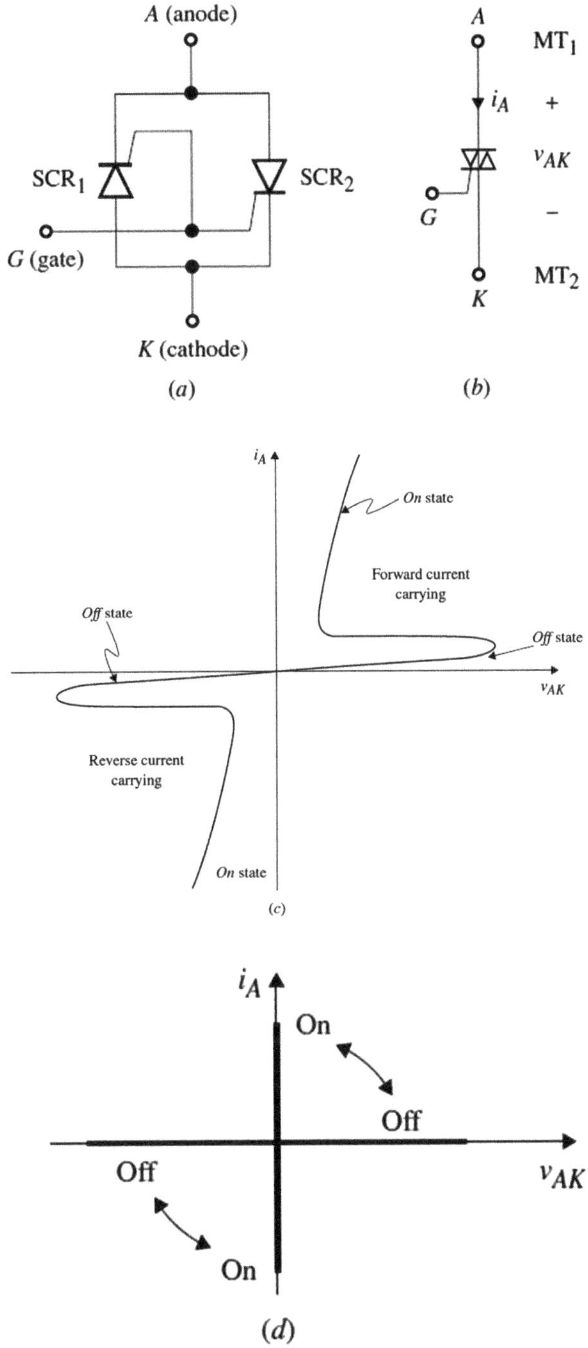

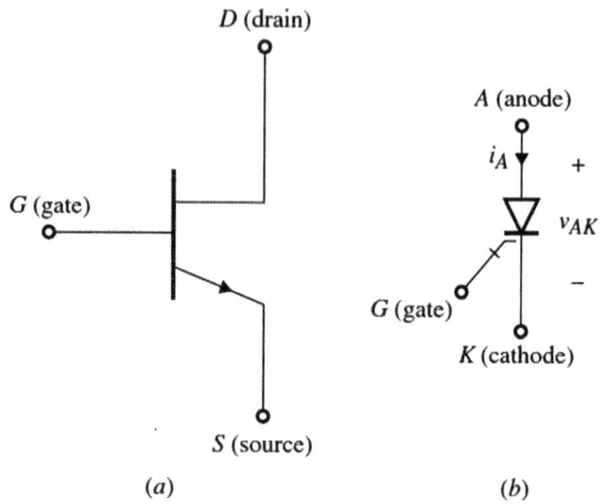

Fig. 2.28 (**a**) SIT symbol and (**b**) SITH symbol

Both SIT and SITH are normally *on* devices and have no reverse voltage blocking capabilities. The SITH device turns *off* the same way as the GTO; by applying a negative gate current, but it has a higher conduction drop than the GTO. Finally, both devices are majority carrier devices with positive temperature coefficients, allowing device paralleling.

Among the SIT's major applications are the audio and VHF/UHF amplifiers, microwaves, AM/FM transmitters, induction heating, and high-voltage low-current power supplies. It has a large forward voltage drop when compared to the MOSFET; hence, it is not normally used in power electronic converter applications. The applications of the SITH include static VAR compensators and induction heating.

2.6.2.6 The MOS-Controlled Thyristor

The simplified equivalent circuit model and the schematic symbol for a p-type MOS-controlled thyristor (MCT) are shown in Fig. 2.29a, b, respectively. Its ideal *i*-*v*-switching characteristic is similar to that of the GTO's as shown in Fig. 2.29c. The device was commercially introduced in 1988. Like the GTO device, it has a high turn-off current gain.

The p-MCT is turned *on* by applying a negative gate voltage (less than −5 V) with respect to the cathode, turning *on* the p-FET and turning *off* the n-FET, initiating the regenerative mechanism in the SCR connected npn and pnp transistors. Similarly, applying a positive gate signal with respect to the cathode initiates the turn-off. The n-MCT has the same device structure, except that the p-FET and n-FET are interchanged; hence, a positive and negative gate signals turn the n-MCT *on* and *off*, respectively. The schematic symbol for the n-MCT is shown in Fig. 2.29d.

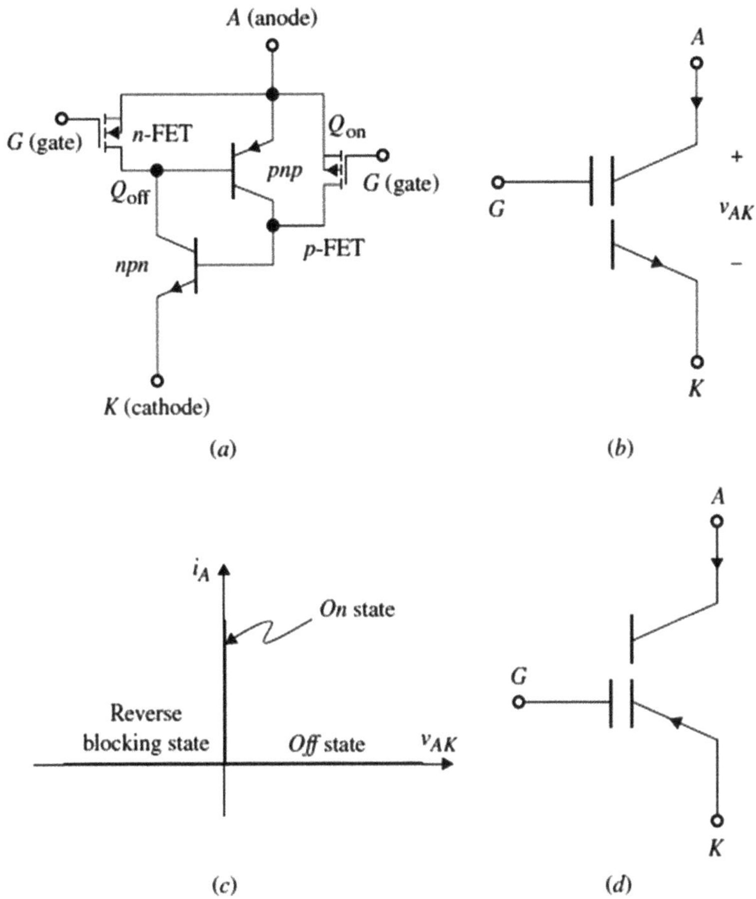

Fig. 2.29 MCT switching characteristics: (**a**) equivalent circuit, (**b**) symbol, (**c**) ideal switching characteristics, (**d**) p-MCT symbol

The MCT's current and voltage ratings exceed 1 kV and 100 A and are continuously being improved. The device can be easily connected in series and in parallel combinations to boost power rating.

This device is a serious competition for the IGBT. It has the same frequency of operation as the IGBT but with a smaller voltage drop and a higher operating temperature. Intensive efforts are under way to introduce a new improvement in the device, and it is expected to have a wider acceptance in medium- and high-power applications.

2.6.2.7 Other Power Devices

Other devices of the thyristor family include the reverse-conducting thyristor (RCT), which is nothing but a built-in antiparallel body diode connected across the SCR to allow current to flow in the opposite direction, and the light-activated SCR (LASCR), which is used in high-voltage and high-current applications such as the HVDC systems. Their power ratings are up to hundreds of kilovolts and hundreds kiloamperes, and they provide complete electrical isolation between the power and control circuits.

2.7 Comparison of Power Devices

Depending on the applications, the power range processed in power electronics is very wide, from hundreds of milliwatts to hundreds of megawatts. Therefore, it is very difficult to find a single switching device type to cover all power electronic applications. Today's available power devices have tremendous power and frequency rating range as well as diversity. Their forward current ratings range from a few amperes to a few kiloamperes, their blocking voltage rating ranges from a few volts to a few kilovolts, and their switching frequency ranges from a few hundreds of hertz to a few megahertz, as illustrated in Table 2.2. This table illustrates the relative comparison between available power semiconductor devices because there is no straightforward technique that gives a ranking for these devices. Devices are still being developed very rapidly with higher current, voltage, and switching frequency ratings. Finally, Fig. 2.30 shows a plot of the frequency versus the power, illustrating the frequency and power ratings of various available power devices.

Table 2.2 Comparison of power semiconductor devices

Device type	Year made available	Rated voltage	Rated current	Rated frequency	Rated power	Forward voltage
Thyristor (SCR)	1957	6 kV	3.5 kA	500 Hz	100'sMW	1.5–2.5 V
Triac	1958	1 kV	100A	500 Hz	100'skW	1.5–2 V
GTO	1962	4.5 kV	3kA	2 kHz	10'sMW	3–4 V
BJT (Darlington)	1960s	1.2 kV	800A	10 kHz	1 MW	1.5–3 V
MOSFET	1976	500 V	50A	1 MHz	100'skW	3–4 V
IGBT	1983	1.2 kV	400A	20 kHz	100'skW	3–4 V
SIT	1950	1.2 kV	300A	100 kHz	10'skW	10–20 V
SITH	1960's	1.5 kV	300A	10 kHz	10'skW	2–4 V
MCT	1988	3 kV	2 kV	20–100 kHz	10'sMW	1–2 V

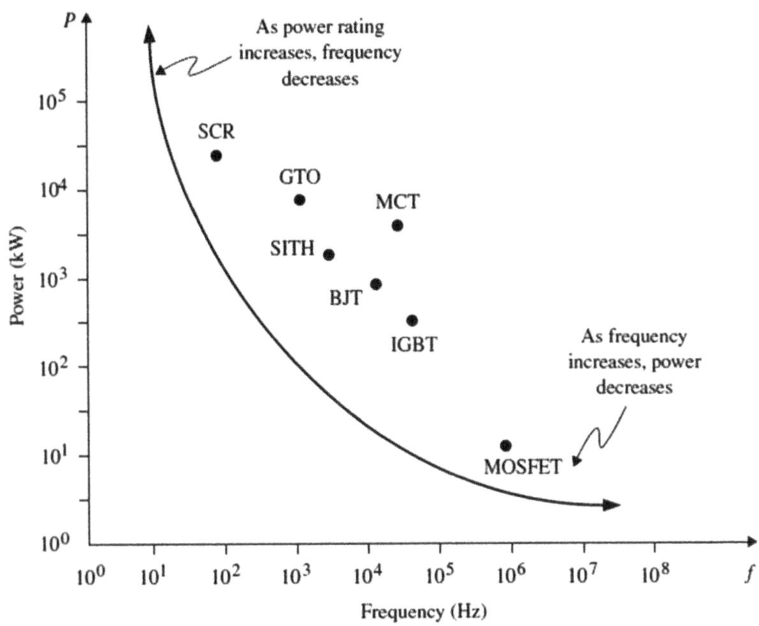

Fig. 2.30 Frequency vs. power rating plot for various power devices

2.8 Future Trends in Power Devices

It is expected that the improvement in power-handling capabilities and increases in the frequency of operation of power devices will continue to drive the research and developments in semiconductor technology. From power MOSFETs to power MOS-IGBTs and to power MOS-controlled thyristors, the power rating has consistently increased by a factor of 5 from one type to another. Major research activities will focus on obtaining new device structures based on the MOS-BJT technology integration to rapidly increase power ratings. It is expected that the power MOS-BJT technology will capture more than 90% of the total power transistor market.

The continuing development of power semiconductor technology has resulted in power systems with driver circuits, logic and control, device protection, and switching devices designed and fabricated on a single chip. Such power IC modules are called "smart power" devices. For example, some of today's power supplies are available as ICs for use in low-power applications. There is no doubt that the development of smart power devices will continue in the near future, addressing more power electronic applications.

2.9 Snubber Circuits

In order to relieve switches from overstress during switching, switching aid circuits, known as *snubber circuits*, are normally added to the power switching device. The objectives of snubber circuits may be summarized as (1) the reduction of the switching power losses in the main power device in the power electronic circuit, (2) avoiding second breakdowns, and (3) controlling the device's *dv*/*dt* or *di*/*dt* in order to avoid latching in pnpn devices. There are wide ranges of turn-*on* and turn-*off* snubber circuits available in today's power electronic circuits. These include dissipative and nondissipative passive snubber circuits and nondissipative active snubber circuits. In dissipative snubber circuits, a capacitor is used in order to slow down the device's voltage rise during turn-*off* or an inductor to slow down the device's current rise during turn-*on*. Figure 2.31a, b shows popular turn-*off* and turn-*on* snubber circuits, respectively. In Fig. 2.31a, a capacitor is used to reduce the voltage rise dv_{sw}/dt across the switch during turn-*off*. In Fig. 2.31b, a snubber inductor, L_s, is used to slow down the rise of the inductor current di_{sw}/dt (the inductor current equals the switch current i_{sw}). Figure 2.32 shows the switching loci for a practical switch (transistor) with and without snubber circuits. For detailed discussion on all types of snubber circuits and their design methods, refer to the references at the end of the textbook.

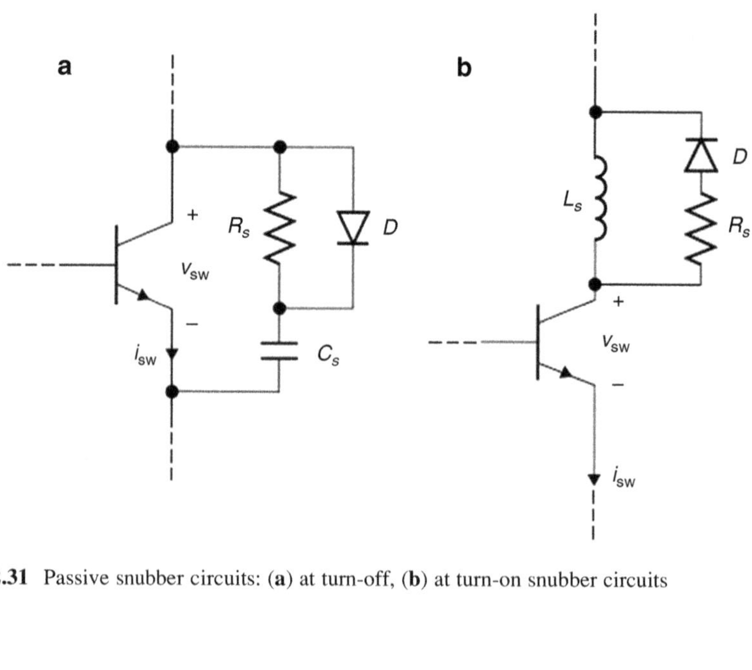

Fig. 2.31 Passive snubber circuits: (**a**) at turn-off, (**b**) at turn-on snubber circuits

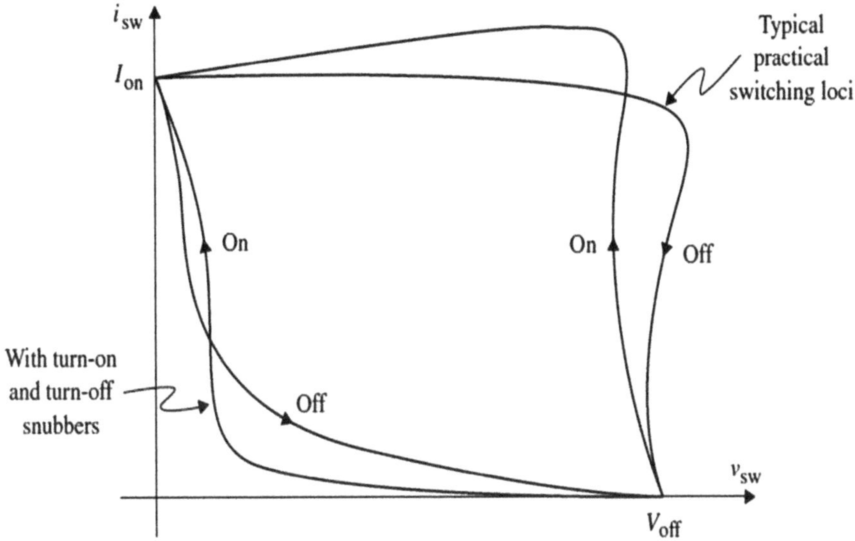

Fig. 2.32 i_{sw} vs. v_{sw} switching loci

2.10 Interest in High-Temperature Power Devices: The Wide Band Gap

The final observation is that while in the past half century, industry has relied heavily on the silicon (Si) as the primary semiconductor material for power devices. In recent years, interest has been steadily shifting toward new material that can handle high temperature and reduced on resistance. Such material is the wide band gap (WBG) semiconductor types such as silicon carbide (SiC) and gallium nitride (GaN). SiC power semiconductors are a relatively new entrant in the commercial marketplace, with the first SiC Schottky diode introduced in 2001. WBG has shown the capability to meet the higher performance demands of the evolving power equipment, operating at higher voltages and temperatures and enabling switching frequencies with greater efficiencies compared with existing Si devices. Along with performance improvements, WBG-based power electronics can be fabricated with a much smaller footprint (reduced volume) compared with comparable Si devices due to decreased cooling requirements and smaller passive components, contributing to overall lower system costs.

The main challenge remains the cost of WBG verses the cost of their counterpart silicon devices. However, in certain power ratings, WBG devices are in parity with the cost of silicon devices. In fact, SiC and GaN are expected to totally replace silicon for high-voltage, high-temperature, and high-frequency applications. Figures 2.33 and 2.34 show a comparison of the Si, SiC, and GaN devices in three areas of operation: high voltage, high frequency, and high temperature.

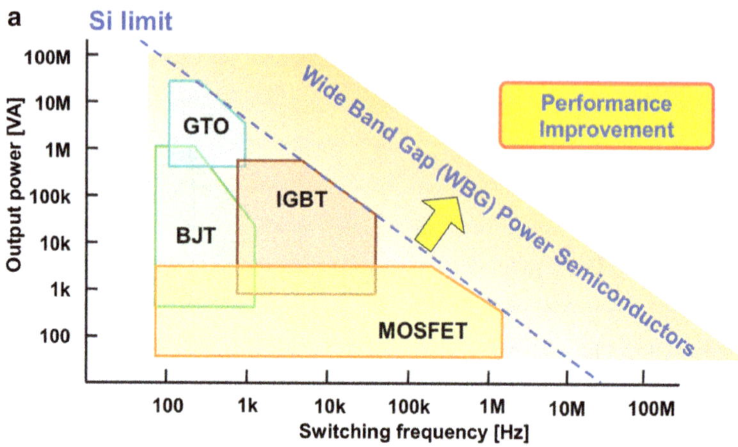

Fig. 2.33 Output power and switching frequency relationship (Courtesy of Professor John Shen of the Illinois Institute of Technology)

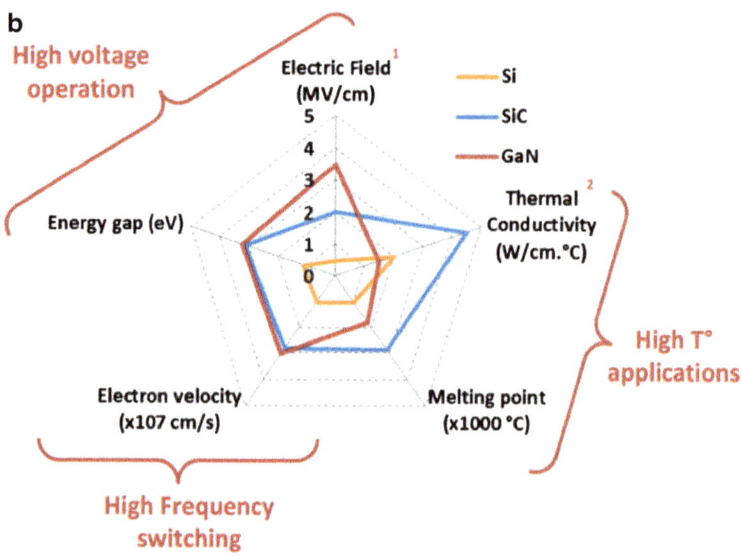

Fig. 2.34 Comparison between Si, SiC, and GaN devices when operated in high voltage, high frequency, and high temperature ranges (Courtesy of Professor John Shen of the Illinois Institute of Technology)

Problems

Ideal Switch Characteristics

2.1 Consider the switching circuit shown in Fig. P2.1 with a resistive load. Assume the switch is ideal and operating at a duty ratio of 40%.

(a) Sketch the waveforms for i_{sw} and v_{sw}.
(b) Determine the average output voltage.
(c) Determine the average output power delivered to the load.
(d) Determine the average output power supplied by the *dc* source.
(e) Determine the efficiency of the circuit.

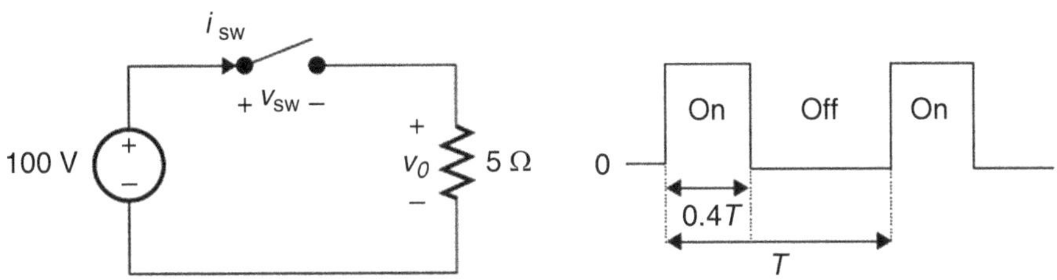

Fig. P2.1

2.2 Consider the two-switch circuit given in Fig. P2.2. Assume ideal S_1 and S_2 with the shown switching sequence. Sketch i_{sw1},v_{sw1},i_{sw2}, and v_{sw2}. Determine (a) the average output voltage, (b) average power delivered to the load, and (c) efficiency.

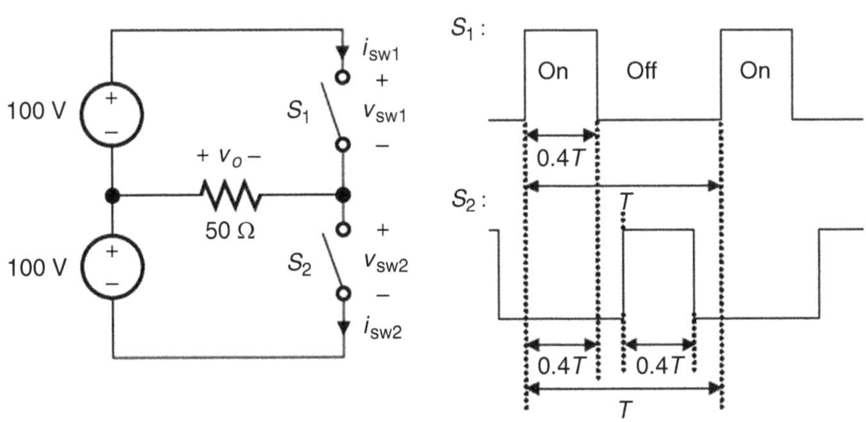

Fig. P2.2

2.3 The circuit shown in Fig. P2.3 is known as a current-driven full-bridge inverter. Assume S_1, S_2, S_3, and S_4 are ideal with their switching waveforms as shown. Sketch the waveforms for i_o, v_o, v_{sw1}, i_{sw1},, v_{sw4}, and i_{sw4}, and derive the expression for the average output voltage.

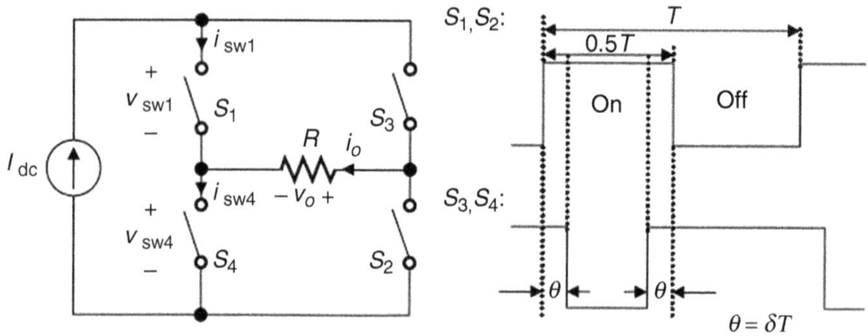

Fig. P2.3

Nonideal Switching Characteristics

2.4 Use Fig. E2.3, and by assuming that the switching waveform has the following parameters, $I_{on} = 1$ A, $V_{off} = 150$ V, $I_{off} = 10$ μA, $V_{on} = 2$ V, $T_s = 10^{-4}$ sec , $t_d = 90$n s, $t_{fall} = 120$ ns, $t_{rise} = 100$ ns, determine the equation for the instantaneous power and find P_{ave}.

2.5 Repeat Problem 2.1 by assuming the switch has a 0.2Ω *on*-state resistance and a 2 V forward drop during conduction.

2.6 Repeat Problem 2.3 by assuming each switch has a 0.2 Ω *on*-state resistance and a 2 V forward voltage drop during conduction.

2.7 Consider the circuit of Fig. P2.3 by assuming each switch has a forward voltage drop, V_F, and an *on*-state resistance r_{on}. Derive the expression for the circuit's efficiency in terms of I_{dc}, R, δ, V_F, and r_{on}. Determine δ for maximum efficiency.

2.8 Determine the conduction and switching average power dissipation for Example 2.2 by using $t_{on} = 5$ μs, $t_{off} = 8$ μs, $T_s = 150$ μs, $V_{off} = 150$ V, and $I_{on} = 15$ A.

2.9 Repeat Problem 2.8 for Fig. E2.3 with $t_d = 4$ μs.

2.10 Consider a power switching device whose current and voltage waveforms are shown in Fig. P2.10.

(a) Derive the expression for the instantaneous power and sketch it.
(b) Determine the average power dissipation.
(c) Calculate part (b) for $t_r = 120$ ns, $t_f = 180$ ns, $t_d = 90$ ns, $f_s = 100$ kHz, $V_{off} = 150$ V, and $I_{on} = 15$ A.
(d) Repeat parts (a)–(c) by assuming $V_{on} = 2$ V and $I_{off} = 10$ μA.

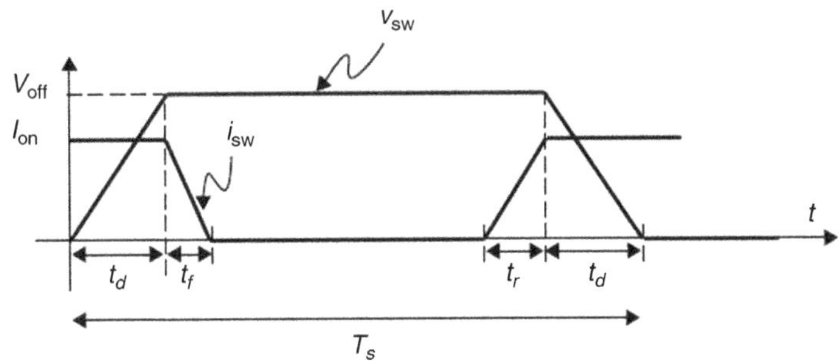

Fig. P2.10

2.11 Use the linear approximated switching current and voltage given in Fig. 2.4 with zero forward voltage and zero leakage (reverse) current. (a) Calculate the average power for $t_{on} = 5$ μs, $t_{off} = 8$ μs, $T_s = 150$ μs, $V_{off} = 150$ V, and $I_{on} = 15$ A. (b) Repeat part (a) by assuming the t_{on} is measured from 10% to 90% of I_{on} and t_{off} is 10% to 90% of V_{off}.

Power Device Switching Characteristics:

2.12 The switching current and voltage waveforms of Fig. 2.3 at turn-*on* are represented mathematically as follows:

$$isw = I_{on}\left(1 - e^{-t/\tau}\right)$$
$$v_{sw} = V_{off} e^{-t/\tau}$$

where τ represents the time constant which is a function of the on-state resistance of the device and its capacitance. Assume negligible I_{off} and V_{on}. Show that the switching power dissipation at turn-*on* is given by:

$$P_{diss} = \frac{V_{off} I_{on}}{2T_s} \tau$$

Assume $t_{on} \gg \tau$ and $t_{off} \gg \tau$

2.13 Assume the switching current and voltage of Fig. E2.1 at the turn-*off* interval ($0 \le t < t_d + t_{off}$) are represented as follows:

$$v_{sw}(t) = V_{off}$$
$$i_{sw}(t) = I_{on} e^{-t/T}$$

Derive the expression for the average switching power dissipation during the *off*-time. Assume $t_d + t_f \approx 5\tau$.

2.14 The diode i-v characteristic curve in the forward region can be mathematically represented by:

$$i_{\mathrm{D}} = I_{\mathrm{s}} e^{v_{\mathrm{D}}/nV_{\mathrm{T}}}$$

where

I_s: reverse saturation current.
V_{T}: thermal voltage equals 25 mV at $20^{\circ}C$ temperature.
n: Empirical constant whose value depends on the semiconductor material and the physical construction of the device. The value is normally between 1 and 2.

(a) Show that a decade change in the forward diode current results in a $2.3nV_{\mathrm{T}}$ change in the forward voltage. Mathematically it is expressed as follows:

$$V_{\mathrm{F2}} - V_{\mathrm{F1}} = 2.3nV_{\mathrm{T}} \log \frac{I_{\mathrm{F1}}}{I_{\mathrm{F2}}}$$

(b) Consider the circuit of Fig. P2.14. Assume the diode has $I_{\mathrm{F}} = 5$ A at $V_{\mathrm{F}} = 1$ V with $n = 1.5$. Determine I_{D} and V_{D} for (i) $R = 10\ \Omega$ and (ii) $R = 5\ \Omega$.

Fig. P2.14

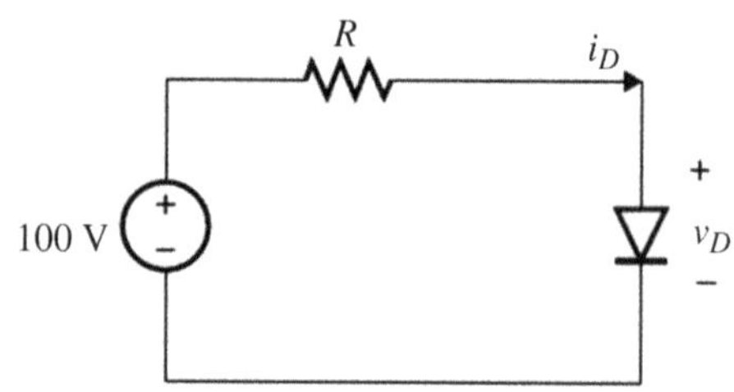

General Device Switching Problems

2.15 Consider Fig. P2.15 that shows an approximated reverse recovery turn-*off* characteristics for a power diode.
Show that the following relation can express the total reverse recovery charge, Q_{rr}:

$$Q_{\mathrm{rr}} = \frac{1}{2} t_{\mathrm{rr}} t_{s1} \frac{di_1}{dt} = \frac{1}{2} t_{\mathrm{rr}} t_{s2} \frac{di_2}{dt}$$

where di_1/dt and di_2/dt are the slope of the diode current during t_{s1} and t_{s2}, respectively.

Fig. P2.15

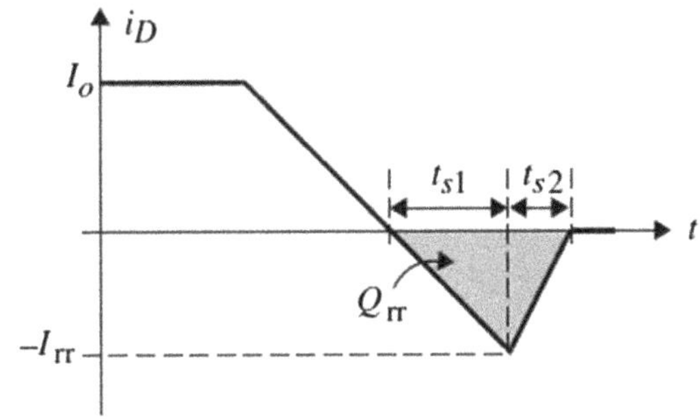

2.16 Consider the switching circuit with the series snubber R_s-C_s network connected across the diode as shown in Fig. P2.16.

Assume the diode switching characteristic curve is given by Fig. P2.16b and the switch is ideal.

(a) Derive the expression for all the branch currents and v_D, and sketch them.
(b) Derive the expression for the capacitor C_s that is needed.

Fig. P2.16

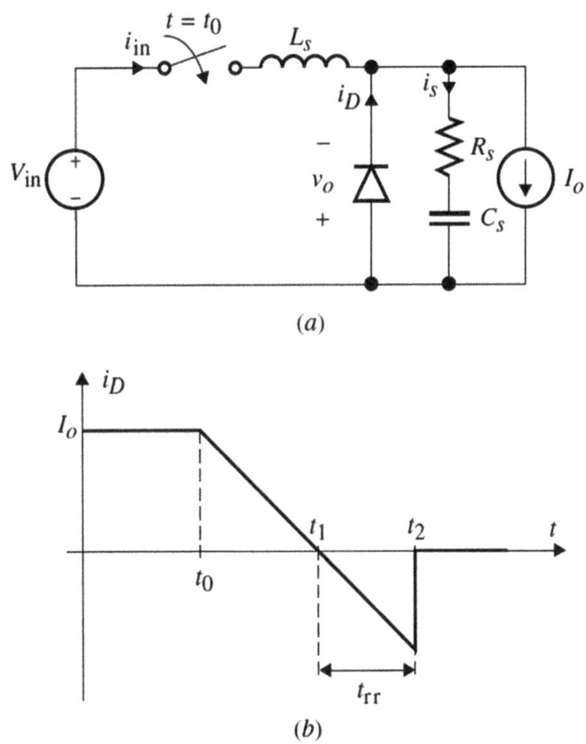

2.17 One way to speed the turn-*off* time of the Darlington connection is to include a diode between the bases of the two transistors as shown in Fig. P2.17. Discuss how the turn-*off* speed is improved by adding the diode *D*.

Fig. P2.17

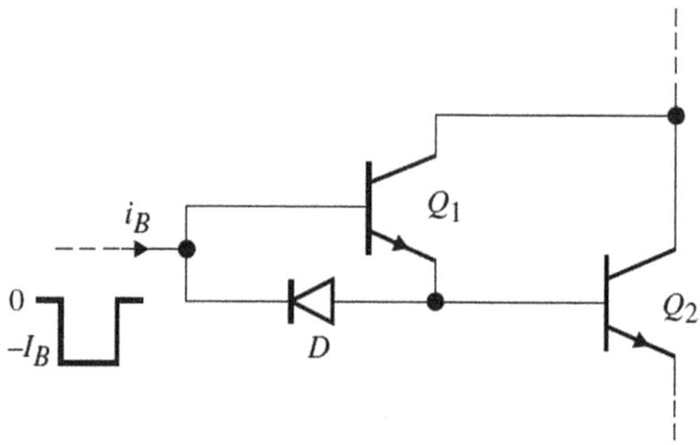

2.18 Consider the transistor switching circuit and its switching waveforms in Fig. P2.18a, b, respectively.

(a) Sketch the waveforms for i_D and v_D.
(b) Calculate the average power dissipated in the transistor.

Use $f_s = 10$ kHz, $t_d = 150$ ns, $t_r = 100$ ns, $t_d = 100$ ns, and $t_f = 250$ ns.

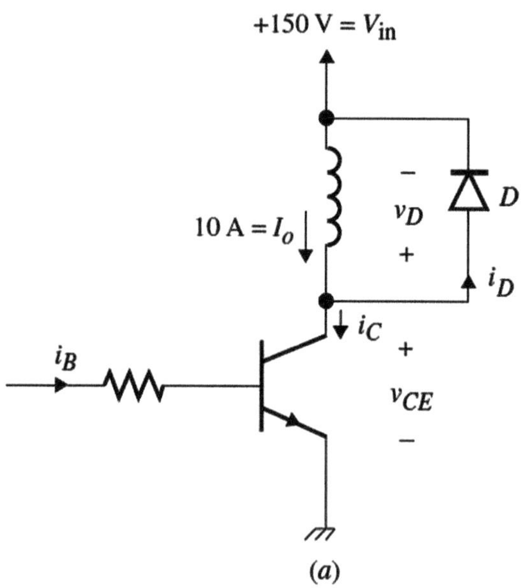

Fig. P2.18

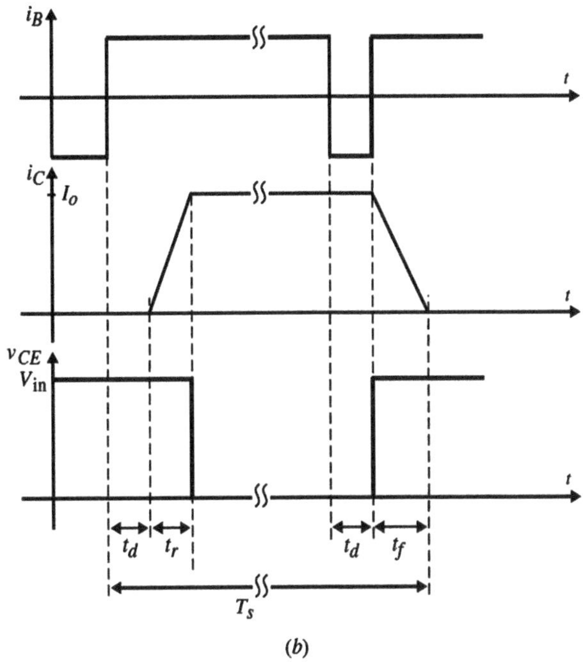

Fig. P2.18 (continued)

2.19 Figure P2.19 shows a BJT switching circuit known as the “baker clamp,” whose objective is to limit how deep in saturation the transistor is permitted. Show that if we assume $v_D = V_{BE}$, then the minimum V_{CE} is clamped to V_Z. If the collector is connected to a 15 V power supply through a collector resistor, R_C, and the base is connected to a 5 V power supply through a base resistor, R_B, design for V_Z, R_C, and R_B so that the transistor is driven into saturation with $V_{CE,\,sat} = 1.8$ V and $\beta_{forced} = 10$.

Fig. P2.19

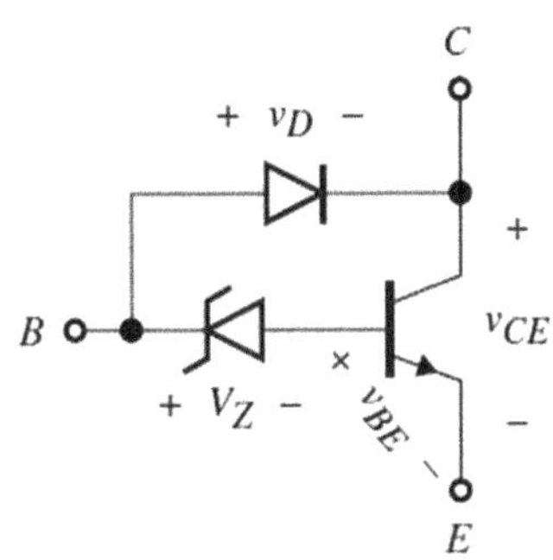

2.20 Derive the SCR forward current relation given in Eqs. (2.1) and (2.2).

2.21 Consider the half-bridge inverter circuits with the unidirectional MOSFET switching a resistive load given in Fig. P2.21. Use $R = 10\ \Omega$:

(a) Determine the average power delivered to the load when S_1 and S_2 are assumed ideal.
(b) Repeat part (a) for $R_{\mathrm{DS(on)}} = 0.2\ \Omega$ for each MOSFET.
(c) Find the efficiency of the circuit in part (b).

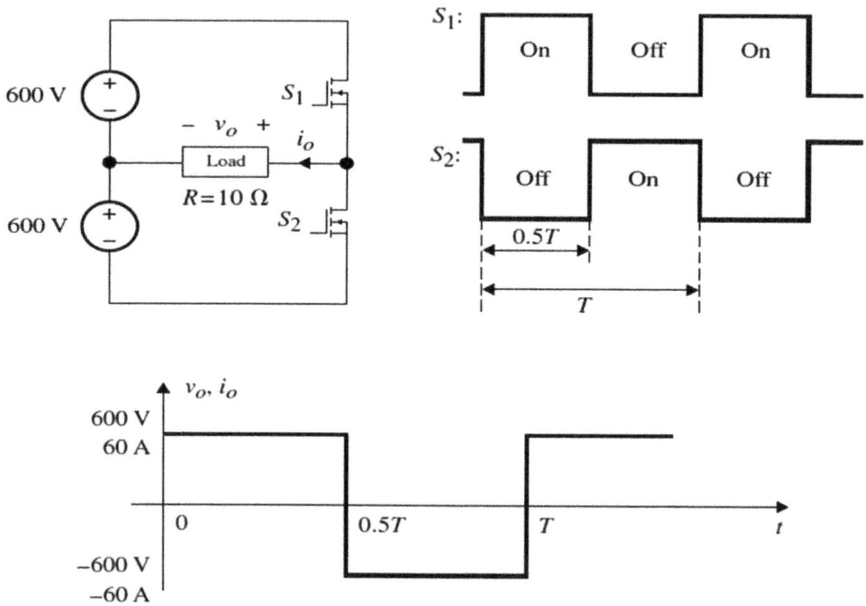

Fig. P2.21

2.22 Calculate β_{forced} in Fig. P2.22a, b for $R_{\mathrm{B}} = 1\ \mathrm{k}\Omega$ and $R_{\mathrm{B}} = 10\ \mathrm{k}\Omega$. Assume $V_{\mathrm{BE}} = 0.75$ V and $V_{\mathrm{CE,\,sat}} = 0.4$ V.

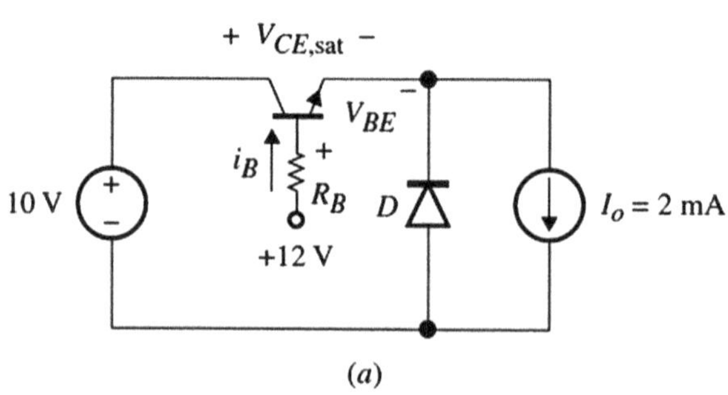

Fig. P2.22

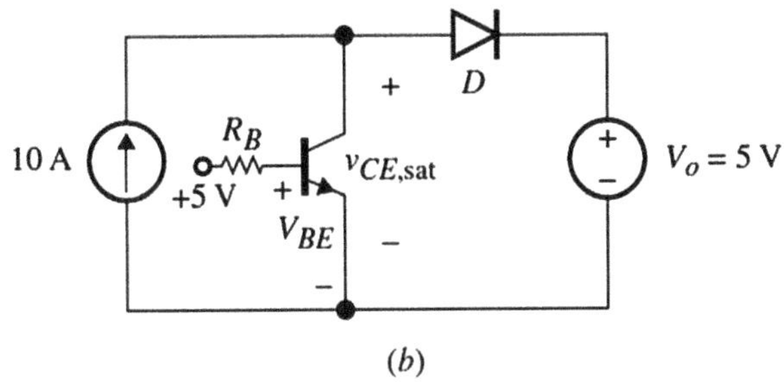

Fig. P2.22 (continued)

2.23 Consider the Darlington transistor pair given in Fig. P2.23. Assume transistor Q_1 is driven into saturation with $V_{\text{CE1, sat}} = 0.3\ V$ and $\beta_{1,\text{forced}} = 0.4$. Calculate $\beta_{2,\text{forced}}$ and $V_{\text{CE2, sat}}$. Use $V_{\text{BE1}} = V_{\text{BE2}} = 0.7\ V$ and $V_{\text{B}} = +5$ V.

Fig. P2.23

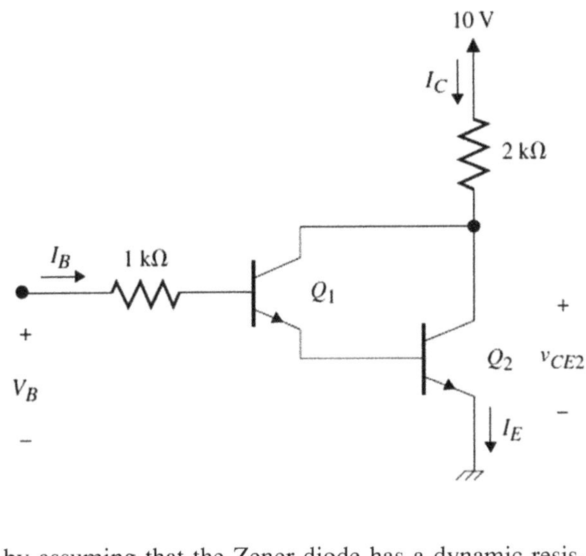

2.24 Repeat Exercise 2.1 by assuming that the Zener diode has a dynamic resistance 45 mΩ, when in regulation mode. What is the efficiency?

2.25 Consider a diode switching circuit shown in Fig. P2.25a with its corresponding diode voltage and current waveforms shown in Fig. P2.25b. At $t = 0^+$ the diode is forced to turn off while it was fully conducting for $t < 0$.

Fig. P2.25

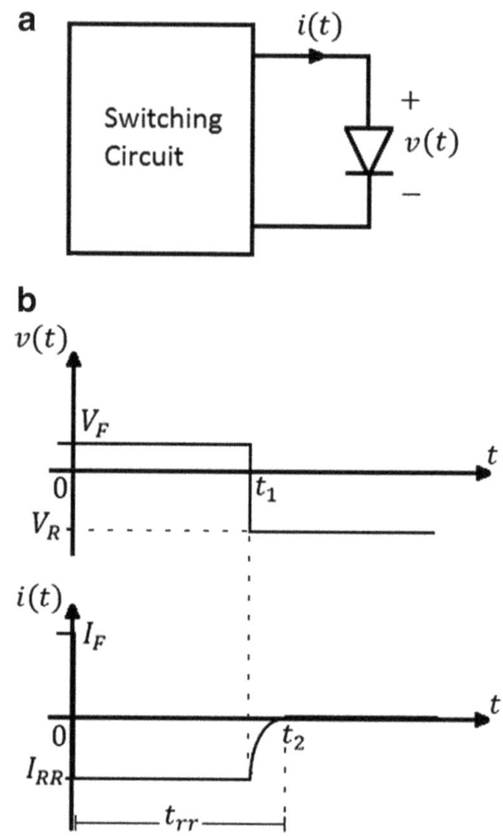

The voltage V_F is the forward voltage drop across the diode when fully conducting, and the voltage V_R is the diode blocking voltage when fully reverse biased. I_{RR} is the reverse recovery current, and $t1$ represents the delay time needed to remove the excess minority carriers from the diode depletion region. The objective of this problem is to estimate the reverse recovery time t_{rr}. Assume that the total excess charge, $q(t)$, within the depletion region is governed by the following equation:

$$\frac{dq(t)}{dt} = i(t) - \frac{q(t)}{\tau}$$

where τ is the time constant of the charge carriers removed within the depletion region.

(a) Draw one possible switching circuit that may give the resultant waveform. Explain the operation of the circuit. Assume $V_F = 1$ V, $V_R = 20$ V, $I_F = 2$ A.

(b) If we assume the time interval $t_2 - t_1 \approx 0$, compare the delay time t_1, i.e., $t_{rr} \approx t_1$. Show that t_{rr} can be approximately given by:

$$t_{\mathrm{rr}} \approx \tau.\ln\left(1+\frac{I_{\mathrm{F}}}{I_{\mathrm{RR}}}\right)$$

[Hint: $q(t_2)=q(t_1)=0$]

(c) Find I_{RR} for a diode where delay time is 100 ns. Assume $\tau=20$.

2.26 Derive the expression for the diode conduction and switching power losses for the circuit and its waveforms shown in Fig. P2.26a, b, respectively.

Fig. P2.26

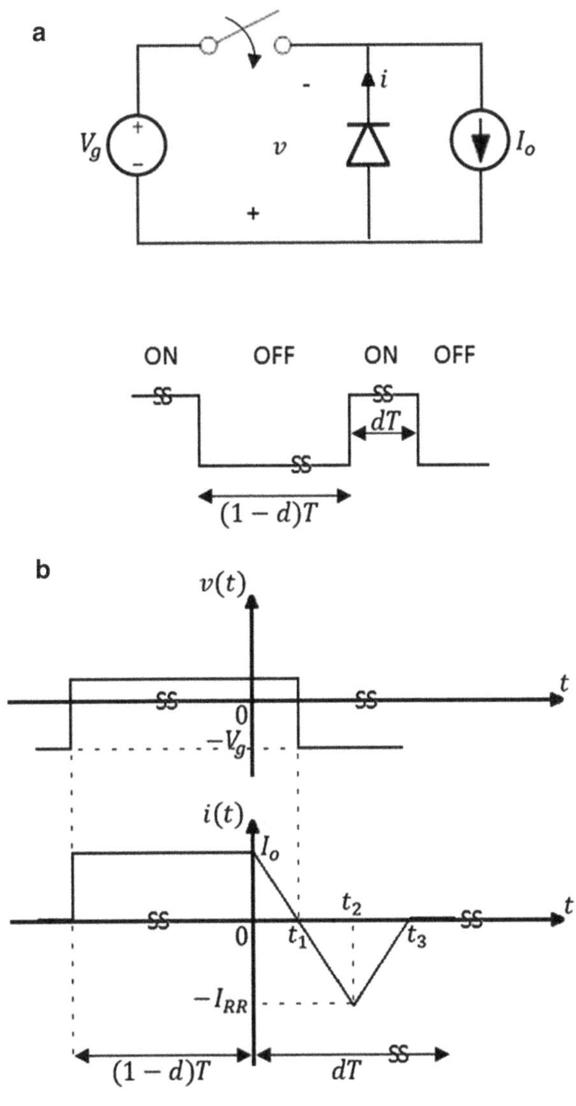

Chapter 3
Switching Circuits, Power Computations, and Component Concepts

3.1 Introduction

In this chapter, several general concepts will be discussed to provide necessary background material in power electronics. The review material will cover switching diode circuits, power computation, harmonic analysis, and component concepts.

3.2 Switching Diode Circuits

Switching power electronic circuits process power more efficiently than linear power electronic circuits. This is why power electronic design engineers must understand the analysis switching circuits. In this section, we will discuss the analysis of switching circuits that include diodes, SCRs, and ideal switches under both *dc* and *ac* excitations.

Diode circuits are nonlinear and their analysis is normally not as straightforward as the analysis of linear circuits. By adding switches to diode circuits, additional nonlinearity is introduced, making the analysis even more complex. These circuits are encountered frequently in power electronic circuits, such as in diode and SCR rectifier circuits, pulse width modulation (PWM), and resonant converters. To simplify the analyses, we will assume throughout this chapter that the diodes, SCRs, and the switches are ideal.

3.2.1 Switching Diode Circuits Under dc *Excitation*

We begin by analyzing diode resonant networks under *dc* excitation, which are often encountered in resonant type *dc-dc* and *dc-ac* power electronic circuits. It will

I. Batarseh, A. Harb, *Power Electronics*,
https://doi.org/10.1007/978-3-319-68366-9_3

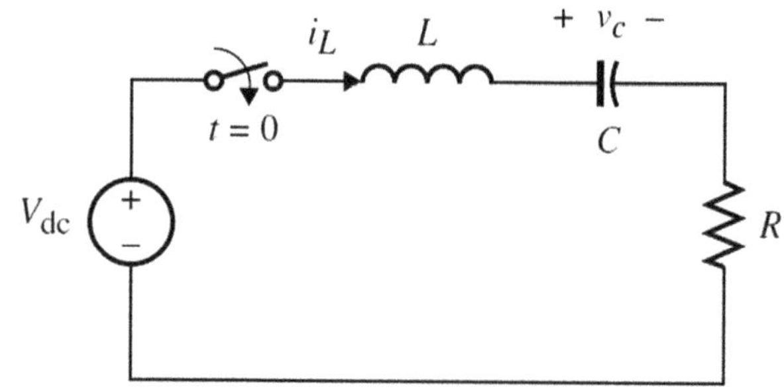

Fig. 3.1 Series resonant RLC circuit with *dc* source

be shown later that resonance is used in power electronics to achieve several important functions including filtering and soft switching. Let us first consider a series RLC circuit as shown in Fig. 3.1. Assume the switch is closed at $t = 0$, and $v_C(0^-)$ and $i_L(0^-)$ are the capacitor and inductor initial values, respectively, just before the switch is closed at $t = 0$.

The analysis of this circuit is straightforward and is given here as a review. Using KVL around the loop yields the following integral-differential equation in terms of the resonant current $i_L(t)$:

$$L\frac{di_L}{dt} + Ri_L + \frac{1}{C}\int_0^T i_L dt = V_{dc}$$

By taking the first derivative of this equation, we obtain a second-order differential equation:

$$\frac{d^2 i_L}{dt^2} + \frac{R}{L}\frac{di_L}{dt} + \frac{1}{LC} i_L(t) = 0 \tag{3.1}$$

Since the excitation is a *dc* source, there exists only a *transient* (natural) solution,[1] whose roots are obtained from the *characteristic equation* given by:

$$s^2 + \frac{R}{L}s + \frac{1}{LC} = 0 \tag{3.2a}$$

The roots of Eq. (3.2a) are given by:

$$s_{1,2} = -\frac{R}{2L} \pm \sqrt{\left(\frac{R}{2L}\right)^2 - \frac{1}{LC}} \tag{3.2b}$$

Depending on the different values of these roots, there exist three different well-understood general solutions for the transient responses:

[1] Also known as the homogeneous solution.

Case I: [Equal Real Roots]
When the roots are real and equal, we obtain what is called a *critically damped* circuit, which occurs when the following circuit condition is met:

$$\left(\frac{R}{2L}\right)^2 = \frac{1}{LC}$$

and the roots of Eq. (3.2b) become:

$$s_{1,2} = -\frac{R}{2L}$$

Under this condition, the general solution for the inductor current is given by:

$$i_L(t) = (A_1 + A_2 t)e^{-(R/2L)t} \tag{3.3}$$

and the capacitor voltage is given by:

$$\begin{aligned} v_c(t) &= V_{dc} - L\frac{di_L}{dt} - Ri_L \\ &= V_{dc} + \left[A_2\left(1 - \frac{R}{2L}t\right) - A_1\frac{R}{2L}\right]e^{-(R/2L)t} \end{aligned} \tag{3.4}$$

The constants A_1 and A_2 are obtained from the given initial conditions.

Case II: [Unequal Real Roots]
Under this condition, the circuit is known as *overdamped* and it occurs when:

$$\left(\frac{R}{2L}\right)^2 > \frac{1}{LC}$$

The general solution is given as follows:

$$i_L(t) = A_1 e^{s_1 t} + A_2 e^{s_2 t} \tag{3.5}$$

and the capacitor voltage is given by:

$$v_C(t) = V_{dc} - (Ls_1 - R)A_1 e^{s_1 t} - (Ls_2 - R)A_2 e^{s_2 t} \tag{3.6}$$

where

$$s_1 = -\frac{R}{2L} + \sqrt{\left(\frac{R}{2L}\right)^2 - \frac{1}{LC}} \tag{3.7a}$$

$$s_2 = -\frac{R}{2L} - \sqrt{\left(\frac{R}{2L}\right)^2 - \frac{1}{LC}} \tag{3.7b}$$

Case III: [Complex Pair of Roots]
Under this case, the circuit is known as *underdamped* and it occurs when:

$$\left(\frac{R}{2L}\right)^2 < \frac{1}{LC}$$

The responses are oscillatory with the general solution for $i_L(t)$ as given in Eq. (3.8):

$$i_L(t) = e^{-\alpha t}(A_1 \cos \omega_d t + A_2 \sin \omega_d t) \tag{3.8}$$

While the voltage equation is given by:

$$\begin{aligned} v_C(t) = V_{dc} - Le^{-\alpha t}[(A_2\omega_d - A_1\alpha)\cos(\omega_d t) - (A_2\alpha + A_1\omega_d)\sin(\omega_d t)] \\ - Re^{-\alpha t}[A_2 \sin(\omega_d t) + A_1 \cos(\omega_d t)] \end{aligned}$$

And the complex roots are given by:

$$\begin{aligned} s_1 &= -\alpha + j\omega_d \\ s_2 &= -\alpha - j\omega_d \end{aligned}$$

The parameters ω_d, α, and ω_0 are given by:

$$\begin{aligned} \omega_d &= \sqrt{\omega_0^2 - \alpha^2} \\ \alpha &= \frac{R}{2L} \\ \omega_0 &= \sqrt{\frac{1}{LC}} \end{aligned}$$

where ω_0 is known as the *resonant frequency*, α is called the *damping factor*, and ω_d is known as the *damped resonant frequency*. The ratio α/ω_0 is defined as the *damping ratio*, δ:

$$\delta \equiv \frac{\alpha}{\omega_0} = \frac{R}{2\sqrt{L/C}}$$

It can be shown that the constants A_1 and A_2 for a given initial capacitor voltage, $v_C(0)$, and an initial inductor current, $i_L(0)$, for the series resonant RLC circuit of Fig. 3.1, can be found using Table 3.1.

For the critically damped case, we have $\delta = 1$, and for a purely capacitive-inductive circuit ($R = 0$), we have $\delta = 0$. In the latter case, the response is purely oscillatory. Such a response is encountered frequently in *dc-dc* soft-switching power electronic circuits. Another parameter that is normally given in the RLC circuit is the *quality factor*, Q_0, which is defined as:

Table 3.1 Three possible cases and their response constants for the series resonant RLC circuit

Circuit type	A_1	A_2
Case I: critically damped	$i_L(0^-)$	$\frac{V_{dc}-v_C(0^-)}{L}+\frac{R}{2L}i_L(0^-)$
Case II: overdamped	$\frac{V_{dc}-v_C(0^-)}{L(s_1-s_2)}+\frac{s_1 i_L(0^-)}{s_1-s_2}$	$-\frac{V_{dc}-v_C(0^-)}{L(s_1-s_2)}+\frac{s_2 i_L(0^-)}{s_1-s_2}$
Case III: underdamped	$i_L(0^-)$	$\frac{V_{dc}-v_C(0^-)}{\omega_d L}-\frac{R}{2\omega_d L}i_L(0^-)$

$s_1=-\frac{R}{2L}+\sqrt{\left(\frac{R}{2L}\right)^2-\frac{1}{LC}}$, $s_2=-\frac{R}{2L}+\sqrt{\left(\frac{R}{2L}\right)^2-\frac{1}{LC}}$, $\omega_d=\sqrt{\omega_0^2-\alpha^2}$, $\alpha=\frac{R}{2L}$, $\omega_0=\sqrt{\frac{1}{LC}}$

$$Q_0 \equiv \frac{\omega_0 L}{R}=\frac{\sqrt{L/C}}{R}=\frac{1}{2\delta} \tag{3.9}$$

The higher the Q_0, the more oscillatory the current response becomes.

Another important parameter of particular interest in power electronics is what is commonly referred to as the circuit *characteristic impedance*, Z_0, which is defined as:

$$Z_0=\sqrt{\frac{L}{C}} \tag{3.10}$$

If R in the RLC circuit represents the load, then Q_0 is known as the *normalized load* and is given by:

$$Q_0=\frac{Z_o}{R} \tag{3.11a}$$

To give the same measure of oscillation in the series case for parallel resonant RLC circuits, the normalized load is defined as:

$$Q_0=\frac{R}{Z_0} \tag{3.11b}$$

The higher the Q_0, the more oscillation the voltage response becomes.

Normalized loads will be studied in *dc-dc* resonant converters in Chap. 6. Notice that Q_0 is defined the same as the quality factor of the resonant circuit. However, one should distinguish between the Q_0 of the resonant circuit in which R represents the losses in the resonant circuit that can be ignored and when R represents the load, where Q_0 becomes a normalized load, i.e., a design parameter.

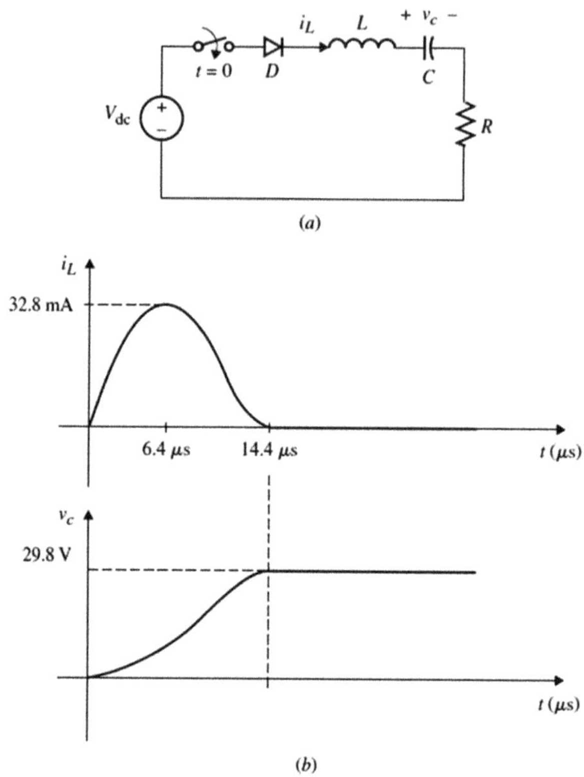

Fig. 3.2 (**a**) Circuit for Example 3.1 and (**b**) its current and voltage waveforms

Example 3.1

Consider the circuit of Fig. 3.2a with $R = 200\ \Omega$, $L = 2$ mH, $C = 0.01$ μF, and $V_{dc} = 20$ V. Derive the expressions for $i_L(t)$ and $v_C(t)$ for $t > 0$. Assume the initial inductor current and the initial capacitor voltage are zeros, i.e., $i_L(0^-) = 0$ and $v_C(0^-) = 0$.

Solution

Here we have $R/L = 100 \times 10^3$ rad/s and $1/LC = 50 \times 10^9$ rad^2/s^2. Since $(R/2L)^2 < 1/LC$, the circuit is *underdamped*; hence the roots of the characteristic equation are $s_1 = (-50 \times 10^3 - j218 \times 10^3)$ rad/s and $s_2 = (-50 \times 10^3 - j218 \times 10^3)$ rad/s.

The general solution for the inductor current is given by:

$$i_L(t) = e^{-50\times 10^3 t}\left(A_1 \cos\left(218 \times 10^3 t\right) + A_2 \sin\left(218 \times 10^3 t\right)\right) A \tag{3.12}$$

The constants A_1 and A_2 can be obtained from Table 3.1. However, for illustration purposes we will show how to find A_1 and A_2.

Since $i_L(0^+) = i_L(0^-) = 0$, then $A_1 = 0$. To solve for A_2, we use the capacitor initial condition. Applying KVL to the circuit, we obtain:

$$-V_{dc} + L\frac{di_L}{dt} + v_C + Ri_L = 0$$

Evaluating this equation at $t = 0^+$, we obtain the first derivative at $t = 0^+$:

$$\frac{di_L(0^+)}{dt} = \frac{V_{dc} - v_C(0^+)}{L} = 10 \text{ A/ms}$$

Taking the first derivative of Eq. (3.12) and evaluating it at $t = 0^+$ and setting it to 10 A/ms, we obtain the following equation:

$$\left.\frac{di_L(t)}{dt}\right|_{t=0^+} = i_L(t) = -50e^{-50\times10^3 t}A_2\left(\sin 218\times10^3 t\right) + 218\times10^3 e^{-50t}A_2\left(\cos 218t\right)\Big|_{t=0}$$
$$= 10 \text{ A/ms}$$

Solving for A_2, we obtain $A_2 = 0.046$.

The inductor current and capacitor voltage expressions for $t > 0$ are given below:

$$i_L(t) = 0.046e^{-50\times10^3 t}\sin\left(218\times10^3 t\right)A$$
$$v_C(t) = V_{dc} - Ri_L - L\frac{di_L}{dt}$$

Substituting for $i_L(t)$, the capacitor voltage is given by:

$$v_C(t) = 20 - e^{-50\times10^3 t}\left(4.6\sin\left(218\times10^3 t\right) - 20.056\cos\left(218\times10^3 t\right)\right) V$$

The diode switches *off* when $i_L(t) = 0$ which occurs when $218\times10^3 t = \pi$, or $t = \pi/(218\times10^3) = 14.4$ μs, at which the capacitor voltage equals 29.76 V. The sketches for i_L and v_C are given in Fig. 3.2b.

Example 3.2

Repeat Example 3.1 by replacing the diode across R as shown in Fig. 3.3a. Again, assume $i_L(0^-) = 0$ and $v_C(0^-) = 0$.

Solution

At $t = 0^+$, the diode begins conducting, because the inductor current just after $t > 0$ is positive (assuming an ideal diode). The equivalent circuit for $t \geq 0$ is shown in Fig. 3.3b. This circuit has a damping ratio of zero, resulting in a purely sinusoidal response with $i_L(t)$ given by:

$$i_L(t) = A_1\cos\omega_0 t + A_2\sin\omega_0 t$$

where $\omega_0 = 2.23\times10^5$ rad/s. Since the initial inductor current is zero, the constant $A_1 = 0$, and from Table 3.1, $A_2 = 45\times10^{-3} = 0.045$, then the inductor current and capacitor voltage become:

$$i_L(t) = 45\sin\omega_0 t \text{ mA}$$
$$v_C(t) = 20(1 - \cos\omega_0 t) \text{ V}$$

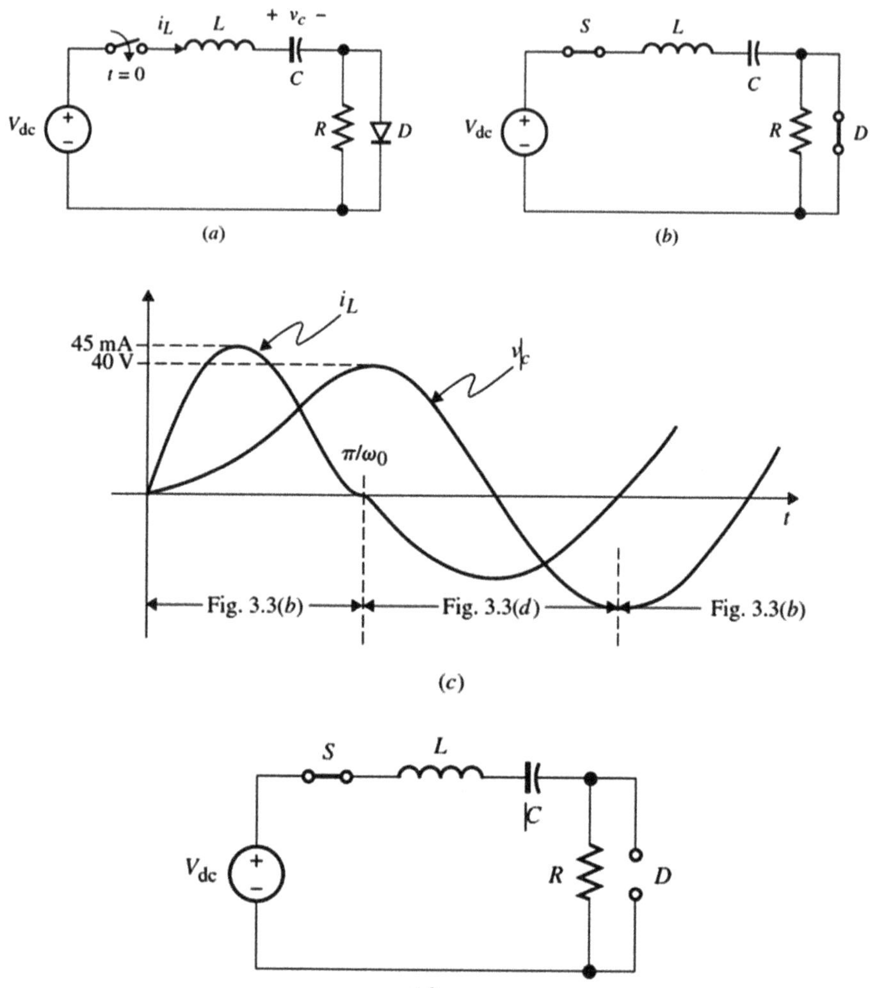

Fig. 3.3 (**a**) RLC diode circuit for Example 3.2. (**b**) Equivalent circuit for $0 \leq t \leq \pi/\omega_0$. (**c**) The inductor and capacitor waveforms. (**d**) Equivalent circuit for $t \geq \pi/\omega_0$

The inductor and capacitor waveforms are shown in Fig. 3.3c.

At $t = \pi/\omega_0$, the diode becomes reverse biased since i_L starts becoming negative. At this time the equivalent circuit model changes to a series RLC circuit with a new capacitor initial value, $v_C(\pi/\omega_0) = 40$ V, and $i_L(\pi/\omega_0) = 0$. Figure 3.3d shows the equivalent circuit for $t > \pi/\omega_0$. The solution for i_L and v_C is similar to that in Example 3.1 except that when i_L becomes zero again, the diode will conduct and the equivalent circuit mode of Fig. 3.3b becomes valid again. This process continues until the diode current decreases to zero.

Exercise 3.1
Repeat Example 3.2 with the diode direction reversed. Sketch the capacitor voltage and inductor current waveforms for $0 \le t < \pi/\omega_d$.

Answer: For $0 \le t < \pi/\omega_d$, $i_L(t) = 0.046e^{-50\times10^3 t} \sin\left(2.18 \times 10^5 t\right)$ A

$$v_C(t) = 20 - \left(4.6 \sin 2.18 \times 10^5 t + 20.1 \cos 2.18 \times 10^5 t\right) e^{-50\times10^3 t} \ v$$

Exercise 3.2
Solve for $i_L(t)$ and $v_C(t)$ after the switch is closed at $t = 0$ in Fig. E3.2. This circuit is frequently encountered in *dc-dc* resonant converters. Assume the initial values are $v_C(0^-) = 0$ and $i_L(0^-) = 1.5\ I_g$. (The circuit that established this initial inductor current is not shown.)

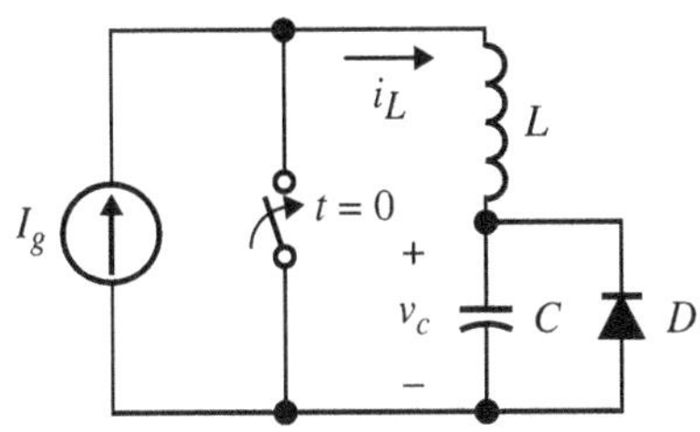

Fig. E3.2 Circuit for Exercise 3.2

Answer: $i_L(t) = 1.5Ig \cos w_0 t$A, $v_C(t) = 1.5I_g L\omega_0 \sin \omega_0 t$ V,
$\omega_0 = \sqrt{1/LC}$ for $\omega_0 t \le \pi$
$i_L(t) = -1.5Ig$, $v_C(t) = 0$, for $\omega_0 t > \pi$

Exercise 3.3
Repeat Exercise 3.2 by assuming the initial conditions are $v_C(0^-) = 0$ V and $i_L(0^-) = 0.5\ I_g$ A.

Answer: $i_L(t) = 0.5Ig \cos w_0 t$ A, $v_C(t) = 0.5I_g L\omega_0 \sin \omega_0 t$ V

$$\omega_0 = \sqrt{1/LC} \text{ for } \omega_0 t \le \pi$$

$i_L(t) = -0.5I_g$, $v_C(t) = 0$, for $\omega_0 t > \pi$

Exercise 3.4
Give switch implementation (unidirectional or bidirectional) for Exercises E3.2 and E3.3.

Exercise 3.5
Consider the resonant circuit in Fig. E3.5. These types of circuits are known as resonant circuits that exhibit "soft-switching" phenomenon to be discussed in Chap. 6. By definition, the soft-switching converter employs resonance to bring voltage across the off switch to zero before the switch is turned on (known as

zero-voltage switching (ZVS)) or to bring the current through the on switch to zero before the switch is turned off (known as *zero-current switching* (ZCS)). For the circuit shown, what is the earliest time switch S1 can be turned off and still achieve ZCS. Assume before $t = 0$, S1 has been closed while S2 has been open. As $t = 0$, S2 closes and S1 remains closed. Given parameters are Lr = 50 μH, Cr = 20 μF, Iin = 10 A, VCr(0-) = −20 V, and ILr(0-) = 0 A, and assume all components are ideal.

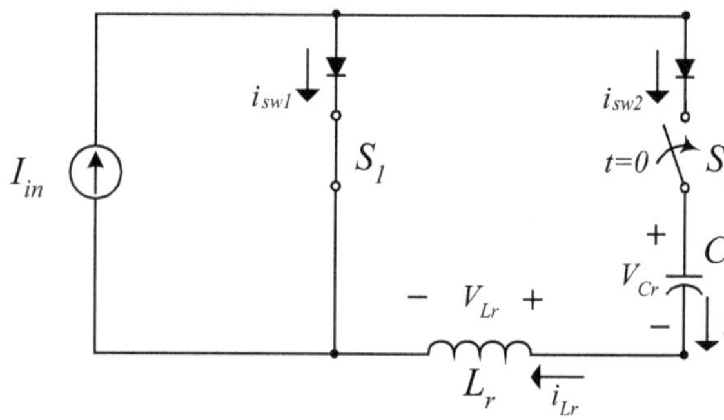

Fig. E3.5 Resonant circuit for Exercise 3.5

Answer: $T = 28.82$ μs

3.2.2 *Switching Diode Circuits with an* ac *Source*

Large classes of switching converter circuits use an *ac* source excitation rather than a *dc* source. The analysis of diode switching circuits with *ac* sources are carried out in two steps: first obtain the transient response (also known as the natural response) by setting the *ac* source to zero, and then obtain the steady-state response (also known as the *forced response*) by converting the circuit to phasor domain. The final solution is the sum of both the natural and the forced responses. Figure 3.4a shows an RLC circuit with an *ac* source, $v_s(t) = V_s \sin \omega t$ V. Assume the switch is turned on at $t = 0$, and we wish to solve for the inductor current.

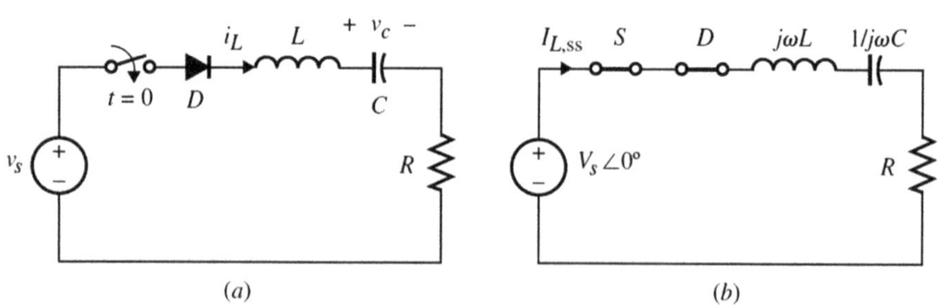

Fig. 3.4 RC circuit with *ac* excitation: (**a**) time-domain circuit and (**b**) equivalent phasor-domain circuit

For $t>0$, KVL yields the following differential-integral equation in terms of $i_L(t)$:

$$-V_s \sin \omega t + L\frac{di_L}{dt} + \frac{1}{C}\int_{-\infty}^{t} i_L(t)dt + Ri_L = 0$$

Taking the first derivative of the above equation we obtain:

$$\frac{d^2 i_L}{dt^2} + \frac{R}{L}\frac{di_L}{dt} + \frac{1}{LC}i_L(t) = \frac{V_s \omega}{L}\cos \omega t \tag{3.13}$$

The right-hand side of Eq. (3.13) represents the forced excitation. First we obtain the natural response component of the complete solution by setting the forced function (source function v_s and its derivatives) to zero, to yield:

$$\frac{d^2 i_L}{dt^2} + \frac{R}{L}\frac{di_L}{dt} + \frac{1}{LC}i_L(t) = 0 \tag{3.14}$$

The natural response of i_L is the same as the response done for the *dc* source RLC circuit shown in Table 3.1 assuming the overdamped case, i.e., the transient response, $i_{L,\text{tran}}$, is given by:

$$i_{L,tran}(t) = A_1 e^{s_1 t} + A_2 e^{s_2 t} \tag{3.15}$$

The steady-state response component is obtained by transferring the circuit into the phasor domain as shown in Fig. 3.4b. The steady-state response, $i_{L,\text{ss}}$, is easily obtained:

$$I_{L,ss} = \frac{V_s \angle 0}{R + j(\omega L - \frac{1}{\omega C})} = \frac{V_s \angle -\theta}{|Z|} \tag{3.16}$$

where

$$|Z| = \sqrt{R^2 + \left(\omega L - \frac{1}{\omega C}\right)^2} \text{ and } \theta = \tan^{-1}\left(\frac{\omega L - \frac{1}{\omega C}}{R}\right)$$

In time domain, $i_{L,\text{ss}}$ is given by:

$$i_{L,ss}(t) = \frac{V_s}{|Z|}\sin(\omega t - \theta) \tag{3.17}$$

The total response is obtained by adding Eqs. (3.15) and (3.17) to yield:

$$\begin{aligned} i_L(t) &= i_{L,tran}(t) + i_{L,ss}(t) \\ &= A_1 e^{s_1 t} + A_2 e^{s_2 t} + \frac{V_s}{|Z|}\sin(\omega t - \theta) \end{aligned} \tag{3.18}$$

The constants A_1 and A_2 are obtained from the initial condition of the inductor current and capacitor voltage at $t = 0^+$.

Example 3.3

Consider the *ac*-diode circuit shown in Fig. 3.5a with a *dc* source in the load side representing either a charged battery or a back electromotive force (emf) to excite the armature circuit in a *dc* motor. Sketch the waveforms for i_o, v_D, and v_o. Assume $v_s = 100 \sin 377t$. What is the average value of v_o?

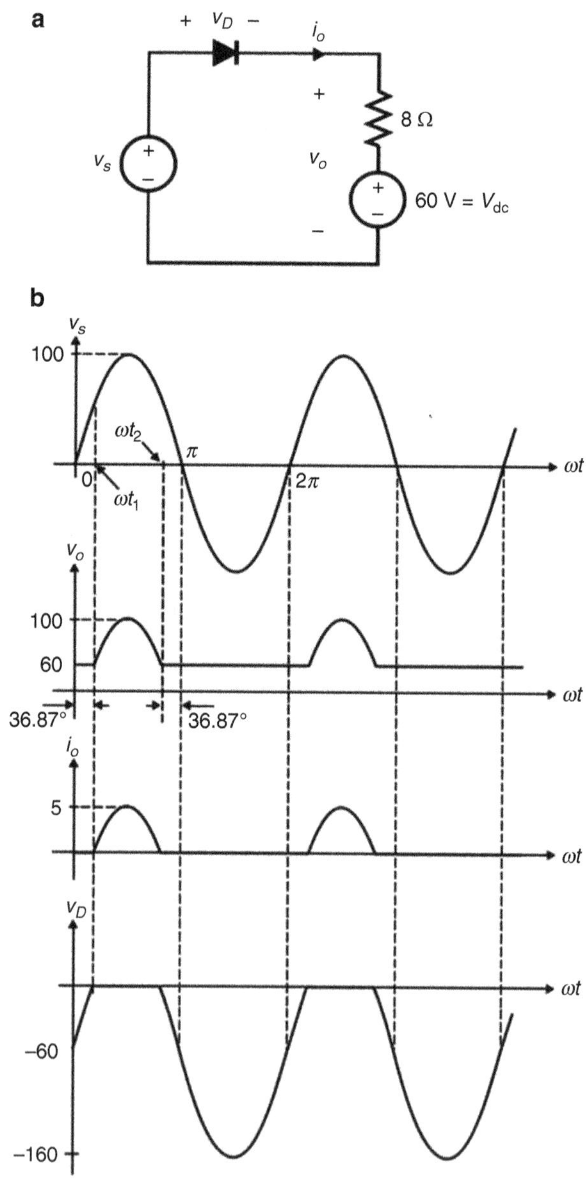

Fig. 3.5 (**a**) *Ac*-diode circuit for Example 3.3. (**b**) The voltage and current waveforms

Solution

The diode will turn *on* when $v_D > 0$ which occurs at $t = t_1$ when $v_s(t_1) = 60$ V, i.e.:

$$100 \sin 377t_1 = 60 \text{ V}$$

resulting in $t_1 = 1.7$ ms. For $0 \le t \le t_1$ the diode is *off* and for the interval $t_1 \le t < t_2$ the diode is *on*, where $t = T/2 - t_1 = 6.63$ ms.

Between $t_1 \le t < t_2$, the output voltage, v_o, equals v_s, and the output current is given by:

$$i_o(t) = \frac{v_s - V_{dc}}{R} = 12.5 \sin 337t - 7.5 \text{ A}$$

For all other times, namely, $0 \le t \le t_1$ and $t_2 \le t < T$, $i_o = 0$ and $v_o = V_{DC} = 60$ V. The average output voltage is obtained from the following equation:

$$V_{o,\text{ave}} = \frac{1}{t}\left[\int_{t_1}^{t_2} V_s \sin \omega t \, dt + V_{dc}(T/2 + 2t_1)\right]$$

$$= \frac{1}{16.67 \text{ ms}}\left[\int_{1.7 \text{ ms}}^{6.63 \text{ ms}} 100 \sin 377t \, dt + 60(8.33 \text{ ms} + 3.4 \text{ ms})\right] = 67.79 \text{ V}$$

The voltage and current waveforms are shown in Fig. 3.5b.

Exercise 3.6

Repeat Example 3.3 by including a freewheeling diode across the load as shown in Fig. E3.6.

Answer: 67.79 V

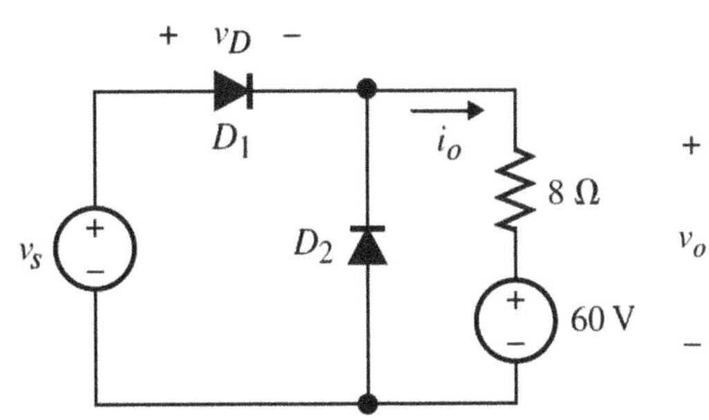

Fig. E3.6 Circuit for Exercise 3.6

Exercise 3.7

Repeat Exercise 3.6 by reversing the polarity of the 60 V *dc* source.

Answer: 31.83 V

Example 3.4
Consider the circuit shown in Fig. 3.6a with a *dc* source in the load side. Assume ideal diodes, zero initial inductor current and $V_{dc} < V_s$. Find the expressions for i_{D1}, i_{D2}, i_o, and v_o for $0 < t < 2T$, where T is the period of v_s given by $T = 2\pi/\omega$. Assume $v_s = V_s \sin \omega t$ with $V_s = 100$ V, $\omega = 377$ rad/s., $T = 16.67$ ms.

Solution
At $t = 0$ the switch closes and v_s is switched into the circuit. However, when $v_s \le V_{dc}$, D_1 remains *OFF* and so does D_2.

At $t = t_1$, $v_s = V_{dc}$, which force D_1 to turn *ON* and D_2 to turn *OFF*, resulting in the equivalent circuit of Mode 1 for $t > t_1$ as shown in Fig. 3.6b. t_1 can be determined by the following equation:

$$t_1 = \frac{1}{\omega} \sin^{-1}\left(\frac{V_{dc}}{V_s}\right) = 0.67 \text{ ms}$$

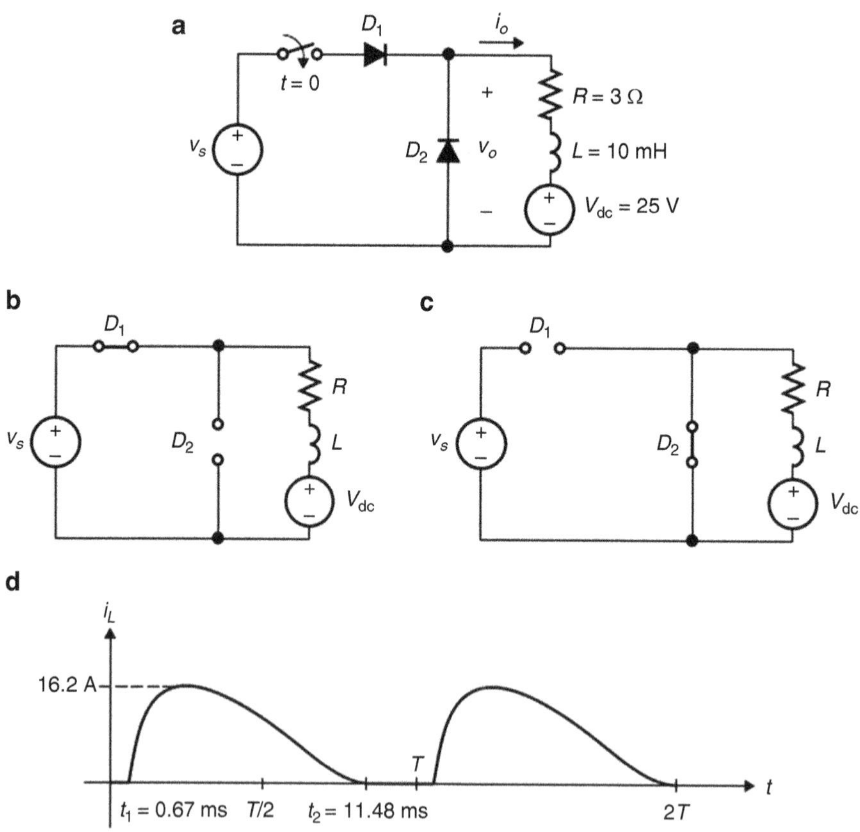

Fig. 3.6 (**a**) Circuit for Example 3.4. (**b**) Mode 1: $0 \le t < T/2$. (**c**) Mode 2: $T/2 < t < T$. (**d**) Sketch for $i_L(t)$ for $0 < t < 2T$

The differential equation for Mode 1 is given by:

$$\frac{di_{\mathrm{L}}}{dt} + \frac{R}{L} i_{\mathrm{L}} = \frac{v_{\mathrm{S}} - V_{dc}}{L}$$

The transient response of $i_{\mathrm{L}}(t)$ is:

$$i_{\mathrm{L,tran}}(t) = I_1 e^{-(t-t_1)/\tau}$$

where $\tau = L/R = 3.33$ ms, and the steady-state response is given by:

$$i_{L,ss}(t) = \frac{V_s}{|Z|} \sin(\omega t + \theta) - \frac{V_{dc}}{R} = 20.75 \sin(377t - 51.5°) - 8.33 \text{ A}$$

where $|Z| = \sqrt{R^2 + (\omega L)^2} = 4.82\ \Omega$ and $\theta = \tan^{-1}(\omega L/R) = 51.5°$.

The overall response is given by:

$$i_L(t) = I_1 e^{-(t-t_1)/\tau} + \frac{V_s}{|z|} \sin(\omega t_1 - \theta) + \frac{V_{dc}}{R} \tag{3.19}$$

The constant I_1 is obtained by setting $i_{\mathrm{L}} = 0$ at $t = t_1$ in Eq. (3.19) which yields:

$$\begin{aligned} I_1 &= -\frac{V_{\mathrm{S}}}{|z|} \sin(\omega t_1 - \theta) - \frac{V_{dc}}{R} \\ &= 20.82 \text{ A} \end{aligned}$$

The general solution for $i_{\mathrm{L}}(t)$ is given by:

$$i_{\mathrm{L}}(t) = 24.75\, e^{-(t-t_1)/\tau} + 20.75 \sin(377t - 51.5°) - 8.33 \quad 0 \le t < T/2 \tag{3.20}$$

where

$$i_L(T/2) = 10.03 \text{ A} \tag{3.21}$$

At $t = T/2$ the source voltage becomes negative, forcing D_2 to turn *ON* and D_1 to turn *OFF*. The equivalent circuit for Mode 2 is shown in Fig. 3.6c. The inductor current at $t = T/2$ becomes the initial inductor current for the next cycle at $t = T/2$.

This mode remains until the inductor current become zero again and the transient response of $i_{\mathrm{L}}(t)$ at Mode 2 is given by:

$$i_{\mathrm{L,tran}}(t) = I_2 e^{-(t-T/2)/\tau}$$

and the steady-state response is given by:

$$i_{L,ss}(t) = -\frac{V_{dc}}{R} = -8.33 \text{ A}$$

resulting in the following expression for $i_L(t)$:

$$i_L(t) = I_2 e^{-(t-T/2)/\tau} - 8.33 \text{ A}$$

The constant I_2 is obtained by setting $i_L = 10.03$ at $t = T/2$ which yields:

$$I_2 = 10.03 + 8.33 = 18.36 \text{ A}$$

and the overall response for Mode 2 can be given by:

$$i_L(t) = 18.36\, e^{-(t-T/2)/\tau} - 8.33 \quad \text{A} \quad T/2 \le t < t_2$$

where t_2 is the time when $i_L(t)$ decreases to zero. It can be found setting $i_L = 0$ at $t = t_2$ which yields:

$$t_2 = -\tau \ln \frac{8.33}{18.36} + T/2 = 10.97 \text{ ms}$$

After that $i_L(t)$ will remain zero until $t = T + t_1$. The sketch for $i_L(t)$ for $0 < t < 2T$ is shown in Fig. 3.6d.

3.3 Controlled Switching Circuits

Consider the SCR circuit shown in Fig. 3.7a driven by an *ac* voltage source $v_s = V_s \sin \omega t$. Assume the switch is turned on at $t = 0$, and the SCR is triggered by applying a gate current, i_g, at $t = t_1$.

At $t = 0$ when the switch is closed, it is still an open circuit and $i_o = 0$. At $t = t_1$, the SCR is triggered and since $v_s(t_1) = V_s \sin \omega t_1 > 0$, the SCR turns *on* resulting in the equivalent circuit shown in Fig. 3.7b.

The circuit is equivalent to the case of the *ac*-diode circuit for $t \geq 0$. The differential equation representing this circuit is given by:

$$\frac{d^2 i_L}{dt^2} + \frac{R}{L}\frac{di_L}{dt} + \frac{1}{LC} i_L(t) = \frac{V_s \omega}{L} \cos \omega t \quad t \ge t_1$$

The general solution for $i_o(t)$ for $t > t_1$ is given by:

$$i_o(t) = \frac{V_s}{|Z|} \sin[\omega t - \theta] + A_1 e^{s_1(t-t_1)} + A_2 e^{s_2(t-t_1)} \tag{3.22}$$

where

$$|Z| = \sqrt{R^2 + \left(\omega L - \frac{1}{\omega C}\right)^2}$$

$$\theta = tan^{-1} \frac{\omega L - \frac{1}{\omega C}}{R}$$

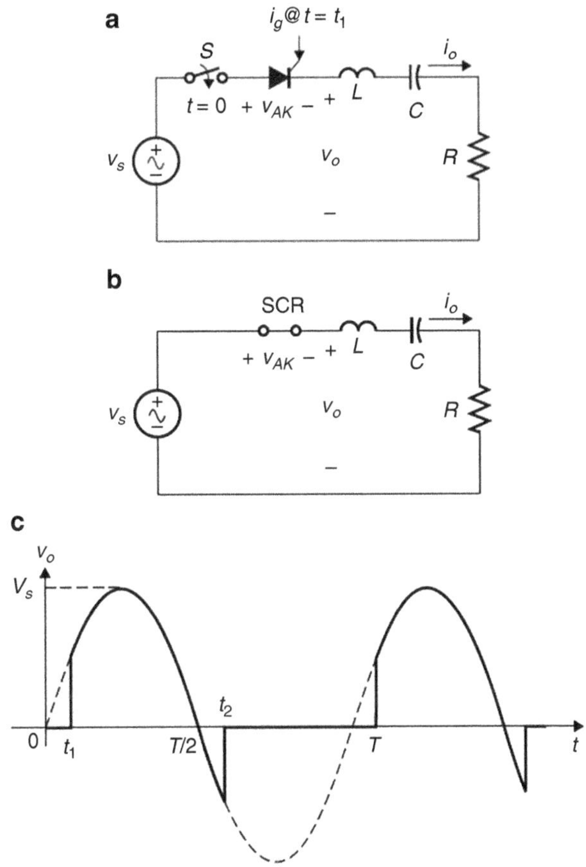

Fig. 3.7 (**a**) SCR circuit. (**b**) Equivalent circuit for $t > t_1$. (**c**) Output voltage waveform

The constants A_1 and A_2 are obtained by applying the initial conditions of the inductor and the capacitor.

The output voltage, v_o, for $t > 0$ is given by:

$$v_o = \begin{cases} 0 & 0 \le t < t_1 \\ V_s \sin \omega t & t_1 \le t < t_2 \\ 0 & t_2 \le t < T \end{cases}$$

where t_1 is the time at which SCR is first turned on and t_2 is when the SCR is turned off because $i_o(t)$ is zero for $t_2 > T/2$. If the SCR gate current is applied at $t = t_1$ into the new cycle, v_o becomes a periodical waveform as shown in Fig. 3.7c whose average value is given by:

$$v_{o,ave} = \frac{1}{T} \int_{t_1}^{t_2} V_s \sin \omega t dt = \frac{V_s}{2\pi} (\cos \alpha - cos\beta) \tag{3.23}$$

where $\alpha = \omega t_1$ and $\beta = \omega t_2$. The angle α is known as the *firing angle*, and $(\beta - \alpha)$ is known as the SCR conduction angle. Notice that by varying α, we can vary the average output voltage. Such circuits will be studied in details in Chap. 8.

Exercise 3.8

Assume the SCR circuit in Fig. 3.7a is triggered at $\alpha = 30^\circ$ (i.e. $t = 1.39$ ms) after the switch is closed at $t = 0$. Derive the expression for $i_L(t)$ and $v_c(t)$ for $0 < \omega t < 2\pi$, assuming zero initial condition. Use $R = 200\ \Omega$, $L = 2$ mH, $C = 0.01\ \mu F$, and $v_s(t) = 20 \sin 377t$.

Answer:

$$i_L(t) = e^{-\alpha(t-t_1)}[A_1 \cos \omega_d(t - t_1) + A_2 \cos \omega_d(t - t_1)] + 75.4 \times 10^{-6} \sin\left(\omega t + 89.9^\circ\right)$$
$t_1 < t < t_2$

where
$\alpha = 50 \times 10^3$, $\omega_d = 217{,}945$, $t_1 = 1.39$ ms, $t_2 = 1.404$ ms, $A_1 = -65.35 \times 10^{-6}$, $A_2 = 22.92 \times 10^{-3}$

Exercise 3.9

Repeat E3.8 by placing an ideal diode as shown in Fig. E3.9.

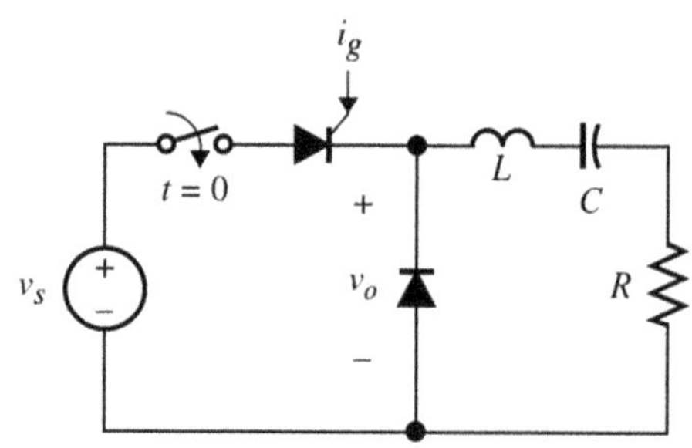

Fig. E3.9 Circuit for Exercise 3.9

Answer:

$$i_L(t) = e^{-\alpha(t-t_1)}[A_1 \cos \omega_d(t - t_1) + A_2 \cos \omega_d(t - t_1)] + 75.4 \times 10^{-6} \sin\left(\omega t + 89.9^\circ\right)$$

where
$\boldsymbol{\alpha} = 50 \times 10^3$, $\boldsymbol{\omega_d} = 217{,}945$, $\boldsymbol{t_1} = 4.27$ **ms**, $\boldsymbol{A_1} = -75.4 \times 10^{-6}$, $\boldsymbol{A_2} = -17.3 \times 10^{-3}$

3.4 Basic Power and Harmonic Concepts

In this section we will review some basic power concepts applied to sinusoidal and non-sinusoidal current waveforms that are of particular importance in power electronic circuits.

3.4.1 Average, Reactive, and Apparent Powers

Power Flow

As stated earlier, the function of the power electronic circuit is to process power by performing some conversion function through a set of switching actions dictated by some control circuit. The direction of power flow in a power electronic circuit is an important concept since it relates to identifying the input and output ports in the power electronic circuit. Normally, the situation exists as shown in Fig. 3.8a where the power flow is from the input side to the output side. Here P_{in} and P_{out} represent

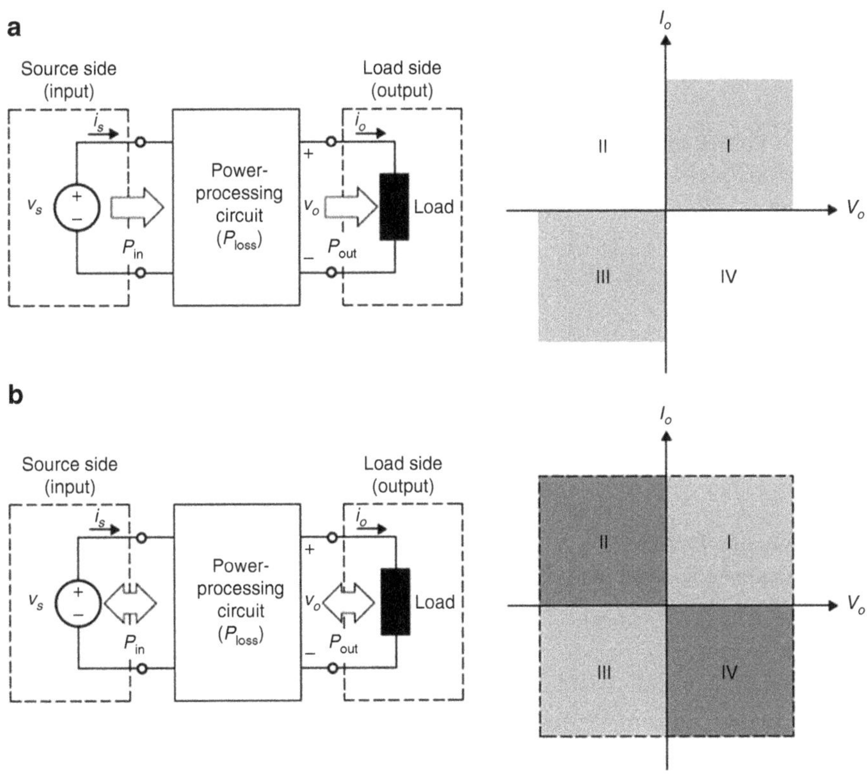

Fig. 3.8 Power flow: (**a**) unidirectional (input-to-output) (**b**) bidirectional

the average input power from the source side (input *ac* or *dc*) and the average output power at the load side (output *ac* or *dc*), respectively. One may conclude from Fig. 3.8a that the average power flow is from the input terminal (source side) to the output terminal (load side); hence, the direction of power flow becomes the basis for defining the power circuit port. However, since some power electronic circuits have a *dc* source in the load side, it is possible to have the power flow in the opposite direction, i.e., the circuit is capable of bidirectional power flow, as shown in Fig. 3.8b. As a result, one has to be careful in identifying the source and load sides. A good discussion on this issue, supported by several examples, is given in Kassakian et al.

The efficiency of the power processing circuit is an extremely important parameter since it has a direct impact on the cost, performance, size, and weight of the system as discussed in Chap. 1. For Fig. 3.8a, the efficiency, η, is defined by:

$$\eta = \frac{P_{out}}{P_{in}} \times 100\%$$
$$= \frac{P_{out}}{P_{out} + P_{loss}}$$

If the power circuit consists of ideal switching devices that operate in either the *on*- or the *off*-states and lossless energy storage elements like capacitors, inductors, and transformers, then the overall efficiency of the power processing circuit is 100%.

Average Values and rms

For a given periodical voltage signal, $v(t)$, with period T, its average value is defined by:

$$V_{ave} = \frac{1}{T}\int_0^T v(t)dt \tag{3.24}$$

and the *root-mean-square* (rms) value is given by:

$$V_{rms} = \sqrt{\frac{1}{T}\int_0^T v^2(t)dt} \tag{3.25}$$

Instantaneous Power

The instantaneous power delivered to a load element that has $v(t)$ across it and $i(t)$ through it, as shown in Fig. 3.9a, is given by:

$$P(t) = v(t)\, i(t) \tag{3.26}$$

The voltage and current waveforms can be sinusoidal, periodical, or constant as shown by the arbitrary waveforms in Fig. 3.9c. Notice that the instantaneous power in this figure can be either positive, zero, or negative. For example,

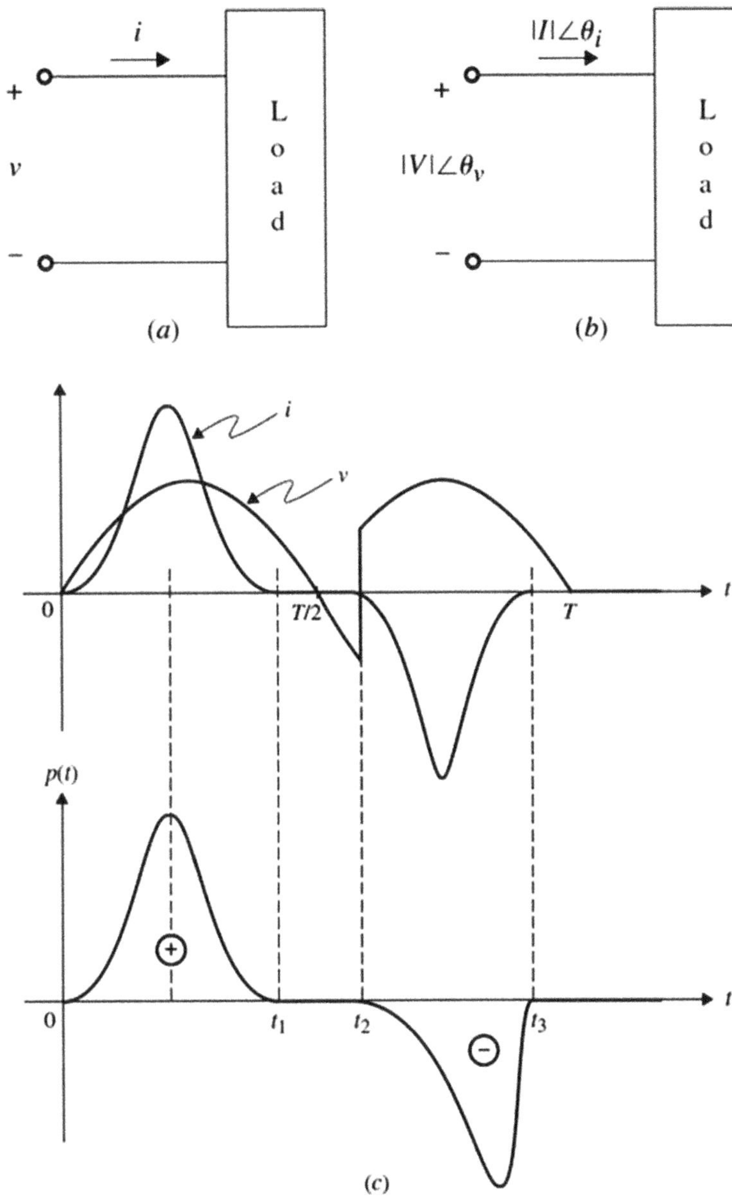

Fig. 3.9 (**a**) Time-domain and (**b**) phasor-domain elements and (**c**) arbitrary current and voltage waveforms

between $0 \leq t \leq t_1$, the circuit element absorbs or dissipates power (positive), between $t_1 \leq t \leq t_2$, the power transfer to the element is zero, and finally between $t_2 \leq t \leq t_3$, the load generates power (positive) or returns power (negative) to the source.

Average Power
If the voltage and current repeat periodically, then we can define the total *average power* either generated or dissipated by the circuit element, which is given by:

$$\begin{aligned} P_{\text{ave}} &= \frac{1}{T}\int_0^T p(t)dt \\ &= \frac{1}{T}\int_0^T i(t)v(t)dt \end{aligned}$$

The average power is also known as real power, which comes from the complex number representation. In terms of phasor representation, the linear time-domain circuit element of Fig. 3.9a is redrawn as a phasor-domain circuit element as shown in Fig. 3.9b.

Apparent Power
The phasor magnitudes $|I_s|$ and $|V_s|$ represent the peak of the source current and voltage, respectively, and θ_i, θ_v represent the current and voltage phase shift, respectively. If we let the total impedance of the element be given by:

$$Z = R + jX = |Z|e^{j\theta}$$

where X is the total reactance of a capacitive circuit, $-1/\omega C$, or a total reactance of an inductive circuit, ωL, then the total input complex power may be defined by:

$$\begin{aligned} P_{\text{T}} &= \frac{VI^*}{2} = \frac{|V||I|}{2}e^{j(\theta_v - \theta_i)} \\ &= V_{rms}I_{rms}e^{j\theta} \\ &= Se^{j\theta} \end{aligned}$$

where I^* is the complex conjugate of I. The parameter S is known as *apparent power*, and θ represents the total phase shift between $i(t)$ and $v(t)$ and is also known as *power factor angle*.

Normally, P_{T} is expressed in terms of real and imaginary complex numbers as follows:

$$P_T = P + jQ \tag{3.27}$$

The real part, P, is the *average power*, which is given by:

$$\begin{aligned} P &= S\cos\theta \\ &= V_{rms}I_{rms}\cos\theta \end{aligned} \tag{3.28}$$

and Q is the *reactive power*, which is given by:

$$\begin{aligned} Q &= S\sin\theta \\ &= V_{rms}I_{rms}\sin\theta \end{aligned} \tag{3.29}$$

From these expressions, S can also be expressed mathematically as:

$$S = \sqrt{P^2 + Q^2} \tag{3.30}$$

The units of P are watts, representing the power being dissipated. Units for Q [2]are volt-amperes reactive (var), representing the reactive power being stored in the inductor or/and capacitor, and the units for S is volt-amp (VA), representing the rms product of the voltage and current values. The reactive power is not a useful parameter, and it is normally desired to make Q equal to zero, which means the total power is equal to the real or average power. Knowing the Q of the circuit helps the designer to compensate so that the load always draws real power. This case corresponds to a unity power factor to be discussed shortly.

Exercise 3.10
Consider the one-port network of Fig. E3.10 with $i(t)$ a triangular waveform. Determine the average and rms current values and the average power absorbed by the network in steady state under the following cases: (i) purely resistive with $R = 0.5\ \Omega$, (ii) purely inductive $L = 1$ mH, (iii) purely capacitive $C = 1\ \mu$F, and (iv) resistive-inductive with $R = 0.5\ \Omega$ and $L = 1$ mH. Assume zero initial conditions.

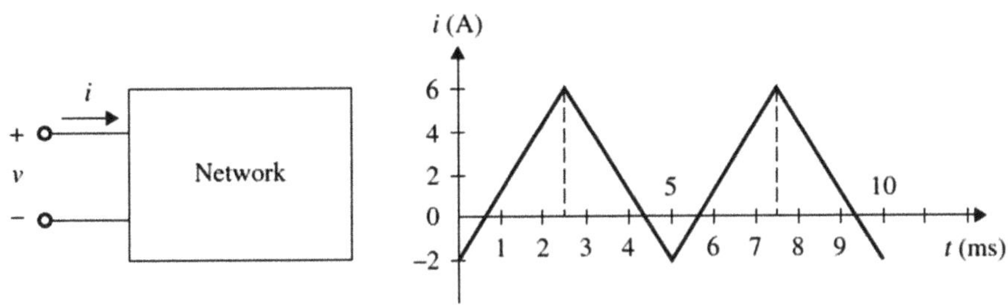

Fig. E3.10 A network and its input current waveform for E3.10

Answer: 2 A, 3.1 A, 4.66 W, 0 W, 0 W, 4.66 W.

3.4.2 *Sinusoidal Waveforms*

3.4.2.1 Instantaneous and Average Powers

First let us consider the case for linear, one-port network shown in Fig. 3.10a. Since the network consists of linear components, its instantaneous source current and voltage expression may be represented as follows:

[2]This Q has nothing to do with resonant circuit parameter Q discussed earlier and the one to be discussed in Chap. 6.

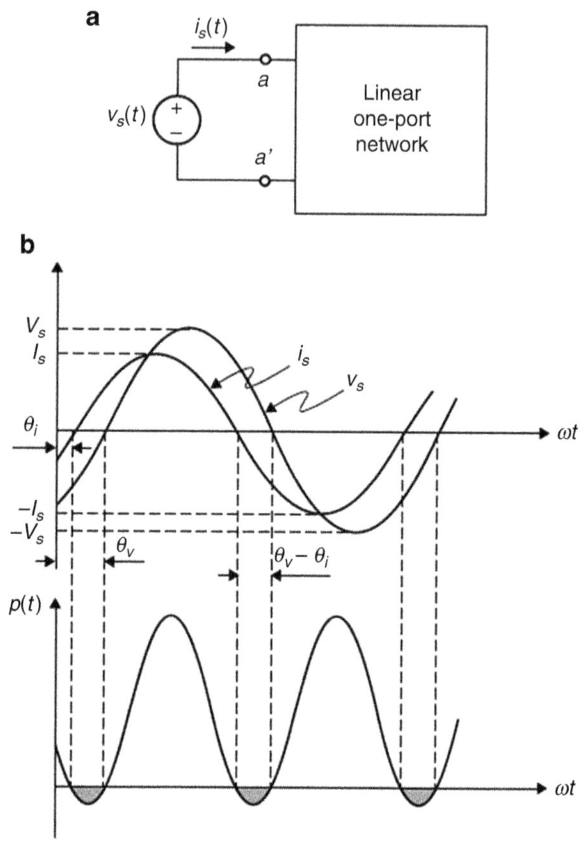

Fig. 3.10 (**a**) Linear one-port circuit. (**b**) Circuit waveforms

$$i_s(t) = I_s \sin(\omega t - \theta_i) \tag{3.31a}$$

$$v_s(t) = V_s \sin(\omega t - \theta_v) \tag{3.31b}$$

where θ_i and θ_v are the port current and voltage phase shift, respectively.

The *instantaneous power* $p(t)$ is given by:

$$\begin{aligned} p(t) &= i_s v_s \\ &= I_s V_s \sin(\omega t - \theta_i) \sin(\omega t - \theta_v) \end{aligned} \tag{3.32}$$

The waveforms for $i_s(t)$, $v_s(t)$, and $p(t)$ are shown in Fig. 3.10b.

The input average power can be calculated from the following integral:

$$\begin{aligned} P_{ave} &= \int_0^T p(t)dt \\ &= \frac{1}{T}\int_0^T i_s(t)v_s(t)dt \end{aligned} \tag{3.33}$$

Substituting for i_s and v_s from Eqs. (3.31) and by using the following trigonometric identities:

$$\cos(\theta_1 \pm \theta_2) = \cos\theta_1 \cos\theta_2 \pm \sin\theta_1 \sin\theta_2$$
$$\sin(\theta_1 \pm \theta_2) = \sin\theta_1 \cos\theta_2 \pm \cos\theta_1 \sin\theta_2$$

the instantaneous power may be expressed as follows:

$$p(t) = \frac{I_s V_s}{2}[\cos(\theta_v - \theta_i) - \cos(2\omega t + \theta_v + \theta_i)] \tag{3.34}$$

Substitute Eq. (3.34) in the integral of Eq. (3.33), and the average power becomes:

$$P_{ave} = \frac{I_s V_s}{2}[\cos(\theta_v - \theta_i)] \tag{3.35}$$

In terms of the rms parameters, the average power is given by:

$$P_{ave} = I_{s,rms} V_{s,rms} \cos(\theta_v - \theta_i) \tag{3.36}$$

3.4.2.2 Power Factor

Power factor is a very important parameter in power electronics because it gives a measure of the effectiveness of the real power utilization in the system. It also represents a measure of distortion of line voltage and line current and the phase shift between them. Let us consider Fig. 3.10a in providing the basic definition of power factor.

Power factor (*pf*) is defined as the ratio of the average power measured at the terminals a-a' of Fig. 3.10a and the rms product of v_s and i_s as given in Eq. (3.37):

$$\text{Power factor} = \frac{\text{Real power (average)}}{\text{Apparent power}} \tag{3.37}$$

For purely sinusoidal current and voltage waveforms, the average power is given in Eq. (3.36) and the apparent power is given by $I_{s,\,rms}V_{s,\,rms}$. As a result, Eq. (3.37) yields:

$$\begin{aligned} \text{Power factor} &= \frac{I_{s,rms} V_{s,rms} \cos\theta}{I_{s,rms} V_{s,rms}} \\ &= \cos\theta \end{aligned} \tag{3.38}$$

Hence in linear power systems, when the line voltage and line currents are purely sinusoidal, the power factor is equal to the cosine of the phase angle between the current and voltage. However, in power electronic circuits, due to the switching of

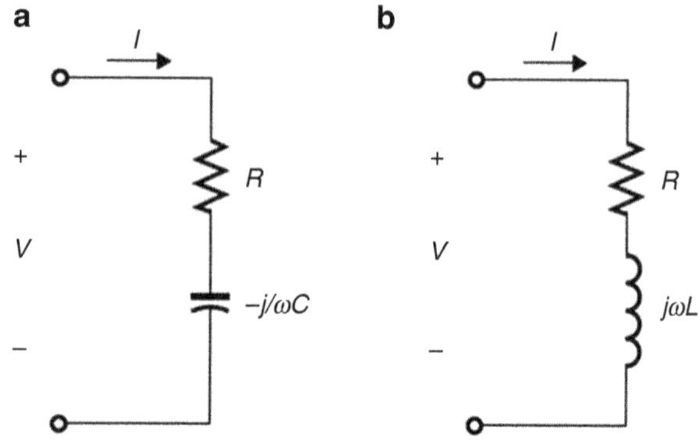

Fig. 3.11 (**a**) Leading power factor. (**b**) Lagging power factor

(a)

$$Z = R - j\frac{1}{\omega C}$$

$$\frac{V}{I} = |Z|\angle\theta$$

$$|Z| = \sqrt{R^2 + \left(\frac{1}{\omega C}\right)^2}$$

$$\theta = \tan^{-1}\frac{-1}{\omega RC} < 0^\circ$$

(b)

$$Z = R + j\omega L$$

$$\frac{V}{I} = |Z|\angle\theta$$

$$|Z| = \sqrt{R^2 + (\omega L)^2}$$

$$\theta = \tan^{-1}\frac{\omega L}{R} > 0^\circ$$

active power devices, the phase angle representation alone is not valid. This is why we will shortly define power factor for terminals, whose currents and/or voltages are non-sinusoidal (distorted).

The angle θ is known as the *power factor angle*; therefore, the power factor varies between 0 and 1, depending on the type of the network. For $\theta > 0$, the current lags the voltage, representing inductive-resistive load as shown in Fig. 3.11b. The network load is said to be having a lagging power factor. Similarly, for $\theta < 0$, the current leads the voltage, representing capacitive-resistive load with leading power factor as shown in Fig. 3.11a.

Let us calculate the power factor for resistive, inductive, and capacitive two-terminal networks:

1. Resistive network

   The voltage and current relation is given by:

$$v_s = i_s R$$

   and the power factor angle is $\theta = \theta_v - \theta_i = 0$, resulting in a power factor equal to 1.

2. Capacitive network

   The capacitor current and voltage relation is given by:

$$i_s = C\frac{dv_s}{dt}$$

In phasor domain, we have:

$$\frac{|V_\mathrm{S}|\angle\theta_\mathrm{v}}{|I_\mathrm{S}|\angle\theta_\mathrm{i}} = -j\frac{1}{\omega C}$$

$$\frac{|V_S|}{|I_s|}\angle\theta_v - \theta_i = \frac{1}{\omega C}\angle -90^\circ$$

Therefore, the power factor angle is $\theta_\mathrm{v} - \theta_\mathrm{i} = \theta = -90^\circ$, resulting in zero power factor. This means the purely capacitive circuit has no average power delivered, (as expected from an ideal capacitor). This is a leading power factor because current leads voltage by 90°.

3. Inductive network

$$\frac{|V_S|\angle\theta_v}{|I_s|\angle\theta_i} = -j\omega L$$

$$\frac{|V_S|}{|I_s|}\angle\theta_v - \theta_i = \omega L\angle 90^\circ$$

The power factor angle is $\theta = +90^\circ$, resulting in lagging power factor because current lags voltage by 90°.

Example 3.5
Determine L in the circuit of Fig. 3.12 so that the power factor becomes unity.

Solution
The total impedance seen by the source is given by:

$$Z_\mathrm{in} = j\omega L - j\frac{1}{\omega C} + R_L$$

For unity power factor, the phase angle of Z_in must be zero, i.e., Z_in is a real number. Setting the imaginary part to zero yields:

$$\omega L - \frac{1}{\omega C} = 0$$

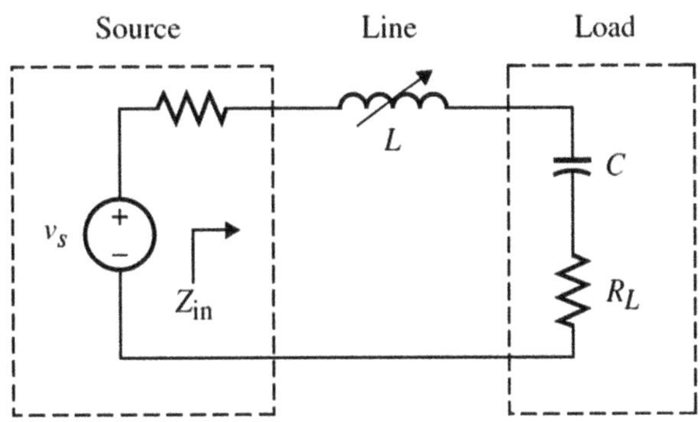

Fig. 3.12 Circuit for Example 3.5

and solving for the inductor value, we obtain:

$$L = \frac{1}{\omega^2 C}$$

Exercise 3.11
Consider the circuit of Fig. E3.11 with 10 kVA load and $v_s = 100 \sin \omega t$.

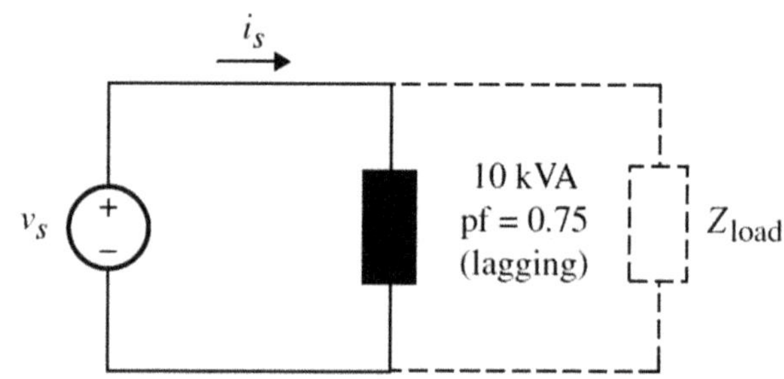

Fig. E3.11 Circuit for Exercise 3.11

It is required to compensate the load to improve power factor to 0.97. Determine the type and value of the load impedance Z_{load} that must be added in parallel to achieve $pf = 0.97$, and repeat the process to achieve $pf = 1.0$.

Answer: For pf = 0.97, $Z_{load} = -j1.056\Omega$ (C = 2.5 mF for ω = 60 Hz) and for $pf = 1.0$, $Z_{load} = -j0.756\Omega$ (C = 3.5 mF for ω = 60 Hz)

3.4.3 Non-sinusoidal Waveforms

Harmonics cause serious problems of interference with sensitivity measurements and communication systems. To reduce or eliminate line voltage and current harmonics, it is necessary to add filters in the *ac* input side. *Ac* and *dc* filter design is a very specialized topic, and the literature is rich in analysis, design, and implementation of such filters. For this reason the topic will not be addressed in this textbook. The voltage source of an ideal electrical power system supplies energy at a constant and single frequency with constant voltage amplitude under all load conditions. However, in practical power systems, single and constant frequency and fixed amplitude voltage sources are not available.

The importance of studying current and voltage harmonics has grown recently because of the widespread use of power switching devices in various power electronics applications.

The topic of studying a system's harmonics is specialized and cannot be fully addressed in a textbook like this. Nevertheless, some important harmonics issues will be addressed here. The reader is encouraged to see the references listed at the end of this chapter

Because of the switching nature of the majority of power electronic circuits, the line (source) current is highly distorted while the line voltage remains nearly sinusoidal. Only in a limited number of circuits we do have almost sinusoidal line voltage and current. In steady state, the non-sinusoidal nature of the line current produces unwanted oscillatory components at different frequencies. Such signals are called harmonics or harmonic components. Under some load conditions, these harmonics have high amplitudes that result in highly undesirable effects. These harmonics must be removed or at least significantly reduced. As a result, it becomes necessary to study the harmonics and power factor values in non-sinusoidal current waveforms. We will assume that the distorted waveforms are in steady state with a given fundamental frequency. The existence of these harmonics affects the overall efficiency, performance, and cost of the power electronic system. It will be shown that because of this distortion, the apparent power rating (volt-amp) of the source must be higher than the actual real power needed by the load.

The best tool available to study harmonics is the use of the Fourier analysis method. The basis for harmonic calculations is the Fourier theorem introduced by the French mathematician Jean Baptiste Joseph Fourier in 1822. First we review the Fourier analysis technique and harmonic components.

3.4.3.1 Fourier Analysis

The Fourier theorem states that, in steady state, for any given periodic function *f*(*t*), one can represent it by the sum of a constant F_0 and infinite sine and cosine functions (f_2,f_2, . . .) defined by the following formula:

$$\begin{aligned} f(t) &= F_0 + f_1(t) + f_2(t) + \ldots + f_n(t) \\ &= F_0 + \sum_{n=1}^{\infty}(a_n \cos n\,\omega t + b_n \sin n\,\omega t) \end{aligned} \tag{3.39}$$

The constant coefficient can be obtained by taking the integral of both sides of Eq. (3.39) from 0 to *T* to give the following expression for F_0:

$$F_O = \frac{1}{T}\int_0^T f(t)dt \tag{3.40}$$

From Eq. (3.40), we see that F_0 represents the average (*dc*) value of *f*(*t*).

Also it can be shown that the coefficient a_n and b_n are evaluated from the following integrals:

$$a_n = \frac{2}{T}\int_0^T f(t)\cos n\omega t\,dt \qquad n = 1, 2, 3, \ldots, \infty \tag{3.41a}$$

$$b_n = \frac{2}{T}\int_0^T f(t)\sin n\omega t\,dt \qquad n = 1, 2, 3, \ldots, \infty \tag{3.41b}$$

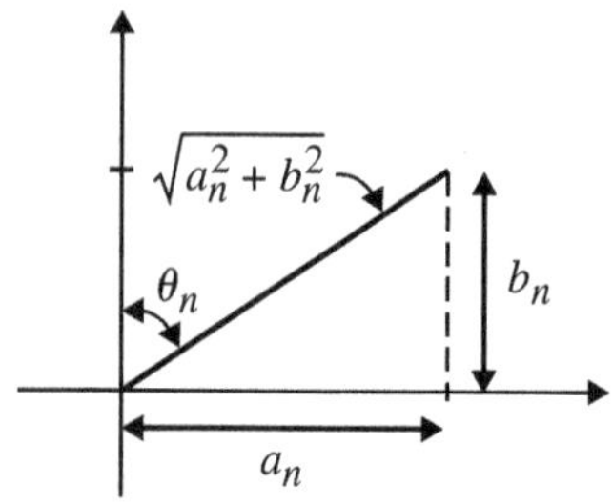

Fig. 3.13 Phase representation for Eq.(3.42)

These equations constitute a frequency domain representation of *f*(*t*)[3] and suggest that any non-sinusoidal waveform with frequency ω, (the waveform repeated every time period $T = 2\pi/w$) can be expressed as the sum of a constant and infinite sinusoidal waveforms at multiples ($n\omega$, with $n = 1, 2, \ldots$) of the original frequency. The frequency of the original waveform (at $n = 1$) is known as the *fundamental frequency*. The values of *f*(*t*) *at* $n = 1, 2, 3, \ldots, \infty$ are known as the Fourier components of the waveform *f*(*t*) or the harmonic components of *f*(*t*). At $n = 1$, the component is usually called the *fundamental component* or fundamental harmonic, denoted by the subscript 1.

By defining the triangle shown in Fig. 3.13, it is also possible to represent *f*(*t*) in the following form:

$$f(t) = F_o + \sum\nolimits_{n=1}^{\infty} F_n \sin(n\,\omega t + \theta_n) \tag{3.42}$$

where F_n is the peak value of the nth harmonic and θ_n is the phase shift, which are given by:

$$F_n = \sqrt{a_n^2 + b_n^2}$$

$$\theta_n = \tan^{-1}\left(\frac{a_n}{b_n}\right)$$

If the waveform has any form of symmetry shown in Table 3.2, calculation of the integrals for a_n and b_n can be significantly reduced. In power electronic circuits, odd symmetry waveforms are more frequently encountered than the even symmetry waveforms.

[3] It is also common to represent a_n and b_n in terms of frequency as follows:

$$a_n = \frac{1}{\pi}\int_0^{2\pi} f(\omega t)\,\sin(n\omega t)\,d(\omega t)\;\; b_n = \frac{1}{\pi}\int_0^{2\pi} f(\omega t)\,\cos(n\omega t)\,d(\omega t)\;\text{ and }\; F_0 = \frac{1}{2\pi}\int_0^{2\pi} f(\omega t)\,d(\omega t)$$

Table 3.2 Simplification of Fourier coefficients due to function symmetry

Symmetry	a_n	b_n
Odd symmetry $f(t) = -f(-t)$	$a_n = 0$ (for all n)	$b_n = \frac{4}{T}\int_0^{T/2} f(t) \sin n\,\omega t\, dt$
Even symmetry $f(t) = f(-t)$	$a_n = \frac{4}{T}\int_0^{T/2} f(t) \cos n\,\omega t\, dt$	$b_n = 0$ (for all n)
Half-wave symmetry $f(t) = -f(t + T/2)$	$a_n = \begin{cases} \frac{4}{T}\int_0^{T/2} f(t) \cos n\,\omega t\, d & n \text{ odd} \\ 0 & n \text{ even} \end{cases}$	$b_n = \begin{cases} \frac{4}{T}\int_0^{T/2} f(t) \sin n\,\omega t\, d & n \text{ odd} \\ 0 & n \text{ even} \end{cases}$
Odd and half-wave symmetry $f(t) = -f(t + T/2)$ $f(t) = -f(-t)$	$a_n = 0$ (for all n)	$b_n = \begin{cases} \frac{8}{T}\int_0^{T/2} f(t) \sin n\,\omega t\, d & n \text{ odd} \\ 0 & n \text{ even} \end{cases}$
Even and half-wave symmetry $f(t) = -f(t + T/2)$ $f(t) = f(-t)$	$a_n = \begin{cases} \frac{8}{T}\int_0^{T/2} f(t) \cos n\,\omega t\, d & n \text{ odd} \\ 0 & n \text{ even} \end{cases}$	$b_n = 0$ (for all n)

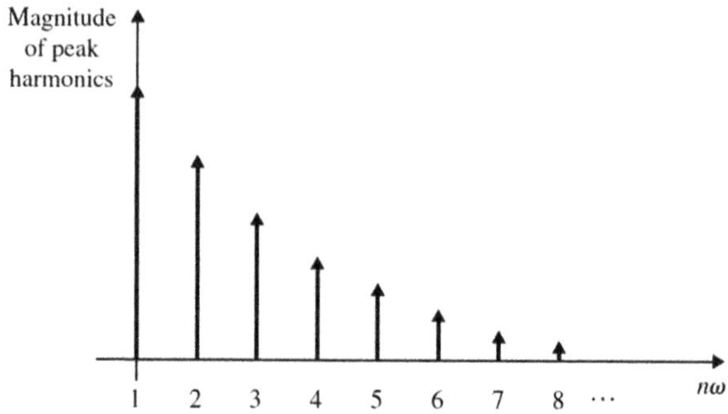

Fig. 3.14 Frequency spectrum representation for f(t)

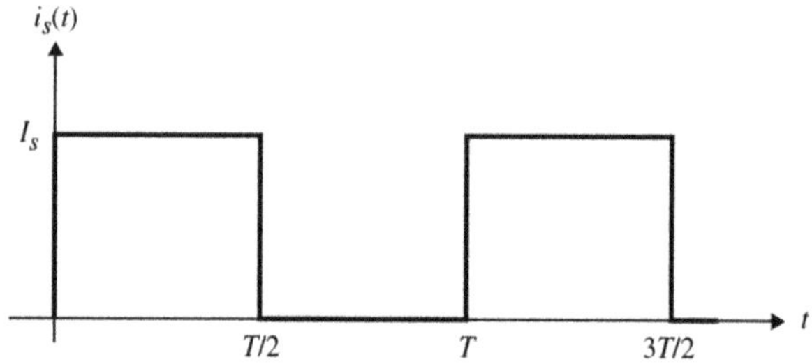

Fig. 3.15 Current waveform for Example 3.6

Another useful representation of $f(t)$ is in terms of its frequency spectrum as shown in Fig. 3.14. It is a plot of the magnitude of the harmonic component against its frequency. As n increases the peak of the higher harmonic components decreases. The peak value at F_1, i.e., the fundamental, is normally referred to as the "wanted" harmonic or component, whereas the higher components represent the "unwanted" components.

Example 3.6
Consider a half-wave power electronic circuit whose line current $i_s(t)$ waveform is shown in Fig. 3.15. Calculate the harmonics of $i_s(t)$.

Solution
The average value is clearly $I_s/2$, so we have:

$$F_0 = \frac{I_s}{2}$$

The coefficients a_n are given by:

$$\begin{aligned} a_n &= \frac{1}{\pi}\int_0^{2\pi} f(\omega t)\cos n\,\omega t d(\omega t) \\ &= \frac{1}{\pi}\int_0^{\pi} I_s \cos(n\,\omega t) dt = \frac{I_s}{n\pi}\sin(n\,\omega t)\Big|_0^{\pi} = 0 \qquad n = 1, 2, \end{aligned}$$

The coefficients b_n are given by:

$$b_n = \frac{1}{\pi}\int_0^{2\pi} f(\omega t)\sin n\,\omega t d(\omega t)$$

$$= \frac{1}{\pi}\int_0^{\pi} I_s \sin(n\,\omega t)dt = \frac{-I_s}{n\pi}\cos(n\,\omega t)\Big|_0^{\pi} = \frac{2I_s}{n\pi} \qquad n = 1, 3, 5$$

Exercise 3.12
Determine the average and rms values for the output voltage waveform for a half-wave rectifier circuit shown in Fig. E3.12:

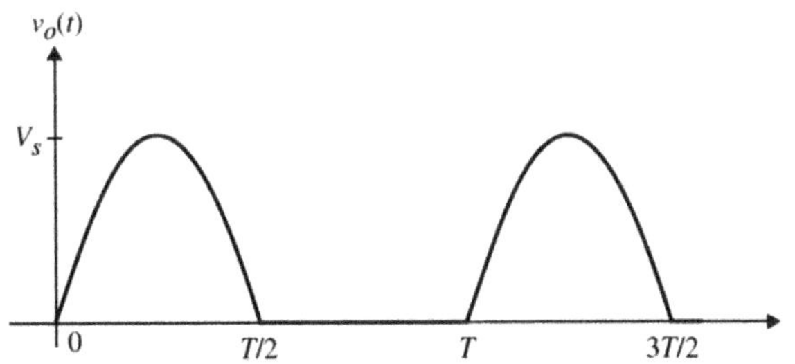

Fig. E3.12 Circuit for Exercise 3.12

Answer: V_s/π, $V_s/2$

3.4.3.2 Line Current Harmonics

If we apply the Fourier equations to the line current i_s and the line voltage v_s, which are periodical non-sinusoidal waveforms with period T and zero dc, then $i_s(t)$ and $v_s(t)$ are given as follows:

$$\begin{aligned} i_s(t) &= I_{dc} + \sum_{n=1}^{\infty} I_{sn}\sin(n\,\omega t + \theta_{ni}) \\ &= I_{dc} + I_{s1}\sin(\omega t + \theta_{1i}) + \sum_{n=2}^{\infty} I_{sn}\sin(n\,\omega t + \theta_{ni}) \end{aligned} \tag{3.43}$$

$$\begin{aligned} v_s(t) &= V_{dc} + \sum_{n=1}^{\infty} V_{sn}\sin(n\,\omega t + \theta_{nv}) \\ &= V_{dc} + V_{s1}\sin(\omega t + \theta_{1v}) + \sum_{n=2}^{\infty} V_{\mathrm{sn}}\sin(n\,\omega t + \theta_{\mathrm{nv}}) \end{aligned} \tag{3.44}$$

The fundamental current and voltage components are:

$$i_{s1}(t) = I_{s1} \sin(\omega t + \theta_{1i}) \tag{3.45a}$$

$$v_{s1}(t) = V_{s1} \sin(\omega t + \theta_{1v}) \tag{3.45b}$$

where I_{s1} and V_{s1} are the peak values of the current and voltage fundamental components, respectively.

The rms values of $i_s(t)$ and $v_s(t)$ can be easily obtained using Eqs. (3.43) and (3.44):

$$\begin{aligned} I^2_{\mathrm{s,rms}} &= \frac{1}{T}\int_0^T i_{\mathrm{s}}^2(t)dt \\ I^2_{\mathrm{s,rms}} &= I^2_{dc} + \left(\frac{I_{\mathrm{s1}}}{\sqrt{2}}\right)^2 + \left(\frac{I_{\mathrm{s2}}}{\sqrt{2}}\right)^2 + \ldots + \left(\frac{I_{\mathrm{sn}}}{\sqrt{2}}\right)^2 \\ &= I^2_{dc} + I^2_{s1,rms} + I^2_{s2,rms} + \ldots + I^2_{sn,rms} \qquad n = 1, 2, \ldots, \infty \end{aligned} \tag{3.46}$$

Similarly,

$$\begin{aligned} V^2_{\mathrm{s,rms}} &= V^2_{dc} + \left(\frac{V_{\mathrm{s1}}}{\sqrt{2}}\right)^2 + \left(\frac{V_{\mathrm{s2}}}{\sqrt{2}}\right)^2 + \ldots + \left(\frac{V_{\mathrm{sn}}}{\sqrt{2}}\right)^2 \\ &= V^2_{dc} + V^2_{s1,rms} + V^2_{s2,rms} + \ldots + V^2_{sn,rms} \qquad n = 1, 2, \ldots, \infty \end{aligned} \tag{3.47}$$

These equations are obtained since the integration of the product of two different frequency components over T is zero.

The instantaneous power is given by:

$$\begin{aligned} p(t) &= i_s(t)v_s(t) \\ &= I_{dc}V_{dc} + I_{dc}\sum_{n=1}^{\infty} V_{sn} \sin(n\,\omega t - \theta_{nv}) + V_{dc}I_{dc}\sum_{n=1}^{\infty} V_{sn} \sin(n\,\omega t - \theta_{nv}) \\ &+ \left(I_{dc}\sum_{n=1}^{\infty} V_{sn} \sin(n\,\omega t - \theta_{nv})\right)\left(\sum_{n=1}^{\infty} V_{sn} \sin(n\,\omega t - \theta_{nv})\right) \end{aligned} \tag{3.48}$$

Evaluating the average value of $p(t)$ of Eq. (3.48) over the fundamental frequency, ω, shows that the second and third terms in Eq. (3.48) are zeros and the fourth term is simply the sum of expressions similar to Eq. (3.46) but evaluated at each harmonic component as shown below:

$$\begin{aligned} P_{ave} &= I_{dc}V_{dc} + I_{s1,rms}V_{s1,rms} \cos(\theta_{v1} - \theta_{i1}) + I_{s2,rms}V_{s2,rms} \cos(\theta_{v2} - \theta_{i2}) + \ldots \\ &= I_{dc}V_{dc} + \sum_{n=1}^{\infty} I_{\mathrm{sn,rms}}V_{\mathrm{sn.rms}} \cos\theta_n \end{aligned} \tag{3.49}$$

where $\theta_n = \theta_{vn} - \theta_{in} (n = 1, 2, \ldots, \infty)$, which represents the phase shift between the nth voltage and current harmonics.

It is clear from the above equation that true *dc* power can be obtained only if both the line current and voltage have *dc* components. The second term represents the average power at the source terminal obtained from the rms value of the harmonic components.

3.4.3.3 Total Harmonic Distortion

Since the wanted portion of the distorted waveform $i_s(t)$ is the fundamental component, then the difference between the "desired" rms value of $i_s(t)$ and the "wanted" value is appropriately called the *distorted* portion of $i_s(t)$ defined as:

$$i_{s,dist} = i_s(t) - i_{s1}(t) = \sum_{n=2}^{\infty} i_{sn}(t) \tag{3.50a}$$

$$v_{s,dist} = v_s(t) - v_{s1}(t) = \sum_{n=2}^{\infty} v_{sn}(t) \tag{3.50b}$$

The relative measure of the distortion is defined through an index called the *total harmonic distortion* (THD), which is the ratio of the rms value of the distorted waveform and the rms value of the fundamental component. The THD in the current and voltage is given in Eqs. (3.51a) and (3.51b), respectively, assuming no *dc* components.

$$\begin{aligned}\mathrm{THD}_i &= \frac{I_{dist,rms}}{I_{s1,rms}} = \frac{\sqrt{I_{s2,rms}^2 + I_{s3,rms}^2 + I_{s4,rms}^2 + \ldots}}{I_{s1,rms}} \\ &= \sqrt{\left(\frac{I_{s2,rms}}{I_{s1,rms}}\right)^2 + \left(\frac{I_{s3,rms}}{I_{s1,rms}}\right)^2 + \ldots}\end{aligned} \tag{3.51a}$$

$$\begin{aligned}\mathrm{THD}_v &= \frac{V_{dist,rms}}{V_{s1,rms}} = \frac{\sqrt{V_{s2,rms}^2 + V_{s3,rms}^2 + V_{s4,rms}^2 + \ldots}}{V_{s1,rms}} \\ &= \sqrt{\left(\frac{V_{s2,rms}}{V_{s1,rms}}\right)^2 + \left(\frac{V_{s3,rms}}{V_{s1,rms}}\right)^2 + \ldots}\end{aligned} \tag{3.51b}$$

In terms of the rms of the original waveform, Eqs. (3.51a) and (3.51b) may be rewritten as:

$$\mathrm{TH}D_i = \sqrt{\left(\frac{I_{s,rms}}{I_{s1,rms}}\right)^2 - 1} \tag{3.52a}$$

$$\mathrm{THD}_v = \sqrt{\left(\frac{V_{s,rms}}{V_{s1,rms}}\right)^2 - 1} \tag{3.52b}$$

It is also common to refer to THD as a percentage.

Exercise 3.13
Calculate the THD_i for i_s(t) given in Example 3.6 and THD_v for v_s of E3.12.

Answer: $121\,\%$, $100\,\%$.

3.4.3.4 Power Factor

The equation to calculate power factor for distorted waveforms are more complex when compared to the sinusoidal case discussed earlier. Applying the definition of the power factor given in Eq. (3.37), to the distorted current and voltage waveforms of Eqs. (3.43) and (3.44) and the average power given in Eq. (3.49) (with zero *dc* components), *pf* may be expressed as:

$$pf = \frac{\sum_{n=1}^{\infty} I_{s,rms} V_{sn,rms} \cos\theta_n}{I_{s,rms} V_{s,rms}} = \frac{\sum_{n=1}^{\infty} I_{s,rms} V_{sn,rms} \cos\theta_n}{\sqrt{\sum_{n=1}^{\infty} I_{sn,rms}^2 V_{sn,rms} \sum_{n=1}^{\infty} v_{sn,rms}^2}} \tag{3.53}$$

The above expression for *pf* can be significantly simplified if we assume the line voltage is purely sinusoidal and distortion is only limited to $i_s(t)$; thus it can be shown that *pf* can be expressed as:

$$pf = \frac{I_{s1,rms}}{I_{s,rms}} \cos\theta_1 \tag{3.54}$$

where θ_1 is the phase angle between the voltage $v_s(t)$ and the fundamental component of $i_s(t)$. This assumption is valid in many power electronics applications. The line voltage is normally undistorted, and the line current is what gets distorted, i.e.:

$$v_s(t) = V_s \sin\omega t \tag{3.55a}$$

$$i_s(t) = \text{distorted}\,(\text{nonsinusoidal}) \tag{3.55b}$$

The current is expressed in terms of the Fourier series as follows:

$$i_s(t) = I_1 \sin(\omega t + \theta_1) + I_2 \sin(\omega t + \theta_2) + \ldots + I_n \sin(\omega t + \theta_n) \tag{3.56}$$

resulting in the average power given by:

$$
\begin{aligned}
P_{\text{ave}} &= \frac{1}{T}\int_0^T v_s i_s dt \\
&= \frac{1}{T}\left[\int_0^T (V_s \sin\omega t) I_1 \sin(\omega t + \theta_1) + I_2 \sin(\omega t + \theta_2) + \ldots + I_n \sin(\omega t + \theta_n) DT\right] \\
&= \frac{1}{T}\left[\int_0^T V_s I_1 \sin\omega t \sin(\omega t + \theta_1) dt\right] \\
&= \frac{V_s I_1}{2} \cos\theta_1 \\
P_{\text{ave}} &= I_{\text{s,rms}} V_{\text{s,rms}} \cos\theta_1
\end{aligned}
$$

Hence the power factor is given by:

$$
\text{Power factor} = \frac{I_{\text{s1,rms}} V_{\text{s,rms}} \cos\theta_1}{I_{\text{s,rms}} V_{\text{s,rms}}} = \frac{I_{\text{s1,rms}}}{I_{\text{s,rms}}} \cos\theta_1 \tag{3.57}
$$

The expression $I_{\text{s1, rms}}/I_{\text{s, rms}}$ is caused by the distortion of the line current and appropriately called the *distortion power factor,* k_{dist}, and the expression $\cos\theta_1$, is caused by the displacement angle between the line voltage and the fundamental current component and commonly known as the *displacement power factor,* k_{disp}. Hence, the power factor in power electronics is more useful if it is represented as a product of the k_{dist} and k_{disp}:

$$
pf = k_{dist}\, k_{disp} \tag{3.58}
$$

where

$$
\begin{aligned}
k_{\text{disp}} &= \cos\theta_1 \\
k_{\text{disp}} &= I_{\text{s1,rms}}/I_{\text{s,rms}}
\end{aligned}
$$

In terms of k_{dist}, it can be shown that the current THD_i can be expressed as:

$$
\text{THD}_i = \sqrt{\frac{1}{k_{dist}{}^2} - 1} \tag{3.59}
$$

Example 3.7

Consider a source with terminal voltage, $v_s(t)$, and with its terminal current, $i_s(t)$, in a nonlinear circuit given by the following expressions:

$$
\begin{aligned}
v_s(t) &= 4\cos(\omega_o t) \\
i_s(t) &= 3\cos(\omega_o t - 30^\circ) + 2 \cdot \cos(3\omega_o t - 50^\circ)
\end{aligned}
$$

(a) Calculate the total rms values of the terminal voltage, $v_s(t)$, and current, $i_s(t)$.
(b) Calculate PAVG (real power) delivered by source.
(c) Calculate the rms values of the fundamental components of the terminal voltage, $v_s(t)$, and current, $i_s(t)$.
(d) Calculate the current THD and voltage THD.
(e) Calculate power factor.

Solution

(a) $V_{S\ \text{rms}} = \frac{4}{\sqrt{2}} = 2.83\ V,\ I_{S\ \text{rms}} = \sqrt{\left(\frac{3}{\sqrt{2}}\right)^2 + \left(\frac{2}{\sqrt{3}}\right)^2} = 2.55\ \text{A}$

(a) $P_{Ave} = \frac{4\times 3}{2}\cos\left(0 - 30^\circ\right) = 5.2\ \text{W}$

(c) $V_{s1\ \text{rms}} = 2.83\ V,\ I_{s1\ \text{rms}} = \frac{3}{\sqrt{2}} = 2.12\ \text{A}$

(d) $\text{THD}_v = \sqrt{\left(\frac{V_{s\ \text{rms}}}{V_{s1\ \text{rms}}}\right)^2 - 1} = 0\%,\ \text{THD}_i = \sqrt{\left(\frac{I_{s\ \text{rms}}}{I_{s1\ \text{rms}}}\right)^2 - 1} = 66.8\%$

(e) Power Factor $= \frac{\text{Real Power}}{\text{Apparent Power}} = \frac{5.2\ \text{W}}{2.83\times 2.55} = 71.8\%$

Or $= k_{\text{disp}} k_{\text{dist}} = \cos 30 \times \frac{I_{s1\ \text{rms}}}{I_{s\ \text{rms}}} = 0.866 \times \frac{2.12}{2.55} = 71.99\%$

Example 3.8
Calculate k_{dist}, k_{disp}, THD_{ii}, and THD_v for the waveforms shown in Fig. 3.16.

Solution
Since the voltage is purely sinusoidal, then $\text{THD}_v = 0\%$.

To obtain the current THD_i, first we obtain the rms value of $i_s(t)$ and its fundamental component:

$$I_{s,\text{rms}}^2 = \frac{1}{2\pi}\int_{-\theta}^{2\pi-\theta} i_s^2\, d\omega t = \frac{1}{2\pi}\left[\int_{-\theta}^{2\pi-\theta} I_o^2\, d\omega t + \int_{\pi-\theta}^{2\pi-\theta} \left(-I_0^2\right) d\omega t\right]$$
$$= I_0^2$$

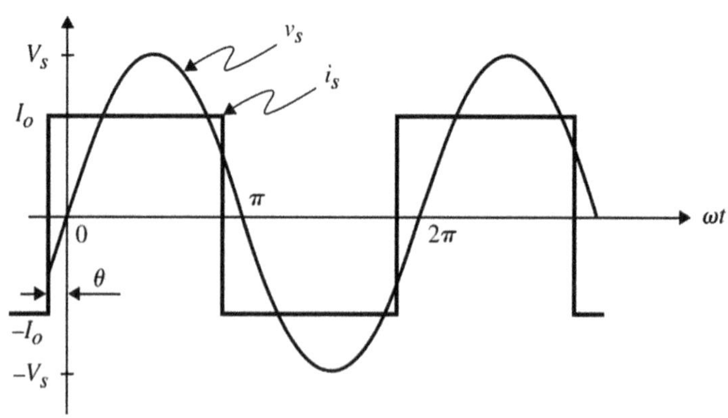

Fig. 3.16 Waveforms for Example 3.8

The fundamental component of $i_s(t)$ is given by:

$$I_{s1} = I'_{s1} \sin \omega t + I''_{s1} \cos \omega t$$

where

$$I'_{s1} = \frac{1}{2\pi}\left[\int_{-\theta}^{2\pi-\theta} I_o \sin \omega t d\omega t + \int_{\pi-\theta}^{2\pi-\theta} -I_o \sin \omega t d\omega t\right]$$

If we let $\omega\tau = \omega t + \theta$, the above integral becomes:

$$\begin{aligned} I'_{s1} &= \frac{1}{2\pi}\left[\int_{\theta}^{\pi} I_o \sin(\omega\tau - \theta) d\omega t - \int_{\pi}^{2\pi} I_o \sin(\omega\tau - \theta) d\omega t\right] \\ &= \frac{I_o}{\pi}[-\cos(\pi-\theta) + \cos(-\theta) + \cos(2\pi-\theta) + \cos(-\theta)] \\ &= \frac{4I_o}{\pi}\cos\theta \end{aligned}$$

Similarly, $I''_{S1} = (4I_o/\pi)\sin\theta$. Therefore, the peak fundamental component is given by:

$$I_{s1} = \sqrt{\left(I'_{s1}\right)^2 + \left(I''_{S1}\right)^2} = \frac{4I_o}{\pi}$$

Hence, the fundamental component of $i_s(t)$ is given by:

$$i_{s1}(t) = I_{s1} \sin(\omega t + \theta)$$

and the rms of I_{s1} is given by:

$$I_{s1,\text{rms}} = \frac{4}{\sqrt{2}}\frac{I_0}{\pi} = \frac{2\sqrt{2}}{\pi} I_o$$

k_{dist}, k_{disp}, and THD are given by the following expressions:

$$\begin{aligned} k_{dist} &= \frac{2\sqrt{2}}{\pi} \\ k_{disp} &= \cos\theta \\ \text{THD} &= \left(\sqrt{\frac{\pi^2}{8} - 1}\right) \times 100\% \\ &= 48.34\% \end{aligned}$$

Example 3.9
The circuit diagram shown in Fig. 3.17a shows a conventional light-dimming scheme used with incandescent lamps (incandescent lamps can be assumed purely

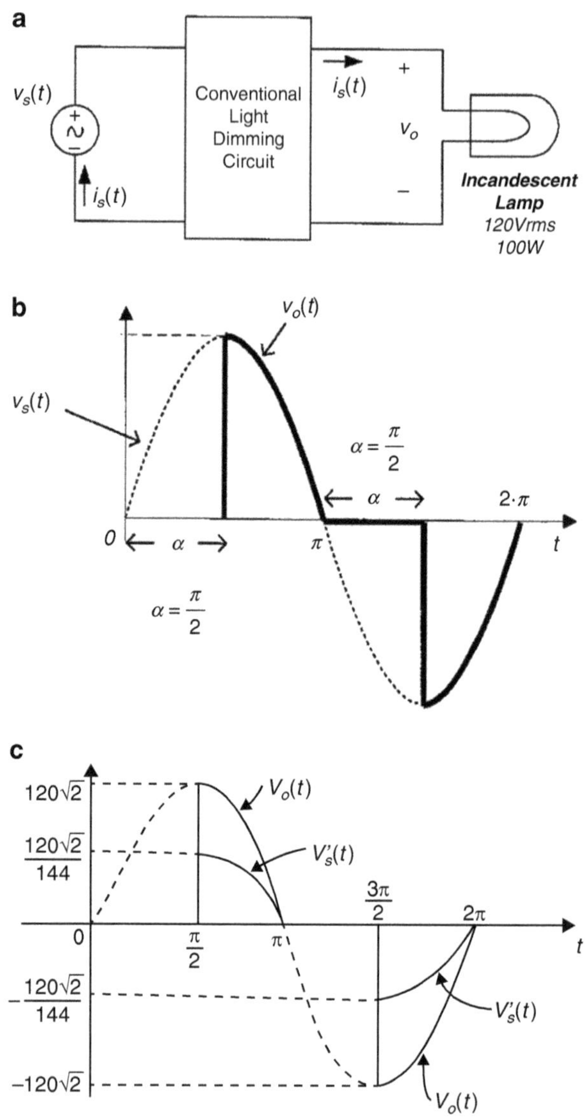

Fig. 3.17 (**a**) Conventional light-dimming circuit for Example 3.9. (**b**) The voltage waveform. (**c**) The current waveform

resistive). The voltage waveform of Fig. 3.17b shows the input and output voltages for the dimming circuit shown at an instant the dimmer is reducing light intensity. For this lighting control application, assuming that the incandescent lamp acts like a pure resister, determine the following:

(a) The Fourier series coefficients for the fundamental component of the source current $i_s(t)$
(b) The THD$_i$ of the source current $i_s(t)$
(c) The power factor as seen by the source

Assume the $v_s(t) = V_{s,\text{peak}} \sin \omega t = 120\sqrt{2} \sin(2.\pi.60t)$ V, and the rated lamp power is 100 W at 120Vrms.

Solution

(a) Since it is assumed that the incandescent lamp acts like a pure resister, the current $i_s(t)$ will be the same shape as the output voltage as shown in Fig. 3.17c and given by:

$$i_s(t) = I_{s,\text{peak}} \sin \omega t$$

with its peak $I_{s,\text{ peak}}$ is given by:

$$I_{s,peak} = \frac{V_{s,peak}}{R_{lamp}} = \frac{120\sqrt{2}}{144} = 1.2 \text{ A}$$

where R_{lamp} is equivalent lamp resistance at the full rated power, which is obtained from:

$$R_{lamp} = \frac{{V_{rms}}^2}{P_o} = \frac{120^2}{100} = 144\ \Omega$$

By inspection, the current $i_s(t)$ is half-wave symmetric with:

$$i_s(t) = -i_s\left(t - \frac{T}{2}\right)$$

Hence, its *dc* value is zero and coefficients $a_n \,\&\, b_n = 0$ for n even and for n odd:

$$a_n = \frac{4}{T}\int_0^{\frac{T}{2}} f(t) \cos n\omega t dt$$

$$b_n = \frac{4}{T}\int_0^{\frac{T}{2}} f(t) \sin n\omega t dt$$

For $n = 1$, the Fourier series coefficients for the fundamental component of $i_s(t)$ are obtained, where $\theta = \omega t$:

$$a_1 = \frac{2}{\pi}\int_{\frac{\pi}{2}}^{\pi} I_{s,\text{peak}} \sin\theta \cos\theta d\theta$$

$$b_1 = \frac{2}{\pi}\int_{\frac{\pi}{2}}^{\pi} I_{s,\text{peak}} \sin^2\theta d\theta$$

Using the trigonometric relation $\sin\theta\cos\theta = \frac{1}{2}\sin 2\theta$ and $\sin^2\theta = \frac{1}{2}(1 - \cos 2\theta)$ and evaluating the above two integrals, it can be shown that the coefficients a_1 *and* b_1 are given by:

$$a_1 = -\frac{I_{s,peak}}{2} = -0.38$$
$$b_1 = \frac{I_{s,peak}}{2} = 0.59$$

Therefore

$$I_{s1,peak} = \sqrt{{a_1}^2 + {b_1}^2} = 0.7 \text{ A}$$
$$I_{s1,rms} = \frac{I_{s1,peak}}{\sqrt{2}} = 0.49 \text{ A}$$

and the fundamental source current, $i_{s1}(t)$, may be expressed by:

$$i_{s1}(t) = I_{s1,\text{peak}} \sin\left(\omega t + \varphi^{\circ}\right)$$
$$i_{s1}(t) = 0.7\ \sin\left(2.\pi.60t - 32.78^{\circ}\right)$$

(b) THD_i for $i_s(t)$ is given by:

$$THD_i = \sqrt{\left(\frac{I_{s,rms}}{I_{s1,rms}}\right)^2 - 1} \times 100\%$$

The rms value for the current $i_s(t)$, $I_{s,\text{rms}}$, is obtained from:

$$I_{s,rms} = \frac{V_{o,rms}}{R_{lamp}} = \frac{84.85}{144} = 0.59 \text{ A}$$

Therefore:

$$THD_1 = \sqrt{\left(\frac{0.59}{0.5}\right)^2 - 1} \times 100\% = 62.6\%$$

(c) The power factor is given by:

$$PF = \frac{I_{s1,rms}}{I_{s,rms}} \cos\left(\theta_v - \theta_i\right) = \frac{0.49}{0.59} \cos\left(32.78^{\circ}\right) = 0.7$$

Therefore, current (fundamental) lags voltage by 32.78°.

Exercise 3.14
Calculate the THD_i and the *pf* for the current waveform shown in Fig. 3.15, by using up to the fifth harmonic.

Answer: 117 % ,0.64

Exercise 3.15

Calculate the THD_i and the *pf* for the current waveform shown in Fig. E3.15.

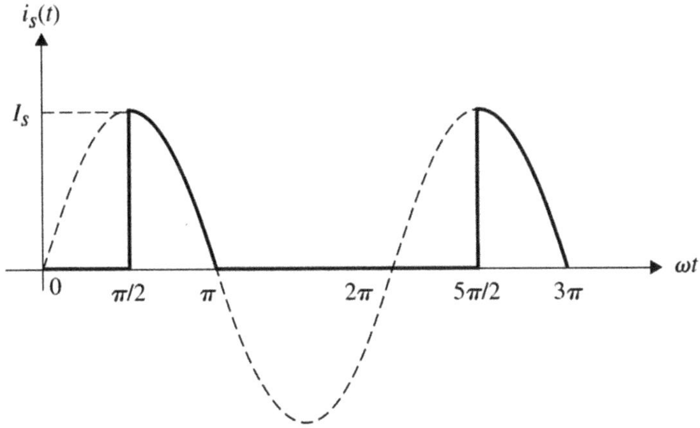

Fig. E3.15 Waveform for Exercise 3.15

Answer: 136 % , 0.51.

3.5 Capacitor and Inductor Responses

The transient and steady-state values of the capacitor and inductor voltage and current are well understood by undergraduate electrical engineering students. However, a brief review of such responses might be useful to some readers at this point.

3.5.1 Capacitor Transient Response

Consider the RC circuit of Fig. 3.18a with a *dc* excitation and an ideal switch. Assume the switch was open for $t < t_0$, and at $t = t_0$, the switch is closed. The capacitor voltage for $t < t_0$ is equal to the *dc* source, V_{dc}.

For $t > t_0$ the time-domain capacitor current is given in terms of its voltage:

$$i_C = C\frac{dv_C}{dt} \tag{3.60}$$

Substituting for $i_C = (0 - v_C(t))/R$ and solving for $v_C(t)$, we obtain the following general solution:

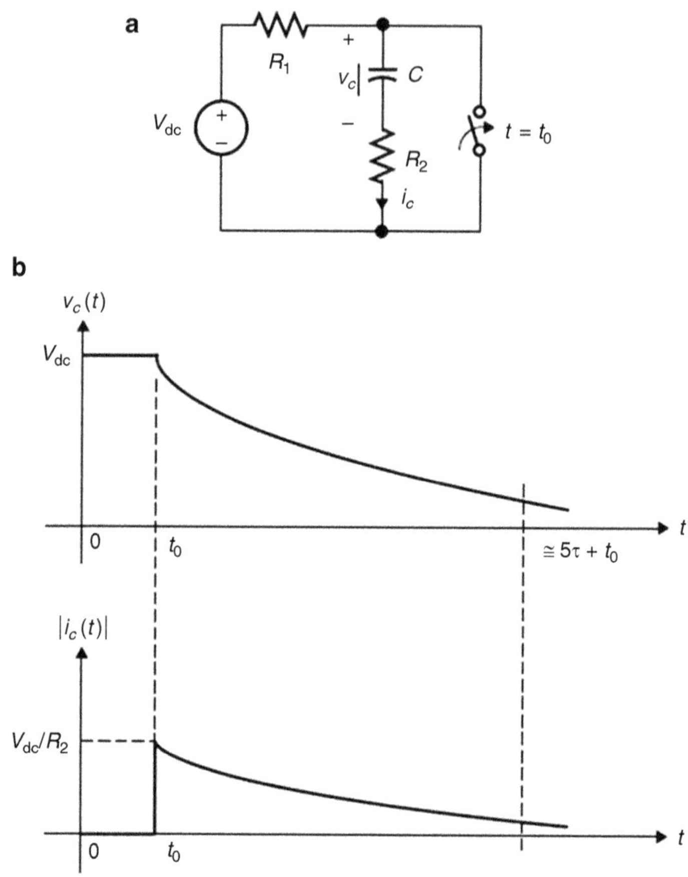

Fig. 3.18 (**a**) RC circuit with dc excitation and an ideal switch. (**b**) The current change through the capacitor

$$v_C(t) = v_{C,f} + \left(v_{C,i} - v_{C,f}\right)e^{-(t-t_0)/\tau} \tag{3.61}$$

where

$v_{C,f}$= final value at $t = \infty$.
$v_{C,i}$= initial value at $t = t_{0^+}$.
τ= circuit time constant with $\tau = R_2C$
Since the capacitor voltage does not change instantaneously, we have:

$$v_C(t_{0^-}) = v_C(t_{0^+})$$

From the circuit diagram, at $t = \infty$ the capacitor becomes fully discharged through R_2 with a time constant R_2C. The capacitor's final value is given by:

$$v_{c,f} = v_c(\infty) = 0$$

The final expression for $v_C(t)$ and $i_C(t)$ for $t > t_0$ is given by:

$$v_C(t) = V_{dc}e^{-(t-t_0)/\tau} \tag{3.62a}$$

$$i_C(t) = \frac{-V_{dc}}{R}e^{-(t-t_0)/\tau} \tag{3.62b}$$

The sketch is shown in Fig. 3.18b. The steady state is reached at approximately 5τ.

The capacitor current shows that at $t = t_{0^-}$ it was zero, but at $t = t_{0^+}$, it suddenly became $-V_{dc}/R_2$. This brings us to another important statement about the capacitor current: *The current through the capacitor can change instantaneously in switching circuits as shown in Fig. 3.18b.*

Example 3.10

Consider the circuit of Fig. 3.19a where the switch has been closed for $t < 0$. At $t = t_{0^+}$, the switch is opened. Sketch the capacitor voltage and current for $t > 0$. Assume the capacitor was initially uncharged.

Solution

The capacitor voltage equation is given by:

$$C\frac{dv_c}{dt} = i_C$$
$$= I_g$$

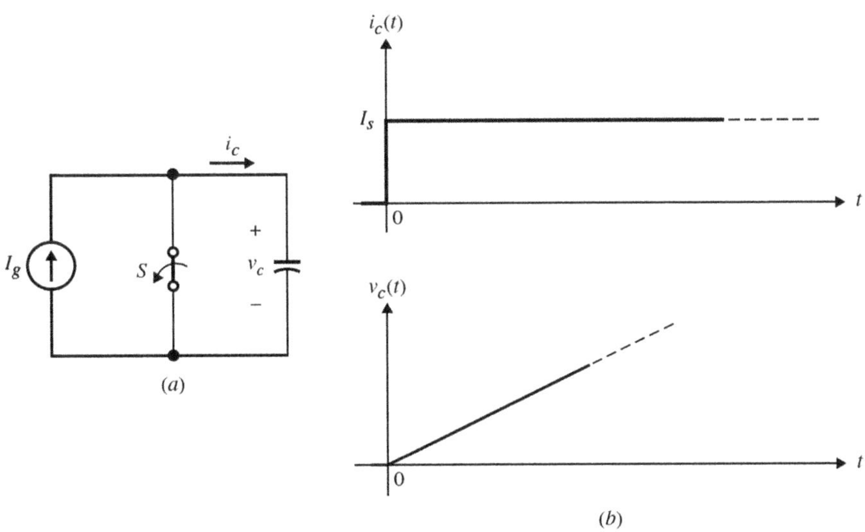

Fig. 3.19 (**a**) Circuit for Example 3.10. (**b**) The capacitor current and voltage waveforms

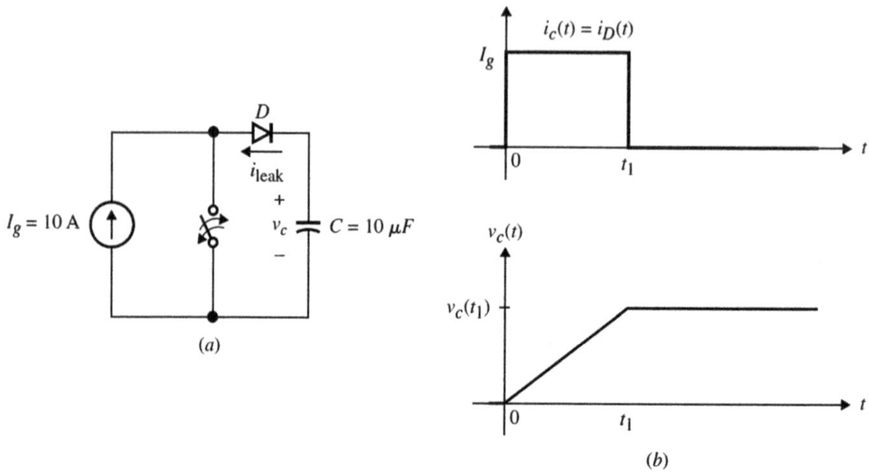

Fig. 3.20 (**a**) Series combination of capacitor and diode. (**b**) Capacitor current and voltage waveforms

Therefore, $v_C(t)$ is:

$$v_C(t) = \frac{I_g}{C}t$$

The capacitor current and voltage waveforms are shown in Fig. 3.19b.

Since the capacitor voltage does not change instantaneously, the switch in this example is not allowed to close again unless a way is found by which the switch voltage is prevented from appearing across the capacitor (i.e., voltage diversion). This can be accomplished by adding a diode in series with the capacitor as shown in Fig. 3.20a. As long as the switch is closed, the capacitor voltage remains constant.

At $t = t_1$, the switch is closed again, forcing *D* to turn *OFF*, since its anode voltage is pulled to the ground. The capacitor voltage at $t = t_1$ is given by:

$$v_c(t_1) = \frac{I_g}{C}t_1$$

In practice, the energy stored in the capacitor dissipates through the capacitor's equivalent resistance and the diode's leakage current, as illustrated in Exercise 3.16.

Exercise 3.16

Consider the circuit of Fig. 3.20a with the switch waveform shown in Fig. 3.21. Assume an ideal switch and ideal diode except it has a 100 μA leakage current. What are the capacitor voltages at $t_1 = 5$ μs, 10 μs, 0.5 s? At what time does the capacitor voltage become zero again?

Answer: 5 V, 10 V, 5 V, 1 s

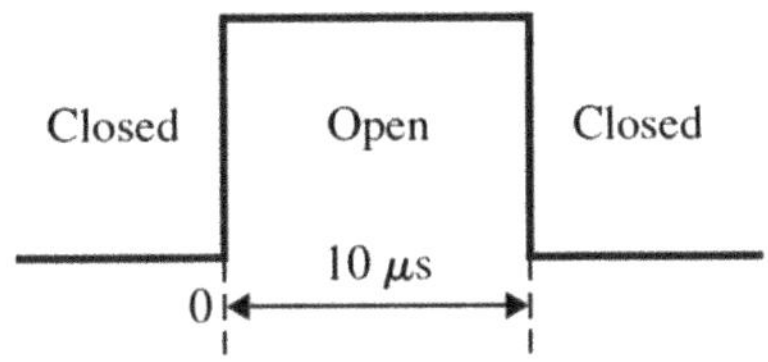

Fig. 3.21 Switch waveform

3.5.2 Capacitor Steady-State Response

Now let us consider the circuit of Fig. 3.18a with the switch turned *ON* and *OFF* repeatedly according to the waveform shown in Fig. 3.22a. After a few switching cycles, the capacitor voltage reaches steady state at which its value at the beginning of each switching cycle is the same. Mathematically, this can be shown as follows:

$$v_C(nT + t_0) = v_C((n+1)T + t_0) \tag{3.63}$$

Figure 3.22b, c show the steady-state capacitor voltage and current waveforms, respectively. Now we come to another important property of the capacitor. In steady state, the average capacitor current is zero.

$$I_{\mathrm{c,ave}} = \frac{1}{T}\int_{nT}^{(n+1)T} i_C dt = \int_{nT}^{(n+1)T} C\frac{dv_C}{dt}dt = \frac{C}{T}[v_C((n+1)T - v_C(nT))] = 0 \tag{3.64}$$

This is illustrated by the equal negative and positive shaded areas for $i_C(t)$ shown in Fig. 3.22c.

3.5.3 Inductor Transient Response

Similarly, let us look at the transient response for the inductor current by considering the dual circuit of Fig. 3.18a, which is shown in Fig. 3.23. Assume the switch was open for a long time before it is closed at $t = t_0$. Assume that initially the inductor current is zero.

For $t > t_0$, the time-domain inductor voltage is given by:

$$v_L = L\frac{di_L}{dt} \tag{3.65}$$

Substituting for $v_L = R(I_{dc} - i_L(t))$, *where* $R = R_1 || R_2$,and solving for $i_L(t)$, we obtain the following general solution for $i_L(t)$:

$$i_L(t) = i_{L,f} + \left(i_{L,i} - i_{L,f}\right)e^{-(t-t_0)/\tau} \tag{3.66}$$

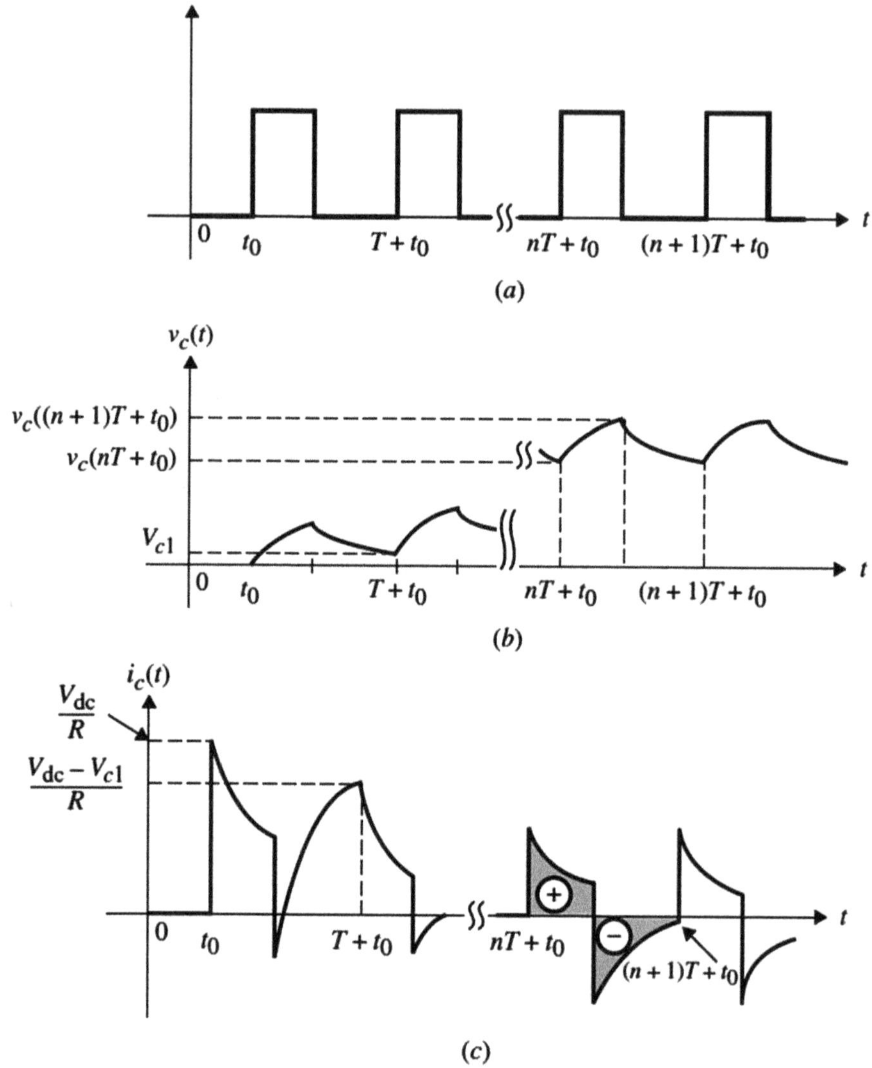

Fig. 3.22 (**a**) Switch waveform, (**b**) steady-state capacitor voltage, (**c**) steady-state capacitor current

Fig. 3.23 Inductor switching circuit

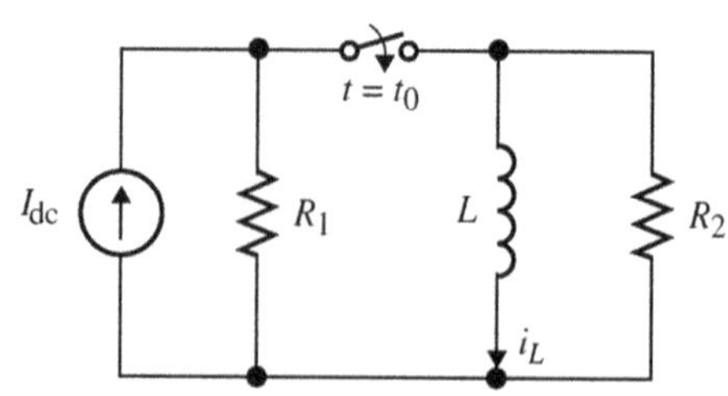

where

$i_{L,f}$ = final value as $t=\infty$
$i_{L,i}$ = initial value at $t=t_{0^+}$
τ = circuit time constant with $\tau = L/R$
Since the inductor current does not change instantaneously, we have:

$$I_L(t_{0^-}) = I_L(t_{0^+}) \tag{3.67}$$

From the circuit, it is clear that at $t=\infty$, $i_L(t=\infty)=i_{L,f}=I_{dc}$, and its voltage becomes zero.

The final expression for $i_L(t)$ is given by:

$$i_L(t) = I_{dc}\left(1 - e^{-(t-t_0)/\tau}\right)$$

and the inductor voltage is given by:

$$v_L(t) = RI_{dc}e^{-(t-t_0)/\tau}$$

Again at $t = t_{0^-}$, $v_L(t_{0^-})$, and at $t = t_{0^+}$, the inductor voltage is given:

$$\mathbf{v_L(t_{0^+}) = RI_{dc}}$$

We conclude that *the voltage across the inductor can change instantaneously under switching action*, as shown in the inductor current and voltage waveforms in Fig. 3.24.

Exercise 3.17

Consider the switching circuit given in Fig. E3.17 with an ideal diode. Assume the initial inductor current is zero and the switch is turned on at $t=0$ and turned off at $t=10$ μs. Sketch the inductor current waveform for $t>0$. If the diode has a 5 Ω forward resistance, what is i_L at $t=0.5$ μs, 10 μs, and 100 μs? At what time does the inductor becomes discharged?

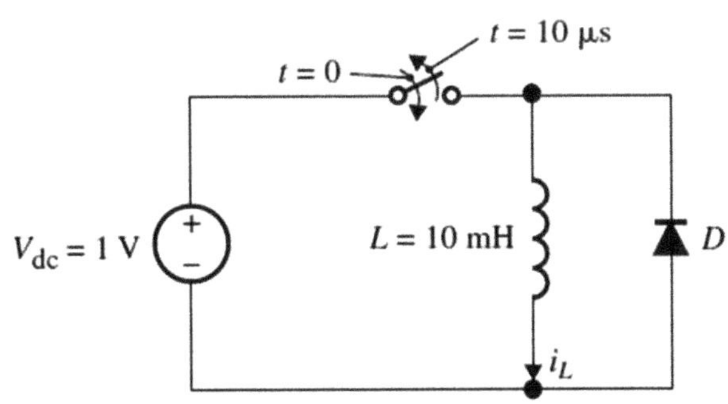

Fig. E3.17 Switching circuit for Exercise 3.17

Answer: 50 μA, 1 mA, 0.95 mA, 10 ms

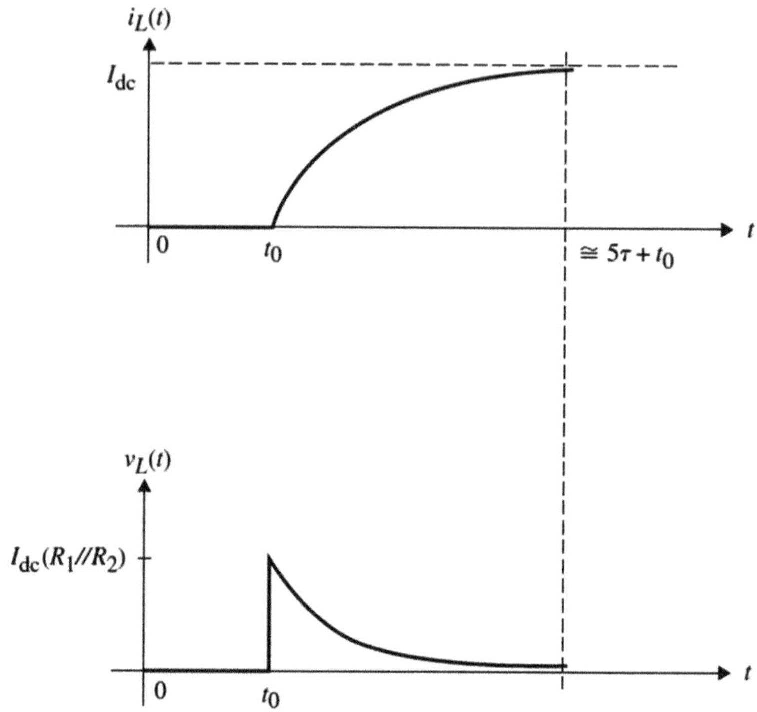

Fig. 3.24 Inductor current and voltage waveforms

Notice in the above exercise that when the switch is opened again at $t = 10$ ms, the inductor current gets trapped and the diode turns on and diverts the current from the switch branch to the diode branch. This is known as *current commutation* and will be used extensively in Chaps. 5 and 6. In practice, trapped energy in the inductor would dissipate through the inductor and diode resistances. To improve the converter's efficiency, normally the trapped energy is allowed to be reconnected by returning it to the source through a feedback circuit along with the use of transformers, to be illustrated in Chap. 5.

3.5.4 Inductor Steady-State Response

Let us reconsider the same circuit of Fig. 3.23 except that the switch turns *ON* and *OFF* repeatedly according to the waveform shown in Fig. 3.25a.

After a few switching cycles, the inductor reaches steady state at which its value at the beginning of each switching cycle is the same. Mathematically, we express the steady-state condition as:

$$i_L(nT + t_0) = i_L((n+1)T + t_0) \tag{3.68}$$

The above condition makes it possible to show that the average value of the steady-state inductor voltage is zero. This is shown as follows:

$$V_{L,ave} = \frac{1}{T}\int_{nT}^{(n+1)T} v_L dt = \frac{1}{T}\int_{nT}^{(n+1)T} L\frac{di_L}{dt}dt = \frac{L}{T}[i_L((n+1)T) - i_L(nt)] \tag{3.69}$$

The steady-state current and voltage waveforms are shown in Fig. 3.25b, c, respectively. The zero average inductor voltage is illustrated by the equal negative and positive shaded areas for $v_L(t)$ shown in Fig. 3.25c.

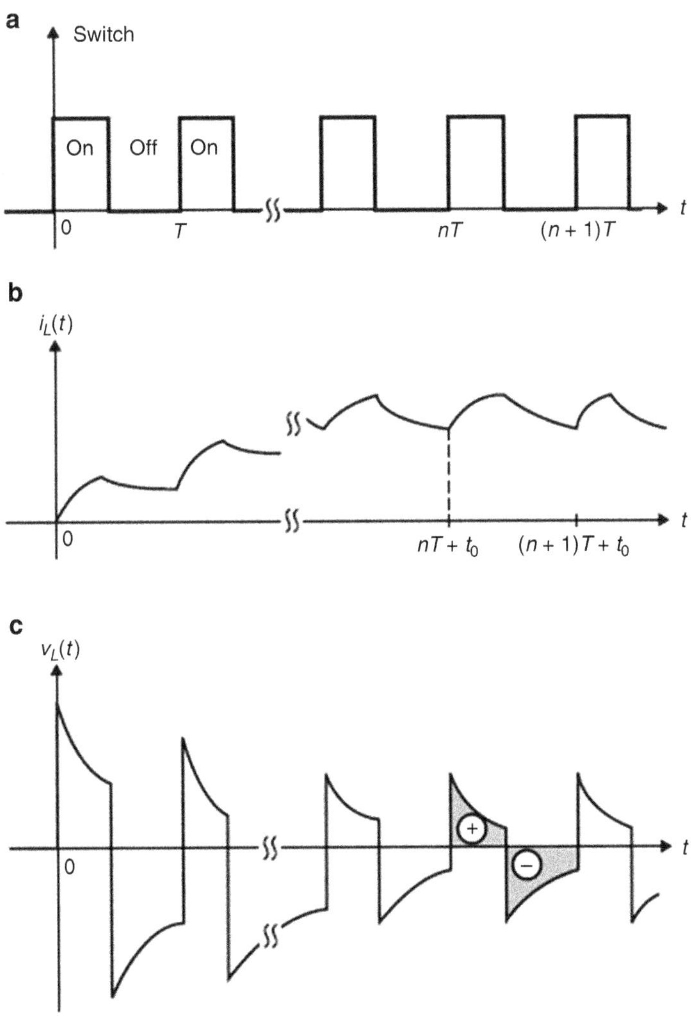

Fig. 3.25 Inductor current and inductor voltage waveforms for Fig. 3.22 when the switch is repeatedly turned on and off according to waveform (**b**) and (**c**), repectively

Problems

In all the problems, assume ideal diodes unless stated otherwise.

3.1 Derive the expressions for $i_L(t)$ and $v_c(t)$ in switching circuits shown in Fig. P3.1. Assume all initial conditions are zero. Use $V_{dc} = 20$ V, $I_{dc} = 2$ A, $R = 200\ \Omega$, $C = 0.01\ \mu$F, and $L = 2$ mH, sketch i_L and v_C.

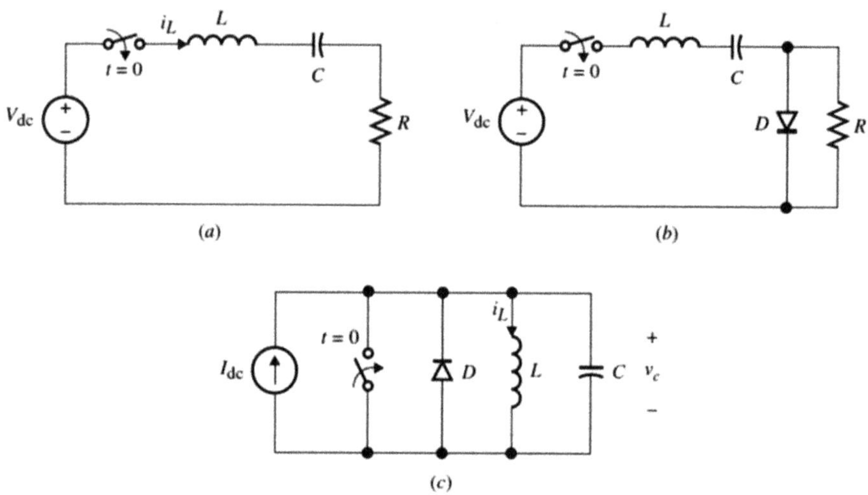

Fig. P3.1

3.2 Assume the switch in Fig. P3.2 is opened at $t = 0^+$, derive the expressions for i_L, v_c and i_{sw} and sketch them for $t > 0$. Assume $i_L(0) = I_g$ and $v_c(0) = V_o$.

Fig. P3.2

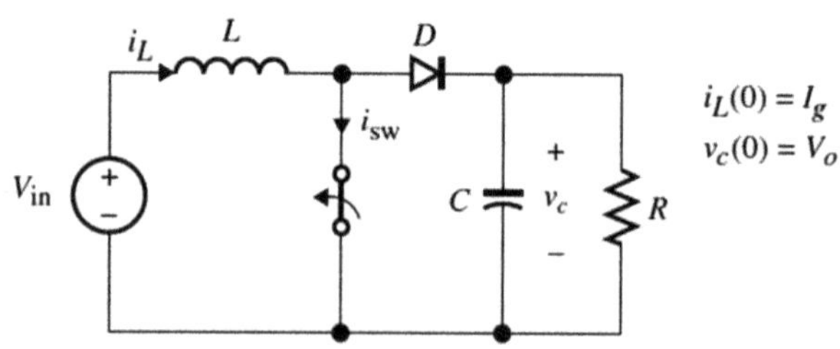

3.3 Consider the switching diode circuits of Fig. P3.3 for $t > 0$ when the switch is turned on. (a) Derive expressions for $i_L(t)$, $i_{sw}(t)$, and $v_c(t)$. Assume $i_L(0) = 1.0I_g$ and $v_c(0) = V_o$. Suggest possible switch implementations (unidirectional or bidirectional).

Fig. P3.3

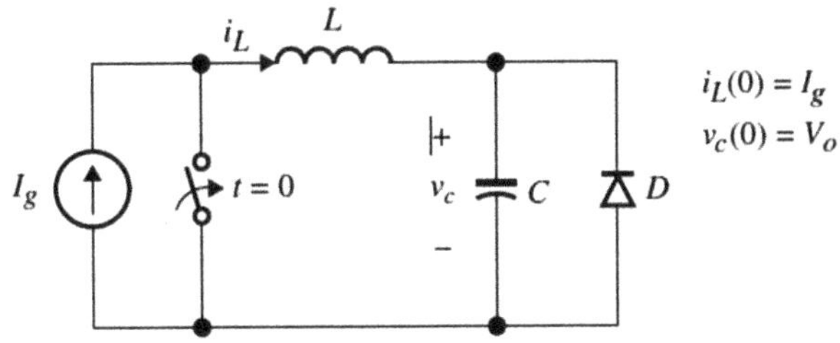

3.4 Consider the transistor switching circuits of Fig. P3.4. At $t = 0$ the transistor is turned on by a signal to the base. Derive and sketch the waveforms for i_Q, i_L, v_c, and v_o. Assume $v_c(0) = -20$ V and $i_L(0) = 0$. Compare the two circuits.

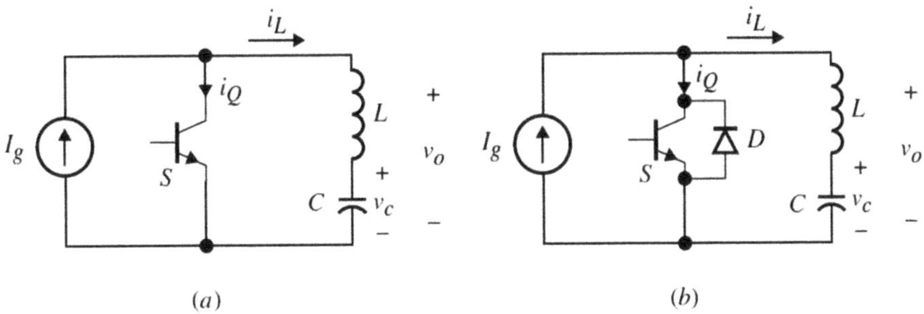

Fig. P3.4

3.5 Consider the diode circuit of Fig. P3.5 with switch S_1 opened and S_2 closed at $t=0$. Assume the initial conditions are given as $v_{c1}(0)=0$ and $i_L(0)= -0.5\ I_g$. Derive the expressions for v_{c1} and i_L and sketch them. Use $I_g=2\ A$, $C_1=C_2=0.01\ \mu F$, and $L=2$ mH.

Fig. P3.5

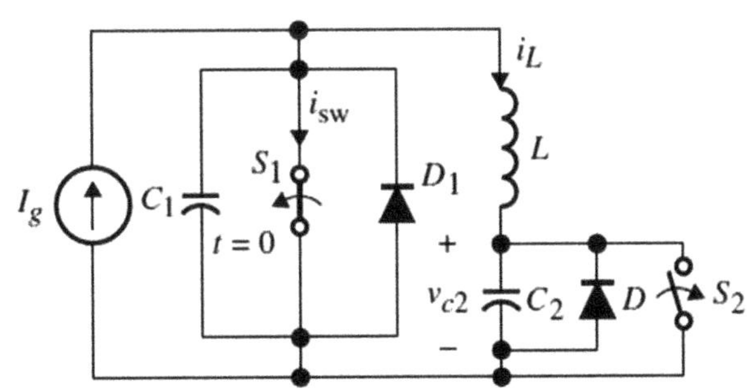

3.6 Assume the switch in Fig. P3.6 is opened at $t = 0^+$. Derive the expressions and sketch the waveforms for v_{c1}, v_{c2}, i_L, and i_D. Assume $v_{c1}(0) = v_{c2}(0)$ and $i_L(0) = 0$.

Fig. P3.6

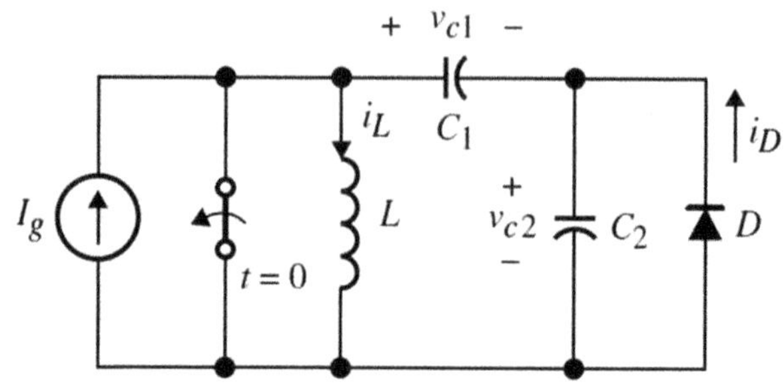

3.7 Consider the capacitor current shown in Fig. P3.7 for a given switch-mode power supply with $C = 1\ \mu F$, and assume $v_c(0) = 100$ V. (a) Sketch v_C showing the peak values and times, and (b) determine the ripple voltage across the capacitor.

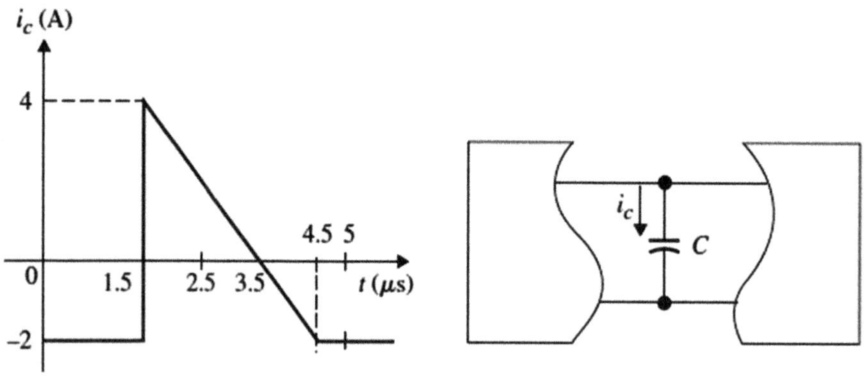

Fig. P3.7

3.8 Consider the SCR-diode switching circuit shown in Fig. P3.8 with the switch closing at $t = 0$. This circuit is a voltage-commutation circuit known as the impulse-commuted chopper used to force the turnoff of the SCR by additional switch S. Assume ideal diode and the capacitor were initially charged to $-V_o$ and the SCR was on for $t < 0$.

(a) Sketch the waveforms of i_{SCR}, v_c, i_D, and v_o.
(b) Show that the time, Δt, it takes for i_C to reach zero again after $t > 0$ is given by:

$$\Delta t = \frac{(V_{dc} + V_o)C}{I_o}$$

Discuss the drawback of such an arrangement.

Fig. P3.8

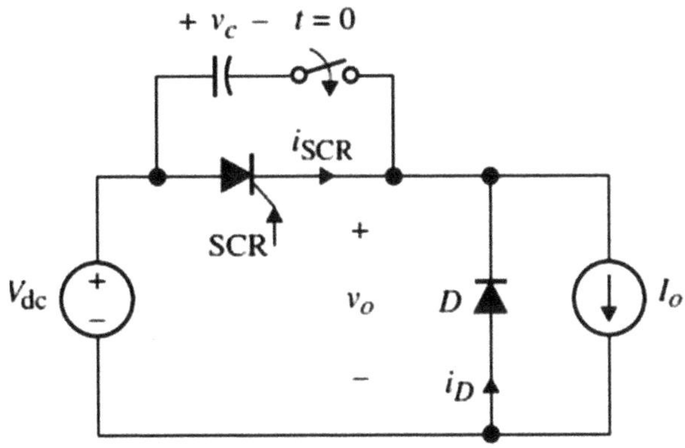

3.9 For the SCR to turn off in the circuit of Fig. P3.8, the capacitor voltage must first be charged to a large negative value. This is done through an external circuit consisting of a diode and an inductance as shown in Fig. P3.9. The purpose of D_1 and L is that when the switch is turned off, the capacitor voltage returns to its original negative value, $-V_o$. Assume S and SCR were open for a long time with a capacitor initial value equal to$+V_o$. At $t=0$, SCR is triggered. Sketch the waveforms for i_{D1}, i_{D0}, i_{SCR}, v_C, and v_o. Compare this circuit with the circuit given in Problem 3.8.

Fig. P3.9

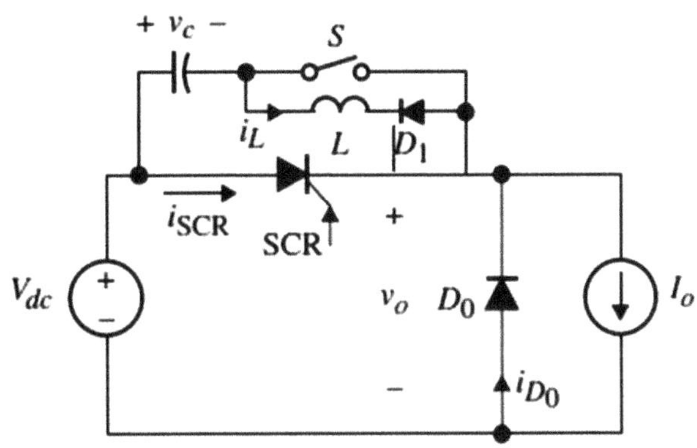

3.10 Consider the SCR circuit shown in Fig. P3.10. Because the SCR turns *OFF* naturally due to the fact that its current becomes zero, the circuit is known as a self-commutated circuit. At $t=0$, the SCR is turned *ON* by applying i_g. Derive the expressions for $v_c(t)$, $i_L(t)$, and the SCR commutation time. Let $L=0.1$ mH, $C=47$ μF, and $V_{dc}=120$ V. Assume the initial capacitor voltage is (i) zero and (ii) $-V_i$ (where $|V_i|<V_{dc}$).

Fig. P3.10

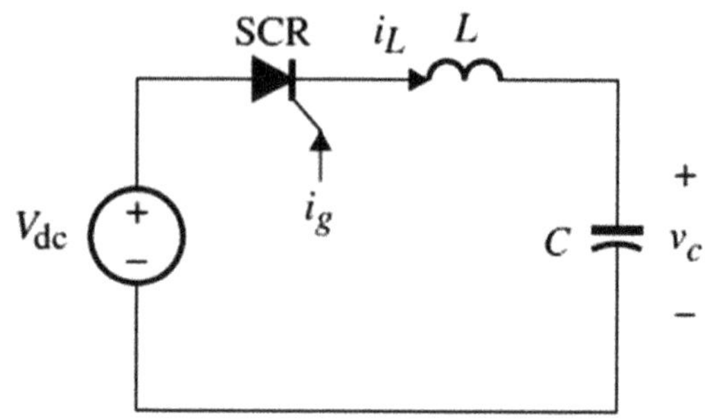

D3.11 Design for L and C in Problem 3.10 so that the SCR commutation time is 100 μs, and the peak capacitor voltage does not exceed 200 V.

3.12 Consider a source terminal connected to a power electronic circuit whose current and voltage waveforms are shown in Fig. P3.12, with $t_1 = 4$ μs and $T = 10$ μs.

(a) Calculate the average and rms source current and source voltage values.
(b) Calculate the average input power.

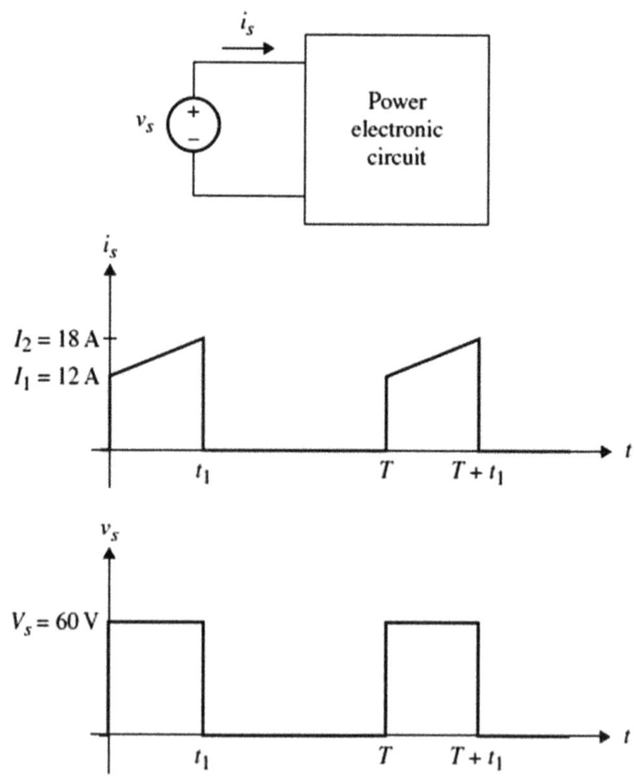

Fig. P3.12

3.13 Consider the circuit of Fig. P3.13 where v_s is a train of pulses as shown. Assume $RC \gg T$ so that V_o is assumed constant and equal to 75 V. Sketch the steady-state inductor current waveform, i_L.

Fig. P3.13

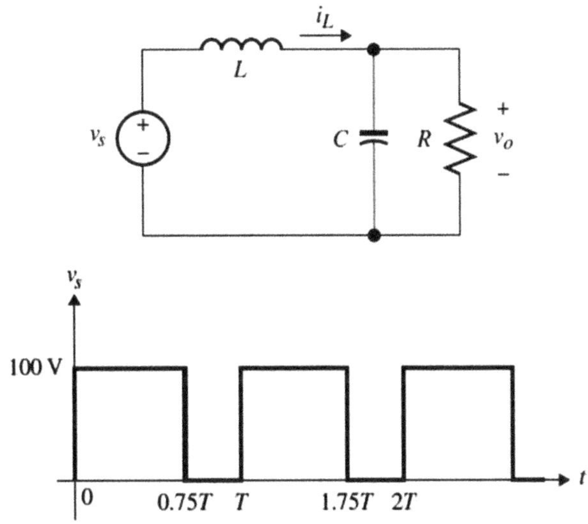

3.14 The switch in Fig. P3.14 is closed at $t = 0$. Assume ideal diodes and all inductor initial conditions are zero. Obtain the expressions for the inductor current and sketch them for $0 \le t \le 2T$. Assume $v_s(t)$ is sinusoidal with $\omega = 377$ rad/s and $V_s = 100$ V. Let $L = 1$ mH, $R = 1\ \Omega$, and $V_{dc} = 20$ V.

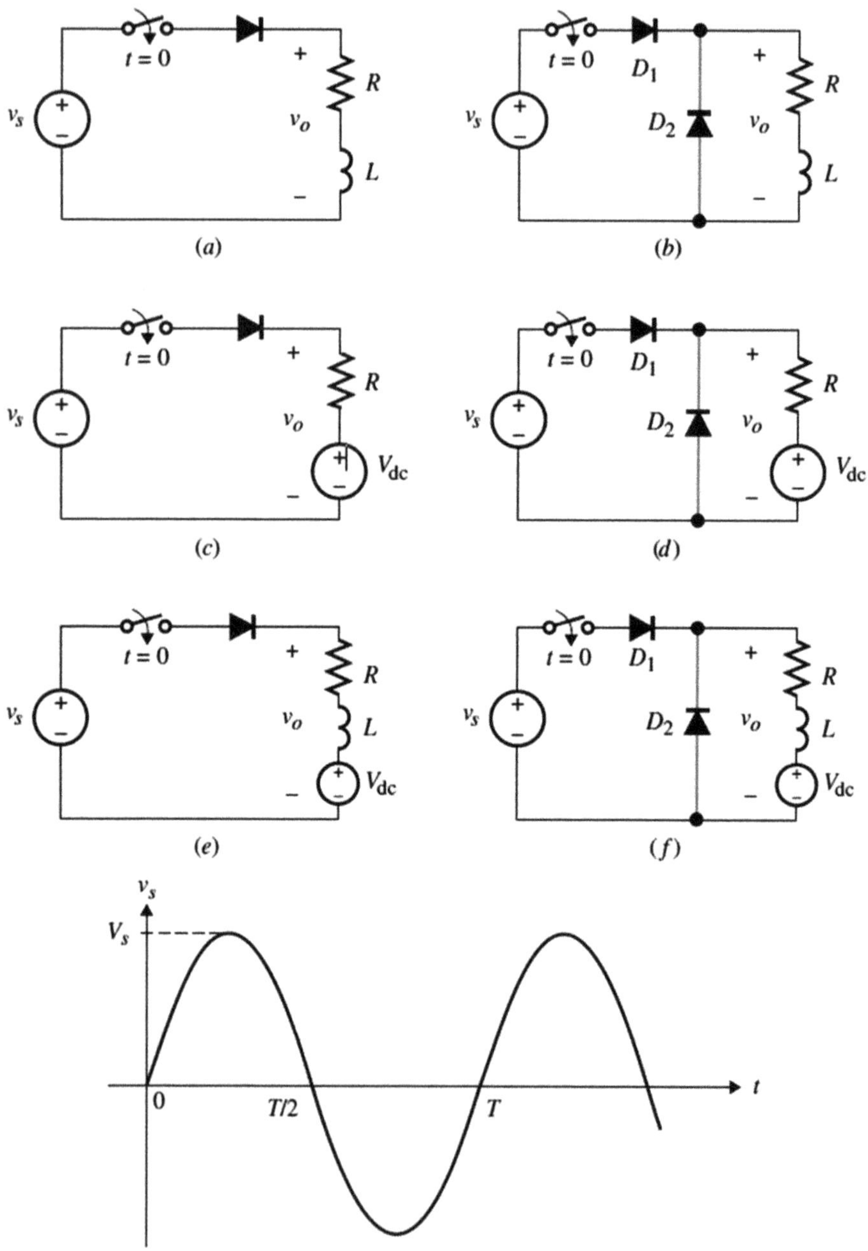

Fig. P3.14

3.15 Consider a power source with its terminal voltage and current given as:

$$v_s(t) = 100 + 80\sin\left(\omega t - 100^\circ\right) + 70\cos\left(2\omega t + 120^\circ\right) + 25\sin 3\omega t \text{ V}$$
$$i_s(t) = 12 + 10\sin\left(\omega t + 25^\circ\right) + 5\sin\left(2\omega t - 30^\circ\right) + 2\cos 3\omega t \text{ A}$$

where ω is the fundamental angular frequency.
Calculate:

(a) The rms value of $i_s(t)$ and $v_s(t)$
(b) The average input power supplied by the source
(c) The rms values of the fundamental components of $i_s(t)$ and $v_s(t)$
(d) THD_i and THD_v
(e) k_{dist}, k_{disp}, and pf

3.16 Consider the phasor circuit of Fig. P3.16 with $Z_L = 20\angle -36^\circ\,\Omega$ and $V_s = 80\angle 0^\circ$ V.

(a) Determine the circuit's real, apparent, and reactive powers and the input power factor.
(b) Determine the type and value of the load needed to be connected between a and a' to achieve unity power factor.

Fig. P3.16

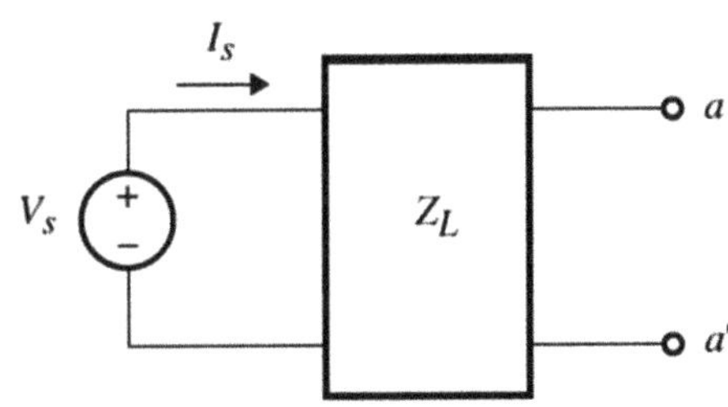

3.17 Prove the following integrals:

$$F_n = \int_0^T \sin n\,\omega t \sin m\,\omega t dt = \int_0^T \cos n\,\omega t \cos m\,\omega t dt = \begin{cases} 0 & n \neq m \\ \dfrac{T}{2} & n = m \end{cases}$$

and

$$F_n = \int_o^T \sin(n\omega t)\cos(m\omega t)dt$$

$$= \begin{cases} 0 & n \neq m \\ -\frac{T}{2} & n = m \end{cases}$$

Where $\omega = 2\pi/T$, and $n = 1, 2, \ldots \infty$; $m = 1, 2, \ldots \infty$.

3.18 Sketch the steady-state i_L and v_L waveforms by assuming the switch in Fig. P3.18 is repeatedly opened and closed as shown. Use $T = 10$ ms.

Fig. P3.18

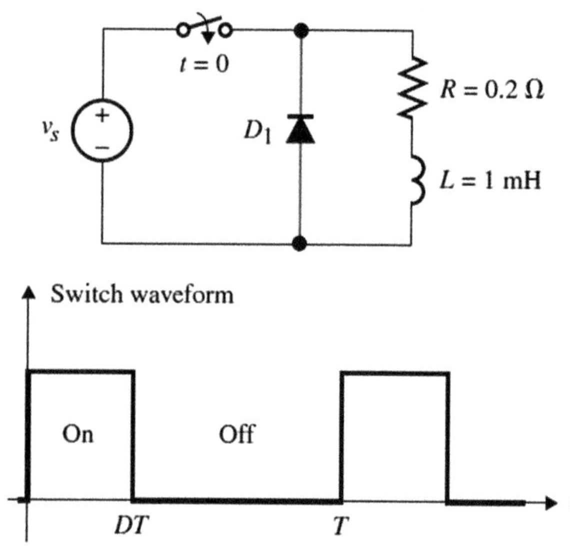

3.19 Show that the Fourier series for the half-wave rectifier output voltage of Fig. P3.19 is given by:

$$v_o(t) = \frac{V_s}{\pi} + \frac{2V_s}{\pi}\sin\omega t - \frac{2V_s}{\pi}\sum_{n=2,4,6\ldots}^{\infty}\frac{1}{n^2-1}\cos\omega t$$

Fig. P3.19

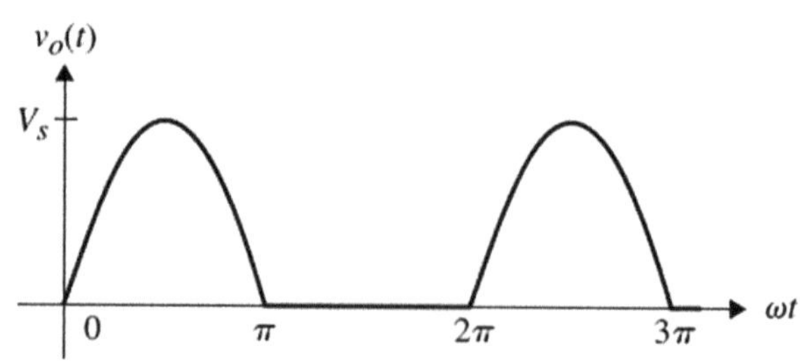

3.20 Show that the Fourier series for the output voltage waveform of a full-wave rectifier of Fig. P3.20 is given by:

$$v_o(t) = \frac{2V_s}{\pi} - \frac{2V_s}{\pi} \sum_{n=2,4,6\ldots}^{\infty} \frac{1}{n^2 - 1} \cos \omega t$$

Fig. P3.20

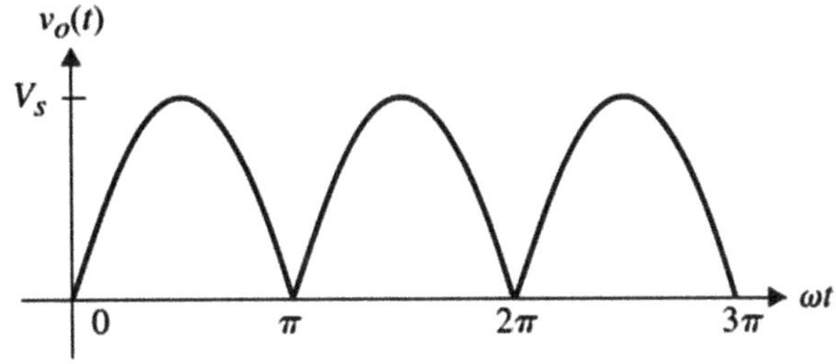

3.21 By considering only the first four terms of the Fourier series given in Problem 3.19, verify the answers given in Exercise 3.11.

3.22 Derive the Fourier coefficient equations a_n and b_n given in Eq. (3.41).

3.23 The phase current waveform in a three-phase full-wave rectifier SCR circuit under highly inductive load is shown in Fig. P3.23. Show that the harmonics of *i*(*t*) are given by the following expression:

$$i(t) = \frac{2\sqrt{3}I_o}{\pi}\left[\cos \omega t - \frac{\cos 5\omega t}{5} + \frac{\cos 7\omega t}{7} - \frac{\cos 11\omega t}{11} + \frac{\cos 13\omega t}{13} + \ldots\right]$$

No triple harmonics in *i*(*t*) and $n = 6k \pm 1 (k = 0, 1, 2, \ldots)$. Show that the rms magnitude of the *nth* harmonic is $\sqrt{6}I_o/\pi n$.

Fig. P3.23

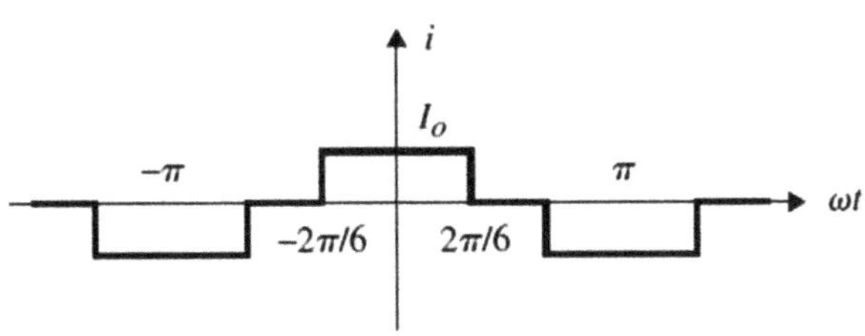

3.24 Show that the Fourier series for the phase current of a six-pulse SCR converter shown in Fig. P3.24 is given by:

$$i(t) = \frac{6I_o}{\pi}\left[\cos\omega t + \frac{\cos 5\omega t}{5} - \frac{\cos 7\omega t}{7} - \frac{\cos 11\omega t}{11} + \frac{\cos 13\omega t}{13} + \frac{\cos 17\omega t}{17} + \ldots + \frac{\cos n\omega t}{n}\right]$$

Fig. P3.24

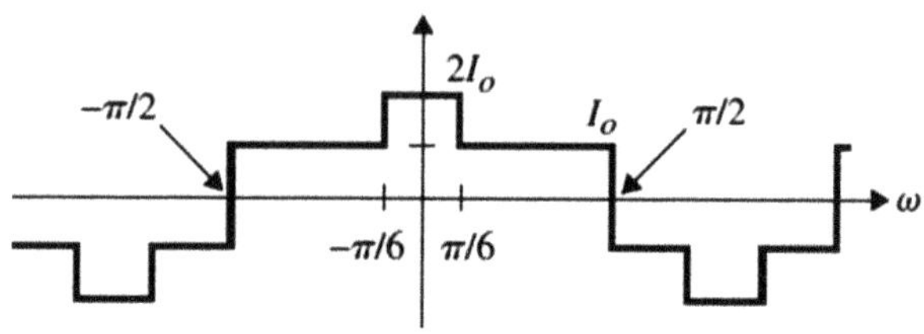

3.25 (a) Figure P3.25a shows a typical *ac* line current waveform in a single-phase, full-wave SCR controlled rectifier circuits under resistance load and sinusoidal *ac* source. Show that the fundamental line current component is given by:

$$i_{s1}(t) = I_{s1}\sin(\omega t + \theta)$$

where

$$\omega = \frac{2\pi}{T}$$

$$\theta = \tan^{-1}\frac{\cos 2\alpha - 1}{2\pi - 2\alpha + \sin 2\alpha}$$

$$I_{s1} = \sqrt{1 + 2(\pi - \alpha)^2 + 2(\pi - \alpha)\sin 2\alpha - \cos 2\alpha} \times \frac{I_s}{\sqrt{2\pi}}$$

(b) Find $i_{s1}(t)$ for the half-wave SCR controlled line current shown in Fig. P3.25b.

Fig. P3.25

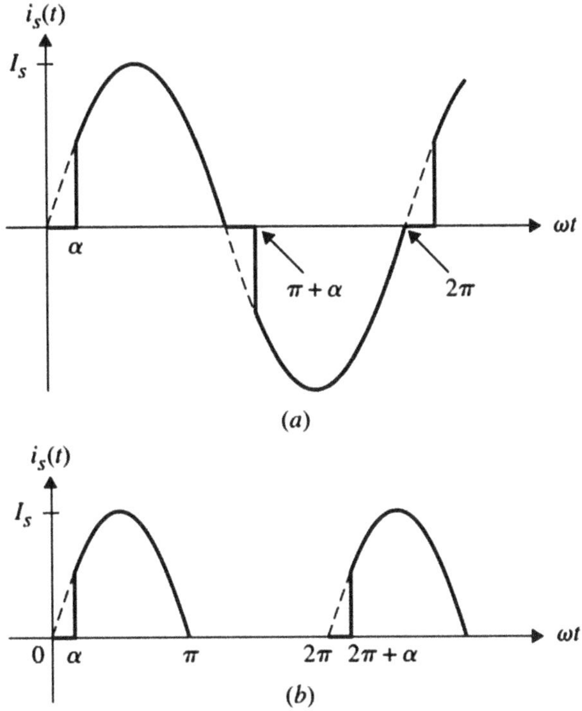

3.26 Show that the high harmonic Fourier coefficients for Fig. P3.26 are given by:

$$a_n = 0$$

$$b_n = \frac{1}{\pi}(2\alpha + \sin 2\alpha) + \sum_{n=3,5,7}^{\infty} \left[\frac{2}{\pi(1-n)} \sin \alpha(1-n)\right.$$

What value of h will achieve 95% input power factor assuming the waveform represents a line current with a sinusoidal line voltage

Fig. P3.26

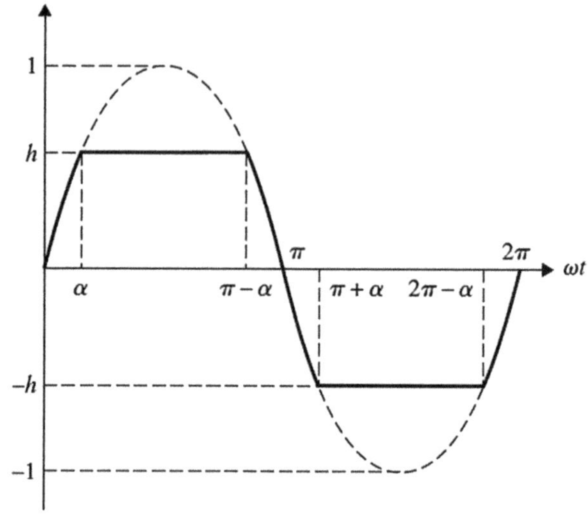

3.27 Figure P3.27 shows the *ac* line current for a full-wave, single-phase SCR controlled converter under highly inductive load. Show that the a_n coefficients are given by:

$$a_n = \frac{4I_o}{n\pi} \sin \frac{n(\pi - \alpha)}{2}$$

for $n = 1, 3, 5, 7, \ldots$

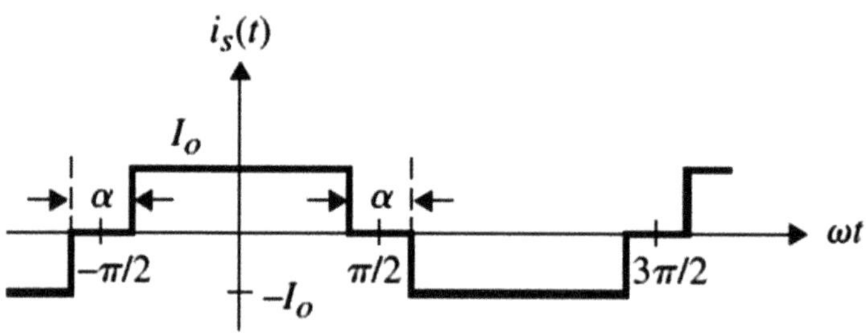

Fig. P3.27

3.28 Show that the Fourier components of $i_s(t)$ of Fig. P3.28 are given by:

$$i_s(t) = \frac{4I_o}{n\pi} \sum_{n=1,3,5,\ldots}^{\infty} \cos n\alpha \sin n\,\omega t$$

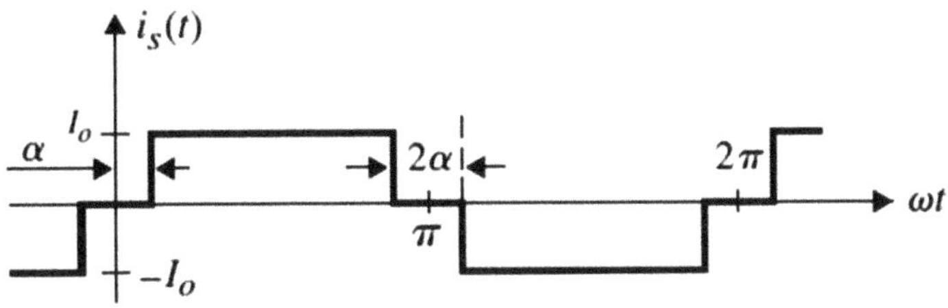

Fig. P3.28

3.29 Determine the Fourier components for the waveforms given in Fig. P3.29.

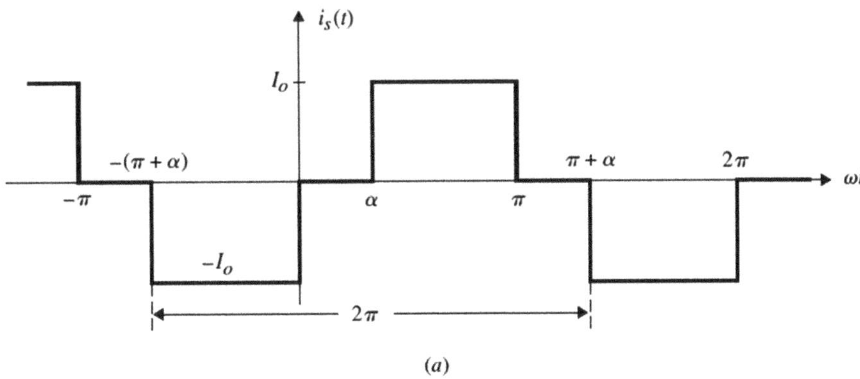

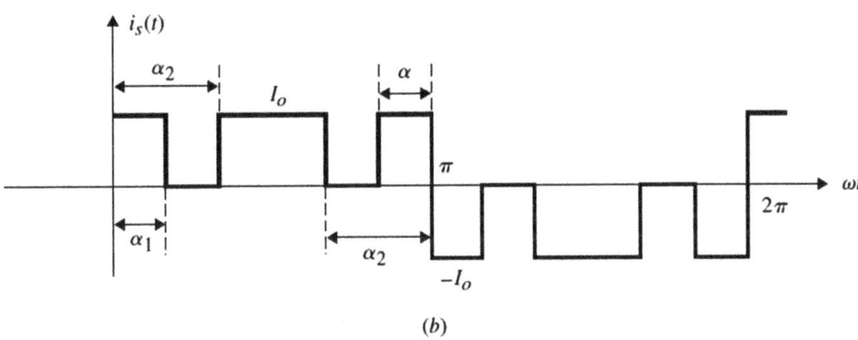

Fig. P3.29

General Problems

3.30 In the circuit of Fig. P3.30, the switch is opened at $t = 0$ after it has been closed for a long time. Derive the expression for $i_L(t)$, $i_D(t)$ and $v_C(t)$ for $t > 0$, and sketch them.

Fig. P3.30

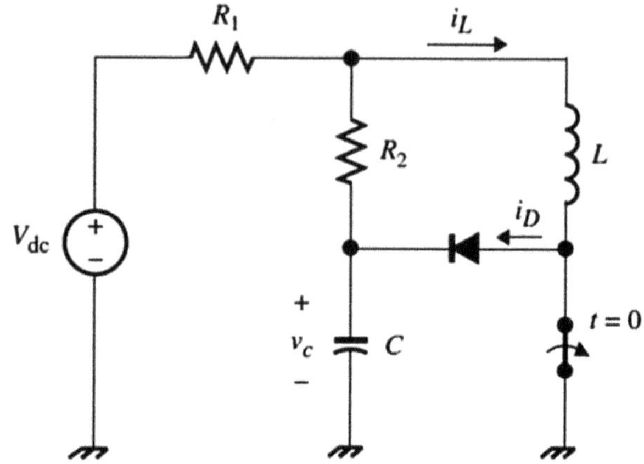

3.31 Show that the load current after the switch is closed at t=0 in the diode circuit given in Fig. P3.31 is given by:

$$i_o(t) = 0 \qquad 0 \le t < t_1$$

$$i_o(t) = \frac{V_s}{|Z|}\sin(\omega t - \theta) - \frac{V_{dc}}{R} + \mathrm{A}e^{-\frac{t}{\tau}} \qquad t_1 \le t < t_2$$

$$i_o(t) = 0 \qquad t_2 \le t < T$$

where

$$\tau = L/R$$

$$\theta = \tan^{-1}\frac{\omega l}{R}$$

$$A = \frac{V_{dc}}{R} - \frac{V_s}{|Z|}\sin(\omega t_1 - \theta)$$

Assuming the inductor is not initially charged, and here t_1 is the time when the diode turns on and is given by $\omega t_1 = \sin^{-1}(V_{dc}/V_s)$, and t_2 is the time at which the diode turns off before the next cycle starts at $t = T$. Find the expression for t_2.

Fig. P3.31

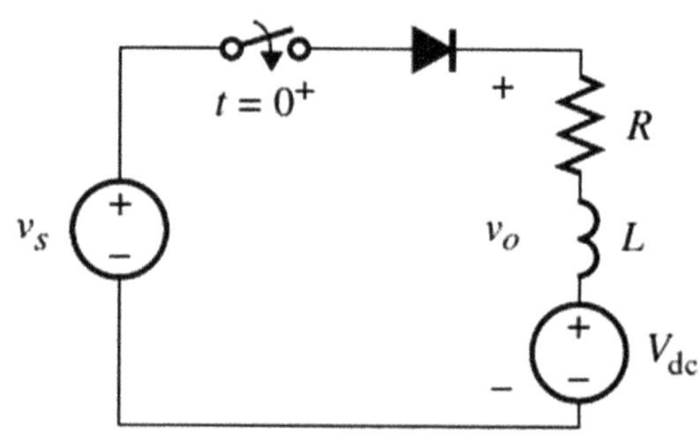

3.32 The circuit of Fig. P3.32a is known as a single-phase bridge inverter whose purpose is to convert the *dc* input voltage, V_{dc}, to an *ac* output voltage, v_o. If the switching sequence of S_1, S_2, S_3, and S_4 is done in such a way that the output voltage is shown in Fig. P3.32b,

(a) show that the Fourier series for v_0 is given by:

$$V_o = V_1 \sin \omega t + V_3 \sin 3\,\omega t + V_5 \sin 5\,\omega t + \ldots + V_n \sin n\,\omega t$$

where

$$V_n = \frac{4V_{in}}{n\pi} \sin\frac{n\alpha}{2} \qquad n = 1, 3, 5, 7,$$

Use the above results to find the Fourier series for $i_o(t)$.

Fig. P3.32

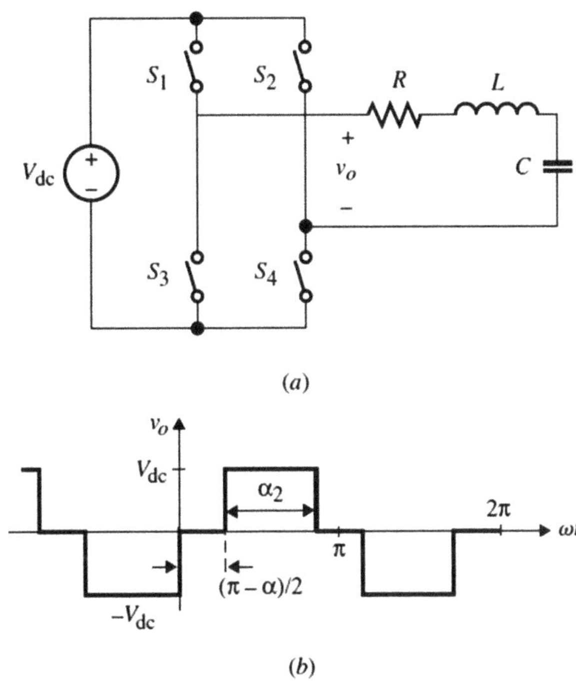

3.33 Consider the two-phase switching commutation circuit shown in Fig. P3.33. Assume S_1 has been on for a long time prior to $\omega t = 150o$ when S_2 is turned on. (a) Sketch the waveforms for $i_a(t)$ and $i_b(t)$, and (b) determine the time during which both D_1 and D_2 were on. Assume $v_a(t) = 100\sin(377t)$ and $v_b(t) = 100\sin(377t - 120^{\circ})$ V. Use $L_S = 1$ mH and $R_s = 2\ \Omega$.

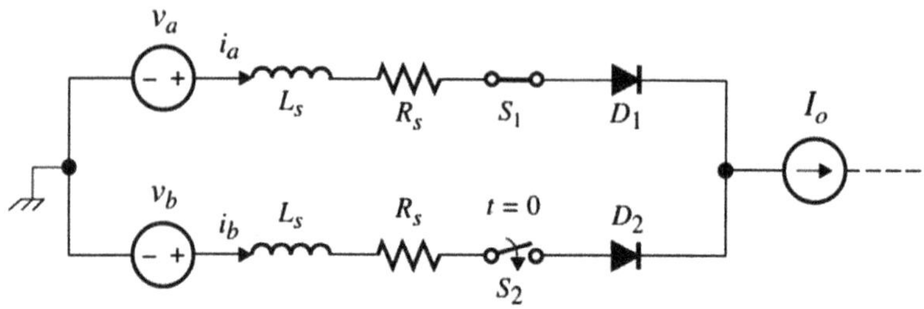

Fig. P3.33

3.34 The circuit shown in Fig. P3.34 is known as a forced commutation circuit whose voltage waveform over one switching period, T_S, is shown for $t > 0$. Derive the expression for $v_C(t)$ over one switching cycle. The conduction states of SCR_1 and SCR_2 are shown on the waveform.

Fig. P3.34

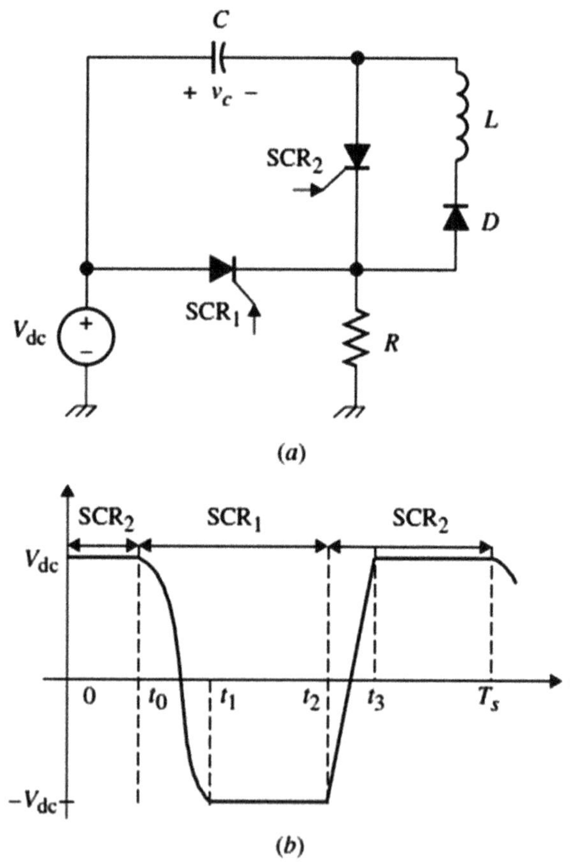

3.35 Figure P3.35 shows a DIAC-TRIAC switching circuit used in a heater controller, motor speed variation and light dimmer applications. Design for R and R_L so that the TRIAC triggers at 30° and 210° during the positive and negative cycles, respectively. Assume $v_s(t) = 110 \sin 2\pi 60t$ and the DIAC breakover voltage is 24 V.

Fig. P3.35

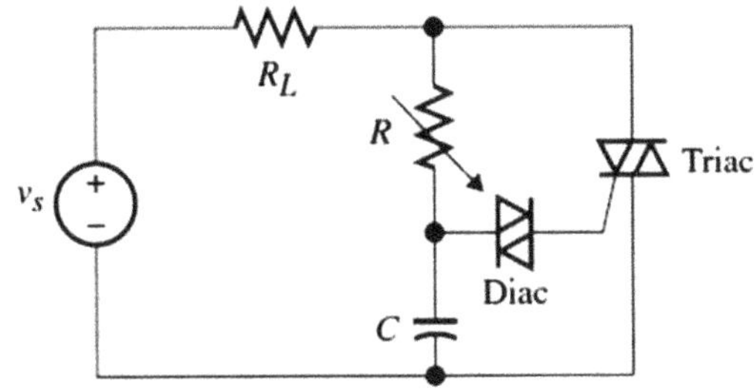

3.36 Assume the switch is turned on at $t = 0$ and the capacitor was initially discharged in the circuit of Fig. P3.36. Derive the expression for $v_C(t)$ and sketch it for $0 < t < 40$ ms, where $v_s(t) = 110 \sin 2\pi 50t$, $R = 10$ kΩ, and $C = 1$ μF.

Fig. P3.36

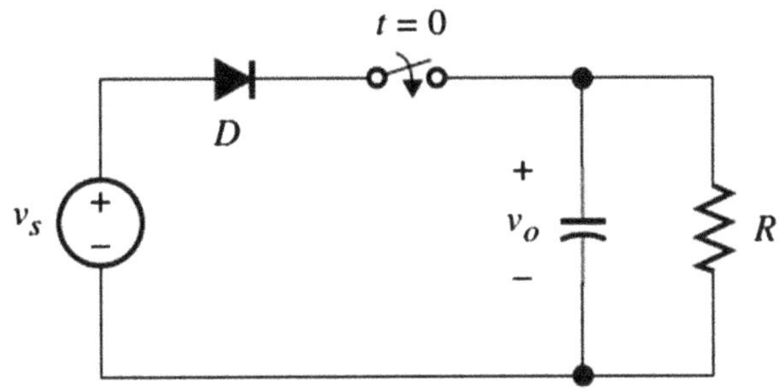

3.37 Figure P3.37 shows a self-oscillating LC circuit that allows the SCR to turn off naturally without using an additional auxiliary SCR. This circuit is known as a series resonant turnoff circuit. (a) Derive the expression for $v_C(t)$ and $i_L(t)$. (b) Determine the power rating of the SCR. Assume the SCR is first triggered at $t = 0$ and repeatedly every $T = 2$ ms. Assume $v_C(0^-) = 30$ V and $i_L(0^-) = 0$.

Fig. P3.37

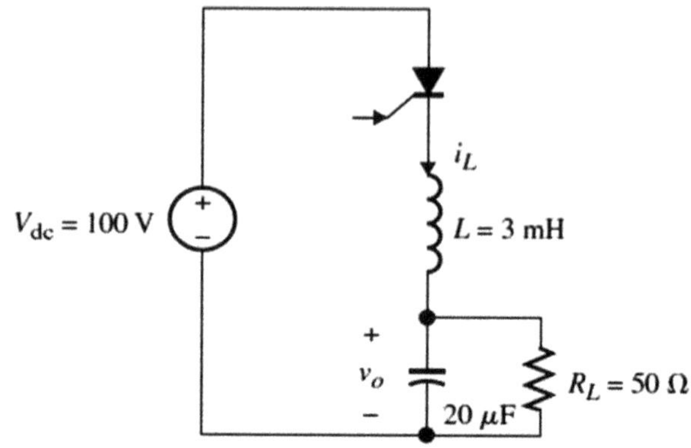

3.38 The circuit given in Fig. P3.38 represents one possible implementation of a family of *dc-dc* converters known as *soft-switching converters.*

(a) If we assume the switch is turned *ON* at $t = 0$, show that the capacitor voltage for $t > 0$ is given by:

$$v_{c1} = V_o + (V_{dc} - V_o)\cos\omega_o t$$

where $\omega_o = \sqrt{C_{eq}L}$ and $C_{eq} = C_1 + C_2$.

(b) In order to turn S back on while the capacitor voltage across C_1 is zero, the voltage across it must be allowed to reach zero again during the off-time of the switch. Show that for zero-voltage switching to occur, the following condition must be met:

$$V_o < \frac{V_{dc}}{2}$$

Fig. P3.38

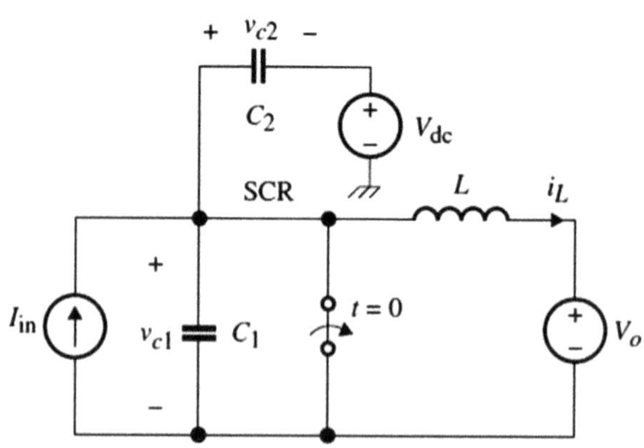

3.39 Figure P3.39 shows output and input waveforms in a cycloconverter-type power electronic circuit, where the output frequency is one-half of the input voltage frequency. Find the Fourier series representation for such waveform.

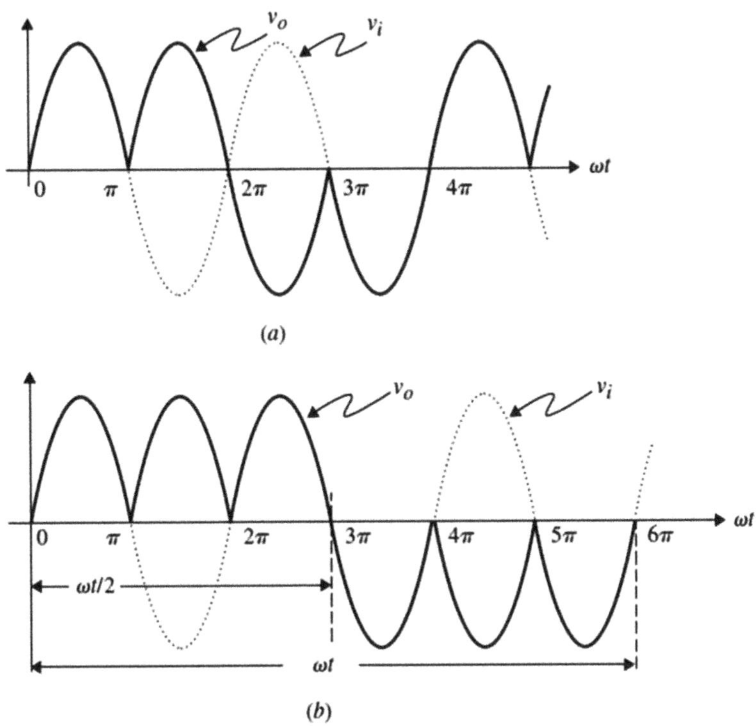

Fig. P3.39

3.40 Figure P3.40 shows a typical inductor current waveform in a switch-mode power supply when operating at the boundary of continuous and discontinuous mode operations. (a) Determine the Fourier components for $i(t)$, (b) calculate its average and rms values, and (c) calculate the THD_i.

Fig. P3.40

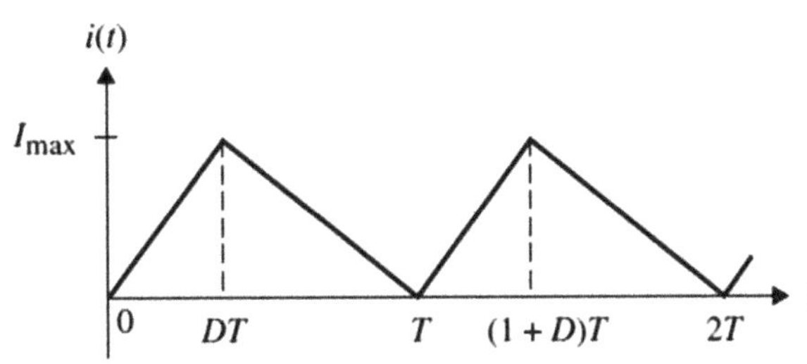

3.41 Repeat Problem 3.40 for the inductor current shown in Fig. P3.41, which represents a discontinuous conduction mode of operation in a switch-mode power supply.

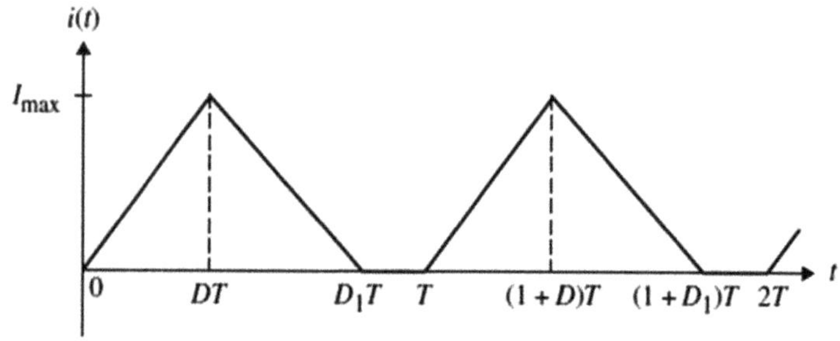

Fig. P3.41

3.42 Figure P3.42 shows a half-wave rectifier waveform with a sine-squared pulse represented mathematically for $0 < t < T/2$ by the following equation:

$$i_s(t) = I_o \sin^2 \omega t$$

(a) Determine the Fourier coefficients.
(b) Calculate the rms and THD_i values.

Fig. P3.42

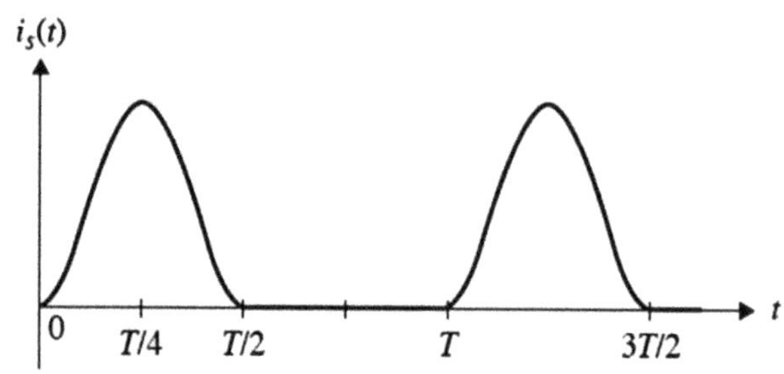

3.43 Consider a sinusoidal line current shown in Fig. P3.43 that can be represented mathematically for $0 < t < T$ by the following equation:

$$i_s(t) = I_o \sin^n \omega t$$

where n is an odd integer. By assuming that the line voltage is given by $v_s(t) = V_s \sin \omega t$ V, derive the expression for the power factor the THD_i.

Fig. P3.43

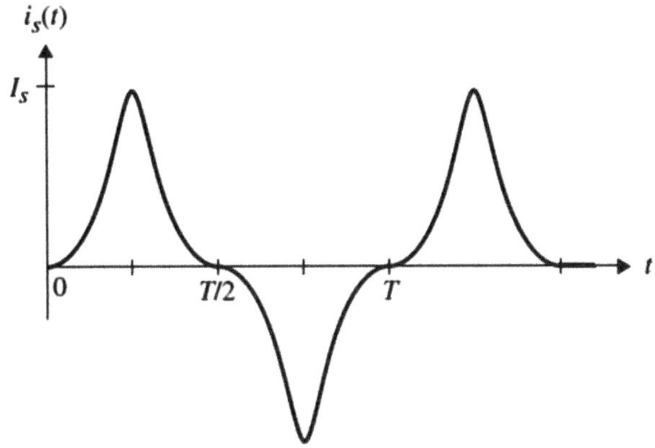

3.44 The simplified equivalent circuit of a *dc-dc* converter known as a buck converter is shown in Fig. P3.44a. Assume the switch is turned on and off according to the waveform shown in Fig. P3.44b Sketch the steady-state waveforms for v_{sw}, i_{sw}, and i_o and derive the expression for the average output power.

Fig. P3.44

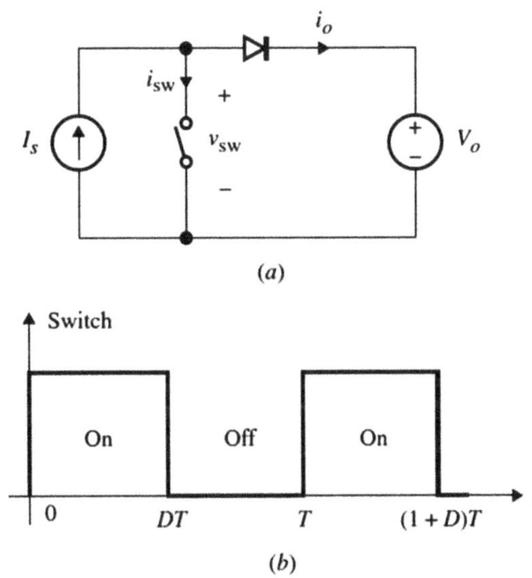

3.45 Figure P3.45 shows eight different switch-mode topologies of which only one topology is valid. Identify such topology and state which circuit law (KVL or KCL) is being violated for each of the other topologies.

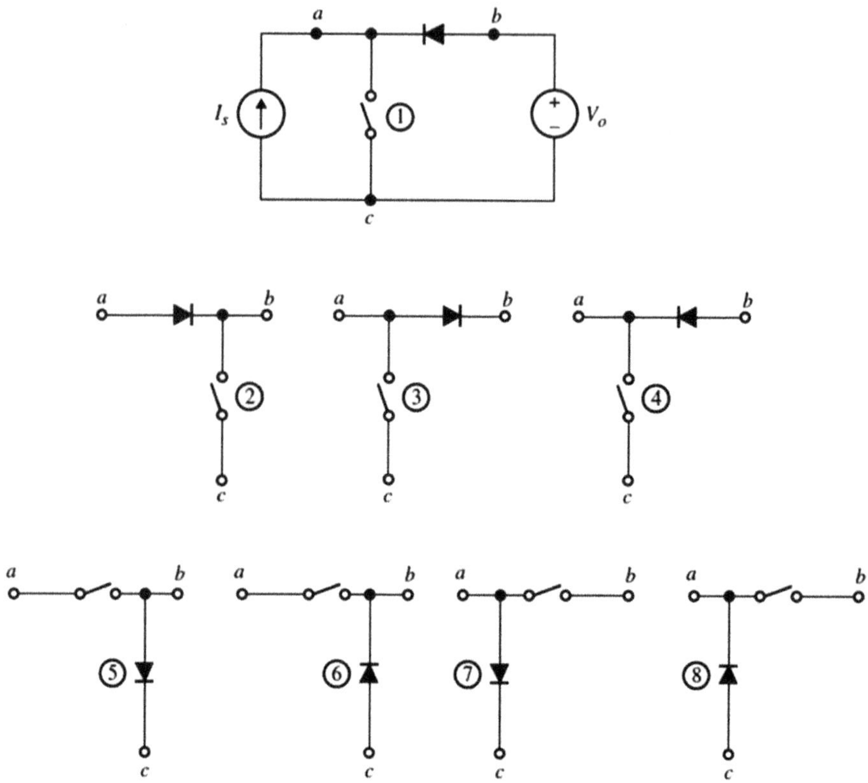

Fig. P3.45

3.46 Sketch the waveforms for i_{D1}, i_{D2}, v_o and find the average output voltage for the circuit of Fig. P3.46a, where $v_{s1}(t)$ and $v_{s2}(t)$are shown in Fig. P3.46b.

Fig. P3.46

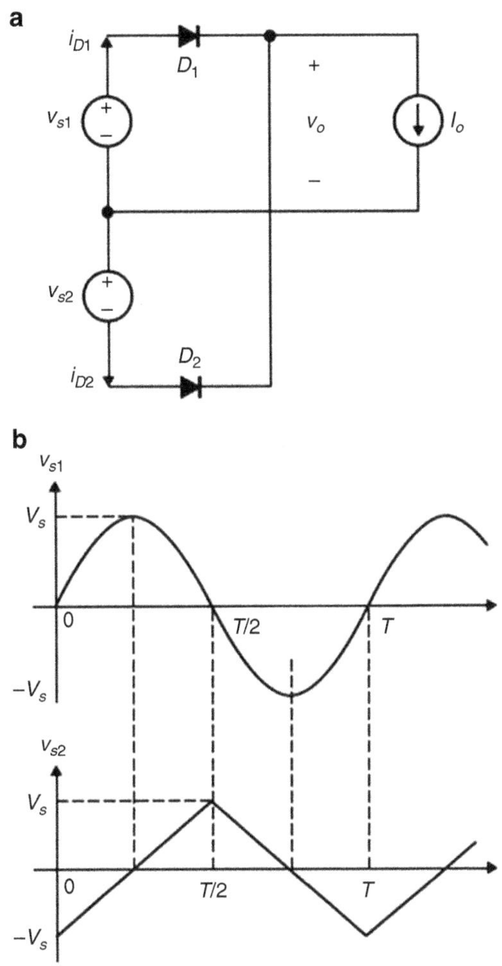

3.47 Repeat Problem 3.46 by replacing D_1 and D_2 by two switches S_1 and S_2 as shown in Fig. P3.47. Assume that S_1 is conducting only when $|v_{s1}(t)| > V_s/3$ and S_2 is the complement of S_1.

Fig. P3.47

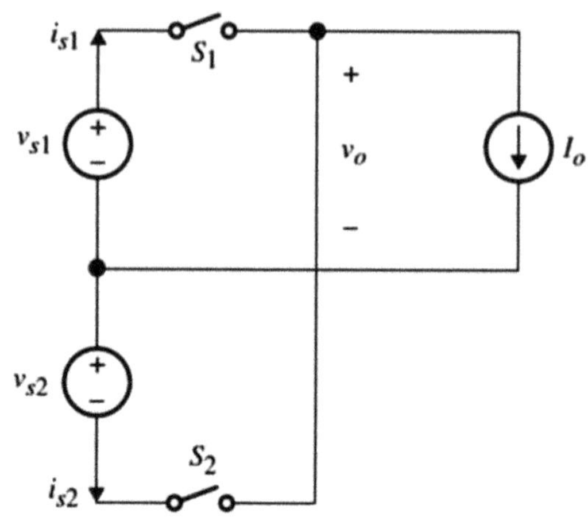

3.48 Consider the full-bridge rectifier of Fig. P3.48 with a current source $i_s(t) = I_s \sin \omega t$ and a constant output voltage, V_o. Sketch v_s and i_o, and find the average output power.

Fig. P3.48

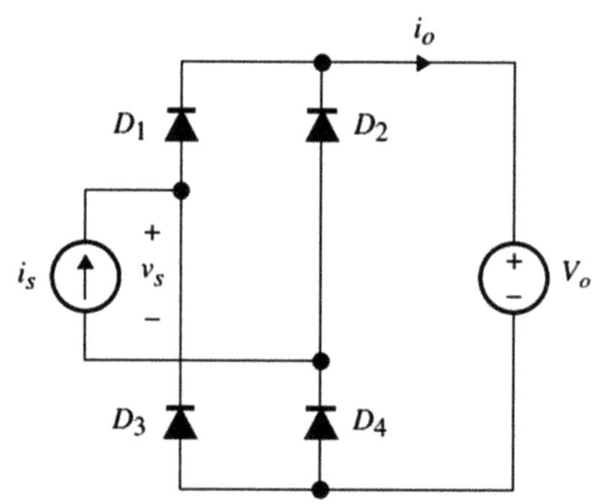

3.49 Consider the full-bridge rectifier of Fig. P3.49 with a voltage source $v_s(t) = V_s \sin \omega t$ and a constant output current, I_o. Sketch i_s and v_o and find the average output power.

Fig. P3.49

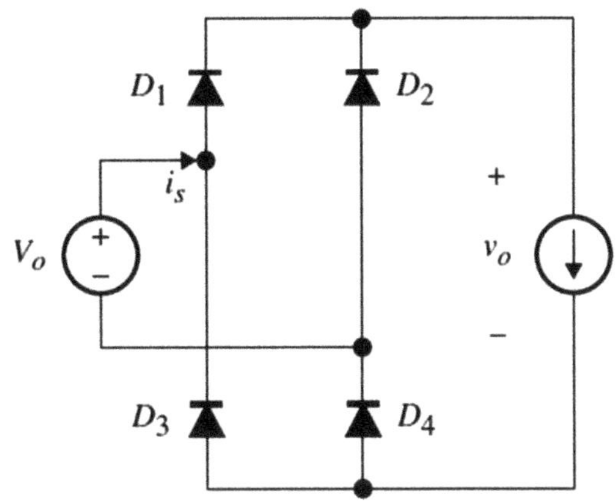

3.50 Figure P3.50 shows four different switch-mode topologies of which only one topology is valid. Identify such topology and state the circuit law (KCL or KVL) that is being violated for each of the other topologies.

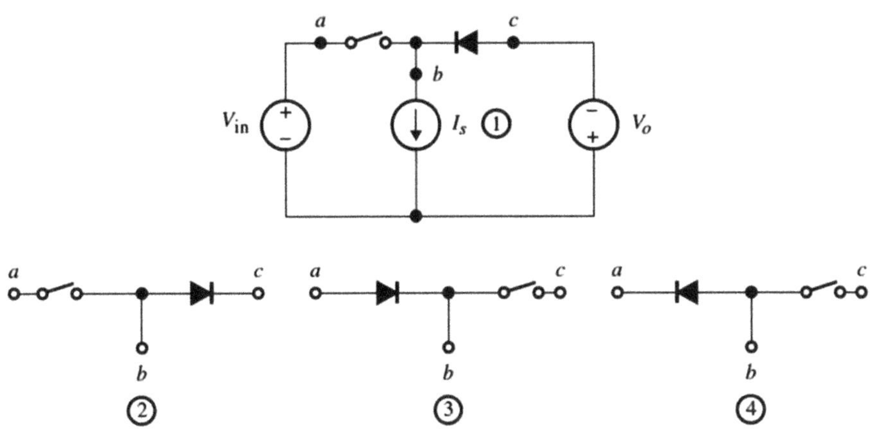

Fig. P3.50

3.51 Figure P3.51a shows an equivalent circuit for a switch-mode converter—Cuk (pronounced chook). Show that the ratio between the average output voltage, V_o, and the average input voltage, V_{in}, is given by the following relation:

$$\frac{V_0}{V_{in}} = \frac{D}{1 - D}$$

The switch waveform is shown in Fig. P3.51b.
Hint: Use the average voltage V_x *as given by* $V_{in} - V_o$.

Fig. P3.51

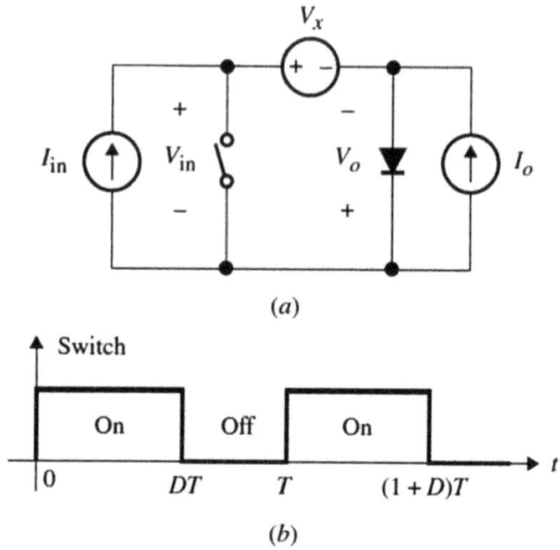

3.52 Figure P3.52a shows a switch-mode power electronic converter known as the buck-boost converter. This circuit is used to convert a DC input voltage to another *dc* level at the load. If we assume that the inductor, L, is large and the average output current is represented by a constant current source, then the equivalent circuit becomes as shown in Fig. P3.52b. Assume the switch is turned on and off as shown in Fig. P3.52c. Figure P3.52a show that the capacitor current is given in Fig. P3.52d, b sketch the steady-state average capacitor voltage.

Fig. P3.52

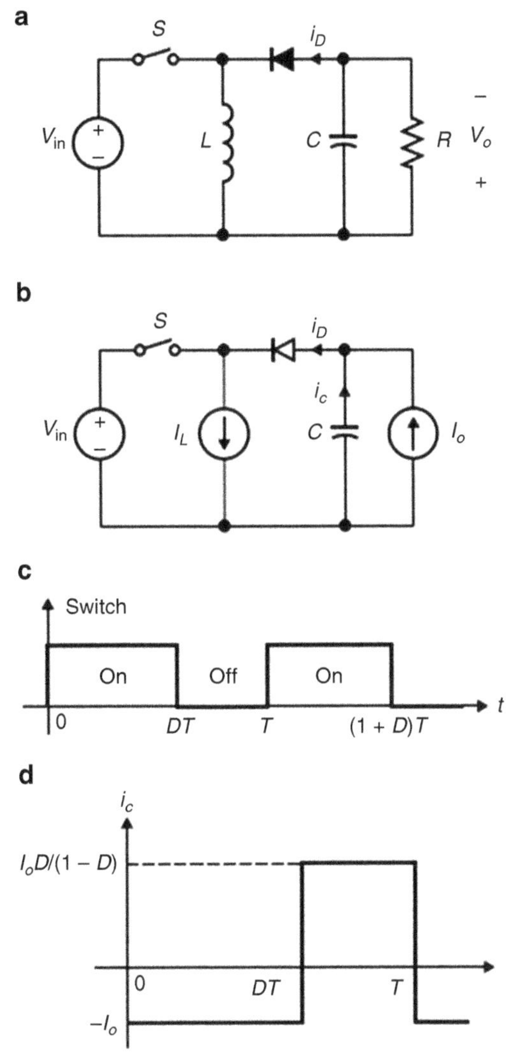

3.53 Determine the power factor for the circuits given in Problems 3.48 and 3.49.

3.54 Consider a transformer with a turn ration $N_1/N_2 = 200$, and rated at 240 kVA (44 kV/220 V) used to step down a 60 Hz voltage in a distributed system.

(a) Determine the rated primary and secondary currents.

(b) Determine the load impedance seen between the primary terminals when the load is fully loaded.

3.55 Consider the periodical current waveform shown in Fig. P3.55.

Determine its RMS expression. If you can only reduce one of the duty ratios D1, D2, and D3 by 10%, determine which one will give you the smallest reduction in RMS value.

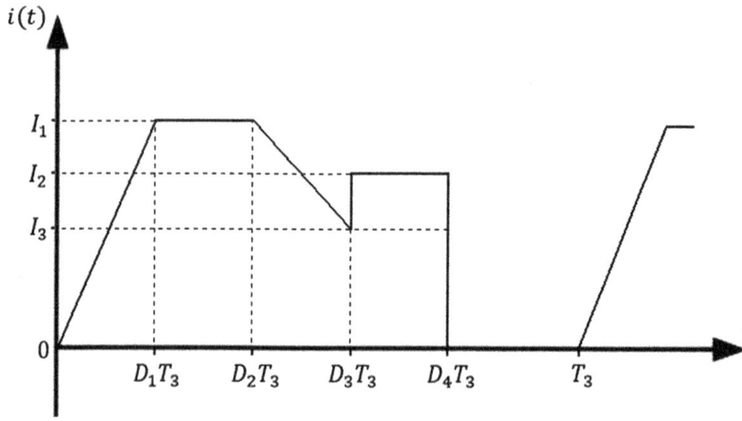

Fig. P3.55

Chapter 4
Non-isolated Switch Mode DC-DC Converters

4.1 Introduction

In this chapter we discuss converter circuits that are used in power electronic circuits and systems to change the voltages from one dc level to another dc level. Once again, switching devices will be used to process energy from the input to the output. Since the input here is dc, which comes from a post-filtering stage, these devices are normally operated at much higher frequencies than the line frequency, reaching as high as a few hundred kilohertz. This is why such converter circuits are known as high-frequency dc-dc switching converters or regulators. The term regulator is used since the circuit's main commercial application is in systems that require a stable and regulated dc output voltage. Depending on whether or not an output transformer is used, high-frequency dc-dc switching converters are classified as isolated or non-isolated. In this chapter and in the next, the emphasis will be on the steady-state analysis and design of several well-known second- and fourth-order dc-dc converters, each having its own features and applications. We will consider those topologies that do not use high-frequency isolation transformers as part of their power stage. Moreover, a large number of applications require output electrical isolation and multiple outputs that cannot be achieved using the basic topologies discussed in this chapter. The isolated and magnetically coupled topologies will be discussed in Chap. 5. Such topologies are the most popular in the power supply industry and are used in various types of electronic equipment whose design requires outputs with electrical isolation and multi-outputs.

I. Batarseh, A. Harb, *Power Electronics*,
https://doi.org/10.1007/978-3-319-68366-9_4

4.2 Power Supply Application

4.2.1 Linear Regulators

A typical block diagram of a linear regulator power supply is shown in Fig. 4.1. The front end of the linear regulator is a 60 Hz transformer, T_1, used to provide input electrical isolation and to step up or step down the line voltage, and this is followed by a full-wave bridge rectifier to convert the ac input to a dc input by adding a large filtering capacitor at the input of the linear regulator. This input to the linear regulator, V_{in}, is unregulated dc and cannot be used to drive the load directly. Using a linear circuit that provides a stable dc output regulates the dc voltage at the output, V_o.

For many years, most power supplies available in the market were of the linear regulator type, in which a series-pass active element is used to regulate the output voltage. In general, the active semiconductor element is used as a variable resistance to dissipate unwanted or excess voltage. Such an arrangement results in large amounts of power being dissipated in the active element, which can cause the efficiency to drop to as low as 40%. Because of this low efficiency, linear regulators have not been used for medium- and high-power applications since the early 1970s, when switch-mode dc-dc converters entered the marketplace. Despite the fact that linear regulators are simple to use and provide tight control, good output voltage ripples, and a low component count, their disadvantages are so numerous that their practical use is limited. Because of their high-power losses, they suffer from high thermal dissipation, resulting in low-power density and low efficiency.

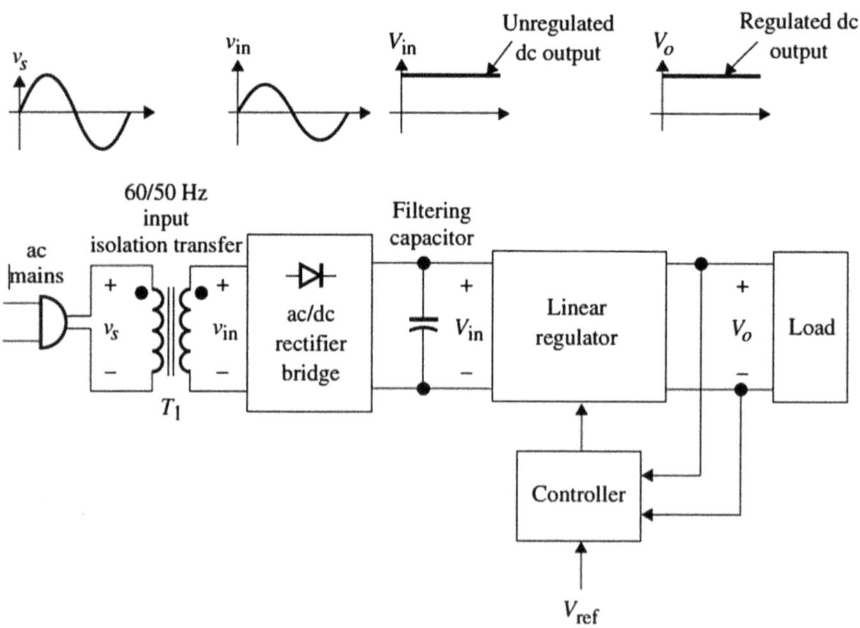

Fig. 4.1 Typical block diagram of a linear regulator power supply

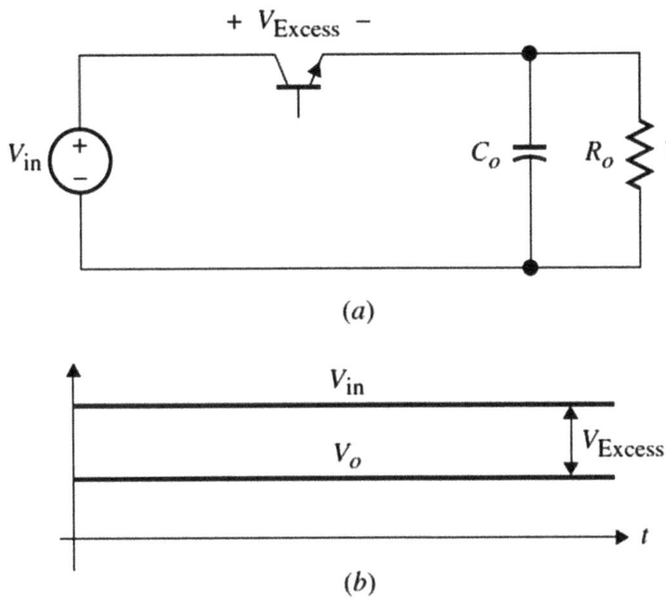

Fig. 4.2 (**a**) Typical configuration for a series active element used in a linear regulator and (**b**) average voltage waveforms

Figure 4.2 illustrates a simplified circuit to show how the series active element is connected in a linear regulator. V_{in} represents the unregulated input voltage, V_{Excess} is the excess voltage to be dissipated across the linear device, and V_{o} is the output voltage. V_{Excess}, which is the difference between V_{in} and V_{o}, must be large enough to keep the transistor in the linear active mode, acting as a variable resistor used only to absorb the difference in the voltage. This is why V_{Excess} is considered one of the key design parameters of linear regulation. To illustrate how the linear resistor works, we present a simplified topology showing the series element regulator in Fig. 4.3, where V_{ref} is generated by a Zener diode and R_1 and R_2 are used as a voltage divider. The magnitude of i_{B} determines how deeply the transistor is driven in the saturation region. The comparator is used to compare the output voltage with a fixed reference voltage. As the output voltage increases, the base current, i_{B}, decreases, and the excess voltage, v_{CE}, increases, hence reducing the output voltage. Similarly, if the output voltage decreases, the result is a reduced v_{CE} and an increased output.

4.2.2 Switch-Mode Power Supplies

The development of the power semiconductor switch made it possible for power electronics engineers to design power supplies with much higher efficiencies compared with linear regulators. Since transistors are used as switching devices,

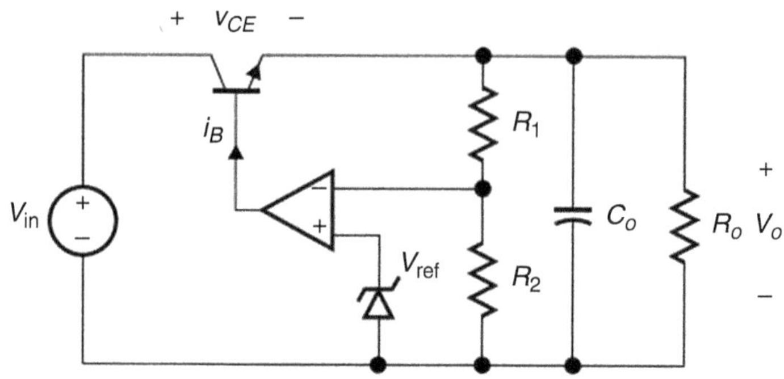

Fig. 4.3 Simplified representation of a simple series element regulator

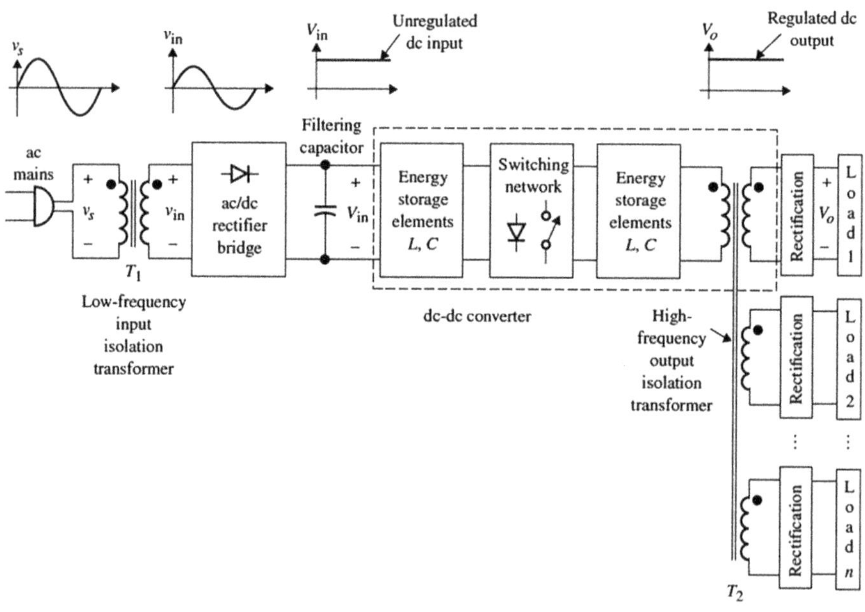

Fig. 4.4 Block diagram of a switch-mode power supply with multiple outputs

such power supplies are known as switch-mode power supplies or simply switching converters. In recent years, switching converters have become very popular due to recent advances in semiconductor technology. Today switching devices are available with very high switching speeds and very high power-handling capabilities. It is possible to design switch-mode power supplies with efficiency greater than 90% with low cost and relatively small size and light weight.

Unlike linear regulators, switching converters use power semiconductor devices to operate in either the on-state (saturation or conduction) or the off-state (cutoff or nonconduction). Since either state will lead to low switching voltage or low switching current, it is possible to convert dc to dc with higher efficiency using a switching regulator. Figure 4.4 shows a simplified block diagram for a switch-mode ac-dc power converter with multi-output application. Compared with the block diagram of Fig. 4.2, a switching network and high-frequency output electrical isolation transformer T_2 are added. The objective is to control the on time of the power devices to regulate the dc output voltage. The post-filtering is used to reduce the output voltage ripple. This chapter will discuss the detailed power stage operation of the dc-dc block shown in Fig. 4.4 without including the high-frequency isolation transformer. Because of the regulation method used, these converters are known as pulse-width-modulation (PWM) converters.

4.3 Continuous Conduction Mode

Steady-state analyses of the basic direct-connected second-order converters such as the buck (step-down), boost (step-up), and buck-boost (step-up/down) and fourth-order converters such as Cuk and SEPIC converters will be presented in this chapter. Both the continuous and discontinuous conduction modes of operation, as well as some non-ideal effects will be included in the analysis. The steady-state analysis of isolated or magnetically coupled converters (which are derived from these basic converters) such as the flyback, forward, push-pull, half- and full-bridge, and Weinberg converters will be discussed in Chap. 5.

First, we introduce the basic concept of a switch-mode circuit and the conversion technique in switching converters. Consider the simplest switching voltage converter, shown in Fig. 4.5. We assume that the switch is ideal and it is turned on at $t = t_0$ and turned off at t_1 alternately as shown in Fig. 4.6a, where $f = 1/T$ is the switching frequency.

The waveforms for the output voltage, v_o, and the output current, i_o, are shown in Fig. 4.6b, c, respectively, where V_{in} is the dc input voltage. The average output voltage is given by:

Fig. 4.5 A simple switching circuit

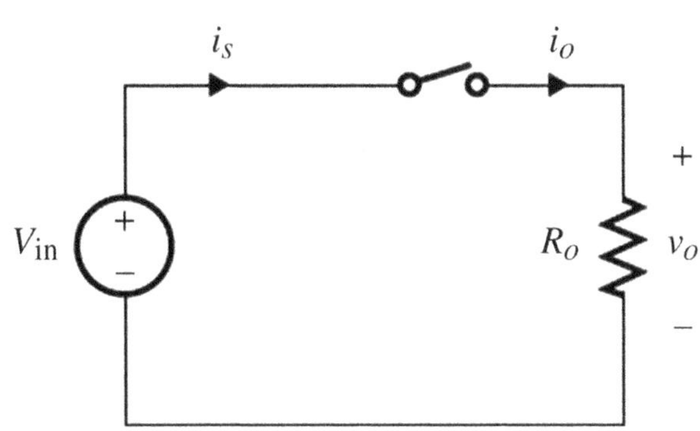

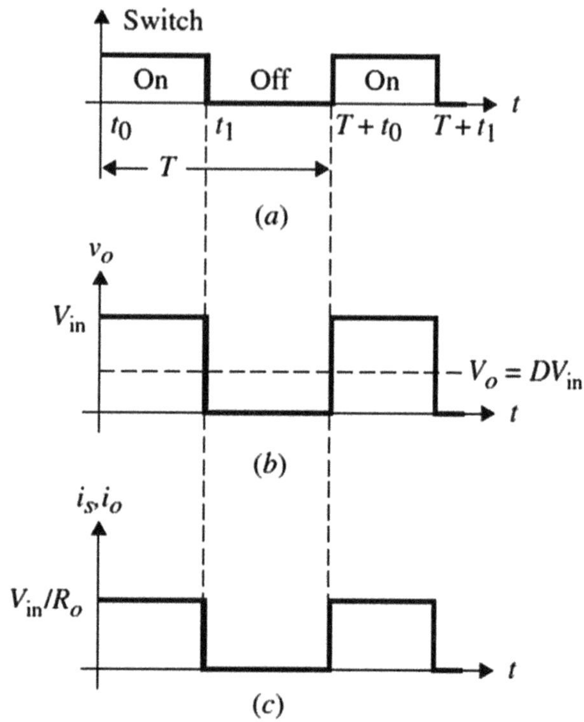

Fig. 4.6 Switching waveform for Fig. 4.5

$$\begin{aligned} V_\mathrm{o} &= \frac{1}{T}\int\limits_{t_0}^{T+t_0} v_o(t)dt \\ &= \frac{1}{T}\int\limits_{t_0}^{t_1} V_\mathrm{in}dt = \frac{t_1 - t_0}{T} V_\mathrm{in} \end{aligned} \tag{4.1}$$

If we let D to be defined as the duty ratio or duty cycle:

$$D = \frac{\text{On time}}{\text{Switching period}} = \frac{t_1 - t_o}{T} \tag{4.2}$$

then, the average output voltage V_o is given by:

$$V_\mathrm{o} = DV_\mathrm{in} \tag{4.3}$$

Exercise 4.1

If the output ripple factor $K_{\text{o, ripple}}$ is defined by the following relation:

$$K_{\text{o,ripple}} = \frac{\sqrt{V_{\text{o,rms}}^2 - V_{\text{o}}^2}}{V_{\text{o}}}$$

where $V_{\text{o, rms}}$ and V_{o} are the *rms* and average values of $v_{\text{o}}(t)$, respectively. Determine the output ripple factor for Fig. 4.5.

Answer: $K_{\text{o, ripple}} = \sqrt{\frac{1}{D} - 1}$

It is clear from Eq. (4.3) that the average output voltage is less than the applied dc voltage. Assuming an ideal switch, theoretically speaking, the efficiency of this converter is 100%. The drawback of this simple switching circuit is that the output voltage is not constant but a chopped dc with high ripple voltage. This results in high harmonics being generated at the load. This will not be acceptable for an electronic load when the output must be regulated to a fixed dc level with little ripple. However, in some application where a precise output voltage is not required, such as heating, light dimming, electroplating, and mechanical applications, this simple arrangement might be used, especially if the frequency is very high.

One approach to smooth the output voltage is to use a low-pass filter at the output of the circuit in order to filter the high switching-frequency components of the output voltage. The filter may consist of a simple capacitor and inductor. The capacitor is used to hold a dc value across the output resistor, and the inductor is used within the circuit to serve as a non-dissipative storage element needed to store energy from the input source and deliver it to the load. It can be argued that one single-pole, single-throw switch will not be sufficient to perform energy processing from the input to the output; rather, two switches or a single-pole, double-throw switch is required. This can be easily justified since the inductor current cannot be instantaneously interrupted. When one switch is switched, resulting in a sudden change in the inductor current, a second switch must be switched so that the continuity of the inductor current is maintained. Hence, a practical representation for a switch-mode converter must include either two switches, as shown in Fig. 4.7a, or a single switch, as shown in Fig. 4.7b. The inductor and capacitor elements are used as energy storage components to allow energy transfer from the input to the output. The output capacitor forms a low-pass filter to produce dc output voltage with little ripple. Most topologies, either isolated or non-isolated, will consist of one inductor and one output capacitor—hence the name second-order voltage converter. Fourth-order voltage converters consist of two inductors and two capacitors, which are considered series or combinations of second-order converters.

In second-order converters, depending on the arrangements of the switches and L, three possible topologies can be obtained, namely, buck, boost, and buck-boost, as shown in Fig. 4.8. Since these switches cannot be turned on and off simultaneously, single-pole, double-throw switches are used, which can be either

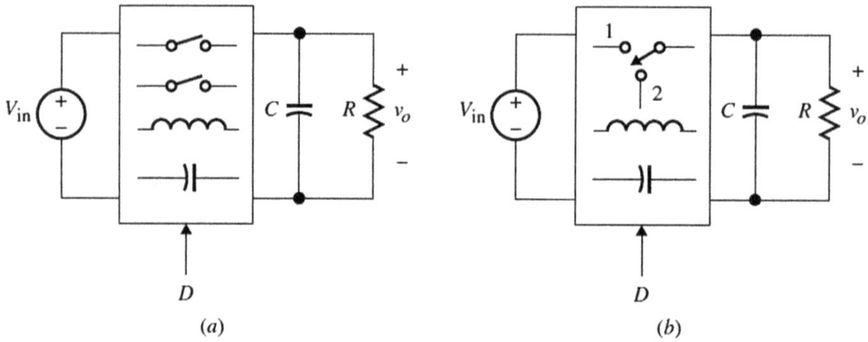

Fig. 4.7 Typical block diagram representation for switching converters with (**a**) two independent switches and (**b**) single-pole double-throw switch

Fig. 4.8 Three possible ways to insert a low-pass LC filter between (**a**) a dc source and a capacitive across the load: (**b**) buck converter, (**c**) boost converter, and (**d**) buck-boost converter

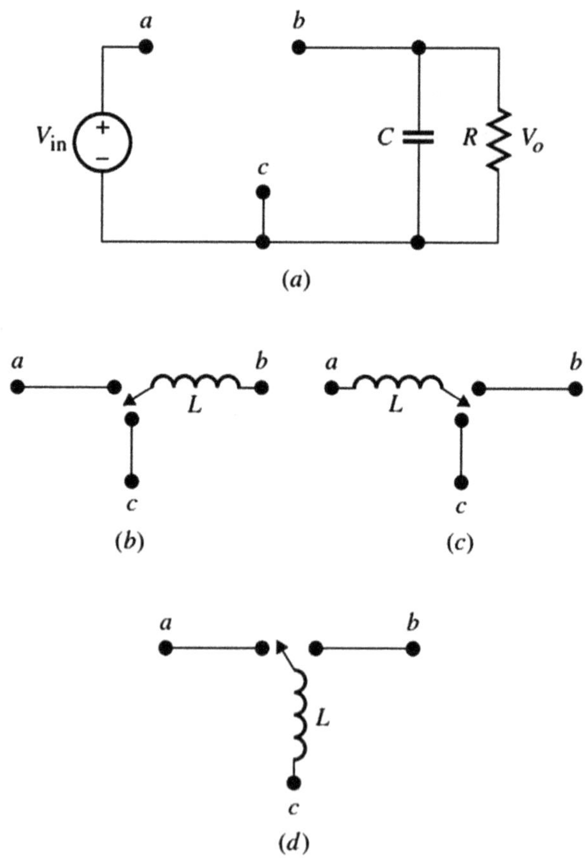

on or off. In the following sections, these converters will be thoroughly analyzed by deriving their steady-state characteristics. Since all switch-mode converters have pulsating currents and voltages, the dc output voltage can be obtained by adding a low-pass filter. It will be shown that in most of these converters an LC-type low-pass filter section is present. In fourth-order circuits like the Cuk and SEPIC, multiple LC low-pass filter sections are added.

Example 4.1
Consider the simple PWM converter of Fig. 4.5 with $V_{in} = 28$ V and $f_s = 50$ kHz. Assume an ideal switch. (a) Design for D and R so that the power delivered to the load is 25 W at average output current of 1.5 A. (b) Find the output ripple factor and (c) find the converter efficiency.

Solution

(a) For 25 W average output power and average output current of 1.5 A, the average output voltage is given by:

$$V_o = \frac{P_o}{I_o} = \frac{25}{1.5} = 16.7 \text{ V}$$

The load resistance:

$$R = \frac{16.7}{1.5} = 11.13\ \Omega$$

For $V_o = 16.7$ V and $V_{in} = 25$ V, the duty cycle:

$$D = \frac{16.7}{25} = 0.6$$

(b) The *rms* value for the output voltage is:

$$V_{o,\mathrm{rms}} = \sqrt{D}V_{in} = 21.69 \text{ V}$$

The ripple factor is given by:

$$K_{o,\mathrm{ripple}} = \sqrt{\frac{21.69^2 - 16.7^2}{16.7^2}}$$

This circuit results in 83% output voltage ripple that is by no means acceptable in many dc power supply applications.

(c) Since the switch is ideal, the converter efficiency is 100%, which can be illustrated as follows:

The average output power is given by:

$$
\begin{aligned}
P_{\mathrm{o}} &= \frac{1}{T}\int_0^{T_{\mathrm{s}}} v_{\mathrm{o}} i_{\mathrm{o}} dt \\
&= \frac{1}{T}\int_0^{T_{\mathrm{s}}} \frac{v_{\mathrm{o}}^2}{R} dt = \frac{1}{RT_{\mathrm{s}}}\int_0^{DT_{\mathrm{s}}} V_{\mathrm{in}}^2\, dt \\
&= \frac{DV_{\mathrm{in}}^2}{R}
\end{aligned}
$$

and the average input power is given by:

$$
\begin{aligned}
P_{\mathrm{in}} &= \frac{1}{T_{\mathrm{s}}}\int_0^{T_{\mathrm{s}}} V_{\mathrm{in}} i_{\mathrm{in}} dt = \frac{V_{\mathrm{in}}}{T_{\mathrm{s}}}\int_0^{DT_{\mathrm{s}}} i_{\mathrm{in}} dt \\
&= \frac{V_{\mathrm{in}}}{T_{\mathrm{s}}}\int_0^{DT_{\mathrm{s}}} i_{\mathrm{o}} dt = \frac{V_{\mathrm{in}}}{T_{\mathrm{s}} R}\int_0^{DT_{\mathrm{s}}} v_{\mathrm{o}} dt \\
&= \frac{DV_{\mathrm{in}}^2}{R}
\end{aligned}
$$

As expected, the average input and output powers are equal.

Exercise 4.2
Repeat Example 4.1 by assuming the switch has a 1.8 **V** voltage drop across it when conducting.

Answer: 0.64, 0.75, 93.6%

4.3.1 *The Buck Converter*

4.3.1.1 Topology and Basic Operation

Figure 4.9a, b shows the circuit configuration for a buck converter with a single- and two-switch implementation. Figure 4.9c shows the transistor-diode implementation. This topology is known as a buck converter because it steps down the average output voltage below the input voltage.

Throughout this chapter to obtain the steady-state characteristic equations, we will assume that power switching devices and the converter components are lossless. Moreover, the exact steady-state analysis of these converters requires solving second-order nonlinear systems. Such analysis is complex, and because of the nature of the output voltage, it is not necessary. Since these converters' function is to produce dc output, the output voltage $v_{\mathrm{o}}(t)$ consists of the desired dc and the undesired ac components. Practically, the output ripple due to switching is very small (less than 1%) compared to the level of the dc output voltage. As a result, we will assume the output ripple voltage is small and can be neglected when evaluating converter voltage gains, i.e., $v_{\mathrm{o}} = V_{\mathrm{o}}$. In other words, the ripple-free

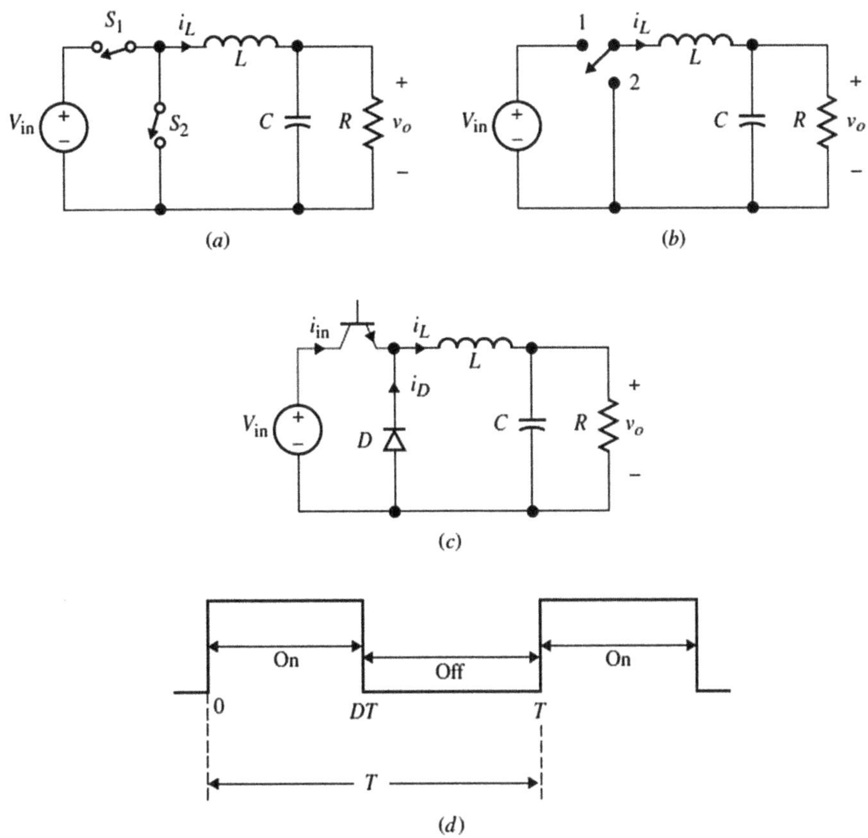

Fig. 4.9 The buck (step-down) converter: (**a**) two-switch implementation, (**b**) single-pole double-throw switch implementation, (**c**) transistor-diode implementation, and (**d**) switching waveform for the power switch

output voltage assumption is made since the output time constant for the filter capacitor and the output resistor, RC, is very large. Moreover, the analysis will be based on the converter operating in the steady-state condition, i.e., the converter currents and voltages have reached their steady-state values. These assumptions can be summarized and represented mathematically as follows:

1. Since we assumed lossless components and ideal switching devices, then the average input power, P_{in}, and the average output power, P_{out}, are equal, as given in Eq. (4.4):

$$P_{in} = P_{out} \tag{4.4}$$

Since we assumed steady-state operation, then the inductor current and the capacitor voltages are periodic over one switching cycle, i.e.:

$$i_L(t_o) = i_L(t_o + T) \tag{4.5a}$$

$$v_c(t_o) = v_c(t_o + T) \tag{4.5b}$$

where t_o is the initial switching time and T is the switching period.

Since we assumed ideal capacitors and inductors, then the average inductor voltage and the average capacitor current are zero:

$$I_c = \frac{1}{T}\int_{t_o}^{T+t_o} i_c(t)dt = 0 \tag{4.6}$$

$$V_L = \frac{1}{T}\int_{t_o}^{T+t_o} v_L(t)dt = 0 \tag{4.7}$$

In fact, Eq. (4.7) is a representation of Faraday's law, which states that voltage time during charging equals voltage time during discharging. This is also known as the volt-second principle. These two relations suggest that the total energy stored in the capacitor or the inductor over one switching cycle is zero. Finally, throughout the analysis in this chapter, the typical switching waveform for the power devices given in Fig. 4.9d will be used to represent the switching action of the power switch. For simplicity we set the initial switching time to zero $t_0 = 0$.

Again, D is known as the duty ratio or duty cycle, defined in Eq. (4.2). The power transistor is turned on for a period of DT and turned off for the remaining time $(1-D)T$. Depending on whether the switch is turned on or off, the inductor current will be either charging through V_{in} or discharging through the diode, respectively. As a result, there are two modes of operation. We first consider Mode 1, when the switch is on, shown in Fig. 4.10a.

As shown in the figure, when the switch is on, the input voltage, V_{in}, forces the diode into the reverse bias region. To determine the voltage conversion ratio, the average input and output currents, and the output voltage, we use the inductor current as a state variable in the following equation:

$$\begin{aligned} V_{in} &= v_L + V_o \\ &= L\frac{di_L}{dt} + V_o \end{aligned}$$

Eq. (4.8) can be rearranged as follows:

$$\frac{di_L}{dt} = \frac{1}{L}(V_{in} - V_o) \tag{4.9}$$

Integrating Eq. (4.9) from $t = 0$ to t with $I_L(0)$ as the initial condition, we obtain:

$$i_L(t) = \frac{1}{L}(V_{in} - V_o)t + I_L(0) \tag{4.10}$$

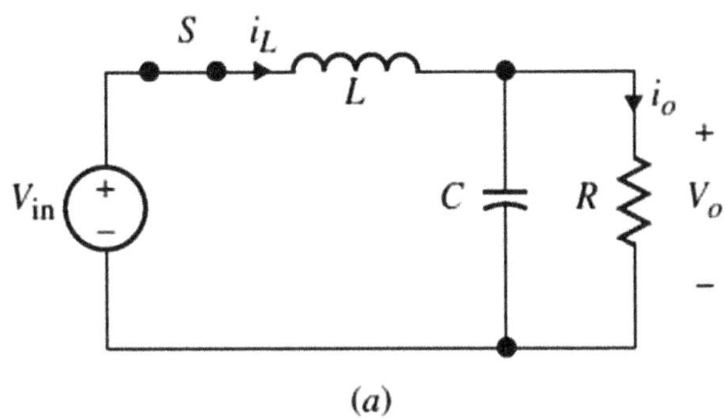

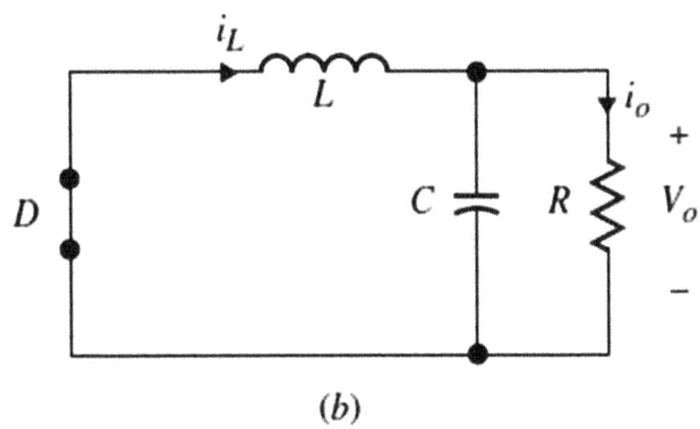

Fig. 4.10 Equivalent circuit modes for the buck converter: (**a**) Mode 1: The power switch is ON, (**b**) Mode 2: The power switch is OFF

Equation (4.10) suggests that the inductor current charges linearly with a slope of $(V_{in} - V_o)/L$, where $I_L(0)$ is the initial inductor current value at $t = 0$ when the switch is first turned *ON*. This equation applies as long as the switch is *ON*. However, the equivalent circuit model changes when the power switch is turned off at $t = DT$, resulting in the equivalent circuit of *Mode 2* shown in Fig. 4.10b during which the diode is conducting.

As shown in Fig. 4.10b, in order for the inductor current to maintain its continuity, the diode is forced to conduct by becoming forward biased so the diode "picks up" the current in the direction shown. The diode is known as "flyback" or "free-wheeling" because of the manner in which it is forced to turn *ON*. The resultant equations that describe *Mode 2* operation are given by:

$$\frac{di_L}{dt} = -\frac{1}{L}V_o \tag{4.11}$$

Integrating both sides of Eq. (4.11) for $t > DT$ with $i_L(DT)$ as an initial condition, we obtain:

$$I_L(t) = -\frac{V_o}{L}(t - DT) + I_L(DT) \tag{4.12}$$

where $I_L(DT)$ is the initial inductor current when the switch first turned *OFF*.

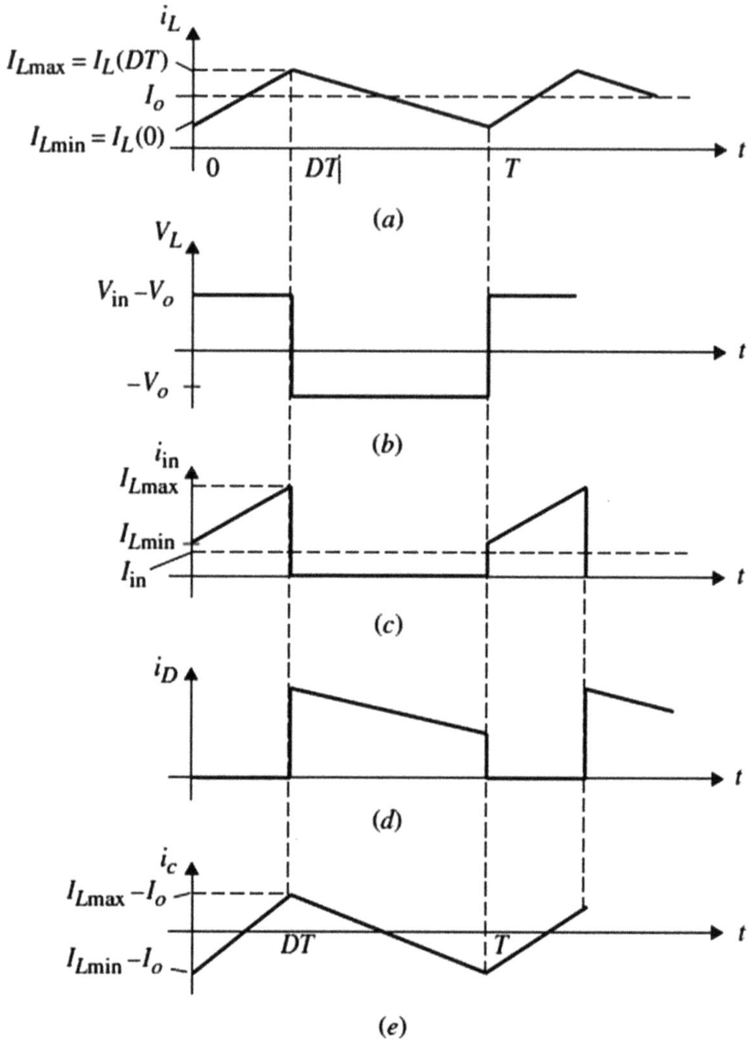

Fig. 4.11 Steady-state waveforms for the buck converter: (**a**) inductor current, (**b**) inductor voltage (**c**) input current, (**d**) diode current, and (**e**) capacitor current

Equation (4.12) suggests that the inductor current starts discharging at $t = DT$ with the slope of $-V_o/L$ as shown in Fig. 4.11a. In the steady-state operation, we have:

$$I_L(0) = I_L(t) \tag{4.13}$$

Evaluating Eq. (4.10) at $t = DT$ and Eq. (4.12) at $t = T$ and using Eq. (4.13), we obtain the following two relations for $I_L(0)$ and $I_L(DT)$:

$$I_{\mathrm{L}}(DT) = \frac{1}{L}(V_{\mathrm{in}} - V_{\mathrm{o}})DT + I_{\mathrm{L}}(0) \tag{4.14a}$$

$$I_{\mathrm{L}}(0) = -\frac{V_{\mathrm{O}}}{L}(1 - D)T + I_{\mathrm{L}}(DT) \tag{4.14b}$$

The steady-state current and voltage waveforms are shown in Fig. 4.11. I_{Lmax} and I_{Lmin} are the inductor current value at the instants the switch is turned OFF and ON, respectively.

Voltage Conversion

Next we use the above relations to derive expressions for the voltage conversion and average input and output currents. From Eq. 4.14, we obtain:

$$\frac{V_{\mathrm{o}}}{V_{\mathrm{in}}} = D \tag{4.15}$$

Hence, the maximum output voltage gain is 1. We should point out that Eq. (4.15) can be obtained easily by using the *volt-second* principle across the inductor which is given as follows:

$$\overbrace{(\text{Inductor Voltage})(\text{time})}^{\substack{\text{Mode 1} \\ (\text{interval } DT)}} + \overbrace{(\text{Inductor voltage})(\text{time})}^{\substack{\text{Mode 2} \\ (\text{interval } (1-D)T)}} = 0$$

$$v_{\mathrm{L}}(t)DT + v_{\mathrm{L}}(t)(1 - D)T = 0 \tag{4.16}$$

where v_{L} equals $(V_{\mathrm{in}} - V_{\mathrm{o}})$ and $-V_{\mathrm{o}}$ during time intervals DT and $(1 - D)T$, respectively.

We can make two observations on the buck voltage gain equation $V_{\mathrm{o}} = DV_{\mathrm{in}}$. First, since all the converter components (L, C, D, Q) are ideal, they do not dissipate any power, resulting in 100% voltage efficiency. Second, the average input and output voltage ratio has a linear control characteristic curve as shown in Fig. 4.12. By varying the value of the duty cycle, D, we can control the average output voltage to the desired level.

Average Input and Output Currents

The input current,i_{in}, as illustrated in Fig. 4.11c, with an average value of I_{in}, is given by:

$$I_{\mathrm{in}} = \frac{1}{T}\int_0^T i_{\mathrm{in}}(t)dt \tag{4.17}$$

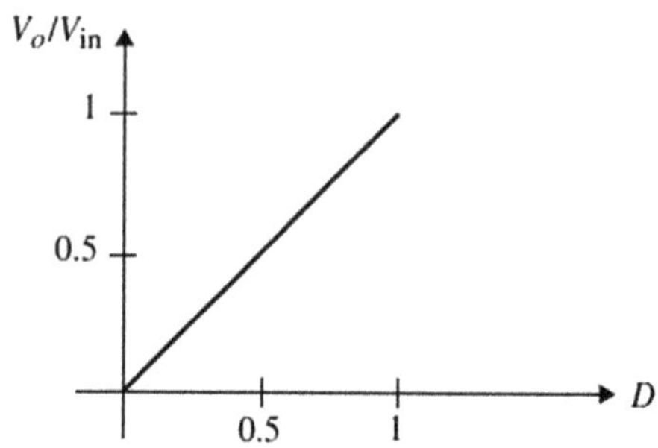

Fig. 4.12 Ideal output control characteristic curve for the buck converter

Since $i_{in} = i_L$ in *Mode 1*, then we substitute for $i_L(t)$ from Eq. (4.10), and by evaluating the integral between $t = 0$ and $t = DT$, we obtain:

$$I_{in} = \frac{1}{2L}(V_{in} - V_o)D^2T + I_L(0)D \tag{4.18}$$

Using Eq. (4.14), we obtain:

$$I_{in} = \frac{1}{2}(I_{Lmax} + I_{Lmin})D \tag{4.19}$$

where I_{Lmax} and I_{Lmin} represent $I_L(DT)$ and $I_L(0)$, respectively.

Similarly, by inspection, the average output current is given by:

$$I_O = I_L = \frac{I_{Lmin} + I_{Lmax}}{2} = \frac{V_o}{R} \tag{4.20}$$

From Eqs. (4.14) and (4.20), we can solve for the maximum and minimum inductor currents to obtain:

$$I_{Lmax} = DV_{in}\left(\frac{1}{R} + \frac{(1-D)T}{2L}\right) \tag{4.21}$$

$$I_{Lmin} = DV_{in}\left(\frac{1}{R} - \frac{(1-D)T}{2L}\right) \tag{4.22}$$

Substituting the above equations in Eqs. (4.19) and (4.20), we obtain:

$$I_o = \frac{DV_{in}}{R}$$

$$I_{in} = \frac{D^2V_{in}}{R}$$

Hence, the current gain is given by:

$$\frac{I_o}{I_{in}} = \frac{1}{D} \tag{4.23}$$

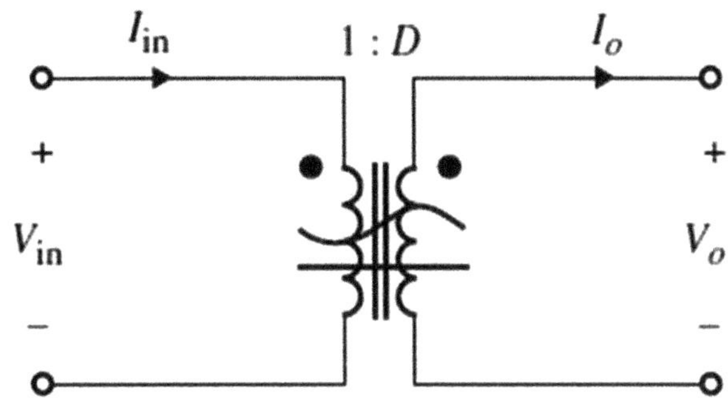

Fig. 4.13 Equivalent circuit representation for the buck converter referred to it as dc-dc transformer

The above relation can be obtained by equating the average input and output power to yield:

$$I_{\text{in}}V_{\text{in}} = I_{\text{o}}V_{\text{o}}$$
$$\frac{I_{\text{in}}}{I_{\text{o}}} = \frac{V_{\text{o}}}{V_{\text{in}}} = D$$

From Eqs. (4.15) and (4.23), it is clear that the current and voltage relations for the converter are equivalent to a dc transformer model with a ratio of D, as shown in Fig. 4.13. The sinusoidal curve and straight line drawn across the transformer windings indicate that the transformer is capable of transferring ac and dc, respectively.

Critical Inductor Value

It is clear that for $L_{\text{min}} \neq 0$ the converter will operate in the *continuous conduction mode (ccm)*. To find the minimum inductor value that is needed to keep the converter in the *ccm*, we set I_{Lmin} to zero and solve for L:

$$I_{\text{Lmin}} = DV_{\text{in}}\left(\frac{1}{R} - \frac{(1-D)T}{2L}\right) = 0$$

$$L_{\text{crit}} = \left(\frac{1-D}{2}\right)TR \tag{4.24}$$

where L_{crit} is the critical inductance minimum value for a given D, T, and R before the converter enters the *discontinuous conduction mode (dcm)* of operation.

Output Voltage Ripple

Since we have assumed that the output voltage has no ripple, the entire ac output current from the inductor passes through the parallel capacitor, and only dc current is delivered to the load resistor. In practice, the value of the output capacitor is an important design parameter since it influences the overall size of the dc-dc converter and how much of the switching frequency ripple is being removed. Having said that, it is design practice to choose a larger output capacitor in order to limit the ac ripple across V_o. Theoretically speaking, if $C \to \infty$,the capacitor acts like a short circuit to the ac ripple, resulting in zero output voltage ripple. If we assume C is finite, then there exists a voltage ripple superimposed on the average output voltage. In order to derive an expression for the capacitor ripple voltage, we first obtain an expression for the capacitor current, which is given by the following relation:

$$i_c(t) = i_L(t) - I_o$$

As a result, the initial capacitor current at $t=0$ is given by:

$$\begin{aligned} I_c(t) &= I_L(t) - I_o \\ &= -\left(\frac{I_{Lmax} - I_{Lmin}}{2}\right) \end{aligned}$$

and at $t=DT$:

$$\begin{aligned} I_c(DT) &= I_L(DT) - I_o \\ &= +\left(\frac{I_{Lmax} - I_{Lmin}}{2}\right) \end{aligned}$$

The resultant capacitor current and voltage are shown in Fig. 4.14.

The instantaneous capacitor current can be expressed in terms of ΔI from Eqs. (4.21) and (4.22) as shown in the following equations:

$$i_c(t) = \frac{I_{Lmax} - I_{Lmin}}{DT}t - \frac{I_{Lmax} - I_{Lmin}}{2} = \frac{\Delta I}{DT}t - \frac{\Delta I}{2} \quad 0 \le t \le DT \tag{4.25a}$$

Fig. 4.14 Capacitor current and voltage waveforms

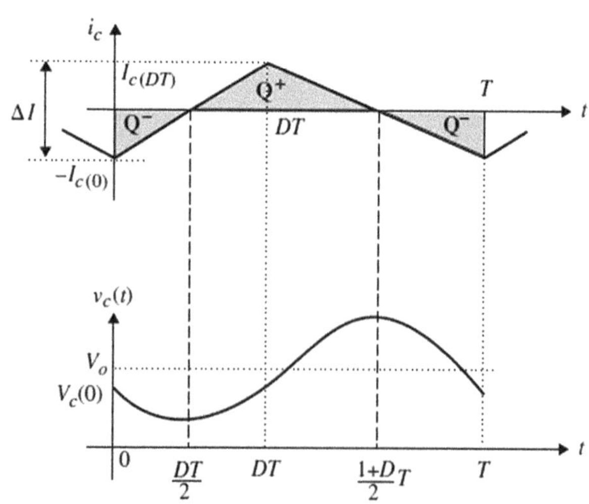

$$i_c(t) = -\frac{I_{\text{Lmax}} - I_{\text{Lmin}}}{(1-D)T}(t - DT) + \frac{I_{\text{Lmax}} - I_{\text{Lmin}}}{2}$$
$$= \frac{-\Delta I}{(1-D)T}(t - DT) + \frac{\Delta I}{2} \qquad DT \leq t < T \tag{4.25b}$$

where $\Delta I = (V_{\text{in}}(1-D)TD)/L$.

From the capacitor voltage-current relation, $i_c = C(dv_c/dt)$, the capacitor voltage, v_{c1} , can be expressed by the following integral for $t \geq 0$:

$$v_{c1} = \frac{1}{c}\int_0^t i_c dt + V_C(0)$$

where $V_C(0)$ is the initial capacitor voltage at $t=0$. Substituting for $i_c(t)$ from Eq. (4.25a), we obtain the following equation:

$$v_{c1}(t) = \frac{1}{C}\int_0^t \left(\frac{\Delta I}{DT}t - \frac{\Delta I}{2}\right)dt + V_C(0)$$

Evaluating this integral yields:

$$v_{c1}(t) = \frac{1}{C}\frac{\Delta I}{DT}\frac{t^2}{2} - \frac{\Delta I}{2C}t + V_c(0) \quad 0 \leq t < DT \tag{4.26}$$

Similarly, for $t \geq DT$, the capacitor voltage is given by:

$$v_{c2}(t) = \frac{1}{C}\int_{DT}^t i_c dt + V_c(DT)$$

where $V_c(DT)$ is the initial capacitor voltage when the switch is turned OFF at $t=DT$. Substitute for $i_C(t)$ from Eq. (4.25b), $v_{c2}(t)$ is given by:

$$v_{c2} = -\frac{\Delta I}{C(1-D)T}\frac{(t-DT)^2}{2} + \frac{\Delta I}{2C}(t - DT) + V_c(DT) \qquad DT \leq t < T \tag{4.27}$$

Since the capacitor voltage is in the steady state, we have $v_{c1}(t=DT) = v_{c2}(t=DT)$ and $v_{c1}(0) = v_{c2}(T)$, resulting in the following boundary conditions for the capacitor voltage:

$$V_c(T) = V_c(DT) = V_c(0)$$

Since the average capacitor voltage is V_o, then we have in general:

$$V_o = \frac{1}{T}\left[\int_0^{DT} v_{c1}(t)dt + \int_{DT}^T v_{c2}dt\right]$$

Substitute for v_{c1} and v_{c2} from Eqs. (4.26) and (4.27) to obtain:

$$V_o = \frac{\Delta I}{12C}(1-2D)T + V_c(0) \tag{4.28}$$

Substitute for $\Delta I = (DV_{in}(1-D)T)/L$ in Eq. (4.28) to yield:

$$V_c(0) = DV_{in}\left[1 - \frac{(1-D)(1-2D)}{12CL}T^2\right] \tag{4.29}$$

Hence, the capacitor initial values at $t=0$ and $t=DT$ are equal, as expected since the capacitor current is symmetrical. Since the peak capacitor voltage occurs when the inductor current is zero, we then have the capacitor minimum voltage occurring at $t=DT/2$ which is obtained from Eq. (4.26):

$$\begin{aligned} V_{c,\min} &= \frac{1}{C}\frac{\Delta I}{DT}\left(\frac{DT}{2}\right)^2 - \frac{\Delta I}{2C}\left(\frac{DT}{2}\right) + V_c(0) \\ &= -\frac{\Delta I}{8C}DT + V_c(0) \end{aligned} \tag{4.30a}$$

And the maximum capacitor voltage occurring at $t=(1+D)T/2$ and is obtained from Eq. (4.27):

$$\begin{aligned} V_{c,\max} &= -\frac{\Delta I}{2C(1-D)T}\left(\frac{(1+D)}{2}T - DT\right)^2 + \frac{\Delta I}{2C}\left(\frac{(1+D)}{2}T - DT\right) + V_C(DT) \\ &= \frac{\Delta I}{8C}(1-D)T + V_c(DT) \end{aligned} \tag{4.30b}$$

Substitute for $V_c(0)$ from Eq. (4.29) and using $\Delta I = (DV_{in}(1-D)T)/L$, it can be shown that $V_{c,\min}$ and $V_{c,\max}$ are expressed as follows:

$$V_{c,\min} = V_o\left[1 - \frac{(1-D)(2-D)}{24CL}T^2\right] \tag{4.31a}$$

$$V_{c,\max} = V_o\left[1 + \frac{(1-D^2)}{24CL}T^2\right] \tag{4.31b}$$

Hence, the variation in the capacitor peak voltage is given by:

$$\Delta V_C = V_{c,\max} - V_{c,\min}$$

Hence, from Eq. (4.31a) and (4.31b), we obtain:

$$\Delta V_c = \frac{V_o}{8LCf^2}(1-D)$$

Sometimes it is useful to express the ratio of the ripple to the output voltage:

$$\frac{\Delta V_{\mathrm{c}}}{V_{\mathrm{o}}} = \frac{1-D}{8LCf^2} \tag{4.32}$$

This term is known as the output voltage ripple that represents the regulation. As expected, when filtering the capacitor and the frequency increase, the voltage ripple decreases.

4.3.1.2 Using Capacitor Charge to EvaluateΔV_{c}

Another useful way to evaluate the expression for ΔV_{c}without actually having to obtain the exact expression for $v_{\mathrm{c}}(t)$ is to use the total charge, Q, deposited on the capacitor current interval. Figure 4.14 shows the waveform for i_{C} and v_{c} with areas of positive charge (+) and negative charge (−). Because of waveform symmetry, $t = DT/2$ and $t = (1+D)T/2$ represent the i_{c} zero crossing times when the capacitor voltage is minimum, $V_{\mathrm{c,\,min}}$, and maximum, $V_{\mathrm{c,\,max}}$, respectively. Hence the capacitor voltage ripple is ΔV_{c}. The total charge stored in the capacitor between $t = DT/2$ and $t = (1+D)T/2$ is obtained from the following equation:

$$\frac{dQ}{dt} = C\Delta V_{\mathrm{C}}$$

So the total charge is Q and v_{c} between i_{c} zero crossings $DT/2 \le t < (1+D)T/2 \;\; t$ is given by:

$$\Delta Q = C\Delta V_{\mathrm{c}} \tag{4.33}$$

However, since the total charge is related to the current according to the following relation:

$$i = \frac{dQ}{DT}$$

then we have:

$$\begin{aligned} \Delta Q &= \frac{1}{T/2}\int_{DT/2}^{(1+D)T/2} idt = \text{Area under the curve} \\ &= \frac{1}{2}\left(\frac{1+D}{2}T - \frac{D}{2}T\right)\frac{1}{2}\Delta I \\ &= \frac{1}{2}\cdot\frac{T}{2}\cdot\frac{1}{2}\cdot\Delta I \end{aligned} \tag{4.34}$$

From Eqs. (4.33) and (4.34), we obtain:

$$\Delta V_c = \frac{T}{8C}\Delta I$$

Substituting for $\Delta I = (DV_{in}(1 - D)T)/L$, we obtain:

$$\begin{aligned}\frac{\Delta V_c}{V_o} &= \frac{1-D}{8LC}T^2 \\ &= \frac{1-D}{8LCf^2}\end{aligned} \tag{4.35}$$

Example 4.2
Consider a buck converter with the following circuit parameters: $V_{in} = 20$ V, $V_o = 15$ V, and $I_o = 5$ A, for $f = 50$ kHz. Determine (a) D, (b) L_{crit}, (c) maximum and minimum inductor current for $L = 100L_{crit}$, (d) average input and output power, and (e) capacitor voltage ripple for $C = 0.47$ μF.

Solution

$$D = 0.75$$

(a) Using $R = 3\ \Omega$ and $T = 20$, the critical inductor value is given by:

$$L_{crit} = \left(\frac{1-D}{2}\right)TR = 7.5\ \mu H$$

(b) For $L = 100\ L_{crit} = 750$ μH $= 0.75$ mH, we have:

$$\begin{aligned} I_{L,min} &= DV_{in}\left(\frac{1}{R} - \frac{(1-D)T}{2L}\right) \\ &= (0.75)(20)\left(\frac{1}{3} - 3.33 \times 10^{-3}\right) \\ I_{L,min} &= 4.95\ \text{A} \\ I_{L,max} &= (0.75)(20)\left(\frac{1}{3} + 3.33 \times 10^{-3}\right) \\ I_{L,max} &= 5.05\ \text{A} \end{aligned}$$

(c) Since it is an ideal converter, the average output and input powers are given by:

$$P_{in} = P_o = V_o I_o = (15)(5) = 75\ \text{W}$$

(d) The capacitor voltage ripple is given by:

$$\frac{\Delta V_o}{V_o} = \frac{1 - D}{8LCf^2}$$

$$= \frac{(1 - 0.75)}{8(0.75\ \text{mH})(0.47\ \mu F)\left(50 \times 10^3\right)^2}$$

$$\frac{\Delta V_o}{V_o} = 0.035 = 3.5\%$$

Example 4.3

Design a buck converter operating in ccm with the following specifications: $(\Delta V_o)/V_o = 0.5\ \%$, $V_{in} = 20$ V, $P_o = 12$ W, $f = 30$ kHz, and $D = 0.4$.

Solution

In order to design this converter, we need to calculate the values for L, C, and R.

The output voltage is given by:

$$V_o = DV_{in} = 8\ \text{V}$$

Hence, the output current is:

$$I_o = \frac{P_o}{V_o} = 1.5\ \text{A}$$

The output resistance is:

$$R = \frac{8}{1.5} = 5.33\ \Omega$$

The critical inductance for ccm is given by:

$$L_{crit} = \frac{1 - D}{2} TR$$

$$= \left(\frac{1 - 0.4}{2}\right)\left(\frac{1}{30 \times 10^3}\right) 5.33$$

$$53.3\ \mu H$$

Let us select $L = 600$ μH. Based on this value, the maximum and minimum inductor currents are given by:

$$I_{L,\max} = DV_{in}\left(\frac{1}{R} + \frac{(1 - D)T}{2L}\right)$$

$$= (0.4)(20)\left(\frac{1}{5.33} + 0.0167\right)$$

$$= 1.63\ \text{A}$$

$$I_{L,\min} = (0.4)(20)\left(\frac{1}{5.33} - 0.0167\right)$$
$$= 1.37 \text{ A}$$

The ripple voltage is given by:

$$\frac{\Delta V_o}{V_o} = \frac{1-D}{8LCf^2} = 0.005$$

Solve for C:

$$C = \frac{1-D}{(8Lf^2)0.005}$$

$$C = 27.78\ \mu\text{F}$$

Exercise 4.3
Redesign Example 4.2 to achieve an output ripple voltage not to exceed 1% and an inductor current ripple not to exceed 10% at the average load current.

Answer: 0.15 mH, 8.33 μF, 3 Ω

Exercise 4.4
Determine the diode and transistor average and *rms* current values, for Exercise 4.3.

Answer: 1.25 A, 3.75 A, 2.5 A, 4.33 A

Exercise 4.5
Show that the expression for the peak capacitor voltage at $t = (1 + D)T/2$.

4.3.2 *The Boost Converter*

4.3.2.1 Basic Topology and Voltage Gain

Other possible switch and transistor-diode arrangements are shown in Fig. 4.15a, b, respectively. This topology is known as a boost converter since the output voltage is higher than the input, as will be shown in this section.

Similar to the case for the buck converter, we assume all the converter components are ideal and the transistor switching waveform is shown in Fig. 4.9d. When the switch is turned *ON*, the equivalent circuit of *Mode 1* is shown in Fig. 4.16a. This is a charging interval and the voltage across the inductor is V_{in}, and $i_L(t)$ is given by:

$$i_L(t) = \frac{1}{L}V_{in}t + I_L(0) \quad 0 \leq t < DT \tag{4.36}$$

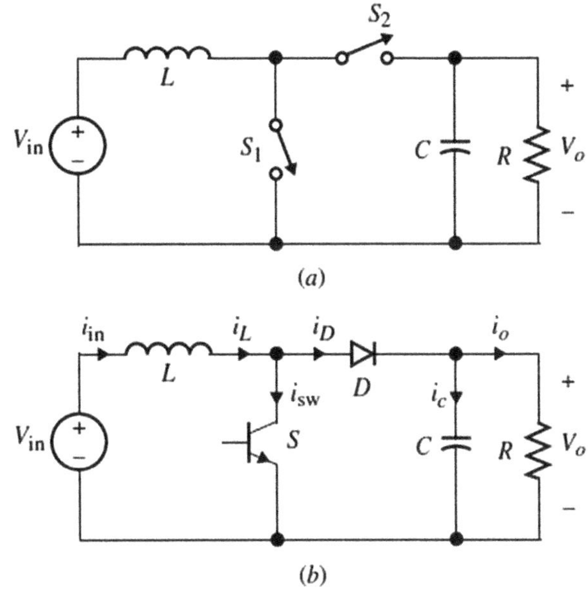

Fig. 4.15 Boost converter: (**a**) two-switch implementation and (**b**) transistor-diode implementation

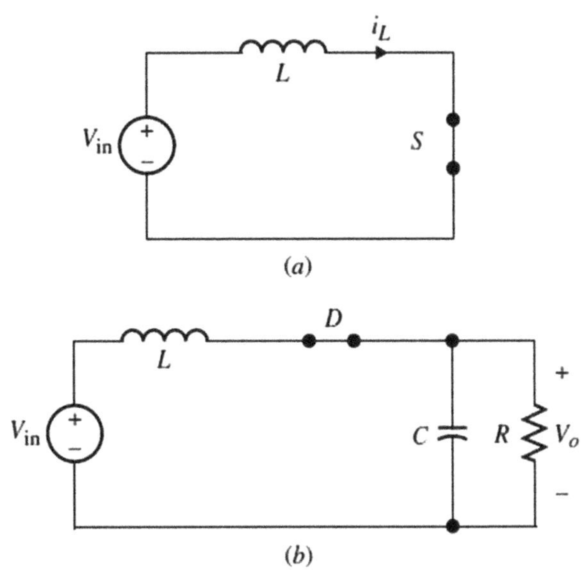

Fig. 4.16 Equivalent circuit modes for the boost converter: (**a**) Mode 1: The switch is ON, (**b**) Mode 2: The switch is OFF

where $I_L(0)$ is the initial inductor current value at $t = 0$.When the switch is turned *off* at $t = DT$, the resultant equivalent circuit *Mode 2* is shown in Fig. 4.16b.

The inductor voltage is $V_{in} - V_o$ and $i_L(t)$ is given by:

$$i_L(t) = \frac{1}{L}(V_{in} - V_o)(t - DT) + I_L(DT) \quad DT \leq t < T \tag{4.37}$$

Evaluating Eqs. (4.36) and (4.37) at $t = DT$ and $t = T$, respectively, and using the fact that $I_L(T) = I_L(0)$, we obtain:

$$I_L(DT) - I_L(0) = \frac{1}{L} V_{in}(DT) \tag{4.38a}$$

$$I_L(DT) - I_L(0) = \frac{1}{L}(V_{in} - V_o)(1 - D)T \tag{4.38b}$$

From Eqs. (4.38a) and (4.38b), the resulting voltage conversion is given by:

$$\frac{V_o}{V_{in}} = \frac{1}{1 - D} \tag{4.39}$$

Hence, the voltage gain is always greater than 1. Also from Eq. (4.38), the inductor ripple current is given by:

$$\begin{aligned} \Delta I &= I_L(DT) - I_L(0) \\ &= I_{L,max} - I_{L,min} \\ &= \frac{1}{L} V_{in} DT \end{aligned} \tag{4.40a}$$

Substituting for V_{in} from Eq. (4.39), we obtain:

$$\Delta I = \frac{1}{L} V_o D(1 - D)T \tag{4.40b}$$

Key current and voltage waveforms are given in Fig. 4.17.

4.3.2.2 Average Input and Output Currents

The input current is the same as the inductor current as shown in Fig. 4.17a. Hence, the average input current by inspection is given by:

$$I_{in} = \frac{I_{L,max} + I_{L,min}}{2} \tag{4.41}$$

whereas the average output current is the same as the average diode current and is given by:

$$I_o = \left(\frac{I_{L,max} + I_{L,min}}{2} \right)(1 - D) = \frac{V_o}{R} \tag{4.42}$$

Since we assume an ideal converter, the average input and output powers must be equal. Using Eqs. (4.41) and (4.42), we get:

$$V_{in} I_{in} = V_o I_o$$

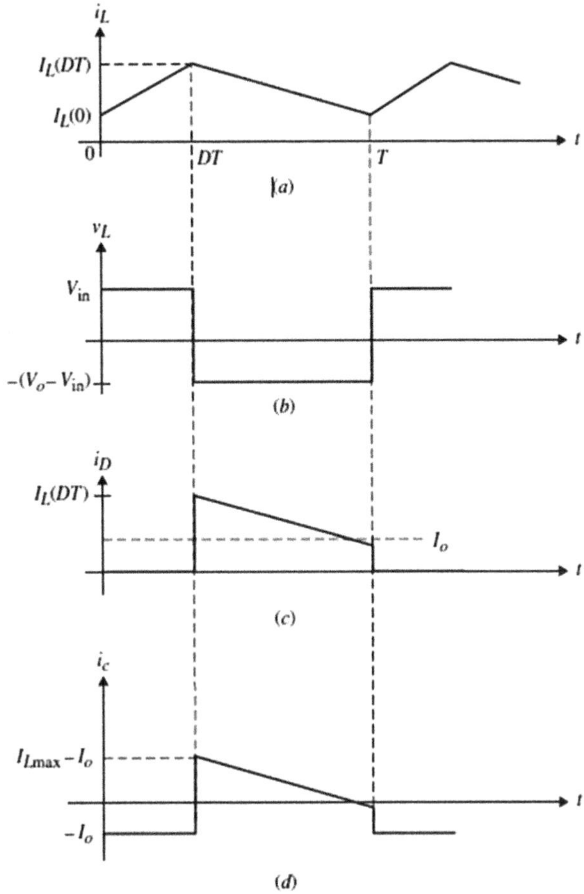

Fig. 4.17 Current and voltage waveforms for the boost converter

resulting in:

$$\frac{I_{\text{in}}}{I_{\text{o}}} = \frac{V_{\text{o}}}{V_{\text{in}}} = \frac{1}{1-D} \tag{4.43}$$

Similar to the buck converter, the input-output current and voltage ratios are equivalent to a dc transformer with a transformer mode ratio equal to $1/(1-D)$, shown in Fig. 4.18.

Using Eqs. (4.38) and (4.42), we can solve for the maximum and minimum inductor current values:

$$I_{\text{L}}(0) = I_{\text{Lmin}} = V_{\text{in}}\left(\frac{1}{R(1-D)^2} - \frac{DT}{2L}\right) \tag{4.44a}$$

$$I_{\text{L}}(DT) = I_{\text{Lmax}} = V_{\text{in}}\left(\frac{1}{R(1-D)^2} + \frac{DT}{2L}\right) \tag{4.44b}$$

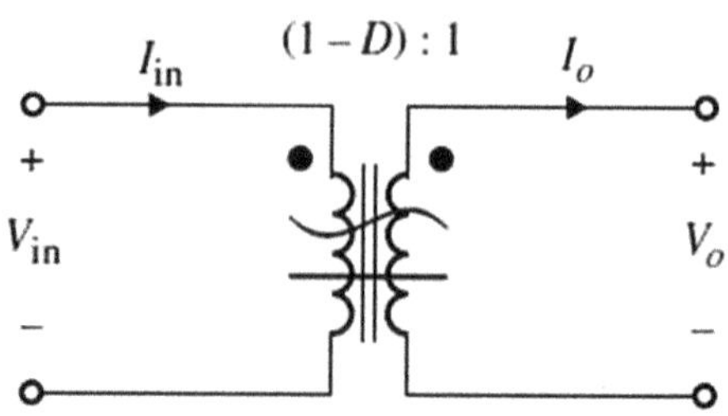

Fig. 4.18 Equivalent transformer circuit representation for the boost converter

For positive values of I_{Lmax} and I_{Lmin}, the converter will operate in the continuous conduction mode. To solve for the minimum critical inductor value that will keep the converter in the ccm, we set I_{Lmin} to zero:

$$I_{\mathrm{Lmin}} = 0$$

Under this boundary condition, the critical inductor value is given by:

$$L_{\mathrm{crit}} = \frac{RT}{2}(1-D)^2 D \tag{4.45}$$

4.3.2.3 Output Ripple Voltage

It is clear from Fig. 4.17 that when the diode is reverse biased, the capacitor current is the same as the load current. Since we assume the load current is purely dc, the capacitor current is given by:

$$\begin{aligned} i_{\mathrm{c}} &= -I_{\mathrm{o}} \qquad 0 \le t < DT \\ i_{\mathrm{c}} &= i_{\mathrm{L}} - I_{\mathrm{o}} \quad DT \le t < T \end{aligned}$$

The capacitor current waveform is shown in Fig. 4.17d and redrawn in Fig. 4.19 along with the capacitor voltage waveform. Mathematical expressions for i_{c} can be obtained directly from this figure.

The current $i_{\mathrm{c}}(t)$ is expressed mathematically as:

$$i_{\mathrm{c}}(t) = -\frac{\Delta I}{(1-D)T}(t-DT) + I_{\mathrm{c}}(DT) \qquad DT \le t < T \tag{4.46}$$

where $I_{\mathrm{c}}(DT)$ is the initial $i_{\mathrm{c}}(t)$ at $t = DT$. The capacitor voltage for $0 \le t < DT$ is given by:

$$\begin{aligned} v_{\mathrm{c}}(t) &= \frac{1}{C}\int_0^t -I_{\mathrm{o}}dt + V_{\mathrm{c}}(0) \\ &= -\frac{I_{\mathrm{o}}}{C}t + V_{\mathrm{c}}(0) \end{aligned} \tag{4.47}$$

where $V_{\mathrm{c}}(0)$ is the initial capacitor voltage at $t = 0$.

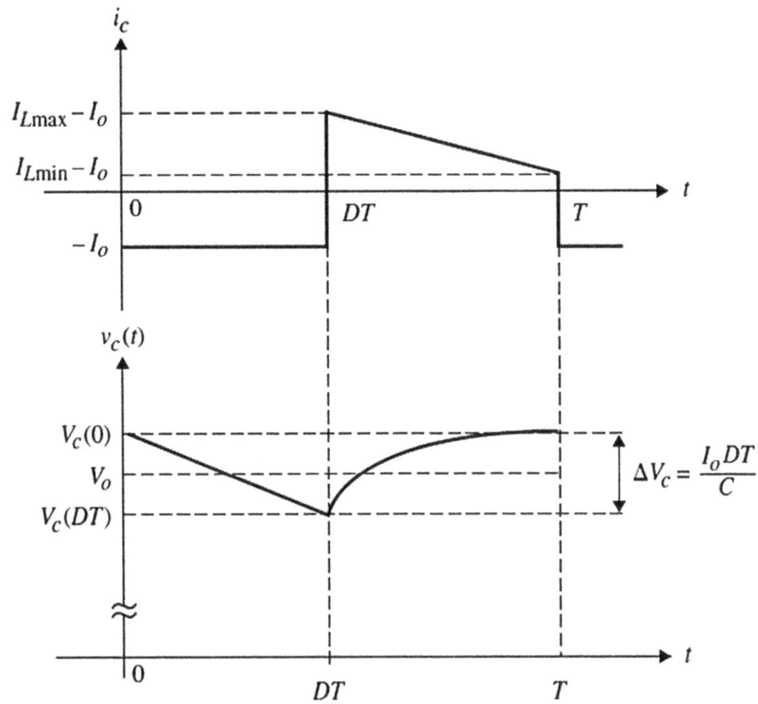

Fig. 4.19 Capacitor current and voltage waveforms for the boost converter, assuming $I_{Lmin} > I_o$

At $t = DT$ we have:

$$V_c(DT) = -\frac{I_o}{C} DT + V_c(0) \tag{4.48}$$

Since the average capacitor voltage is V_o, we can solve for $V_c(0)$ and $V_c(DT)$ as follows:

$$V_c(0) = V_o + \frac{I_o DT}{2C} \tag{4.49}$$

$$V_c(DT) = V_o - \frac{I_o DT}{2C}$$

and the capacitor voltage variation is given by:

$$\Delta V_c = V_c(0) - V_c(DY) = \frac{I_o DT}{C} \tag{4.50}$$

For $DT \leq t < T$ the capacitor voltage is given by:

$$\begin{aligned} v_c(t) &= \frac{1}{c}\int_{DT}^{T}\left[\frac{-\Delta I}{(1-D)T}(t-DT)+I_c(DT)\right]dt + V_c(DT) \\ &= -\frac{\Delta I}{2C(1-D)T}(t-DT)^2 + \frac{I_c(DT)(1-D)T}{C} + V_c(DT) \end{aligned} \tag{4.51}$$

The output ripple voltage is given by:

$$|\Delta V_o| = |\Delta V_c| = I_o\frac{DT}{C} = \frac{V_o DT}{RC}$$

Then the voltage ripple is given by:

$$\begin{aligned} \frac{\Delta V_o}{V_o} &= \frac{DT}{RC} \\ &= \frac{D}{RCf} \end{aligned} \tag{4.51}$$

Example 4.4
Sketch the current waveforms for i_L, i_{in}, i_D, i_o, and i_c for the boost converter with the following parameters: $=1.8$ mH, $V_{in} = 50$ V, $V_o = 120$ V, $R = 20\ \Omega$, $C = 147\ \mu$F, and $f = 15$ kHz.

Also sketch the voltage waveform v_L, v_{sw}, v_c, and v_D.

Solution
In order to sketch the waveforms, we need to find D, the maximum and minimum inductor currents, and the average output current.

The duty cycle is given by:

$$\frac{V_o}{V_{in}} = \frac{1}{1-D} = \frac{120}{50}$$

which yields $D = 0.58$

Using $R = 20\ \Omega$ and $T = 66.67\ \mu$s, the maximum and minimum inductor currents are given by:

$$I_{Lmax} = V_{in}\left(\frac{1}{(1-D)^2 R} + \frac{DT}{2L}\right) = 14.94\ \text{A}$$

and

$$I_{Lmin} = V_{in}\left(\frac{1}{(1-D)^2 R} - \frac{DT}{2L}\right) = 13.86\ \text{A}$$

The average input and output currents are given by:

$$I_{in} = \frac{I_{Lmax} + I_{Lmin}}{2} = 14.4 \text{ A}$$
$$I_o = I_{in}(1 - D) = 6 \text{ A}$$

The capacitor peak currents are given by:

$$I_{cmax} = I_{Lmax} - I_o = 8.94 \text{ A}$$
$$I_{cmin} = I_{Lmin} - I_o = 7.86 \text{ A}$$

Hence:

$$\Delta I_c = (I_{cmax} - I_{cmin}) = 1.074 \text{ A}$$

Notice that this value must be equal to ΔI_L.
The capacitor voltage is given by:

$$v_c(t)|_{t=0} = 120 \text{ V}$$
$$v_c(t)|_{t=DT} = -\frac{I_o}{C}DT + 120 \text{ V}$$
$$= \frac{-5.95 \text{ A}}{147\ \mu F}(0.58)(66.67\ \mu s) + 120 \text{ V} = 118.43 \text{ V}$$

Hence, the ripple is 1.57 V.
Figure 4.20 shows the sketch for the boost waveforms of Example 4.4.

Example 4.5
Design a boost converter with the following specifications: $P_o = 27$ W, $V_o = 40$ V, $V_{in} = 28$ V, $\Delta V_o/V_o = 2\,\%$, and $f_s = 35$ kHz.

Solution
First let's determine the duty cycle, D:

$$D = 1 - \frac{V_{in}}{V_o} = 1 - \frac{28}{40} = 0.3$$

For continuous conduction mode, the inductance minimum value is given by:

$$L_{crit} = \frac{RT}{2}(1 - D)^2 D$$

where

$$T = 28.57\ \mu s$$

$$R = \frac{V_o^2}{P_o} = \frac{(40)^2}{27} = 59.26\ \Omega$$

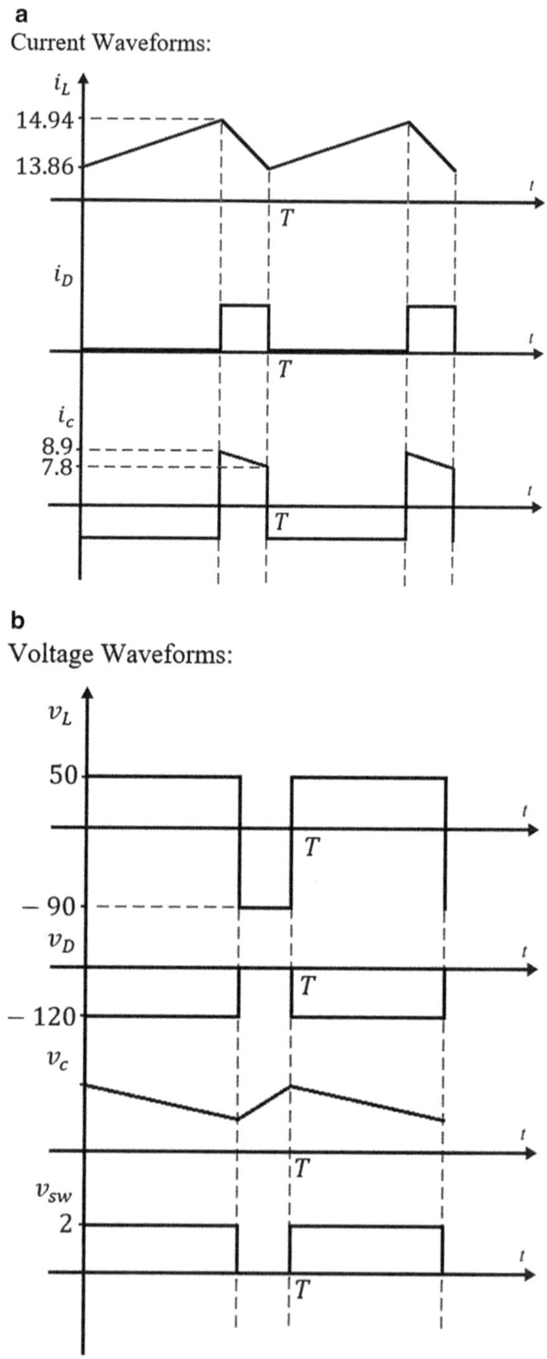

Fig. 4.20 (**a**) Current waveforms. (**b**) Voltage waveforms

We choose $L = 200\ \mu H$, wherein L should be greater than L_{crit} for ccm operation. The output ripple voltage:

$$\frac{\Delta V_o}{V_o} = \frac{D}{RCf}$$

$$0.02 = \frac{0.3}{(59.26)C(35\ \text{kHz})}$$

$$C = 7.23\ \mu F$$

Exercise 4.6
Determine the average and *rms* current values for the diode and transistor in Example 4.4.

Answer: 8.4 A, 11 A, 6 A, 9.28 A

Exercise 4.7
Design a boost converter and its associated voltage mode control loop that meets the following specifications:

- Vo = 12 V
- Vin = 3.6 V nominally, $V_{in,max} = 6$ V, $V_{in,min} = 3$ V
- Maximum output power = 20 W, minimum output power = 10 W
- ccm operation between max and min loading for the entire input voltage range
- Switching frequency = 100 kHz
- Output voltage ripple <0.5% (max allowable sustained ripple)

Answer: L > 9 μH and C > 209 μF

4.3.3 The Buck-Boost Converter

The third possible converter is obtained by interchanging the diode and the inductor of the buck converter to realize the design of Fig. 4.21. This converter is known as a *buck-boost converter* since its voltage gain can be less than, equal to, or greater than 1. Unlike the buck and boost converters, this converter gives a negative output voltage when used without isolation.

In this analysis, we follow assumptions made in performing the steady-state analysis for the buck and boost topologies. When the switch is *ON*, the diode is reverse biased and the equivalent circuit of *Mode 1* as shown in Fig. 4.22a, whereas Fig. 4.22b gives the equivalent circuit for *Mode 2* when the transistor is *OFF*.

When the transistor is turned on, the inductor current starts charging from the source voltage, V_{in}, while the diode D is reverse biased. The voltage across the inductor is V_{in} and i_L is given by:

$$i_L(t) = \frac{1}{L} V_{in}(t) + I_L(0) \tag{4.53}$$

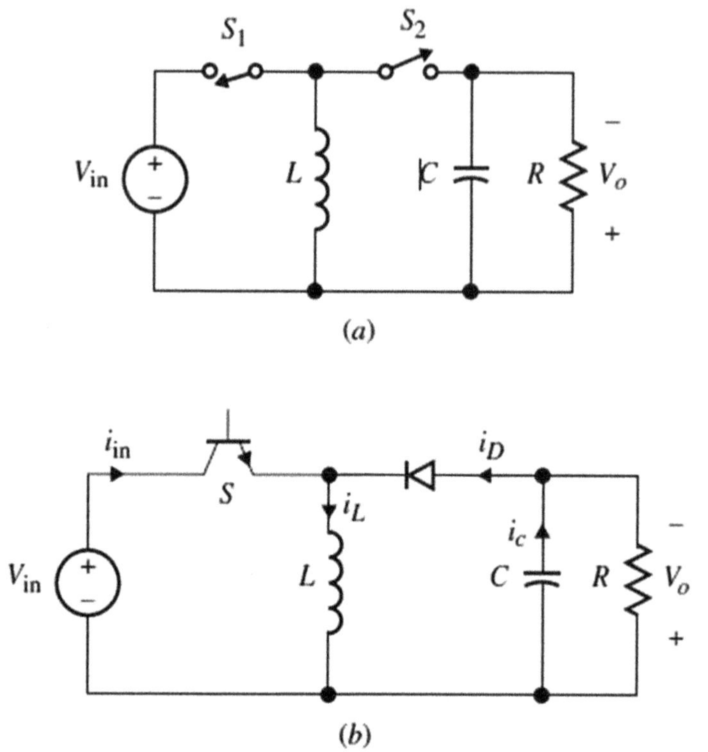

Fig. 4.21 Buck-boost converter: (**a**) switch implementation and (**b**) transistor-diode implementation

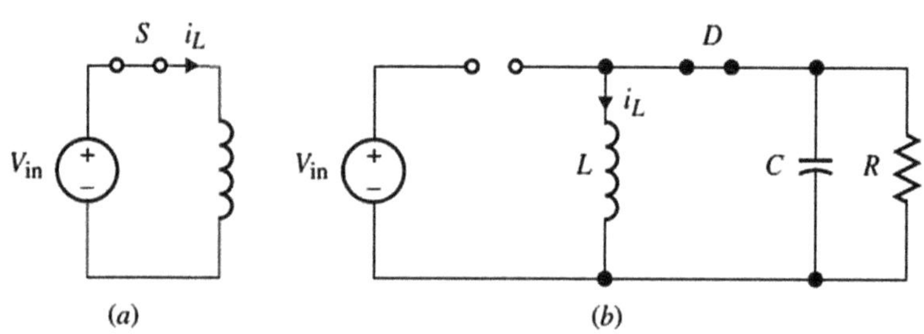

Fig. 4.22 Equivalent circuits: (**a**) Mode 1: when the transistor is conducting (**b**) Mode 2: when the transistor is not conducting

where $I_L(0)$ is the initial inductor value that corresponds to the minimum inductor current value. Evaluating Eq. (4.53) at $t = DT$, when the switch is turned *OFF*, we obtain the maximum inductor current. This yields the following relation:

$$I_{Lmax} - I_{Lmin} = \frac{1}{L} V_{in} DT \tag{4.54}$$

Similarly, in *Mode 2* the inductor current is given by:

$$i_{\mathrm{L}}(t) = -\frac{V_{\mathrm{o}}}{L}(t - DT) + I_{\mathrm{L}}(DT) \tag{4.55}$$

Evaluating Eq. (4.54) at $t = T$ and since we assume steady-state operation, we use $I_{\mathrm{L}}(0) = I_{\mathrm{L}}(T)$, to obtain:

$$I_{\mathrm{L}}(0) = \frac{-V_{\mathrm{o}}}{L}(T - DT) + I_{\mathrm{L}}(DT) \tag{4.56a}$$

$$I_{\mathrm{Lmax}} - I_{\mathrm{Lmin}} = \frac{V_{\mathrm{o}}}{L}(1 - D)T \tag{4.56b}$$

Equating Eqs. (4.54) and (4.54b), we obtain the following voltage conversion ratio:

$$\frac{V_{\mathrm{o}}}{V_{\mathrm{in}}} = \frac{D}{1 - D} \tag{4.57}$$

It is clear from this equation that the output voltage can be either smaller or greater than the input voltage:

$D > 0.5$	Boost
$\mathrm{D} < 0.5$	Buck
$D = 0.5$	Unity gain

Figure 4.23 gives typical current and voltage waveforms for the buck-boost converter.

From Fig. 4.23, it is clear that the average input current is given by:

$$I_{\mathrm{in}} = \frac{I_{\mathrm{L,max}} + I_{\mathrm{L,min}}}{2} D \tag{4.58a}$$

And the average output current is given by:

$$I_{\mathrm{o}} = \frac{I_{\mathrm{L,max}} + I_{\mathrm{L,min}}}{2}(1 - D) \tag{4.58a}$$

Since the conservation of power must hold, we use $I_{\mathrm{in}}V_{\mathrm{in}} = I_{\mathrm{o}}V_{\mathrm{o}}$ to obtain:

$$\frac{I_{\mathrm{in}}}{I_{\mathrm{o}}} = \frac{V_{\mathrm{o}}}{V_{\mathrm{in}}} = \frac{D}{1 - D} \tag{4.59}$$

To solve for $I_{\mathrm{L,max}}$ and $I_{\mathrm{L,min}}$ in terms of the converter components, we substitute for $I_{\mathrm{o}} = V_{\mathrm{o}}/R$. Hence, from Eqs. (4.56) and (4.59), we obtain:

$$I_{\mathrm{Lmax}} = V_{\mathrm{in}}\left[\frac{D}{R(1 - D)^2} + \frac{DT}{2L}\right] \tag{4.60a}$$

Fig. 4.23 Current and voltage waveforms for the buck-boost converter

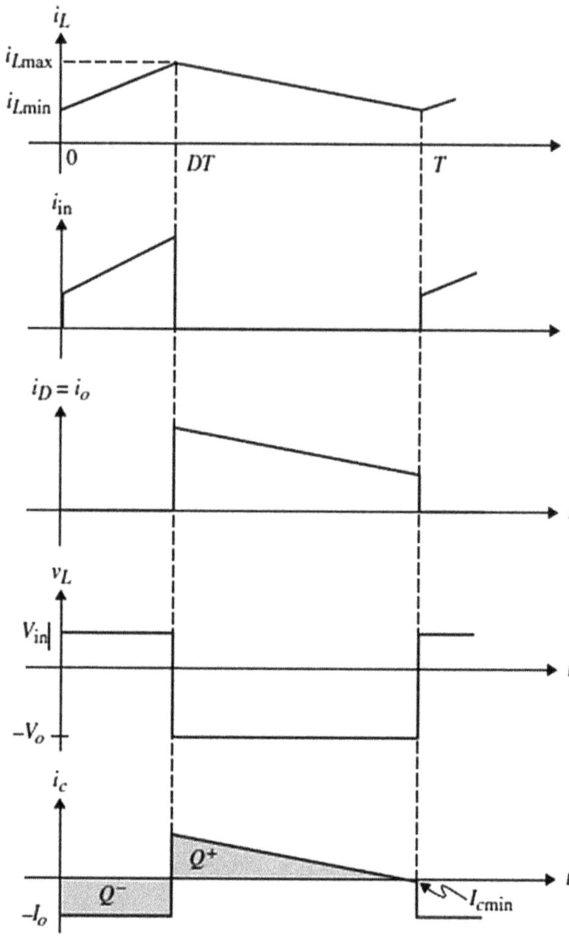

$$I_{\text{Lmin}} = V_{\text{in}} \left[\frac{D}{R(1-D)^2} - \frac{DT}{2L} \right] \tag{4.60b}$$

The critical inductor value that keeps the converter in the *ccm* mode can be obtained by setting $I_{\text{Lmin}} = 0$ in the above equation to yield:

$$L_{\text{crit}} = \frac{RT(1-D)^2}{2} \tag{4.61}$$

We notice that for the same frequency of operation and load resistance, the buck converter has the highest critical inductor limit when compared to the boost and buck-boost, whereas the boost converter has the smallest L_{crit}, resulting in a wider range of inductor design.

4.3.3.1 Output Voltage Ripple

It can be shown that the capacitor current of the buck-boost converter is the same as for that of the boost. Hence, the capacitor voltage is given by:

$$\Delta V_c(t) = \frac{V_o DT}{RC}$$
$$\frac{\Delta V_o}{V_o} = \frac{DT}{RC} = \frac{D}{RCf} \tag{4.62}$$

Example 4.6

(a) Sketch the current and voltage waveforms for a buck-boost converter with the following parameters: $V_{in} = 40$ V, $V_o = 60$ V, $D = 0.6$, $R = 20$ Ω, and $L = 750$ μH.
(b) Find the *rms* value for i_c.

Solution

(a) Let $T = 200$ μs:

$$D = \frac{1}{1 + \frac{V_{in}}{V_o}} = \frac{1}{1 + \frac{40}{60}} = 0.6$$

$$I_{Lmax} = \frac{V_{in}D}{R(1-D)^2} + \frac{V_{in}DT}{2L}$$
$$= \frac{40(0.6)}{20(1-0.6)^2} + \frac{40(0.6)(200\,\mu s)}{2(750\,\mu H)} = 10.7 \text{ A}$$

$$I_{Lmin} = \frac{V_{in}D}{R(1-D)^2} - \frac{V_{in}DT}{2L}$$
$$= \frac{40(0.6)}{20(1-0.6)^2} - \frac{40(0.6)(200\,\mu s)}{2(750\,\mu H)} = 4.3 \text{ A}$$

$$I_{in} = \frac{I_{Lmax} + I_{Lmin}}{2} D$$
$$= \frac{10.7 + 4.3}{2}(0.6) = 4.5 \text{ A}$$

Having found P_o, we can now solve for the average output current:

$$P_o = V_{in} I_{in}$$
$$= (40)(4.5) = 180 \text{ W}$$

Hence, we can calculate the values for I_{cmax} and I_{cmin}:

$$I_{cmax} = I_{Lmax} - I_o = 7.7\ \text{A}$$

$$I_{cmin} = I_{Lmin} - I_o = 1.3\ \text{A}$$

$$I_{crms} = \sqrt{\frac{1}{T}\left[\int_0^{DT} 9dt + \int_{DT}^{T}\left(\frac{-V_o}{L}(t - DT) + I_{cmax}\right)dt\right]}$$

Exercise 4.8

Consider a buck-boost converter that supplies 75 W at $I_o = 5$ A from a 37 V dc source. Let $T = 130$ μs and $L = 250$ μH. Determine (a) the duty ratio D, (b) I_{Lmax} and I_{Lmin}, (c) average input current, (d) average diode and transistor currents, and (e) the *rms* value of the capacitor current.

Answer: 0.29, 9.8 A, 4.25 A, 2.03 A, 3.464 A

Exercise 4.9

Derive the output voltage ripple Eq. (4.61) for a buck-boost converter.

Exercise 4.10

Determine D, I_{Lmax}, I_{Lmin}, I_{in}, I_D, I_{sw}, and I_{Crms} for a buck-boost converter whose capacitor current waveform is shown in Fig. E4.10. Assume $L = 120$ μH.

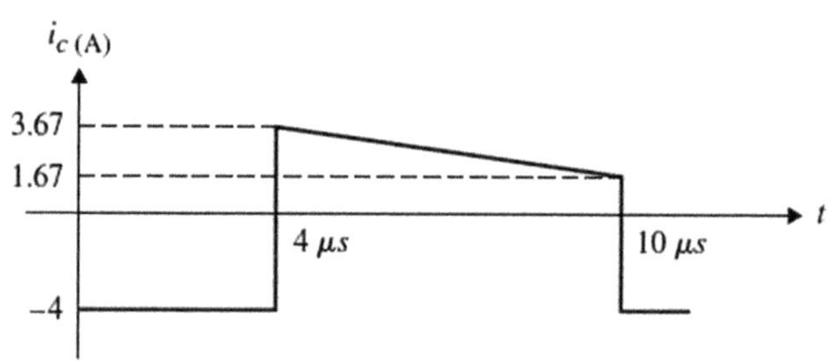

Fig. E4.10 Waveform for Exercise 4.10

Answer: 0.4, 7.67 A, 5.67 A, 2.67 A, 4 A, 2.67 A, 3.3 A

Under certain input and output voltage conditions, none of the preceding three basic converter topologies is suitable. For example, suppose the input voltage varies between 8 and 18 V while the output voltage is desired to be maintained fixed at +12 V. It is clear that the voltage gain varies between 2/3 and 3/2 while the output is constant at +12 V. Even though the buck-boost converter allows the voltage gain to be less or greater than 1, it provides only a negative output voltage polarity. One topology that is capable of providing a positive output voltage is known as the single-ended primary-inductance converter (SEPIC), shown in Fig. 4.24.

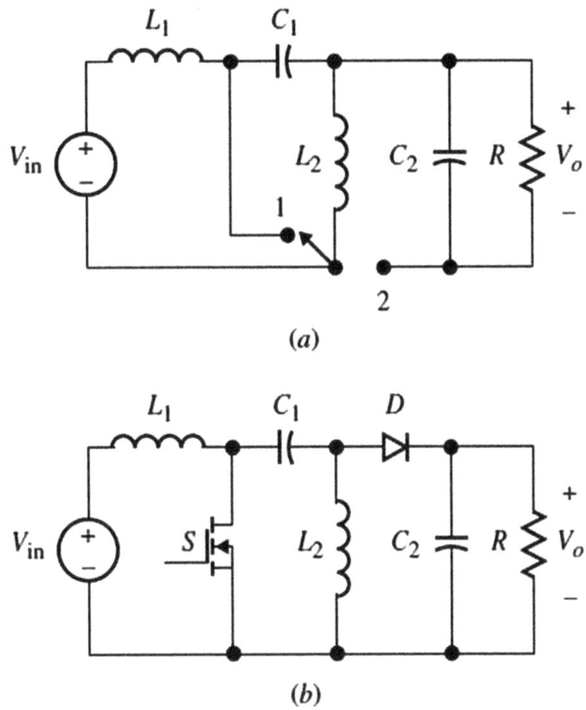

Fig. 4.24 Single-ended primary-inductance converter (SEPIC): (**a**) single-switch double-throw implementation and (**b**) MOSFET-diode implementation

4.3.4 The Fourth-Order Converters

It is possible to cascade or cascode more than one of the basic topologies, buck, boost, and buck-boost, to form new topologies that have new attractive features that the single topology does not. Figure 4.25a, b shows a block diagram of two converters connected in series (cascade) and in parallel (cascode), respectively. The parallel connection is also known as a differential configuration.

The overall voltage gain for Fig. 4.25a, b converter arrangements is given in Eqs. (4.63) and (4.64), respectively:

$$M = \frac{V_o}{V_{in}} = M_1 M_2 \tag{4.63}$$

$$M = \frac{V_o}{V_{in}} = M_1 - M_2 \tag{4.64}$$

where, in Fig. 4.25a, M_1 and M_2 are given by:

$$M_1 = \frac{V_1}{V_{in}} \quad M_2 = \frac{V_o}{V_1}$$

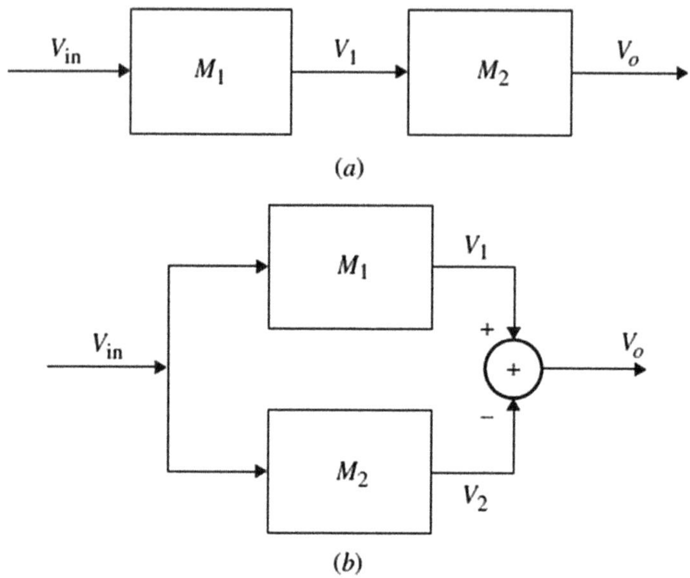

Fig. 4.25 Combination of basic converter topologies: (**a**) cascade and (**b**) cascode

Fig. 4.26 Block diagram representation for the cascade of a boost and buck converters: (**a**) boost-buck and (**b**) buck-boost

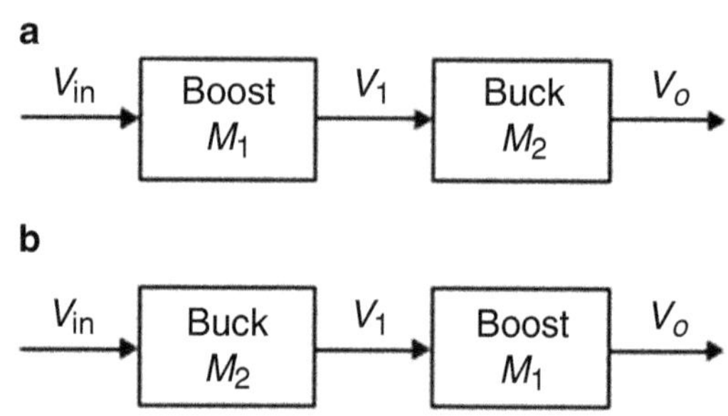

In Fig. 4.25b we have:

$$M_1 = \frac{V_1}{V_{in}} \quad M_2 = \frac{V_2}{V_{in}}$$

Figure 4.26a, b show the block diagram and circuit implementation for the two possible series cascading of the boost converter with a buck converter. Regardless of the sequence, the voltage gain for both converters should be $M_1M_2 = D/(1-D)$.

The equivalent circuit representation for the cascode of the boost and buck converters is shown in Fig. 4.27a. After careful circuit manipulation, it can be shown that the equivalent one-switch implementation of Fig. 4.27a is as shown in Fig. 4.27b.

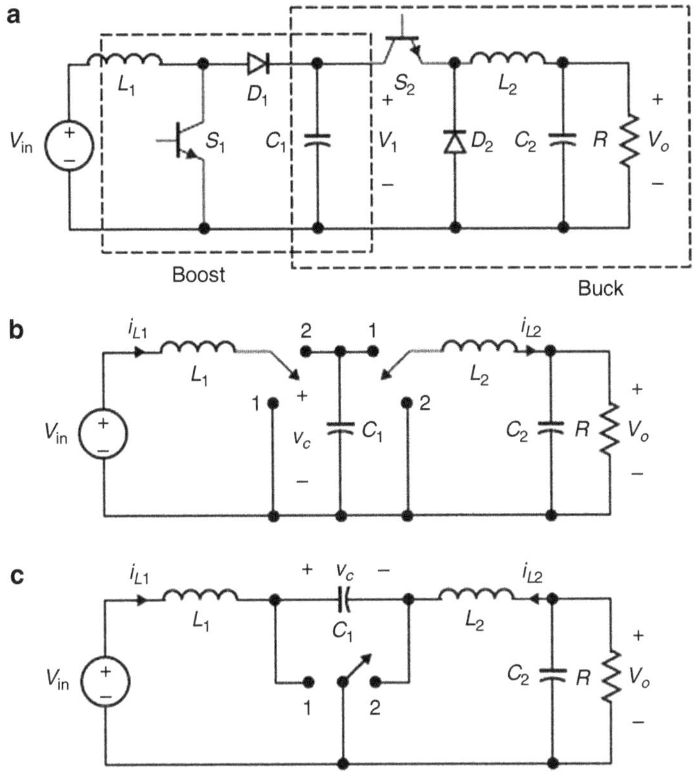

Fig. 4.27 (**a**) Boost-buck cascade, (**b**) two-switch implementation, and (**c**) one-switch equivalent circuit implementation

To illustrate how Fig. 4.27b is obtained from Fig. 4.27a, we assume i_{L1} and i_{L2} are constant current sources as shown in Fig. 4.28a. Figure 4.28b shows the equivalent circuit when S is in positions 1 and 2. Next we redraw the portion of the output circuit as shown in Fig. 4.28c, which is redrawn in Fig. 4.28d. The two modes of Fig. 4.28d are represented in the equivalent circuit shown in Fig. 4.28e.

Similarly, the buck and boost cascade is shown in Fig. 4.29a. The circuit can be simplified through several straightforward steps as shown in Fig. 4.29b–d. Notice that the capacitor in Fig. 4.29a is removed since regardless of the state of S_1 and S_2, the capacitor average current is always zero. In Fig. 4.29c we combine L_1 and L_2 in series. Finally, Fig. 4.29d is obtained using similar steps for the buck-boost cascade.

Example 4.7

Figure 4.30a shows a cascade of a boost and a buck converter and shows that the voltage gain is given by:

$$\frac{V_{\mathrm{o}}}{V_{\mathrm{in}}} = \frac{D}{1 - D} \tag{4.65}$$

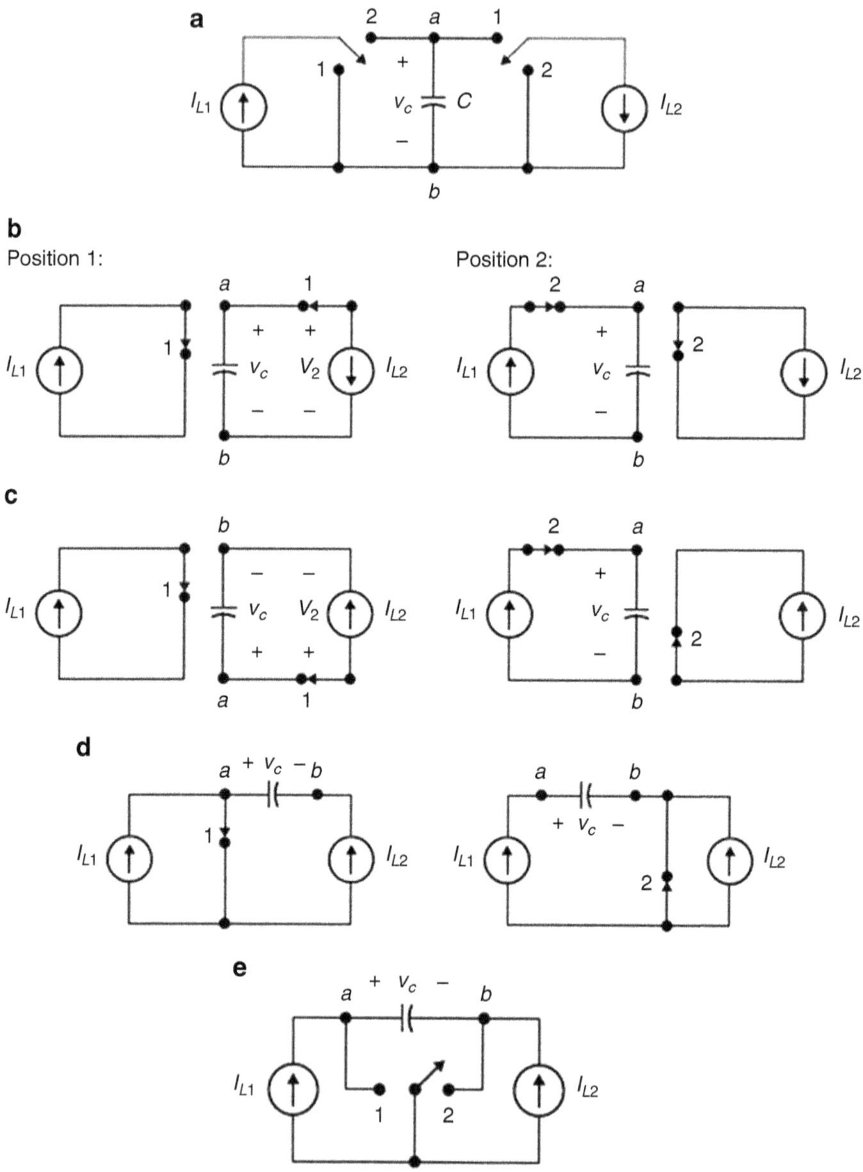

Fig. 4.28 Steps illustrating how Fig. 4.26 (**b**) was derived

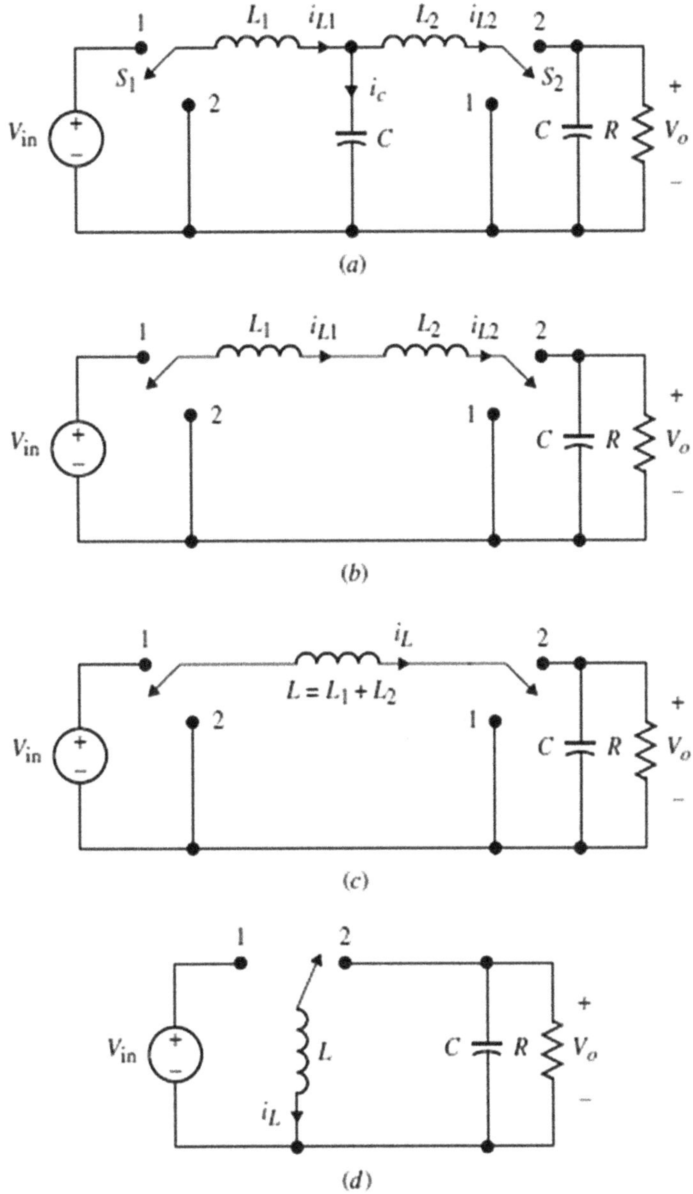

Fig 4.29 Steps illustrating how the buck-boost converter is obtained from the cascade configuration

Solution

During Mode 1 when S_1 and S_2 are in position 1 and during Mode 2 when S_1 and S_2 are in position 2, the inductor voltages are given by:

Mode 1 (DT):

$$v_{L1} = V_{in}$$
$$v_{L2} = -V_o + V_c$$

Mode 2 ($(1 - D)T$):

$$v_{L1} = V_{in} - V_c$$
$$v_{L2} = -V_o$$

Applying the volt-second balance principle to v_{L1} and v_{L2}, we obtain:

$$v_{L1}: \quad DV_{in} + (1 - D)(V_{in} - V_c) = 0$$
$$M_1 = \frac{V_c}{V_{in}} = \frac{1}{1 - D}$$

$$v_{L2}: \quad D(-V_o + V_c) - (1 - D)V_c = 0$$
$$M_2 = \frac{V_o}{V_c} = D$$

Therefore, the total voltage gain is:

$$\frac{V_o}{V_c} = M_1 M_2 = \frac{D}{1 - D}$$

Another straightforward way to find the above gain equation is to realize that the average voltage between a and a' is DV_{in} and use this voltage as an input to the Cuck converter. The switch implementation for Fig. 4.30a is shown in Fig. 4.30b

Exercise 4.11
Show that the cascoded buck and boost converters of Fig. 4.31c produce the same voltage gain conversion of Fig. 4.31a, i.e., $V_o/V_{in} = D/(1 - D)$, when S_1 and S_2 are single-pole double-throw switches, assume S_1 and S_2 are synchronized in position 1 for DT interval and in position 2 for $(1 - D)T$ interval.

Figure 4.27b is known as the Cuk converter whose voltage gain is $D/(1 - D)$, as will be discussed in details shortly. This concept can be extended to other cascaded topologies. The generalized fourth-order switch mode voltage-to-voltage converter is given in Fig. 4.31. Components 1, 2, 3, 4, and 5 consist of a switch, a diode, two inductors, and one capacitor. By disallowing capacitor loops and inductor cut-sets, the total number of physical realizable topologies can be reduced. For example, all possible topologies with components 1 representing a switch are shown in Fig. 4.32.

The topology of Fig. 4.32e is a buck converter with an additional output LC filter. Figure 4.32c is a buck-boost converter with an additional output LC filter. The topologies in Fig. 4.32b, f are physically unrealizable since the average output current in each of the capacitors is zero; hence, no power is delivered to the load. On the other hand, Fig. 4.32d does not have an average input current.

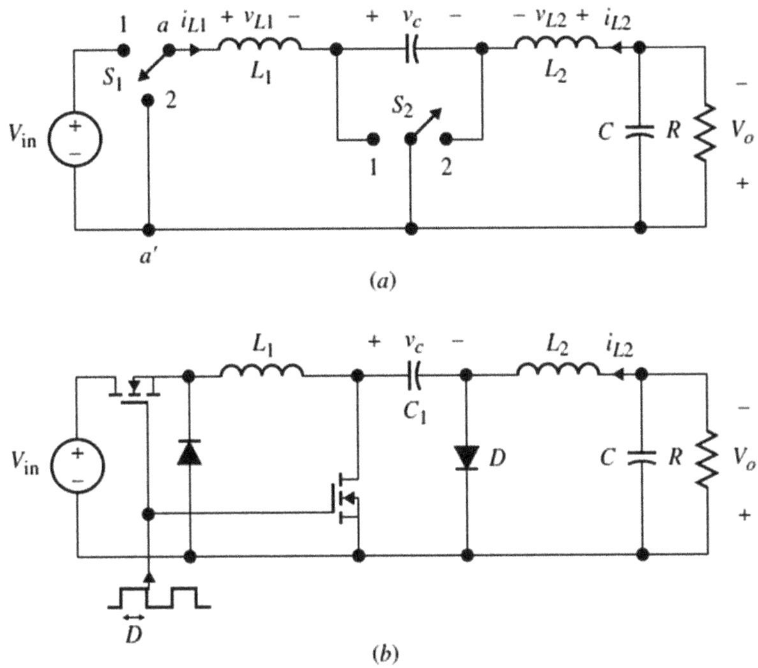

Fig. 4.30 (**a**) Cascade configuration of buck and Cuk converters and (**b**) diode-switch implementation.

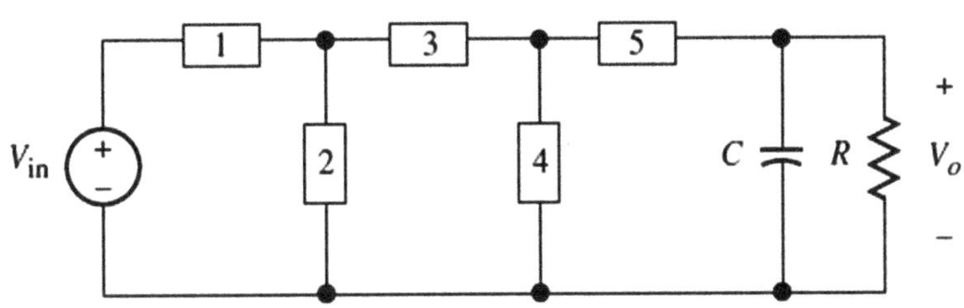

Fig. 4.31 Generalized representation of fourth-order voltage-to-voltage converter

4.3.4.1 The Cuk Converter

It has been shown that other converter topologies can be obtained by combining some of the three topologies discussed earlier. One cascade combination of a buck and a boost converter is known as a Cuk converter given in Fig. 4.27b and redrawn in Fig. 4.33, named after its inventor, Slobodan Cuk from California Institute of Technology. Figure 4.33c shows the Cuk converter with a magnetically coupled inductor representation

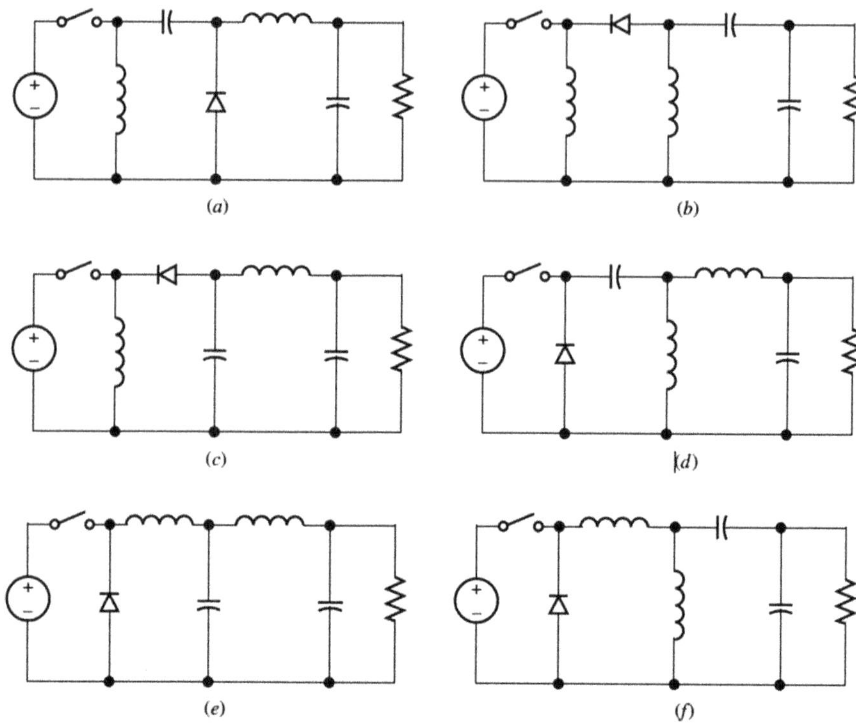

Fig. 4.32 All possible fourth-order topologies with a switch used in Component #1 in Fig. 4.31

The front end of the converter is a boost, and the back end of it is a buck. Hence, we may refer to the Cuk converter as a boost-buck converter. Unlike the previous converters, this converter requires two switches and uses two inductors and a capacitor to store and transfer energy from the input to the output, resulting in a higher level of complexity. Like the buck-boost, the gain of the Cuk converter can be less than, equal to, or greater than 1, with a positive output polarity. The major advantage of this converter is that the input inductor, L_1, and the output inductor, L_2, can be coupled on one magnetic core structure such that with the proper core gap design, the input and output switching currents can be made zero. Other possible combinations of converters are shown in Fig. 4.34a, b, which represents a boost with a LC-output-filter and a buck cascade with a LC-input-filter cascade, respectively.

The analysis of the Cuk converter can be carried out the same way as for the other converter topologies. There are two modes of operations: *Mode 1* when the switch is *ON* and *Mode 2* when the switch is *OFF*.

Fig. 4.33 Cuk converter with magnetically coupled inductors: (a) two-switch implementation, (b) transistor-diode implementation, and (c) core implementation

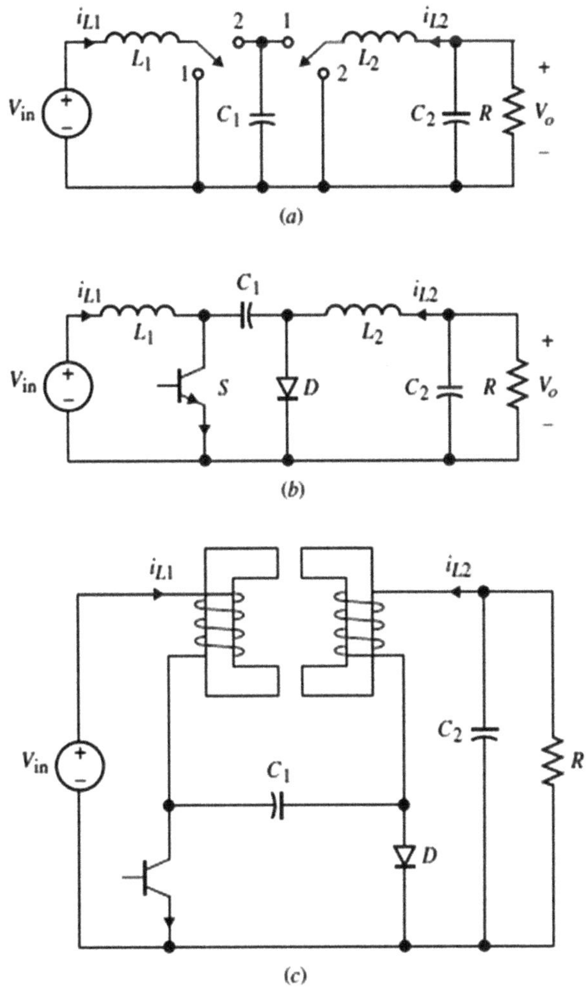

Fig. 4.34 (a) Boost cascade with LC-output-filter. (b) Buck cascade with LC-input-filter

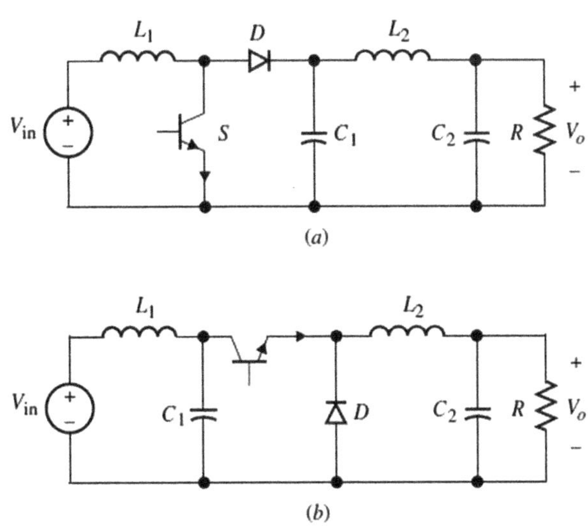

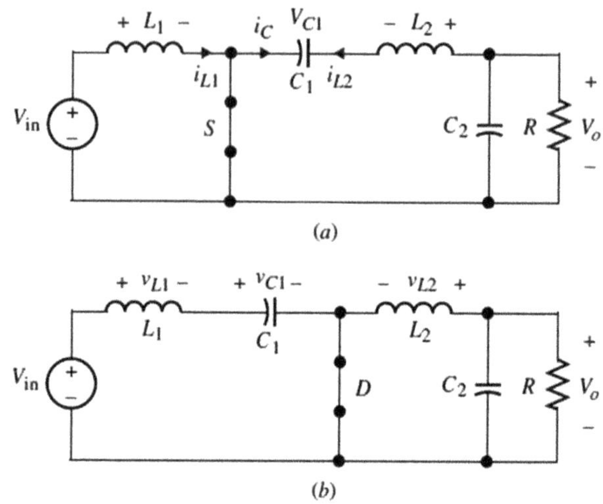

Fig. 4.35 (**a**) Mode 1: transistor is turned ON. (**b**) Mode 2: transistor is turned OFF

Mode 1

Mode 1 starts when the transistor is turned *ON* at $t = 0$ (Fig. 4.35a).

The inductor $L_1(0)$voltage is given by:

$$v_{\mathrm{L1}} = V_{\mathrm{in}}$$

$$L_1 \frac{di_{\mathrm{L1}}}{dt} = V_{\mathrm{in}} \tag{4.66}$$

resulting in the following current relation:

$$i_{\mathrm{L1}}(t) = \frac{V_{\mathrm{in}}}{\mathrm{L_1}} t + I_{\mathrm{L1}}(0) \tag{4.67}$$

where $I_{\mathrm{L1}}(0)$ is the initial current value.

The voltage across L_2 is given by:

$$\begin{aligned} v_{\mathrm{L2}} &= v_{\mathrm{c1}} + V_o \\ &= L_2 \frac{di_{\mathrm{L2}}}{dt} \end{aligned}$$

Since $i_{\mathrm{L2}} = -i_{\mathrm{c}} = -C(dv_{\mathrm{c1}}/dt)$, the above equation yields:

$$L_2 C_1 \frac{d^2 V_{\mathrm{c1}}}{dt^2} + v_{\mathrm{c1}} = -V_{\mathrm{o}}$$

If we assume the average voltage across C_1 has no ripple, then its average value, V_{C1} , is given by:

$$V_{\mathrm{c1}} = V_{\mathrm{in}} - V_{\mathrm{o}} \tag{4.68}$$

The above equation is obtained by writing KVL from the output through C_1, L_1, L_2, and V_{in} and setting the average inductor voltages to zero:

$$v_{L2} = V_{in}$$
$$i_{L2} = \frac{V_{in}}{L_2} t + I_{L2}(0)$$

At $t = DT$:

$$I_{L1} = \frac{V_{in}}{L_1} DT + I_{L1}(DT) \tag{4.69a}$$

Similarly:

$$I_{L2} = \frac{V_{in}}{L_2} DT + I_{L2}(DT) \tag{4.69b}$$

Mode 2

When the transistor is turned *OFF* at $t = DT$, *D* turns *ON*, and the equivalent circuit is given by Fig. 4.35b. Similar analysis shows:

$$v_{L1} = V_o$$

and the inductor currents is given by:

$$i_{L1}(t) = \frac{V_o}{L_1}(t - DT) + I_{L1}(DT)$$
$$i_{L2}(t) = \frac{V_o}{L_2}(t - DT) + I_{L2}(DT)$$

At $t = T$:

$$i_{L1}(T) = \frac{V_o}{L_1}(1 - D)T + I_{L1}(DT) \tag{4.70a}$$
$$i_{L2}(T) = \frac{V_o}{L_2}(1 - D)T + I_{L2}(DT) \tag{4.70b}$$

Since $I_{L1}(T) = I_{L1}(0)$ and $I_{L2}(T) = I_{L2}(0)$, from Eqs. (4.69) and (4.70), we obtain:

$$\frac{-V_{in}}{L_1} DT = \frac{V_o}{L_1}(1 - D)T$$

$$\frac{V_o}{V_{in}} = \frac{-D}{1 - D} \tag{4.71}$$

The average input current is the same as the average inductor current $i_{L1}(T)$, given by:

$$I_{\text{in}} = \frac{I_{\text{L1max}} + I_{\text{L1min}}}{2} \tag{4.72}$$

and the average output current is the same as the average inductor current $i_{L2}(T)$ as follows:

$$I_o = \frac{I_{\text{L2max}} + I_{\text{L2min}}}{2} \tag{4.73}$$

Since the average input power and output power are equal, we can obtain from the above equations the following equations:

$$I_{\text{L2}}(DT) = \left[\frac{D}{(1-D)R} + \frac{DT}{2L_2}\right] V_{\text{in}} \tag{4.74a}$$

$$I_{\text{L2}}(0) = \left[\frac{D}{(1-D)R} - \frac{DT}{2L_2}\right] V_{\text{in}} \tag{4.74b}$$

Similarly, we obtain:

$$I_{\text{L1}}(DT) = \left[\frac{D^2}{(1-D)^2 R} + \frac{DT}{2\text{L}_1}\right] V_{\text{in}} \tag{4.75a}$$

$$I_{\text{L1}}(0) = \left[\frac{D^2}{(1-D)^2 R} - \frac{DT}{2L_1}\right] V_{\text{in}} \tag{4.75b}$$

For continuous input current, $i_{\text{L1}}(t)$, we set $I_{\text{L1}}(0) = 0$ to obtain the critical value of L_1 as follows:

$$L_{\text{1crit}} = \frac{(1-D)^2 RT}{2D} \tag{4.76}$$

Similarly, for continuous current in L_2, the minimum value is given by:

$$L_{\text{2crit}} = \frac{(1-D)RT}{2D} \tag{4.77}$$

The ripple voltage across C_1 and C_2 is given by:

$$\frac{\Delta V_{\text{c1}}}{V_{\text{o}}} = \frac{D}{RC_1 f}$$

$$\frac{\Delta V_{\text{c2}}}{V_{\text{o}}} = \frac{D}{8L_2 C_2 f^2} \tag{4.78}$$

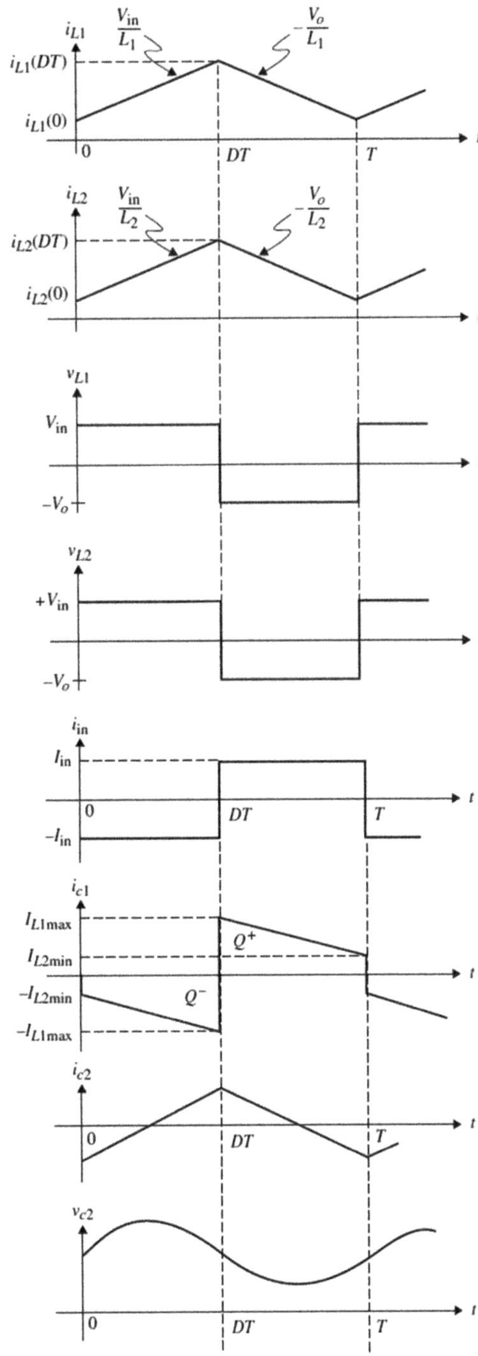

Fig. 4.36 Current and voltage waveforms for the Cuk converter

Exercise 4.12

The converters shown in Fig. E4.12a, b are for a SEPIC and a converter known as Zeta. Derive the voltage gain for both converters.

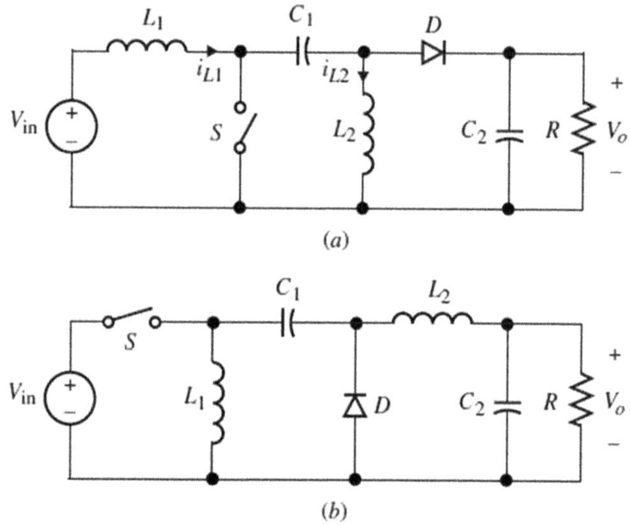

Fig. E4.12 (**a**) SEPIC and (**b**) Zeta

Answer: $\frac{D}{1-D}$

Exercise 4.13

Consider the SEPIC converter in Fig. E4.10a operating in ccm with the following parameters:

$$D = 0.52, \qquad V_{in} = 112\ V, \qquad f_s = 110\ \text{KHz}$$
$$R = 12\ \Omega, \qquad L_1 = L_2 = 50\ \mu H, \qquad C_1 = C_2 = C = 147\ \mu F$$

(a) Sketch the waveforms for $i_{L1},\ i_{L2},\ i_D,\ i_{C1}$, and i_{C2}.
(b) Determine the capacitor ripple voltage $|\Delta V_{C1}/V_o|$ and $|\Delta V_{C2}/V_o|$.

Answer: $\frac{\Delta V_{c1}}{V_o} = \frac{\Delta V_{c2}}{V_D} = \frac{D}{RCf}$

Exercise 4.14

Consider the PWM converter shown in Fig. E4.14 with S_1 and S_2 turning ON and OFF simultaneously. Derive the expression for V_o/V_{in}.

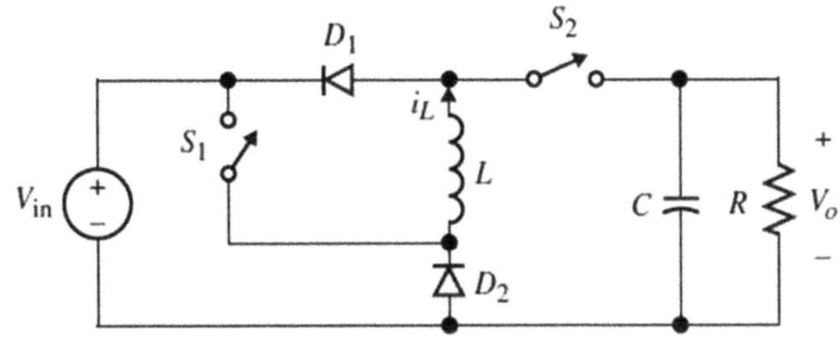

Fig. E4.14 PWM converter for Exercise 4.14

Answer: $(2D - 1)/D$

4.3.5 Bipolar Output Voltage Converter

All single-switch converter topologies discussed so far produce unipolar average output voltage and a unipolar average output current as shown in Fig. 4.37 resulting in unidirectional average power flow from the source to the load.

In some converter topologies like bridge converters, the output voltage can be of two polarities, depending on the duty cycle. Figure 4.38a, b shows two ways to implement a converter that produces bipolar output voltage.

Furthermore, depending on the switching sequence, the average output voltage and current can be unidirectional. Figure 4.39c shows the switch/diode implementation for Fig. 4.39b. It can be shown that the voltage gain is given by:

Fig. 4.37 Average output current vs. output voltage

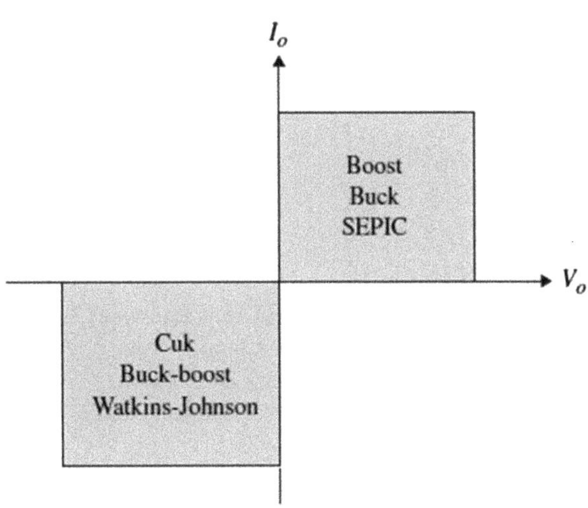

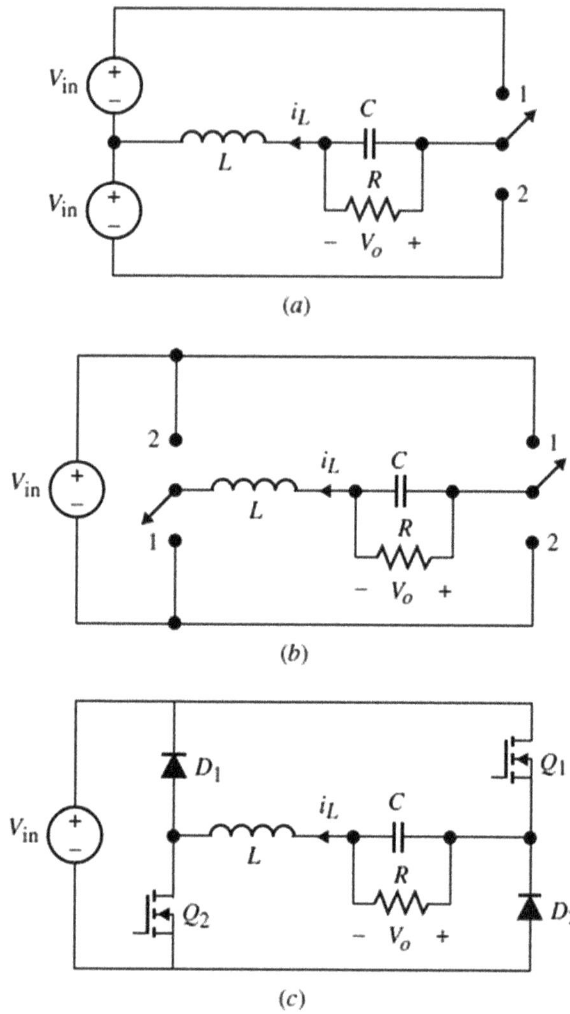

Fig. 4.38 Bipolar output voltage converters: (**a**) half-bridge, (**b**) full-bridge, and (**c**) diode transistor implementation of (**b**)

$$\frac{V_o}{V_{in}} = 2D - 1 \tag{4.79}$$

Exercise 4.15

The converter shown in Fig. E4.15 is known as Watkins-Johnson converter and is capable of producing a bipolar output voltage. Assume S_1 and S_2 are thrown in positions 1 and 2 simultaneously for DT and $(1 - D)T$ time intervals, respectively. Derive the expression for the voltage gain V_o/V_{in}.

Fig. E4.15 Watkins-Johnson converter

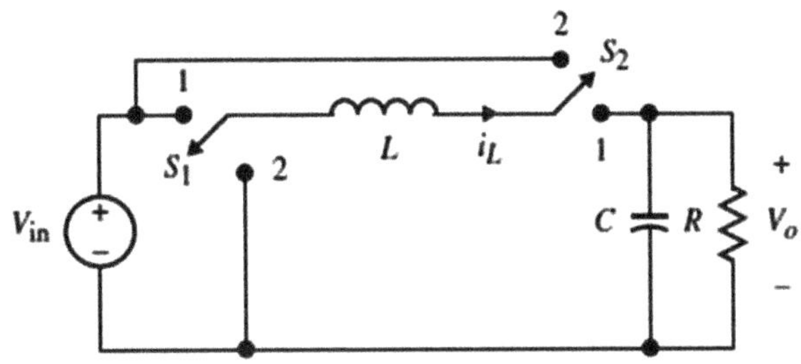

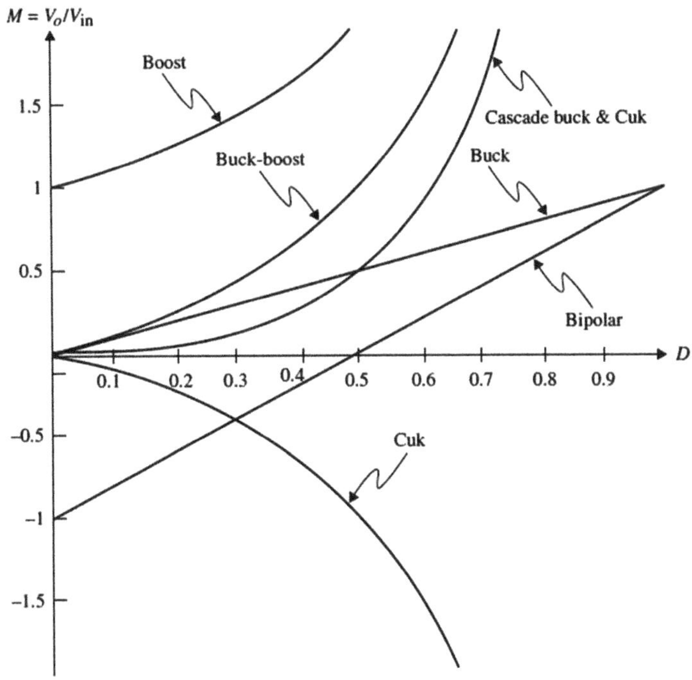

Fig. 4.39 Shows a plot of M vs. D for several converters

Answer: $\frac{V_o}{V_{in}} = \frac{2D-1}{D}$

4.4 Discontinuous Conduction Mode

Unlike the continuous conduction mode (ccm), in the discontinuous conduction mode (dcm), the minimum inductor current in each of the three basic topologies is zero. Hence, there exists a short interval in which the inductor current is zero, i.e., discontinuous. The dcm operation is frequently encountered in dc-dc converters since these converters normally operate under open load conditions. Both the steady-state conversion ratio and the closed-loop dynamic charge significantly change. It will be shown shortly that under dcm the converter voltage gain is a function of not only D but also the load, switching frequency, and circuit components.

As in the ccm case, the steady-state condition will be assumed when it comes to deriving the expression for the voltage gain. In this section, we will present the steady-state analysis for the buck, boost, and buck-boost topologies operated in dcm. Unlike the steady-state analysis in ccm, the analysis for dcm requires solving for the time interval during which the inductor current becomes zero. The basic analysis procedure for each converter is the same.

4.4.1 The Buck Converter

The inductor current waveform under the dcm operation is shown in Fig. 4.40.

Recall the inductor current equations for the buck converter:

$$i_{\mathrm{L}}(t)=\frac{1}{L}(V_{\mathrm{in}}-V_{\mathrm{o}})t+i_{\mathrm{L}}(0) \qquad 0\leq t<DT \tag{4.80a}$$

$$i_{\mathrm{L}}(t)=\frac{-1}{L}V_{\mathrm{o}}(t-DT)+i_{\mathrm{L}}(DT) \qquad DT\leq t<T \tag{4.80b}$$

The edge of discontinuity occurs when $i_{\mathrm{L}}(T)=i_{\mathrm{L}}(0)=0$. This value occurs when $L=L_{\mathrm{crit}}$. The waveform for $i_{\mathrm{L}}(t)$ at the edge of discontinuity is shown in Fig. 4.41. These waveforms correspond to the equivalent circuits given in Fig. 4.42a–c when the switch(es) are on, off, and both off, respectively.

If the converter inductor becomes less than L_{crit}, there exists a dead time greater than zero, as shown in Fig. 4.40. It is shown that D_1 is the duty ratio at which the inductor current becomes zero.

At $t=DT$, Eq. (4.80a) gives:

$$i_{\mathrm{L}}(DT)=\frac{1}{L}(V_{\mathrm{in}}-V_{\mathrm{o}})(DT) \tag{4.81}$$

Notice that the initial inductor current at $t=0$ is zero, $i_{\mathrm{L}}(0)=0$.

At $t=D_1T$, Eq. (4.80b) gives:

$$i_{\mathrm{L}}(D_1T)=\frac{-1}{L}V_{\mathrm{o}}(D_1T-DT)+i_{\mathrm{L}}(DT) \tag{4.82}$$

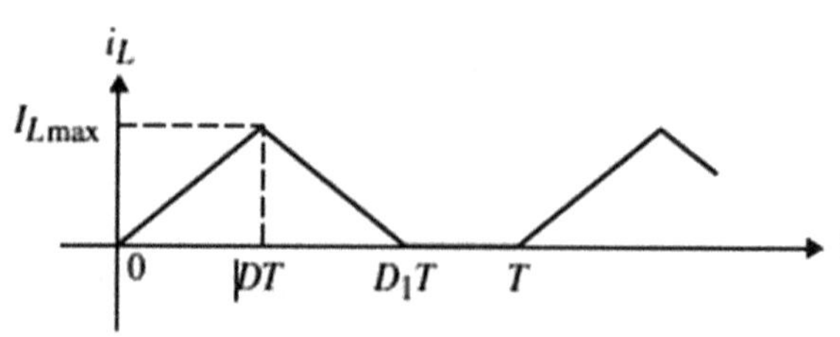

Fig. 4.40 Inductor current waveform under the dcm operation

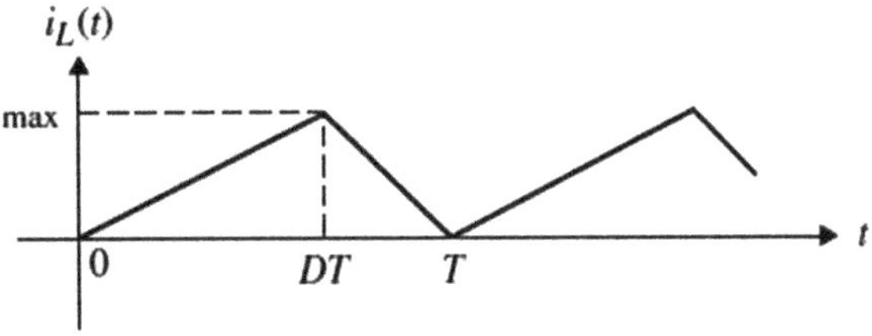

Fig. 4.41 Inductor current waveform at the edge of the discontinuity

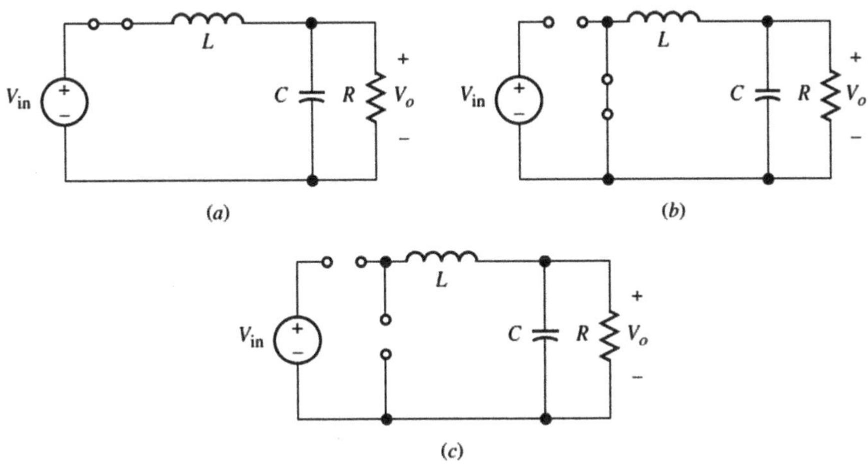

Fig. 4.42 Equivalent circuit modes for the buck converter operating in dcm: (**a**) switch ON, (**b**) switch OFF, and (**c**) when $i_L = 0$

From the above two equations, we obtain:

$$\frac{V_o}{V_{in}} = \frac{D}{D_1} \tag{4.83}$$

Recall from setting $I_{in}V_{in} = I_oV_o$ with $I_{in} = \frac{1}{2}i_L(DT)$, $i_L(DT)$ can also be expressed by:

$$i_L(DT) = \frac{2V_o^2}{DV_{in}R} \tag{4.84}$$

Solve for D_1 from Eqs. (4.82) and (4.84) to obtain:

$$D_1 = \frac{2V_oL}{DRTV_{in}} + D \tag{4.85}$$

Thus, the voltage gain in terms of D and the circuit parameters is given by:

$$\frac{V_o}{V_{in}} = \frac{D}{\frac{2V_oL}{DRTV_{in}} + D} \tag{4.86}$$

Let the voltage gain M be equal to V_o/V_{in}. We must solve for the voltage gain in terms of the circuit components. From Eq. 4.86, we solve for M to yield:

$$M = \frac{D^2RT}{4L}\left[\sqrt{\frac{8L}{D^2RT}+1}-1\right] \tag{4.87}$$

Also we can express D_1 as follows:

$$D_1 = \frac{1}{\frac{D^2RT}{4L}\left[\sqrt{\frac{8L}{D^2RT}+1}-1\right]} \tag{4.88}$$

The voltage conversion characteristics of the buck converter can be expressed in terms of the normalized time constant, τ_n, which is given by:

$$\tau_n = \frac{\tau}{T} \tag{4.89}$$

where $\tau = L/R$ and T is the switching period.

The voltage gain can be expressed as:

$$M = \frac{D^2}{4\tau_n}\left[\sqrt{\frac{8\tau_n}{D^2}+1}-1\right] \tag{4.90}$$

The characteristic curves for M vs. D under different values of normalized time constants are shown in Fig. 4.43.

The normalized relation for D_1 is given by:

$$D_1 = \frac{1}{\frac{D^2}{4\tau_n}\left[\sqrt{\frac{8\tau_n}{D^2}+1}-1\right]} \tag{4.91}$$

The characteristic curves for M vs. D_1 under different values of normalized time constants are shown in Fig. 4.44.

The maximum inductor current of Eq. (4.81) can be expressed in terms of τ_n:

$$I_{\mathrm{Lmax}} = -\frac{V_{in}TD}{L}\left[\frac{D^2}{4\tau_n}\left[\sqrt{\frac{8\tau_n}{D^2}+1}-1\right]-1\right] \tag{4.92}$$

Normalize the current by V_{in}/R:

$$I_{\mathrm{nLmax}} = \frac{I_{\mathrm{Lmax}}R}{V_{in}}$$

$$I_{\mathrm{nLmax}} = -\frac{D^3}{4\tau_n^2}\left[\sqrt{\frac{8\tau_n}{D^2}+1}-1\right]+\frac{D}{\tau_n} \tag{4.93}$$

The plot for I_{nLmax} vs. D under different normalized time constants is given in Fig. 4.45.

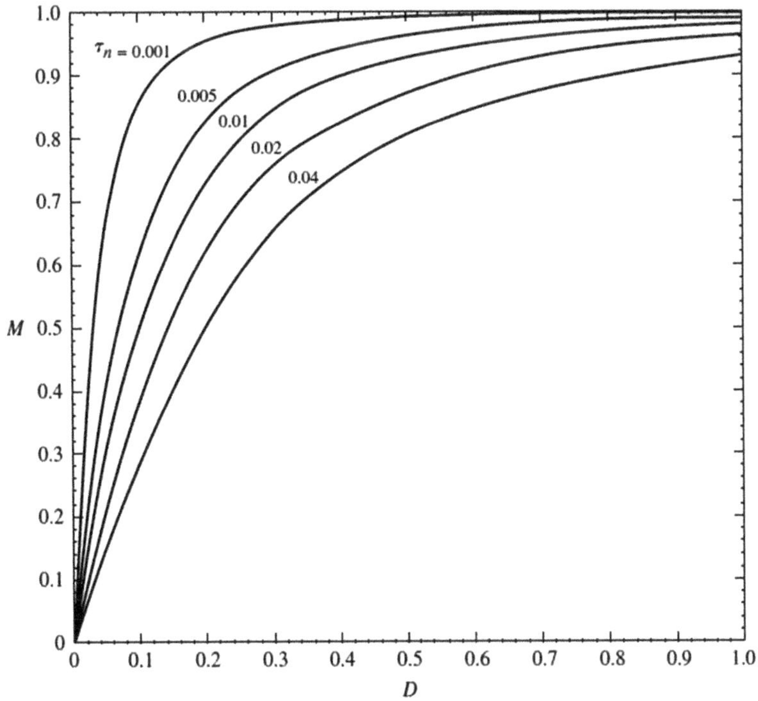

Fig. 4.43 Characteristic curves for M vs. D under different normalized time constants

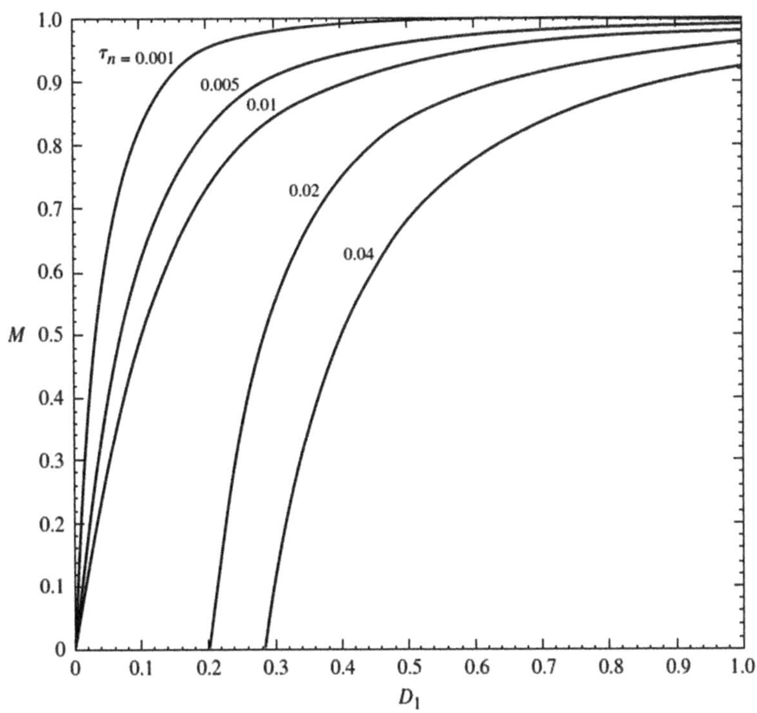

Fig. 4.44 Characteristic curves for M vs. D_1 under different normalized time constants

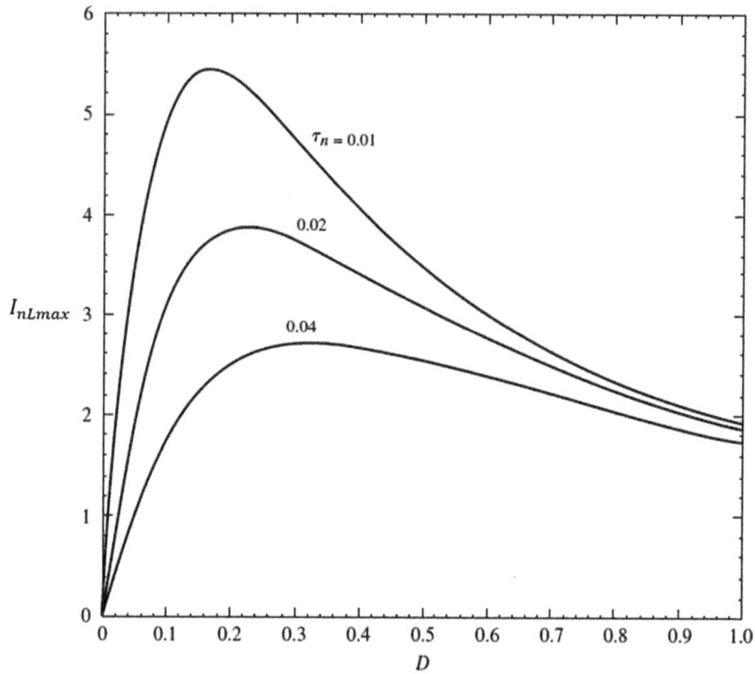

Fig. 4.45 Normalized maximum inductor current vs. D under different normalized time constants

Example 4.8

Consider a buck converter with a dc voltage source of 80 V and a load resistance equal to 18 Ω. It is required that this converter needs to deliver at least 100 W to the load. Assume the switching frequency is 150 kHz. Determine (a) the inductor critical value, L_{crit}, (b) the voltage gain for $L = 0.1L_{\text{crit}}$ and $10L_{\text{crit}}$, (c) D_1 for $L = 0.1\ L_{\text{crit}}$, and (d) the maximum inductor current at $t = DT$.

Solution

(a) The output voltage, V_{o}, is given by:

$$V_{\text{o}} = \sqrt{PR} = \sqrt{(100)(18)} = 42.43\ \text{V}$$

For the given $V_{\text{in}} = 80\text{V}$, the duty ratio under *ccm* is 0.53. Under this condition, with $T = 6.67\ \mu\text{s}$, the inductor critical value is:

$$L_{\text{crit}} = \frac{RT}{2}(1 - D)$$

(b) If the inductor is chosen to be less than $L = 0.1L_{\text{crit}} = 2.812\ \ \mu\text{H}$, we have:

$$M = \frac{V_{\text{o}}}{V_{\text{in}}} = \frac{D^2RT}{4L}\left[\sqrt{\frac{8L}{D^2RT} + 1} - 1\right] = 0.873$$

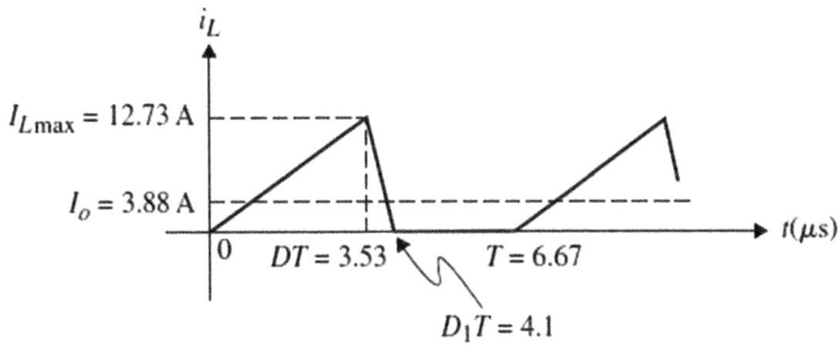

Fig. 4.46 Inductor current waveform for Example 4.8 under dcm

For $L = 10\ L_{crit}$:

$$M = \frac{V_o}{V_{in}} = \frac{42.43}{80} = 0.53$$

The new output voltage is $V_o = 0.873(80) = 69.8$ V.

(c) D_1 is given by:

$$D_1 = \frac{D}{M} = \frac{0.53}{0.873}$$

(d) The maximum inductor current at t = DT is:

$$I_{Lmax} = -\frac{V_{in}TD}{L}(M-1) = 12.73 \text{ A}$$

The time intervals are given as $DT = 3.53$ μs and 4.2 μs with:

$$I_o = I_L = \frac{V_o}{R} = \frac{69.8 \text{ V}}{18} = 3.88 \text{ A} \qquad P_o = (3.88)(69.3) = 270.8\,\text{W}$$

$$I_{in} = \frac{I_L(DT)}{2}D = \frac{(12.73)(0.53)}{2} = 3.37 \text{ A} \quad P_{in} = (3.37)(80) = 269.8\,\text{W}$$

The plot for i_L is shown in Fig. 4.46.

Example 4.9

Consider a buck converter with the following parameters: $V_{in} = 80$ V, $R_o = 18\ \Omega$, $P_o = 100$ W, $L = 0.4$ mH, and $f_s = 150$ kHz. (a) Determine the mode of operation. (b) Determine the range R_o for the converter to remain in the ccm.

Solution

(a) $V_o = \sqrt{PR} = 42.43$ V

Assuming in continuous mode of operation:

$$D = \frac{V_o}{V_{in}} = \frac{42.43 \text{ V}}{80 \text{ V}} = 0.53$$

The critical inductor value is given by:

$$L_{\text{crit}} = \frac{R_{\text{o}}T}{2}(1 - D) = 28.21\ \mu H$$

Since $L > L_{\text{crit}}$, the converter is operating in the ccm.

(b) Notice that the inductor current tends to decrease faster when the load resistance increases.

Recall:

$$I_{\text{L}}(DT) = I_{\text{Lmax}} = \left[\frac{1}{R} + \frac{(1 - D)T}{2L}\right] DV_{\text{in}}$$

$$I_{\text{L}}(0) = I_{\text{Lmin}} = \left[\frac{1}{R} - \frac{(1 - D)T}{2L}\right] DV_{\text{in}}$$

For I_{Lmin}, we have:

$$0 = \left[\frac{1}{R} - \frac{(1 - 0.53)6.67\ \mu s}{2(0.4\ \text{mH})}\right](0.53)(80)$$

$$R = \frac{2(0.4\ \mu H)}{(1 - 0.53)(6.67\ \mu s)} = 255.2\ \Omega$$

for $R < 255.2\ \Omega$	ccm (heavy load)
for $R = 255.2\ \Omega$	dcm (boundary)
for $R > 255.2\ \Omega$	dcm (light load)

4.4.2 The Boost Converter

Similarly, the boost converter can be analyzed to obtain the voltage gain, M, the time at which the inductor current reaches zero, D_1, and the maximum inductor current, I_{Lmax}. For the *ON*-state in the time interval, $0 \le t \le DT$, we have:

$$i_{\text{L}} = \frac{V_{\text{in}}}{L}t + I_{\text{L}}(0) \tag{4.94}$$

In the dcm, $I_{\text{L}}(0)$ equals zero, and the maximum inductor current occurs at $t = DT$, which is given by:

$$I_{\text{Lmax}} = I_{\text{L}}(DT) = \frac{V_{\text{in}}}{L}DT \tag{4.95}$$

For the *OFF*-state in the time interval, $DT \leq t$, we have:

$$i_{\mathrm{L}}(t) = \frac{V_{\mathrm{in}} - V_{\mathrm{o}}}{L}[t - DT] + i_{\mathrm{L}}(DT)$$

In the dcm, $i_{\mathrm{L}}(D_1T) = 0$; hence:

$$0 = \frac{V_{\mathrm{in}} - V_{\mathrm{o}}}{L}(D_1 - D)T + i_{\mathrm{L}}(DT) \tag{4.96}$$

From Eqs. (4.95) and (4.96), we obtain the voltage gain:

$$\frac{V_{\mathrm{o}}}{V_{\mathrm{in}}} = M = \frac{D_1}{D_1 - D} \tag{4.97}$$

To solve for D_1 in terms of the circuit parameters, we use the conservation of power to obtain another relation. The average input power, P_{in}, equals $I_{\mathrm{in}}V_{\mathrm{in}}$, where $I_{\mathrm{in}} = \frac{1}{2}I_{\mathrm{Lmx}}D_1$; thus:

$$P_{\mathrm{in}} = \frac{1}{2}I_{\mathrm{Lmx}}V_{\mathrm{in}}D_1 \tag{4.98}$$

and the average output is given by:

$$P_{\mathrm{o}} = \frac{V_{\mathrm{o}}^2}{R} \tag{4.99}$$

Equating P_{in} and P_{o}, form Eqs. (4.98) and (4.99), we obtain:

$$I_{\max} = \frac{2V_{\mathrm{o}}^2}{RV_{\mathrm{in}}D_1} \tag{4.100}$$

From Eqs. (4.97) and (4.100) and using $I_{\mathrm{in}} = \frac{1}{2}I_{\mathrm{Lmax}}D_1$, we obtain:

$$M = \frac{1}{2}\left[1 + \sqrt{1 + \frac{2RTD^2}{L}}\right] \tag{4.101}$$

In terms of τ_{n}::

$$M = \frac{1}{2}\left[1 + \sqrt{1 + \frac{2D^2}{\tau_{\mathrm{n}}}}\right] \tag{4.102}$$

The duty ratio D_1 is obtained from Eq. (4.97) in terms ofτ_n:

$$D_1 = \frac{\tau_{\mathrm{n}}}{D} + D + \sqrt{\frac{{\tau_{\mathrm{n}}}^2}{D^2} + 2\tau_{\mathrm{n}}} \tag{4.103}$$

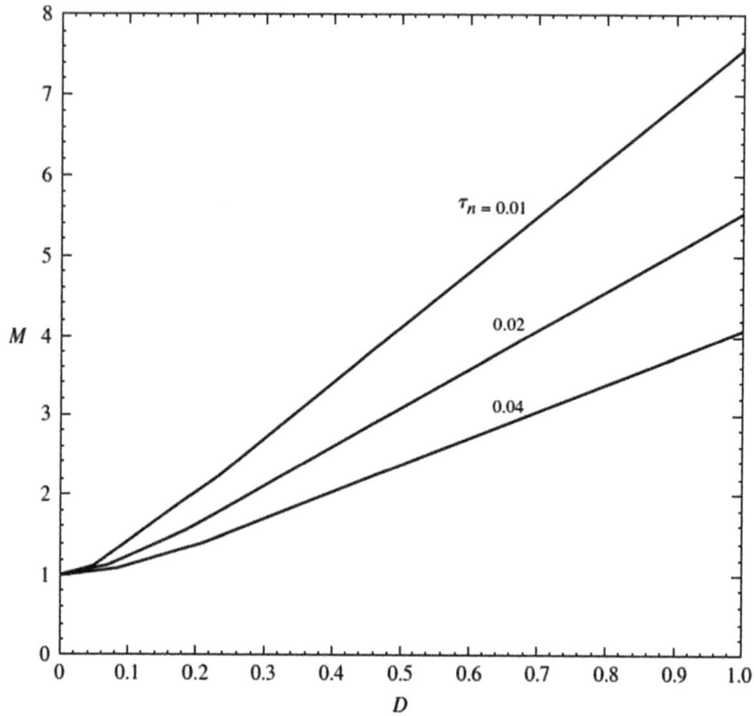

Fig. 4.47 *M* vs. *D* under different normalized values of time constant for the boost converter

Set $D_1 = 1$ and solve for L, we obtain:

$$L = L_{\text{crit}} = \frac{RTD}{2}(1 - D)^2$$

The normalized maximum inductor current is given by:

$$I_{\text{nLmax}} = \frac{D}{\tau_\text{n}} \tag{4.104}$$

The characteristic curves for M vs. D, M vs. D_1, and D vs. I_{nLmax} under different values of normalized time constants are shown in Figs. 4.47, 4.48, and 4.49, respectively.

Example 4.10
Consider a boost converter that supplies 4 A to an 80 V output with $V_{\text{in}} = 60$ V and $L = 67$ μH. Assume the inductor current is discontinuous and $f_s = 100$ kHz. Determine the operation mode and the maximum inductor that can be used to achieve the dcm mode of operation.

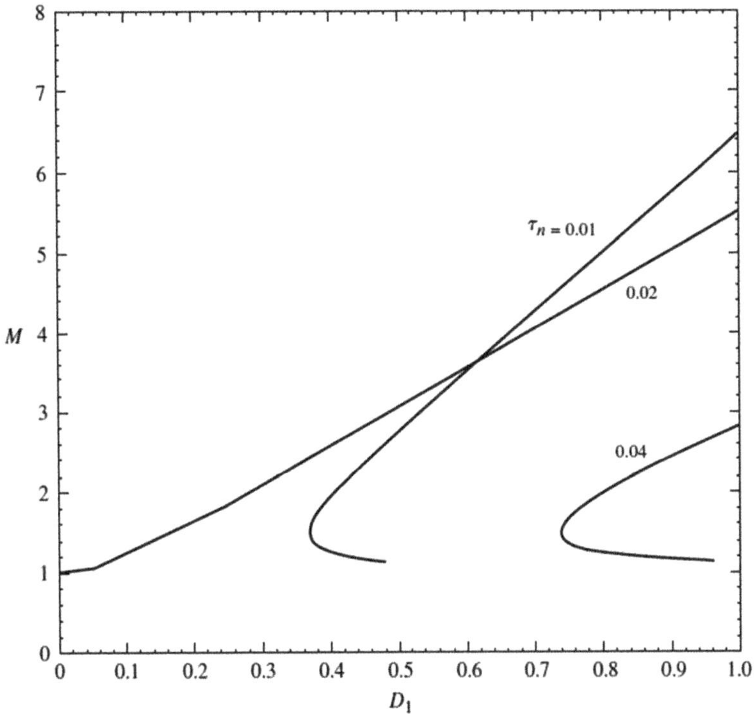

Fig. 4.48 M vs. D_1 under different values of time constant for the boost converter

Solution

The voltage gain is 1.33 and $\tau = L/R = 3.35 \times 10^{-6}$ s and $\tau_n = 0.335$. The duty ratio:

$$D = \frac{M-1}{M} = \frac{1.33-1}{1.33}$$

thus, we obtain the value for D_1:

$$D_1 = \frac{\tau_n}{D} + D + \sqrt{\frac{{\tau_n}^2}{D^2} + 2\tau_n}$$

$$D_1 = \frac{0.335}{0.248} + 0.248 + \sqrt{\frac{(0.335)^2}{(0.248)^2} + 2(0.335)} = 3.18$$

Since $D_1 > 1$, then the converter must be operating in the ccm under the specified values.

The maximum inductor value to maintain:

$$L < L_{crit} = \frac{RTD}{2}(1-D)^2$$

$L \leq 0.125$ mH for dcm

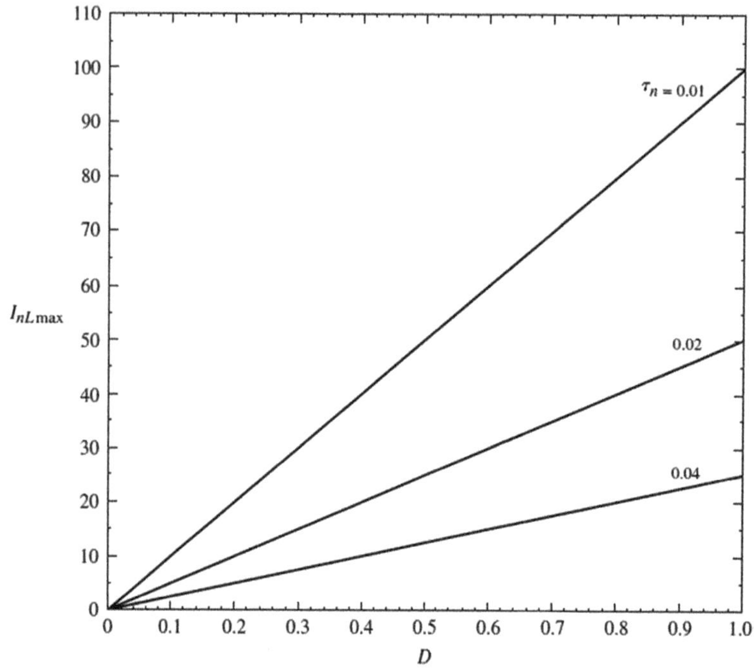

Fig. 4.49 I_{nLmax} vs. D under different normalized values of time constant for the boost converter

4.4.3 The Buck-Boost Converter

Like the other two converters, the derivation of the gain equation for the buck-boost operating in dcm is straightforward. Similar analysis shows the following relations for M and D_1:

$$D_1 = D\left(1 + \frac{1}{M}\right) \tag{4.105}$$

$$M = \frac{D}{\sqrt{2\tau_{\mathrm{n}}}} \tag{4.106}$$

The characteristic curves for M vs. D under different values of normalized time constants are shown in Fig. 4.50. For the M vs. D_1 curve, refer to Fig. 4.51.

And the normalized maximum inductor current is given by:

$$I_{\mathrm{nLmax}} = \frac{MD}{\tau_{\mathrm{n}}} \tag{4.107}$$

Figure 4.52 shows I_{nLmax} vs. D under different normalized time constants.

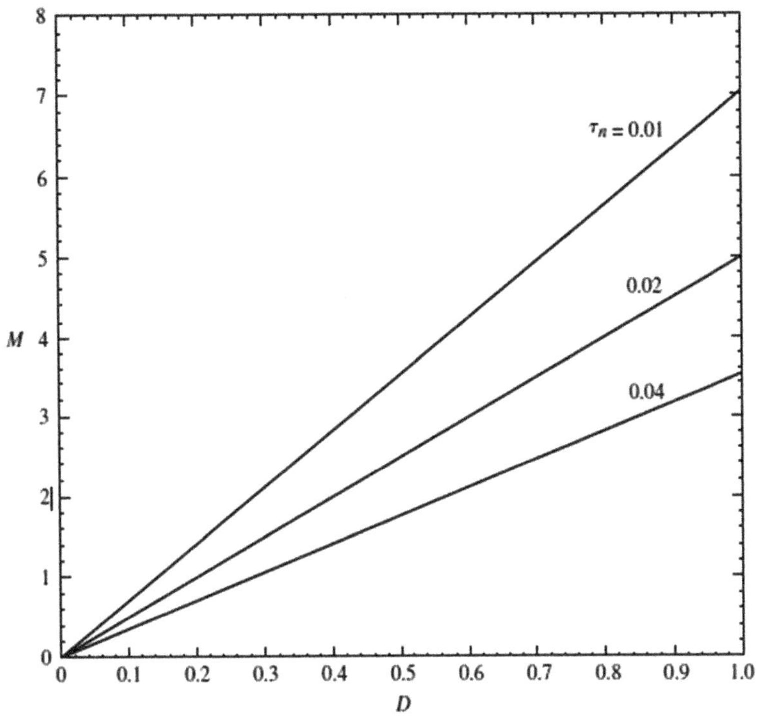

Fig. 4.50 *M* vs. *D* under different values of normalized time constant for the buck-boost converter

Example 4.11

Consider a buck-boost converter with the following values: $V_o = 12$ V, $P_{out} = 25$ W, $V_{in} = 20$ V, and f = 100 kHz. (a) Design the above converter so that it will operate in *ccm*. (b) Repeat part (a) for dcm. (c) Find the maximum inductor current under both ccm and dcm. (d) If the load resistance increases by 50% (i.e., the load current changes 2.08 A to 1.39 A), determine the mode of operation for the two converters and then the maximum inductor current. Sketch the new inductor currents.

Solution

(a) The load resistance at 25 W and 12 V output is given by:

$$R = \frac{V_o^2}{P_o} = \frac{12^2}{25} = 5.76\ \Omega$$

The duty ratio, *D*, under ccm is 0.375.
For *ccm*, the inductor must be larger than the critical value given by:

$$L_{crit} = \left(\frac{RT}{2}\right)(1 - D)^2 = 0.01125\ \text{mH}$$

For ccm, we choose $L = 0.1$mH.

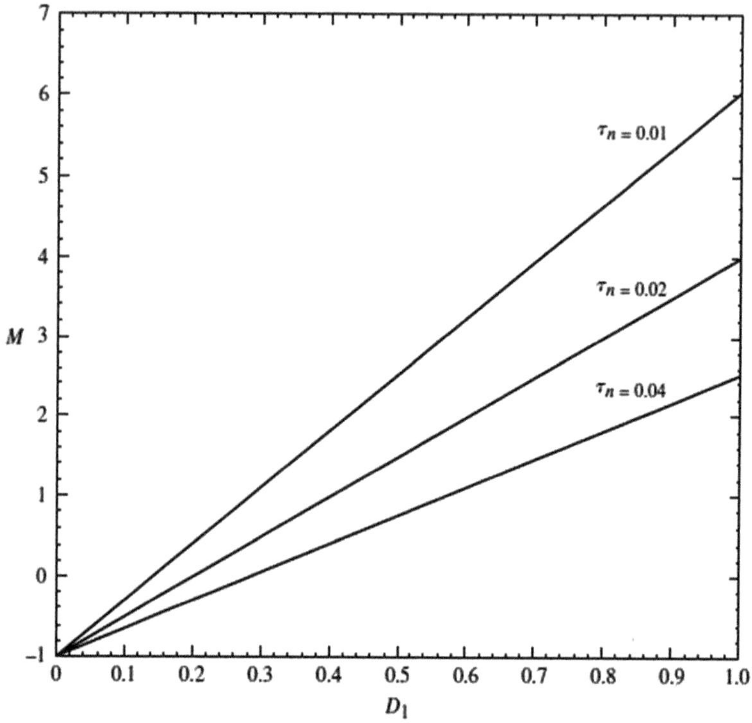

Fig. 4.51 *M* vs. *D1* under different values of normalized time constant for the buck-boost converter

(b) For dcm, we choose L smaller than L_{crit}. Let $L = 0.005\text{mH}$:

$$\tau_n = \frac{\tau}{T} = \frac{0.868\ \mu s}{10\ \mu s} = 0.087$$

Hence, for $M = 0.6$ and $\tau_n = 0.087$, we obtain $D = 0.25$.
Solve for D_1 to yield $D_1 = 0.67$.

(c) The maximum inductor current for ccm is given by:

$$I_{\text{Lmax}} = \frac{DV_{\text{in}}}{R(1-D)^2} + \frac{V_{\text{in}}DT}{2L} = 3.71\ \text{A}$$

The minimum inductor current is given by:

$$I_{\text{Lmin}} = 3.33 - 0.371 \approx 3\ \text{A}$$

For dcm, we have:

$$I_{\text{Lmax}} = \frac{1}{L} V_{\text{in}} DT = 10\ \text{A}$$

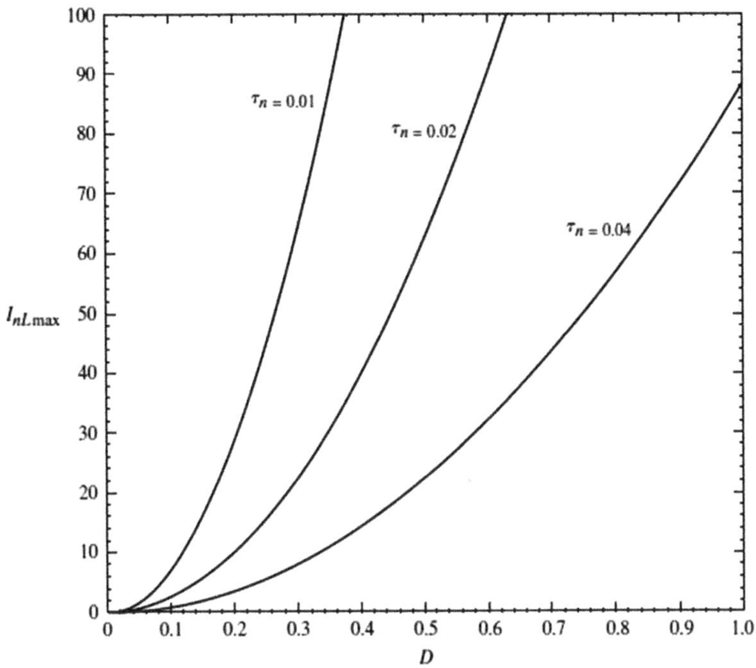

Fig. 4.52 $I_{\mathrm{nL,max}}$ vs. D under different values of normalized time constant for the buck-boost converter

Table 4.1 shows the summary of equations for the three basic converters

	ccm	*dcm*		
Converter type	$\frac{V_o}{V_{in}}$	$\frac{V_o}{V_{in}}$	D_1	$\frac{V_o}{V_{in}}$
Buck	D	$\frac{D}{D_1}$	$\frac{1}{\frac{D}{4\tau_n}\left[\sqrt{\frac{8\tau_n}{D}+1}-1\right]}$	$\frac{D^2}{4\tau_n}\left[\sqrt{\frac{8\tau_n}{D^2}+1}-1\right]$
Boost	$\frac{1}{1-D}$	$\frac{D_1}{D_1-D}$	$D_1=\frac{\tau_n}{D}+D+\sqrt{\frac{8\tau_n}{D}+1}$	$\frac{1}{2}\left[1+\sqrt{1+\frac{2D^2}{\tau_n}}\right]$
Buck-boost	$\frac{D}{1-D}$	$\frac{D}{D_1-D}$	$D+\sqrt{2\tau}$	$D\sqrt{\frac{1}{2\tau}}$

(d) If the load resistance changes by +50% for the designs, the new modes of operations can be obtained by finding the new value of τ_n or determining the new L.

(i). For the ccm:

$$R_{\mathrm{new}}=5.76+\frac{1}{2}(5.76)=8.64\ \Omega$$

The new L_{crit} is given by:

$$L_{crit} = 0.017\ \text{mH}$$

Hence, the converter will remain in the *ccm*:

$$I_{L,max} = 2.6\ \text{mA}$$

$$I_{Lmin} = 1.85\ \text{mA}$$

(ii). For the dcm the new τ is 0.005/8.64 = 0.58μs, then the normalized τ_n is given by:

$$\tau_n = \frac{\tau}{T} = \frac{0.58}{10} = 0.058$$

For $M = 0.6$ and $\tau_n = 0.058$, we have $D = 0.204$ and D_1 is obtained from:

$$D_1 = D + \sqrt{2\tau_n} = 0.544$$

The maximum inductor current is given by:

$$I_{Lmax} = \frac{1}{L} V_{in} DT = 8.165\ \text{A}$$

Exercise 4.16
Consider a boost converter that delivers power to the load with the inductor voltage waveform shown in Fig. E4.16. Assume $R = 176\ \Omega$. Determine L.

Fig. E4.16

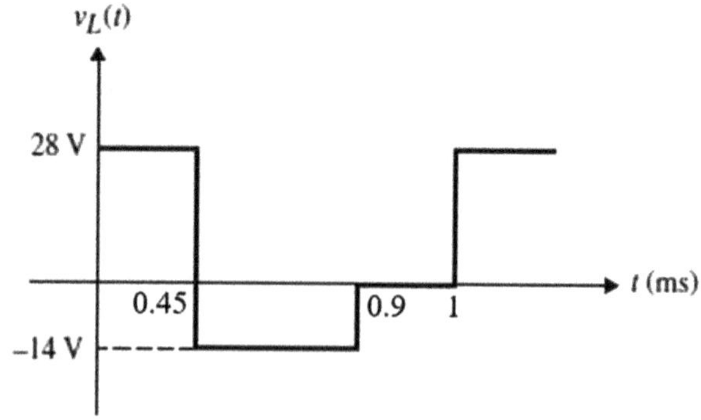

Answer: 12.78 mH

4.5 The Effects of Converter Non-idealities

The analysis thus far has been based on the assumption that all components, switching devices, and diodes are ideal. Depending on the application and power levels, the inclusion of some parasitic effects of both components and devices is very important to the design of acceptable performance. In this section, we are going to study the second-order non-ideal effects on the output voltage and efficiency. The following non-ideal characteristics will be investigated:

1. Inductor resistance (r_L)
2. Transistor and diode voltage drops (V_Q, V_D)
3. Switching and conduction losses (r_{sw})

Other non-idealities include the equal series resistance of the output capacitor, r_{ESR}, which is important to include since its presence affects the design of the closed-loop compensator to stabilize the converter.

4.5.1 *Inductor Resistance*

4.5.1.1 The Buck Converter

To study the effect of inductor resistance on the buck converter performance, we assume the inductor has a finite resistance, r_L, in series as shown in Fig. 4.53. The source of this resistive loss for a practical inductor is from the core and copper losses.

When the switch is *ON*, the inductor voltage is given by:

$$V_{in} = r_L i_L + L\frac{di_L}{dt} + V_o$$

The first-order differential equation is obtained:

$$\frac{di_L}{dt} + \frac{i_L}{\tau} = \frac{1}{L}(V_{in} - V_o)$$

where $\tau = L/r_L$.

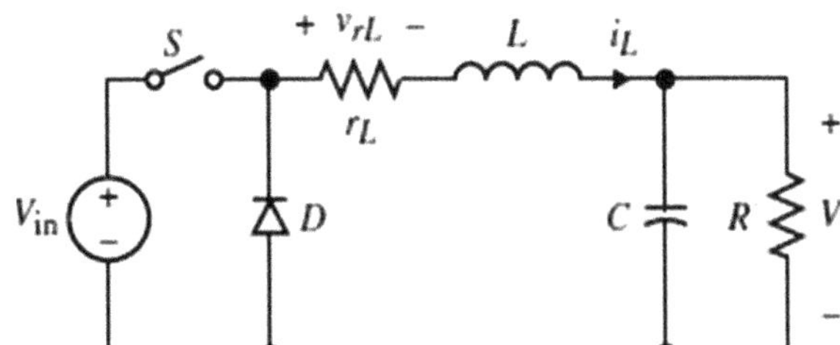

Fig. 4.53 Buck converter with inductor resistance

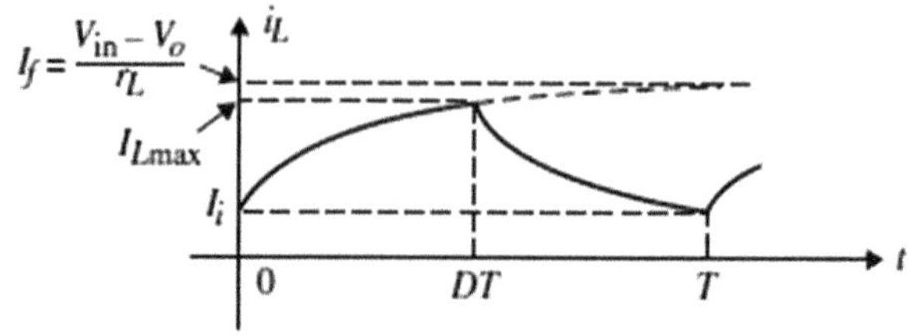

Fig. 4.54 Inductor current including r_L

The general solution for i_L is given by:

$$i_L(t) = (I_i - I_f)e^{-t/\tau} + I_f \tag{4.108}$$

The initial, I_i, and final, I_f, values must be determined in order to solve for $i_L(t)$. At $t = DT$ we have:

$$i_L(DT) = (I_i - I_f)e^{-DT/\tau} + I_f$$

and the final value, I_f, is given by:

$$I_f = \frac{V_{in} - V_o}{r_L}$$

A typical sketch of $i_L(t)$ including inductor resistance is shown in Fig. 4.54.

Let $\boldsymbol{I_i}$ be the minimum inductor current at $\boldsymbol{t} = 0$ and $\boldsymbol{I_{Lmin}}$ and $\boldsymbol{I_{Lmax}}$ be the maximum at $\boldsymbol{t} = \boldsymbol{DT}$. Hence, using the above value for $\boldsymbol{I_f}$, we obtain the following relation:

$$I_{Lmax} = \left[I_{Lmin} - \left(\frac{V_{in} - V_o}{r_L}\right)\right]e^{-DT/\tau} + \frac{V_{in} - V_o}{r_L} \tag{4.109}$$

When the diode is conducting, the inductor equation is given by:

$$\frac{di_L}{dt} + \frac{r_L}{L}i_L = -\frac{V_o}{L}$$

The solution for i_L is given by:

$$i_L(t) = (I_i - I_f)e^{(t-DT)/\tau} + I_f$$

The final value, I_f, is given by:

$$I_f = \frac{-V_o}{r_L}$$

and the initial value, I_i, is $I_{Lmax.}$ Hence, $i_L(t)$ is given by:

$$I_L(t) = \left(I_{Lmax} + \frac{V_o}{r_L}\right)e^{-(t-DT)/\tau} - \frac{V_o}{r_L} \tag{4.110}$$

Evaluating $i_L(t)$ at $t = T$, we obtain:

$$I_{\text{Lmin}} = \left(I_{\text{Lmax}} + \frac{V_o}{r_L}\right)e^{-T(t-DT)/\tau} - \frac{V_o}{r_L} \tag{4.111}$$

where $i_L(T) = I_{\text{Lmin}}$.

To solve for $i_L(t)$, expressions for I_{Lmin} and I_{Lmax} must be obtained from Eqs. (4.109) and (4.111) in terms of the circuit components.

By replacing the exponential function by its first two linear terms, as follows:

$$e^{-DT/\tau} \approx 1 - \frac{DT}{\tau} \tag{4.112a}$$

$$e^{-T(t-DT)/\tau} = 1 - \frac{T}{\tau} + \frac{DT}{\tau} \tag{4.112b}$$

Equations (4.109) and (4.111) may be written as follows:

$$I_{\text{Lmin}} = \left(I_{\text{Lmax}} + \frac{V_o}{r_L}\right)\left(1 - \frac{T}{\tau} + \frac{DT}{\tau}\right) - \frac{V_o}{r_L} \tag{4.113a}$$

$$I_{\text{Lmax}} = \left[I_{\text{Lmin}} - \left(\frac{V_{\text{in}} - V_o}{r_L}\right)\right]\left(1 - \frac{DT}{\tau}\right) + \frac{V_{\text{in}} - V_o}{r_L} \tag{4.113b}$$

Further, it can be shown that the following equation may be obtained:

$$I_{\text{Lmaax}} \approx I_{\text{Lmin}} \approx \frac{(V_{\text{in}}D - V_o)}{r_L} \tag{4.114}$$

Equation (4.114) suggests that the inductor ripple is approximated to zero.

Using the relation:

$$I_o = \frac{I_{\text{Lmin}} + I_{\text{Lmax}}}{2} = \frac{V_o}{R}$$

we obtain:

$$\frac{V_o}{V_{\text{in}}} = \frac{D}{1 + \frac{r_L}{R}} \tag{4.115}$$

Notice that when $r_L \to 0$, the gain becomes D.

Another way to solve for the voltage gain is to assume that the time constant τ is very large when compared to the switching period and the voltage across r_L is small compared to the input voltage $V_{\text{in};}$ hence, an approximation V_{rL} is calculated using the average value of inductor current. Plot for M vs. D for different r_L/R values as shown in Fig. 4.55.

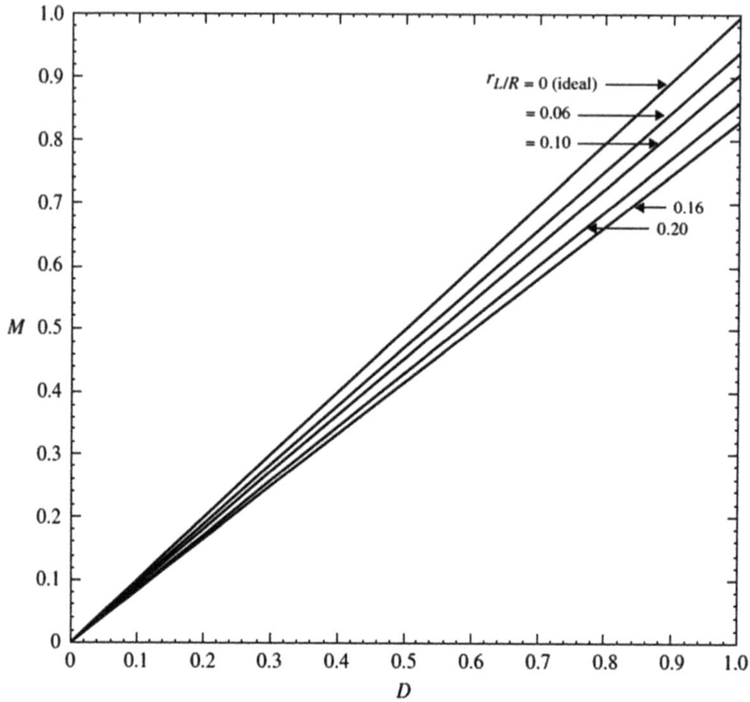

Fig. 4.55 *M* vs. *D* under different inductor resistance values for the buck converter

Fig. 4.56 Boost converter with inductance resistance

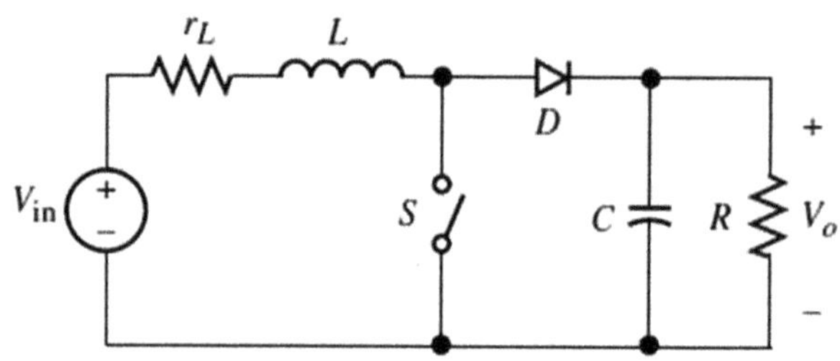

4.5.2 *The Boost Converter*

To study the effect of the inductor resistance on the boost converter performance, we again assume the inductor has a finite resistance, r_L, in series as shown in Fig. 4.56. The inductor current waveform is shown in Fig. 4.57.

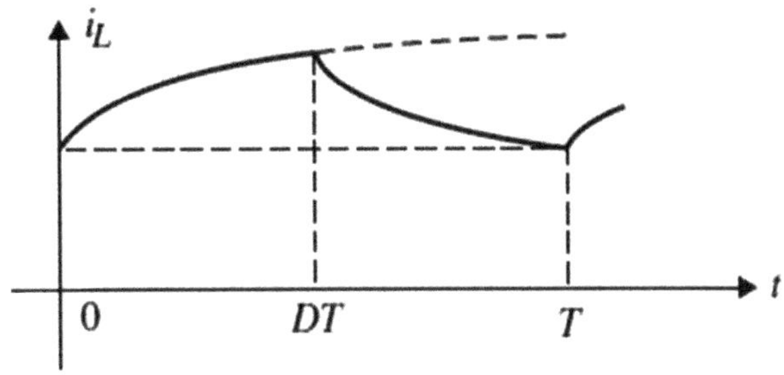

Fig. 4.57 Boost inductor current

Similar analysis gives the following inductor current equations:

$$i_{\mathrm{L}}(t)=\left(I_{\mathrm{Lmin}}+\frac{V_{\mathrm{in}}}{r_{\mathrm{L}}}\right)e^{-t/\tau}+\frac{V_{\mathrm{in}}}{r_{\mathrm{L}}} \qquad 0\le t<DT$$

$$i_{\mathrm{L}}(t)=\left(I_{\mathrm{Lmax}}+\frac{V_{\mathrm{in}}-V_{\mathrm{o}}}{r_{\mathrm{L}}}\right)e^{-(t-DT)/\tau}-\frac{V_{\mathrm{in}}-V_{\mathrm{o}}}{r_{\mathrm{L}}} \qquad DT\le t<T$$

where $i_{\mathrm{L}}(0)=I_{\mathrm{Lmin}}$ and $i_{\mathrm{L}}(DT)=I_{\mathrm{Lmax}}$ Using the same approximation we have applied for the buck converter, the maximum and minimum inductor currents can be approximated as follows:

$$I_{\mathrm{Lmax}}\approx I_{\mathrm{Lmin}}\approx\frac{1}{r_{\mathrm{L}}}[V_{\mathrm{in}}-(1-D)V_{\mathrm{o}}]$$

Using the average output current which is given by:

$$I_{\mathrm{o}}=\left(\frac{I_{\mathrm{Lmax}}+I_{\mathrm{Lmin}}}{2}\right)(1-D)$$

we obtain the equation for the voltage gain:

$$\frac{V_{\mathrm{o}}}{V_{\mathrm{in}}}=\frac{1}{(1-D)+\frac{r_{\mathrm{L}}}{R}\frac{1}{1-D}} \tag{4.116}$$

The plot of Eq. (4.116) is shown in Fig. 4.58.

Exercise 4.17

Consider this buck-boost with inductor resistance shown in Fig. E4.17. Show that the voltage gain equation can be approximated by the following relation:

$$\frac{V_{\mathrm{o}}}{V_{\mathrm{in}}}=\frac{1}{(1-D)+\frac{r_{\mathrm{L}}}{R(1-D)}} \tag{4.117}$$

Fig. E4.17 Buck-boost converter with inductance resistance

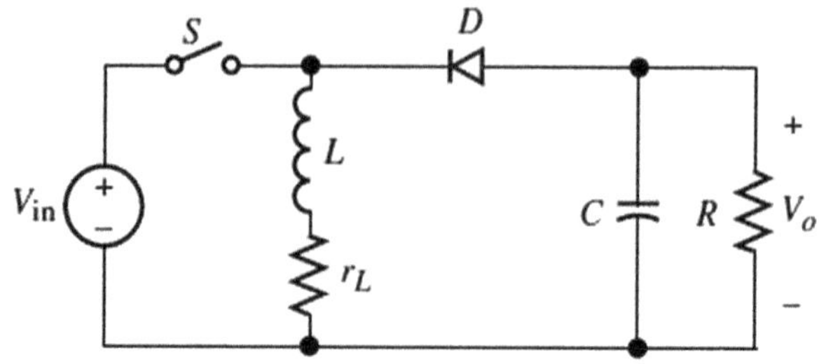

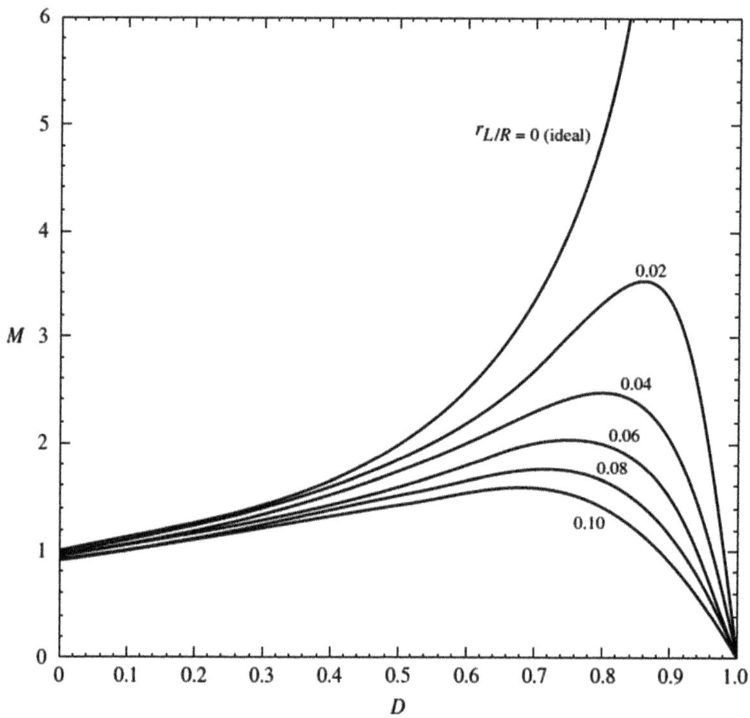

Fig. 4.58 *M* vs. *D* under different inductor resistance values for the boost converter

whose plot is shown in Fig. 4.59.

4.5.3 *Transistor and Diode Voltage Drops*

As an illustration, let us consider the buck converter. Let us assume that when the transistor is on, a nonzero voltage drop across it, V_Q, is present and, when the diode is turned *ON*, a voltage V_D appears across it. Hence, the voltage across the inductor while the transistor and diode are conducting is given by:

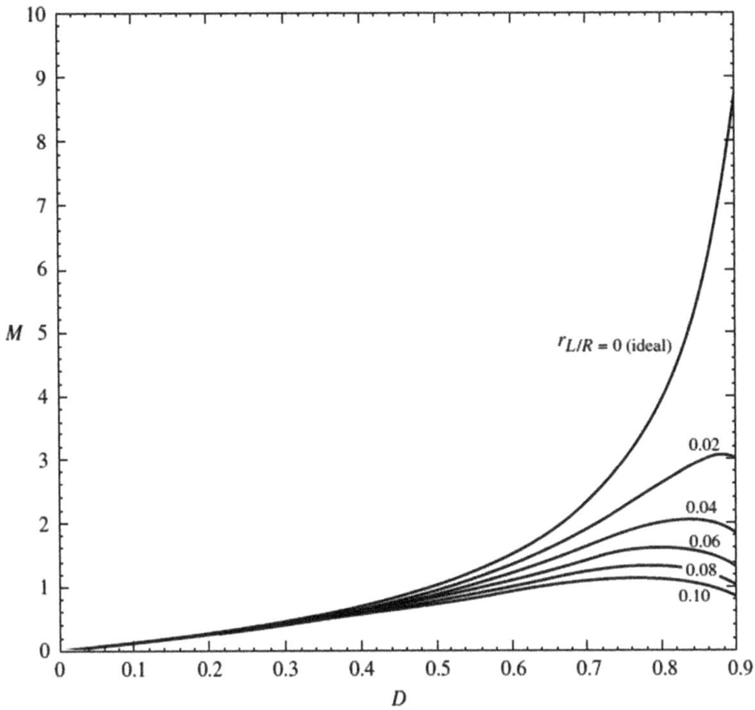

Fig. 4.59 *M* vs. *D* under different inductor resistance values for the buck-boost converter

Switch *ON*: $v_L = V_{in} - V_o - V_Q$ $0 \le t < DT$

Switch *OFF*: $v_L = -V_o - V_D$ $DT \le t < T$

Hence, the inductor currents are given by:

$$i_L(t) = \frac{1}{L}(V_{in} - V_o - V_Q)t + I_{Lmin} \qquad 0 \le t \le DT$$

$$i_L(t) = \frac{1}{L}(-V_o - V_D)(t - DT) + I_{Lmax} \quad DT \le t \le T$$

Evaluating the above equations at $t = DT$ and T, we obtain:

$$i_{Lmax} = \frac{1}{L}(V_{in} - V_o - V_Q)DT + I_{Lmin}$$
$$i_{Lmin} = \frac{1}{L}(-V_o - V_D)(1 - D)T + I_{Lmax}$$

Using this input and output average power, the voltage conversion is given by:

$$\frac{V_o}{V_{in}} = D - D\frac{V_Q}{V_{in}} - \frac{V_D}{V_{in}}(1 - D)$$

If we normalize voltages by V_{in}, we obtain:

$$M = D\left(1 - V_{nQ} - V_{nD}\left(\frac{1}{D} - 1\right)\right) \tag{4.118}$$

where V_{nQ} and V_{nD} are normalized transistor and diode voltage drops.

Exercise 4.18

Derive the normalized voltage gain expression for the boost converter by including V_{nQ} and V_{nD}.

Answer: $M = 1 - V_{nD} - \frac{D}{1-D}V_{nQ} + \frac{D}{1-D}$

Exercise 4.19

Show that gain for the buck-boost is given by the following equation when including the diode and voltage drops for the buck-boost:

$$M = \frac{(1 - V_{nQ})D - (1 - D)V_{nD}}{1 - D} \tag{4.119}$$

4.5.4 *The Effect of Switch Resistance*

To study the effect of the switching resistance on the converter's performance, we assume the switch has a finite resistance, r_{sw}, as shown in Fig. 4.60.

The resultant inductor current is shown in Fig. 4.61.

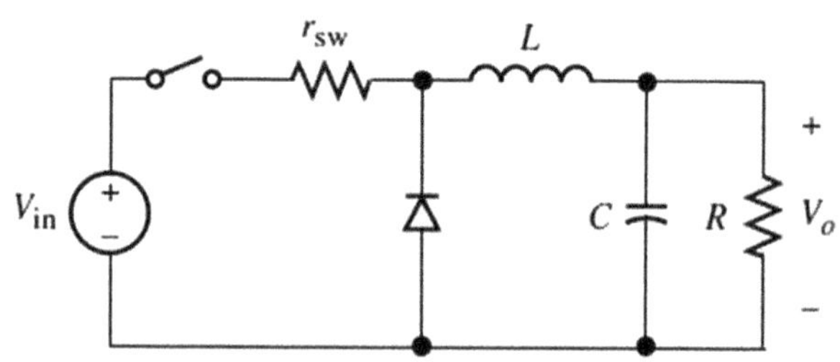

Fig. 4.60 Buck converter with switching resistance

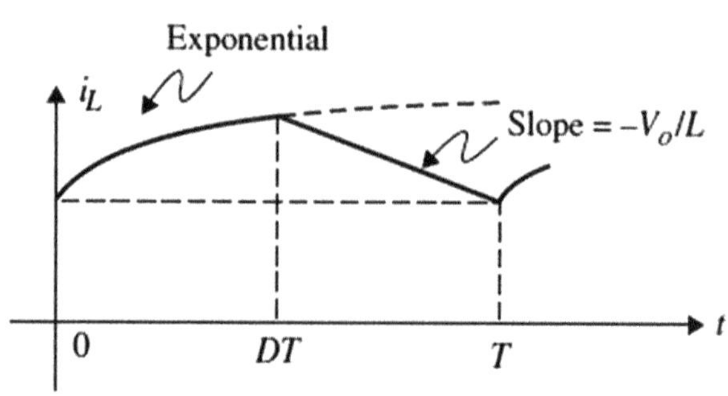

Fig. 4.61 Inductor current for Fig. 4.60

The inductor current equations are given for the ON and OFF times of the switch as follows:

$$i_L(t) = \left(I_{Lmin} - \frac{V_{in} - V_o}{r_{sw}}\right)e^{-t/\tau} + \frac{V_{in} - V_o}{r_{sw}} \qquad 0 \le t < DT$$

$$i_L(t) = \frac{V_o}{L}(t - DT) + I_{Lmax} \qquad DT \le t < T$$

where τ is the time constant and equals to L/r_{sw}.

Using the same approximation applied to the inductor resistance, we obtain the following expression for I_{Lmax} and I_{Lmin}:

$$I_{Lmax} \approx \frac{1 - DV_{in} - V_o}{Dr_{sw}}$$

$$I_{Lmin} = -\frac{V_o}{L}(1 - D)T + \frac{DV_{in} - V_o}{Dr_{sw}}$$

$$I_o = \frac{V_o}{R} = \frac{I_{Lmax} + I_{Lmin}}{2}$$

Hence, the voltage gain is given by:

$$M = \frac{1}{\frac{1}{D} + \frac{1-D}{2\tau/T} + \frac{r_{sw}}{R}} \tag{4.120}$$

Again assuming $2\tau/T \gg 1 - D$, we obtain:

$$M = \frac{1}{1 + D\frac{r_{sw}}{R}} \tag{4.121}$$

Typically, for a MOSFET, $r_{sw} = r_{DS(on)} = 0.15\ \Omega$.

A plot of M vs. D for the buck converter under different values of r_{sw}/R is shown in Fig. 4.62. Figures 4.63 and 4.64 show the plots for M vs. D under different values of r_{sw}/R for the boost and buck-boost, respectively.

Exercise 4.20

Show that the gain expressions for the boost and buck-boost converters when including the switch resistance, r_{sw}, are as shown below:

$$\frac{V_o}{V_{in}} = \frac{1}{1 - D + \left(\frac{D}{1-D}\right)\frac{r_{sw}}{R}} \tag{4.122}$$

$$\frac{V_o}{V_{in}} = \frac{D}{1 - D + \left(\frac{D}{1-D}\right)\frac{r_{sw}}{R}} \tag{4.123}$$

Example 4.12

A buck converter is modeled by including a switch resistance, r_{sw} an inductor resistance, r_L, and a diode voltage drop, V_D. Assume $V_{in} = 50$ V, $V_D = 0.9$ V,

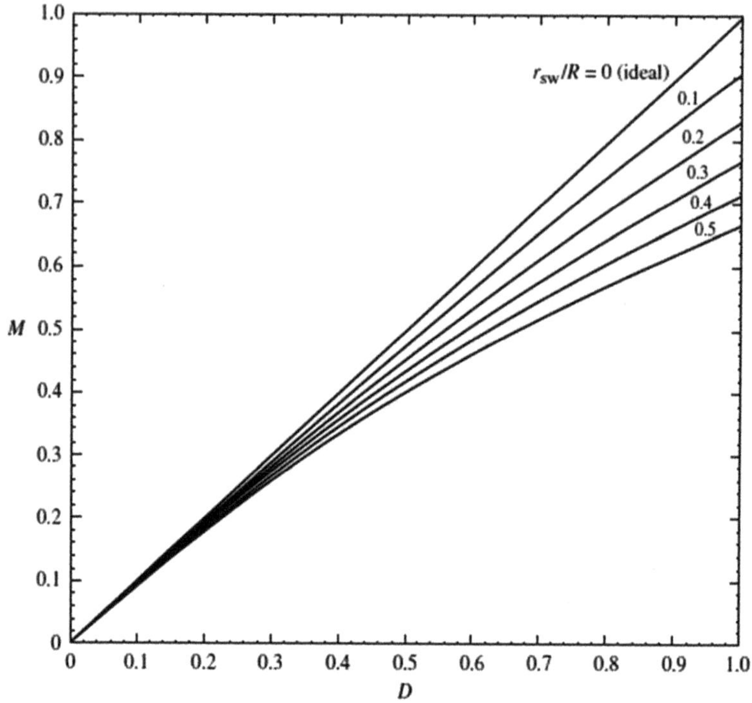

Fig. 4.62 M vs. D for the buck converter under different values of r_{sw}/R

$V_o = 20$ V, $R = 4\ \Omega$, $r_{sw} = 0.08\ \Omega$, and $r_L = 0.06\ \Omega$. (a) Derive the relation for V_o/V_{in} that includes the above effects, (b) find the duty cycle, D, and (c) find the efficiency $\eta = P_o/P_{in}$.

Solution

(a) The inductor current relations are given by:

$$I_{Lmax} - I_{Lmin} = \left(\frac{V_{in} - I_L(r_L + r_{sw}) - V_o}{L}\right) DT$$

$$I_{Lmax} - I_{Lmin} = \left(\frac{V_{in} - I_L r_L - V_o}{L}\right)(1 - D)T$$

$$I_L = \frac{V_o}{R}$$

These relation result in:

$$\frac{V_o}{V_{in}} = \frac{D\left(1 + \frac{V_D}{V_{in}}\right) - \frac{V_D}{V_{in}}}{1 + \frac{r_L}{R} + D\frac{r_{sw}}{R}} \tag{4.124}$$

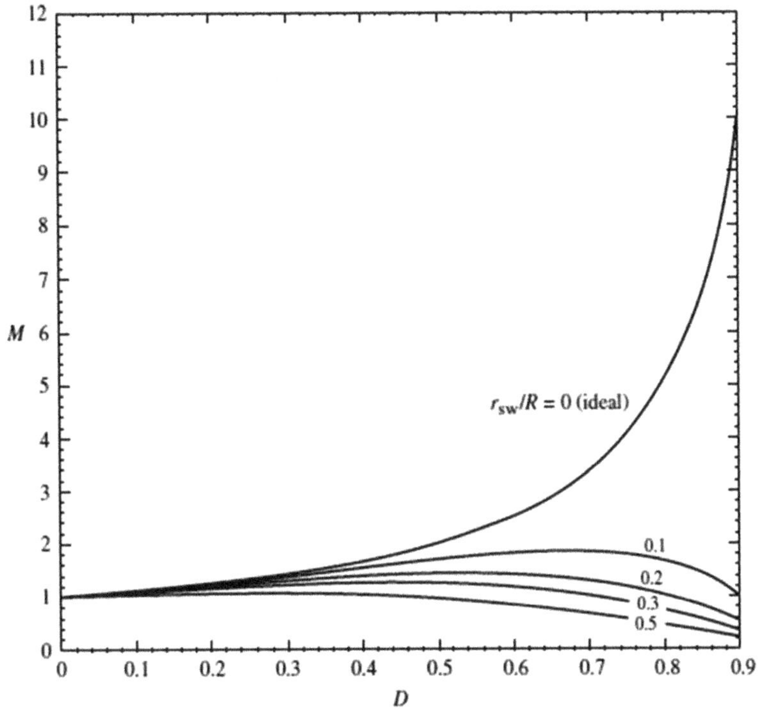

Fig. 4.63 M vs. D under different values of r_{sw}/R for the boost converter

(b) Substituting for $V_{in} = 50$ V, $V_D = 0.9$ V, $V_o = 20$ V, $R = 4\ \Omega$, $r_{sw} = 0.08\ \Omega$, and $r_L = 0.06\ \Omega$, we obtain $D = 0.42$.

(c) The power loss in the inductor and switch resistors are given by:

$$P_{loss} = (I_{in})^2 r_{sw} + (I_L)^2 r_L$$

$$I_{in} = DI_o = D\frac{V_o}{R} = (0.42)\frac{20}{4} = 2.1$$

$$I_o = 5\ \text{A}$$

$$P_{loss} = (2.1)^2(0.08) + (5)^2(0.06) = 0.353 + 1.5 = 1.835\ \text{W}$$

$$P_{in} = V_{in}I_{in} = (50)(2.1) = 105\ \text{W}$$

$$P_o = V_oI_L = (20)(5) = 100\ \text{W}$$

Total loss is 5 W, and the efficiency, η, is given by:

$$\eta = \frac{100}{105}100\% = 98.2\%$$

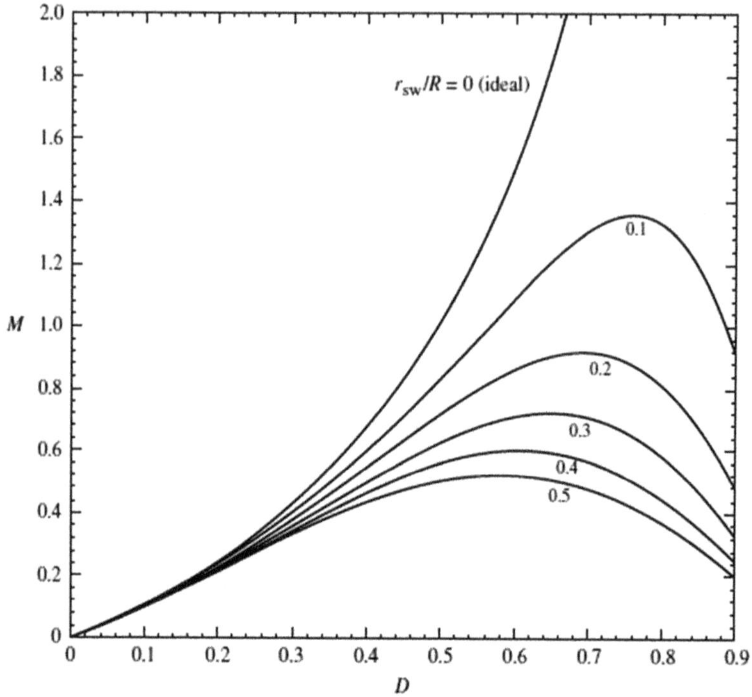

Fig. 4.64 M vs. D under different values of r_{sw}/R for the buck-boost converter

Exercise 4.21

Derive the voltage gain equation for the buck-boost converter by including the switch resistance, r_{SW}, and the inductor resistance, r_L.

Answer: $\frac{V_o}{V_{in}} = \frac{D}{(1-D)+\frac{r_L}{R(1-D)}+\left(\frac{D}{1-D}\right)\frac{r_{SW}}{R}}$

4.5.4.1 DC Transformer Model

We can develop the dc transformer equivalent circuit model by including the converter non-idealities discussed above. Such linear circuit models are very useful when it comes to solving for voltage conversion, average current gain, and efficiency. This is because linear circuit analysis techniques that are well understood by design engineers can be employed in solving these models.

By inspection, it can be seen that the dc transformer equivalent circuit for a buck converter of Fig. 4.65 is given in Fig. 4.66. The derivation of the model is quite simple. All we need to do is to write the dc equation relating V_{in}, V_o, I_{in}, D, and I_o during Modes 1 and 2. For example, for the buck converter of Fig. 4.67, we have the following relations for $v_L(t)$:

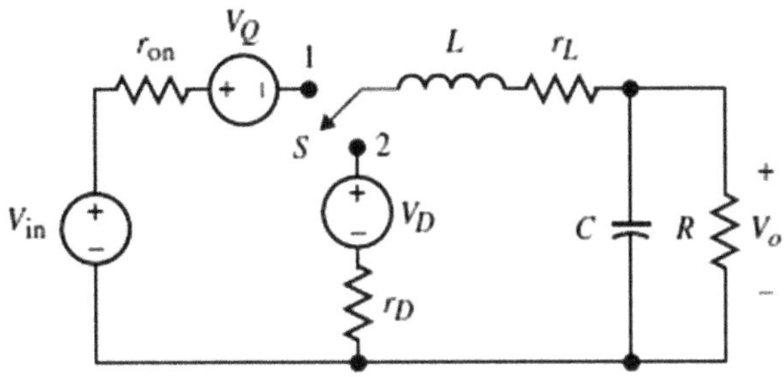

Fig. 4.65 Buck converter with component non-idealities

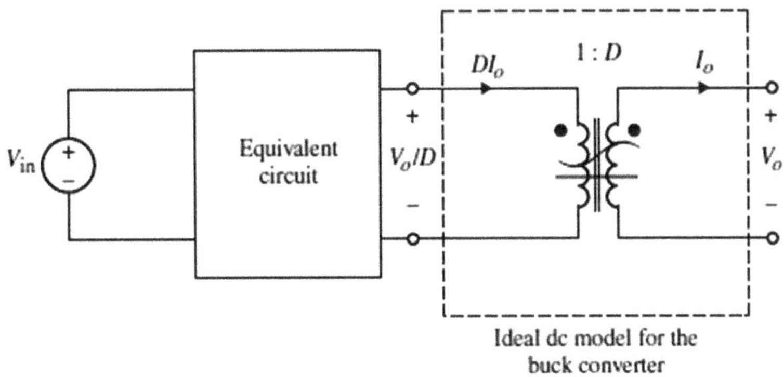

Fig. 4.66 Transformer equivalent circuit diagram for a buck converter

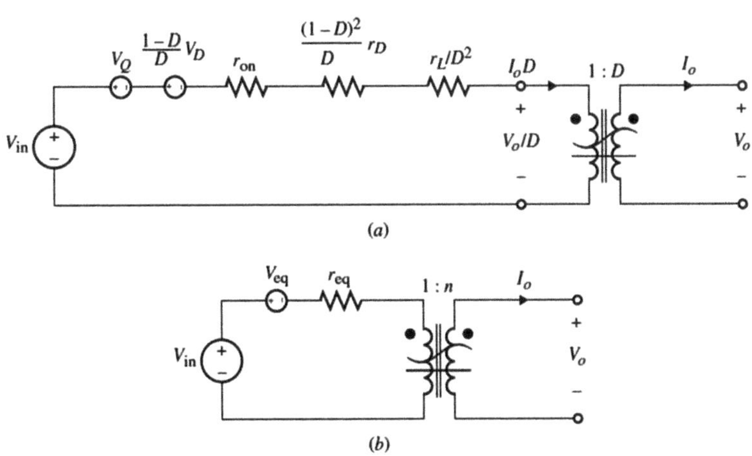

Fig. 4.67 (**a**) Dc equivalent circuit for buck converter of Fig. 4.65 and (**b**) simplified equivalent circuit

$$v_L(t) = V_{in} - V_Q - (r_L + r_{on}D)I_o - V_o \quad 0 \le t < DT \tag{4.125a}$$

$$v_L(t) = -V_D - V_o - (r_D(1 - D) + r_L)I_o \quad DT \le t < T \tag{4.125b}$$

Under dc condition, the average inductor voltage is zero:

$$\frac{1}{T}\int_0^T v_L(t)dt = 0$$

The above integral yields the following dc equation:

$$D[V_{in} - V_o - V_Q - (r_L + r_{on}D)I_o] + (1 - D)[-V_D - V_o - (r_D(1 - D) + r_L)I_o] = 0$$

Rearranging the terms in terms of V_{in}, V_o/D, and DI_o, we obtain:

$$V_{in} = \frac{V_o}{D} + V_Q + \frac{1 - D}{D}V_D + \left[r_{on}D + \frac{(1 - D)^2}{D}r_D + \frac{r_L}{D^2}\right]DI_o \tag{4.126}$$

It can be easily seen that the above equation can be represented by the dc equivalent model shown in Fig. 4.67a.

The simplified equivalent circuit is shown in Fig. 4.67 where:

$$r_{eq} = r_{on}D + \frac{(1 - D)^2}{D}r_D + \frac{r_L}{D^2}$$

$$V_{eq} = V_Q + \frac{1 - D}{D}V_D$$

$$n = D$$

Similarly, we can obtain the dc equivalent circuit for the boost and buck-boost. Table 4.2 shows the values for V_{eq}, R_{eq} and transformer ratio for the buck, boost, and buck-boost.

Table 4.2 Dc equivalent circuit model

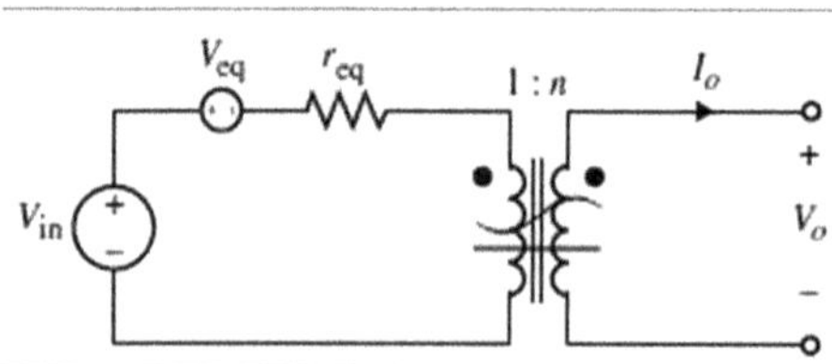

	Buck	Boost	Buck-boost
n	D	$\frac{1}{1-D}$	$\frac{D}{1-D}$
r_{eq}	$r_{on}D + \frac{(1-D)^2}{D}r_D + \frac{r_L}{D^2}$	$r_{on}D + r_D(1-D) + r_L$	$\frac{r_{on}}{D} + \left(\frac{1-D}{D^2}\right)r_D + \frac{r_L}{D^2}$
V_{eq}	$V_Q + \frac{1-D}{D}V_D$	$(1-D)V_D + DV_Q$	$V_Q + \frac{1-D}{D}V_D$

Exercise 4.22
Show that the transformer equivalent circuit model for a boost converter that includes r_L, r_{on}, V_D, and $V_Q$is shown in Fig. 4.67a with

$n = 1/(1-D)$, $r_{eq} = (1-D)V_D + DV_Q$, and $V_{eq} = (1-D)V_D + DV_Q$.

4.6 Switch Utilization Factor

It is possible to evaluate the use of the power switch in the buck, boost, buck-boost, and Weinberg converters by investigating the average switch power-handling capability with respect to the average power delivered to the load. Figure 4.68 shows a block diagram representation for any switch-mode power converter.
where

I_o = average load current.
V_o = average load voltage.
$I_{sw,max}$ = maximum current through the switch when turned ON.
$V_{sw,max}$ = maximum voltage across the switch when turned OFF.
Let us define the switch maximum power as:

$$P_{sw,max} = V_{sw.max} I_{sw,max}$$

and the average output power is given by:

$$P_o = V_o I_o$$

then, we define the switch utilization factor (K_{sw}) as the ratio of $P_{sw,max}$ and P_o:

$$K_{sw} = \frac{P_o}{P_{sw,max}} \tag{4.127}$$

Example 4.13
Determine K_{sw} for the buck converter .

Solution
The maximum switch current and voltage are given by Eqs. (4.128) and (4.129), respectively:

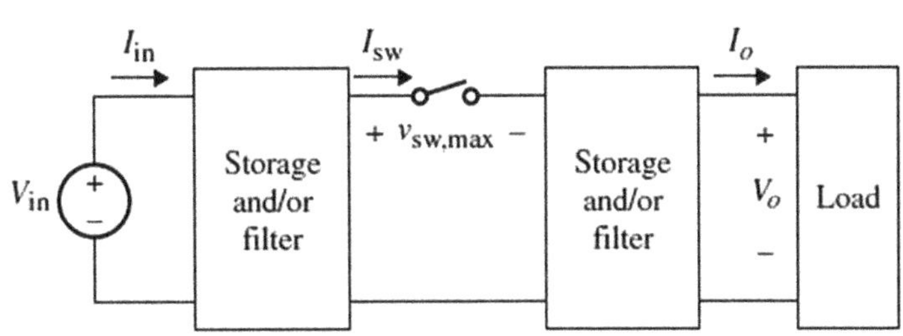

Fig. 4.68 Block diagram for switch-mode power converter

$$I_{sw,max} = DV_{in}\left[\frac{1}{R} + \frac{(1-D)T}{2L}\right] \tag{4.128}$$

$$V_{sw,max} = V_{in} \tag{4.129}$$

The maximum switch power is given by:

$$\begin{aligned} P_{sw,max} &= V_{sw,max} I_{sw,max} \\ &= DV_{in}^2\left[\frac{1}{R} + \frac{(1-D)T}{2L}\right] \end{aligned}$$

The average output power is given by:

$$P_o = \frac{V_o^2}{R}$$

Therefore, the switch utilization factor, K_{sw}, is given by:

$$\begin{aligned} K_{sw} &= \frac{V_o^2/R}{DV_{in}^2\left[\frac{1}{R} + \frac{(1-D)T}{2L}\right]} \\ &= \frac{D}{1 + \frac{(1-D)RT}{2L}} \end{aligned}$$

Using normalized time constant, $\tau_n = L/RT$, K_{sw} can be expressed as:

$$K_{sw} = \frac{D}{1 + \frac{(1-D)}{2\tau_n}}$$

Figure 4.69 shows a plot of K_{sw} vs. D under different values of τ_n for the buck converter.

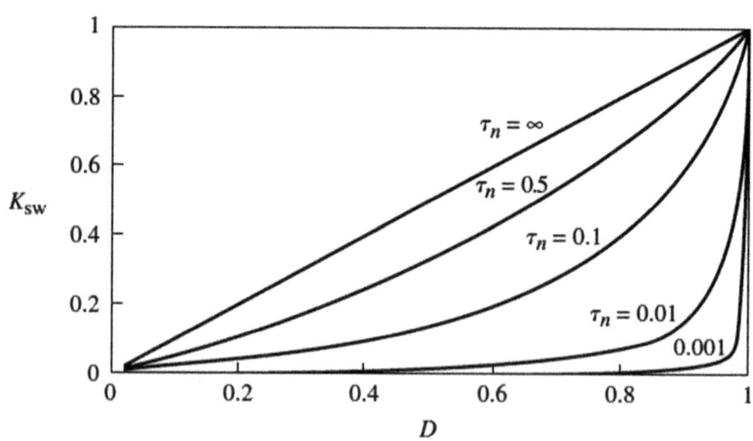

Fig. 4.69 K_{sw} vs. D under different values of τ_n for the buck converter

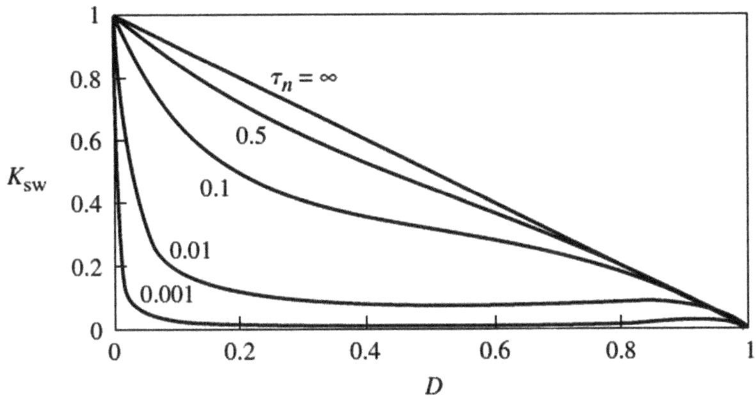

Fig. 4.70 K_{sw} vs. D under different values of τ_n for the boost

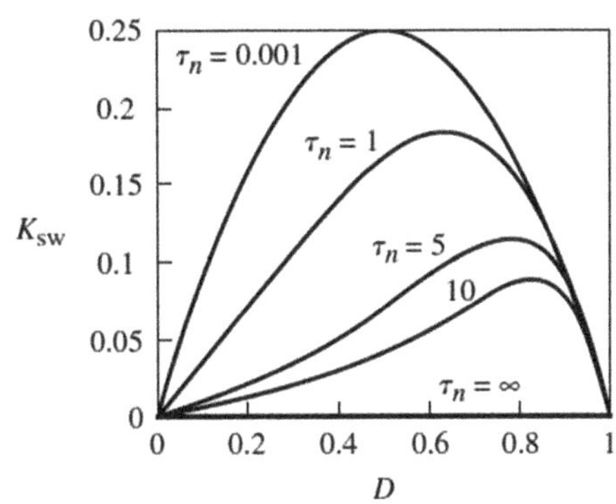

Fig. 4.71 K_{sw} vs. D under different values of τ_n for the buck-boost

Similarly, it can be shown that the switch utilization factor for the buck and the boost are given in Eqs. (4.131) and (4.132), respectively.

$$K_{sw} = \frac{D}{1 + \frac{(1-D)D}{2\tau_n}} \tag{4.131}$$

$$K_{sw} = \frac{D}{\frac{1}{1-D} + \frac{(1-D)}{2\tau_n}} \tag{4.132}$$

Figures 4.70 and 4.71 show the plots of K_{sw} vs. D under different values of τ_n for the boost and buck-boost, respectively.

Exercise 4.23
Show that the switch utilization factor for the boost and the buck-boost are given in Eqs. (4.131) and (4.132).

Problems

Continuous Conduction Mode

Buck Converter

4.1 Consider the following specifications for a buck converter:

$V_{in} = 80$ V
$R_o = 9\ \Omega$
$P_O = 100$ W
$f_S = 150$ KHz

Determine:

(a) The inductor value at the boundary condition (critical value L_{crit})
(b) The maximum inductor current value for $L = 10L_{crit}$
(c) The diode rms and average current values

4.2 Derive the expressions for $V_C(0)$ and $V_C(DT)$ for a buck converter operating in ccm whose current waveform is redrawn in Fig. P4.2.

Fig. P4.2

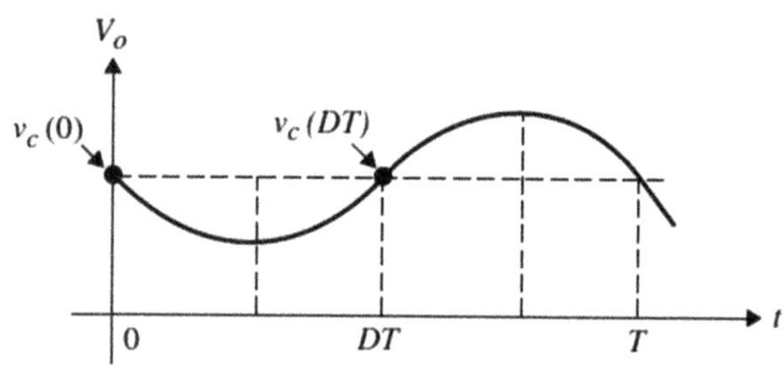

4.3 Consider the buck converter with $V_{in} = 25$ V, $V_o = 12$ V at $I_o = 2$ A, $f_s = 50$ KHz. Determine:

1. The duty cycle D
2. L_{crit}
3. I_{Lmin}, I_{Lmax}, $I_{o,ave}$, $L_{in,ave}$ for $L = 100\ L_{crit}$
4. Output voltage ripple for $C = 0.47\mu$F
5. C for $|V_c| = 2\%$ of V_o

4.4 Derive the expressions for the inductor and capacitor rms currents for the buck. Find these values for Problem 4.3.

D4.5 (a) Design a buck converter for the following given specifications:

$$\Delta V_o/V_o = 0.1\%, \;\; V_{in} = 40 \text{ V}, \;\; P_{o,ave} = 20 \text{ W}, \;\; f_s = 75 \text{ kHz}, \;\; D = 0.5.$$

(b) What is the inductor current's peak-to-peak value?

(c) What is the diode and switch's rms current values?

D4.6 Consider a buck converter with $V_{in} = 48$ V, $V_o = 27$ V at $I_o = 8$ A, and $f_s = 50$ kHz. Design for L and C such that the peak value of the inductor current does not exceed 15% of its average value and the capacitor peak voltage does not exceed 2% of its average value.

D4.7 Consider a buck converter of Problem 4.6 by assuming the input voltage varies by ±20% and the output current varies between 10 A to 6 A. Redesign for L and C using the same specifications with V_{in}(nominal) = 48 V, $V_O = 27$ V, and $f_s = 50$ kHz.

4.8 Determine the diode and switch rms and peak current and voltage ratings for the design of Problem 4.6.

4.9 Determine the circuit parameters including V_{in}, D, f_S, C, L and I_{Lmax}, and I_{Lmax} for a buck converter whose capacitor voltage is given in Fig. P4.9.

Fig. P4.9

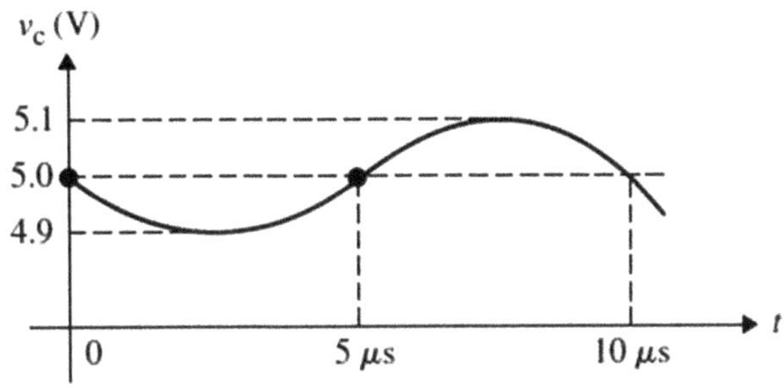

4.10 Design the buck converter to operate in ccm with the following specifications: $V_{in,\,min} = 22$ V, $V_{in,\,max} = 48$ V, $V_{in,\,nom} = 32$ V, $V_o = 12$ V, $I_{o,\,max} = 4$ A, $I_{o,\,min} = 0.5$ A, $f_s = 50$ kHz, and $\Delta V_O = 120$ mV.

Boost Converter

D4.11 Design the boost converter with the following specifications: $V_{in} = 28$ V, $V_o = 48$ V, $P_o = 100$ W, $f_s = 110$ KHz.

4.12 Sketch the inductor current for the boost converter shown in Fig. P4.12. Assume ccm operation with D = 0.4.

Fig. P4.12

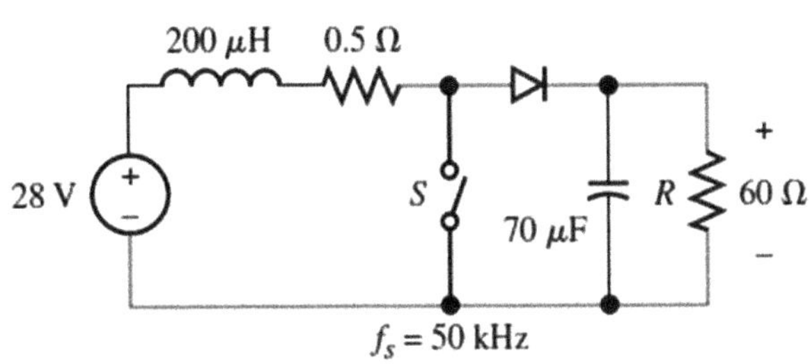

4.13 Design a boost converter to deliver 80 V at 4A from a 60V source with an output ripple voltage not to exceed 1%. Assume $f_s = 20$ KHz.

4.14 The source of a boost converter has an internal resistance R_s as shown in Fig. P4.14. Derive expressions for:

Fig. P4.14

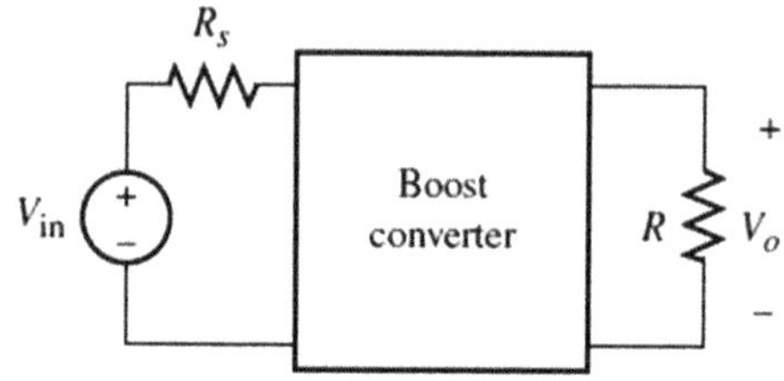

(a) V_o/V_{in}

(b) The efficiency $\eta = \frac{P_{o,\,ave}}{P_{in,\,ave}}$

(c) The duty cycle at which the output voltage is maximized

Your expressions should be given as a function of R/R_s.

4.15 Design a boost converter for the following requirements:
$P_o = 20$ W, $V_{in} = 20$ V, $D = 0.35$, and 1% output voltage ripple.

4.16 Determine ΔVo for a boost converter with $V_{in} = 12$ V, $V_o = 15$ V *at* $I_o = 250$ mA. Use $L = 150$ μH, $C = 470$ μF, and $f_s = 20$ kHz. Sketch the capacitor voltage and inductor current.

4.17 Derive the capacitor and inductor current rms values for Problem 4.16.

4.18 Derive the expressions for V_o/V_{in} for the boost by including r_L and r_{sw}.

4.19 Derive the expressions for V_o/V_{in} for the boost by including V_D and V_Q.

D4.20 Design a boost converter to operate in ccm with the following specifications:

$V_{in,\,min} = 90\ V,\ \ V_{in,\,max} = 150\ V,\ \ V_{n,\,max} = 120\ V,\ \ V_o = 152\ V,$
$I_{o,\,max} = 2\ A,\ I_{o,\,min} = 0.2\ A,\ \ f_s = 50\ kHz,\ \ and\ \ \Delta V_o/V_o = 1\%.$

Buck-Boost Converter

4.21 Consider a buck-boost converter with the following specifications:
$V_o = 12$ V, $P_o = 25$ W, $V_{in} = 24$ V, $f_s = 100$ KHz

(a) Design the above converter so it operates in the ccm.
(b) Find the maximum inductor current under both operating modes.
(c) If the load changes by 50% (lighter load R increases), determine the new mode of operation.
(d) Sketch $i_C(t)$ and $v_c(t)$ for parts (a) and (b).

4.22 Consider the following values for a buck-boost converter:
$V_{in} = 30$ V, $D = 0.25$, $R = 2\ \Omega$, $L = 330$ μH, $f = 10$ KHz

(a) Determine the mode of operation.

(b) Sketch i_L.
(c) Determine V_O.
(d) Determine C for a 1% output ripple.
(e) At what value of R does the transition between ccm and dcm occur?

4.23 Consider a buck-boost converter that supplies 100 W to $V_O = 30$ V from $50 - 30$ V dc source. Let $T = 100$ μs and $L = 800$ μH.

(a) Determine the range of D for the given range of source voltage.
(b) Determine the average input current, diode current, $I_{L\max}$, and $I_{L\min}$, under $V_{in} = 35$V.
(c) Determine the rms value of the diode current and the capacitor current under $V_{in} = 35$V.
(d) Design for C so that the output ripple voltage is limited to 2% of V_o under $V_{in} = 30$V.
(e) If the load resistor changed so that the converter load current is decreased by 10%, what is the new duty cycle when $V_{in} = 35$V?

4.24 For stability consideration, sometimes it is required that the equivalent series resistance (ESR) of the output capacitor be included when modeling the power stage of a PWM converter. Figure P4.24 shows a buck-boost converter by including r_{ESR}. Derive the expression for the output ripple.

Fig. P4.24

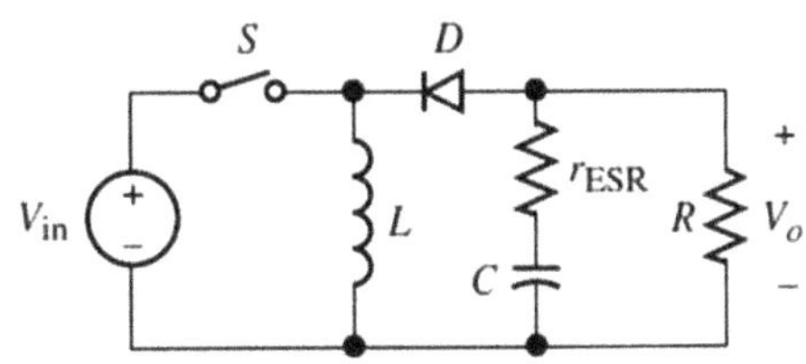

4.25 Repeat Problem 4.4 for a buck-boost converter.
D4.26 Redesign Problem 4.5 for a buck-boost converter.
D4.27 Redesign Problem 4.13 for a buck boost converter.
4.28 Derive the expressions for V_o/V_{in} for the buck-boost converters by including r_L.
4.29 Derive the expressions V_o/V_{in} for the buck-boost by including V_D and V_Q.
4.30 Derive the expression for V_o/V_{in} for the buck-boost converter by including r_{sw}.

Fourth-Order Converters

4.31 Derive the expression for V_o/V_{in} for the zeta converter by including r_L.
4.32 Derive the expression for V_o/V_{in} for the Cuk converter by including V_D and V_Q.
4.33 Derive the expression for V_o/V_{in} for the Cuk converter by including r_{sw}.
4.34 Determine the output voltage ripple equation for the converters Cuk and Zeta.
4.35 Consider the Cuk converter of Fig. P4.35.

(a) Sketch the waveforms for i_{L1}, i_{L2}, i_D and i_s assuming constant voltages across C_1 and C_2.
(b) Determine the rms current values in the diode and switch.
(c) What is the ripple voltage across C_2 and C_2.

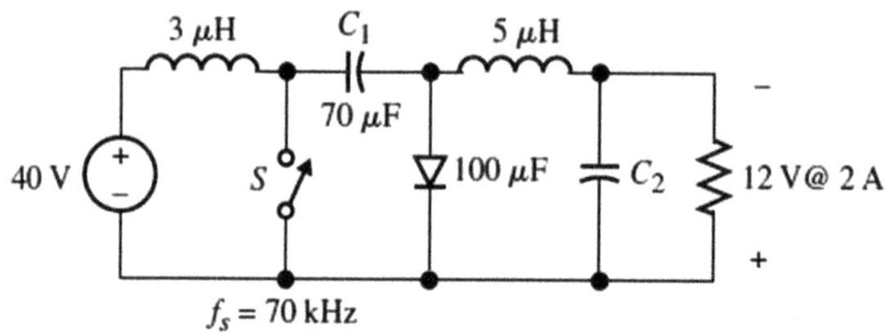

Fig. P4.35

4.36 Derive the voltage gain equation for the single-ended primary-inductance converter. (SEPIC) shown in Fig. 4.23a.

4.37 Derive the *rms* expression for the inductor and diode currents for the Cuk converter.

4.38 (a) Derive the voltage gain expression for the SEPIC converter when operating in the dcm. (b) Design the converter for the following specifications:

$V_{in} = 45$ V, $V_o = 15$ V, $P_o = 25$ W, $f_s = 50$ kHz

The maximum inductor ripple current should not exceed 15% of their average values.
The maximum voltage ripple across C_1 should not exceed 5% of its average value.
The maximum output voltage ripple is less than 0.1%.

4.39 Develop an equivalent circuit for cascaded boost and t buck-boost converters.

4.40 Derive the voltage gain expression for the cascaded arrangement in Problem 4.39.

Discontinuous Conduction Mode

4.41 Consider the buck converter of Problem P4.1.

(a) Determine the voltage gain for $L = 0.1\ L_{crit}$.
(b) Determine D_1.

4.42 Derive the ripple voltage for the buck converter operating in the dcm.

4.43 Derive the ripple voltage for the buck-boost converter operating in the dcm.

4.44 Consider a boost converter with $L = 67$ μH, $V_{in} = 60$ V, $f_s = 20$ kHz, $C = 10$ μF, $V_o = 80$ V at $I_o = 4$ A.

(a) Determine the mode of operation.
(b) Repeat part (a) for $I_o = 2$ A.
(c) Sketch i_e, V_D,and V_c for parts (a) and (b).
(d) Determine D and D_1 for $V_o = 80$ V at (i) $I_o = 4$ A and (ii) $I_o = 2$ A.
(e) Repeat part (d) for $V_{in} = 65$ V.

4.45 Determine the expressions for the diode and transistor currents the buck-boost when operating in the dcm.

4.46 Design the buck-boost converter to operate in the dcm with the following specifications:
$V_{in} = 18$ V, $V_o = -38$ V, $I_{o,max} = 2$ A, $I_{o,min} = 0.2$ A, $f_s = 95$ kHz, and $\Delta V_o = 180$ mV.

4.47 Derive the following expressions for the buck-boost converter operating in dcm:

$$M = \frac{D}{\sqrt{2\tau_n}}$$
$$D_1 = D + \sqrt{2\tau_n}$$

where

$$\tau_n = \frac{\tau}{T}$$
$$\tau = L/R$$

4.48 Derive the expression for the voltage ripple for the boost converter when operating in dcm. Also give an expression for the capacitor voltage at $t = 0$ and $t = DT$ in terms of the circuit parameters.

4.49 Repeat Problem 4.21 under dcm operation.

4.50 Derive the voltage gain for the Cuk converter when operating in the dcm.

4.51 Derive the expression for M and D_1 for the Cuk converter when it operates in the dcm. (dcm operation for the Cuk converter is assumed only when both inductor currents are discontinuous.)

4.52 (a) Derive the expression for ΔV_o (peak-to-peak) for a buck-converter operating in the dcm.
(b) Determine ΔV_o for a boost converter for $V_{in} = 12$ V, $V_o = 15$ V, and $I_o = 250$ mA; use $L = 150$ μH, $C = 470$ μF, and $f_s = 20$ kHz.

Other Converter Topologies

4.53 Derive the expression for the voltage gain for the following two-switch PWM converter shown in Fig. P4.53. S_1 and S_2 are switched simultaneously. Assume $RC \gg T$. Compare this converter to the single-switch buck-boost converter.

Fig. P4.53

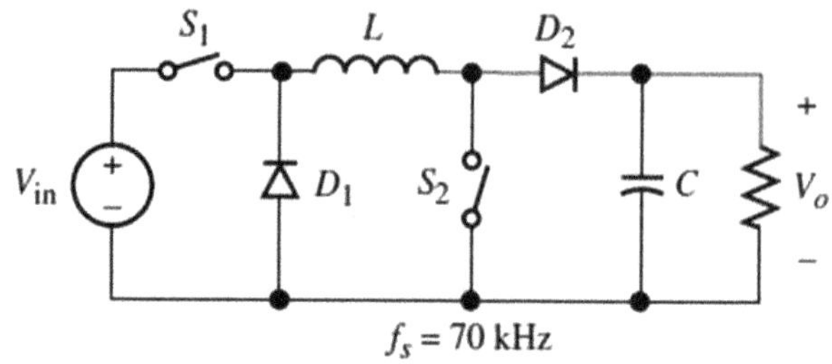

4.54 It is possible to develop basic switch-mode dc converter by using constant input and output current sources. Such converters are called *current inverters*. Figure P4.54 shows the three basic current converters: buck, boost, and buck-boost. Assume *L*/*R* is very large when compared to the switching frequency so that I_o is assumed constant. Show that the current gain for these topologies are given by, respectively:

(a) $\frac{I_o}{I_{in}} = (1 - D)$
(b) $\frac{I_o}{I_{in}} = \frac{1}{D}$
(c) $\frac{I_o}{I_{in}} = \frac{1-D}{D}$

Fig. P4.54

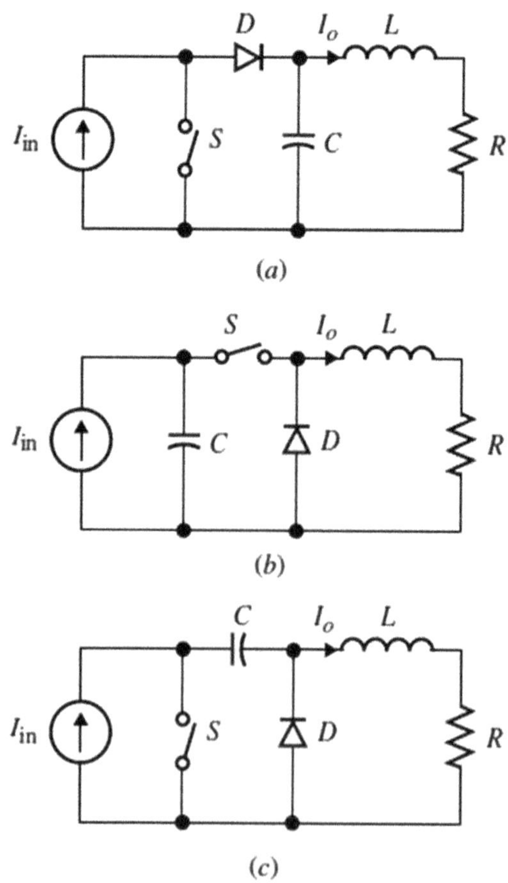

4.55 Show that the critical capacitor value that will produce a continuous capacitor voltage for the buck converter in Fig. P4.54 a is given by:

$$C_{\text{crit}} > \frac{D^2 T}{2R}$$

4.56 Consider the circuits shown in Fig. P4.56 with V_o and I_o are constants, i.e., $RC \gg T$ and $L/R \gg T$. The input current and voltage square waves are as shown. Derive the expressions for the inductor and capacitor ripples in terms of the circuit parameters.

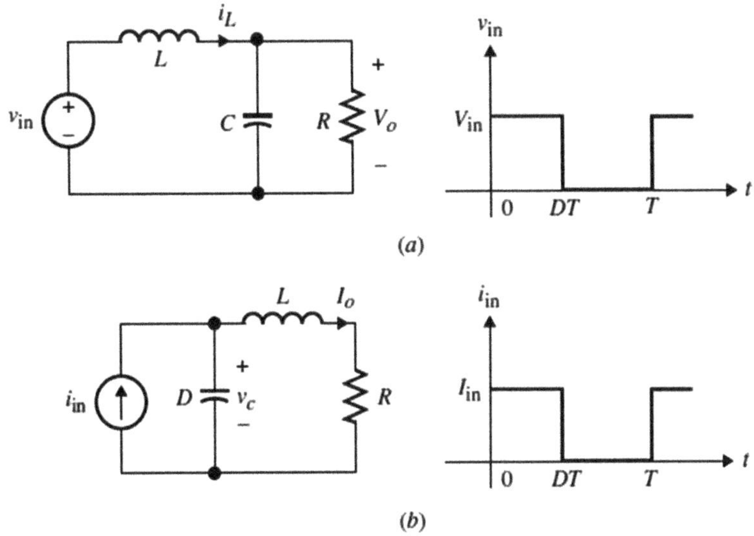

Fig. P4.56

4.57 Repeat Problem 4.56 for Fig. P4.57, where $1 < \beta < 2$

Fig. P4.57

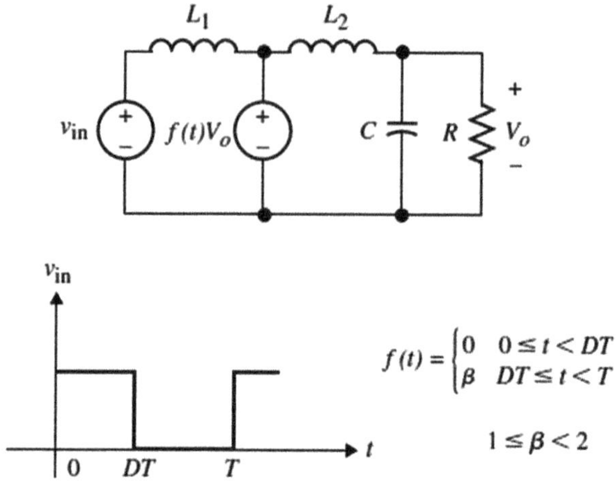

4.58 Derive the dc and the fundamental components of the Fourier series for the input current in the buck converter.

Additional General Problems

4.59 Derive the *rms* expression for the inductor and diode currents for the buck converter.

4.60 Derive the *rms* expression for the inductor and diode currents for the boost converter.

4.61 Derive the *rms* expression for the inductor and diode currents for the buck-boost converter.

4.62 Consider the bidirectional double-switch buck converter shown in Fig. P4.62. Assume S_1 and S_2 are turned on and off simultaneously at duty ratio D. Show that the voltage gain is given by:

$$\frac{V_o}{V_{in}} = 2D - 1$$

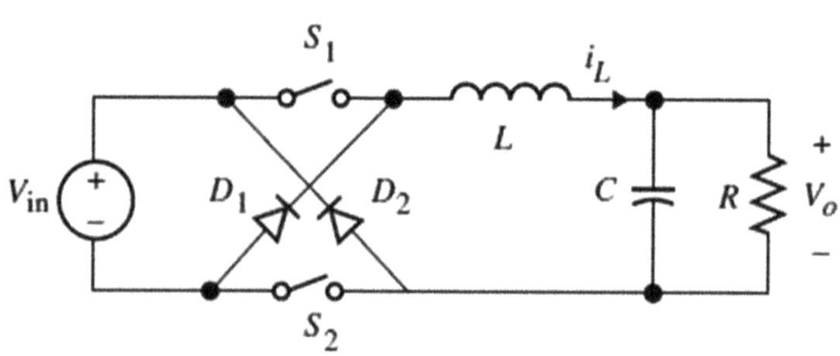

Fig. P4.62

4.63 Derive the ripple voltage expression for the bidirectional converter of Problem 4.62.

4.64 Design the three buck-boost converters needed to be used in the DC distributed power system in Fig. P4.64.

Fig. P4.64

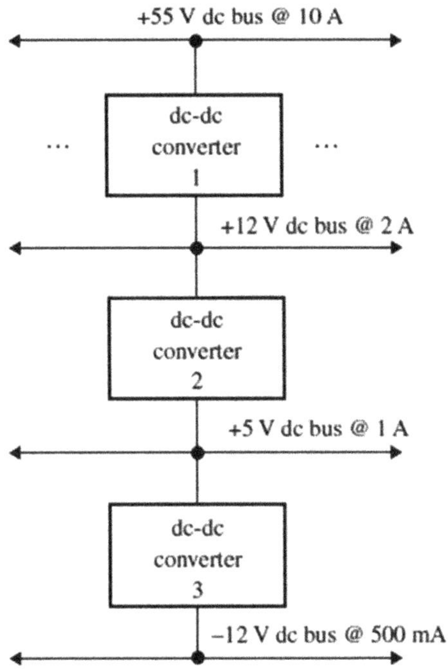

4.65 Show that the nth harmonic peak value for the diode voltage of the buck converter is given by:

$$V_{D,n} = \frac{V_{\text{in}}\sqrt{2}}{n\pi}\sin nD\pi$$

4.66 The converter shown in Fig. P4.66 is known as a non-inverting buck-boost converter. Derive the expression for V_o/V_{in}. Assume S_1 and S_1 are turned ON and OFF simultaneously.

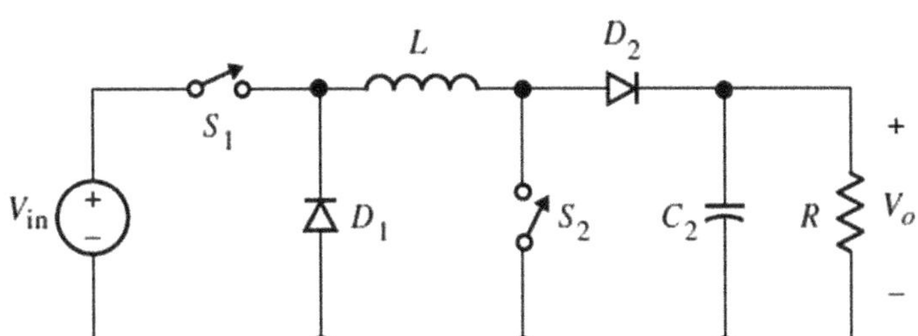

Fig. P4.66

D4.67 Design a switch-mode power supply that is capable of supplying +12 V output voltage with an output power between 25 W and 375 W from a 56 V dc source. Assume the dc input comes from a variable dc source supply that changes by ±25%. The converter should not produce more than ±2% output ripple, given the peak and *rms* rating values for the power devices. Let $f_s = 75$ kHz.

D4.68 Design a switch-mode converter that delivers 25 W output power to a regulated output voltage of +12V. Assume the dc source is obtained from a ac-dc SCR rectifier whose dc unregulated output varies between 8 V and 18 V. It is desired to limit the output voltage ripple to 1% and the switching frequency to 100 kHz.

D4.69 Figure P4.69 shows another way to represent a buck-boost converter using a single-pole double-throw switch. Assume C_1 and C_2 are large enough so that their voltages are constant. Derive the expression for V_o/V_{in} and I_o/I_{in}. Assume in steady state the switch is in position 1 for the interval DT and in position 2 for the interval $(1 - D)T$.

Fig. P4.69

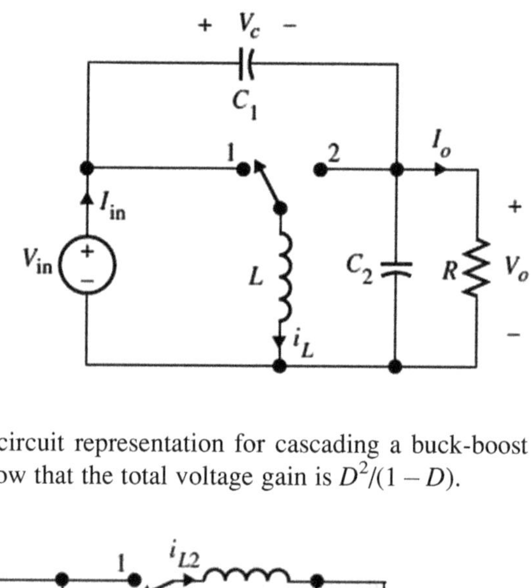

4.70 Figure P4.70 shows the circuit representation for cascading a buck-boost and a buck converter. Show that the total voltage gain is $D^2/(1 - D)$.

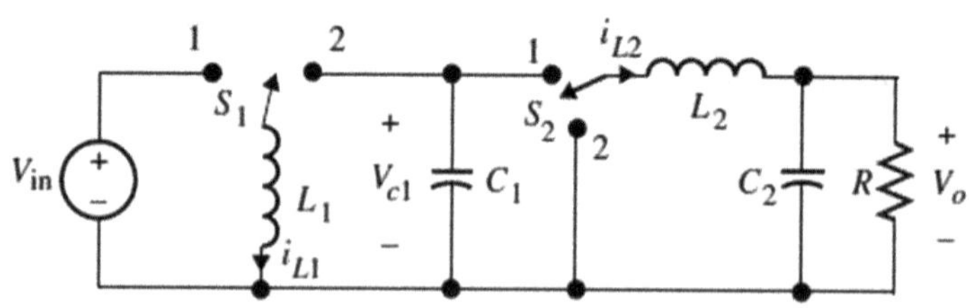

Fig. P4.70

4.71 Determine the voltage gain for the circuit shown in Fig. P4.71. What converters are put in cascade to produce this topology?

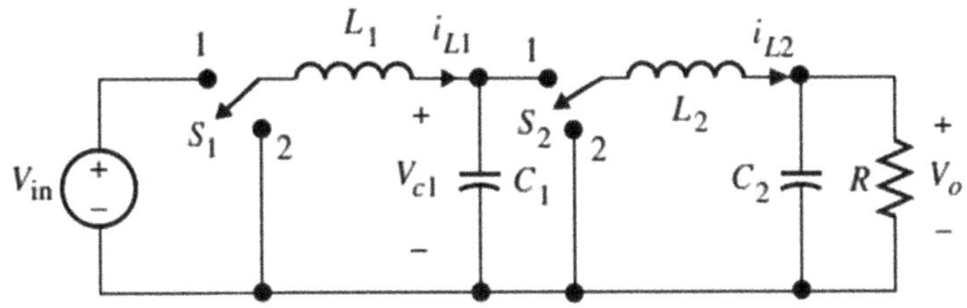

Fig. P4.71

4.72 Consider a boost converter with its dc source coming form an unregulated 124 V dc input that varies by ±18% and delivers power to the load between 75 W and 225 W. It is desired to regulate the output voltage to 72 V. Assume the power switch has an on resistance of 0.5 Ω and the power diode has a 0.7 V voltage drop with a 0.25 Ω forward resistance. Also assume the inductor loss can be modeled by a discrete resistance of a 0.1 Ω. Determine the range of the duty cycle D needed to be employed to maintain a constant output voltage. Assume ccm of operation.

4.73 Figure P4.73 shows a two-quadrant boost converter topology that produces bipolar output voltage. Derive the expression for V_o/V_{in} assuming Q_1 and Q_2 are synchronized switches during turn on for DT and turn off for $(1 - D)T$.

Fig. P4.73

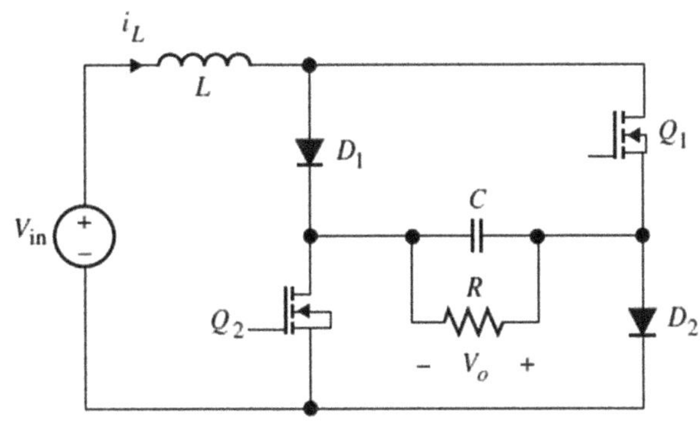

4.74 The circuit shown in Fig. P4.74 is known as an inverse SEPIC converter. Derive the voltage gain equation V_o/V_{in}. The switch is in position 1 during the DT interval and in position 2 during the $(1 - D)$T interval.

Fig. P4.74

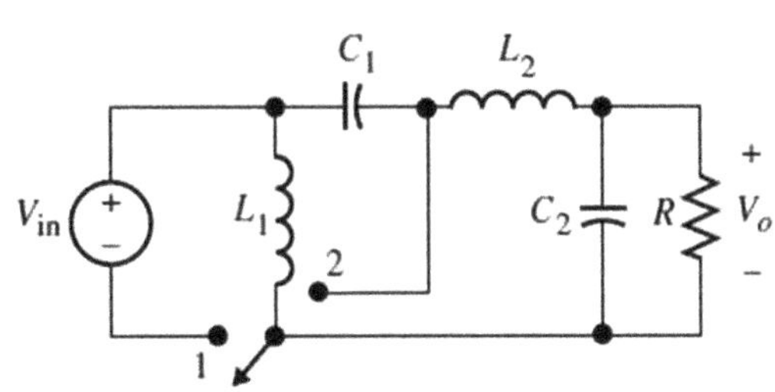

Inverse SEPIC

Chapter 5
Isolated Switch-Mode DC-DC Converters

5.1 Introduction

The discussion in Chap. 4 showed that the four basic converter topologies have their output conversions determined by the duty ratio, D, and that they consist of a single input and a single output with a common reference point. For applications in which the output voltage does not differ from the input voltage by a large factor, those topologies can be used. However, for many applications, the input voltage is normally derived from an off-line half- or full-wave rectifier, and the output voltage is normally a very small fraction of the rectified dc input voltage. As a result, transformers must be added between the power stage and the output for the purpose of voltage scaling. Unlike line-frequency transformers, these transformers are high frequency and much smaller in size and weight. In addition, transformers are used in switching mode converters for electrical isolation between the input and output; reduction of stresses in switching devices; and provision multi-output connections. With isolation transformers, the output voltage polarity reversal does not become a design restriction. Also in some application, system isolation may be required by the certain regulatory body. However, the benefits obtained by adding isolation transformers come with a price. The major drawbacks include high converter volume and weight, reduced efficiency, and added circuit complexity to limit the effect of leakage inductance and avoid core saturation.

In this chapter, we will discuss some widely used high-frequency switch-mode dc-dc converters: the flyback, forward, push-pull, half-bridge, and full-bridge converters. It will be shown that the flyback converter is based on the boost converter, and the forward converter is based on the buck converter. In analyzing these converters, we will use the transformer model by including the magnetizing inductance. In all the dc-dc converters discussed in Chap. 4, the role of the inductor is to store energy as it comes from the source in one portion of the switching cycle and release it to the load in the other portion of the cycle. As the converter reaches the steady state, the mechanism of energy storage and release reaches equilibrium,

I. Batarseh, A. Harb, *Power Electronics*,
https://doi.org/10.1007/978-3-319-68366-9_5

and the net energy stored in the inductor over one-switching cycle is zero. In other words, energy in the inductor does not build up from one cycle to the next. Otherwise, if energy buildup in the inductor were allowed, soon the inductor would reach saturation. The issue of saturation is important not only for the inductor but also for the transformer. Unlike converter topologies presented in Chap. 4, the topologies in this chapter require more than one switch, since the transformer must be magnetized and demagnetized repeatedly within a switching cycle, to avoid saturation. The conversion can be done unidirectionally, in which the magnetization current is positive, or bidirectionally, in which the magnetization current is positive and negative. Finally, we consider only the continuous conduction mode of operation for the isolated converters, since their dcm analysis is very similar to the dcm analysis presented in the previous chapter.

5.2 Transformer Circuit Configurations

5.2.1 Transformer Model

An ideal transformer has no leakage inductances, an infinite magnetizing inductance, no copper and core losses, the ability to pass all signal frequencies without any power loss, and the ability to provide any level of current and voltage ratio transformation. Figure 5.1a shows an ideal transformer with its current and voltage relations. Figure 5.1b shows the equivalent transformer circuit including only the magnetic inductance, L_m, reflected in the primary side.

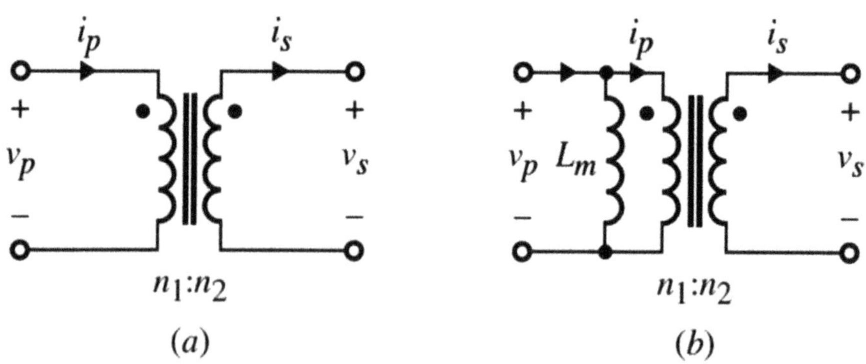

$$\frac{v_p}{v_s} = \frac{n_1}{n_2}, \quad \frac{i_p}{i_s} = \frac{n_2}{n_1}$$

Fig. 5.1 (**a**) Ideal transformer model, (**b**) transformer equivalent circuit including the magnetizing inductance

5.2.2 Circuit Configurations

Throughout this chapter, as far as the primary side is concerned, two configurations of transformers are used: non-center-tap and center-tap transformers. The non-center-tap configuration has three possible connections known as single-ended, half-bridge, and full-bridge, as shown in Fig. 5.2a–c, respectively. The center-tap configuration is known as *push-pull* and is shown in Fig. 5.3.

As far as the transformer's secondary side is concerned, two configurations of transformer-rectifier connections are normally used: center-tap full-wave and bridge full-wave, as shown in Fig. 5.4a and b, respectively. The center-tap arrangement results in one diode forward voltage drop when compared to the two diode voltage drops for the bridge configuration.

These configurations are the most popular ones in today's switch-mode power supply design. Since we have a dc source voltage across S_1 and S_2 of Figs. 5.2b and 5.3, the two switches are not allowed to close simultaneously. Hence, in a normal operation, S_1 and S_2 in these two configurations switch alternatively, each

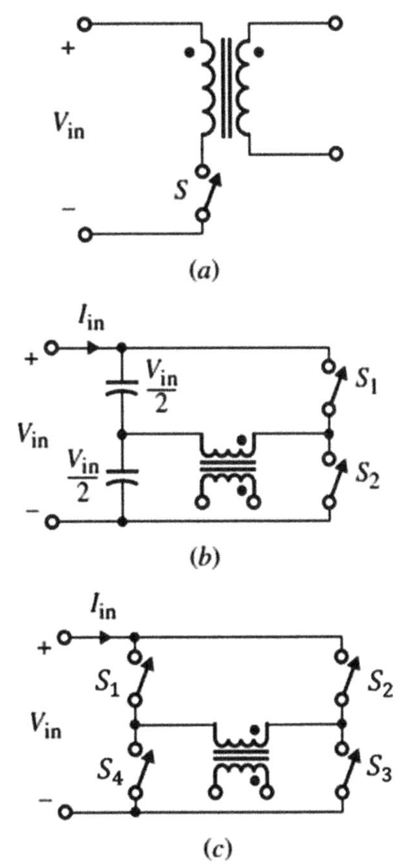

Fig. 5.2 Transformer configurations: (**a**) single-ended, (**b**) half-bridge, (**c**) full-bridge

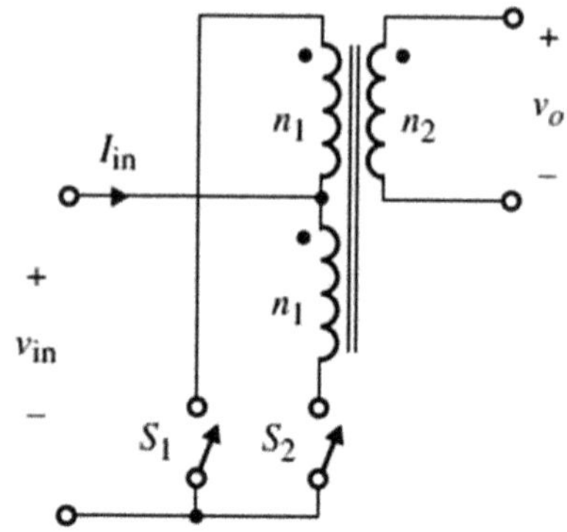

Fig. 5.3 Push-pull transformer configuration

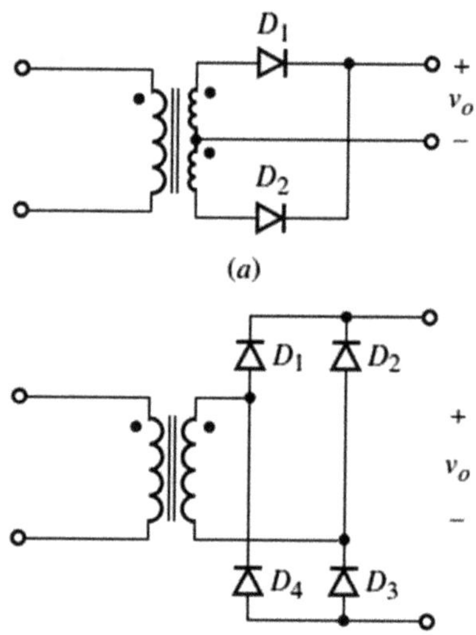

Fig. 5.4 Transformer configuration in the secondary (load) side: (**a**) center-tap full-wave, (**b**) full-wave bridge

having the same duty cycle. This will prevent transformer core saturation. In the full-wave bridge configuration, S_1 and S_3 are closed simultaneously in the first half cycle, and in the second half cycle, switches S_2 and S_4 are closed simultaneously.

The difference between the two-switch transformer connections of Figs. 5.2b and 5.3 is that the switches in the earlier connection must be able to sustain a maximum of V_{in} voltage when compared to $2V_{in}$ for the latter connection. However, in the push-pull configuration, each switch carries the average input current, I_{in}, whereas, in the half-bridge configuration, each switch must carry twice the average input current, I_{in}. This is why the push-pull configuration is chosen when the source voltage is low. In the full-bridge case, each switch must be able to block V_{in} and allow a peak current equal to I_{in}. This makes it attractive to high-power applications. The only disadvantage of the full-bridge configuration is that it requires two floating driving circuits.

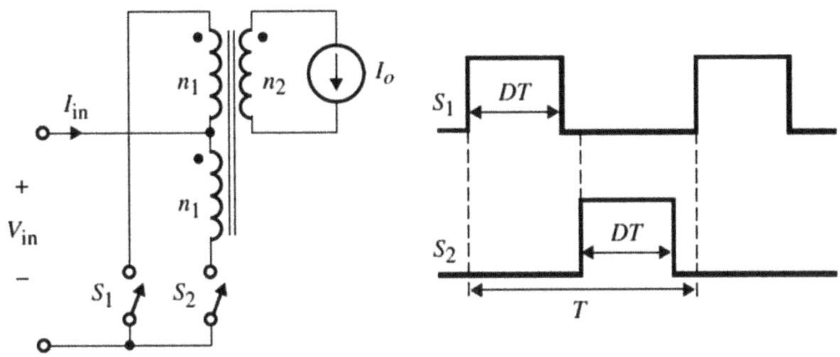

Fig. 5.5 Circuit for Example 5.1

Example 5.1
Consider the push-pull converter with a current source load as shown in Fig. 5.5. If S_1 and S_2 are turned on and off according to the waveforms shown, determine the current and voltage stresses for S_1 and S_2 in terms of the average input current, I_{in}, and voltage V_{in}.

Solution
Consider the first half cycle when S_1 is closed. The input voltage is applied to the upper primary winding of the transformer. An equal amount of voltage is also induced on the lower primary winding. The polarity of this voltage is in the same direction of the input voltage. As a result, S_2 is subjected to twice of the input voltage, i.e., $V_{stress} = 2V_{in}$. S_1 also subjects the same voltage stress when S_2 is turned on.

Neglecting the transformer's magnetizing current, the maximum switch current can be obtained as $I_{max} = I_{stress} > I_{in,ave}/2D = n_2 I_o / n_1 D$ $(0 < D < 0.5)$. Depending on the value of D, the current stress on each switch can be much larger than the average input current.

The comparison between primary-switch connections for the isolated switching mode converters are summarized in Table 5.1.

5.3 Buck-Derived Isolated Converters

Depending on the location at which the isolation transformer is inserted in the power stage of the basic buck converter, and on the type of isolation transformer configuration, several buck-derived converter topologies can be derived. These topologies vary in complexity and features. In this section we will consider some of the more popular buck-derived converters: single-ended forward converter, half- and full-bridge, push-pull, and the Weinberg converters. Other topologies are given as exercises for students at the end of this chapter.

Table 5.1 Comparison between transformer configurations

Features of transformer configuration	Maximum voltage stress across each switch	Maximum current stress through each switch	Core excitation and reset mechanism	Power level applications
Single-ended: forward	$\geq 2V_{in}$[a] $= (1 + n_3/n_1)V_{in}$	$\geq 2I_{in}$[a] $\geq I_{in}/D$	Unidirectional complex reset circuit	Low power <500 W
Flyback	$\geq 2V_{in} = V_{in}(1-D)$	$2I_{in}$ $\geq I_{in}/D$	Unidirectional and core reset is not necessary	<200 W
Half-bridge	V_{in}	$2I_{in}$ $\geq I_{in}/D$[b]	Bidirectional and core reset is not necessary	High input voltage medium power with low source current <800 W
Full-bridge	V_{in}	$2I_{in}$ $\geq I_{in}/D$[b]	Bidirectional and core reset is simple	High power >800 W
Push-pull	$2V_{in}$	$2I_{in}$ $\geq I_{in}/D$[b]	Bidirectional and core reset is simple	Medium power with low source voltage

[a] I_{in} and V_{in} are average values
[b] $D < 0.5$

AC Transformer Insertion

Let us consider isolating the input and output voltages of the buck converter by inserting an *ac* transformer. Figure 5.6a shows four places where physical transformer insertion is possible. Locations AA′ and DD′ are not possible since the transformer's primary and secondary voltages would have been dc V_{in} and V_o, respectively. The obvious location is either BB′ or CC′ since the voltage at either locations is an ac square wave voltage.

It is clear to see that the isolated buck-derived converter of Fig. 5.6b will not function properly in steady state. This is because the average voltage across the transformer is positive (nonzero) which results in continuous increase in the magnetization current that will eventually lead to transformer saturation. Let us further investigate the problem of transformer magnetizing inductance saturation.

When the switch is turned on at $t = 0$, the diode becomes reverse biased as shown in Fig. 5.7a. The magnetizing and output inductor currents are given by:

$$i_{L_m} = \frac{V_{in}}{L_m} t + I_{L_m}(0)$$

$$i_L(t) = \frac{n_1}{n_2} i_p = \frac{\frac{n_2}{n_1} V_{in} - V_o}{L} t + I_L(0)$$

where $I_L(0)$ and $I_{L_m}(0)$ are the initial output and magnetizing inductor currents, respectively.

Fig. 5.6 (**a**) Buck converter and (**b**) isolated buck converter

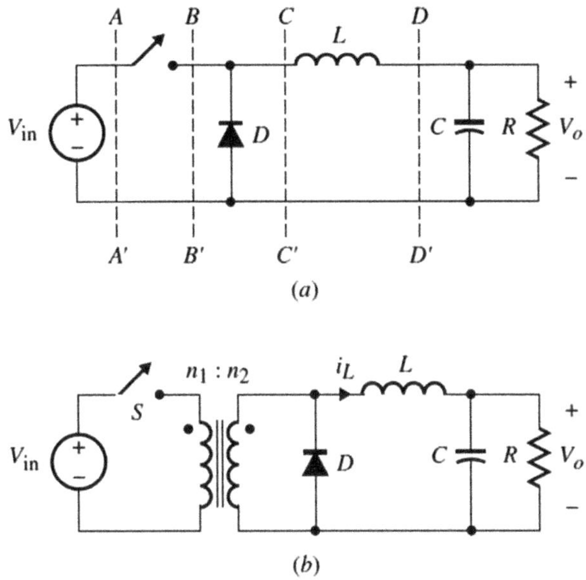

Fig. 5.7 Equivalent circuit model (**a**) switch ON, (**b**) switch OFF, (**c**) magnetizing inductor current

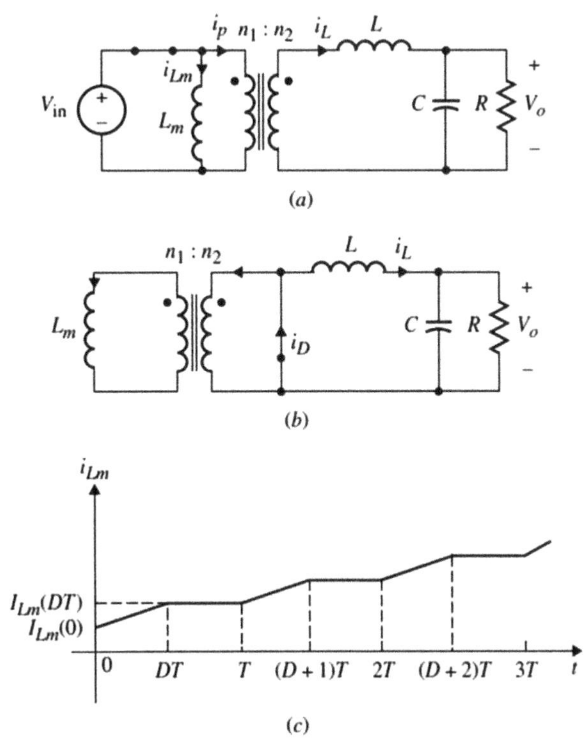

At $t = DT$, S is switched off (Fig. 5.7a), and D turns on. Hence, the magnetizing current stays constant at $I_{L_m}(DT)$ which is given by:

$$I_{L_m}(DT) = \frac{DT}{L_m} V_{in} + I_{L_m}(0)$$

The diode current is given by:

$$i_D(t) = i_L(t) + \frac{n_1}{n_2} I_{L_m}(t)$$

The inductor current $i_L(t)$ for $t \geq DT$ is given by:

$$i_L(t) = -\frac{V_o}{L}(t - DT) + I_L(DT)$$

where

$$i_L(DT) = \frac{\frac{n_2}{n_1} V_{in} - V_o}{L} DT + I_L(0)$$

At $t = T$, when the switch is turned on again, it causes i_{L_m} to increase starting at a higher initial value, $I_{L_m}(DT)$ as shown for three cycles in Fig. 5.7c.

One way to avoid the problem of transformer saturation is to add another diode as shown in Fig. 5.8 to produce a negative voltage across the primary side when the switch is switched off. However, this converter, as shown, provides a path for the magnetizing current to reset to zero when the switch is open. As a result, a third winding, known as "catch" winding, is added to allow i_{L_m} to discharge to zero, preventing magnetic flux buildup from one cycle to the next. This converter is known as a single-ended forward converter, to be discussed in the next section.

The isolation of the buck-boost converter can be also carried out in a similar manner. Figure 5.9a shows the buck-boost converter with a transformer insertion as shown in Fig. 5.9b. Depending on the transformer windings, negative or positive output polarities can be obtained as shown in Fig. 5.9b and c, respectively. Figure 5.9c is known as a single-ended flyback converter which will be discussed later in the chapter.

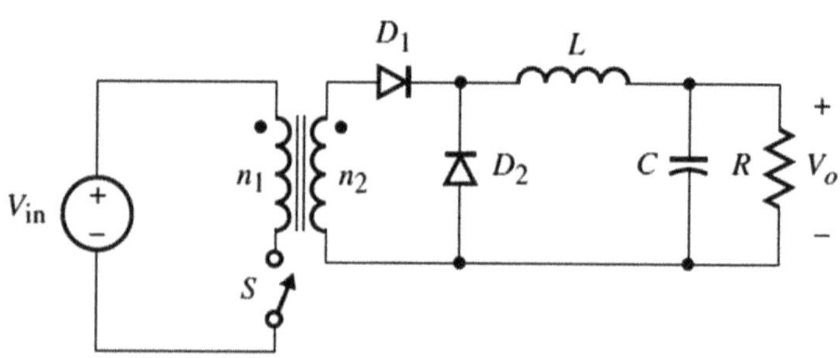

Fig. 5.8 Isolated buck-derived converter, known as single-ended forward converter

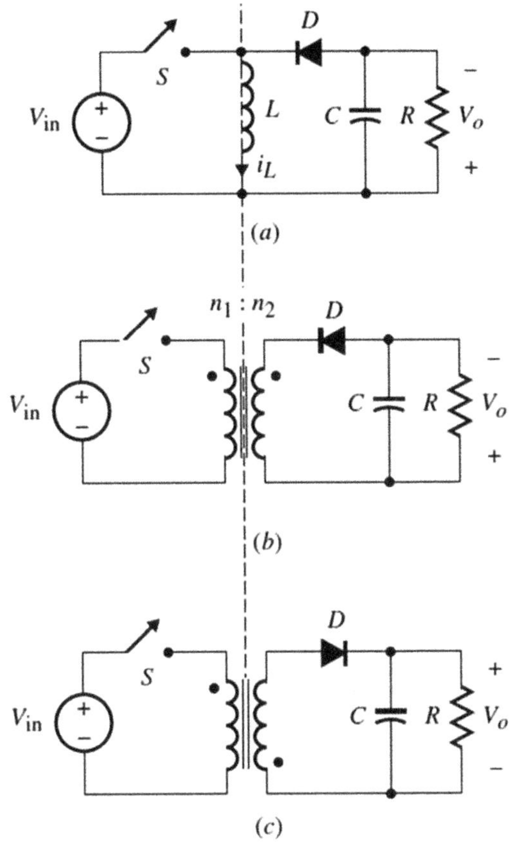

Fig. 5.9 (**a**) Buck-boost converter, (**b**) isolated converter with negative output voltage, (**c**) positive output voltage

5.3.1 Single-Ended Forward Converter

In this section we will analyze Fig. 5.8, which shows the simplest isolated dc-dc converter utilizing one switch and two diodes. As stated earlier, this circuit is commonly referred to as a forward converter. It can be shown that the design of Fig. 5.8 will not work properly since its magnetizing current will not be allowed to reset to zero, causing the magnetizing current to continuously increase linearly until it finally saturates the core. A more practical forward converter must include a transformer core resetting circuit as shown in Fig. 5.10a. The additional winding n_r is known as tertiary winding. Figure 5.10b and c shows two alternative ways to draw Fig. 5.10a.

- To allow for a zero average voltage across the transformer primary winding, the maximum duty cycle is 50%. However, if the winding ratio $n_r/n_1 < 1$, then it is possible to have a duty cycle that exceeds 50%. This will result in a voltage stress across the switch that exceeds $2V_{in}$.

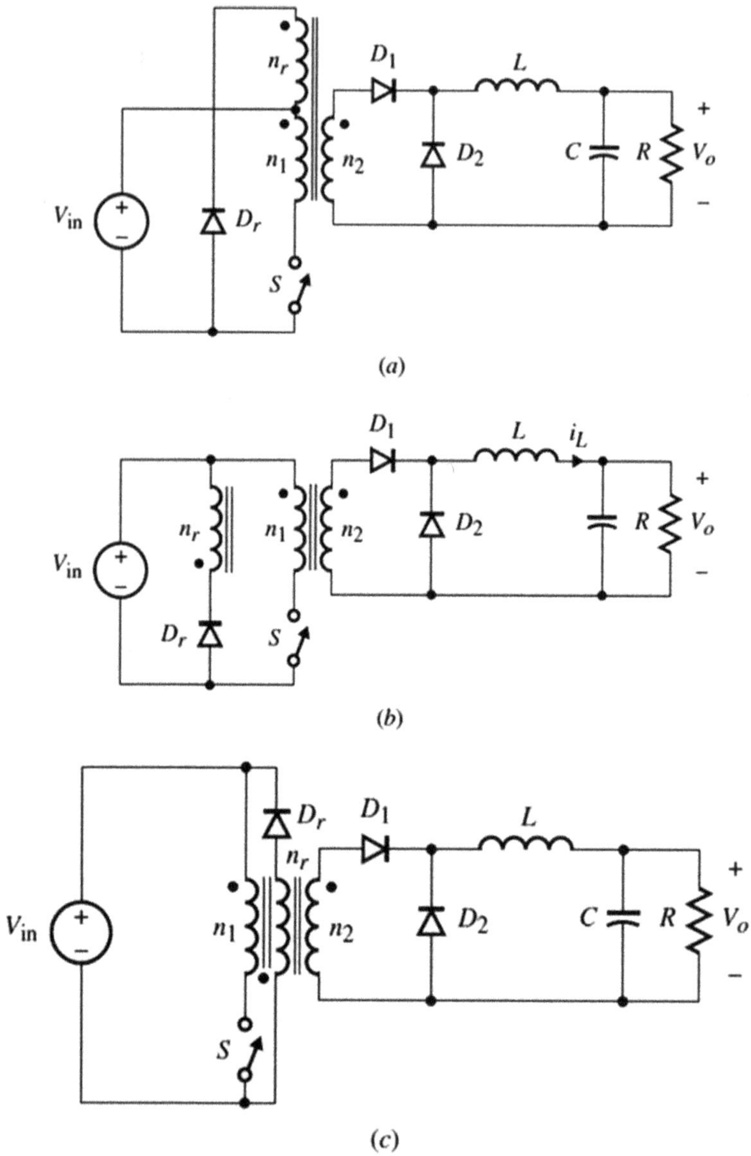

Fig. 5.10 Three ways to draw a single-ended forward converter with a core resetting circuit

- For illustration purposes, we will analyze the converters of Figs. 5.8 and 5.10a.
- If an ideal transformer is assumed in Fig. 5.8, then the steady state is quite simple. Assume the switch is turned on for the period DT and off for the period $(1 - D)T$, resulting in the two modes of operation shown in Fig. 5.11a and b, respectively.

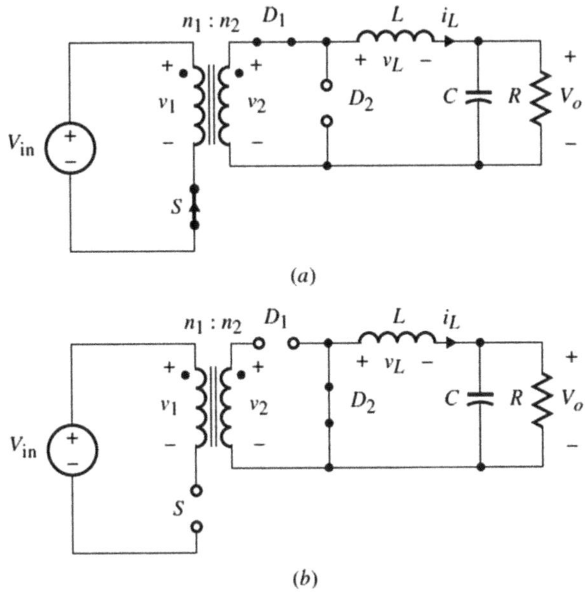

Fig. 5.11 Modes of operation: (**a**) Mode 1, S is ON; (**b**) Mode 2, S is OFF

When the switch is turned on initially at $t = 0$, the initial inductor current is $I_L(0)$, and v_L is given by:

$$v_L = \frac{n_2}{n_1}V_{in} - V_o$$
$$= L\frac{di_L}{dt}$$

The inductor current is given by:

$$i_L(t) = \frac{\frac{n_2}{n_1}V_{in} - V_o}{L}t + I_L(0) \qquad 0 \le t < DT$$

At $t = DT$, the switch turns OFF forcing D_1 to reverse bias and D_2 to become forward biased, resulting in the inductor current given by:

$$i_L(t) = -\frac{V_o}{L}(t - DT) + I_L(DT)$$

where $I_L(DT)$ is the inductor current at $t = DT$ when the switch is turned OFF. From the above equations, we obtain the following voltage gain relation:

$$\frac{V_o}{V_{in}} = \frac{n_2}{n_1}D$$

Sketch of the waveforms for v_1, v_2, v_L, i_L, i_{D_1}, and i_{D_2} is shown in Fig. 5.12.

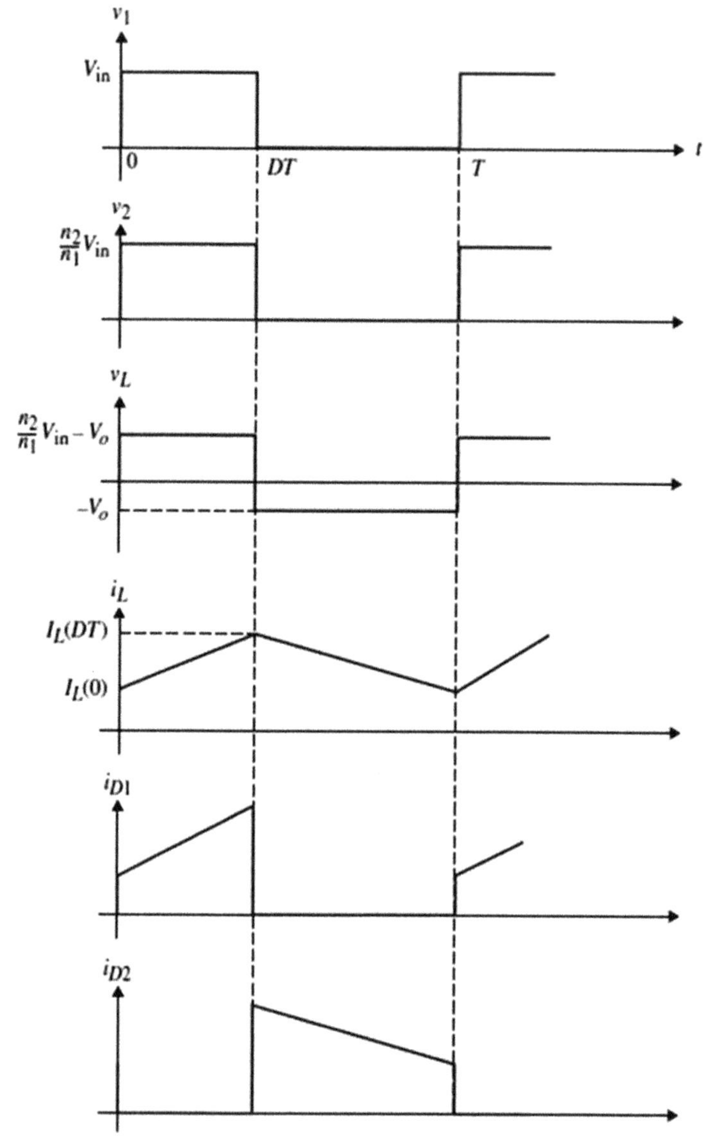

Fig. 5.12 Current and voltage waveforms for the forward converter

For a proper operation that will allow the inductor current to reach steady state, the relation $(n_2/n_1)V_{in} > V_0$ must hold, which provides a step-down operation. The capacitor voltage ripple is similar to the non-isolated buck converter.

Next we carry out the analysis by assuming the transformer has a finite magnetizing inductance, L_m, as shown in the equivalent circuit given in Fig. 5.13. With careful investigation of the above circuit, it is clear that the magnetizing current

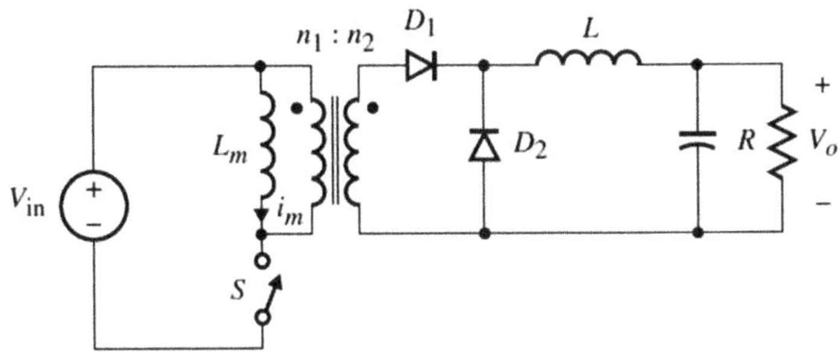

Fig. 5.13 Single-ended forward converter with the magnetizing inductance

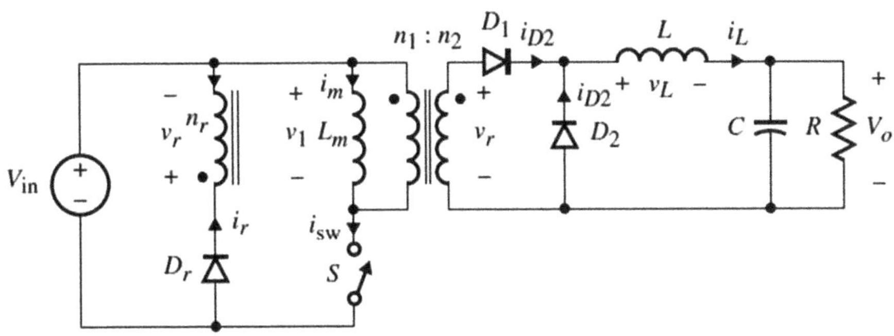

Fig. 5.14 Forward converter with core resetting circuit $n_r - D_r$ including the magnetizing inductance

$i_m(t)$ has no place to discharge its value when the switch is OFF. This will cause the converter to fail. As a result, a core resetting mechanism mentioned above must be used. Figure 5.14 shows the equivalent circuit for the forward converter by including the core resetting circuit given in Fig. 5.10b.

The operation of this converter can be easily explained by assuming that before the switch is turned ON again in a new cycle, the magnetizing current, i_m, has reached zero, i.e., the transformer core is being reset. The energy is delivered to the load during the period the switch is ON, and the core resetting takes place during the OFF time. It will be shown that $D \le 50\,\%$ for $n_r \ge n_1$ to allow time for the magnetic core flux to reset during the OFF switch time. In steady state, this converter has three modes of operation discussed as follows:

The first mode starts when S is turned ON at $t = 0$, casing the voltage across the primary equals to V_{in}. This will force D_1 to turn *ON* and D_2 to reverse bias. Since the reset winding has the opposite polarity of the primary, the diode D_r becomes reverse biased as shown in Fig. 5.15a.

From Fig. 5.15a, the following voltage equations are obtained:

$$v_L = v_2 - V_o, \quad v_1 = V_{in}, v_2 = \frac{n_2}{n_1} v_1, \quad v_r = \frac{n_r}{n_1} V_{in}, \text{ and } v_{Dr} = \left(1 + \frac{n_r}{n_1}\right) V_{in} \quad (5.1)$$

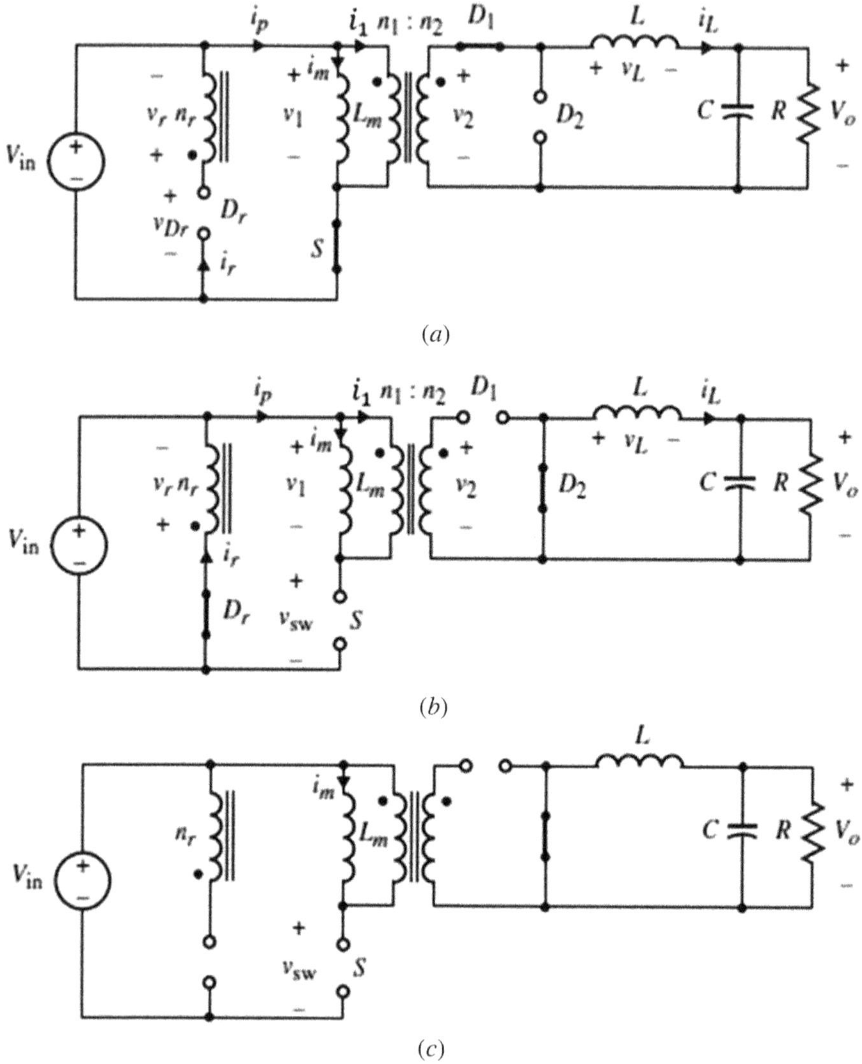

Fig. 5.15 Modes of operation: (**a**) Mode 1, $0 < t < DT$; (**b**) Mode 2, $DT < t < D_1T$; (**c**) Mode 3, $D_1T < t < T$

The current equations are given by:

$$L_{\mathrm{m}} \frac{di_{\mathrm{m}}}{dt} = V_{in} \tag{5.2a}$$

$$L \frac{di_L}{dt} = \frac{n_2}{n_1} V_{\mathrm{in}} - V_o \tag{5.2b}$$

$$i_r = 0 \tag{5.2c}$$
$$i_1 = \frac{n_2}{n_1} i_L \tag{5.2d}$$
$$i_p = i_m + i_1 \tag{5.2e}$$

From Eqs. (5.2a) and (5.2b), the following relations are obtained:

$$i_m(t) = \frac{V_{in}}{L_m} t \tag{5.3}$$
$$i_L(t) = \frac{\frac{n_1}{n_2} V_{in} - V_o}{L} t + I_L(0) \tag{5.4}$$

where $I_L(0)$ is the initial output inductor current and $I_m(0) = 0$ since the core has been reset prior to turning ON the switch. Since $L_m \gg L$, the slope of the magnetizing current is much smaller than the slope of i_L as shown in Fig. 5.13. At $t = DT$, the switch is turned OFF, and the circuit enters Mode 2 as shown in Fig. 5.15b. At the instance S opens, the transformer primary current, i_p, becomes zero, turning OFF D_1 and forcing i_L to go through D_2. At this point, i_m now is forced to flow in the n_1 winding, which forces D_r to turn ON to carry the reflected current through n_r.

The voltage equations in this mode are given by:

$$\begin{aligned} &v_L = -V_o, v_r = -V_{in}, v_1 = -\frac{n_1}{n_2} v_1, v_r = \frac{n_2}{n_1} V_{in}, \\ &v_{Dr} = 0, \text{ and } v_{sw} = \left(1 + \frac{n_1}{n_r}\right) V_{in} \end{aligned} \tag{5.5}$$

and the current equations as follows:

$$L_m \frac{di_m}{dt} = -\frac{n_1}{n_r} V_{in} \tag{5.6a}$$
$$L \frac{di_L}{dt} = -V_o \tag{5.6b}$$
$$i_r = \frac{n_1}{n_r} i_1 \tag{5.6c}$$
$$i_m = -i_1 \tag{5.6d}$$

The waveforms are shown in Fig. 5.16. It is shown that at $t = D_1T$, the magnetizing inductor current becomes zero, since it represents the smaller portion of i_p. The magnetizing and inductor currents in this time interval are given by:

$$i_m(t) = -\frac{n_1}{n_r} \frac{V_{in}(t - DT)}{L_m} + I_m(DT) \tag{5.7}$$
$$i_L(t) = -\frac{V_o}{L}(t - DT) + I_L(DT) \tag{5.8}$$

where $I_m(DT) = (V_{in}/L_m)DT$. Setting Eq. (5.7) to zero at $t = D_1$, we obtain:

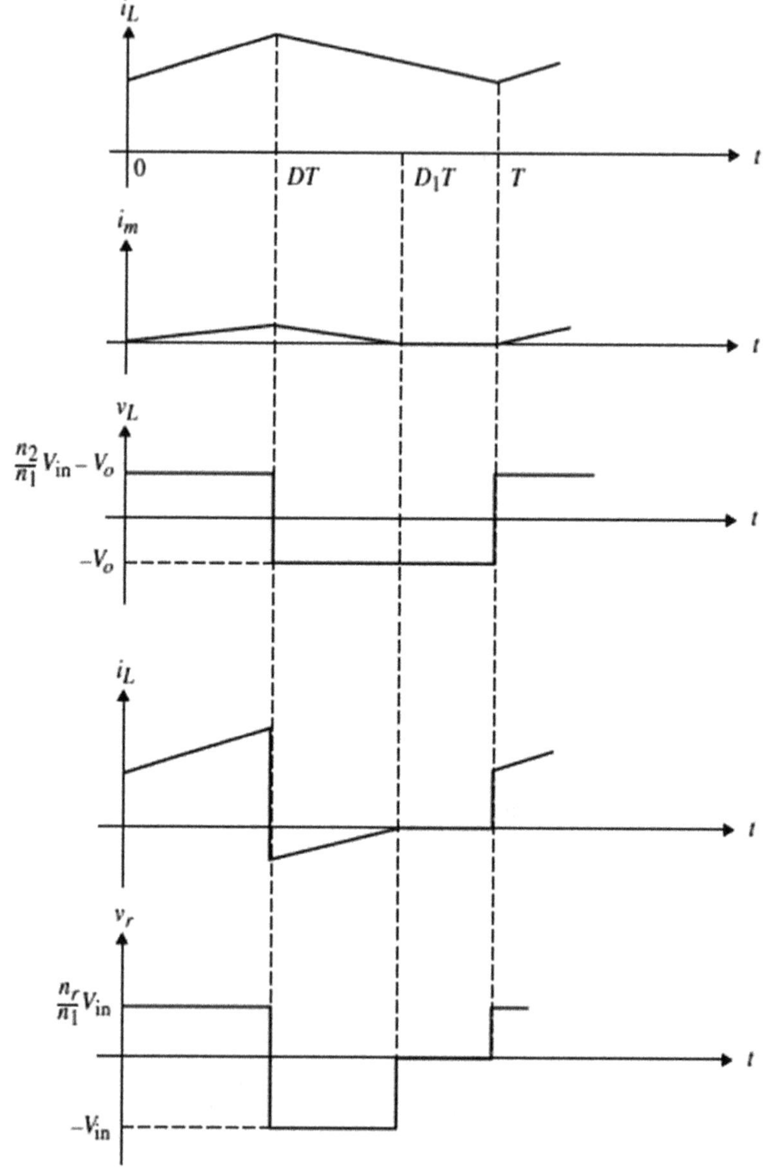

Fig. 5.16 Typical voltage and current waveforms

$$D_1 = \left(1 + \frac{n_r}{n_1}\right)D \tag{5.9}$$

At $t = D_1T$, $i_m = 0$, causing D_r to become zero as shown in Fig. 5.12c.

The voltage and current equations are given by:

$v_L = -V_o$, $v_r = v_1 = v_2 = 0$, $v_{Dr} = V_{in}$, $v_{sw} = V_{in}$. $i_m = i_r = i_p = i_L = 0$, and $i_L(t)$s is the same as that given in Eq. (5.8).

For the core to be fully demagnetized, i_m must reach zero; therefore, D_1 must not be greater than 1. Hence:

$$D_1 \leq 1$$

Hence, from Eq. (5.9), we restrict D by the following relation:

$$D \leq \frac{1}{1 + \frac{n_r}{n_1}} = \frac{n_1}{n_1 + n_r} \tag{5.10}$$

The maximum duty cycle of 50% occurs when $n_r = n_1$.

Example 5.2

Consider the forward converter of Fig. 5.14 with an input voltage of 50 V, an output voltage of 35 V, and $n_1/n_2 = 1$ and $n_1/n_r = 0.25$. With a frequency of 35 kHz, an inductance of 180 μH, and a minimum inductor current of 1.1 A, calculate the duty cycle and the maximum inductor current in this converter.

Solution

The duty cycle is given in the following equation:

$$D = \frac{V_o}{V_{in}} = \frac{35\ V}{50\ V} = 0.7$$

$$T = \frac{1}{f} = \frac{1}{35\ \text{kHz}} = 2.86 \times 10^{-5}\ \text{s}$$

Calculating for the maximum inductor current:

$$\begin{aligned} i_{L,\max} &= \frac{V_{in} - V_o}{L} DT + i_{L,\min} \\ &= \frac{50\ V - 35\ V}{180\ \mu H} (0.7)\left(2.86 \times 10^{-5}\ \text{s}\right) + 1.10\ \text{A} \\ &= 2.77\ \text{A} \end{aligned}$$

Exercise 5.1

Figure E5.1 shows a two-switch forward converter.

(a) Show that the voltage gain expression is given by:

$$\frac{V_o}{V_{in}} = \frac{n_2}{n_1} = D$$

S_1 and S_2 are switched simultaneously.

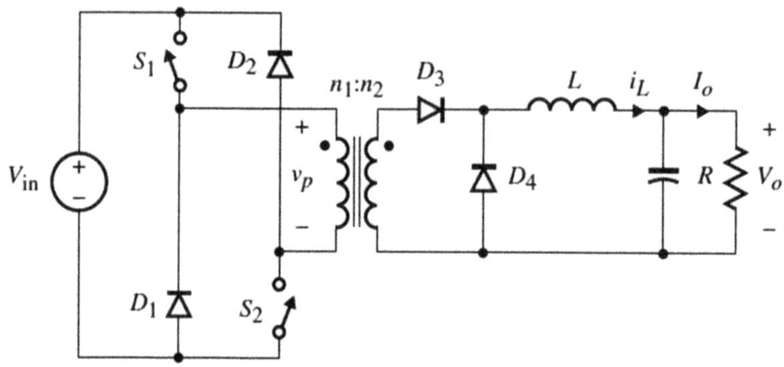

Fig. E5.1 Two-switch forward converter

(b) Find the critical inductor value (L_{crit}) to maintain a continuous conduction mode of operation. Assume $n_1 = n_2 = 1$, $V_{in} = 80$ V, $V_o = 45$ V at $I_o = 5$ A, $f_s = 50$ kHz.

(c) Compare the one-switch, single-ended topology with this two-switch topology.

Answer: 39.4 μH

Exercise 5.2
Find the rms current values for diodes D_3 and D_4 in Fig. E5.1 using the values of part (b) of Ex 5.1 with $L = 10\ L_{crit}$.

Answer: 3.76 A, 3.31 A

Exercise 5.3
Determine the efficiency of the converter described in Exercise 5.1 by assuming that the output rectifier diodes have a 1 V forward voltage drop and a 15 Ω conduction resistance.

Answer: 84.1%

Exercise 5.4
Determine the rms value of the inductor current i_L for Example 5.2.

Answer: 1.99 A

5.3.2 *Half-Bridge Converters*

Another way to avoid the transformer saturation problem is to use half- and full-bridge converter topologies to generate symmetrical *ac* waveforms at the primary side of the transformer. In this way the core flux is excited bidirectionally, resulting in a better utilization of the core, which in turn results in an increased power rating.

- Figure 5.17a shows the circuit topologies for the half-bridge converter with the center-tap output rectifier configuration. The filtering capacitors are relatively large and used as voltage dividers, resulting in a $V_{in}/2$ applied voltage across each primary winding. As stated before, in the half-bridge converter, S_1 and S_2 are switched on and off in complementary fashion but with equal conduction periods. Since the input voltage is not allowed to be shorten, S_1 and S_2 are normally designed such that there exists a dead time during which both switches are off. This results in a duty ratio less than 50%. Key current and voltage waveforms are shown in Fig. 5.17b.
- The voltage gain for the half-bridge converter is the same as the gain for the push-pull converter, to be discussed in a later section. The filtering capacitors C_1 and C_2 are used to divide the input voltage, so each has $V_{in}/2$. Unlike the push-pull converter, the maximum blocking voltage for each switch of the half-bridge converter is V_{in}, rather than $2V_{in}$.
- Finally, the switches S_1 and S_2 are implemented using bidirectional semiconductor devices (i.e., MOSFETs with antiparallel diodes) to provide conduction paths for the inductor leakage currents that exist due to the non-ideal transformers.

Exercise 5.5

(a) Show that the voltage gain for the half-bridge converter on Fig. 5.17a is given by:

$$\frac{V_o}{V_{in}} = \frac{n_2}{n_1} D$$

(b) Find the expression for the maximum current, $i_{D_1 max}$, in D_1.

Answer: $i_{D_1 max} = I_o + \frac{V_o(1-D)T_s}{2L}$

Exercise 5.6
Determine the average load voltage and current and the *rms* inductor current for a half-bridge converter of Fig. 5.17a with the following specifications:
$V_{in} = 135$ V, $V_o = 12$ V, $f_s = 100$ kHz, $R = 2\ \Omega$,
$n_2 = 13$, $n_1 = 39$, $L = 20\ \mu$H

Answer: 6 A, 6.01 A

5.3.3 *Full-Bridge Converter*

Figure 5.18a shows the buck-derived, full-bridge converter with a full-wave center-tap output transformer. The switch pairs S_1/S_3 and S_2/S_4 are switched complementary at a given duty ratio. This is the typical PWM control technique that has been

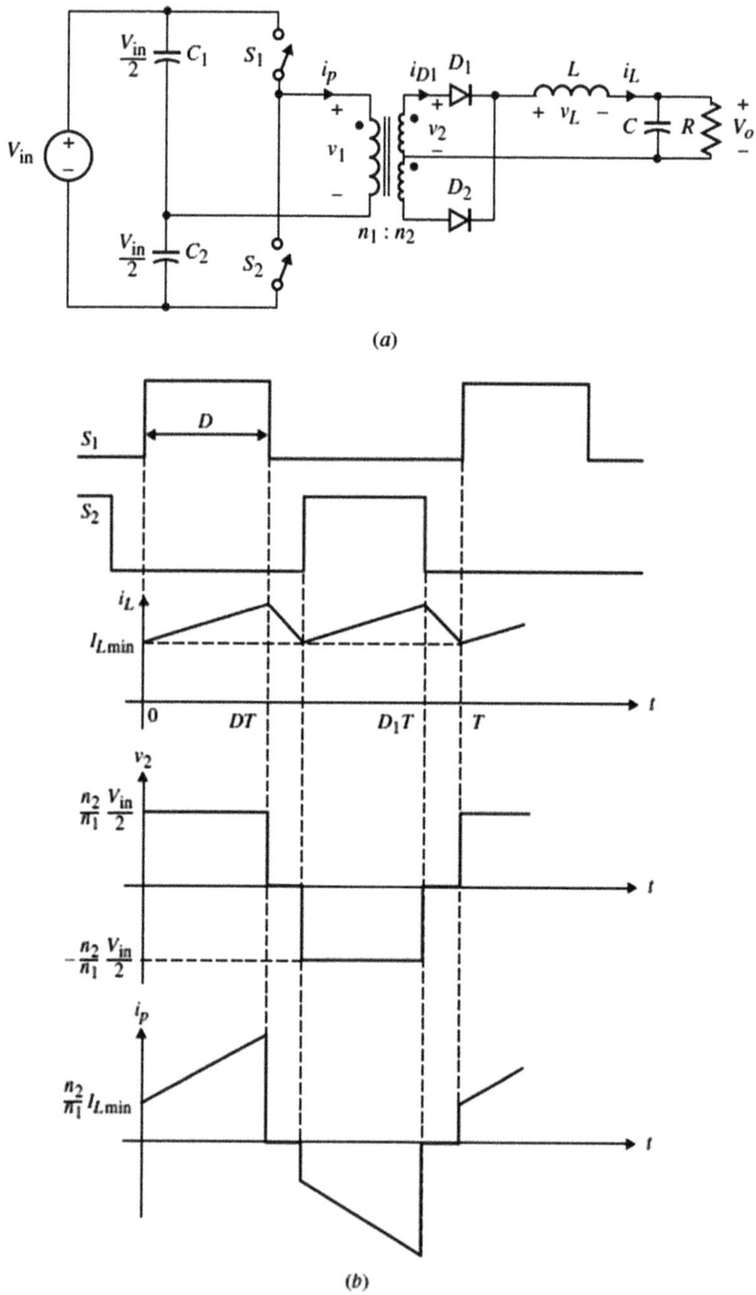

Fig. 5.17 (**a**) Half-bridge buck-derived converter (**b**) current and voltage waveforms

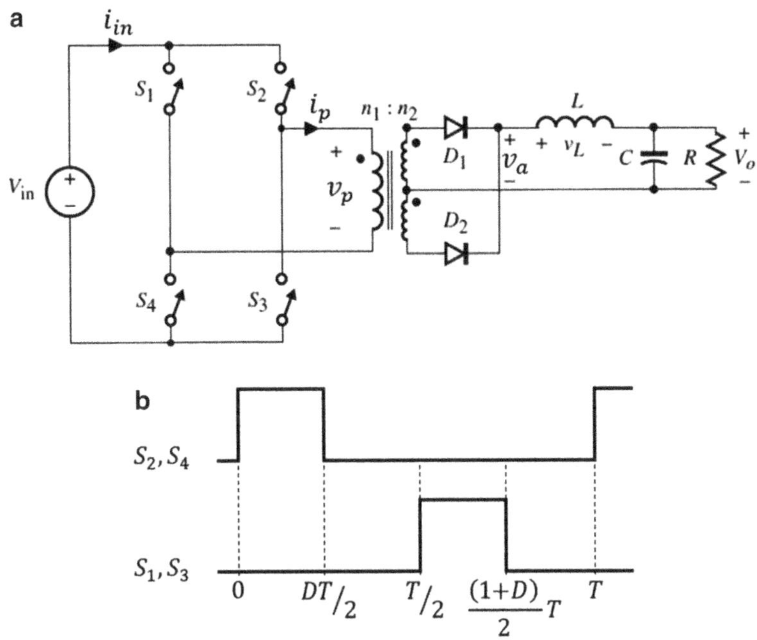

Fig. 5.18 (**a**) Full-bridge converter (**b**) Control signals for S_2/S_4 switches for the full-bridge converter of (a)

discussed in Chap. 4 and in this chapter. It can be shown that the output can be also regulated by controlling the phase shift between the switches.

Figure 5.18b shows the typical driving signals for S_2/S_4 under the PWM control method, with the correspondingly waveforms for $v_p(t)$, v_a, and $i_L(t)$. It can be easily shown that during the periods when either pair S_2/S_4 or S_1/S_3 is on, only one output diode of D_1 or D_2 will be on. Whereas, when all switches are off, both output diodes are on, resulting in $v_a = 0$, with each diode currying 0.5 $i_L(t)$.

It is straight forward to show that the voltage gain for the full-bridge convertor of Fig. 5.18a is given by:

$$\frac{V_o}{V_{in}} = 2\frac{n_2}{n_1}D \tag{5.11}$$

with $D \leq 0.5$, and the inductance current ripple is given by:

$$\Delta I_L = I_L(DT) - I_L(0) = \frac{\left(\frac{n_2}{n_1}V_{in} - V_o\right)DT}{L} \tag{5.12}$$

And since $I_o = \frac{1}{2}(I_L(DT) + I_L(0))$, hence, the value $I_L(DT)$ and $I_L(0)$ are given by:

$$I_L(DT) = I_o + \frac{\Delta I_L}{2} \tag{5.13}$$

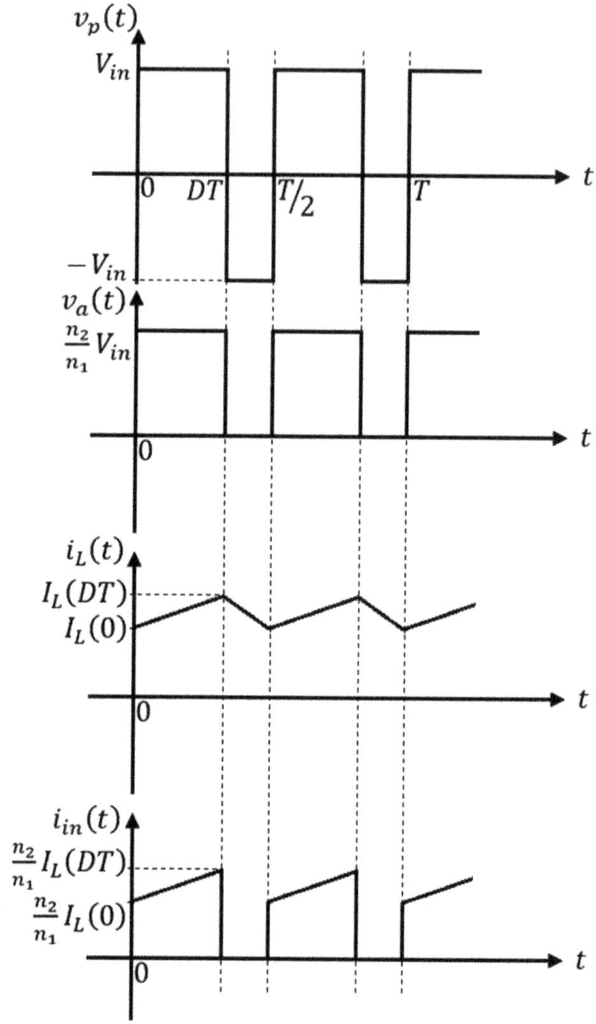

Fig. 5.19 Waveforms for the full-bridge converter of Fig. 5.18a based on drive signals of Fig. 5.18b

$$I_L(0) = I_o - \frac{\Delta I_L}{2} \tag{5.14}$$

The current and voltage waveforms for the full-bridge converter of Fig. 5.18a are shown in Fig. 5.19.

Finally, we must note that the push-pull, half-, and full-bridge converters can use the full-bridge rectifier at the output side. Unlike the half-bridge converter, the full-bridge converter is used in high input voltage applications, since the power switching devices are required to block only V_{in}.

Example 5.3
Design the full-bridge dc-dc converter with the center-tap output transformer as shown in Fig. 5.18a with the following specifications: $V_{in} = 480$ V, $V_o = 600$ V at $I_o = 10$ A, $f_s = 50$ kHz.

Solution
$R = V_o/I_o = 600/10 = 60\ \Omega$. If choosing $n_2 = 2n_1$, then:

$$D = \frac{1}{2}\frac{V_o}{V_{in}}\frac{n_1}{n_2} = \left(\frac{1}{2}\frac{600}{480}\frac{1}{2}\right) = 0.3125$$

since L_{crit} for the buck converter is expressed as:

$$L_{crit} = \left(\frac{1-D}{4}\right)RT$$

we have $L_{crit} = 112.5\ \mu$H. Choose $L > 10\ L_{crit} = 1.125$ mH. If output ripple is less than 1%, according to:

$$\frac{\Delta V_o}{V_o} = \frac{1-D}{8LC(2f)^2}$$

C can be determined as $C = 0.42\ \mu$F. We can choose $C = 10\ \mu$F.

Example 5.4
Determine and sketch all the steady-state waveforms for the full-bridge converter of Fig. 5.18a with the following specifications: $V_{in} = 12$ V, $V_o = 1.2$ V, $(n_1/n_2) = 5$, $P_o = 75$ W, $f_s = 180$KHz, and $L = 0.15\ \mu$H.

Solution
From Eq. (5.11), the duty cycle is given by:

$$D = \frac{1}{2}\frac{n_1}{n_2}\frac{V_o}{V_{in}} = \frac{1}{2}(5)\left(\frac{1.2}{12}\right) = 0.25$$

The on time for each pair of switches is $DT = 0.25 \times 5.56 = 1.39\ \mu$s.
The primary voltage peak is 12 V and the peak of v_a is $(12/5) = 2.4$ V.
The average output current $I_o = \frac{P_o}{V_o} = \frac{75\ \text{W}}{1.2\ \text{V}} = 62.5$, and the output inductor ripple is obtained from Eq. (5.12):

$$\Delta I_L = \frac{\left(\frac{12}{5} - 1.2\right)0.25 \times 5.56}{0.15} = 11.12\ \text{A}$$

The maximum and minimum inductor values are:

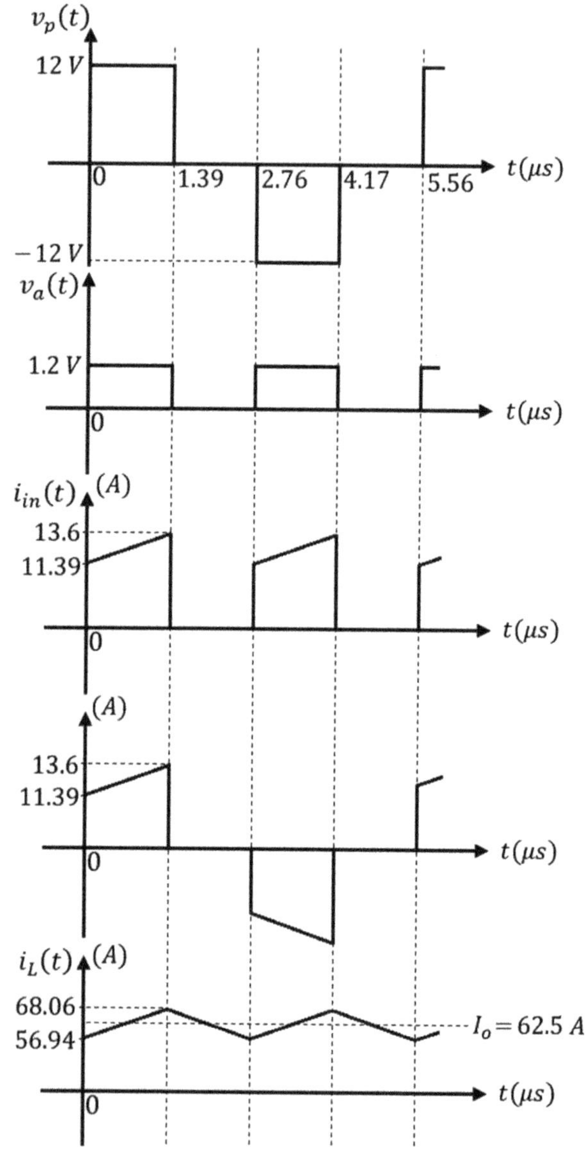

Fig. 5.20 Waveforms for Example 5.4

$$I_{\mathrm{L}}(DT) = I_{\mathrm{o}} + \frac{\Delta I_{\mathrm{L}}}{2} = 62.5 + \frac{11.2}{2} = 68.06\ \mathrm{A}$$

$$I_{\mathrm{L}}(0) = I_{\mathrm{o}} - \frac{\Delta I_{\mathrm{L}}}{2} = 62.5 - \frac{11.2}{2} = 56.94\ \mathrm{A}$$

Therefore, the resulted maximum and minimum input current are 13.6 A and 11.39 A, respectively (Fig. 5.20).

5.3.4 Push-Pull Converter

The circuit configuration for the push-pull converter is shown in Fig. 5.21a. The circuit uses two active switches. It uses the transformer for voltage scaling and electrical isolation, and the output inductor is used for energy storage. Hence, unlike the design of the transformer for the single-ended converter, where care must be taken in selecting the core material and geometry to design for proper magnetizing inductance, in the push-pull converter, the transformer is used as an ideal element. Since S_1 and S_2 share the current, the push-pull converter is used for higher-power applications compared to the single-ended converters. Figure 5.21b illustrates the switching waveforms for S_1 and S_2 with a dead time during which both switches are open.

During this dead time, the load current is carried by the two output diodes, D_1 and D_2. The maximum duty cycle for both switches is 0.5. As can be noticed, the switching frequency of the converter is twice the switching frequency of each switch acting alone. The converter's basic operation is straightforward and similar to the analysis of the half-bridge converter. When S_1 is on, the possible primary voltage causes D_1 to conduct and D_2 to turn off, resulting in the equivalent circuit shown in Fig. 5.22a.

The converter voltages are given by:

$$v_{S_1} = v_{S_2} = \frac{n_2}{n_1} V_{\text{in}} \tag{5.15}$$

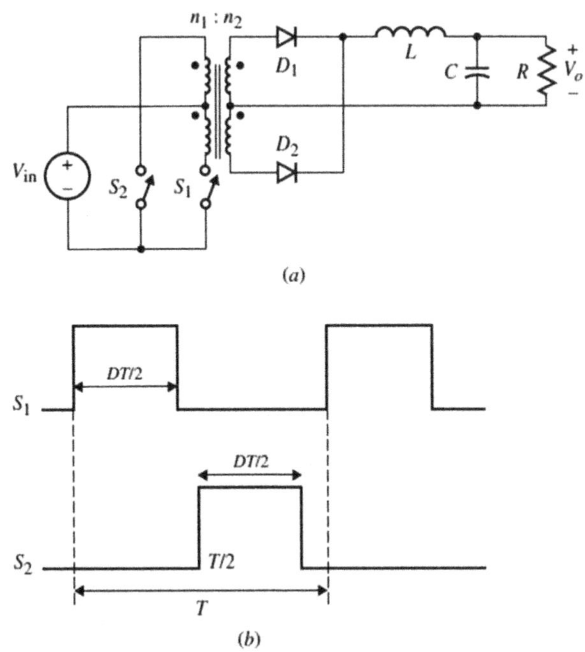

Fig. 5.21 (**a**) Push-pull converter, (**b**) switching waveforms for S_1 and S_2

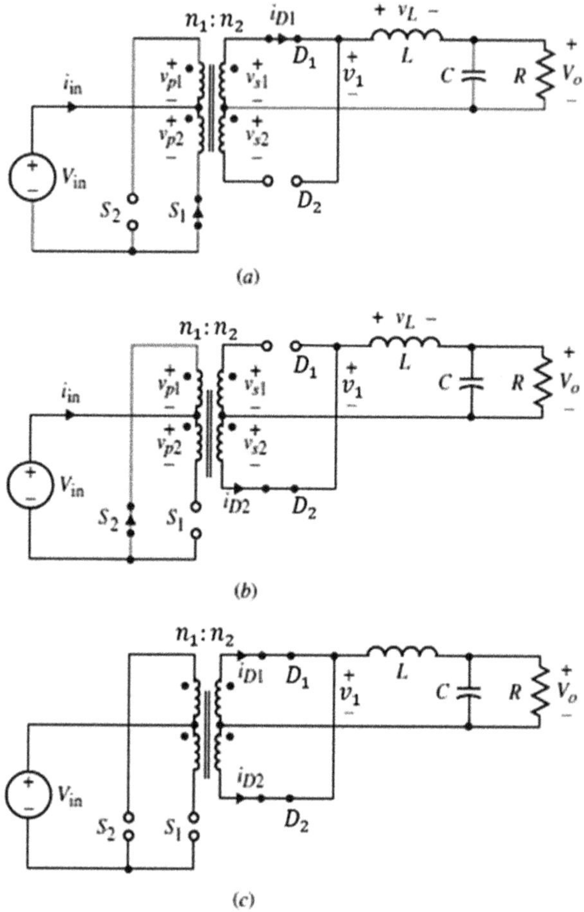

Fig. 5.22 (**a**) Equivalent circuit when S_1 is ON and S_2 is OFF, (**b**) equivalent circuit when S_1 is OFF and S_2 is ON, (**c**) equivalent circuit when S_1 is OFF and S_2 is OFF

$$v_{p_1} = v_{p_2} = +V_{in} \tag{5.16}$$

$$v_L = \frac{n_2}{n_1} V_{in} - V_o \tag{5.17}$$

This circuit is equivalent to a buck converter with a dc input of $(n_2/n_1)V_{in}$. All voltage and current waveforms are similar to the buck converter.

Similarly, when S_1 is OFF and S_2 is ON, the equivalent circuit is shown in Fig. 5.21b.

The voltages are given by:

$$v_{S_1} = v_{S_2} = -\frac{n_2}{n_1} V_{in} \tag{5.18}$$

$$v_{p_1} = v_{p_2} = +V_{in} \tag{5.19}$$

The voltage across the inductor is given by:

$$v_L = -v_{S_2} - V_o = \frac{n_2}{n_1} V_{in} - V_o \tag{5.20}$$

Notice that for both modes, whether S_1 *or* S_2 is *ON*, the circuit is similar to the buck converter when the main switch is ON.

Finally, consider the case when both S_1 and S_2 are OFF. The equivalent circuit is shown in Fig. 5.22c. The voltages v_{S_1} and v_{S_2} are both zero, and the inductor voltage is $-V_o$. Hence, the inductor current starts discharging with a slope of $-V_o/L$. This mode is similar to the buck converter when the main switch is off. Therefore, the voltage gain of the push-pull converter is given by

$$\frac{V_o}{V_{in}} = \frac{n_2}{n_1} D \tag{5.21}$$

where D is the duty cycle for either switch, which range between 0 and 0.5.

We observe that because of the presence of the transformer, each of S_1 and S_2 should be able to withstand a reverse voltage of at least $2V_{in}$. Also, a peak reverse diode voltage for each of D_1 and D_2 is $2(n_2/n_1)V_{in}$. One disadvantage of the push-pull converter is the existing of the imbalance of the voltages applied across the transformer primaries, resulting in an unequal switch current. This in turn results in a nonzero magnetizing inductance current at the end of each switching cycle. This eventually will lead to a transformer saturation problem. This problem is caused by a mismatch in the transistor characteristics, such as switching times and voltage drops. To avoid this problem, push-pull converters are designed with not only a voltage control loop (duty cycle control) but also using a current loop (current-programmed control) that prevents the transformer from saturation.

Key waveforms for the push-pull converter are shown in Fig. 5.23a with no magnetizing inductor being included.

Example 5.5

Another boost-derived converter that uses an isolation transformer is the push-pull arrangement shown in Fig. 5.24a.

The input inductor, L, is very large so that the input current, I_{in}, is assumed constant. These types of converters are useful in high output voltage applications. Since I_{in} is continuous, there should exist an overlap in the conduction time of S_1 and S_2 as shown in the switching wave forms Fig. 5.21b, where δ is the overlapping conduction time. Derive the voltage gain expression for the converter and compare it with the push-pull converter shown earlier. Figure 5.24c shows the waveforms for the primary voltage and input current.

Solution

Let us assume Mode 1 begins at $t = 0$ when both S_1 and S_2 are ON, resulting in the transformer primary voltage equaling to zero since D_1 and D_2 are OFF. The voltage across L equals to the input voltage:

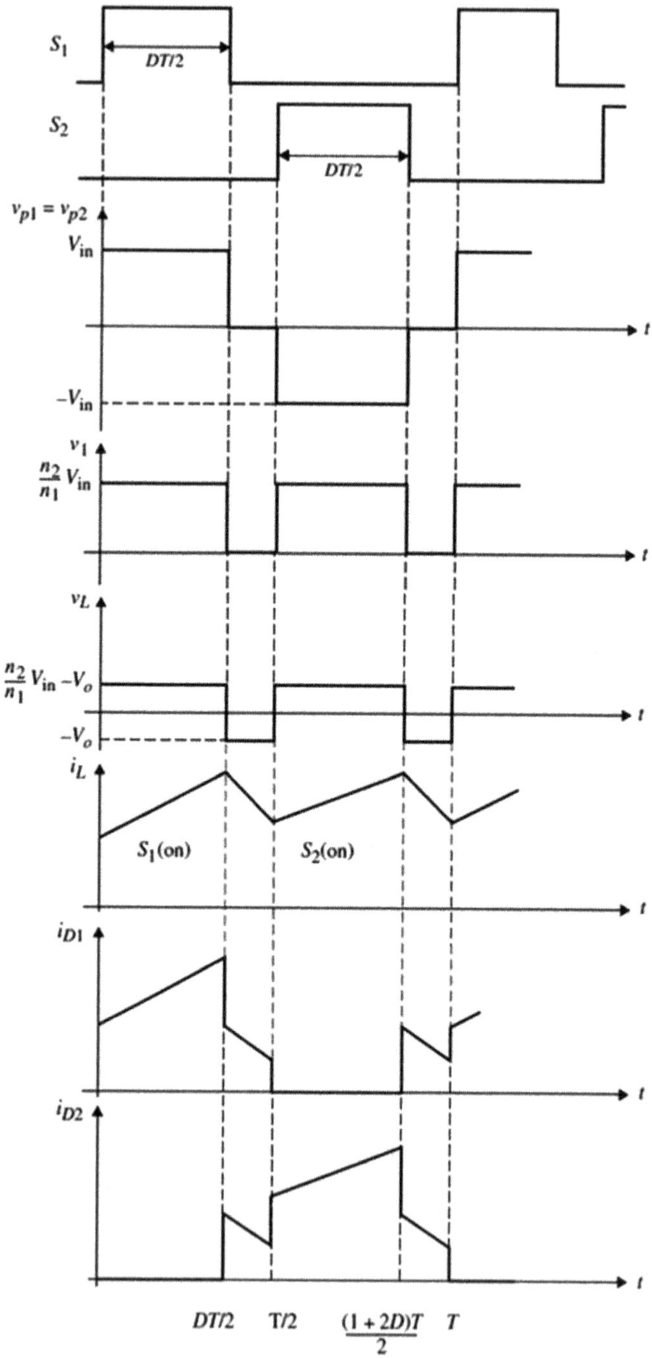

Fig. 5.23 Voltage and current waveforms for the push-pull converter of Fig. 5.21a

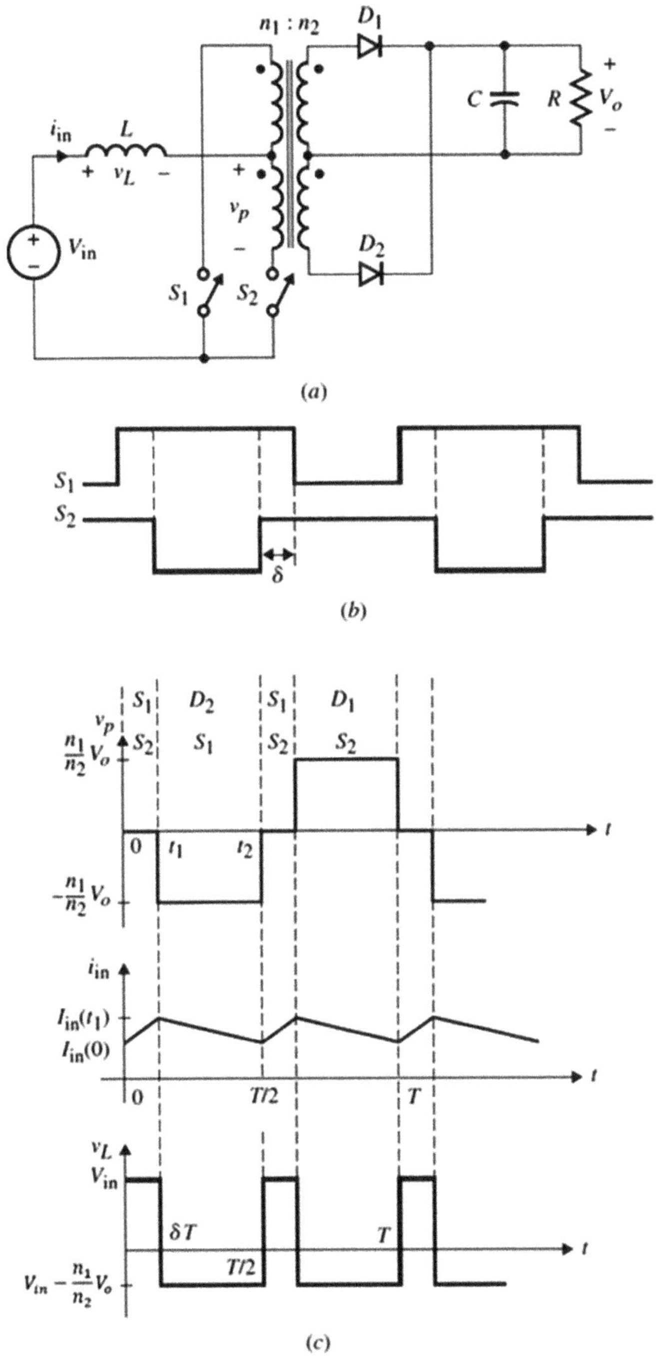

Fig. 5.24 (**a**) Current-fed push-pull converter, (**b**) switching waveforms, and (**c**) converter wave forms

$$v_L = V_{in} = L\frac{di_{in}}{dt}$$

While i_{in} is given by:

$$i_n(t) = \frac{V_{in}}{L}t + I_{in}(0)$$

$I_{in}(0)$ is the initial inductor current value in L.

At $t = t_1$, S_2 is turned off, forcing D_2 to conduct. The voltage across the inductor is given by:

$$v_L = V_{in} - \frac{n_1}{n_2}V_o = L\frac{di_{in}}{dt}$$

The inductor current is given by:

$$i_{in}(t) = \frac{V_{in} - \frac{n_1}{n_2}V_o}{L}(t - t_1) + I_{in}(t_1)$$

The inductor current discharges at a rate of $(V_{in} - (n_1/n_2)V_o)/L$. At $t = t_2$, S_2 turns ON again, resulting in a $v_p = 0$ as shown under Mode 3 in Fig. 5.22a. The next two modes are similar to the first two modes except when S_2 is ON, D_1 conducts.

The gain equation is obtained by applying the volt-second balance to L as follows:

$$V_{in}t_1 = -\left(V_{in} - \frac{n_1}{n_2}V_o\right)(t_2 - t_1)$$

Since $t_1 = \delta T$ and $t_2 = T/2$:

$$V_{in}\delta = \left(\frac{n_1}{n_2}V_o - V_{in}\right)\left(\frac{1}{2} - \delta\right)$$

Solving for V_o/V_{in}, we obtain:

$$\frac{V_o}{V_{in}} = \frac{n_2}{n_1}\frac{1}{1 - 2\delta}$$

If we let DT represent the ON, $(1 - D)T$ is the OFF time of S_1, then we have:

$$\delta = D - \frac{1}{2}$$

Hence, the gain is given by:

$$\frac{V_o}{V_{in}} = \frac{n_2}{n_1} \frac{1}{2(1-D)}$$

Exercise 5.7
Calculate the diode rms and average current values for a push-pull converter of Fig. 5.21a whose current i_L is given in Fig. E5.7. Determine the average output power if the load resistance is 10 Ω.

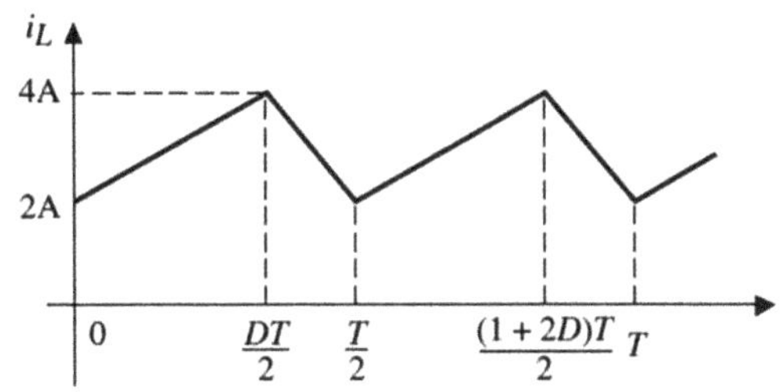

Fig. E5.7 Waveform for i_L of Fig. 5.19a

Answer: $\sqrt{6D + \frac{13}{4}}\ A$, $\frac{3}{4}(1+2D)$, 90 W

5.4 Boost-Derived Isolated Converters

In this section we will discuss several well-known boost-derived isolated converters. All of these converters have the same function of the basic boost converter discussed in Chap. 4. We begin with the simplest: the single-ended one-switch boost-derived converter known as the flyback converter.

5.4.1 Single-Ended Flyback Converter

The circuit topology for the flyback converter is shown in Fig. 5.25a. This converter is one of the most common isolated switch-mode converters.

- The transformer shown in this topology serves as a step-up/step-down to the input voltage, reverses output voltage polarity, provides electrical isolation, and provides energy storage during the operation. Since all the energy obtained from

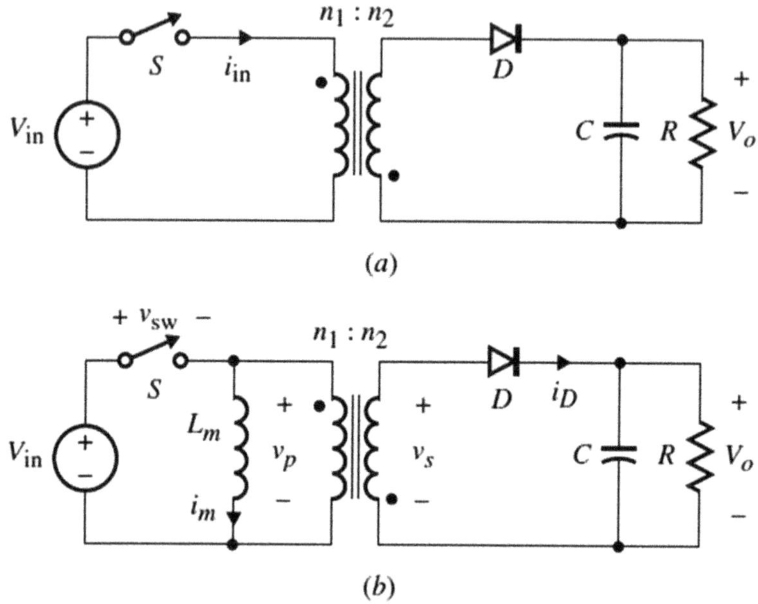

Fig. 5.25 (**a**) Single-ended basic flyback converter, (**b**) flyback converter with magnetizing inductor

the source is first stored in the transformer and then passed on to the load, this converter is also known as an energy storage converter. Commercial flyback converters are normally designed with several multi-coil output transformers. During the turn-on period, energy is stored in the magnetic inductor and transferred to the output side during the turn-off period. In order for the diode to conduct only during the off period in which energy is transferred to the output, the polarities of the transformer windings are reversed, as shown in the figure. One popular application for the flyback converter is in television screens, in which high output voltage is required. This can be obtained by using a high transformer turn ratio, n_2/n_1. Unlike the forward converter, the flyback converter does not need an output inductor; it uses only one diode and does not suffer from a core saturation problem. When operated in dcm, it uses a relatively small, magnetizing inductance. To understand the role of the magnetizing inductance, we replace the transformer by a simple model that includes the magnetizing inductor, as shown in Fig. 5.25b.

- It might be beneficial to students to show how the flyback converter is derived from the non-isolated conventional buck-boost converter that was studied in Chap. 4. In the next section, we discussed how the flyback converter of Fig. 5.25 is derived.

5.4.1.1 Derivation of the Flyback Convertor

Recall that the buck-boost converter was attractive since its output voltage can be smaller and larger than the input voltage. However, the negative output voltage typically limits the applications of the buck-boost converter. Fortunately, output voltage polarity reversed is possible by using a transformer that also serves as isolation if the application calls for it. To illustrate the steps of reversing the output voltage of Fig. 4.21 of the buck-boost converter of Chap. 4, we redraw it again in Fig. 5.26a for convenience. Now imagine that with your both hands you hold the portion of the circuit at points "a" and "b" and then pull the circuit at point "a" toward you and the circuit at the point "b" away from you, in the process rotating the right side portion of the counter 108° as shown in Fig. 5.26b. Figure 5.26a shows the rearrangement of the circuit of Fig. 5.26b by placing the output diode in the top branch of the circuit. Notice when the switch is ON in Fig. 5.26c, the diode D is off, and the voltage across the inductance is + V_{in}, causing the inductance current to increase with slope + V_{in}/L, with the voltage at the terminal $V_{ab} = -V_L = -V_{in} < 0$. It is clear that for the diode to be off in the circuit of Fig. 5.26c, the voltage between terminals "b" and "a" must be the inverse polarity of the voltage between terminals "a" and "b," $V_{ba} = -V_L$. Now when the switch is turned off, the diode must turn ON, changing voltage $V_{ba} = V_o$. Again, the only way for the inductance to demagnetize during the switch off interval is to have $V_{ba} = -V_L = -V_o$. Once again the voltage across terminals "b" and "a" is the reverse of the voltage across the terminals "a" and "b," i.e., V_L.

So, in order for terminals "b" and "a" become the negative polarities of terminals "a" and "b," one can show that one way to accomplish this is to have transformer with its primary and secondary windings done such a way to produce 180° phase shift between them. This means the primary current and secondary current both flow toward the dots of the reverse polarity transformer as shown in Fig. 5.26d of the physical implementation of the primary and secondary windings.

Typical waveforms for the flyback converter of Fig. 5.25b operating in the continuous conduction mode are shown in Fig. 5.27. When the transistor is turned on, the primary voltage v_D becomes equal to the source voltage V_{in}, and the diode D is turned off by the negative polarity of $(n_2/n_1)V_{in} + V_o$. The magnetizing inductance, L_m, starts charging linearly with slope V_{in}/L_m. Note that this mode of operation is like the boost converter, in which the inductor L stores energy during the on time. When S is turned off, the diode is forced to carry the magnetizing inductor current through the secondary winding. The drawback of this topology is the transformer's primary-side leakage inductance, whose energy must be dissipated when the transistor is turned off. For this reason, in order to reduce the stress on the switching transistor, a snubber circuit is normally added.

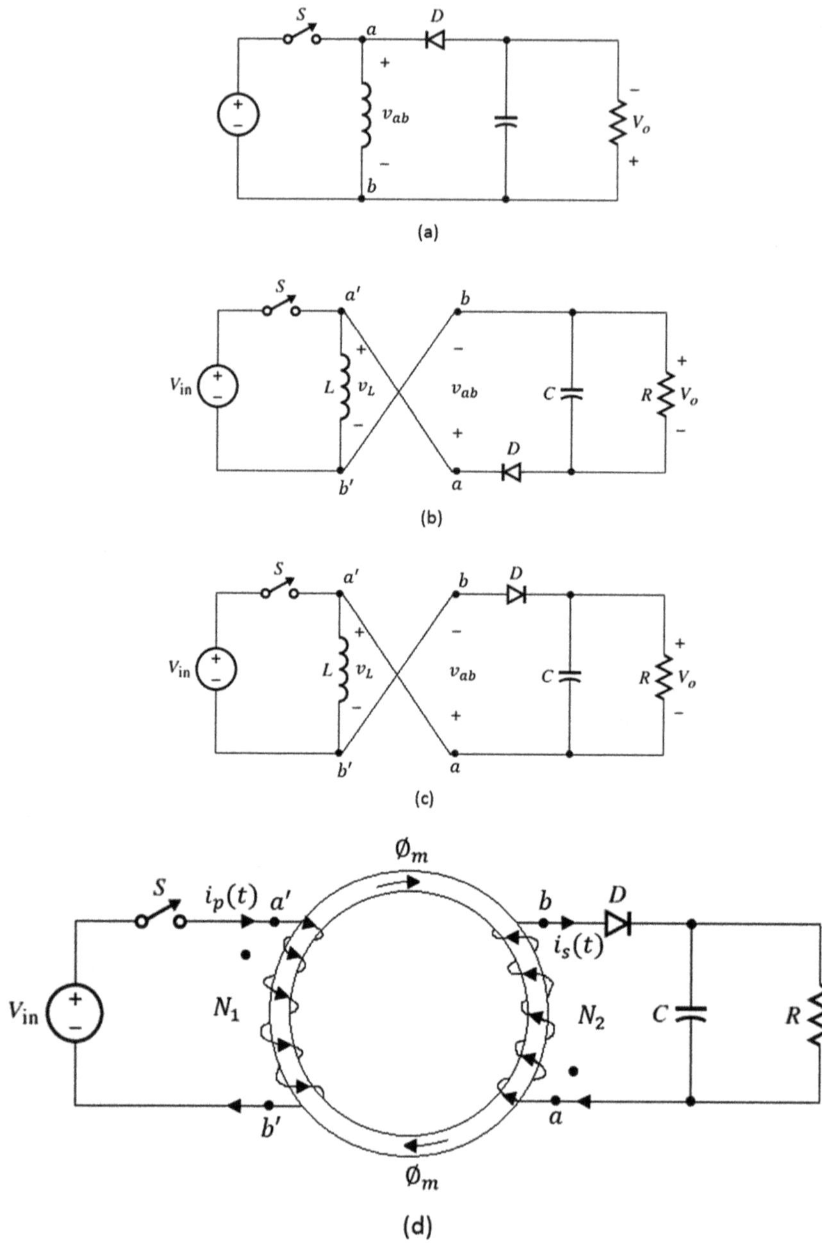

Fig. 5.26 Derivation of the flyback converter from the (**a**) conventional buck-boost, (**b**, **c**) circuit rearrangements, and (**d**) implementation using transformer to achieve output voltage polarity reversal

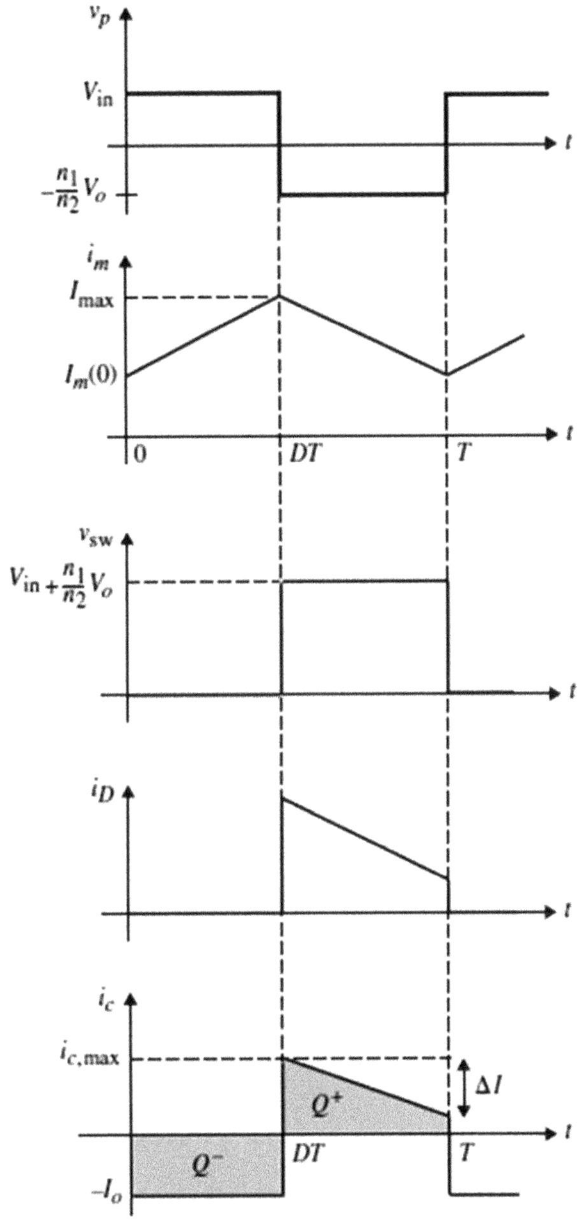

Fig. 5.27 Current and voltage waveforms for the flyback converter

The analysis of this circuit consists of two modes of operation as follows:

Mode 1: [S is turned ON at $t = 0$ and D is OFF].

The *on*-state equivalent circuit model is shown in Fig. 5.28a.

The voltage across L_m is V_{in}, yielding, the following equation for $i_m(t)$:

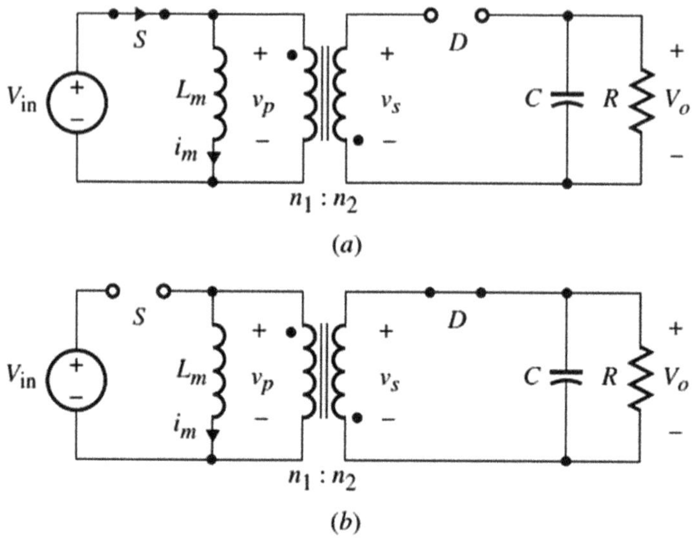

Fig. 5.28 Modes of operation: (**a**) Mode 1, S is ON and D is OFF; (**b**) Mode 2, S is OFF and D is ON

$$i_{\mathrm{m}}(t) = \frac{V_{\mathrm{in}}}{L_{\mathrm{m}}} t + I_{\mathrm{m}}(0) \tag{5.22}$$

where $I_m(0)$ is the initial value of the magnetizing current when the transistor is turned ON. From this equation, we notice that the inductor current, i_m, will linearly charge to $t = DT$ when the transistor is turned OFF to enter *Mode 2*.

Mode 2: [*S* is turned OFF and *D* is ON].

This mode of operation starts at $t = DT$ when the transistor is turned OFF. To maintain the continuity of i_m, the diode *D* turns ON. The equivalent circuit model is shown in Fig. 5.28b.

The voltage across the magnetizing inductance is nV_o, where n is n_1/n_2. In terms of i_m, we have:

$$v_{\mathrm{p}} = -nV_{\mathrm{o}} = L_{\mathrm{m}} \frac{di_{\mathrm{m}}}{dt}$$

Integrating this equation from DT to t, we obtain:

$$i_{\mathrm{m}}(t) = \frac{-V_{\mathrm{o}}\mathrm{n}}{L_{\mathrm{m}}}(t - DT) + I_{\mathrm{m}}(DT) \tag{5.23}$$

where $I_m(DT)$ is the magnetizing current value at $t = DT$, when the diode *D* starts conducting. Evaluating Eq. (5.15) at $t = DT$ and at $t = T$, respectively, and using $i_m(T) = i_m(0)$, we obtain the following relation:

$$I_{\mathrm{m}}(DT) - I_{\mathrm{m}}(0) = \frac{nV_{\mathrm{o}}}{L_{\mathrm{m}}}(1-D)T \tag{5.24a}$$

$$I_{\mathrm{m}}(DT) - I_{\mathrm{m}}(0) = \frac{V_{\mathrm{in}}}{L_{\mathrm{m}}}DT \tag{5.24b}$$

Equating Eqs. (5.24a) and (5.24b), we obtain the following conversion ratio:

$$\frac{V_{\mathrm{o}}}{V_{\mathrm{in}}} = \frac{D}{n(1-D)} \tag{5.25}$$

This gain equation is similar to the buck-boost converter gain when $n=1$. All relevant current and voltage waveforms are shown in Fig. 5.27. The average output current is the same as the average diode current which is given by the following relation:

$$I_{\mathrm{o}} = \frac{n(I_{\mathrm{m}}(DT) + I_{\mathrm{m}}(0))}{2}(1-D) \tag{5.25}$$

and the average input current is given by:

$$I_{\mathrm{in}} = \frac{I_{\mathrm{m}}(DT) + I_{\mathrm{m}}(0)}{2}D \tag{5.26}$$

Hence, the current conversion ratio is given by:

$$\frac{I_{\mathrm{o}}}{I_{\mathrm{in}}} = \frac{n(1-D)}{D} \tag{5.27}$$

This equation can also be obtained by equating the average input and output powers by replacing $I_{\mathrm{o}} = V_{\mathrm{o}}/R$, thus we obtain the following relation:

$$I_{\mathrm{m}}(DT) + I_{\mathrm{m}}(0) = \frac{2V_{\mathrm{o}}}{nR(1-D)} \tag{5.28}$$

From the above relation, we obtain the minimum and maximum current values of i_{m} as follows:

$$I_{\mathrm{m}}(0) = \frac{V_{\mathrm{in}}D}{n^2R(1-D)^2} - \frac{V_{\mathrm{in}}DT}{2L_{\mathrm{m}}} \tag{5.29a}$$

$$I_{\mathrm{m}}(DT) = \frac{V_{\mathrm{in}}D}{n^2R(1-D)^2} + \frac{V_{\mathrm{in}}DT}{2L_{\mathrm{m}}} \tag{5.29b}$$

When setting $I_{\mathrm{m}}(0) = 0$, we obtain the critical value of the magnetizing inductance for the continuous conduction mode of operation:

$$L_{\text{crit}} = \frac{n^2(1-D)^2RT}{2} \tag{5.30}$$

If the inductor current is allowed to reach zero, i.e., $L_m < L_{crit}$, then the converter will operate in dcm, and the core becomes fully demagnetized in each cycle.

Example 5.6
For the flyback converter of Fig. 5.28, consider the case when the convertor is required to deliver 500 W a + 48 output voltage bus from a dc input voltage bus of 400 V while operating at switching frequency of 250 kHz. It is desired to operate the converter between 40% and 60% duty ratio and its magnetizing inductor ripple not to exceed 10% of its average value. Design for transfer ratio and its magnetizing inductor value.

Solution
For $D = 0.4$, $V_o = 48$ V, and $V_{in} = 400$ V, from Eq. (5.17), the transfer ratio is given by:

$$n = \frac{n_1}{n_2} = \frac{D}{\frac{V_o}{V_{in}}(1-D)} = \frac{0.4}{\frac{48}{400}(1-0.4)} = 5.57$$

For $D = 0.6$, we have:

$$n = \frac{n_1}{n_2} = \frac{D}{\frac{V_o}{V_{in}}(1-D)} = \frac{0.6}{\frac{48}{400}(1-0.6)} = 12.5$$

Let us select $n = \frac{n_1}{n_2} = 8$, this in $D = 0.49$.
From Eq. (5.19), the average magnetizing inductor value is given by:

$$I_{L_m}\text{Ave} = \frac{V_{in}}{nR(1-D)}$$

Where the load resistance is given by:

$$R = \frac{(V_o)^2}{P_o} = \frac{(48)^2}{500} = 4.61\ \Omega$$

Then

$$I_{L_m}\text{Ave} = \frac{400}{8 \times 4.61(1-0.49)} = 21.27\ \text{A}$$

The ripple should be set to 10% of I_{L_m}Ave, i.e., should not exceed 2.13 A.
From Eq. (5.16), the magnetizing inductor ripple is given by:

$$\Delta I_m = I_m(DT) - I_m(0) = \frac{V_{in}}{L_m}DT$$

$$L_{\mathrm{m}} = \frac{V_{\mathrm{in}}}{\Delta I_{\mathrm{m}}} DT = \frac{400}{2.13} 0.49 \times \frac{1}{250\ \mathrm{kHz}} = 0.37\ mH$$

Let us check if the inductor is operating in ccm. Recall that the critical value of the inductance is given by:

$$L_{\mathrm{crit}} = \frac{n^2(1-D)^2 RT}{2} = \frac{8^2(1-0.49)^2 4.61}{2 \times 250\ \mathrm{kHz}} = 0.15\ \mu H$$

Yes, the mode is ccm.

Example 5.7
Sketch the waveform for i_{c} and determine the expression for the output voltage ripple for the flyback converter of Fig. 5.25b.

Solution
The capacitor current equals the *ac* ripple in the diode current i_{D}. So i_{c} is the same as the diode current subtracting the average output current:

$$i_{\mathrm{c}} = i_{\mathrm{D}} - I_{\mathrm{o}}$$

The current i_{c} is obtained by simply shifting i_{D} down by I_{o} as shown in Fig. 5.27. It is clear that the waveform is similar to the buck-boost capacitor current. The total charge, ΔQ, during DT interval is given by:

$$\begin{aligned} \Delta Q^{-} &= (DT)(I_{\mathrm{o}}) \\ C\Delta V_{c} &= (DT)(I_{\mathrm{o}}) \\ \therefore \Delta V_{\mathrm{c}} &= \frac{DTI_{\mathrm{o}}}{C} \\ \frac{\Delta V_{\mathrm{c}}}{V_{\mathrm{o}}} &= \frac{D}{RCf} \end{aligned}$$

The peak capacitor current is given by:

$$I_{\mathrm{c,max}} = I_{\mathrm{L_m}}(DT) - I_{\mathrm{o}}$$

Substitute for $I_{\mathrm{L_m}}(DT)$ from Eq. (5.19) and simplify to obtain:

$$I_{\mathrm{c,max}} = I_{\mathrm{o}} \left[\frac{1 - n(1-D)}{n(1-D)} + \frac{nTR(1-D)}{2L_{\mathrm{m}}} \right]$$

Exercise 5.8
Determine the voltage gain $V_{\mathrm{o}}/V_{\mathrm{in}}$ for the flyback converter when operating in the discontinuous conduction mode. Compare this expression with the voltage gain for the continuous conduction mode.

Answer: $\frac{V_{\mathrm{o}}}{V_{\mathrm{in}}} = D\sqrt{\frac{RT}{2nL_{\mathrm{m}}}}$

Exercise 5.9
Consider the flyback converter of Fig. 5.25b with $V_o = 48$ V at $I_o = 1$ A, $V_{in} = 18$ V, $f = 150$ kHz, $n = 0.3$. Find the maximum L_m so that the converter operates in dcm.

Answer: 4.44 μH.

5.4.2 *Half-Bridge Converter*

Because of the presence of an input inductance in series with the dc voltage source, the circuit implementation of the half-bridge converter requires a center-tap transformer. Figure 5.29a, b shows a discrete inductor and a center-tap transformer implementation of the half-bridge boost-derived converter.

- Because switches S_1 and S_2 are switched alternately, the topology of Fig. 5.29a is not suitable due to the abrupt change in the primary current. This problem does not exist for Fig. 5.29b, in which a center-tap transformer is added. It can be shown that the voltage gain is given by:

$$\frac{V_o}{V_{in}} = \frac{1}{2}\left(\frac{n_2}{n_1}\right)\frac{1}{1-D} \tag{5.31}$$

5.4.3 *Full-Bridge Converter*

Figure 5.30 shows the full-bridge boost-derived converter with a full-wave rectifier in the output side. This converter is also known as a double-ended converter.

The main advantage of the full-bridge configuration is that its core material is better utilized when compared to the single-ended case in which the magnetic field density changes from 0 to $+B_{sat}$, whereas the core excitation varies between $+B_{sat}$ and $-B_{sat}$ for the full-bridge configuration. There are several switching sequences for Fig. 5.31 that will result in a duty cycle above 50%. Figure 5.31 shows two possible switching sequences.

- In Fig. 5.31a, each of the two diagonal pairs of switches are switched simultaneously with $D > 0.5$ for each switch. With this switching sequence, there exist two intervals during which all four switches are on at the same time. During these intervals, the inductor current is divided almost equally between the switch legs $S_1 - S_4$ and $S_2 - S_3$, reducing their rms current values. In the switching sequence of Fig. 5.31b, two of the switches operate with a fixed duty cycle and the other two switches operate with a varying duty cycle. Of course, the driving circuit for Fig. 5.31a is more complex since unlike Fig. 5.31b, where only two switches are controlled, the switching sequence of Fig. 5.31a is required to control four switches. Next we consider the analysis of the circuit shown in

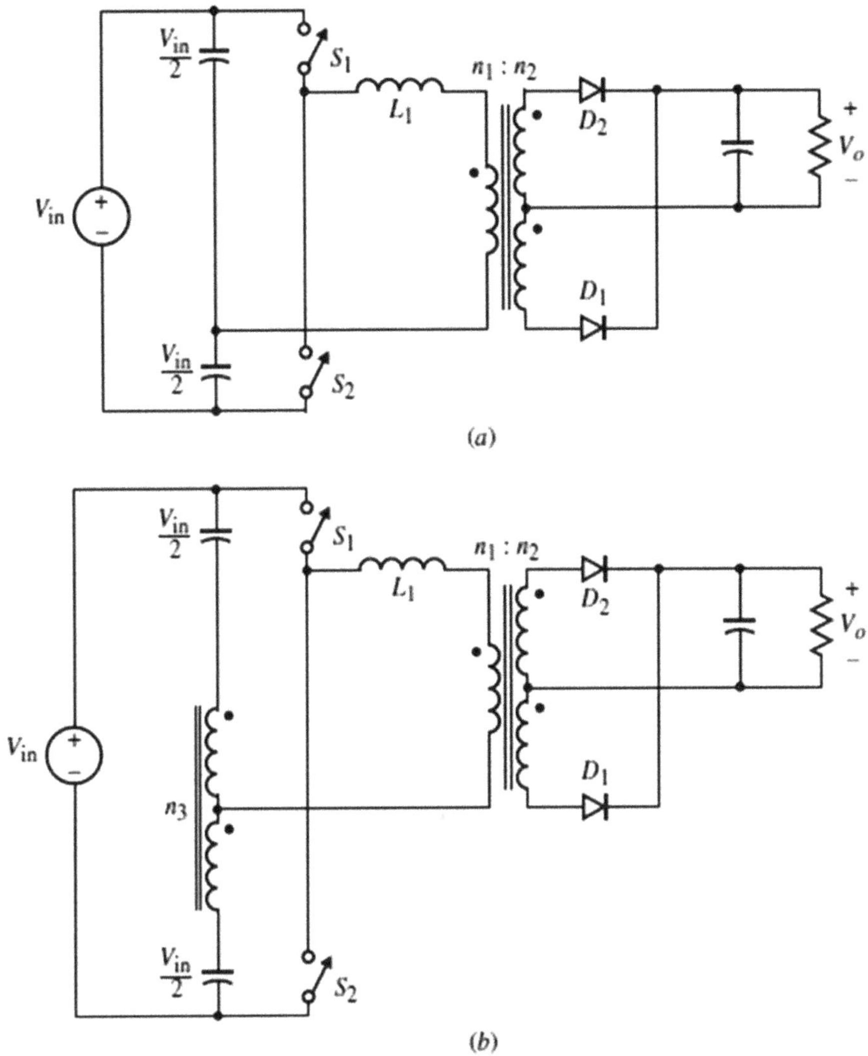

Fig. 5.29 Half-bridge boost-derived converter with (**a**) single inductor implementation, (**b**) center-tap transformer implementation

Fig. 5.30 when operated under the switching sequence of Fig. 5.31a. It can be shown that there are four modes of operation during one-switching period, as shown in Fig. 5.32.

The inductor current during Mode 1 is given by:

$$i_L(t) = \frac{V_{in}}{L} + I_L(0) \tag{5.32}$$

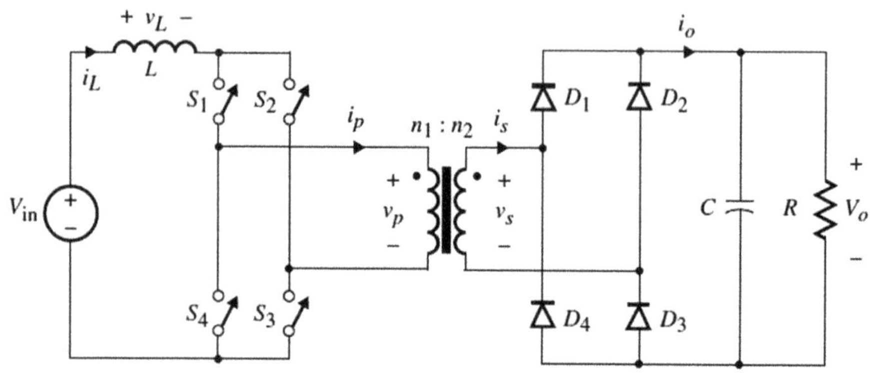

Fig. 5.30 Full-bridge boost-derived converter with full-wave output rectifier

Fig. 5.31 Two possible switching sequences for the full-bridge converter of Fig. 5.30

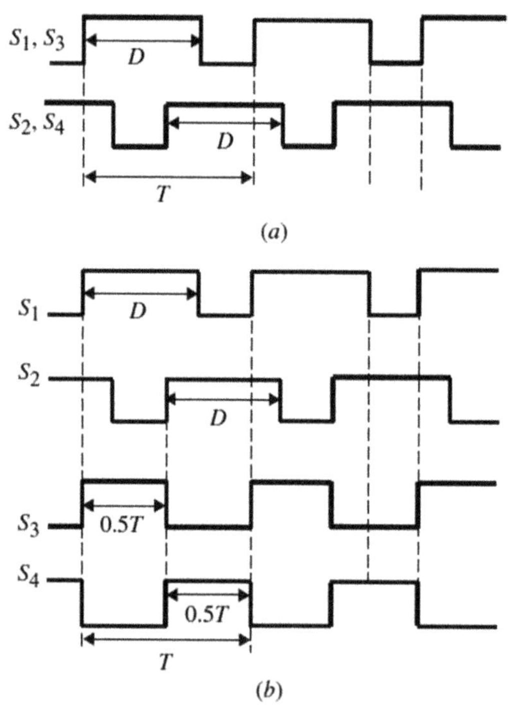

where $I_L(0)$ is the initial inductor current when S_1 and S_3 are turned on, while S_2 and S_4 were on initially. This current is divided equally through $S_1 - S_4$ and $S_2 - S_3$, with $v_s = v_p = 0$ and $i_o = 0$.

- At $t = t_1$, S_2 and S_4 are turned off simultaneously, forcing $i_L(t)$ to flow through the primary, through diodes D_1 and D_3, and then to the load. The equivalent circuit is shown in Fig. 5.32b.

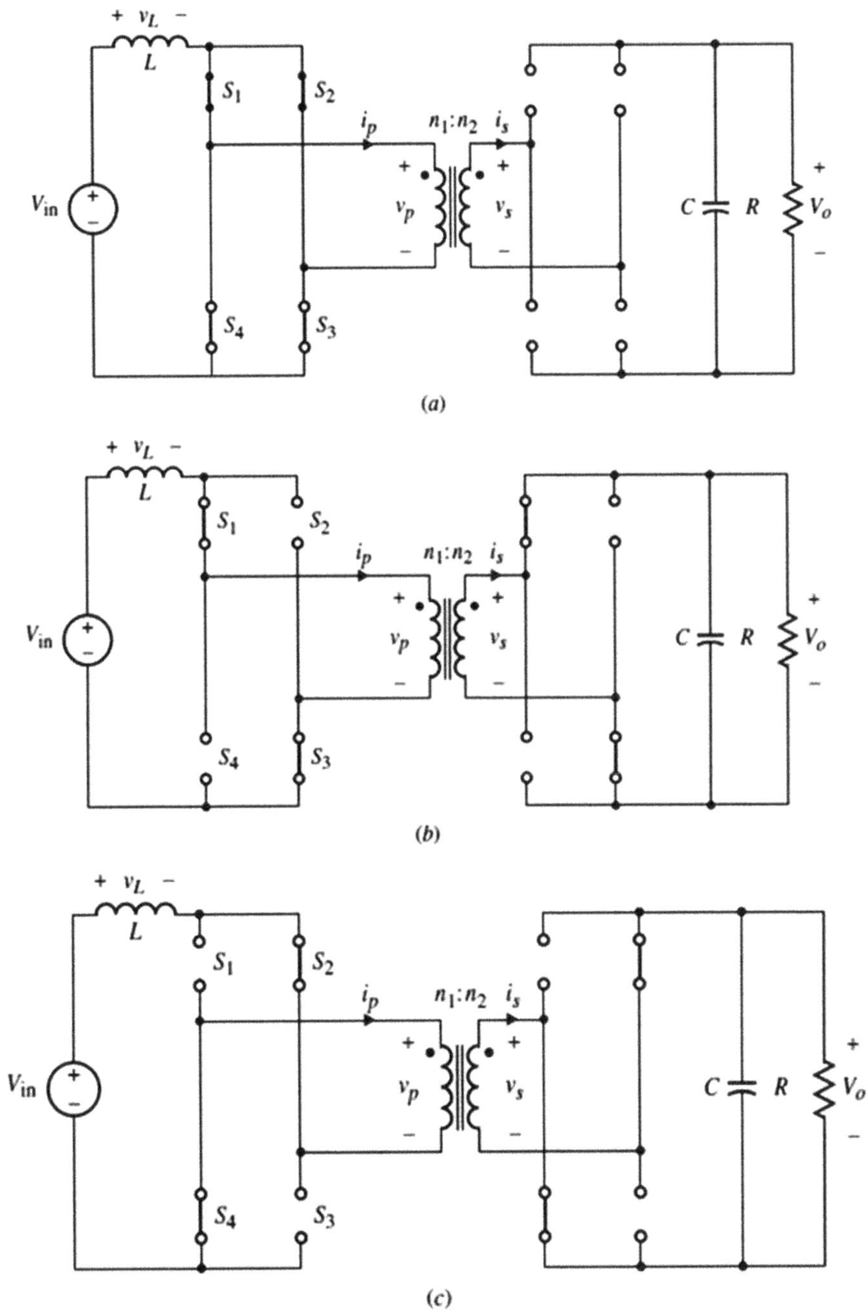

Fig. 5.32 Modes of operation: (**a**) Mode 1 and 3, all switches are ON; (**b**) Mode 2, S_1 and S_3 are ON; (**c**) Mode 4, switches S_2 and S_4 are ON

- The inductor current is obtained from v_L:

$$v_L(t) = V_{\text{in}} - \frac{n_1}{n_2}V_{\text{o}} \qquad t_1 \leq t < t_2$$

Therefore the inductor current is given by:

$$i_{\text{L}}(t) = \frac{V_{\text{in}} - \frac{n_1}{n_2}V_{\text{o}}}{L}(t - t_1) + I_{\text{L}}(t_1) \tag{5.33}$$

In this mode v_{p} and v_{s} are given by:

$$v_{\text{s}} = V_{\text{o}} \quad v_{\text{p}} = \frac{n_1}{n_2}V_{\text{o}}$$

Mode 3 starts at $t = t_2$, when S_2 and S_4 are turned on again, resulting in a similar equivalent circuit to that in Mode 1. The current equation is given by:

$$i_{\text{L}}(t) = \frac{V_{\text{in}}}{L}(t - t_2) + I_{\text{L}}(t_2) \tag{5.34}$$

At $t = t_3$, we have:

$$i_{\text{L}}(t_3) = \frac{V_{\text{in}}}{L}(t_3 - t_2) + I_{\text{L}}(t_2) \tag{5.35}$$

Finally, Mode 4 starts at $t = t_3$ when S_1 and S_3 are turned OFF simultaneously. The voltages are given as follows:

$$v_{\text{L}} = V_{\text{in}} + v_{\text{p}} \qquad v_{\text{p}} = \frac{n_1}{n_2}v_{\text{s}} \qquad v_{\text{s}} = -V_{\text{o}}$$

Hence, the currents are given as:

$$i_{\text{L}}(t) = \frac{V_{\text{in}} - \frac{n_1}{n_2}V_{\text{o}}}{L}(t - t_1) + I_{\text{L}}(t_1) \tag{5.36}$$

$$i_{\text{p}}(t) = -i_{\text{L}} \tag{5.37}$$

$$i_{\text{s}}(t) = \frac{n_1}{n_2}i_{\text{p}} \tag{5.38}$$

The current and voltage waveforms are shown in Fig. 5.33.
From *volt-second* principle, we have:

$$t_1 V_{\text{in}} = (t_1 - t_2)\left(V_{\text{in}} - \frac{n_1}{n_2}V_{\text{o}}\right)$$

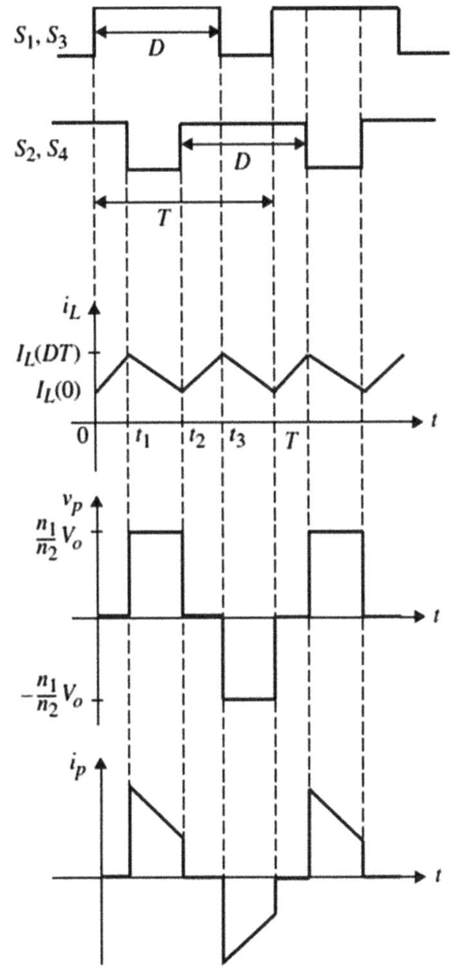

Fig. 5.33 Current and voltage waveforms for Fig. 5.27

From Fig. 5.33, $t_1 = \left(D - \frac{1}{2}\right)$ and $(t_1 - t_2) = (1 - D)T$; hence, we have:

$$\left(D - \frac{1}{2}\right)TV_{\text{in}} = -(1 - D)T\left(V_{\text{in}} - \frac{n_1}{n_2}V_{\text{o}}\right) \tag{5.39}$$

Solving Eq.(5.39) for the voltage gain, we obtain:

$$\frac{V_{\text{o}}}{V_{\text{in}}} = \frac{n_1}{n_2}\frac{1}{2(1 - D)} \tag{5.40}$$

Example 5.8

(a) Derive the voltage gain expression for Fig. 5.30 by using the switching sequence given in Fig. 5.31b. (b) What are the rms currents in each switch?

Solution

(c) Although there is a change in the switching sequence between Fig. 5.31a, b in switches S_3 and S_4, careful investigation of the circuit operation modes indicated that conditions governing the charging and discharging of the inductor are exactly the same for both switching sequences. Therefore, the expression according to the volt-second balance as obtained above also applies here, and the voltage gain expression for the switching sequence given in Fig. 5.31b is the same as that given in Eq. (5.40).

$$I_{\mathrm{s,rms}} = \sqrt{\frac{I_{\mathrm{L}}^{2}}{2} + \frac{\Delta I_{\mathrm{L}}^{2}}{24}} \tag{5.41}$$

where

$$I_{\mathrm{L}} = \frac{n_2}{n_1} \frac{I_{\mathrm{o}}}{2(1-D)}$$

is the average inductor current and

$$\Delta I_{\mathrm{L}} = \frac{V_{\mathrm{in}}}{L}\left(D - \frac{1}{2}\right)T \tag{5.42}$$

is the inductor current ripple.

Exercise 5.10
Derive the expressions for $i_{\mathrm{L}}(0)$ and $i_{\mathrm{L}}(DT)$ and determine the inductor's critical value.

Answer: $\mathrm{I_L(0)} = \frac{\mathrm{n_2}}{\mathrm{n_1}} \frac{\mathrm{I_o}}{2(1-\mathrm{D})} - \frac{\mathrm{V_{in}}}{2\mathrm{L}}\left(\mathrm{D} - \frac{1}{2}\right)\mathrm{T}$, $\mathrm{I_L(DT)} = \frac{\mathrm{n_2}}{\mathrm{n_1}} \frac{\mathrm{I_o}}{2(1-\mathrm{D})} + \frac{\mathrm{V_{in}}}{2\mathrm{L}}\left(\mathrm{D} - \frac{1}{2}\right)\mathrm{T}$
$\mathrm{L_{crit}} = 2\left(\frac{\mathrm{n_2}}{\mathrm{n_1}}\right)^2 \mathrm{R}(1-\mathrm{D})^2\left(\mathrm{D} - \frac{1}{2}\right)\mathrm{T}$

Exercise 5.11
Show that the primary rms current of Fig. 5.30 is given by the following equation:

$$I_{\mathrm{p,rms}} = \sqrt{2(1-D)\left(I_{\mathrm{L}}^{2} + \frac{1}{12} DI^{2}\right)}$$

5.5 Other Isolated Converters[1]

In addition to the circuits presented thus far, there are many other isolated converters that are covered in the open literature. Here we will consider two examples: the Cuk and the Weinberg converters.

5.5.1 *Isolated Cuk Converter*

Figure 5.34 shows the isolated Cuk converter. The converter storage capacitor is split into two series capacitors C_1 and C_2 as shown. The output voltage remains negative since the transformer's primary and secondary windings have the same polarities. Let us derive the expression for the voltage gain. Assume the converter operates in the continuous conduction mode. Like the push-pull and bridge converters, the entire core B-H loop is utilized in both directions.

Figures 5.35a, b show the two equivalent circuits for Mode 1 when S is ON and Mode 2 when S is off.

Mode 1 (during DT interval):

$$v_{\mathrm{L}_1} = V_{\mathrm{in}}$$

$$v_{\mathrm{L}_2} = -V_{\mathrm{o}} - V_{\mathrm{c}_2} - \frac{n_2}{n_1} V_{\mathrm{c}_1}$$

Mode 2 (during $(1 - D)T$ interval):

$$v_{\mathrm{L}_1} = V_{\mathrm{in}} + V_{\mathrm{c}_1} + \frac{n_1}{n_2} V_{\mathrm{c}_2}$$

$$v_{\mathrm{L}_2} = -V_{\mathrm{o}}$$

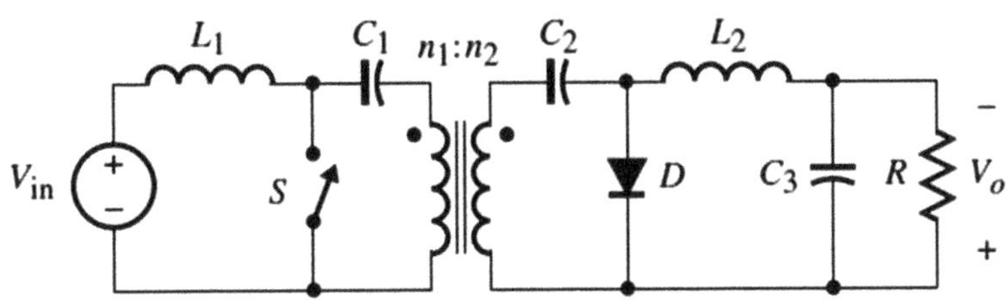

Fig. 5.34 Isolated Cuk converter

[1]This section may be omitted without the loss of continuity.

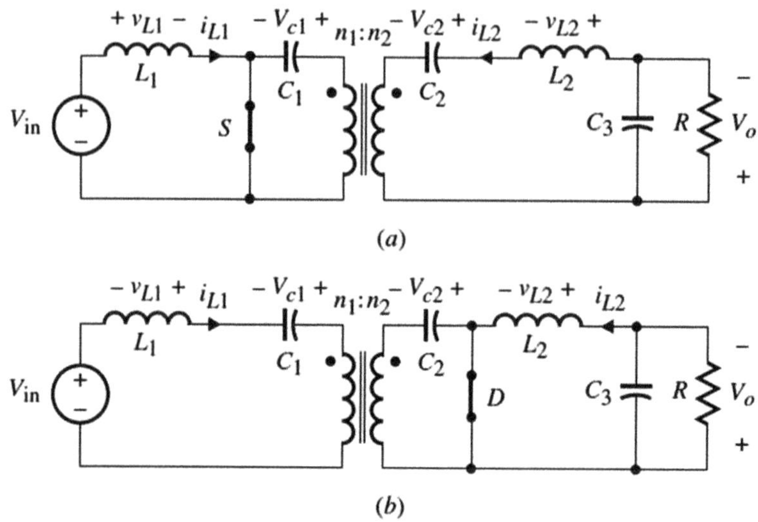

Fig. 5.35 Isolated Cuk converter: (**a**) Mode 1, (**b**) Mode 2

Applying the volt-second principle to v_{L_1} and v_{L_2},we obtain:

v_{L_1}:

$$DV_{in} = -(1-D)\left[V_{in} + V_{c_1} + \frac{n_2}{n_1}V_{c_2}\right]$$

$$V_{in}\frac{-D}{1-D} = V_{in} + V_{c_1} + \frac{n_2}{n_1}V_{c_2}$$

$$V_{in}\frac{-1}{1-D} = V_{c_1} + \frac{n_2}{n_1}V_{c_2} \tag{5.43}$$

v_{L_2}:

$$D\left[-V_o - V_{c_2} - \frac{n_2}{n_1}V_{c_1}\right] = +(1-D)V_o$$

$$-D\left(V_{c_2} + \frac{n_2}{n_1}V_{c_1}\right) = [+(1-D) + D]V_o$$

$$V_{c_2} + \frac{n_2}{n_1}V_{c_1} = \frac{-V_o}{D}$$

$$V_{c_2} + \frac{n_2}{n_1}V_{c_1} = -\frac{n_2}{n_1}\frac{V_o}{D} \tag{5.44}$$

From Eqs. (5. 43) and (5. 44), we obtain:

$$\frac{V_o}{V_{in}} = \frac{D}{1-D}\frac{n_2}{n_1}$$

Exercise 5.12

Consider the Cuk converter shown in Fig. E5.12 that provides a positive output voltage. Assume ideal components to derive the voltage gain equation V_o/V_{in} and sketch the waveforms for i_{L_1}, i_{L_2}, i_D, v_{sw}, v_D, and v_{c_1}.

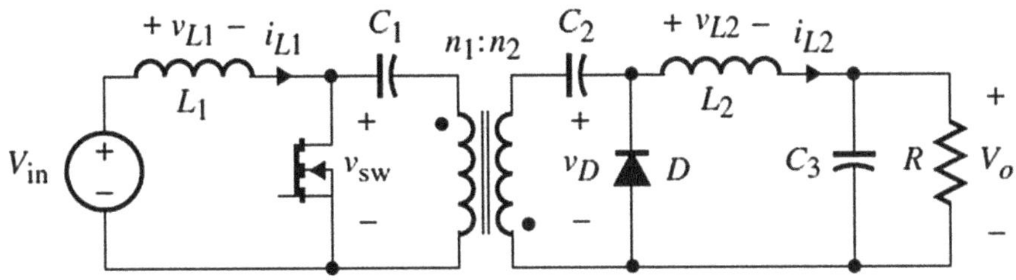

Fig. E5.12

5.5.2 *The Weinberg Converter*

- In this section, the basic topology and operation of another buck-derived push-pull converter, known as the Weinberg converter, will be discussed. There are several versions of this topology that perform the same function. The steady-state analysis will be presented in this section for the converter operating in ccm. The original proposed Weinberg dc-dc topology is shown in Fig. 5.36a. A more simplified equivalent circuit can be obtained. In this analysis we will assume the n-turn-ratio transformer has a magnetizing inductance, L_m, and the m-turn-ratio transformer is ideal.
- The switching waveforms are similar to those for the push-pull converter and are shown in Fig. 5.36b. When switches S_1 and S_2 are turned on and off alternately, the following cases are investigated:

1. Mode 1: S_1 is ON; S_2 is OFF
2. Mode 2: S_1 is OFF; S_2 is ON
3. Mode 3: S_1 is OFF; S_2 is OFF

1. *Mode 1: S_1 is ON; S_2 is OFF $[0 < t \leq DT]$*

   In this case, since the current from V_{in} flows through the magnetizing inductance of the input transformer, causing $v_{s_2} > 0$, hence D_1 and D_4 are turned OFF. With S_1 ON, $v_{s_1} > 0$ and $v_{s_1} < 0$, causing D_3 to reverse bias and D_2 to be forward biased. The equivalent circuit model is shown in Fig. 5.37a.

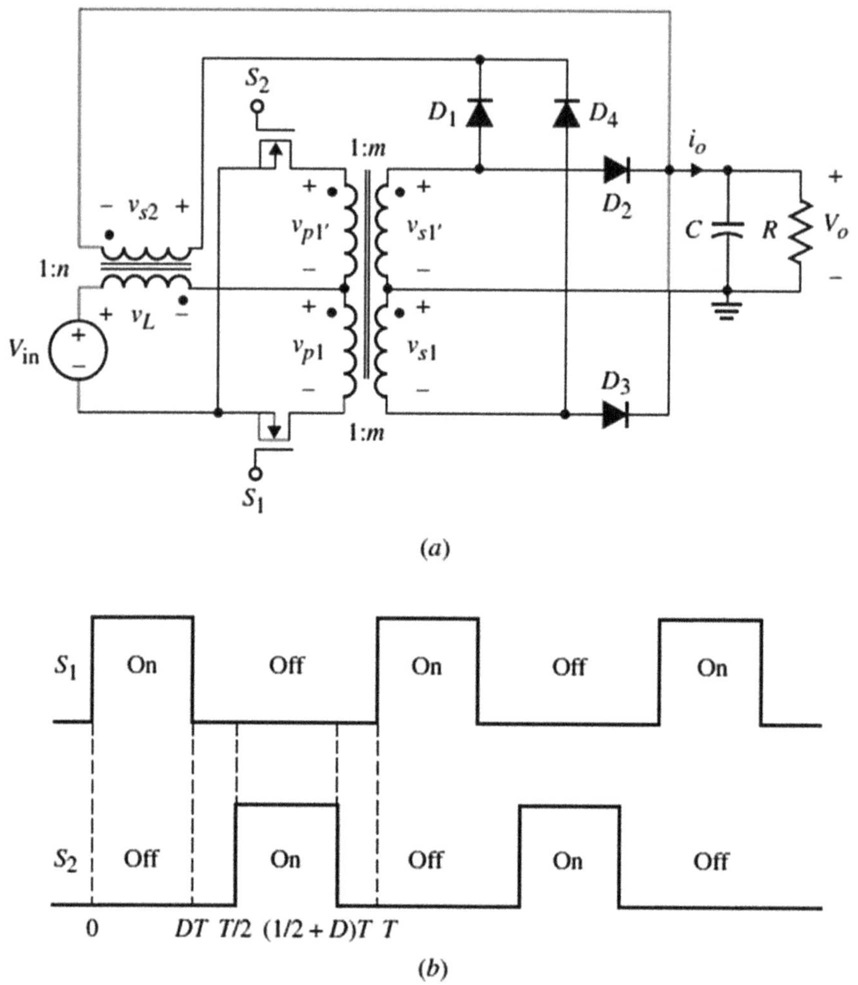

Fig. 5.36 The Weinberg dc-dc converter: (**a**) circuit topology and (**b**) switching waveforms for the power switches S_1 and S_2

From Fig. 5.37 we have $v_{\mathrm{L}} = V_{\mathrm{in}} - V_{\mathrm{o}}/m$, yielding the following equation for i_{L}:

$$i_{\mathrm{L}}(t) = \frac{V_{\mathrm{in}} - V_{\mathrm{o}}/m}{L} t + I_L(0) \tag{5.45}$$

where $I_{\mathrm{L}}(0) = I_{\mathrm{L_{min}}}$ is the initial inductor current at $t = 0$. Since the center-tap transformer has the same ratio, m, diode D_3 blocks twice the output voltage, i.e.:

$$v_{\mathrm{D}_3} = -2V_{\mathrm{o}}$$

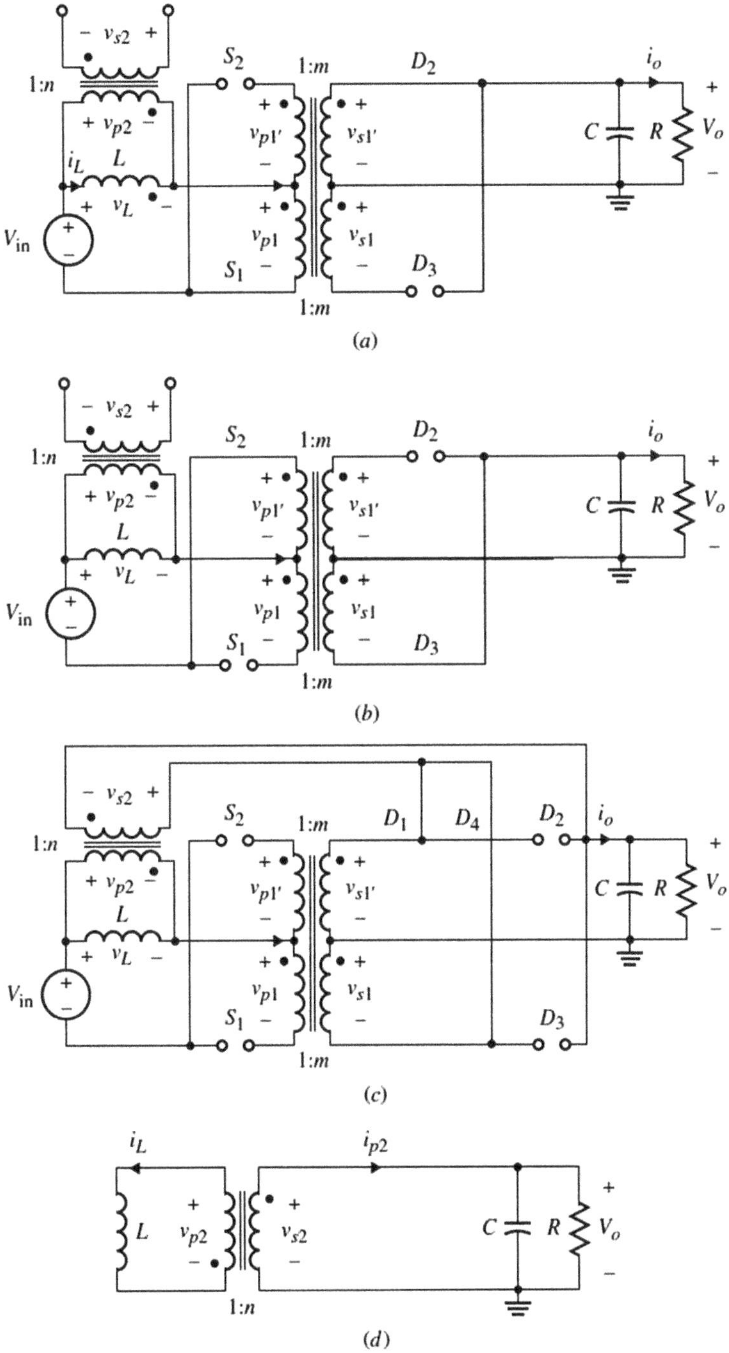

Fig. 5.37 Modes of operation: (**a**) equivalent circuit for Mode 1 when S_1 is ON and S_2 is OFF, (**b**) equivalent circuit for Mode 2 when S_2 is ON and S_1 is OFF, (**c**) equivalent circuit for Mode 3 when both S_1 and S_2 are OFF, (**d**) simplified equivalent circuit of Mode 3

2. *Mode 2: S_1 is OFF; S_2 is ON* ($T/2 < t \leq (0.5 + D)T$)

   At $t = T/2$, S_2 is turned ON, while S_1 is kept OFF (S_1 is turned OFF in the previous interval at $t = DT$). In this case since the current from V_{in} flows through the magnetizing inductance, diodes D_1 and D_4 are turned OFF. The output diode, D_2, is OFF because S_2 is ON; D_3 is conducting to carry the inductor current.

   From Fig. 5.37b, the inductor voltage is $v_L = V_{in} - V_o/m$, and i_L is given by:

$$i_L(t) = \frac{V_{in} - V_o/m}{L}(t - T/2) + I_L(T/2) \tag{5.46}$$

   The blocking voltage of D_2 is $-2V_o$, and this case is identical to the case in Mode 1. Since we have a symmetric circuit operation, the above two modes are similar when either switch is ON or the other one is OFF.

3. *Mode 3: S_1 is OFF; S_2 is OFF [$DT < t \leq T/2$ and $(1/2 + D)T < t \leq T$]*

   In this case, since the two switches are turned OFF, D_1 and D_4 are turned ON, while D_2 and D_3 are turned OFF. The equivalent circuit is shown in Fig. 5.34c, and its simplified equivalent circuit is shown in Fig. 5.37d.

   From Fig. 5.37d, the inductor voltage is $-V_o/n$, and the inductor current for the first interval is given by:

$$i_L(t) = -\frac{V_o}{nL}(t - (DT)) + I_L(DT) \tag{5.47}$$

where $I_L(DT) = I_{L_{max}}$ is the initial inductor current at $t = DT$.

Exercise 5.13

Show that the average input and output currents for the Weinberg converter are given by Eqs. (5.48) and (5.49), respectively.

$$I_{o.ave} = \frac{(I_{L_{max}} + I_{L_{max}})}{2}\left(\frac{D}{m} + \frac{1 - D}{n}\right) \tag{5.48}$$

$$I_{in,ave} = \frac{(I_{L_{max}} + I_{L_{max}})D}{2} \tag{5.49}$$

Example 5.9

It can be shown that the two-switch Weinberg converter of Fig. 5.36a can be simplified by using a one-switch equivalent circuit model as shown in Fig. 5.38. Derive the inductor current and voltage gain equation, and draw the key current and voltage waveforms for the simplified converter of Fig. 5.38.

Solution

Figure 5.39a and b shows the two equivalent circuit modes. Figure 5.39c shows the switching waveforms for the simplified circuit of Fig. 5.38.

The inductor current for the two modes are given by:

$$i_L(t) = \frac{V_{in} - V_o/m}{L}t + I_L(0) \qquad 0 < t \leq \text{DT} \tag{5.50a}$$

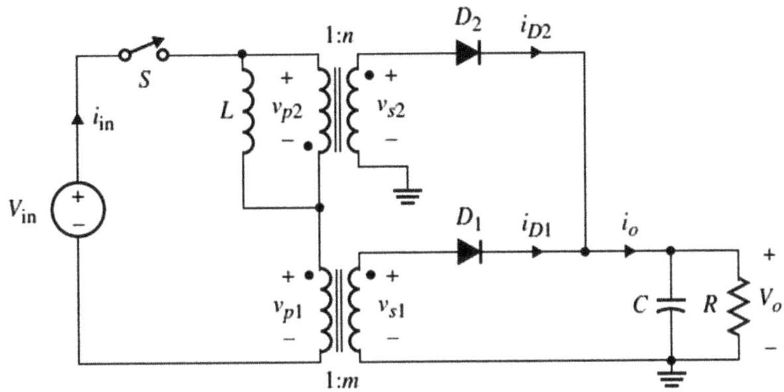

Fig. 5.38 Simplified one-switch equivalent circuit model

$$i_{\mathrm{L}}(t) = \frac{-V_{\mathrm{o}}}{L}(t - (\mathrm{DT})) + I_{\mathrm{L}}(\mathrm{DT}) \qquad \mathrm{DT} < t < T \tag{5.50b}$$

It can be shown that the voltage gain equation is:

$$\frac{V_{\mathrm{o}}}{V_{\mathrm{in}}} = \frac{D}{\frac{D}{m} + \frac{1-D}{n}} \tag{5.51}$$

If $m = n$, then the voltage gain becomes the gain of the buck converter. Therefore:

$$\frac{V_{\mathrm{o}}}{V_{\mathrm{in}}} = D \tag{5.52}$$

It can be also shown that $I_{\mathrm{L_{max}}}$ and $I_{\mathrm{L_{min}}}$ are given by:

$$I_{\mathrm{L_{max}}} = \frac{V_{\mathrm{o}}^2}{RDV_{\mathrm{in}}} + \frac{V_{\mathrm{o}}}{2nL}(1 - D)T \tag{5.53a}$$

$$I_{\mathrm{L_{min}}} = \frac{V_{\mathrm{o}}^2}{RDV_{\mathrm{in}}} - \frac{V_{\mathrm{o}}}{2nL}(1 - D)T \tag{5.53b}$$

For ccm operation, $I_{\mathrm{L_{min}}} > 0$ and to obtain the critical inductor set $I_{\mathrm{L_{min}}} = 0$, then:

$$L_{\mathrm{crit}} > \frac{RDV_{\mathrm{in}}}{2nV_{\mathrm{o}}}(1 - D)T \tag{5.54}$$

The normalized output capacitor ripple voltage is given by:

$$\Delta V_{\mathrm{nc}} = \frac{n^2\tau_{\mathrm{n}}}{2\tau_{\mathrm{nRC}}}\left[\left(\frac{I_{\mathrm{n,max}}}{n} - 1\right)^2 + \left(\frac{mD}{n(1-D)}\right)\left(1 - \frac{I_{\mathrm{n,min}}}{m}\right)^2\right] \tag{5.55}$$

where, $V_{\mathrm{nc}} = V_{\mathrm{c}}/V_{\mathrm{in}}$, $\tau_{\mathrm{n}} = \tau/T = L/RT$, $\tau_{\mathrm{nRC}} = RC/T$, $I_{\mathrm{n,max}} = I_{\mathrm{max}}/I_{\mathrm{o}}$, and $I_{\mathrm{n,min}} = I_{\mathrm{min}}/I_{\mathrm{o}}$.

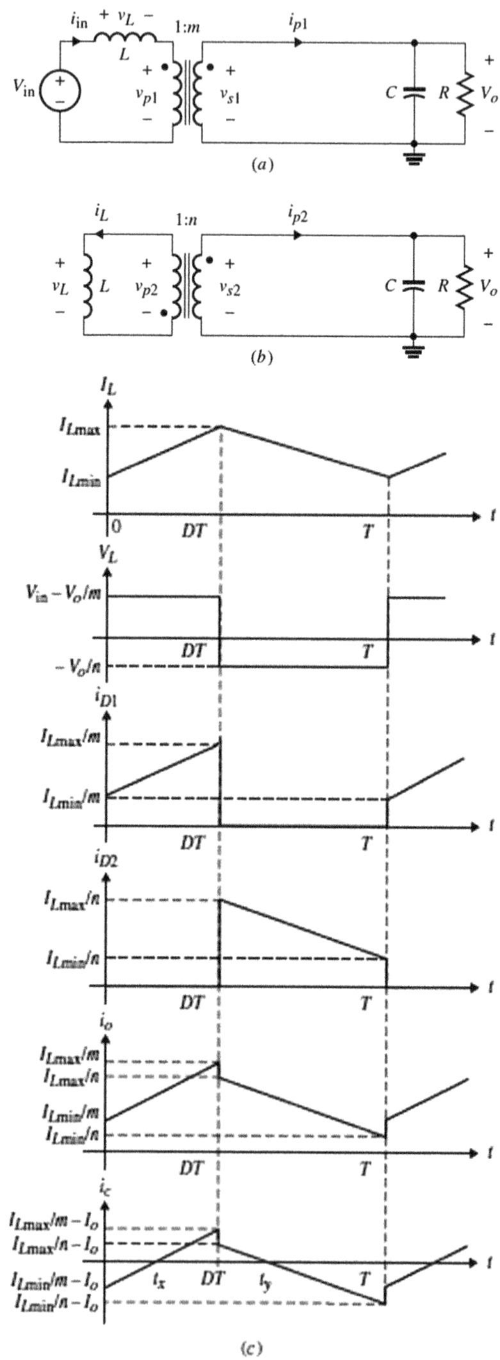

Fig. 5.39 Equivalent circuits for Fig. 5.38: (**a**) Mode 1, when the switch is ON, (**b**) Mode 2, when the switch is OFF, (**c**) converter waveforms

Example 5.10

Consider a dc-dc converter unit system that uses the Weinberg converter with an input dc equal to 125 V and an output of 48 V. The design has the following specifications:

Output power $P_o = 2$ kW

Switching frequency $f = 40$ kHz

Output ripple voltage $\Delta V_c/V_o = 0.2\%$

Determine the following:

(a) The value of n where $D = 0.7$ and $m = 0.5$
(b) The critical inductor value
(c) The maximum and minimum inductor current
(d) The average input and output current
(e) The capacitor current value

Solution

(a) From the gain relation, we have:

$$\frac{V_o}{V_{in}} = \frac{D}{\frac{D}{m} + \frac{1-D}{n}}$$

$$\frac{48}{125} = \frac{0.7}{\frac{0.7}{0.5} + \frac{1-0.7}{n}}$$

$$n \approx 0.7$$

(b) The minimum inductor value is given by:

$$L_{crit} > \frac{RDV_{in}}{2nV_o}(1-D)T$$

$R = V_o^2/P = 1.152\ \Omega$, and $T_s = 1/f = 25\ \mu s$; therefore:

$$L_{crit} = \frac{(1.125)(0.7)(125)}{2(48)(0.7)}(1-0.7)(25\ \mu s)$$

$$L_{crit} = 11.25\ \mu H$$

Select $L = 30\ \mu$H.

(c) The maximum inductor current is:

$$I_{L_{max}} = \frac{V_o^2}{RDV_{in}} + \frac{V_o}{2nL}(1-D)T_s$$

$$I_{L_{max}} = \frac{(48)^2}{(1.125)(0.7)(125)} + \frac{48}{2(0.7)(30\ \mu H)}(1-0.7)(25\ \mu s)$$

$$I_{L_{max}} = 31.4\ \text{A}$$

And the minimum value is given by:

$$I_{L_{max}} = \frac{V_o^2}{RDV_{in}} - \frac{V_o}{2nL}(1-D)T_s$$

$$I_{L_{max}} = 14.3 \text{ A}$$

(d) From Eq. (5.34), we have:

$$I_{o.ave} = \frac{(I_{L_{max}} + I_{L_{max}})}{2}\left(\frac{D}{m} + \frac{1-D}{n}\right)$$

$$I_{o.ave} = \frac{(31.4 \ + 14.3\ )}{2}\left(\frac{0.7}{0.5} + \frac{1-0.7}{0.7}\right)$$

$$I_{o.\ ave} = 41.78\ A$$

And from Eq. (5.35) we have:

$$I_{in,ave} = \frac{(I_{L_{max}} + I_{L_{max}})D}{2}$$

$$I_{in,ave} = \frac{(31.4 \ + 14.3)}{2}0.7$$

$$I_{in,\ ave} = 16\ A$$

(e) The times at which the capacitor current becomes zero are given by:

$$t_x = \frac{I_o - I_{L_{min}}/m}{\frac{1}{L}\left(\frac{V_{in}}{m} - \frac{V_o}{m^2}\right)}$$

$$t_x = \frac{41.78 - 14.3/0.5}{\frac{1}{30\ \mu H}\left(\frac{125}{0.5} - \frac{48}{0.5^2}\right)}$$

$$t_x = 6.83\ \mu s$$

and

$$t_y = \frac{nL}{V_o}(I_{L_{max}} - nI_o) + DT_s$$

$$t_y = \frac{(0.7)(30\ \mu H)}{48}(31.4 - 0.7 \times 41.78) + (0.7)(25\ \mu s)$$

$$t_y = 18.44\ \mu s$$

Recall, $\tau_n = \frac{\tau}{T_s} = \frac{L/R}{T_s}$ and $\tau_{nRC} = \frac{RC}{T_s}$

Therefore

$$\Delta V_{nc} = \frac{n^2 \tau_n}{2\tau_{nRC}}\left\{\left(\frac{I_{n,max}}{n} - 1\right)^2 + \left(\frac{mD}{n(1-D)}\right)\left(1 - \frac{I_{n,min}}{m}\right)^2\right\}$$

The value for the capacitor is obtained as:

$$C = 476.6\ \mu F$$

5.6 Multi-output Converter

As stated earlier, one of the advantages of using an isolated transformer is to allow for multiple outputs at the secondary side of the output transformer. This is very common in today's power supplies used in components to provide several voltage levels at different currents and polarities. Figure 5.40a, b shows three-output examples of a half-bridge buck-derived converter and a single-ended flyback converter, respectively.

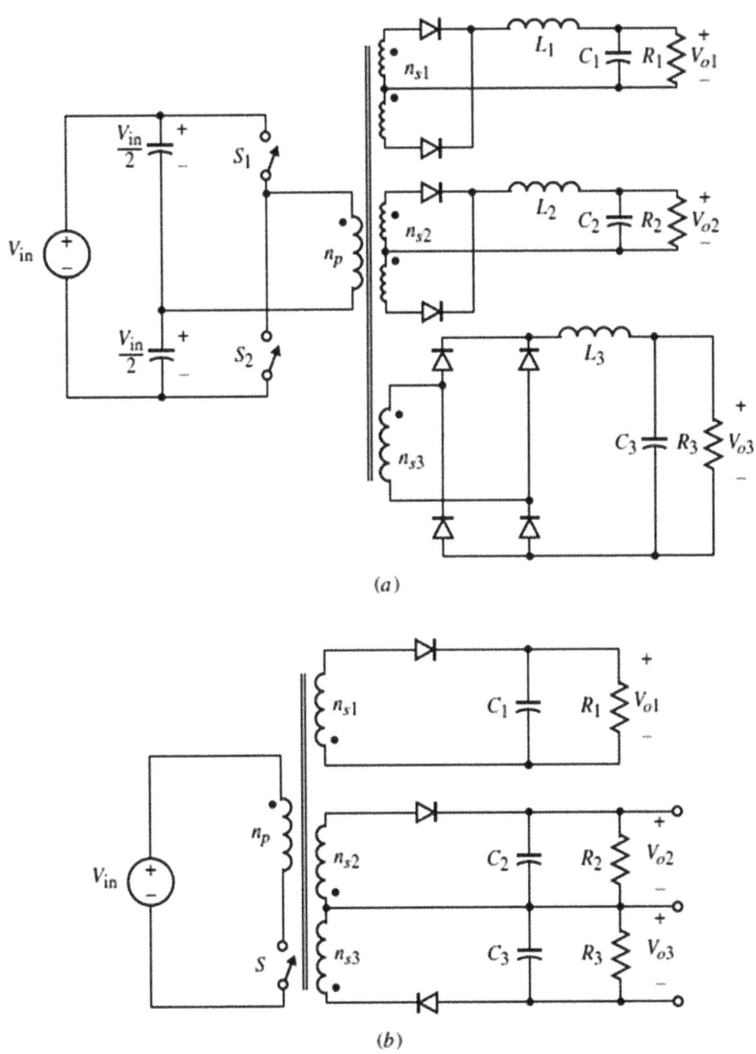

Fig. 5.40 (**a**) Half-bridge three-output buck-derived converter, (**b**) three-output flyback converter

Exercise 5.14

It is required to design the half-bridge converter of Fig. 5.40a to provide three output voltages with $V_{o1} = 5$ V at 6 A, $V_{o2} = 37$ V at 0.5 A, and $V_{o3} = 12$ V at 2 A. Design for transformer turns ratio and the converter components such that the voltage ripple for any output does not exceed 0.1%. Assume $V_{in} = 160$ VDC and a switching frequency of 100 kHz.

Exercise 5.15

Determine the rms currents for the power switches and transformer windings of Example 5.8.

Answer: 0.877 A, 1.24 A, 3.24 A, 0.534 A, 1.34 A.

Exercise 5.16

Design the multi-output converter of Fig. 5.40b for the following output specifications: $V_{o1} = +5$ V at 4 A, $V_{o2} = +12$ V at 0.5 A, and $V_{o3} = -12$ V at 0.3 A.

The ripple voltage does not exceed 100 mV peak-to-peak for each output. Use $V_{in} = 185$ V, $f_s = 50$ kHz, and $D = 0.5$.

Answer: $n_{s_1}/n_p = 1/37, n_{s_2}/n_p = n_{s_3}/n_p = 12/185, L_m > 2.925$ mH, $C_1 > 400$ μF, $C_2 = 50$ μF, $C_3 > 30$ μF

In practice, normally the output voltage with the highest power is controlled with the other outputs using linear regulators to stabilize their voltages. This is why the open-loop outputs are designed for a higher voltage than specified.

Problems

Forward Converter

5.1 Consider the single-ended forward converter of Fig. P5.1 by including a leakage inductance of the power transformer. Discuss the effect of L_k on the circuit operation.

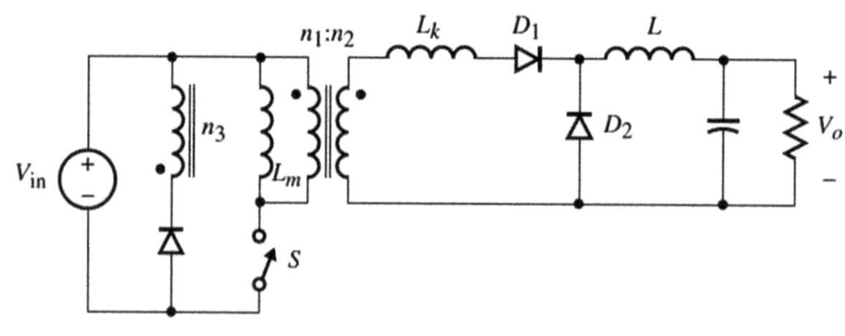

Fig. P5.1

D5.2 Design a forward converter of Fig. 5.11 that needs to supply 400 W at 15 A load current. The output ripple should not exceed 1%.

5.3 Derive the expression for the output ripple voltage for a forward converter of Fig. 5.11.

5.4 The circuit given in Fig. P5.4 is another type of a forward converter. Sketch the primary voltage and diode currents waveforms for the circuit.

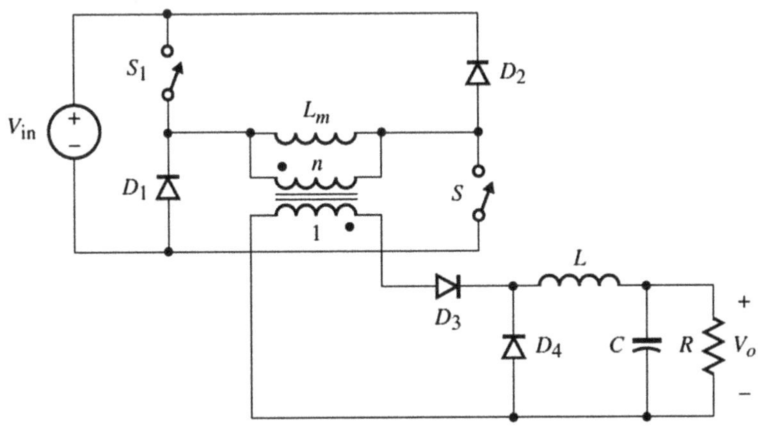

Fig. P5.4

5.5 Replace D_1 and D_2 by S_3 and S_4, respectively, in Problem 5.45. Derive the voltage gain if the four switches have the switching waveform in Fig. P5.5.

Fig. P5.5

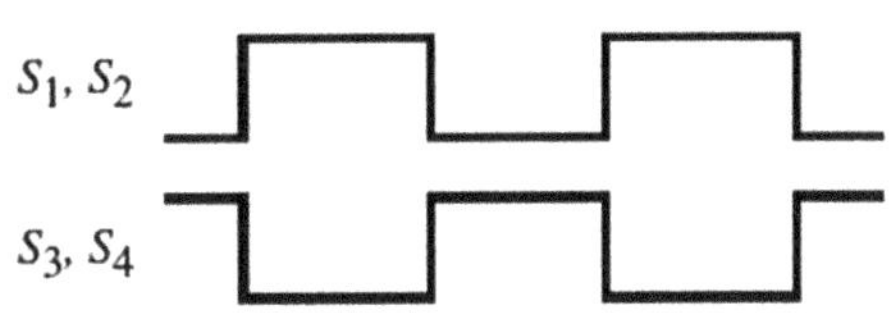

5.6 Consider the single-ended forward converter given in Fig. P5.6 that uses a regular and a Zener diode to provide a negative voltage damp across the primary in order to reset the core to avoid transformer saturation. (a) Discuss the operation of the circuit, and (b) sketch the current waveform i_{D_1}, i_{D_2}, i_L, i_{D_3}, i_{sw}, and i_D.

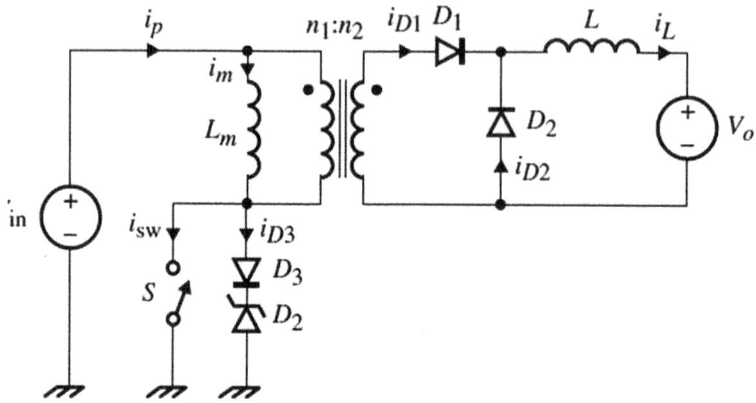

Fig. P5.6

Full-Bridge

D5.7 Design the full-bridge dc-dc converter with a center-tap output transformer with the following specifications: $V_{in} = 480$ V, $V_o = 240$ V at $I_o = 20$ A, and $f_s = 75$ kHz.

5.8 Sketch the labeled current and voltage waveforms for the double-ended buck-derived converter shown in Fig. P5.9. Assume the switching sequence for $S_1 - S_3$ and $S_2 - S_4$ is the same as that in Fig. 5.18b. Assume $RC \gg T$.

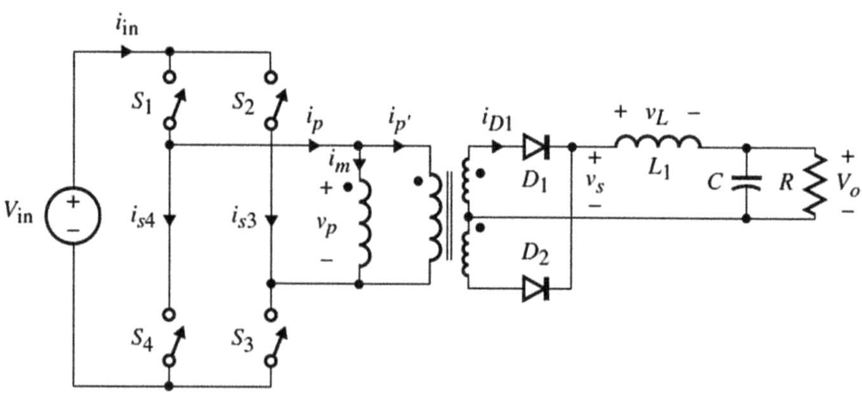

Fig. P5.8

5.9 Repeat Example 5.6 by using the switching sequence in Fig. P5.10. Derive the voltage gain in terms of the phase shift period D_1.

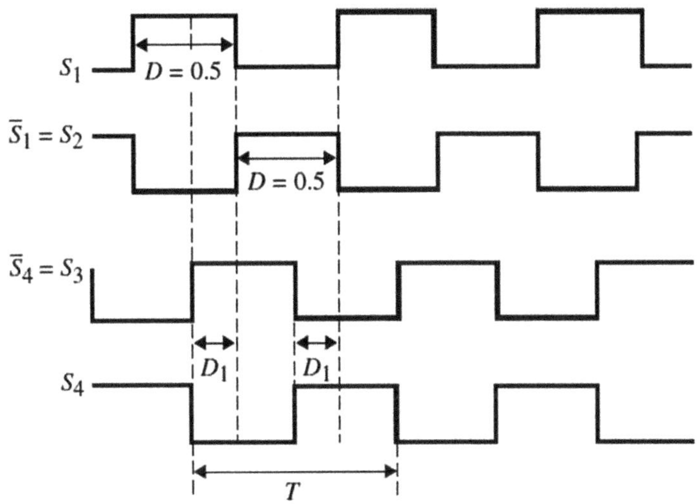

Fig. P5.9

Push-Pull Converter

5.10 Derive the output voltage ripple for the push-pull converter.

Isolated SEPIC Converter

5.11 Derive the voltage gain equation for the isolated SEPIC converter shown in Fig. P5.11.

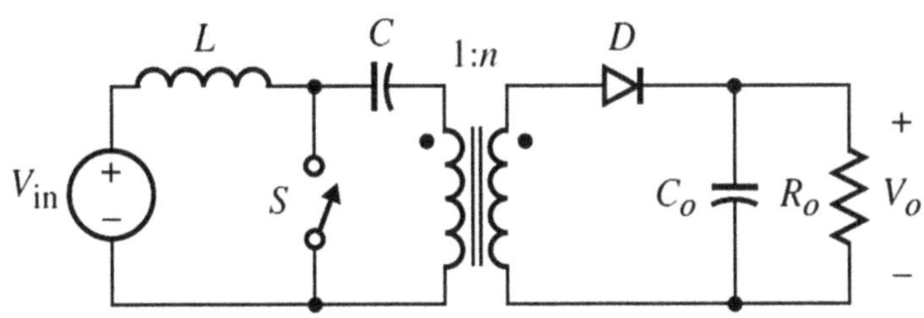

Fig. P5.11

5.12 Derive the voltage gain expression for the isolated Cuk converter shown in Fig. P5.12.

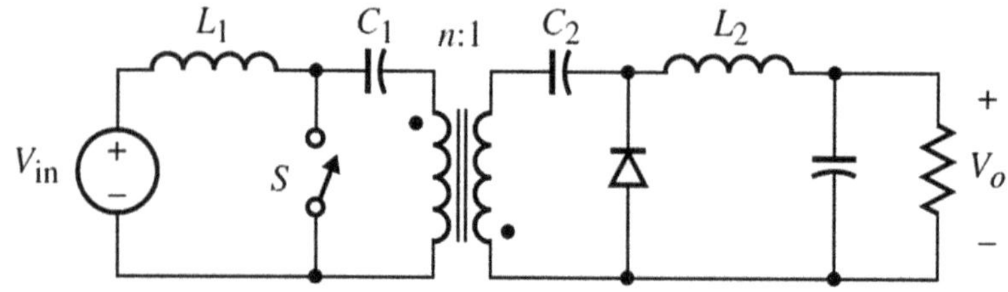

Fig. P5.12

Weinberg Converter

5.13 Repeat Problem 5.3 for the Weinberg converter.

5.14 Show that the voltage gain equation for the Weinberg converter of Fig. 5.35 is given by:

$$\frac{V_o}{V_{in}} = \frac{D}{\frac{D}{m} + \frac{1-D}{n}}$$

5.15 Show that t_x and t_y of Fig. 5.35 are given by:

$$t_x = \frac{I_o - I_{L_{min}}/m}{\frac{1}{L}\left(\frac{V_{in}}{m} - \frac{V_o}{m^2}\right)}$$

$$t_y = \frac{nL}{V_o}(I_{L_{max}} - nI_o) + DT$$

For a special case when $I_{n,max}/n = 1$, show that $t_y = DT_s$, t_x is the same as above, and ΔV_{nc} is given by:

$$\Delta V_{nc} = \frac{n^2 \tau_n}{2\tau_{nRC}}\left(\frac{mD}{n(1-D)}\right)\left(1 - \frac{I_{n,min}}{m}\right)^2$$

Design Problems

Multi-output Converters

D5.16 Design the four-output buck-boost converter shown in Fig. P5.16 to operate in ccm with the following specifications:

$V_{in,min} = 28$ V, $V_{in,max} = 48$ V, $V_{in,nom} = 36$ V, $V_{o1} = 5$ V at $I_o = 10$ A, $V_{o2} = 5$ V at $I_o = 1$ A, $V_{o3} = 12$ V at $I_o = 2$A, $I_o = 2$ A, $V_{o4} = -12$ V at $I_o = 3$ A, $f_s = 70$ kHz, and $\Delta V_o/V_o = 1\%$ for each output.

Fig. P5.16

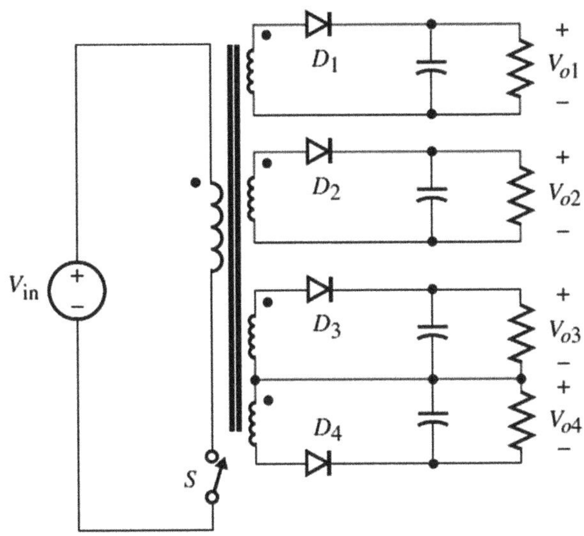

D5.17 Design the two-output forward converter shown in Fig. P5.17 to operate in ccm with the following specifications:

$V_{\text{in, min}} = 78$ V, $V_{\text{in, max}} = 110$ V, $V_{\text{in, nom}} = 92$ V, $V_{o1} = 5$ V at $I_o = 15$ A, $V_{o2} = 12$ Vat $I_o = 5$ A $f_s = 70$ kHz, and $\Delta V_o/V_o = 1.5\%$ for each output.

Fig. P5.17

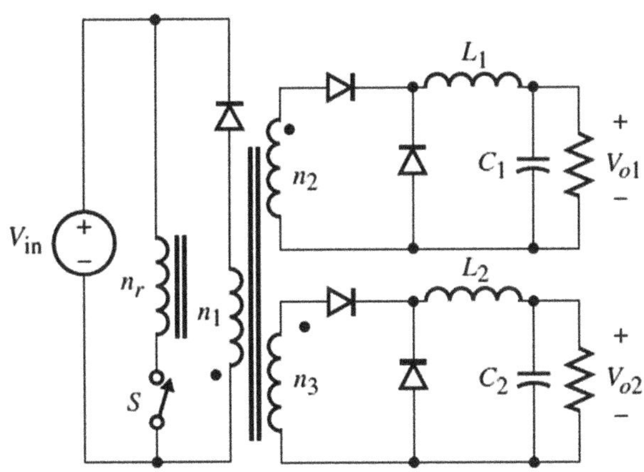

D5.18 Design the two-output, two-switch forward converter shown in Fig. P5.18 to operate in ccm with the following specifications:

$V_o = 210$ V, $V_{\text{in, max}} = 480$ V, $V_{\text{in, nom}} = 370$ V, $V_{o1} = 5$ V at $I_o = 1.5$ A, $V_{o2} = 28$ V at $I_o = 5$ A, $f_s = 85$ kHz, $\Delta V_{o1}/V_{o1} = 1\,\%$, $\Delta V_{o2}/V_{o2} = 5\%$, $\Delta I_{o1}/I_{o1} = 15\%$, $\Delta I_{o2}/I_{o2} = 10\%$.

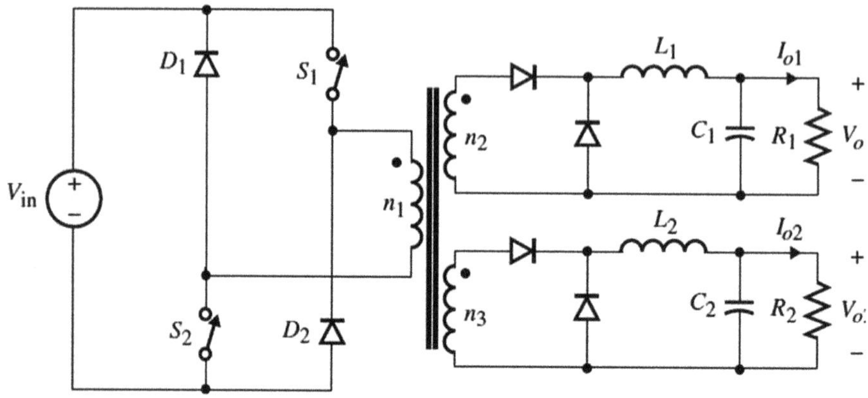

Fig. P5.18

Single-Output Converters

D5.19 Design the simplified Weinberg converter shown in Fig. P5.19 to operate in ccm with the following specifications:

$V_{in,\,nom} = 48$ V, $V_o = 22$ Vat $I_o = 3$ A to 0.3 A, $f_s = 120$ kHz, and $\Delta V_o/V_o = 1\%$

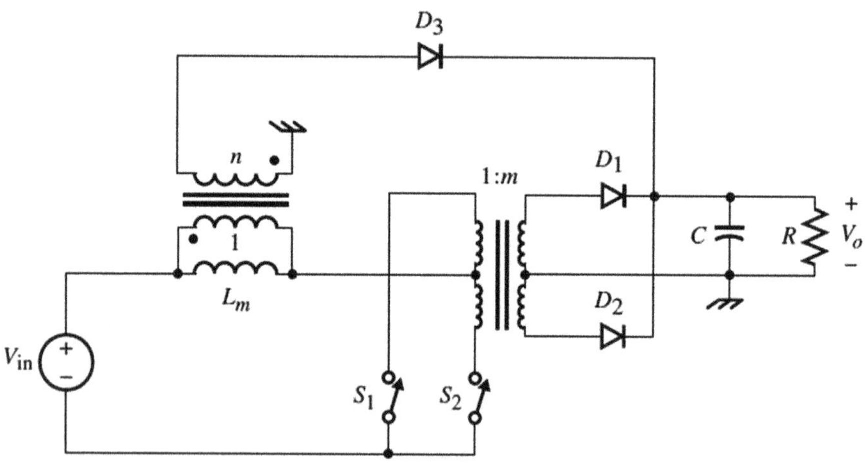

Fig. P5.19

D5.20 Design the single-ended forward converter of Fig. 5.8 to operate in ccm with the following specifications:

$V_{in,\,nom} = 28$ V, $V_o = 12$ V at $I_o = 4$ A to 1 A, $f_s = 125$ kHz, and $\Delta V_o/V_o = 1\%$. Assume diode forward voltage drop is 1 V.

D5.21 Redesign Problem 5.20 using a push-pull converter of Fig. 5.19a.

D5.22 Design the full-bridge converter of Fig. 5.18 to operate in ccm with the following specifications:

$V_{\text{in, nom}} = 150$ V, $V_o = 12$ V at $I_o = 2$ A, $f_s = 100$ kHz, $\Delta V_o/V_o = 1\%$, $\Delta I_o/I_o = 5\%$

D5.23 Design the half-bridge converter of Fig. 5.17a to operate in ccm with the following specifications:

$V_{\text{in, nom}} = 185$ V, $V_o = 28$ V at $I_o = 2$ A, $f_s = 100$ kHz, $\Delta V_o/V_o = 0.1\%$, $\Delta I_o/I_o = 5\%$.

D5.24 Design the forward converter that uses a tertiary winding shown in Fig. 5.14 with the following specifications:

$V_{\text{in, nom}} = 285$ V $\pm 20\%$, $V_o = 12$ Vat $I_o = 2$ A to 200 mA, $f_s = 120$ kHz, $\Delta V_o = 100$ mV.

Determine the diode and transistor voltage and current ratings.

D5.25 Repeat Problem 5.26 with the same specifications except that it is desired to add an additional output voltage with the following specifications:

$V_{o2} = 5$ Vat 4 A to 500 mA and the output voltage ripple does not exceed 50 mV. The circuit configuration is shown in Fig. P5.25.

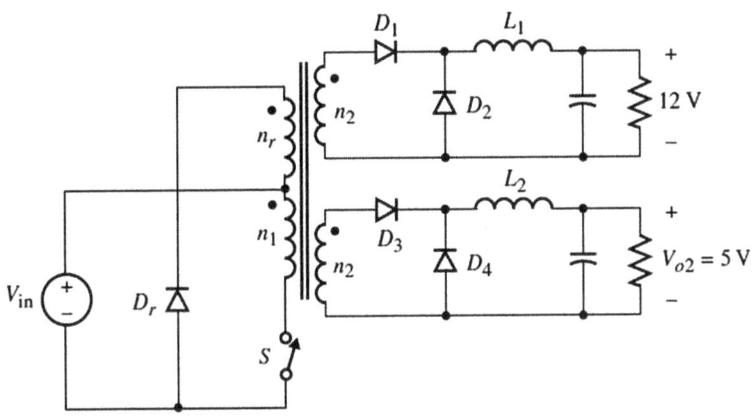

Fig. P5.25

Additional Topologies

5.26 For the three circuits given in Fig. P5.26a–c, the voltage gains are given by the following equations, respectively. Assume S_1 and S_2 are switched simultaneously.

(a) $\frac{V_o}{V_{in}} = \frac{D}{(1-D)^2}\frac{n_2}{n_1}$

(b) $\frac{V_o}{V_{in}} = \frac{D^2}{(1-D)}\frac{n_2}{n_1}$

(c) $\frac{V_o}{V_{in}} = \frac{-D}{(1-D)(1-2D)}\frac{n_2}{n_1}$

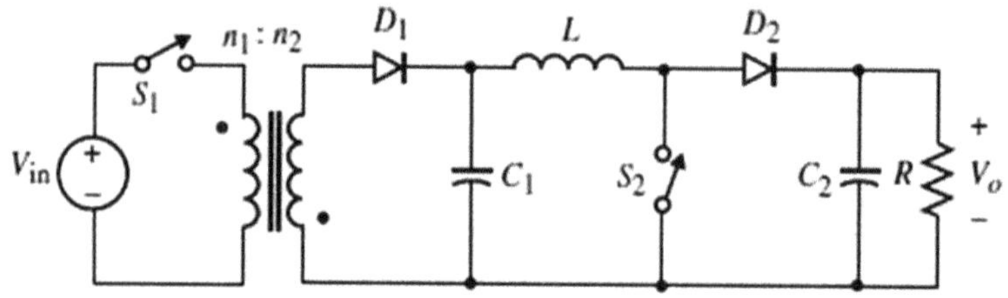

(a)

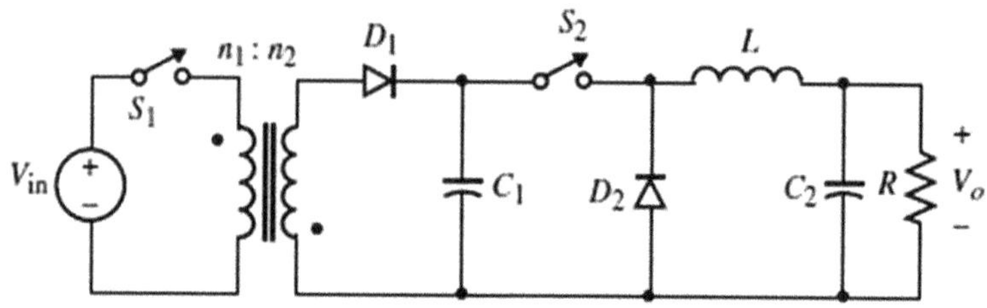

(b)

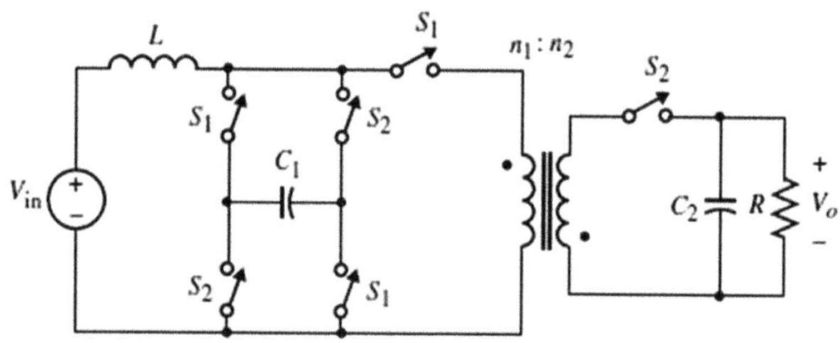

(c)

Fig. P5.26

5.27 The converter shown in Fig. P5.27a is another version of the forward converter. Assume ideal diodes and L_m is the transformer magnetizing inductance. The switching waveforms for S_1, S_2, S_3, and S_4 are shown in Fig. P5.27. Assume $RC \gg T$.

(a) Sketch the current and voltage waveforms labeled on the figure.
(b) Derive the voltage gain expression V_o/V_{in}.

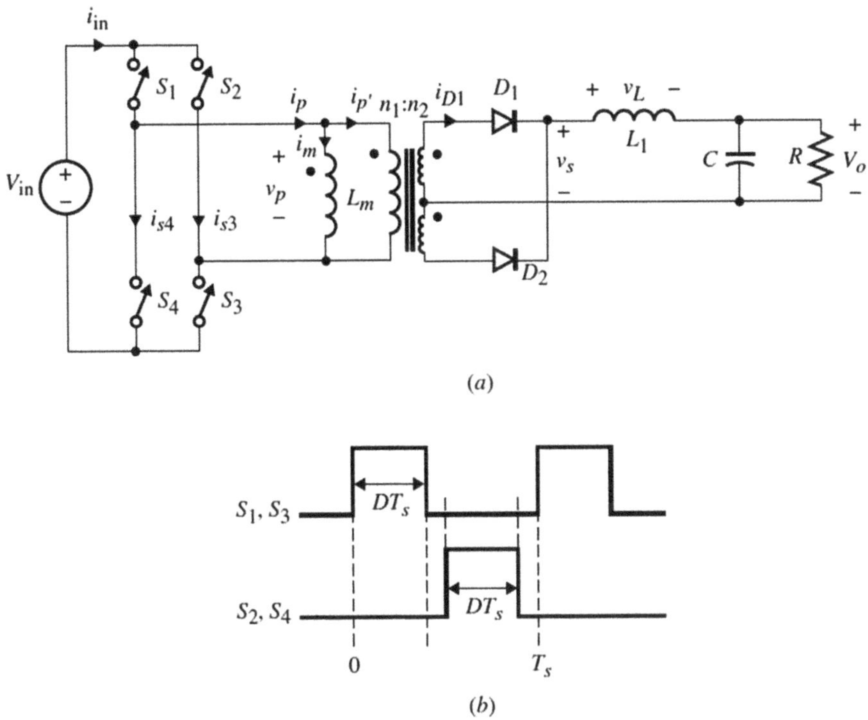

Fig. P5.27

5.28 Draw the waveforms for the branch currents and voltages for the push-pull converter by including a finite magnetic inductance as shown in Fig. P5.28.

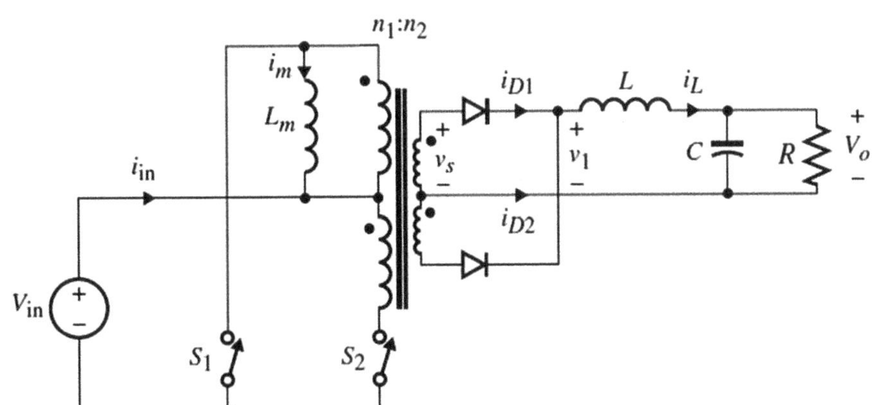

Fig. P5.28

5.29 (a) Show that if the single-ended flyback converter of Fig. 5.23b is operating in dcm (magnetizing current is discontinuous), then the voltage conversion ratio is given by:

$$\frac{V_o}{V_{in}} = \frac{n_2}{n_1} D \sqrt{\frac{1}{2\tau}}$$

where

$$\tau = L/RT$$

5.30 Consider the circuit given in Fig. P5.30 with the switching sequence given in Fig. P5.30:

(a) Sketch the waveforms for i_L, v_s, i_{D_1}, and i_{D_4}.
(b) Derive the expression for V_o/V_{in}.
(c) What is the expression for the critical value of L to maintain ccm of operation?

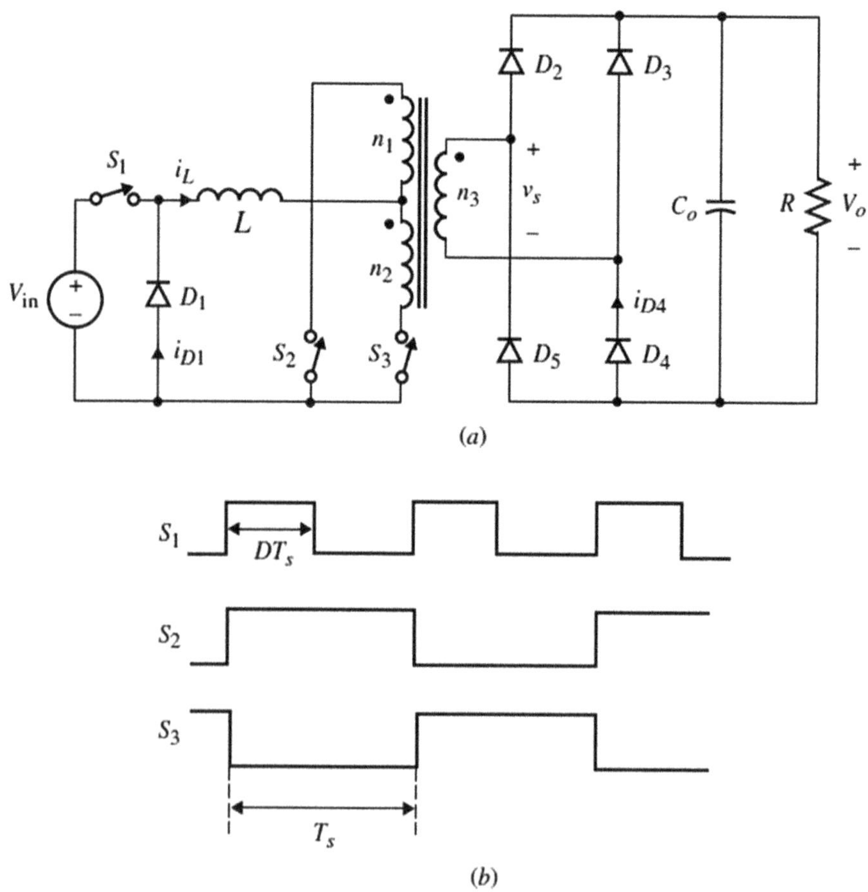

Fig. P5.30

5.31 There are several core reset schemes available in today's dc-dc converters. Figure P5.31 shows one possible core resetting circuit used in high-power applications. The winding n_r is the reset winding. (a) Discuss the operation of each circuit and compare them to each other and (b) draw the waveforms for i_{in}, v_{sw}, i_o, v_o, and i_{D_1}.

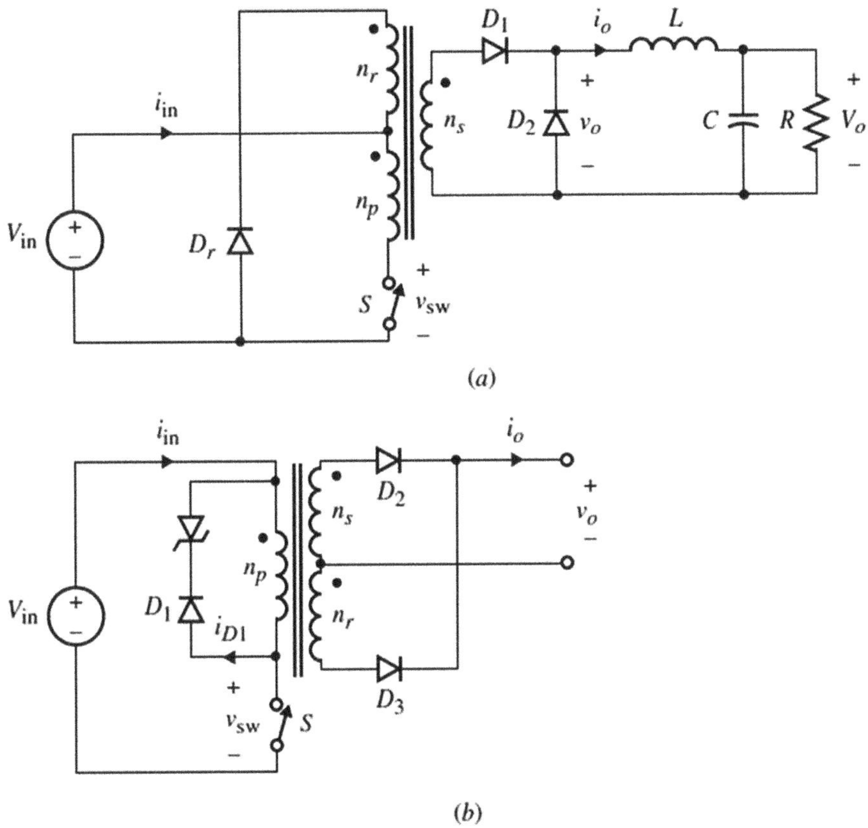

Fig. P5.31

5.32 Derive the voltage gain equation for the isolated double-switch, single-ended dc-dc converter in Fig. P5.32. S_1 and S_2 are switched simultaneously.

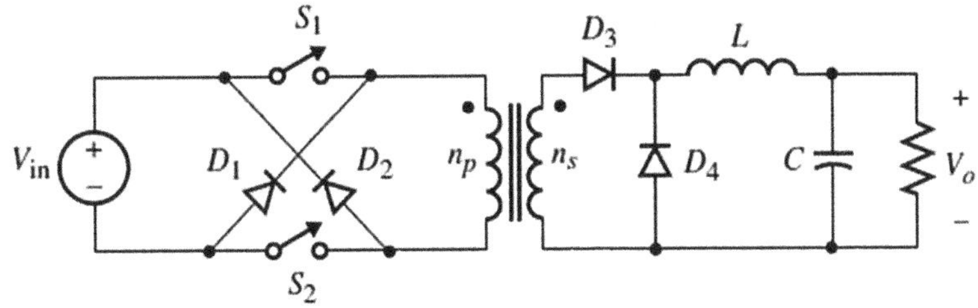

Fig. P5.32

5.33 Consider the current-fed push-pull converter given in Fig. P5.33. Derive the current gain expression I_o/I_{in}.

Fig. P5.33

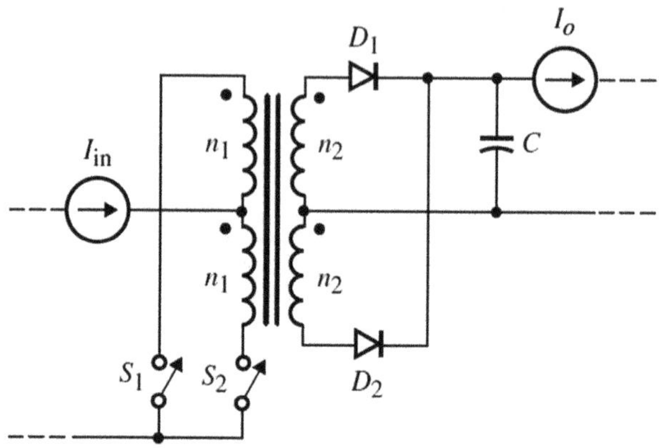

5.34 The converter shown in Fig. P5.34 is an isolated version of the converter given in Chap. 4. Assume the magnetizing inductor is finite, L_m. Drive the expression for V_o/V_{in}.

Fig. P5.34

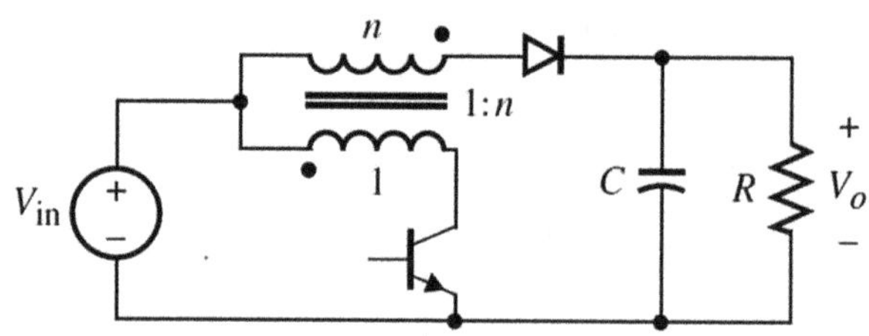

5.35 Figure P5.35 shows the isolated version of the SEPIC converter, where a transformer is inserted in place of L_2. Assume the transformer has a magnetizing inductance, L_m, and a turns ratio of $n_1 : n_2$.

(a) Sketch the waveforms for i_{in}, v_{sw}, i_o, v_o, and i_{D_1} and v_{sw}.
(b) Show that the voltage gain is given by:

$$\frac{V_o}{V_{in}} = \frac{n_2}{n_1} \frac{D}{(1-D)}$$

where D is duty cycle of the switch.

Fig. P5.35

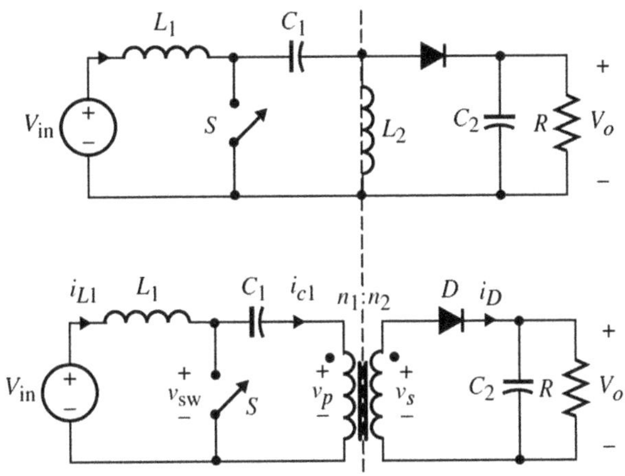

5.36 The converter in Fig. P5.36a, b is known as the isolated Watkins-Johnson converter. Assume S_1 and S_2 are switched on and off as shown in Fig. P5.36c. Assume finite magnetizing inductance. Derive the expressions for V_o/V_{in} for each converter.

Fig. P5.36

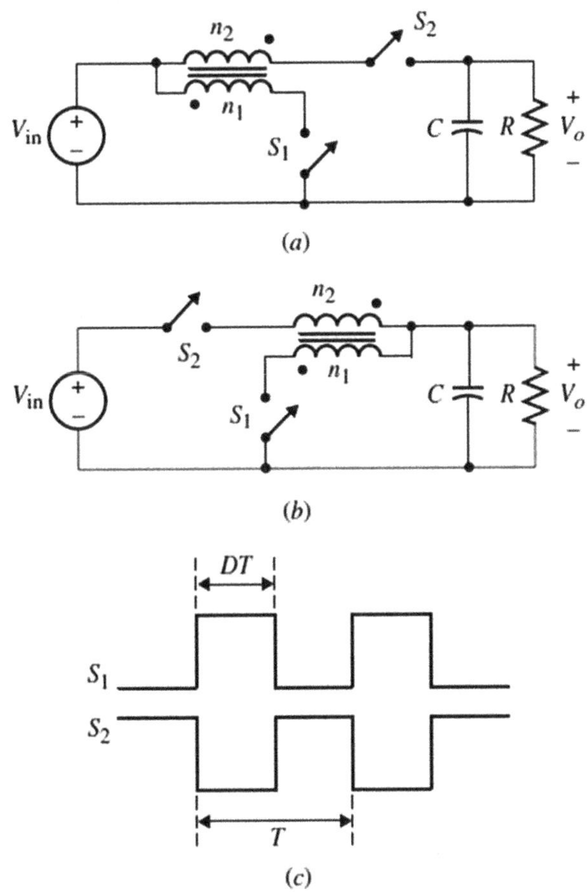

5.37 Draw the waveforms for the voltage and current variables in the push-pull converter shown in Fig. P5.37. Unlike Fig. 5.16a, this converter assume finite magnetizing inductance.

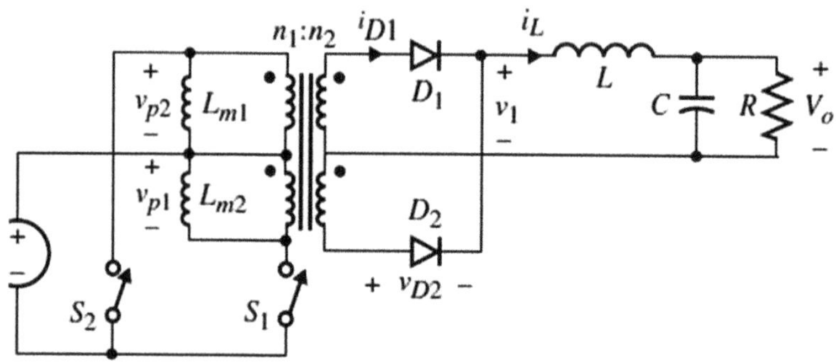

Fig. P5.37

D5.38 Consider a flyback converter that draws its input current from a dc voltage source that varies between 185 V and 275 V and produces an output voltage of +12 V at an average load current of 10 A.

(a) Determine the value of the transformer ratio and magnetizing inductance L_m.
(b) Determine the range of the duty cycle needed to maintain a constant output voltage. Assume $f = 85$ KHz and a maximum duty cycle of 0.6.

D5.39 Repeat the design in Problem P5.38 by assuming the power switch and diode have 3 V and 0.7 V voltage drops when conducting, respectively.

5.40 Determine the efficiency of the flyback converter of Fig. 5.28 of Example 5.6 by assuming its output diode has a voltage drop of 0.75 V, and the on resistance of the switch is 48 mΩ.

5.41 The Fig. P5.41a shows a forward convertor with resent winding and magnetizing inductor whose current waveforms are shown in Fig. P5.41b. Assume the convertor is operating in ccm with its output current and assume constant with ripple free. Based on the values shown, determine the convertor parameters and the primary-side magnetizing inductance.

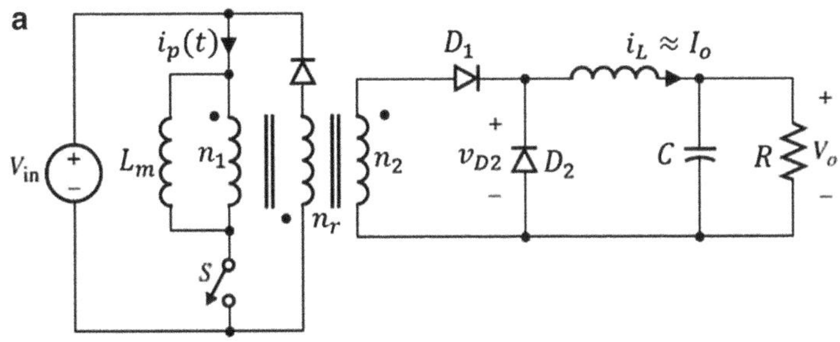

Fig. P5.41a

Fig. P5.41b

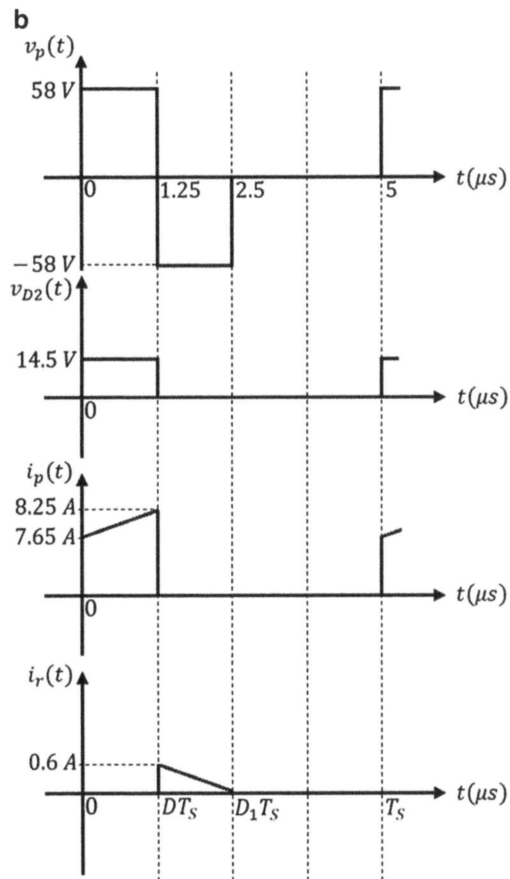

Chapter 6
Soft-Switching dc-dc Converters

A class of dc-dc converters, known in the literature as soft-switching resonant converters, has been thoroughly investigated in recent years for its various attractive features. Soft switching means that one or more power switches in a dc-dc converter have either the turn-on or turn-off switching losses eliminated. This is in contrast to hard switching, where both turn-on and turnoff of the power switches are done at high current and high voltage levels. One approach is to create a full-resonance phenomenon within the converter through series or parallel combinations of resonant components. Such converters are generally known as resonant converters. Another approach is to use a conventional PWM buck converter, boost, buck-boost, Cuk, and SEPIC and replace the switch with a resonant switch that accomplishes the loss elimination. Because of the nature of the PWM circuit, resonance occurs for a shorter time interval compared to the full-resonance case. This class of converters, combining resonance and PWM, is appropriately known as quasi-resonance converters. In this chapter, our focus will be on the latter method, mainly using the resonance PWM switch to achieve soft switching. For simplicity, here we use the term soft switching to refer to dc-dc converters, quasi-resonance converters, and other topologies that employ resonance to reduce switching losses. Two major techniques are employed to achieve soft switching: zero-current switching (ZCS) and zero-voltage switching (ZVS). This chapter will focus only on ZCS and ZVS types of PWM dc-dc resonant switches and their steady-state analyses.

6.1 Types of dc-dc Converters

As shown in previous chapters, linear-mode and switch-mode converters have been used widely in the design of commercial dc-dc power supplies. Linear power supplies offer the designer four major advantages: simplicity in design, no electrical noise in the output, fast dynamic response time, and low cost. Their applications,

I. Batarseh, A. Harb, *Power Electronics*,
https://doi.org/10.1007/978-3-319-68366-9_6

however, are limited due to several disadvantages: (1) The input voltage is at least 2 or 3 V higher than the output voltage because the circuit can only be used as a step-down regulator, (2) each regulator is limited to only one output, and (3) efficiency is low compared to other switching regulators (30–60% for an output voltage less than 20 V).

High-frequency pulse-width modulation (PWM) switching regulators overcome all the linear regulators' shortcomings: (1) They have higher efficiency (>95%); (2) power transistors operate at their most efficient points—cutoff and saturation—allowing for power densities of around 100'*s* W/in^3; (3) multi-output applications are possible; and (4) the size and the cost are much lower, especially at high power levels.

However, PWM switching converters still have several limitations, among them (1) greater circuit complexity compared to the linear power supplies, (2) high electromagnetic interference (EMI), and (3) switching speeds below 100 kHz because of high stress levels on power semiconductor devices.

A third generation of power converters is known as soft-switching resonant converters. Compared with linear and switch-mode converters, the potential advantage of the soft-switching resonant converters is reduced power losses, thus achieving high switching frequency and high power density while maintaining high efficiency. Moreover, due to the higher switching frequency, such converters exhibit faster transient responses. Today's soft-switching techniques are used in the design of both high-frequency dc-dc conversion and high-frequency dc-ac inversion. Only the first application is discussed in this text.

6.1.1 The Resonant Concept

Like switch-mode dc-dc converters, resonant converters are used to convert dc-dc through an additional stage, the resonant stage, in which the dc signal is converted to a high-frequency ac signal. The advantages of the resonant converter include the natural commutation of power switches, resulting in low switching power dissipation and reduced component stresses, which in turn results in increased power efficiency and increased switching frequency, and higher operating frequencies, resulting in reduced size and weight of equipment and in faster response and hence a possible reduction in EMI problems.

Since the size and weight of the magnetic components (inductors and transformers) and the capacitors in a converter are inversely proportional to the converter's switching frequency, many power converters have been designed for progressively higher frequencies in order to reduce the size and weight and to obtain fast converter transients. In recent years, the market demand for wide applications that require variable-speed drives, highly regulated power supplies, and uninterruptible power supplies, as well as the desire for smaller size and lighter weight, has increased.

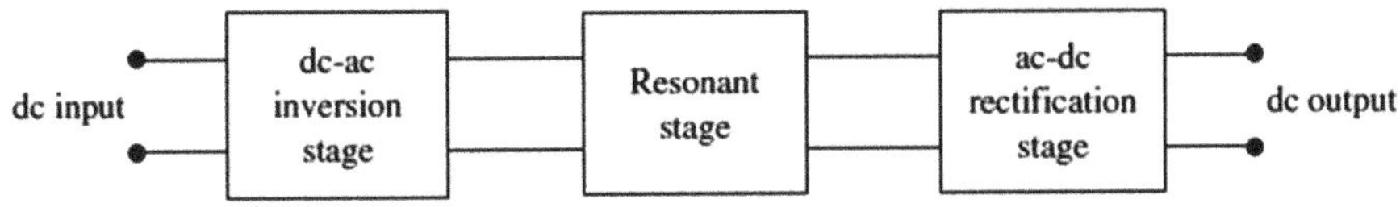

Fig. 6.1 Typical block diagram of soft switching dc-dc converter

There are many soft-switching techniques available in the literature to improve the switching behavior of dc-dc resonant converters. At the time of this writing, intensive research in soft switching is under way to further improve the efficiency through increased switching frequency of power electronic circuits.

From a circuit standpoint, a dc-dc resonant converter can be described by three major circuit blocks as shown in Fig. 6.1: the dc-ac input inversion circuit, the resonant energy buffer tank circuit, and the ac-dc output rectifying circuit. Typically, the dc-ac inversion is achieved by using various types of switching network topologies. The resonant tank, which serves as an energy buffer between the input and the output, is normally synthesized by using a lossless frequency-selective network. The purpose of that network is to regulate the energy flow from the source to the load. Finally, the ac-dc conversion is achieved by incorporating rectifier circuits at the output section of the converter.

6.1.2 Resonant Versus Conventional PWM

For many years, high-efficiency power processing circuits have been achieved by operating power semiconductor devices in the switching mode, whereby switching devices are operated in either the on or off state, as in the PWM method. In PWM converters, the switching of semiconductor devices normally occurs at high current levels. Therefore, when switching at high frequencies, these converters are associated with high power dissipation in their switching devices. Furthermore, PWM converters suffer from EMI caused by high-frequency harmonic components associated with their quasi-square switching current and/or voltage waveforms. Unfortunately, even though the technological advancements of PWM switch mode converters have resulted in faster switching, their operating frequency is limited by the factors mentioned above.

In the resonant technique, switching losses in the semiconductor devices are avoided due to the fact that the current through or voltage across the switching device at the switching point is equal to or near zero. This reduction in switching losses allows the designer to attain a higher operating frequency without sacrificing the converter's efficiency. Compared to the PWM converters, the resonant converters show the promise of achieving the design of small-size and low-weight converters. Currently, resonant power converters operating in the range of a few megahertz are available. Another advantage of resonant converters over PWM converters is the decrease in the harmonic content in the converter voltage and current waveforms. Therefore, when the resonant and PWM converters are operated at the same power level and frequency, the resonant converter can be expected to have lower harmonic emission.

6.2 Classification of Soft-Switching Resonant Converters

The literature is very rich with resonant power electronic circuits used in applications such as dc-dc and dc-ac resonant converters. To date, there exists no general classification of resonant converter topologies. In fact, more and more resonant topologies continue to be introduced in the open literature. Several types of dc-dc converters employing additional resonant stages that have been explored thus far may be summarized as follows:

- *Quasi-resonant converter (single ended)*
  - Zero-current switching (ZCS)
  - Zero-voltage switching (ZVS)
- *Full-resonant converters (conventional)*
  - Series resonant converter (SRC)
  - Parallel resonant converter (PRC)
- *Quasi-square wave (QSW) converters*
  - Zero-current switching (ZCS)
  - Zero-voltage switching (ZVS)
- *Zero-clamped topologies*
  - Zero-clamped-voltage (CV)
  - Zero-clamped-current (CC)
- *Class E resonant converter*
- *Dc- link resonant inverters*
- *Multi-resonant converters*
  - Zero-current switching (ZCS)
  - Zero-voltage switching (ZVS)
- *Zero transition topologies*
  - Zero-voltage transition (ZVT)
  - Zero-current transition (ZCT)

Many other variations of soft-switching topologies that exist today are beyond the scope of this text. Since the scope of this text targets undergraduate electrical engineering students, we focus mainly on the quasi-resonant type of PWM converters that will provide good insight into the principles of soft switching.

6.3 Advantages and Disadvantages of ZCS and ZVS

The major advantage of ZCS and ZVS quasi-resonant converters is that the power switch is turned on and off at zero voltage and zero current, respectively. In ZCS topologies, the rectifying diode has ZVS, whereas in ZVS topologies, the rectifying diode has ZCS. A second advantage is that both ZVS and ZCS converters utilize transformer leakage inductors and diode junction capacitors and the output parasitic capacitor of the power switch.

The major disadvantage of the ZVS and ZCS techniques is that they require variable- frequency control to regulate the output. This is undesirable since it complicates the control circuit and generates unwanted EMI harmonics, especially under wide load variations. In ZCS, the power switch turns off at zero current, but at turn-on the converter still suffers from the capacitor turn-on loss caused by the output capacitor of the power switch.

6.3.1 Switching Loci

Most regulator converter switches need to turn on or turn off the full load current at a high voltage, resulting in what is known as hard switching. Figure 6.2a, b shows typical switching loci for a hard-switching converter without and with a snubber circuit, respectively.

In a soft-switching converter topology, an LC resonant network is added to shape the switching device voltage or current waveform into a quasi-sine wave in such a way that a zero voltage or a zero current condition is created. This technique eliminates the turn-on or turn-off loss associated with the charging or discharging of the energy stored in the MOSFET's parasitic junction capacitors. Figure 6.3a, b shows typical switching loci for the ZVS at turn-on and the ZCS at turn-off cases.

6.3.2 Switching Losses

As the frequency of operation increases, the switching losses also increase. As shown in Chap. 3, there are two types of switching losses:

1. At turn *off*, the power transformer leakage inductance produces high *di*/*dt*, which results in a high voltage spike across it.
2. At turn-*on*, the switching loss is mainly caused by the dissipation of energy stored in the output parasitic capacitor of the power switch.

Figure 6.4a, b show typical switching waveforms at turn *off* and turn-*on*, without and with snubber circuit, respectively.

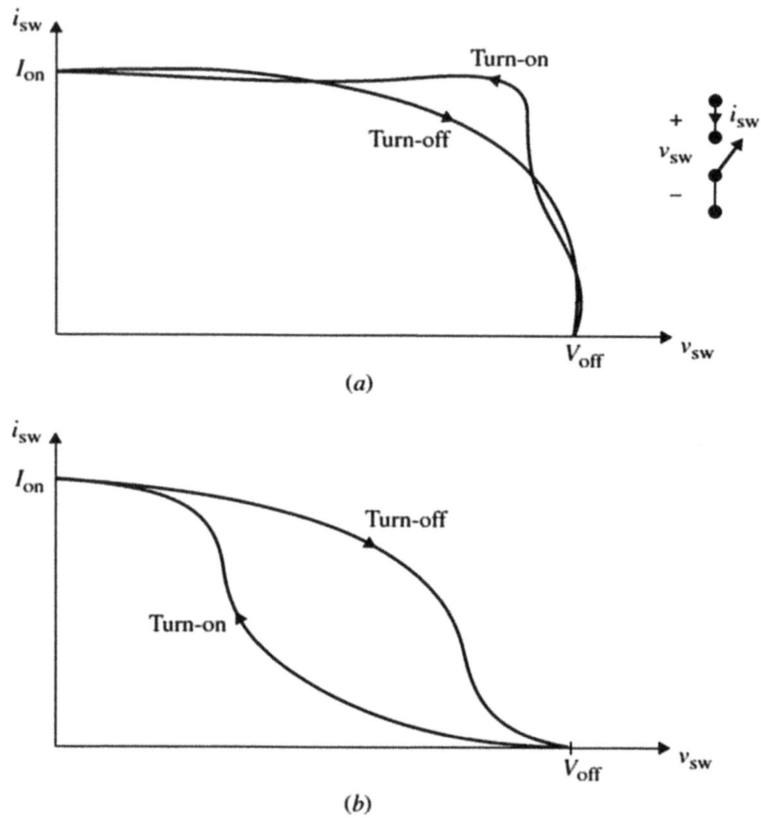

Fig. 6.2 Switching loci: (**a**) without snubber circuit and (**b**) with snubber circuit

6.4 Zero-Current Switching Topologies

6.4.1 The Resonant Switch

In this section, we will present one class of PWM converters that was introduced in the open 1980s based on the concept of using the conventional PWM switching along with an LC tank circuit. Depending on the inductor-capacitor arrangements, there are two possible resonant switch configurations. The switch is either an L-type or an M-type[1] and can be implemented as half-wave or full-wave, which corresponds to whether the switch current can be unidirectional or bidirectional, respectively. Figure 6.5 shows the L-type in both the half- and full-wave implementation. Similarly, the M-type switch is shown in Fig. 6.6.

[1]The notations L-type and M-type were first used by the original authors.

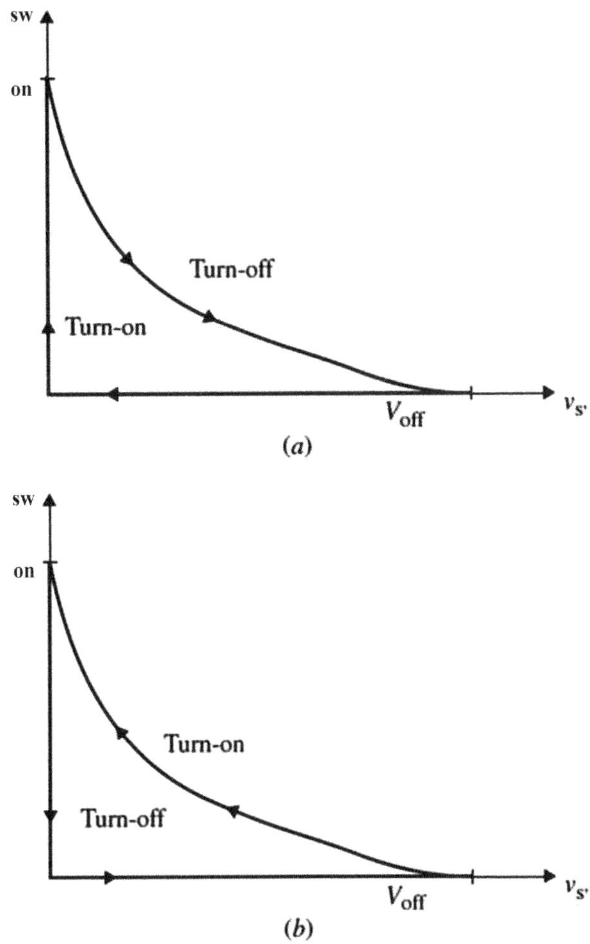

Fig. 6.3 (**a**) ZVS at turn-on and (**b**) ZCS at turnoff

The three conventional converter topologies, buck, boost, and buck-boost, will be analyzed here. In all of these topologies, the LC tank forms the resonant tank that causes ZCS to occur. The buck, boost, and buck-boost converters are shown in Fig. 6.7a–c, respectively.

The detailed steady-state analysis of these conventional converters was presented in Chap. 4. There exist two ccm modes of operation (as far as energy transfer is concerned): one mode during which energy is transferred from the source to the storage inductor and a second mode during which energy is transferred from the storage inductor to the load. A low-pass LC filter is used in all three topologies to filter the fundamental frequency from the harmonics. Using the two types of switch arrangements, it is possible to convert the three topologies into quasi-resonant converters.

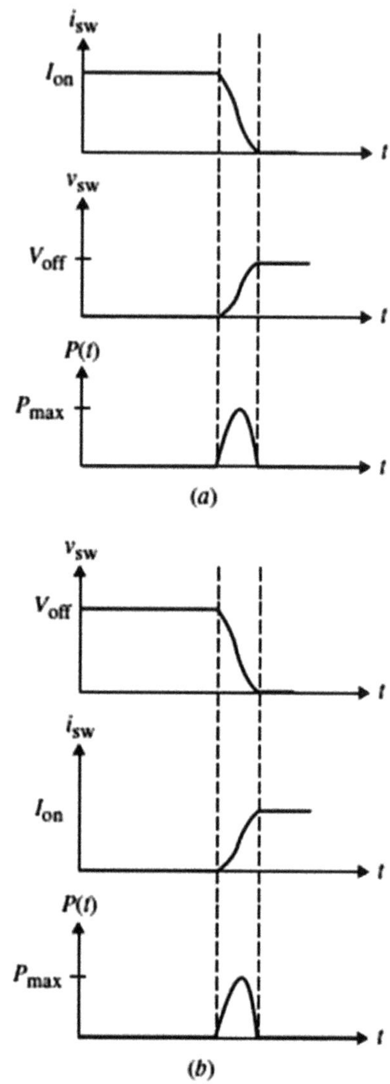

Fig. 6.4 Typical switching current, voltage, and power losses waveforms at (**a**) turn-on and (**b**) turnoff

6.4.2 Steady-State Analysis

To simplify the steady-state analysis of the steady-state condition for the above converters, there are some assumptions need to be made:

1. The filtering components $L_o, L_{in}, L_F,$and C_o are very large when compared to the resonant components L and C.
2. The output filter $L_o - C_o - R$ is treated as a constant current source, I_o.

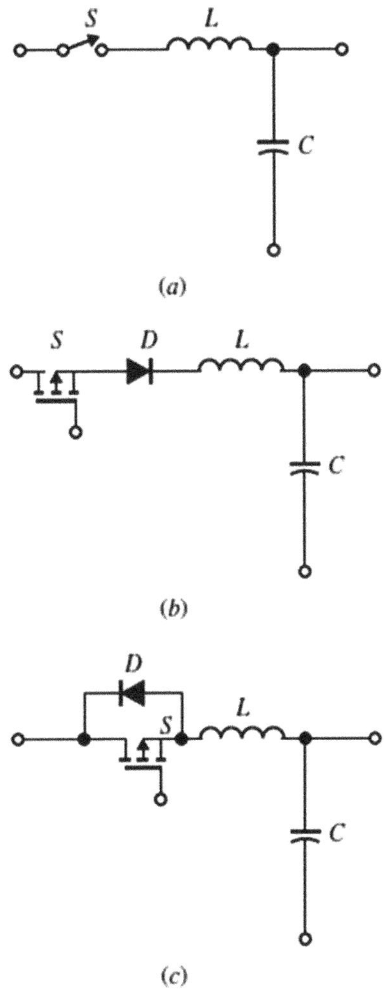

Fig. 6.5 Resonant switch: (**a**) L-type switch, (**b**) half-wave implementation, and (**c**) full-wave implementation

3. The output filter $C_o - R$ is treated as a constant voltage source, V_o.
4. Ideal switching devices and diodes.
5. Ideal reactive circuit components.

6.4.2.1 The Buck-Resonant Converter

Replacing the switch in Fig. 6.7a by the resonant-type switch of Fig. 6.5a, we obtain a quasi-resonant PWM buck converter, as shown in Fig. 6.8a. Its simplified equivalent circuit is shown in Fig. 6.8b. It can be shown that there are four modes of operation under the steady-state condition.

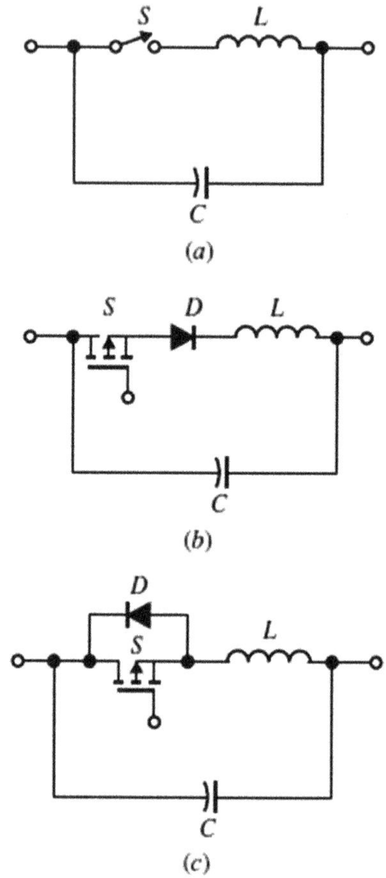

Fig. 6.6 Resonant switch: (**a**) M-type switch, (**b**) half-wave implementation, and (**c**) full-wave implementation

Mode 1 ($0 \le t < t_1$)
Mode 1 starts at $t = 0$, when S is turned on. Since the switch was off prior to $t = 0$, it is clear that the diode must have been on for $t < 0$ to carry the output inductor current. Hence, we assume for $t > 0$, both S and D are on. The output current is equal to the constant current source, I_o, as shown in Fig. 6.9a. In this mode, the capacitor voltage, v_C, is zero, and the input voltage is equal to the inductor voltage as given by:

$$V_{in} = L\frac{di_L}{dt} \tag{6.1}$$

Integrating Eq. (6.1) from 0 to t, the inductor current i_L is given by:

$$i_L(t) = \frac{V_{in}}{L}t \tag{6.2}$$

Equation (6.2) assumes zero initial condition for i_L. The current and voltage waveforms are given in Fig. 6.10.

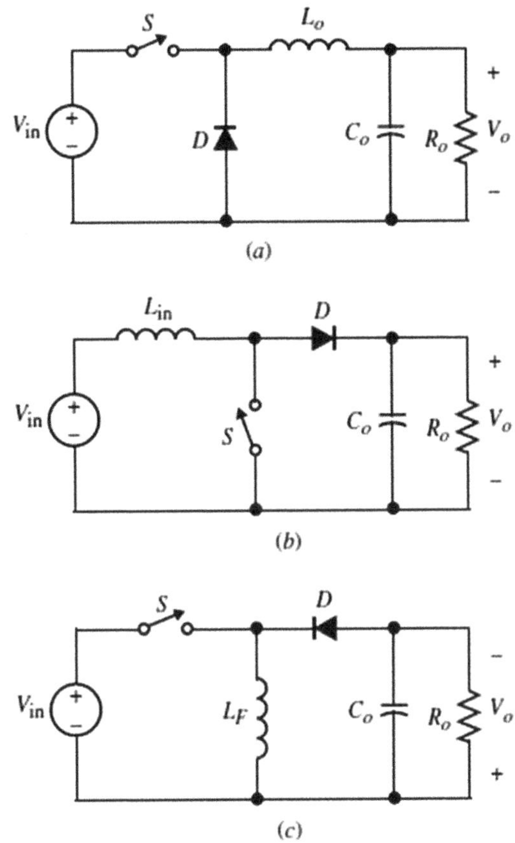

Fig. 6.7 Conventional converters (**a**) buck, (**b**) boost, and (**c**) buck-boost

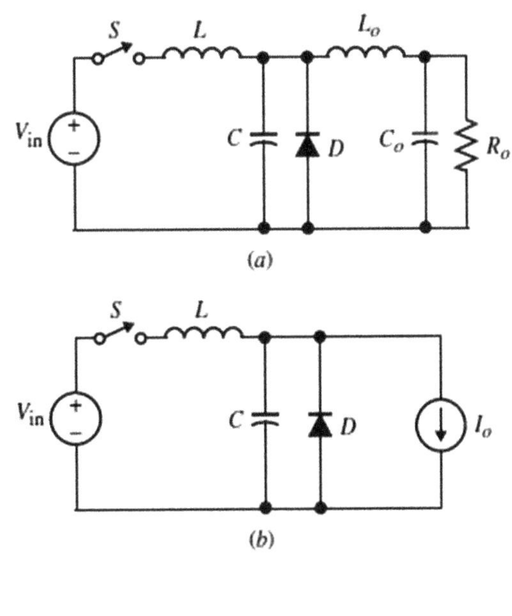

Fig. 6.8 (**a**) Buck conventional with L-type resonant switch and (**b**) simplified equivalent circuit

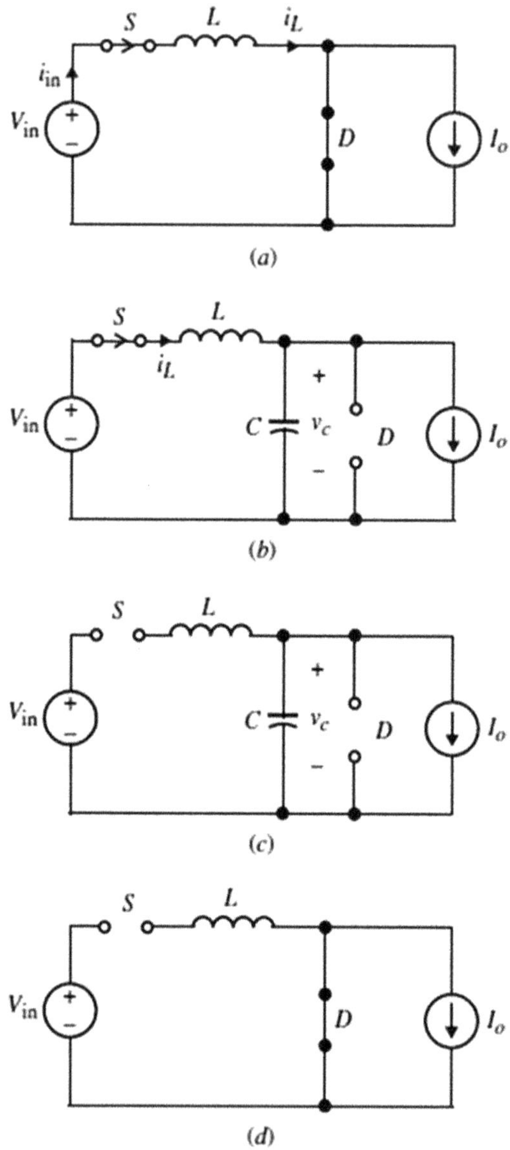

Fig. 6.9 (**a**) Equivalent circuit for Mode 1 (**b**) Equivalent circuit for Mode 2 (**c**) Equivalent circuit for Mode 3 (**d**) Equivalent circuit for Mode 4

As long as the inductor current is less than I_o, the diode will continue conducting, and the capacitor voltage remains at zero. At time t_1, the inductor current becomes equal to I_o, the diode stops conducting, and the circuit enters Mode 2. Evaluating Eq. (6.2) at $t = t_1$, we have:

$$I_o = \frac{V_{in}}{L} t_1 \tag{6.3}$$

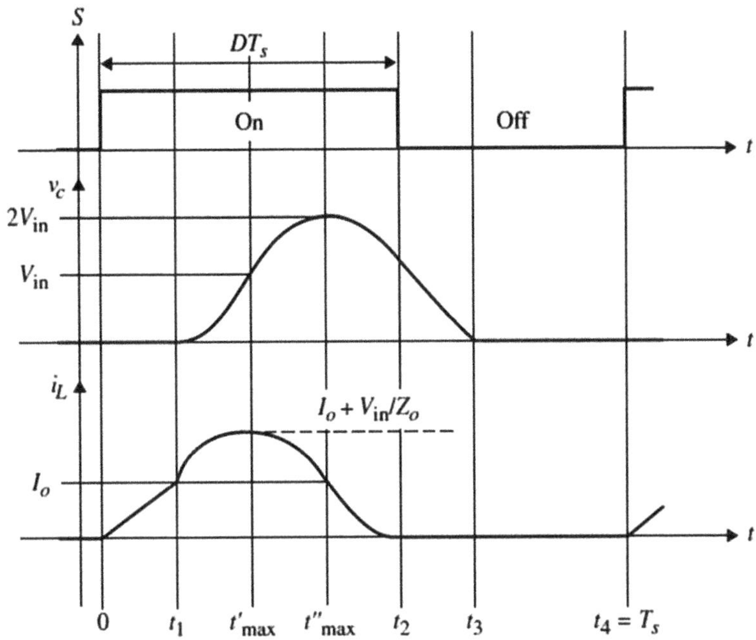

Fig. 6.10 Characteristic waveform of buck L-type

Hence, the time interval $\Delta t_1 = t_1$ is given by:

$$\Delta t_1 = t_1 = \frac{LI_o}{V_{in}} \tag{6.4}$$

This is the inductor current charging state.

Mode 2 ($t_1 \le t < t_2$)
Mode 2 starts at t_1, when the diode is open circuited as shown in Fig. 6.9b, resulting in a resonant stage between L and C. During the time between t_1 and t_2, the switch remains on, but the diode is off. The initial capacitor voltage remains zero, but the initial inductor current changes to I_o.

The first-order differential equations that represent this mode are given by Eq. (6.5):

$$C\frac{dv_C}{dt} = i_L - I_o \tag{6.5a}$$

$$L\frac{di_L}{dt} = V_{in} - v_C \tag{6.5b}$$

Next we express the inductor current in a second-order differential equation for $t \ge t_1$ as given in Eq. (6.6):

$$\frac{d^2 i_L}{dt^2} = \frac{1}{LC} i_L = \frac{I_o}{LC} \tag{6.6}$$

From Eq. (6.6), the general solution for $i_L(t)$ is given by:

$$i_L(t) = A_1 \sin \omega_o (t - t_1) + A_2 \cos \omega_o (t - t_1) + A_3 \tag{6.7}$$

where the resonant angular frequency is defined by:

$$\omega_o = \sqrt{\frac{1}{LC}}$$

Now we evaluate the constants A_1, A_2, and A_3. Recall at $t = t_1$, $i_L(t_1) = I_o$. Using this value, Eq. (6.7) becomes:

$$A_2 + A_3 = I_o$$

The other relation in terms of the constants is obtained from the following equation:

$$L \frac{di_L(t)}{dt} = V_{in} - v_C(t)$$

Substituting for i_L from Eq. (6.7), and since the capacitor voltage is zero at t_1, we have:

$$A_1 = \frac{V_{in}}{L\omega_o}$$

By taking the derivative of i_L again where $L(di_L{}^2(t_1)/dt^2) = 0$, we find A_2 to equal zero. Therefore, $A_3 = I_o$.

By solving the above equations, i_L and v_c are given by:

$$i_L(t) = I_o + \frac{V_{in}}{Z_o} \sin \omega_o (t - t_1) \tag{6.8}$$

$$v_C(t) = V_{in}[1 - \cos \omega_o (t - t_1)] \tag{6.9}$$

where $Z_o = \sqrt{L/C}$ is known as the characteristic impedance.

This mode will last until $t = t_2$, when the transistor turns off because the inductor current reaches zero. The peak inductor current occurs at $t = t'_{max}$, where $\omega_o (t'_{max} - t_1) = \pi/2$. When $v_C(t) = V_{in}$, the peak inductor current is $I_o + V_{in}/Z_o$. Moreover, the peak capacitor voltage occurs at $t = t''_{max}$, where $\omega_o (t''_{max} - t_1) = \pi$. When $i_L(t) = I_o$, the peak capacitor voltage is $2V_{in}$.

The time interval in this mode can be derived at $t = t_2$ by setting $i_L(t_2) = 0$:

$$\begin{aligned} i_L(t_2) &= I_o + \frac{V_{in}}{Z_o} \sin \omega_o (t_2 - t_1) \\ &= 0 \end{aligned} \tag{6.10}$$

therefore:

$$\Delta t_2 = t_2 - t_1 = \frac{1}{\omega_o} \sin^{-1}\left(\frac{-Z_o I_o}{V_{in}}\right) \tag{6.11}$$

It should be noted that the angle $\omega_o(t_2 - t_1)$ is in the third quadrant; hence, Eq. (6.11) may be written as:

$$t_2 - t_1 = \frac{T_S}{2} + \frac{1}{\omega_o} \sin^{-1}\left(\frac{Z_o I_o}{V_{in}}\right) = \frac{1}{\omega_o}\left(\pi + \sin^{-1}\frac{Z_o I_o}{V_{in}}\right)$$

Mode 3 starts at $t = t_2$, when the switch is turned *off*.

Mode 3 ($t_2 \leq t < t_3$)
At t_2 the inductor current becomes zero, and the capacitor linearly discharges from $v_C(t_2)$ to zero during t_2 to t_3. The diode remains off since its voltage is negative, as shown in Fig. 6.9c.

Now the initial value of $i_L(t_2)$ is zero, and the initial value of the capacitor at $t = t_2$ is $V_C(t_2)$. The inductor has no current going through it when the switch is off, so the capacitor current equals I_o, as given by:

$$i_C = C\frac{dv_C}{dt} = -I_o \tag{6.12}$$

The capacitor voltage $v_C(t)$ is obtained by integrating Eq. (6.12) from t_2 to t with $V_C(t_2)$ as the initial value, resulting in:

$$v_C(t) = \frac{-I_o}{C}(t - t_2) + V_C(t_2) \tag{6.13}$$

The initial value $V_C(t_2)$ is obtained from Eq. (6.9) in the previous mode, to yield:

$$V_C(t_2) = V_{in}[1 - \cos\omega_o(t_2 - t_1)] \tag{6.14}$$

Substituting Eq. (6.14) into Eq. (6.13), we obtain:

$$v_C(t) = \frac{-I_o}{C}(t - t_2) + V_{in}[1 - \cos\omega_o(t_2 - t_1)] \tag{6.15}$$

At $t = t_3$, the capacitor voltage becomes zero, and the equation for this time interval is given by:

$$\Delta t_3 = t_3 - t_2 = \frac{C}{I_o}V_{in}[1 - \cos\omega_o(t_2 - t_1)] \tag{6.16}$$

At this point, the diode turns *on*, and the circuit enters Mode 4. In this mode, the capacitor voltage and inductor current remain zero until the switch is turned *on* again to repeat Mode 1.

Mode 4 ($t_3 \leq t < t_4$)

At this mode the switch remains *off*, but the diode starts conducting at $t = t_3$. Mode 4 will continue as long as the switch is *off*, and the output current starts the freewheeling stage through the diode. The inductor current and the capacitor voltage are zero when the switch closes:

$$i_L(t) = 0$$
$$v_C(t) = 0$$

Therefore, there will be no power transfer during this mode. By turning on the switch at $t = T_S$, the cycle will repeat these four modes. The dead timeΔt_4 is given by:

$$\Delta t_4 = T_S - \Delta t_1 - \Delta t_2 - \Delta t_3 \tag{6.17}$$

Figure 6.10 shows the steady-state waveforms for v_C and i_L for the buck converter with L-type switch.

6.4.2.2 Voltage Gain

In this section, we will derive the expression for the voltage gain, $M = V_o/V_{in}$, in terms of the circuit parameters.

The average output voltage, V_o, can be obtained by evaluating the following integral:

$$V_o = \frac{1}{T_S}\int_0^{T_S} v_C(t)dt \tag{6.18}$$

Substitute for $v_C(t)$ from Eqs. (6.9) and (6.13) for intervals $(t_2 - t_1)$ and $(t_3 - t_2)$, respectively, into Eq. (6.18) to yield:

$$V_o = \frac{1}{T_S}\left[\int_{t_1}^{t_2} V_{in}(1 - \cos\omega_o(t - t_1))dt + \int_{t_2}^{t_3}\left(\frac{-I_o}{C}(t - t_2) + V_C(t_2)\right)dt\right]$$

The voltage gain ratio is given by:

$$\frac{V_o}{V_{in}} = \frac{1}{T_S}\left[(t_2 - t_1) - \frac{\sin\omega_o(t_2 - t_1)}{\omega_o} - \frac{I_o}{V_{in}C}\frac{(t_3 - t_2)^2}{2} + \frac{V_C(t_2)}{V_{in}}(t_3 - t_2)\right] \tag{6.19}$$

Substitute for $(t_2 - t_1)$, $(t_3 - t_2)$, and $V_C(t_2)$ from Eqs. (6.11), (6.16), and (6.14), respectively, into Eq. (6.19) to yield a closed-form expression for M in terms of the circuit parameters. However, for illustration purposes, we will show that the same expression can be obtained from the law of power conservation. The conservation of energy per switching cycle states that since the converter is assumed to be ideal, then the average input and output powers should be equal. Next we evaluate the average input and output powers and then equate them since the converter is assumed to be ideal.

The total input energy over one switching cycle is given by;

$$E_{in} = \int_{0}^{T_S} i_{in} V_{in} dt \tag{6.20}$$

Since i_{in} is equal to $i_L(t)$, Eq. (6.20) is rewritten as:

$$E_{in} = \int_{0}^{t_1} i_L(t) V_{in} dt + \int_{t_1}^{t_2} i_L(t) V_{in} dt \tag{6.21}$$

Substituting for $i_L(t)$ from Eqs. (6.2) and (6.8) into the above integrals, respectively, Eq. (6.21) becomes:

$$E_{in} = V_{in} \left\{ \frac{V_{in}}{2L} t_1{}^2 + I_o(t_2 - t_1) - \frac{V_{in}}{Z_o \omega_o} [\cos \omega_o (t_2 - t_1) - 1] \right\} \tag{6.22}$$

Substituting for $\cos \omega_o (t_2 - t_1) = 1 - [I_o(t_3 - t_2)]/CV_{in}$ from Eq. (6.16) in Mode 3, Eq. (6.22) becomes:

$$E_{in} = V_{in} \left\{ \frac{t_1}{2} I_o + I_o(t_2 - t_1) + \frac{V_{in}}{Z_o \omega_o} \left[\frac{I_o(t_3 - t_2)}{CV_{in}} \right] \right\} \tag{6.23}$$

With $Z_o \omega_o = 1/C$, Eq. (6.23) becomes:

$$E_{in} = V_{in} I_o \left[\frac{t_1}{2} + (t_2 - t_1) + (t_3 - t_2) \right] \tag{6.24}$$

The output energy over one cycle is obtained by evaluating Eq. (6.25):

$$E_o = \int_{0}^{T_S} I_o V_o dt = I_o V_o T_S \tag{6.25}$$

From the conservation of energy theory, equating the input and output energy expressions from Eq. (6.24) and (6.25), we have:

$$I_o V_o T_S = V_{in} I_o \left[\frac{t_1}{2} + (t_2 - t_1) + (t_3 - t_2) \right] \tag{6.26}$$

From Eq. (6.26) the voltage gain is expressed by:

$$\frac{V_o}{V_{in}} = \frac{1}{T_S} \left[\frac{t_1}{2} + (t_2 - t_1) + (t_3 - t_2) \right] \tag{6.27}$$

Substituting for t_1, $(t_2 - t_1)$, and $(t_3 - t_2)$ from Eqs. (6.4), (6.11), and (6.16), respectively, into Eq. (6.27), the voltage gain becomes:

$$\frac{V_o}{V_{in}} = \frac{1}{T_S}\left\{\frac{LI_o}{2V_{in}} + \frac{1}{\omega_o}\sin^{-1}\frac{-Z_o I_o}{V_{in}} + \frac{CV_{in}}{I_o}[1 - \cos\omega_o(t_2 - t_1)]\right\} \tag{6.28}$$

To simplify and generalize the gain equation, the following normalized parameters are defined:

$$M = \frac{V_o}{V_{in}} \quad \text{normalized output voltage} \tag{6.29a}$$

$$Q = \frac{R_o}{Z_o} \quad \text{normalized load} \tag{6.29b}$$

$$I_o = \frac{V_o}{R_o} \quad \text{average output current} \tag{6.29c}$$

$$f_{ns} = \frac{f_s}{f_o} \quad \text{normalized switching frequency} \tag{6.29d}$$

By substituting Eq. (6.29) into Eq. (6.28), the final voltage gain is simplified into:

$$M = \frac{f_{ns}}{2\pi}\left[\frac{M}{2Q} + \alpha + \frac{Q}{M}(1 - \cos\alpha)\right] \tag{6.30}$$

where:

$$\alpha = \sin^{-1}\left(\frac{-M}{Q}\right) \tag{6.31}$$

A plot of the control characteristic curve of M vs. f_{ns} under various normalized loads is given in Fig. 6.11.

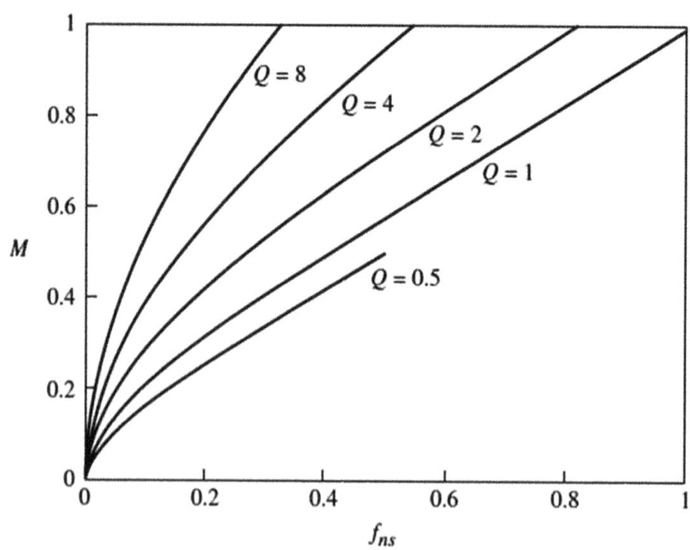

Fig. 6.11 Control characteristic curve of M vs. f_{ns} for the ZCS buck converter

Example 6.1

Consider the following specifications for a ZCS buck converter of Fig. 6.8a. Assume the parameters are $V_{in} = 25$ V, $V_o = 12$ V, $I_o = 1$ A, and $f_S = 250$ kHz.

Design for the resonant tank parameters L and C, and calculate the peak inductor current and peak capacitor voltage. Determine the time interval for each mode.

Solution

The voltage gain is $M = V_o/V_{in} = 12/25 = 0.48$. Let us select $f_{ns} = 0.4$. Next we determine Q from either the control characteristic curve of Fig. 6.11 or from the gain equation of Eq. (6.30). This results in Q approximately equaling 1. Since $R_o = V_o/I_o$, the characteristic impedance is given by:

$$\begin{aligned} Z_o &= \frac{R_o}{Q} = 12\ \Omega \\ &= \sqrt{\frac{L}{C}} = 12\ \Omega \end{aligned} \tag{6.32}$$

The second equation in terms of L and C is obtained from f_O. From the normalized switching frequency, f_O may be given by:

$$\begin{aligned} f_o &= \frac{f_S}{f_{ns}} \\ &= \frac{f_S}{0.4} = 625\ \text{kH} \end{aligned}$$

In terms of the angular frequency, ω_o, we have:

$$\omega_o = 2\pi f_o = \sqrt{\frac{1}{LC}} \tag{6.33}$$

Solving Eqs. (6.32) and (6.33) for L and C, we obtain:

$$\begin{aligned} L = \frac{Z_o}{\omega_o} &= \frac{12\ \ \Omega}{2\pi \times 625 \times 10^3\ \ \text{rad/s}} \\ &= 3.06 \times 10^{-6} \approx 3\ \mu H \end{aligned}$$

$$\begin{aligned} C = \frac{1}{Z_o\omega_o} &= \frac{1}{12 \times 2 \times \pi \times 625 \times 10^3} \\ &\approx 0.02\ \mu F \end{aligned}$$

The peak inductor current is given by:

$$\begin{aligned} I_{L,\text{peak}} &= I_o + \frac{V_{in}}{Z_o} \\ &\approx 3\ \text{A} \end{aligned}$$

and the peak capacitor voltage is:

$$v_{c,\text{peak}} = 2V_{\text{in}} \\ = 50\text{ V}$$

The time intervals are calculated from the following expressions:

$$t_1 = \frac{I_o L}{V_{\text{in}}} = \frac{1\text{ A} \times \left(3 \times 10^{-6}\, H\right)}{25\text{ V}} \approx 0.12\,\mu s$$

$$t_2 = t_1 + \frac{1}{\omega_o}\sin^{-1}\left(\frac{-Z_o I_o}{V_{\text{in}}}\right)$$

$$= 0.12 + \frac{1}{2\pi \times 625 \times 10^3}\sin^{-1}\left(\frac{-12 \times 1}{25}\right)$$

$$\approx 1.05 \quad \mu s$$

$$t_3 = t_2 + \frac{CV_{\text{in}}(1 - \cos\omega_o(t_2 - t_1))}{I_o}$$

$$= 1.05 + \frac{\left(0.02 \times 25 \times 10^{-6}\right)(1 - \cos(2\pi \times 0.625 \times 0.67))}{1\text{ A}}$$

$$\approx 1.99\,\mu s$$

For $t'_{\max}$ we have:

$$\omega_o\left(t'_{\max} - t_1\right) = \frac{\pi}{2}$$

$$t'_{\max} = \frac{\pi/2}{\omega_o} + t_1$$

$$= 0.4\,\mu s + 0.12\,\mu s$$

$$= 0.52 \quad \mu s$$

And finally, $t_4 = 4\ \mu\text{s} = T_S$.

Exercise 6.1

Consider the ZCS buck converter with the following parameters: $V_{\text{in}} = 40$ V, $V_o = 28.7$ V @ $I_o = 0.6$ A, $f_S = 100$ kHz, $L = 15$ μH, and $C = 60$ nF. Determine the peak inductor current and the time at which the capacitor reaches its maximum voltage of 80 V.

Answer: 3.1 A, 3.2 μs.

Exercise 6.2

Figure 6.12a shows another type of quasi-resonant buck converter that uses an M-type resonant switch arrangement. Its simplified circuit is shown in Fig. 6.12b. Derive the voltage gain equation, $M = V_o/V_{\text{in}}$, for the steady-state waveforms shown in Fig. 6.12c.

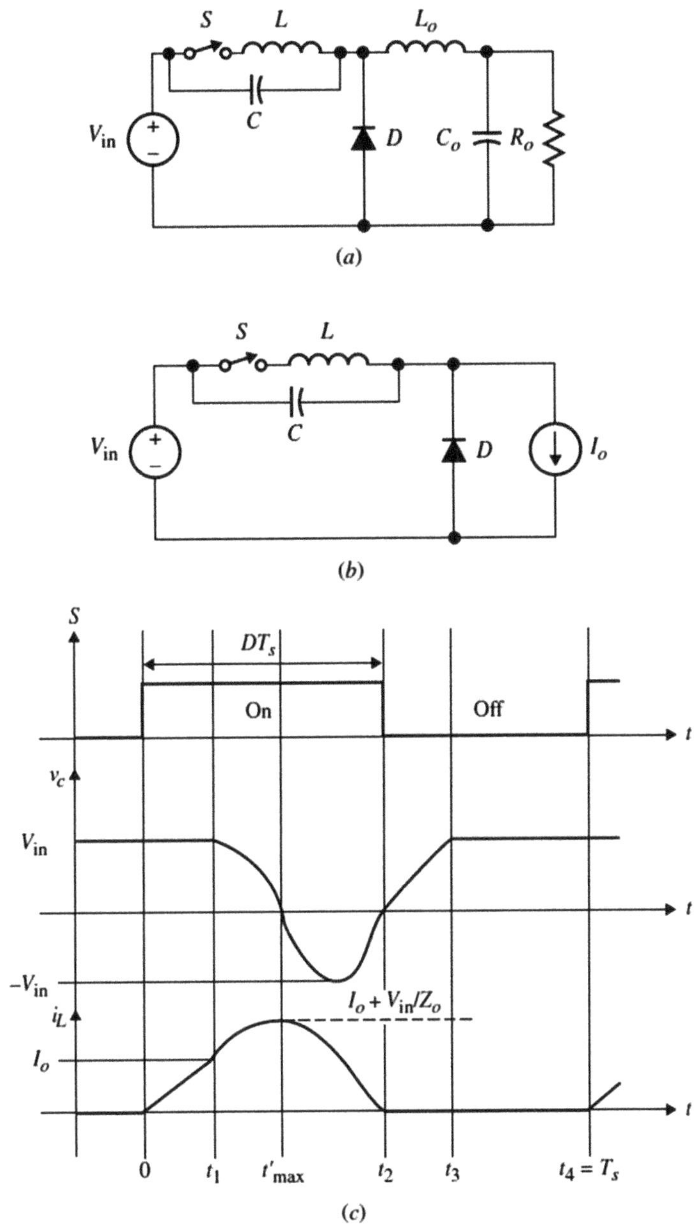

Fig. 6.12 (**a**) ZCS buck converter with M-type, (**b**) simplified equivalent circuit, and (**c**) steady-state waveforms

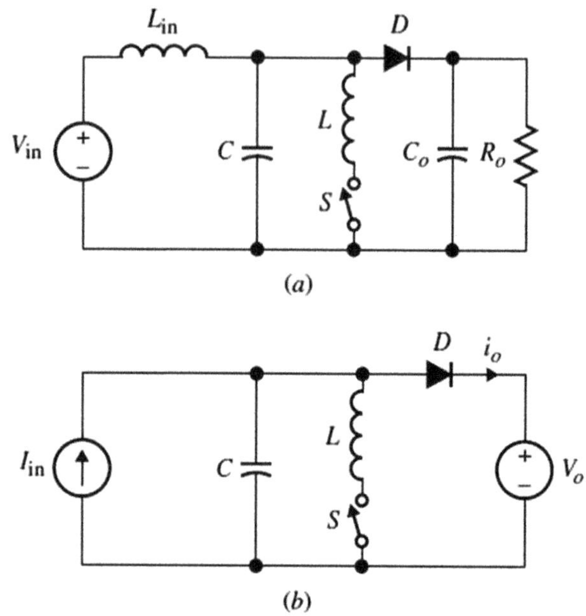

Fig. 6.13 (**a**) ZCS converter with boost M-type switch and (**b**) simplified equivalent circuit

6.4.2.3 The ZCS Boost Converter

Let us consider the boost quasi-resonant converter with an M-type switch as shown in Fig. 6.13a, with its equivalent circuit shown in Fig. 6.13b. Here we assumed the input current is constant, and the load voltage is constant.

The following operation assumes half-wave operation.

Mode 1 ($0 \le t < t_1$)

We first assume that the switch and the diode are both *on* in Mode 1 as shown in Fig. 6.14a.

The output voltage is given by:

$$V_{\mathrm{o}} = L\frac{di_{\mathrm{L}}}{dt} \tag{6.34}$$

The initial inductor current and capacitor voltage are given by:

$$i_{\mathrm{L}}(0) = 0$$

$$v_{\mathrm{C}}(0) = V_{\mathrm{o}}$$

Integrating Eq. (6.34), the inductor current becomes:

$$i_{\mathrm{L}}(t) = \frac{V_{\mathrm{o}}}{L}t + i_{\mathrm{L}}(0) = \frac{V_{\mathrm{o}}}{L}t \tag{6.35}$$

Fig. 6.14 (**a**) Equivalent circuit for Mode 1 (**b**) Equivalent circuit for Mode 2 (**c**) Equivalent circuit for Mode 3 (**d**) Equivalent circuit for Mode 4

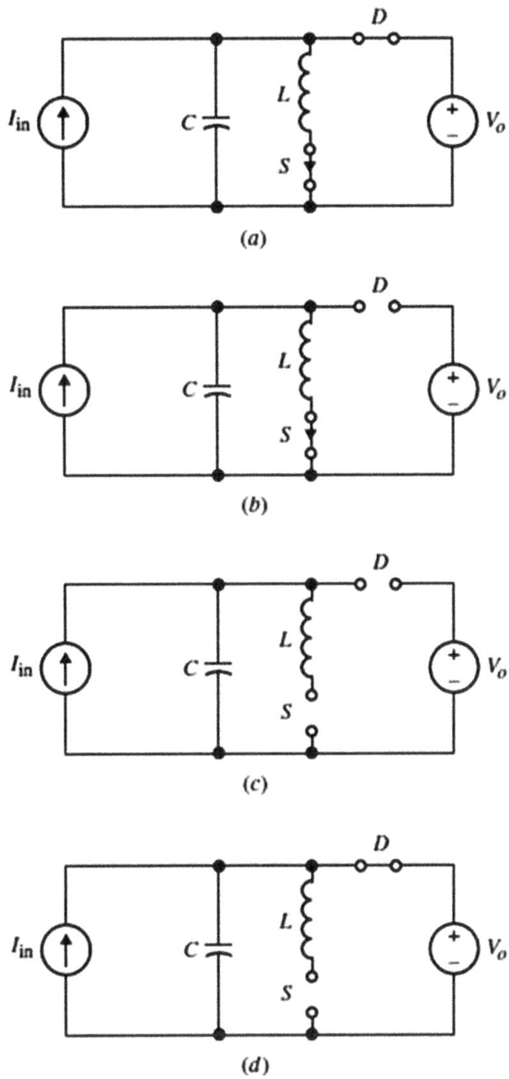

When the resonant inductor current reaches the input current, I_{in}, the diode turns *off*, hence we have:

$$\frac{V_{\mathrm{o}}}{L} t_1 = I_{\mathrm{in}}$$

with t_1 given by:

$$t_1 = \frac{I_{\mathrm{in}} L}{V_{\mathrm{o}}} \tag{6.36}$$

At $t = t_1$, the diode turns off since $i_{\mathrm{L}} = I_{\mathrm{in}}$, and the converter enters Mode 2.

Mode 2 ($t_1 \leq t < t_2$)

The switch remains closed, but the diode is off at t_1 in Mode 2 as shown in Fig. 6.14b. This is a resonant mode during which the capacitor voltage starts decreasing resonantly from its initial value of V_o. When $i_L = I_{in}$, the capacitor reaches its negative peak. At $t = t_2$, i_L equals zero, and the switch turns off (i.e., switching at zero current).

The initial conditions are given by:

$$v_C(t_1) = V_o \quad \text{and} \quad i_L(t_1) = I_{in}$$

From Fig. 6.14b, the first derivatives for i_L and v_C are given by:

$$L\frac{di_L}{dt} = v_C(t)$$
$$C\frac{dv_C}{dt} = I_{in} - i_L(t)$$

Next the two first-order differential equations need to be solved in this mode. Using the same solution technique used in the buck converter to solve the above differential equations, the expression for $i_L(t)$ and $v_C(t)$ are given by:

$$i_L(t) = I_{in} + \frac{V_o}{Z_o}\sin\omega_o(t - t_1) \tag{6.37}$$

$$v_C(t) = V_o\cos\omega_o(t - t_1) \tag{6.38}$$

where $\omega_o = 1/\sqrt{LC}$

At $t = t_2$, $i_L(t_2) = 0$ and the time interval can be obtained from evaluating Eq. (6.37) at $t = t_2$, to yield:

$$\begin{aligned}(t_2 - t_1) &= \frac{1}{\omega_o}\sin^{-1}\left(-\frac{I_{in}Z_o}{V_o}\right) \\ &= \frac{1}{\omega_o}\left[\pi + \sin^{-1}\left(\frac{I_{in}Z_o}{V_o}\right)\right]\end{aligned} \tag{6.39}$$

Mode 3 ($t_2 \leq t < t_3$)

Mode 3 starts at t_2, and the switch and the diode are both open, as shown in Fig. 6.14c. Since v_C is constant, the capacitor starts charging up by the input current source. The capacitor voltage is given by:

$$\begin{aligned}v_C(t) &= \frac{1}{C}\int_{t_2}^{t} I_{in}dt \\ &= \frac{I_{in}}{C}(t - t_2) + v_C(t_2)\end{aligned} \tag{6.40}$$

The diode begins conducting at $t = t_3$ when the capacitor voltage is equal to the output voltage, i.e., $v_C(t_3) = V_o$. Equation (6.40) becomes:

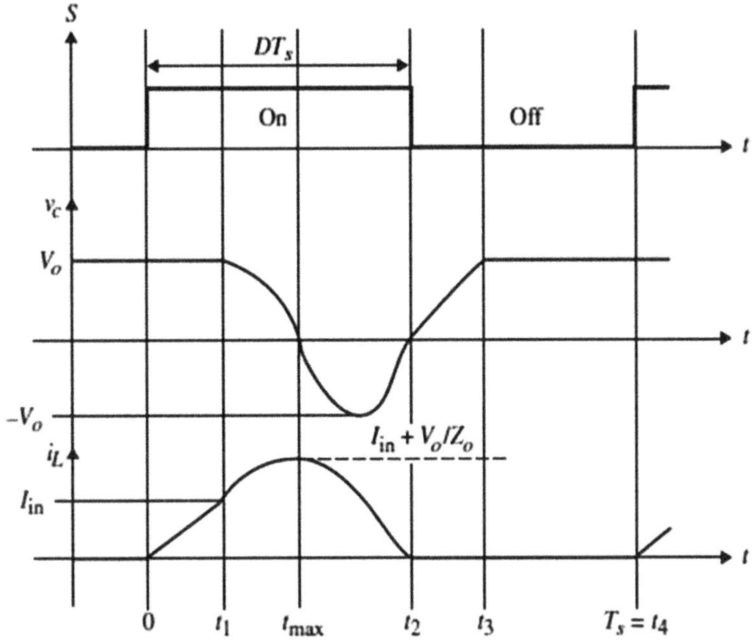

Fig. 6.15 Steady-state waveform of the boost M-type switch converter

$$v_C(t_3) = V_o = \frac{I_{in}}{C}(t_3 - t_2) + v_C(t_2)$$

so the time interval in this period can be expressed as in Eq. (6.41):

$$t_3 - t_2 = \frac{C}{I_{in}}[V_o - v_C(t_2)] \tag{6.41}$$

where $v_C(t_2)$ may be obtained from Eq. (6.38).

Mode 4 ($t_3 \le t < t_4$)
At t_3, the capacitor voltage is clamped to the output voltage, and the diode starts conducting again, but the switch remains open. This condition remains as long as the switch is open. The cycle of the mode will repeat again at the time of T_s when S is turned *on* again.

Typical steady-state waveforms are shown in Fig. 6.15.

6.4.2.4 Voltage Gain

As we did for the buck converter, we apply the conservation of energy per switching cycle to express the voltage gain, $M = V_o/V_{in}$, in terms of the circuit parameter.

The input energy is given by:

$$E_{\text{in}} = V_{\text{in}} I_{\text{in}} T_S \tag{6.42}$$

and the output energy:

$$E_{\text{o}} = \int_0^{T_S} i_{\text{o}} V_{\text{o}} dt \tag{6.43}$$

The output current equals $i_o = I_{\text{in}} - i_L$ and $i_o = I_{\text{in}}$ for intervals $0 \le t < t_1$ and $t_3 \le t < T_S$, respectively. Therefore, E_o becomes:

$$E_{\text{o}} = \int_0^{t_1} (I_{\text{in}} - i_L) V_{\text{o}} dt + \int_{t_3}^{T_S} I_{\text{in}} V_{\text{o}} dt \tag{6.44}$$

The input current is obtained from the conservation of output power as:

$$I_{\text{in}} = \frac{V_{\text{o}}^2}{V_{\text{in}} R_{\text{o}}}$$

Substituting for the input current and by evaluating Eq. (6.44), the output energy becomes:

$$\begin{aligned} E_{\text{o}} &= V_{\text{o}} \int_0^{t_1} \left(I_{\text{in}} - \frac{V_{\text{o}}}{L} t \right) dt + I_{\text{in}} V_{\text{o}} (T_S - t_3) \\ &= V_{\text{o}} \left(I_{\text{in}} t_1 - \frac{1 V_{\text{o}}}{2 L} t_1^2 \right) + I_{\text{in}} V_{\text{o}} (T_S - t_3) \end{aligned} \tag{6.45}$$

If we substitute for $t_1 = I_{\text{in}} L / V_o$ and $(T_S - t_3) = T_S - [t_1 + (t_2 - t_1) + (t_3 - t_2)]$ and use the equations for $(t_2 - t_1)$ and $(t_3 - t_2)$ from Eqs. (6.39) and (6.41), Eq. (6.45) becomes:

$$E_{\text{o}} = -\frac{1}{2} I_{\text{in}}^2 L + V_{\text{o}} I_{\text{in}} \left[T_S - \frac{\alpha}{\omega_o} - \frac{C}{I_{\text{in}}} V_{\text{o}} (1 - \cos \alpha) \right] \tag{6.46}$$

Following similar steps as in the quasi-resonant buck converter, it can be shown that the voltage gain expression is given by:

$$\frac{M-1}{M} = \frac{f_{\text{ns}}}{2\pi} \left[\frac{M}{2Q} + \alpha + \frac{Q}{M} (1 - \cos \alpha) \right] \tag{6.47}$$

where α, M, I_o, and f_{ns} are given in Eq. (6.29).

Figure 6.16 shows the characteristic curve for M vs. f_{ns} as a function of the normalized load.

Example 6.2

Design a ZCS boost converter for the following parameters: $V_{\text{in}} = 20$ V, $V_o = 40$ V, $P_o = 20$ W, and $f_S = 250$ kHz.

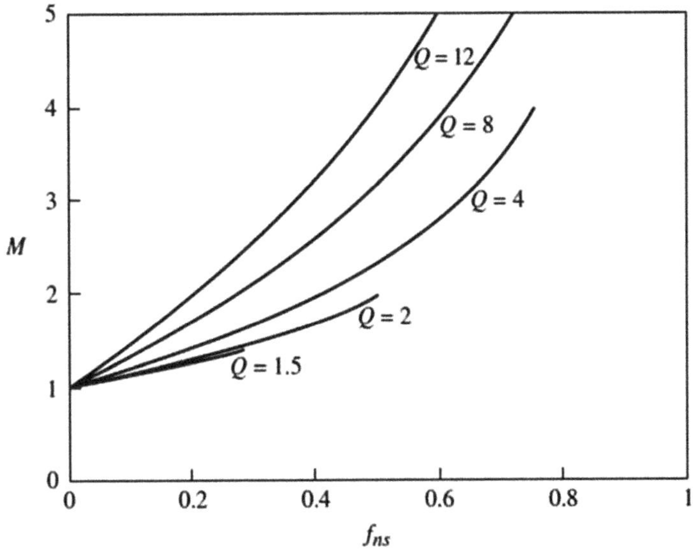

Fig. 6.16 Characteristic curve for M vs.f_{ns} for the ZCS boost converter

Solution

The voltage gain is $M = V_o/V_{\mathrm{in}} = 40/20 = 2$. Let us select $f_{\mathrm{ns}} = 0.38$. From the characteristic curve of Fig. 6.16, Q can be approximated as 6.0.

The characteristic impedance is given by:

$$Z_o = \frac{R_o}{Q} = \frac{V_o^2/P_o}{Q} = \frac{80\ \Omega}{6} = 13.33\ \Omega \tag{6.48}$$

and the resonant frequency is:

$$\begin{aligned} f_o &= \frac{f_S}{f_{\mathrm{ns}}} \\ &= \frac{250\ \mathrm{kHz}}{0.38} = 657.89\ \mathrm{kHz} \end{aligned} \tag{6.49}$$

Solve Eqs. (6.48) and (6.49) for L and C:

$$L = \frac{Z_o}{2\pi f_o} = \frac{13.33\ \Omega}{2\pi \times 657.89 \times 10^3} = 3.22 \times 10^{-6}\ \mathrm{H}$$

$$C = \frac{1}{Z_o \omega_o} = \frac{1}{(13.33)\left(2\pi \times 657.89 \times 10^3\right)} = 18.14\ \mathrm{nF}$$

To limit the input ripple current and the output voltage, we set:

$$L_o = 100L = 322 \times 10^{-6}\ \mathrm{H}$$

$$C_o = 100C = 1.8 \times 10^{-6}\ \mathrm{F}$$

Example 6.3
Design a boost converter with ZCS, with the following design parameters: $V_{in} = 25$ V, $P_o = 30$ W, at $I_o = 0.5$ A, and $f_S = 100$ kHz. Assume the output voltage ripple $\Delta V_o/V_o$ is 0.2%.

Solution
The load resistance, $R_o = P_o/I_o^2 = 30/(0.5)^2 = 120\ \Omega$:

$$M = \frac{V_o}{V_{in}} = \frac{60}{25} = 2.4$$

From the characteristic curve of Fig. 6.16, we approximate Q to 6 when we assume $f_n = 0.58$. Hence, $f_o = 100$ kHz/0.58 = 172.4 kHz.

From Q and R_o, the characteristic impedance is obtained from:

$$Q = \frac{R_o}{Z_o} = \frac{120}{Z_o} = 6 \qquad \text{and} \qquad Z_o = 20\ \Omega$$

hence, $\sqrt{L/C} = 20\ \Omega$

$$\omega_o = 2\pi\left(172 \times 10^3\right) = 1080.7 \times 10^3\ \text{rad/s}$$

$$\sqrt{\frac{1}{LC}} = 1080.7 \times 10^3\ \text{rad/s}$$

Solving for C and L:

$$C = 46.27\ \text{nF}$$

$$L = 18.51\ \mu H$$

From $\Delta V_o/V_o = 0.2\%$, C_o can be obtained from the ripple voltage equation for the conventional boost converter, which is given by:

$$\frac{\Delta V_o}{V_o} = \frac{D}{f_S R_o C_o}$$

Solving the above equation for C_o, we obtain:

$$C_o = \frac{D}{f_S R_o (\Delta V_o/V_o)}$$

$$= \frac{\left(\frac{10 - 4.43 - 1.91}{10}\right)}{\left(100 \times 10^3\right)(120)\left(\frac{0.2}{100}\right)} = 15.25\ \mu F$$

where DT is the *on* time of the switch:

$$t_1 = \frac{LI_{\text{in}}}{V_{\text{o}}} = \frac{18.51 \times (10^{-6}) \times 1.2}{60} = 0.370\ \mu s$$

$$\begin{aligned} t_2 - t_1 &= \frac{1}{\omega_o}\left[\pi + \sin^{-1}\left(\frac{Z_{\text{o}} I_{\text{in}}}{V_{\text{o}}}\right)\right] \\ &= \frac{1}{1080.7 \times 10^3}\left[\pi + \sin^{-1}\left(\frac{20 \times 1.2}{60}\right)\right] = 3.29\ \mu s \end{aligned}$$

$$\begin{aligned} t_3 - t_2 &= \frac{1}{\omega_{\text{o}}}\frac{V_{\text{o}}}{Z_{\text{o}} I_{\text{in}}}(1 - \cos\alpha) \\ &= \frac{1}{1080.7 \times 10^3}\frac{60}{20 \times 1.2}(1 - \cos 3.553) \\ &= 4.43\ \mu s \end{aligned}$$

$$\begin{aligned} t_4 - t_3 &= T - t_1 - (t_2 - t_1) - (t_3 - t_2) \\ &= 10 - 0.370 - 3.29 - 4.43 = 1.91\ \mu s \end{aligned}$$

The output inductor is obtained, with $D = 0.366$, from:

$$\begin{aligned} L_{\text{crit}} &= \frac{R_{\text{o}}}{2f_{\text{S}}}(1 - D)^2 D \\ &= 88.8\ \mu H \end{aligned}$$

Hence, we select $L_{\text{o}} = 890\ \mu\text{H}$.

Figure 6.17a shows the quasi-resonant boost converter by using the L-type resonant switch, and the simplified circuit and its steady-state waveforms are shown in Fig. 6.17b, c, respectively. The reader is invited to verify these waveforms.

Exercise 6.3

Consider a quasi-resonant boost converter of Fig. (*6.13*) with the following design parameters: $V_{in} = 25$ V, $P_o = 30$ W *at* $I_o = 0.5$ A, and $f_S = 100$ kHz. Determine the resonant tank capacitor and inductor components.

Answer: $C = 42.4\ \mu\text{F}$, $L = 9.55\ \mu\text{H}$

6.4.2.5 ZCS Buck-Boost Converter

Let us consider the quasi-resonant buck-boost converter by using the L-type switch as shown in Fig. 6.18a, b shows the simplified equivalent circuit.

Like the buck and the boost converters, the buck-boost converter also leads to four modes of operations.

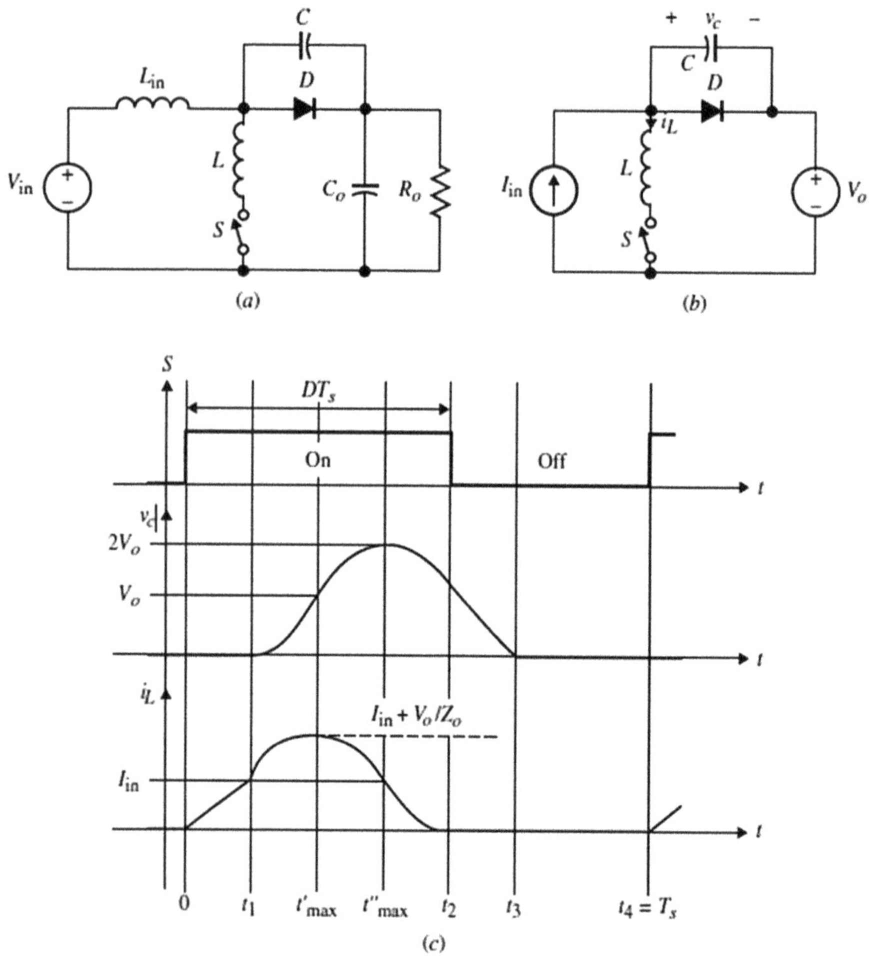

Fig. 6.17 (**a**) ZCS boost converter with L-type switch, (**b**) simplified equivalent circuit, and (**c**) steady-state waveforms

Mode 1 ($0 \leq t < t_1$)

Mode 1 starts at $t = 0$; the switch and the diode are both conducting. According to Kirchhoff's law, the voltage equation can be written as:

$$L\frac{di_{\mathrm{L}}(t)}{dt} = V_{\mathrm{in}} + V_{\mathrm{o}} \tag{6.50}$$

By integrating both sides of Eq. (6.50) with the initial condition of $i_{\mathrm{L}}(0) = 0$, $i_{\mathrm{L}}(t)$ is given by:

$$i_{\mathrm{L}}(t) = \frac{V_{\mathrm{in}} + V_{\mathrm{o}}}{L}t \tag{6.51}$$

and $v_{\mathrm{C}}(t) = 0$

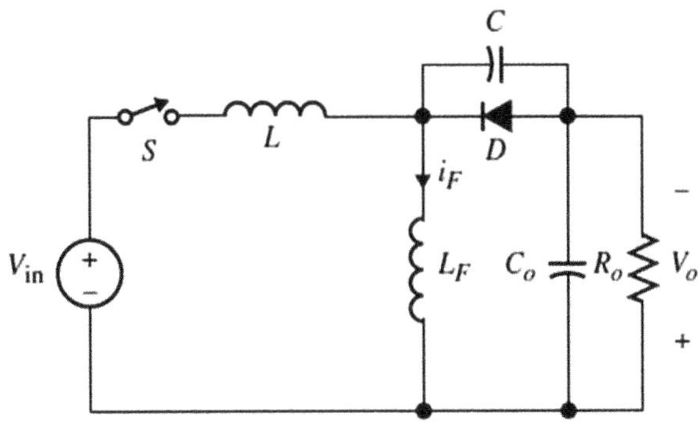

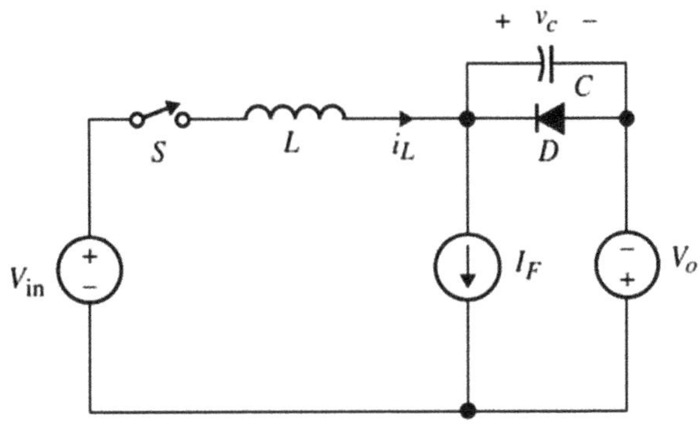

Fig. 6.18 (**a**) ZCS buck-boost with L-type switch and (**b**) simplified equivalent circuit

At $t = t_1$, the inductor current reaches I_F, forcing the output diode to stop conducting, so t_1 can be express as:

$$t_1 = \frac{LI_F}{V_{in} + V_o} \tag{6.52}$$

The steady-state waveforms are shown in Fig. 6.20.

Mode 2 ($t_1 \leq t < t_2$)
This is a resonant stage between L and C with the initial conditions given by:

$$v_C(t_1) = 0$$
$$i_L(t_1) = I_F$$

Applying Kirchhoff's law, in Fig. 6.19b, the inductor current and capacitor voltage equations may be given as:

$$L\frac{di_L}{dt} = V_{in} + V_o - v_C(t) \tag{6.53a}$$

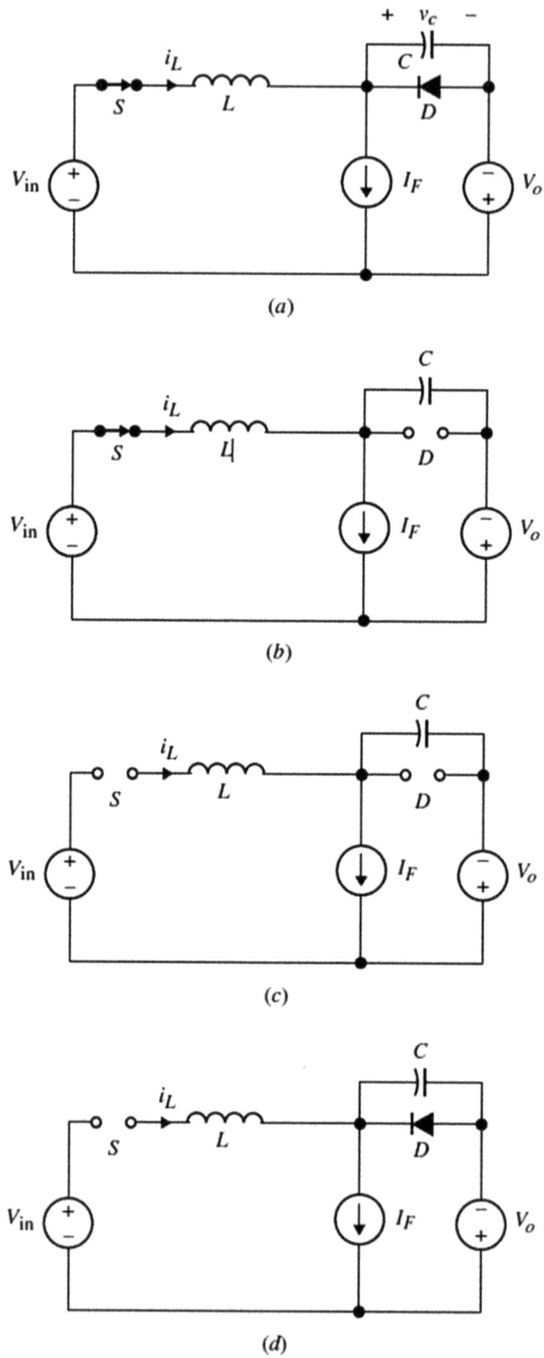

Fig. 6.19 (**a**) Equivalent circuit for Mode 1 (**b**) Equivalent circuit for Mode 2 (**c**) Equivalent circuit for Mode 3 (**d**) Equivalent circuit for Mode 4

$$C\frac{dv_C}{dt} = i_L(t) - I_F \tag{6.53b}$$

Solving Eq. (6.53) for $t > t_1$, we obtain:

$$i_L(t) = I_F + \frac{V_{in} + V_o}{Z_o}\sin\omega_o(t - t_1) \tag{6.54}$$

$$v_C(t) = (V_{in} + V_o)[1 - \cos\omega_o(t - t_1)] \tag{6.55}$$

At $t = t_2$, the inductor current reaches zero, $i_L(t_2) = 0$, and the switch stops conducting. The time interval $(t_2 - t_1)$ is given by:

$$(t_2 - t_1) = \frac{1}{\omega_o}\sin^{-1}\left(-\frac{I_F Z_o}{V_{in} + V_o}\right) \tag{6.56}$$

Mode 3 ($t_2 \leq t < t_3$)

Mode 3 starts at $t = t_2$, when the inductor current reaches zero. The switch and the diode are both off. The capacitor starts to discharge until it reaches zero, and the diode will start to conduct again at $t = t_3$. During this period, the inductor current is zero:

$$v_C(t) = \frac{-1}{C}\int_{t_2}^{t} I_F dt = \frac{-I_F}{C}(t - t_2) + V_C(t_2) \tag{6.57}$$

The diode begins to conduct at the end of this mode, $t = t_3$, because the capacitor voltage is equal to zero:

$$0 = \frac{-I_F}{C}(t_3 - t_2) + V_C(t_2)$$

where $V_C(t_2)$ may be obtained from Eq. (6.55) by evaluating it at $t = t_2$. The expression from Eq. (6.57) for the time between t_2 and t_3 is:

$$(t_3 - t_2) = \frac{C}{I_F}V_C(t_2) \tag{6.58}$$

Mode 4 ($t_3 \leq t < t_4$)

Between t_3 and t_4, the switch remains off, but the diode is on. At the end of the cycle, the switch is closed again when the current becomes zero. The cycle of the modes will repeat at T_S.

The steady-state waveforms shown in Fig. 6.20 are the characteristic waveforms for the switch, v_C, and i_L.

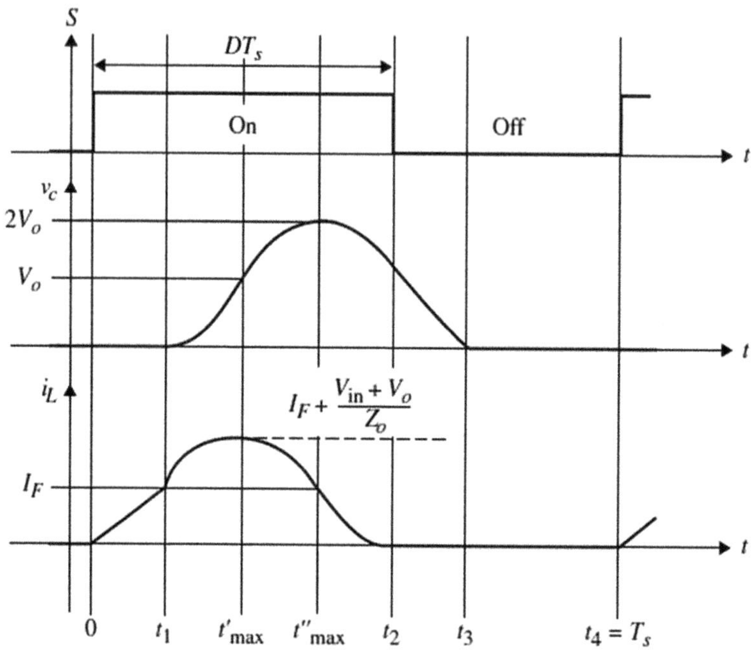

Fig. 6.20 Steady-state waveform of buck-boost L-type switch

6.4.2.6 Voltage Gain

As before, the conservation of energy per switching cycle is used to express the voltage gain, $M = V_o/V_{in}$, in terms of the normalized circuit parameter. It can be shown that M for the ZCS buck-boost converter is given by:

$$\frac{M}{1+M} = \frac{f_{ns}}{2\pi}\left[\frac{M}{2Q} + \alpha + \frac{Q}{M}(1 - \cos\alpha)\right] \tag{6.59}$$

Figure 6.21 shows the characteristic curve for M vs. f_{ns} for the ZCS buck-boost converter.

Example 6.4
Consider a QRC-ZCS buck-boost converter with the following specifications:

$V_{in} = 40$ V, $P_o = 80$ Wat $I_o = 4$ A, $f_S = 250$ kHz, $L_o = 0.1$ mH, and $C_o = 6$ μF. Design values for L and C and determine the output ripple voltage, assuming $D = 0.5$.

Solution
The output voltage and load resistance are given by:

$$V_o = \frac{80}{4} = 20 \text{ V} \quad \text{and} \quad R_o = \frac{20}{4} = 5\ \Omega$$

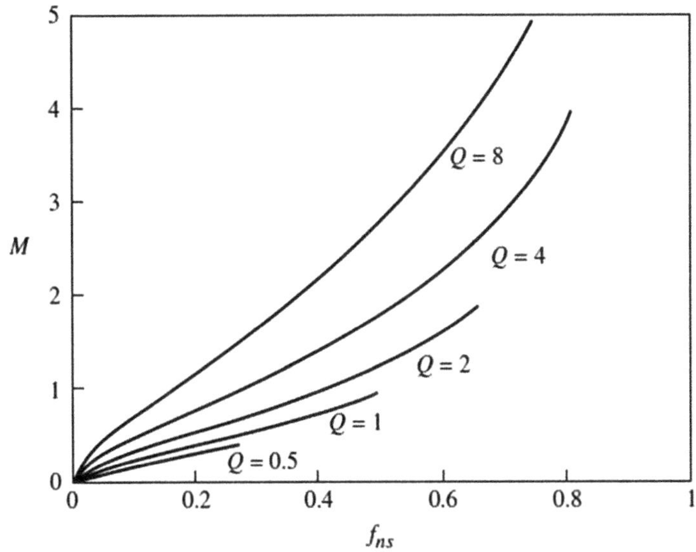

Fig. 6.21 Characteristic curve for M vs. f_{ns} for the ZCS buck-boost converter

The voltage gain is given by:

$$\frac{V_o}{V_{in}} = \frac{20}{40} = 0.5$$

With $M = 0.5$ and $f_{ns} = 0.17$, we have $Q = 3$, resulting in $f_o = 250$ kHz/0.17 = 1470.6 kHz.

From Q and Z_o, we have:

$$Q = \frac{R_o}{Z_o} = \frac{5}{Z_o} \quad \text{and} \quad Z_o = \frac{5}{3} = 1.667\ \Omega$$

Hence:

$$\sqrt{L/C} = 1.667\ \Omega$$

and

$$\sqrt{1/LC} = 2\pi f_o = 2\pi \times 1470.6 \times 10^3\ \text{rad/s}$$
$$\frac{1}{C} = 1.667 \times 2\pi \times 1470.6 \times 10^3$$

From the above equation, C and L are given by:

$$C = 64.93\ \text{nF}$$
$$L = 1.667^2 \times C = 0.1804\ \mu H$$

The duty cycle D is approximately 33% since the voltage gain for the buck-boost is 0.5. Hence, the voltage ripple is:

$$\frac{\Delta V_o}{V_o} = \frac{D}{R_o C_o f} = \frac{0.33}{5 \times 6 \times 10^{-6} \times 250 \times 10^3} = 4.4\%$$

Example 6.5

Derive the steady-state waveform for the ZCS buck-boost topology of Fig. 6.22.

Solution

It is clear that there are four modes of operation in steady state as shown in Fig. 6.22. Mode 1 starts when the switch turns on at $t = t_0$, forcing $i_L(t)$ to increase linearly until it reaches the same value as the filter inductor current I_F. Here Mode 2 starts as resonant condition between L and C since D turns off when $i_L(t) = I_F$. Resonant Mode 2 ends when $i_L(t)$ becomes zero. Mode 3 begins during which the capacitor linearly charge to $(V_o + V_{in})$ at which the diode conducts during Mode 4 until the cycle repeats at $t = T_s + t_0$ when S is turned on again.

Exercise 6.4

For the buck-boost converter with an M-type switch shown in Fig. 6.23, show that the voltage gain, $M = V_o/V_{in}$, is given by:

$$\frac{M}{1+M} = \frac{f_{ns}}{2\pi}\left[\frac{M}{2Q} + \alpha + \frac{Q}{M}(1 - \cos\alpha)\right] \tag{6.60}$$

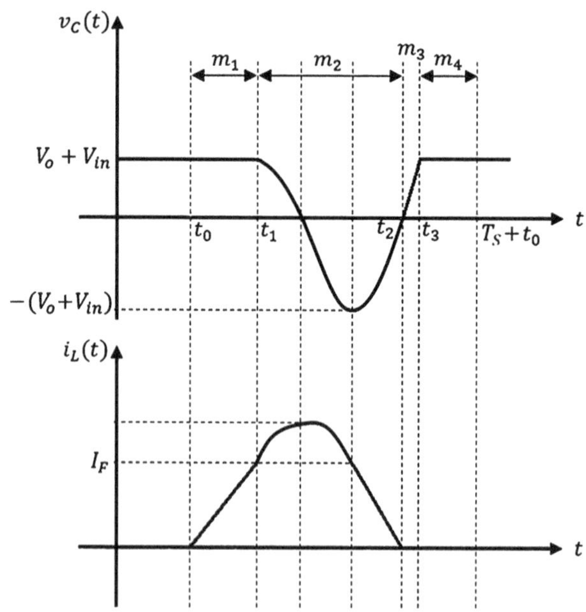

Fig. 6.22 Steady-state waveform for the M-type ZCS buck-boost converter

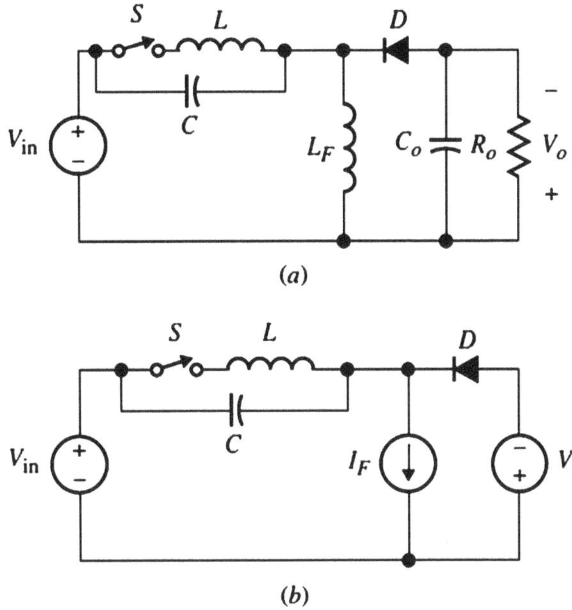

Fig. 6.23 (**a**) ZCS buck-boost converter with an M-type switch and (**b**) simplified equivalent circuit

6.5 Zero-Voltage Switching Topologies

In this section, we will investigate the zero-voltage switching (ZVS) quasi-resonant converter family. Like the ZCS topologies, M-type or L-type switch arrangements can be used. In those topologies, the power switch is turned on at zero voltage (of course, turnoff also occurs at zero voltage). While the switch is off, a peak voltage will appear across it, causing the stress to be higher than in the hard-switching PWM case. In a ZVS topology, a flyback diode across the switch (body diode) is used to damp the voltage across the capacitor, which results in a zero voltage across the switch. We should point out that the capacitor across the switch can be the same as the switch's parasitic capacitor, and the flyback diode could be the same as the internal body diode of the power semiconductor switch.

Figure 6.24a shows a MOSFET switch implementation including an internal body diode and a parasitic capacitor.

We will assume C_{gd} and C_{gs} are too small to be included. If the body diode, D_S, is not fast enough for the designed application or has limited power capabilities, it is possible to block it and use an external, fast flyback diode as shown in Fig. 6.24b. D_1 is used to block D_S, and D_2 is the diode actually used to carry the reverse switch current. Both the current- and voltage-mode control methods are used in conjunction with a direct duty ratio PWM control approach to vary the on or off time of the power switch.

Over the last 15 years, many different zero-voltage resonant converter topologies have been introduced. Only the quasi-resonant soft-switching ZVS topologies

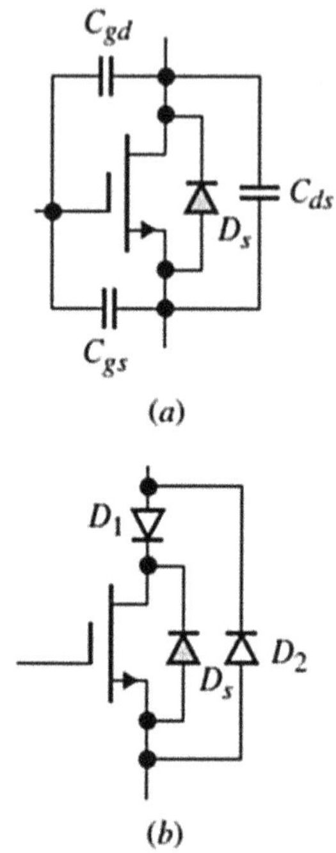

Fig. 6.24 (**a**) MOSFET implementation and (**b**) MOSFET switch with fast flyback diode

will be analyzed and their control characteristic curves studied here. Regardless of their topological variations, many of these converters have several common features.

6.5.1 Resonant Switch Arrangements

Next we investigate the ZVS buck, boost, and buck-boost topologies using L-type and M-type resonant switches. Figure 6.25a shows the two possible implementations using L- and M-type resonant switches. The half-wave L-type and M-type MOSFET implementations are shown in Fig. 6.25b, c shows the full-wave implementations for L- and M-type switches.

6.5.2 Steady-State Analyses

As the ZCS case, to simplify the steady-state analysis, here we make the same assumptions made in Sect. 4.2.

6.5.2.1 The Buck Converter

Replacing the switch in Fig. 6.7a by the M-type switch of Fig. 6.25a, we obtain a new ZVS buck converter as shown in Fig. 6.26a. The simplified equivalent circuit is given in Fig. 6.26b.

As the switch and diode are *on* and/or *off* at the same time, it can be shown that under steady-state conditions, there are four modes of operation. Unlike ZCS topologies, the switching cycle in ZVS starts with the main switch first being in the nonconduction state. This is in order to establish a zero-voltage condition across the switch during the resonant stage while it is open. Figure 6.27 shows the

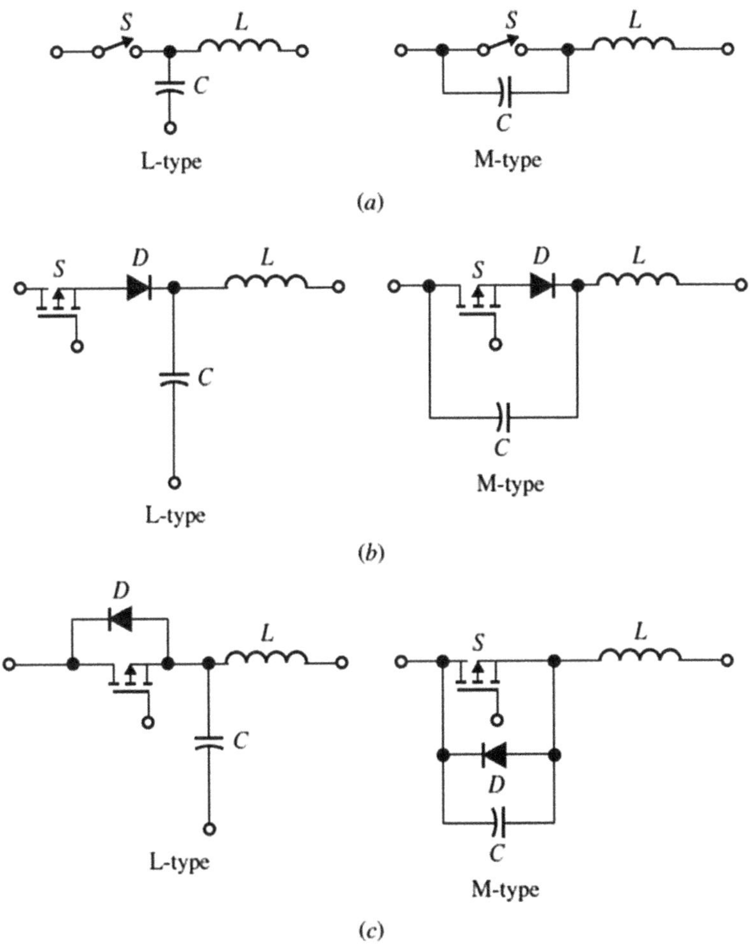

Fig. 6.25 (**a**) Resonant switch arrangement types for ZVS operation (**b**) Half-wave MOSFET implementation (**c**) Full-wave MOSFET implementation

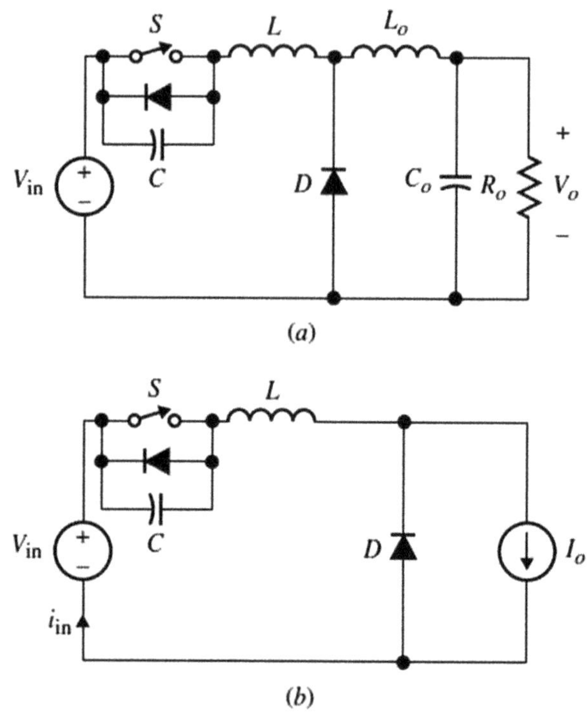

Fig. 6.26 (**a**) Quasi-resonant buck converter with M-type switch and (**b**) simplified equivalent circuit

equivalent circuit modes under the steady-state condition. The four modes are discussed as follows.

Mode 1 ($0 \leq t < t_1$)

Assume that initially the power switch is conducting and the diode is off. Mode 1 starts at $t = 0$, when the switch is turned *off*. In this mode, since S has been closed for $t < 0$, the initial capacitor voltage, v_C, is zero, and the inductor current is I_o as shown in Fig. 6.27a:

$$v_C(0) = 0$$
$$i_L(0) = I_o$$

Applying KCL to Fig. 6.27a, we have:

$$C\frac{dv_C}{dt} = i_L(t)$$

Since $i_L(t) = I_o$, the capacitor starts to charge according to Eq. (6.61):

$$v_C(t) = \frac{1}{C}I_o t \tag{6.61}$$

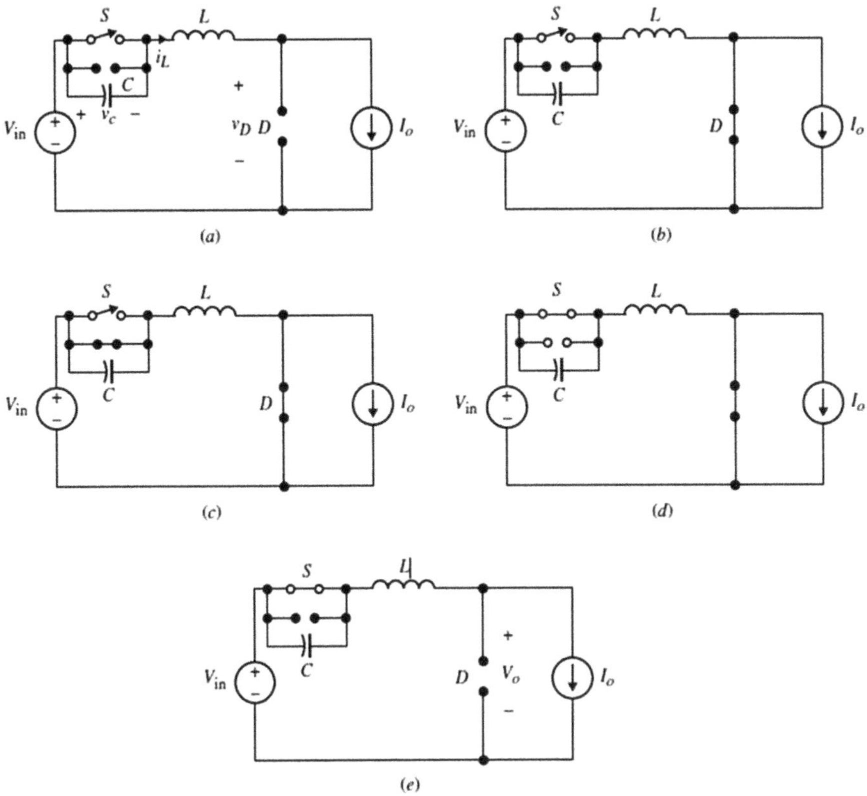

Fig. 6.27 Equivalent circuits for (**a**) Mode 1 (**b**) Mode 2, (**c**) Mode 3, (**d**) Mode 4, and (**e**) Mode 5

The voltage across the output diode is given by:

$$v_D(t) = V_{in} - v_C(t)$$

As long as $v_C < V_{in}$, the diode remains *off*.

The capacitor voltage reaches the input voltage, V_{in}, at $t = t_1$, causing the diode to turn *on*. Hence, at $t = t_1$, we have:

$$V_C(t_1) = V_{in}$$

and t_1 can be expressed as:

$$t_1 = \frac{CV_{in}}{I_o} \tag{6.62}$$

At $t = t_1$, the circuit enters Mode 2. The current and voltage waveforms are shown in Fig. 6.28.

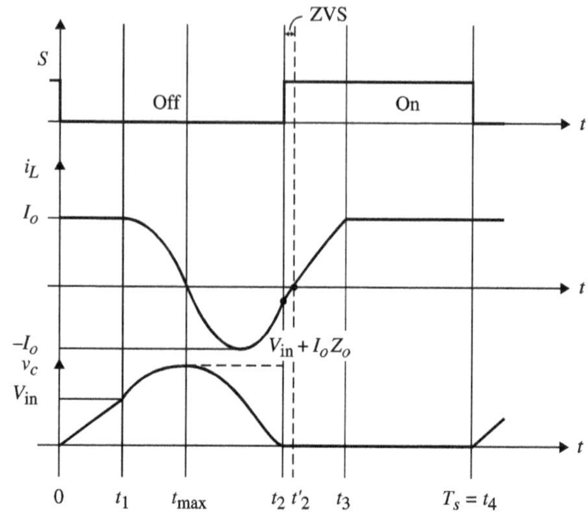

Fig. 6.28 Steady-state waveforms for the ZVS buck converter

Mode 2 ($t_1 \leq t < t_2$)
Mode 2 starts at t_1, when the diode turns on, and the circuit enters the resonant stage. During the time between t_1 and t_2. The switch remains off. At $t = t_2$, the capacitor voltage tends to go negative, forcing the diode across S to turn on. The initial capacitor voltage and inductor current in this mode are given by:

$$V_C(t_1) = V_{in}$$

$$I_L(t_1) = I_o$$

The expressions of the current and the voltage in the time domain are given in Eqs. (6.63) and (6.64), respectively:

$$i_L(t) = I_o \cos \omega_o (t - t_1) \tag{6.63}$$

$$v_C(t) = V_{in} + I_o Z_o \sin \omega_o (t - t_1) \tag{6.64}$$

where the resonant frequency and the characteristic impedance are defined as before.

The inductor current is zero when the capacitor voltage reaches the peak, and the capacitor starts discharging while the inductor current is a negative value. The inductor current reaches the peak when the capacitor drops to the input voltage, and at the end of the mode, i.e., at $t = t_2$, the capacitor equals zero, $v_C(t_2) = 0$.

The period between t_2 and t_1 is given by:

$$t_2 - t_1 = \frac{1}{\omega_o} \sin^{-1}\left(\frac{-V_{in}}{I_o Z_o}\right) = \frac{\alpha}{\omega_o} \tag{6.65}$$

and the inductor current at $t = t_2$ is:

$$I_L(t_2) = I_o \cos \alpha$$

where:

$$\alpha = \omega_o(t_2 - t_1) \tag{6.66}$$

Mode 3 ($t_2 \leq t < t_3$)
At t_2 the capacitor voltage becomes zero, and the inductor current starts to charge linearly and reaches the output current at $t = t_3$. The body diode of the switch turns on at $t = t_2$ to maintain inductor current continuity, and the output diode also remains on at this point, as shown in Fig. 6.27c. As long as the inductor current is less than I_o, the output diode will stay on. The switch may be turned on at 2 V any time after t_2 and before t_2', when the inductor current reverses polarity.

The initial value of the capacitor voltage in Mode 3 is zero:

$$V_c(t_2) = 0$$

The inductor voltage is equal to the input voltage:

$$L\frac{di_L}{dt} = V_{in} \tag{6.67}$$

Integrating Eq. (6.67) from t_2 to t, the inductor current can be expressed as:

$$i_L(t) = \frac{V_{in}}{L}(t - t_2) + I_o \cos \alpha \tag{6.68}$$

At $t = t_3$, the inductor current reaches the output current, $i_L(t_3) = I_o$, forcing the diode to turn off. The time interval from t_3 to t_2 is:

$$t_3 - t_2 = \frac{I_o L}{V_{in}}(1 - \cos \alpha) \tag{6.69}$$

Therefore, there will be no power transfer and no charge or discharge intervals when the switch turns on again in the next mode.

Mode 4 ($t_3 \leq t < t_4$)
In this mode, as shown in Fig. 6.27d, the inductor current is trapped and held constant at $i_L = I_o$, with $v_c = 0$:

$$i_L = 0$$
$$v_c = 0$$

Mode 4 will continue as long as the switch remains on. By turning off the switch at $t = t_4 = T_s$, the switching cycle repeats. The dead time $t_4 - t_3$ is given by:

$$t_4 - t_3 = T_s - t_1 - (t_2 - t_1)(t_3 - t_2) \tag{6.70}$$

6.5.2.2 Voltage Gain

We follow the same approach as in the ZCS by using the energy balance concept. The input energy is given by:

$$E_{in} = \int_0^{T_s} i_{in} V_{in} dt$$

i_{in} is the current which is equal to $i_L(t)$. Hence, we have:

$$E_{in} = \int_0^{t_1} i_L(t) V_{in} dt + \int_{t_1}^{t_2} i_L(t) V_{in} dt + \int_{t_2}^{t_3} i_L(t) V_{in} dt + \int_{t_3}^{T_s} i_L(t) V_{in} dt \tag{6.71}$$

The inductor current equals the output current in Mode 1 and 4, and for $t_1 \leq t < t_2$ and $t_2 \leq t < t_3$, i_L is given in Eqs. (6.63) and (6.68), respectively. Substitute the inductor currents into Eq. (6.71) to yield Eq. (6.72):

$$\begin{aligned} E_{in} = V_{in} \Big[& I_o t_1 + I_o \sqrt{LC} \sin \omega_o (t_2 - t_1) + \frac{V_{in}}{2L} (t_3 - t_2)^2 \\ & + I_o (t_3 - t_2) \cos \alpha + I_o (T_s - t_3) \Big] \end{aligned} \tag{6.72}$$

Substituting for the time intervals t_1, $(t_2 - t_1)$, $(t_3 - t_2)$, and $(T_s - t_3)$ from Eqs. (6.62), (6.65), (6.69), and (6.70), respectively, and using the normalized parameters M, Q, and ω_o, Eq. (6.72) becomes:

$$E_{in} = V_{in} I_o \left[\frac{-Q}{M \omega_o} - \frac{ML}{2R_o} + T_s - \frac{\alpha}{\omega_o} + \frac{ML}{R_o} \cos \alpha - \frac{ML}{2R_o} \cos^2 \alpha \right] \tag{6.73}$$

The output energy is expressed by:

$$E_o = \int_0^{T_s} I_o V_o dt = I_o V_o T_s \tag{6.74}$$

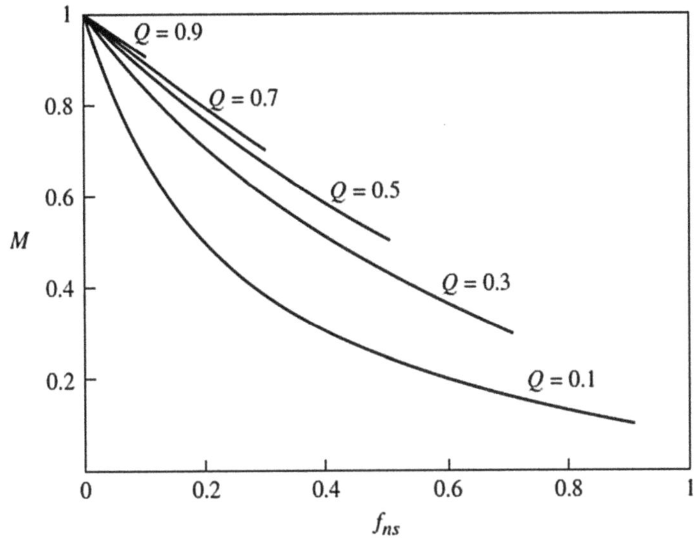

Fig. 6.29 Control characteristic curve of M vs. f_{ns} for ZVS buck converter

Equating the input and output energy in Eqs. (6.73) and (6.74), the voltage gain expression becomes:

$$M = 1 - \frac{f_{ns}}{2\pi}\left[\frac{Q}{2M} + \alpha + \frac{M}{Q}(1 - \cos\alpha)\right] \tag{6.75}$$

where:

$$\alpha = \sin^{-1}\left(\frac{-Q}{M}\right) + \pi$$

A plot of the control characteristic curve of M vs. f_{ns} is shown in Fig. 6.29.

6.5.2.3 The Boost Converter

In this section, we consider the quasi-resonant boost converter using the M-type switch as shown in Fig. 6.30a, with its simplified circuit shown in Fig. 6.30b. The four circuit modes of operation are shown in Fig. 6.31.

Mode 1 ($0 \leq t < t_1$)
Assume for $t < 0$, the switch is closed, while D is open. At $t = 0$, the switch is turned *off*, allowing the capacitor to charge by the constant current I_{in} as given by:

$$I_{in} = i_L(t) = i_c(t) = C\frac{dv_c}{dt} \tag{6.76}$$

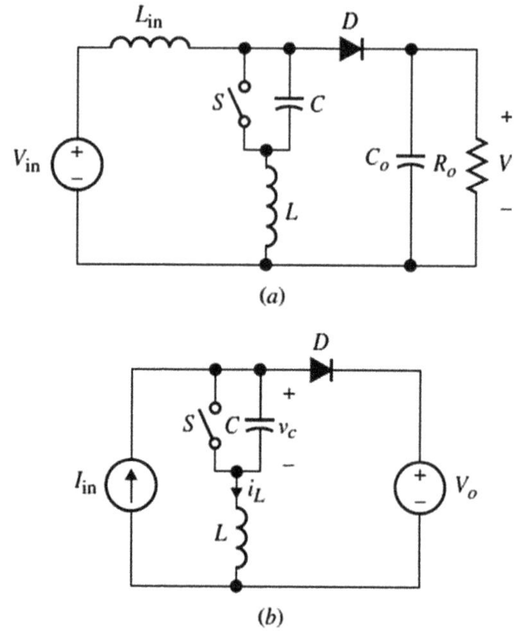

Fig. 6.30 (**a**) Quasi-resonant boost converter with M-type switch and (**b**) equivalent circuit

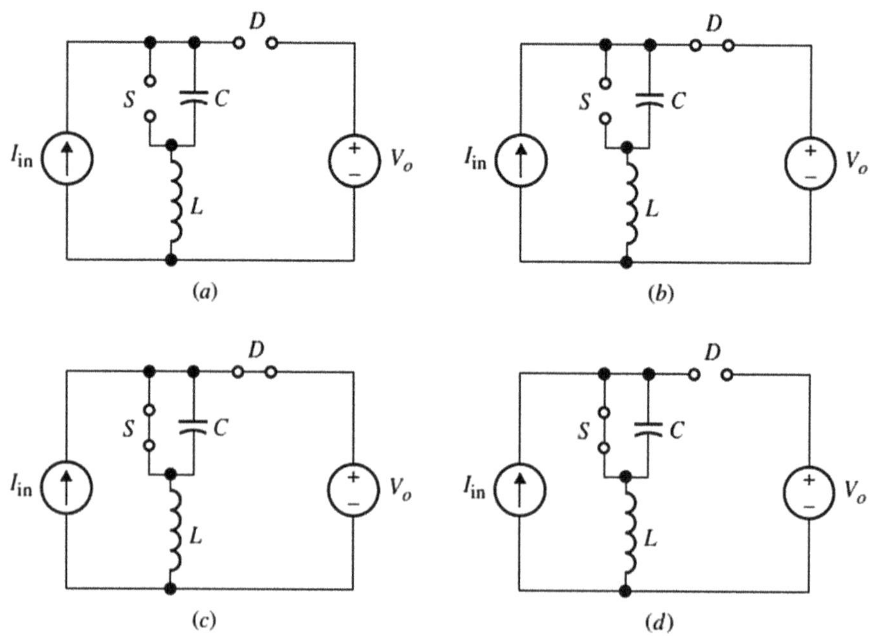

Fig. 6.31 Equivalent circuit modes: (**a**) Mode 1, (**b**) Mode 2, (**c**) Mode 3, and (**d**) Mode 4

Since the initial capacitor voltage equals zero, Eq. (6.76) gives the following expression for $v_c(t)$:

$$v_c(t) = \frac{I_{in}}{C}t \tag{6.77}$$

The capacitor voltage reaches the output voltage at $t = t_1$, i.e., $v_{c1}\ (t_1) = V_o$, resulting in:

$$t_1 = \frac{CV_o}{I_{in}} \tag{6.78}$$

At $t = t_1$, the diode starts conducting since $v_c = V_o$, and the converter enters Mode 2.

Mode 2 ($t_1 \leq t < t_2$)
At $t = t_1$, the resonant stage begins since D is *on* and S is *off* as shown in Fig. 6.31b. When the capacitor voltage reaches the output voltage, i_L reaches the negative peak.

The initial conditions are $v_c(t_1) = V_o$ and $i_L(t) = I_{in}$.

The expression for $v_c(t)$ is given by Eq. (6.79):

$$v_c(t) = V_o + I_{in}Z_o \sin \omega_o(t - t_1) \tag{6.79}$$

and the inductor current is:

$$i_L(t) = I_{in} \cos \omega_o(t - t_1) \tag{6.80}$$

Evaluating Eq. (6.79) at $t = t_2$ with $v_c(t_2) = 0$, the time interval between t_1 to t_2 can be found to be:

$$(t_2 - t_1) = \frac{1}{\omega_o} \sin^{-1}\left(\frac{-V_o}{I_{in}Z_o}\right) \tag{6.81}$$

Mode 3 ($t_2 \leq t < t_3$)
Mode 3 starts at t_2, when v_c reaches zero and S turns on at ZVS. The switch and the diode are both conducting, and the inductor current linearly increases to I_{in} as shown in Fig. 6.32. At $t = t_2$, the diode (antiparallel diode) turns on, clamping the voltage across C to zero.

The initial conditions at $t = t_2$ are:

$$V_c(t_2) = 0 \tag{6.82a}$$

$$I_L(t_2) = I_{in} \cos \omega_o(t_2 - t_1) \tag{6.82b}$$

Because the capacitor voltage is zero, the inductor voltage is equal to the output voltage:

$$L\frac{di_L}{dt} = V_o \tag{6.83}$$

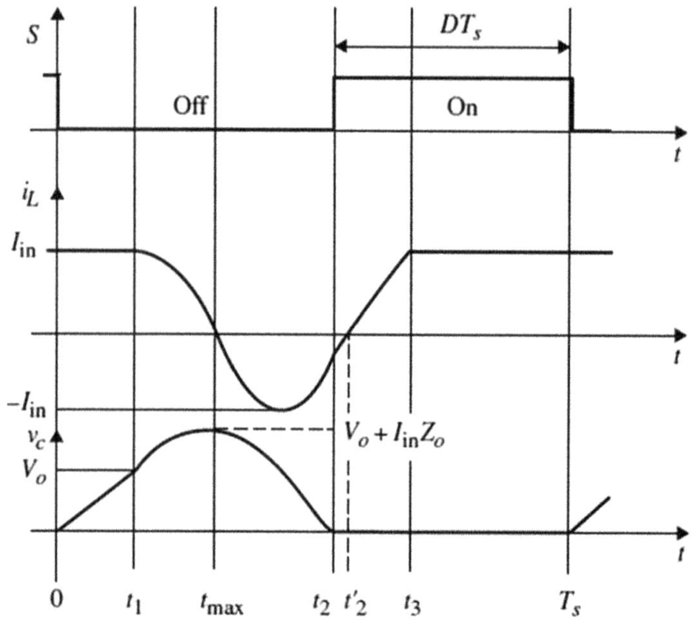

Fig. 6.32 Steady-state waveforms for ZVS boost converter

By integrating Eq. (6.83), the inductor current becomes:

$$i_L(t) = \frac{V_o}{L}(t - t_2) + I_L(t_2) \tag{6.84}$$

To achieve ZVS, the switch can be turned on anytime after t_2 and before t_2'. At $t = t_2'$, the inductor current reverses polarity, and the switch picks up the current. At $t = t_3$, i_L reaches I_{in}, resulting in the time interval given in Eq. (6.85):

$$(t_3 - t_2) = \frac{L}{V_o}[I_{in} - i_L(t_2)] \tag{6.85}$$

Substituting the initial condition into the equation, we get:

$$(t_3 - t_2) = \frac{L}{V_o} I_{in}(1 - \cos \omega_o(t_2 - t_1)) \tag{6.86}$$

At $t = t_3$, the output diode turns *off*, and the entire I_{in} current flows in the transistor and the inductor.

Mode 4 ($t_3 \le t < t_4$)
At time t_3, the inductor current reaches I_{in}, and the output diode turns *off*, but the switch remains closed. The cycle of the mode will repeat again at $t = T_s$.

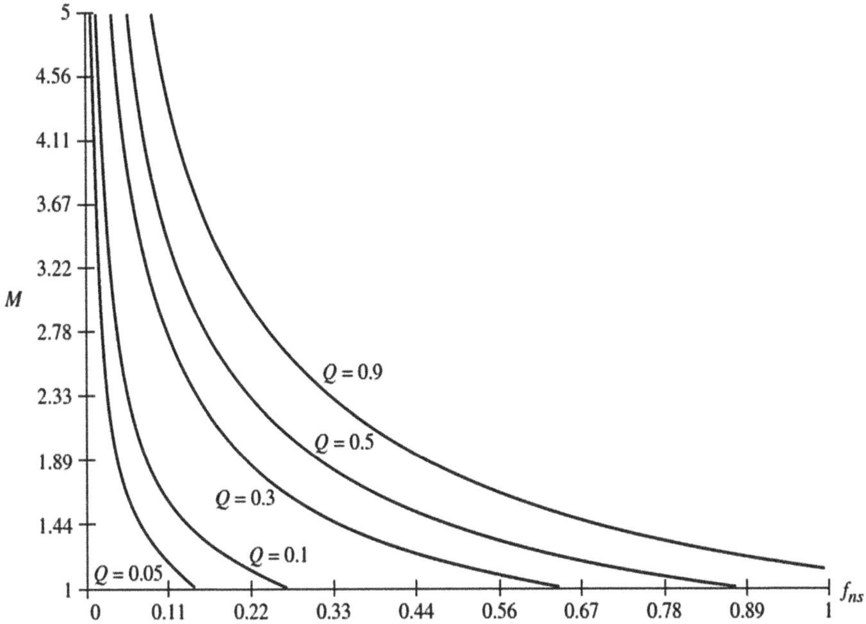

Fig. 6.33 Control characteristic curve of M vs. f_{ns} for ZVS boost converter

6.5.2.4 Voltage Gain

As before, we use the conservation of energy per switching cycle to express the voltage gain. It can be shown that the voltage gain in terms of the normalized parameter is given in Eq. (6.87), and its plot is given in Fig. 6.33.

A plot of the control characteristic curve of M vs. f_{ns} is shown in Fig. 6.33:

$$\frac{1}{M} = \frac{f_{ns}}{2\pi}\left[\frac{Q}{2M} + \alpha + \frac{M}{Q}(1 - \cos\alpha)\right] \tag{6.87}$$

Example 6.6

Design a ZVS quasi-resonant boost converter for the following design parameters: $V_{in} = 30$ V, $P_o = 30$ W at $V_o = 38$ V, $f_{ns} = 0.4$, and $T_s = 4$ μs. Assume the output voltage ripple is limited to 2% at $D = 0.4$.

Solution

The voltage gain is $M = V_o/V_{in} = 1.3$, and with $f_{ns} = 0.4$, we obtain $Q = 0.2$.

Using the switching frequency $f_s = 1/T_s = 1/4$ μs $= 250$ kHz and $f_o = f_s/0.4 = 250/0.4 = 625$ kHz, the resonant frequency is obtained from:

$$\omega_o = \frac{1}{\sqrt{LC}} = (2\pi)(625) \times 10^3 \text{ rad/s}$$

The second equation in terms of L and C is obtained from:

$$Q = \frac{R_o}{Z_o} = \frac{R_o}{\sqrt{L/C}} = 0.2$$

The load resistance is:

$$R_o = \frac{38^2}{30} = 48.13\ \Omega$$

Substituting the above relation for Q, we obtain:

$$\sqrt{\frac{L}{C}} = \frac{48.13}{0.2} = 240.65\ \Omega$$

Solving the above two equations for C and L, we obtain:

$$C = \frac{1}{(2\pi)\,(625)\,\left(10^3\right)\,(240.65)} = 1.06\ \text{nF}$$

To calculate L_o and C_o, using the voltage ripple to be 2 % , we use the following relation:

$$\frac{D}{f_s R_o C_o} = 0.02$$

where:

$$\begin{aligned}
D &= 0.4 \\
C_o &= \frac{0.4}{f_s R_o}\frac{100}{0.2} \\
&= \frac{40}{0.2 \times 250 \times 10^3 \times 48.13} = 16.6\ \mu F
\end{aligned}$$

The critical inductor value is given by:

$$\begin{aligned}
L_{\text{crit}} &= \frac{R_o}{2f_s}(1-D)^2 D \\
&= \frac{48.13}{2 \times 250 \times 10^3}(1-0.4)^2(0.4)
\end{aligned}$$

To achieve a limited ripple current, it is recommended that L_o be set to be about a 100 times the critical inductor value. So we select $L_o = 1.4$ mH.

Exercise 6.5

The quasi-resonant boost converter using the L-type switch is shown in Fig. 6.34a, and its simplified circuit is shown in Fig. 6.34b. Derive the expression for the voltage gain.

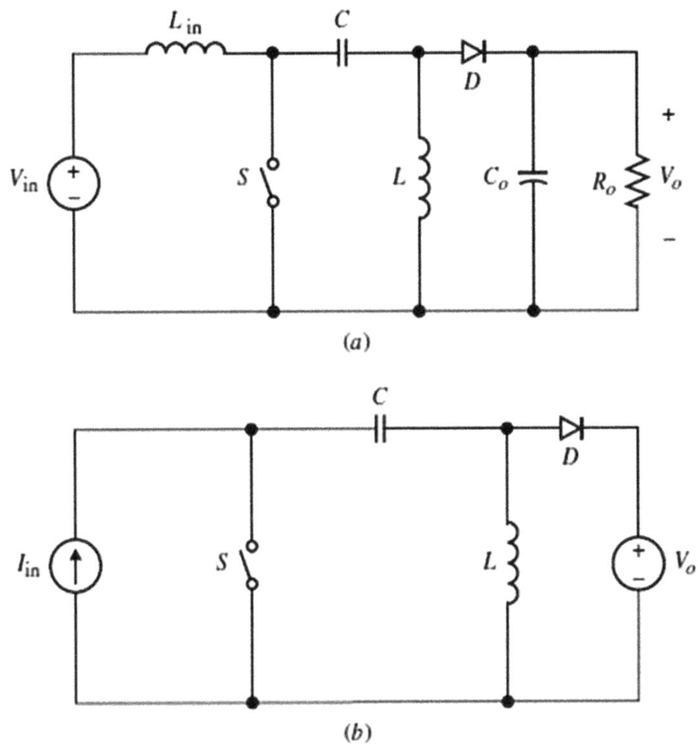

Fig. 6.34 (**a**) Quasi-resonant boost converter with L-type switch and (**b**) simplified equivalent circuit

6.5.2.5 The Buck-Boost Converter

The ZVS buck-boost converter with an M-type switch is shown in Fig. 6.35a with its equivalent circuit shown in Fig. 6.35b.

Like the buck and the boost converters, the buck-boost converter also leads to four modes of operation as shown in Fig. 6.36. Figure 6.36e shows the typical steady-state waveforms for v_c and i_L.

Following a similar analysis as before, it can be shown that the voltage gain in terms of M, Q, and f_{ns} is given in Eq. (6.88):

$$M = \frac{1}{\frac{f_{ns}}{2\pi}\left[\alpha + \frac{Q}{2M} + \frac{M}{Q}(1 - \cos\alpha)\right]} - 1 \qquad (6.88)$$

Figure 6.37 shows the control characteristic curve for M vs. f_{ns}.

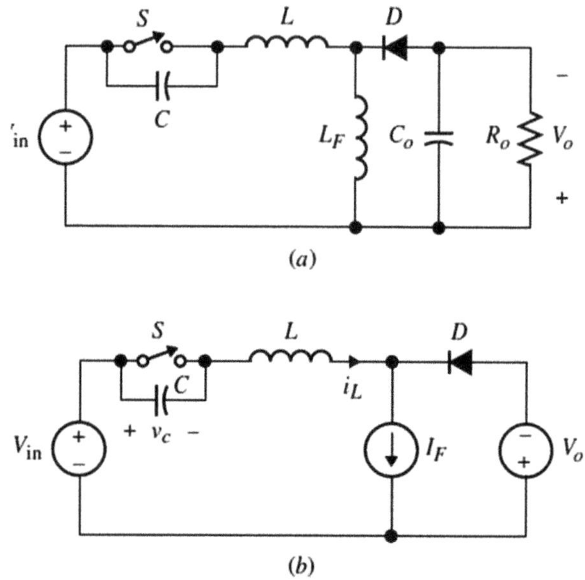

Fig. 6.35 (**a**) ZVS buck-boost M-type and (**b**) simplified equivalent circuit

6.6 Zero-Voltage and Zero-Current Transition Converters

Traditional converters operate with a sinusoidal current through the power switches, which results in high peak and rms currents for the power transistors and high-voltage stresses on the rectifier diodes. Furthermore, when the line voltage or load current varies over a wide range, quasi-resonant converters are modulated with a wide switching-frequency range, making the circuit design difficult to optimize. As a compromise between the PWM and resonant techniques, various soft-switching PWM converter techniques have been proposed lately to combine the desirable features of the conventional PWM and quasi-resonant techniques without a significant increase in the circulating energy.

6.6.1 Switching Transition

To overcome the limitations of the quasi-resonant converters, zero-voltage transition (ZVT) or zero-current transition (ZCT) is the solution. Instead of using a series resonant network across the power switch, a shunt resonant network is used across the power switch. A partial resonance is created by the shunt resonant network to achieve ZCS or ZVS during the switching transition. This retains the advantages of a PWM converter because after the switching transition is over, the circuit reverts to the PWM operation mode.

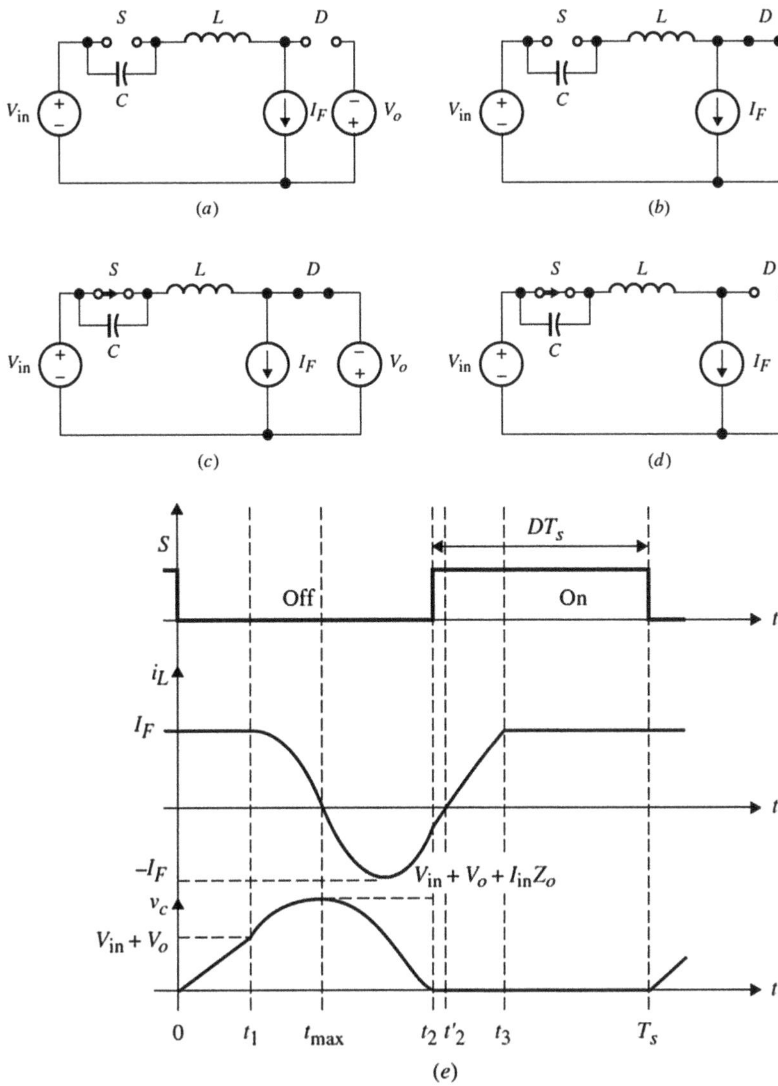

Fig. 6.36 Equivalent circuit for (**a**) Mode 1, (**b**) Mode 2, (**c**) Mode 3, (**d**) Mode 4, and (**e**) steady-state waveforms for v_C and i_L

The features of the ZCT-PWM and ZVT-PWM soft-switching converters are summarized as follows:

- Zero-current/voltage turnoff/on for the power switch
- Low-voltage/current stresses of the power switch and rectifier diode
- Minimal circulating energy
- Constant frequency operation
- Soft switching for a wide line and load range

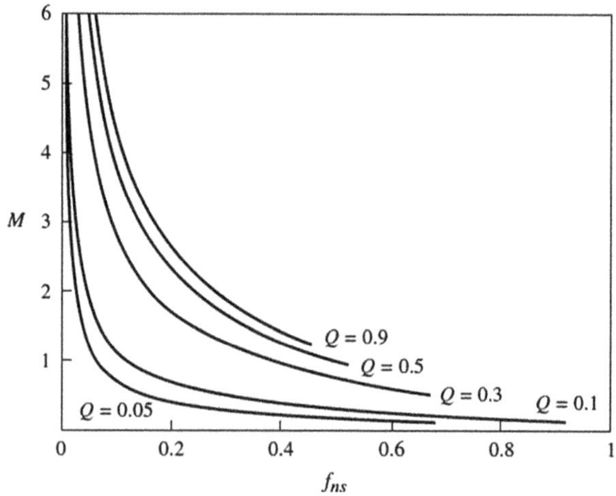

Fig. 6.37 Control characteristic curve of M vs. f_{ns} for ZVS buck-boost converter

One disadvantage is that the auxiliary switch does not operate with soft switching; it is hard switching, but the switching loss is much lower than that of a PWM converter. Another disadvantage is that the transformer leakage inductance is not utilized, which is similar to the quasi-resonant converters. Therefore, the transformer should be designed with minimum leakage inductance.

The ZVT and ZCT converters differ from the conventional PWM converters by the introduction of a resonant branch shown in Fig. 6.38. Figure 6.38a shows the ZVT-PWM switching cell, and Fig. 6.38b shows the ZCT-PWM switching cell. L is a resonant inductor, C is a resonant capacitor, S_1 is an auxiliary switch, and D_1 is an auxiliary diode.

6.6.2 *The ZVT Buck Converter*

In this section, we consider the ZVT buck PWM shown in Fig. 6.39 by replacing the ZVT-PWM switching cell of Fig. 6.38a, into the conventional buck converter. The ZVT buck converter equivalent circuit is shown in Fig. 6.39b. The resonant elements are L_r, C_{r1}, and C_{r2}. The capacitors C_{r1} and C_{r2} represent the switch $S1$ and output D junction capacitors, respectively. The main switch is implemented by $S1$ and its associated body diode D_s. The auxiliary switch is S_a and its body diode is D_a.

6.6.2.1 Steady-State Analysis

Let us start the steady-state analysis by first assuming that for $t < t_o$, the main switch S_1 and the auxiliary switch S_a were turned off as shown in Fig. 6.40 The output

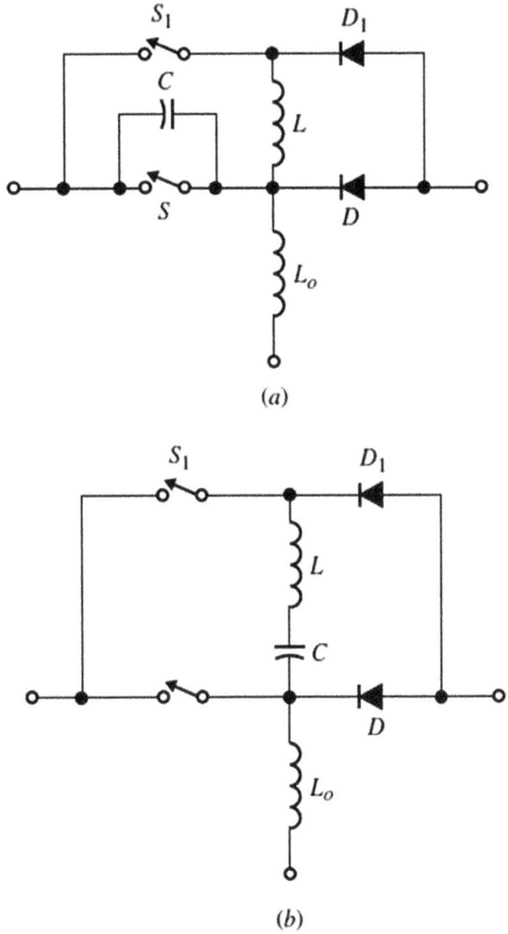

Fig. 6.38 (**a**) ZVT-PWM switching cell (**b**) ZCT-PWM switching cell

current I_o flows through the output diode D, and the capacitor voltage v_{cr1} is charged to V_{in}. It will be shown next that the switching cycle is divided into seven modes as shown in Fig. 6.41.

These modes are discussed as follows:

Mode 1 ($t_0 \le t < t_1$)

Mode 1 begins when the auxiliary switch S_a is turned on at $t = t_o$ resulting in the equivalent circuit as shown in Fig. 6.41a.

The voltage across L_r is clamped to V_{in}. Hence, $i_{Lr}(t)$ is obtained from:

$$L_r \frac{di_{Lr}(t)}{dt} = V_{in}$$

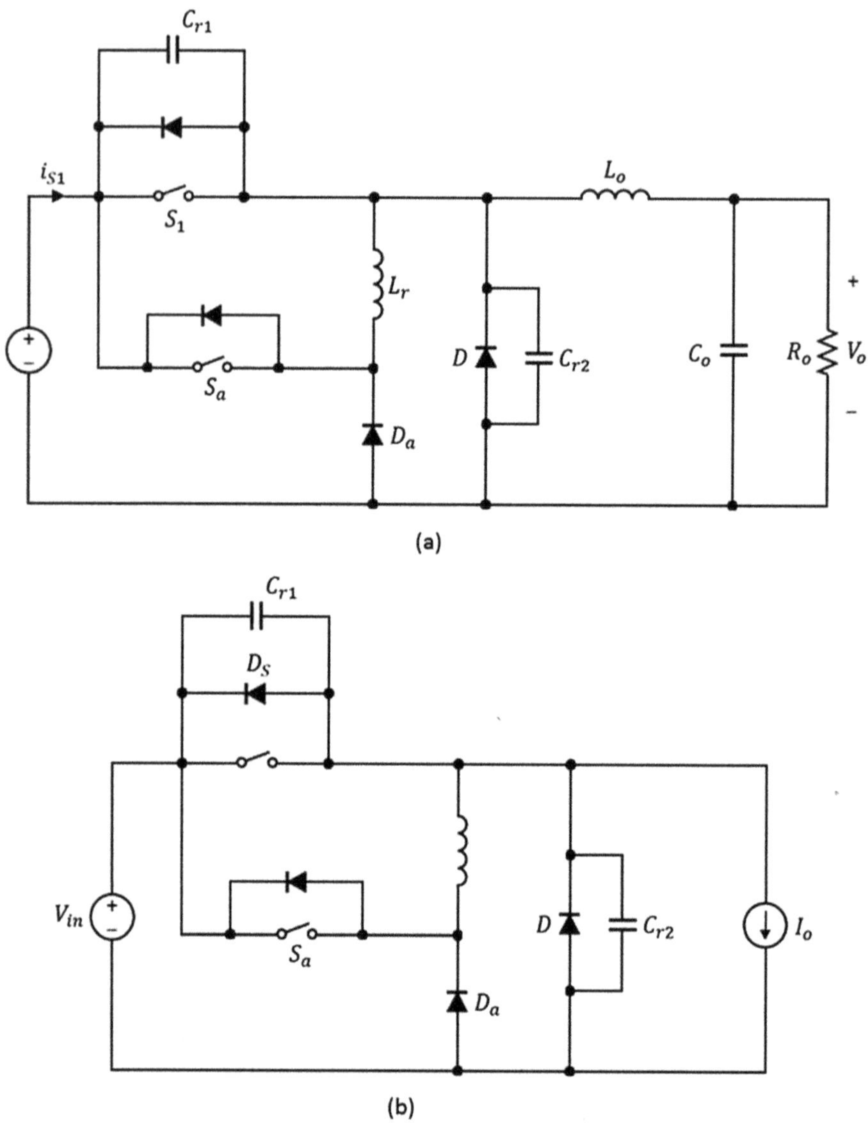

Fig. 6.39 (**a**) ZVT boost converter and its (**b**) simplified equivalent circuit

For $t > t_{o,}$ $i_{Lr}(t)$ is given by:

$$i_{Lr}(t) = \frac{V_{in}}{L_r}(t - t_0)$$

The current $i_{Lr}(t)$ linearly charges until $t = t_1$ when it becomes equal to I_o, $i_{Lr}(t_1) = I_0$, which gives us the following time intervals for $(t_1 - t_o)$ for Mode 1:

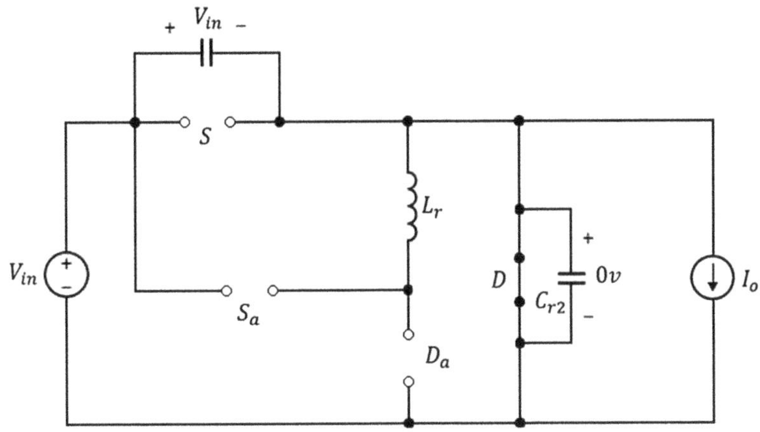

Fig. 6.40 The steady-state equivalent circuit for Fig. 6.39b for $t < t_o$

$$t_1 - t_o = \frac{L_r I_o}{V_{in}}$$

At $t = t_1$, D turns off at zero current, and the circuit enters Mode 2 as shown in Fig. 6.41b with the following inductor current and capacitor voltage values:

$$i_{Lr}(t_1) = I_0,$$
$$v_{Cr1}(t_1) = V_{in}, \text{ and}$$

$$v_{Cr2}(t_1) = 0.$$

Mode 2 ($t_1 \leq t < t_2$)

This mode is formed as a result of turning off of the diode D. The equivalent circuit is as shown in Fig. 6.41b.

The differential equations in this mode are given by:

$$v_{Cr1}(t) = \frac{i_{Lr} di_{Lr}(t)}{dt} \tag{6.89}$$

$$I_o - i_{Lr}(t) = c_{r1}\frac{dv_{cr1}}{dt} - c_{r2}\frac{dv_{cr2}}{dt} \tag{6.90}$$

Applying KVL to obtain:

$$V_{in} = v_{Cr1} + v_{Cr2}$$

Hence Eq. 6.90 becomes after replacing $v_{cr2} = V_{in} - v_{cr1}$

$$I_o - i_l(t) = (C_{r1} + C_{r2})\frac{dv_{Cr1}}{dt}$$

$$= C_r\frac{dv_{Cr1}}{dt}$$

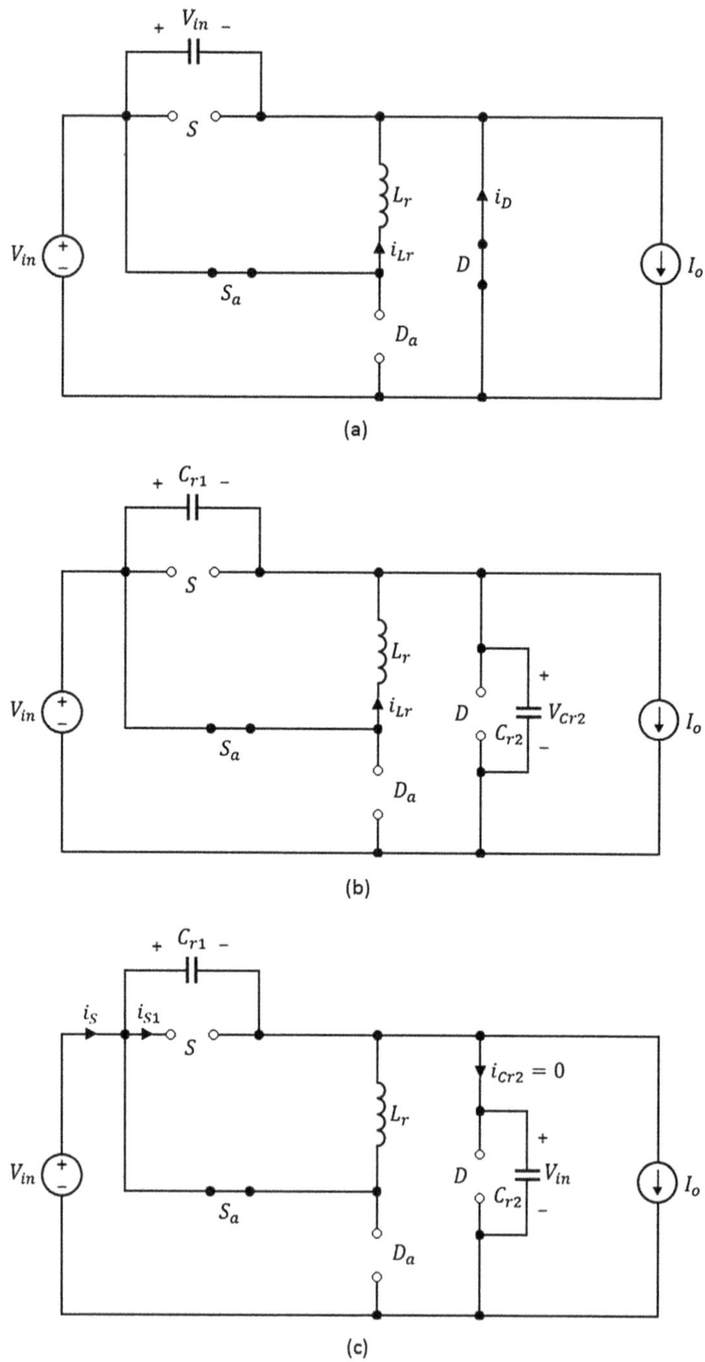

Fig. 6.41 Equivalent circuits for the seven modes of operation: (**a**) Mode 1, (**b**) Mode 2, (**c**) Mode 3, (**d**) Mode 4, (**e**) Mode 5, (**f**) Mode 6, and (**g**) Mode 7

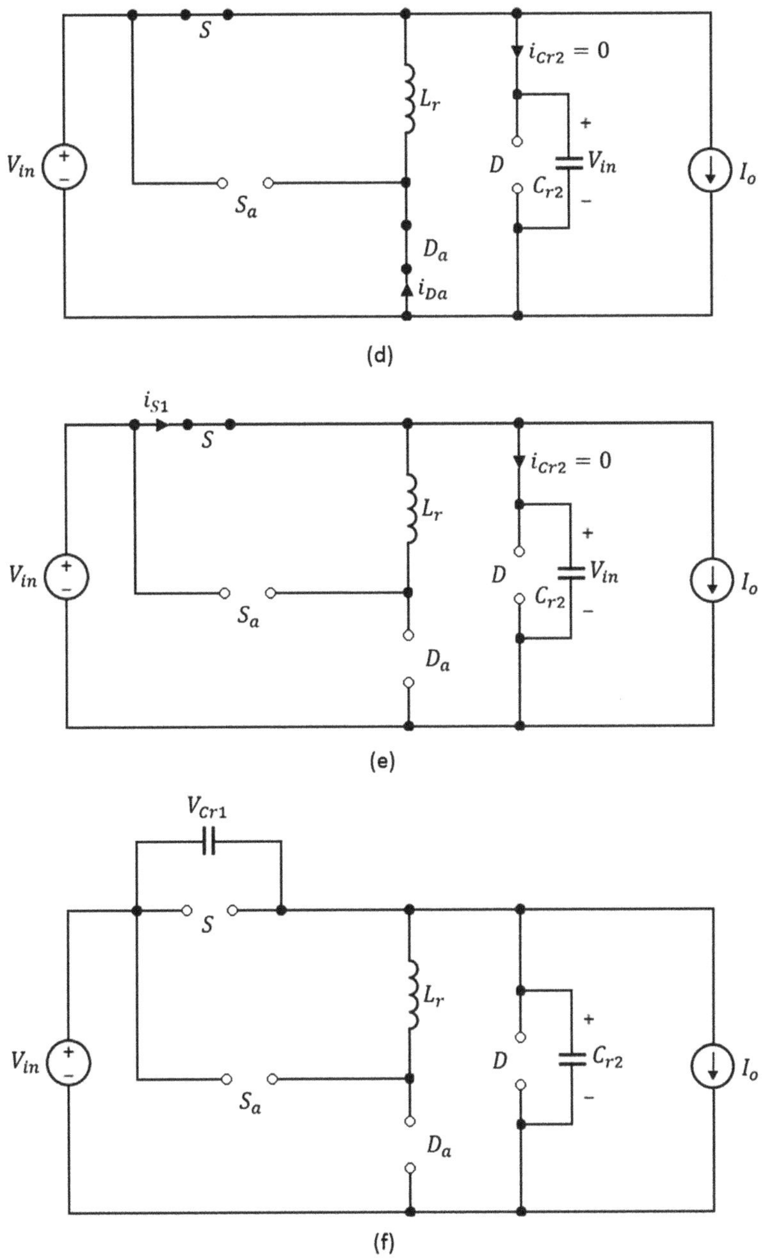

Fig. 6.41 (continued)

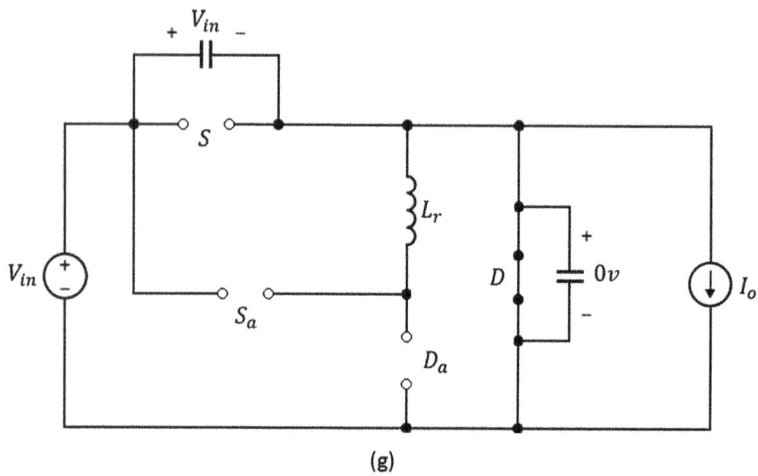

Fig. 6.41 (continued)

With the above initial condition, the solution for $i_{Lr(t)}$ and $v_{cr1}(t)$:

$$i_{Lr}(t) = \frac{V_{in}}{Z_0} \sin \omega_0(t - t_1) + I_0 \tag{6.91}$$

$$v_{cr1}(t) = V_{in} \cos \omega_0(t - t_1) \tag{6.92}$$

where:

$$\omega_0 = \sqrt{\frac{1}{L_r C_r}}; Z_0 = \sqrt{\frac{L_r}{C_r}}; \qquad \text{where } C_r = C_{r1} + C_{r2}$$

Applying KVL again, we have:

$$i_{Lr}(t_2) = I_{Lr2}$$

$$v_{cr2}(t) = V_{in} - v_{cr1}(t)$$

$$= V_{in}(1 - \cos \omega_0(t_2 - t_1))$$

At $t = t_2$, $v_{cr2}(t_2) = V_{in}$, $v_{cr1}(t_2) = 0$, and $i_{Lr}(t_2) = I_{Lr2}$. The time interval $(t_2 - t_1)$ is obtained from Eq. (6.92) to yield:

$$(t_2 - t_1) = \frac{T_s}{4}$$

At $t = t_2$, the resonant stage ends, and the circuit enters Mode 3, where the body diode of S_1 turns on as shown in Fig. 6.41c.

Mode 3 ($t_2 \le t < t_3$)
This mode serves to generate delay so that, at $t = t_3$, the voltage across the main switch is zero to achieve ZVS, and the switch S_1 can be turned on at any time $t > t_2$. The circulating current in S_a, D_s, and the inductor L_r is fixed by the trapped current of $i_{Lr}(t_2) = I_{Lr2}$. This current remains as long as S_a is turned off.

The delay time T_{delay} is kept as small as possible and must be:

$$t_2 + T_{delay} \ge (t_1 - t_0) + (t_2 - t_1)$$

$$\ge \frac{\pi}{2\omega_0} + \frac{L_r I_0}{V_{in}}$$

Let us set $T_{delay} = 0.1\,\%$ of T_s.

Mode 4 ($t_3 \le t < t_4$)
At $t = t_3$, we turn on $S_1$with ZVS, and we turn off S_a, to enter Mode 4 as shown in Fig. 6.41d with the initial conditions at $t = t_3$ as follows:

$$i_{Lr}(t_3) = i_{Lr}(t_2) = I_{Lr2},$$

$$v_{cr1}(t_3) = 0, \quad v_{cr2}(t_3) = V_{in},$$

$$i_{DS} = i_{Lr}(t_2) - I_0 = I_{Lr2} - I_0, \text{ and}$$

$$i_s = -i_{DS}$$

Note that when S_a is turned off, the current i_{Lr} forces itself into the auxiliary diode D_a.

The voltage across L_r is $-V_{in}$. Hence $i_L(t)$ is given by:

$$L_r \frac{di_{Lr(t)}}{dt} = -V_{in}$$

With $i_{Lr}(t_3) = I_{Lr2}$, the current $i_{Lr}(t)$ is given by:

$$i_{Lr}(t) = -\frac{V_{in}}{L_r}(t - t_3) + I_{Lr2}$$

The resonant inductor discharges with slopes $-\frac{V_{in}}{L_r}$ until it becomes zero at $t = t_4$, giving the following expression for the time interval $(t_4 - t_5)$:

$$(t_4 - t_3) = \frac{I_{Lr2} L_r}{V_{in}}$$

The current $i_{s1} = I_o - i_{Lr}(t)$, and at $t = t_4$, $i_{s1}(t_4) = I_0$. Forcing D_a off and the current i_{s1} become clamped to I_o, the circuit enters Mode 5 at $t = t_4$ as shown in Fig. 6.41e.

Mode 5 ($t_4 \leq t < t_5$)
This mode is formed when the diode D_a is turned off with the equivalent circuit as shown in Fig. 6.41f. In this period, the circuit remains in the same mode until the main S_1 is turned off at $t = t_5$.

Mode 6 ($t_5 \leq t < t_6$)
When S_1 is turned off, a new circuit between V_{in}, c_{r1}, and c_{r2} is formed as shown in the equivalent circuit of Mode 6 in Fig. 6.41f.

The expressions for the currents and voltages are as follow:

$$i_{cr1} = I_0 = Cr_1 \frac{d_{v_{cr1}}}{dt},$$

where $v_{cr1}(t_5) = 0$. Since $V_{in} = V_{cr1} + V_{cr2}(t)$, we have:

$$v_{cr1}(t) = \frac{I_o}{C_{r1}}(t - t_5) \quad \text{and}$$

$$v_{cr2}(t) = V_{in} - \frac{I_0}{C_{r1}}(t - t_5)$$

Mode 7 ($t_6 \leq t < t_7$)
At $t = t_6$, the converter enters Mode 7 when the diode D starts conducting the output current I_0. At $t = t_6$ we have:

$$v_{cr1}(t_6) = V_{in} \text{ and } v_{cr2}(t_6) = 0.$$

Mode 7 remains in effect until the auxiliary switch S_a is turned on again at $t = T_s + t_0$ to repeat the switching cycle.

In Fig. 6.42, all the relevant waveforms are plotted.

6.6.3 *The ZVT Boost Converter*

In this section, we consider the boost ZVT converter by placing the ZVT-PWM switching into the conventional boost topology. Figure 6.43a, b shows the circuits for the ZVT boost converter topologies based on ideal switches and non-ideal device implantation, respectively. In the following two sections, we will analyze both circuits of Fig. 6.43.

6.6.3.1 The Ideal Boost ZVT Converter

Figure 6.44 shows the simplified equivalent circuit for the ZVT boost converter of Fig. 6.43a. It will be shown that the switching cycle is divided into seven modes of operation as discussed below.

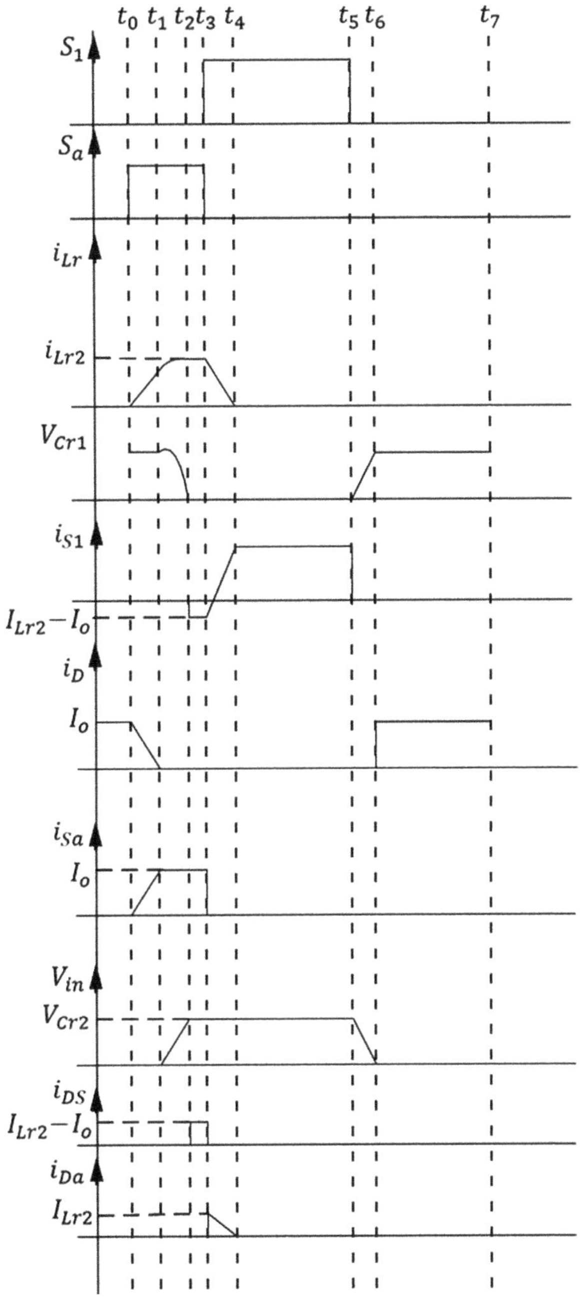

Fig. 6.42 Steady-state waveforms for the ZVT buck converter of Fig. 6.39b

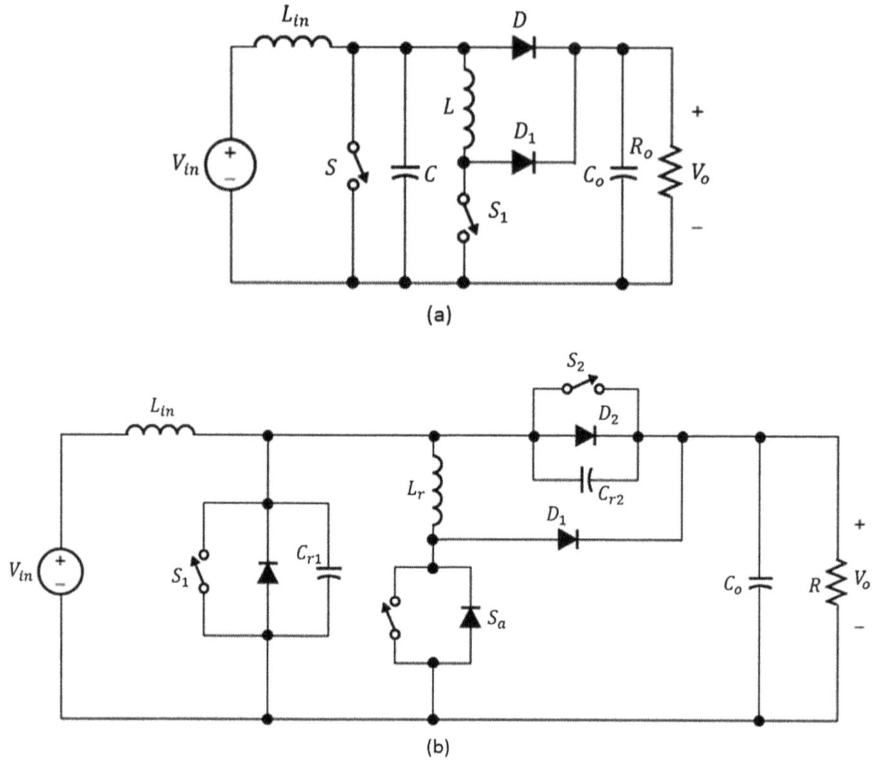

Fig. 6.43 The ZVT boost converter: (**a**) Ideal devices and (**b**) including device junction capacitors

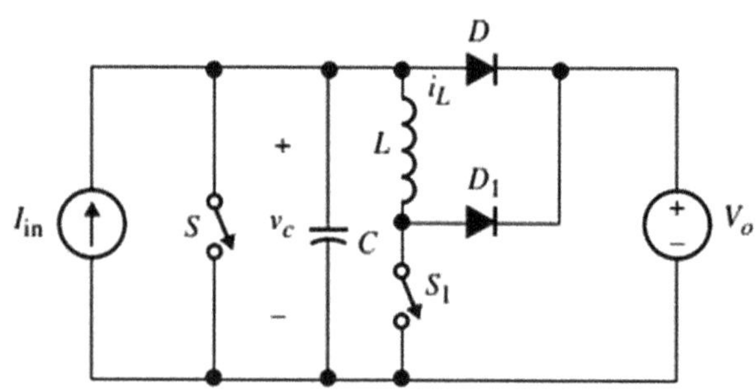

Fig. 6.44 The simplified ZVT boost converter of Fig. 6.43a

Mode 1 ($t_0 \leq t < t_1$)

Mode 1 as shown in Fig. 6.45a starts at $t = t_0$ when the auxiliary switch S_1 is turned *on*. Since the main switch, S, and the auxiliary switch S_1 were *off* prior to $t = t_0$, it is clear that the diode, D, must have been *on* for $t < t_0$ to carry the output current. Hence, we assume D is *on* and D_1 is *off* at $t = t_0$. So for $t > t_0$, S_1 is *on*. The diode current starts to decrease, and it reaches zero when the inductor current i_L increases

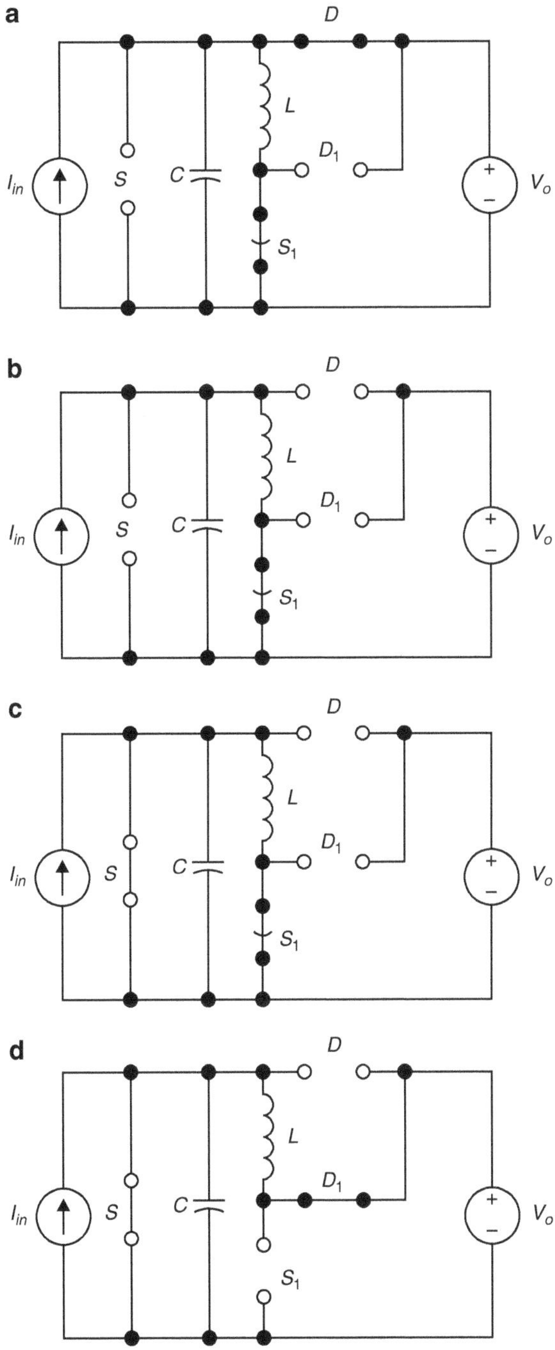

Fig. 6.45 Equivalent circuits for the seven modes of operation: (**a**) Mode 1, (**b**) Mode 2, (**c**) Mode 3, (**d**) Mode 4, (**e**) Mode 5, (**f**) Mode 6, and (**g**) Mode 7

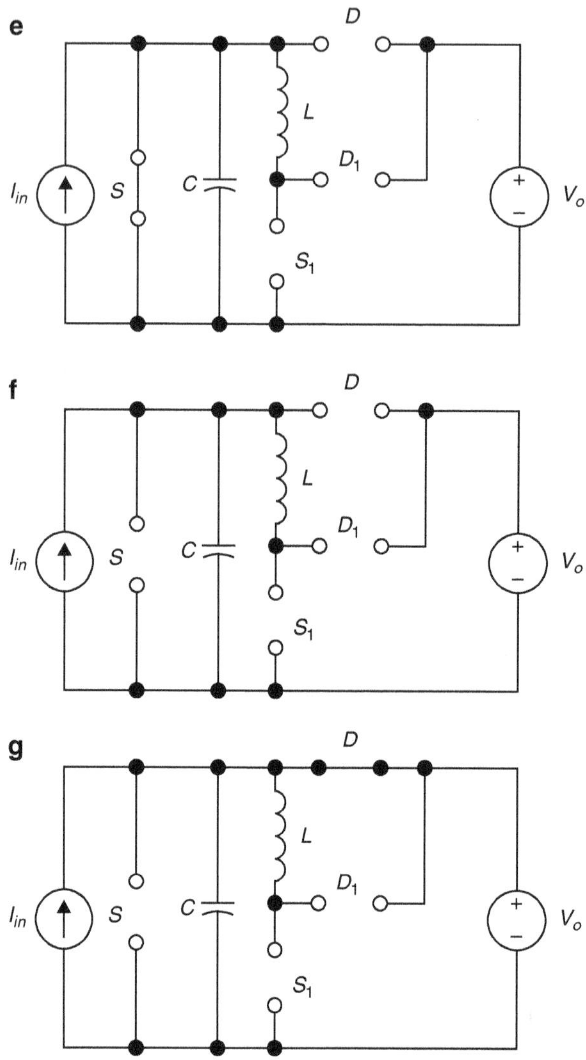

Fig. 6.45 (continued)

and reaches the constant current source, I_{in}, at t_1. In this mode, the capacitor voltage, v_c, is equal to the output voltage V_o and also equal to the inductor voltage as given by:

$$V_o = L\frac{di_L}{dt} = v_c(t)$$

From the above equation, the inductor current i_L is given by:

$$i_L(t) = \frac{V_o}{L}(t - t_0)$$

The above equation assumes zero initial condition for i_L.

As long as the inductor current is less than I_{in}, the diode will stay conducting, and the capacitor voltage remains at V_o. At time $t=t_1$, the inductor current becomes equal to I_{in}, D stops conducting, and the circuit enters Mode 2. From the above equation, we have:

$$I_{in} = \frac{V_o}{L}(t - t_0)$$

The time interval is given by:

$$(t_1 - t_0) = \frac{LI_{in}}{V_o}$$

This is the inductor current charging state.

Mode 2 ($t_1 \leq t < t_2$)

Mode 2 starts at $t_1$when D is *off*; the circuit is shown in Fig. 6.45b, resulting in a resonant stage between L and C. During the time between t_1 and t_1, the main switch, S, remains *off*, and S_1 is still *on*, but both diodes are *off*. The initial capacitor voltage is still V_o, but the initial i_L has changed to I_{in}. The first-order differential equations that represent this mode are given by:

$$C\frac{dv_c}{dt} = I_{in} - i_L(t)$$
$$L\frac{di_L}{dt} = v_c(t)$$

Equation (6.93) is obtained from the above two equations:

$$\frac{d^2 i_L}{dt^2} - \frac{1}{LC} i_L(t) = \frac{1}{LC} I_{in} \tag{6.93}$$

The solution for i_L and v_c is given by:

$$i_L(t) = \frac{V_o}{Z_o}\sin\omega_o(t - t_1) + I_{in}$$
$$v_c(t) = V_o\cos\omega_o(t - t_1)$$

The time interval between t_1 and t_2 is given by:

$$(t_2 - t_1) = \frac{1}{\omega_o}\frac{\pi}{2}$$

The diode voltage starts to charge up due to the decreasing capacitor voltage:

$$v_D(t) = V_o - v_c(t)$$

Substituting for v_c, the diode voltage becomes:

$$v_D(t) = V_o(1 - \cos\omega_o(t - t_1))$$

Mode 3 ($t_2 \leq t < t_3$)
Mode 3 starts when the capacitor discharging to zero. The body diode of S turns on to clamp v_c to zero. In this mode, the main switch, S, remains *off*, the auxiliary switch, S_1, is still *on*, and both diodes are *off*.

Now:

$$v_c(t) = 0$$
$$T_{delay} = t_3 - t_2$$

Mode 4 ($t_3 \leq t < t_4$)
Mode 4 starts at $t = t_3$, when the main switch, S, is turned *on* and the auxiliary switch, S_1, is turned *off*. At t_3, the initial capacitor voltage is zero, and the inductor starts linearly discharging from $i_L(t_2)$ to zero during t_3 to t_4. The diode D remains *off* since its voltage is negative, but D_1 turns *on* at $t = t_3$ to carry the inductor current.

The input voltage is equal to the inductor voltage, and the output voltage is equal to the negative of inductor voltage, v_L:

$$v_L(t) = L\frac{di_L}{dt} = -V_o$$

The inductor current for $t > t_2$ is given by:

$$i_L(t) = -\frac{V_o}{L}(t - t_2) + I_L(t_2)$$

At $t = t_4$, $i_L(t_4) = 0$, hence, the interval $(t_4 - t_3)$ is given by:

$$t_4 - t_3 = \frac{I_L(t_2)L}{V_o} = \left(\frac{V_o}{Z_o}\sin\omega_o(t_2 - t_1) + I_{in}\right)\frac{L}{V_o}$$

Mode 5 ($t_4 \leq t < t_5$)
This mode starts at $t = t_4$, when the diode *D1* is turned off after inductor current drops to zero. The operation of the circuit in this mode is identical to that of the PWM boost converter.

Mode 6 ($t_5 \leq t < t_6$)
In this mode, at $t = t_5$, both switches are *off*, and also both diodes are *off*. The inductor current is zero, and the input current is only going through the capacitor:

$$I_{in} = C\frac{dv_c}{dt}$$

The capacitor voltage can be expressed as:

$$v_c(t) = \frac{1}{C} I_{in}(t - t_5)$$

The capacitor is charging up from zero and will reach the output voltage at $t = t_6$. The time interval is:

$$(t_6 - t_5) = \frac{V_o C}{I_{in}}$$

The circuit enters Mode 7 at this point.

Mode 7 ($t_6 \le t < t_7$)
When the capacitor reaches the output voltage, D starts conducting, but in this mode, both switches are still *off*. The diode current will equal the input current immediately. At $t = t_6$, the capacitor voltage is equal to the output voltage until the auxiliary switch is turned *on* again, and then the cycle will repeat from Mode 1. The waveforms for the seven modes of operation are shown in Fig. 6.46.

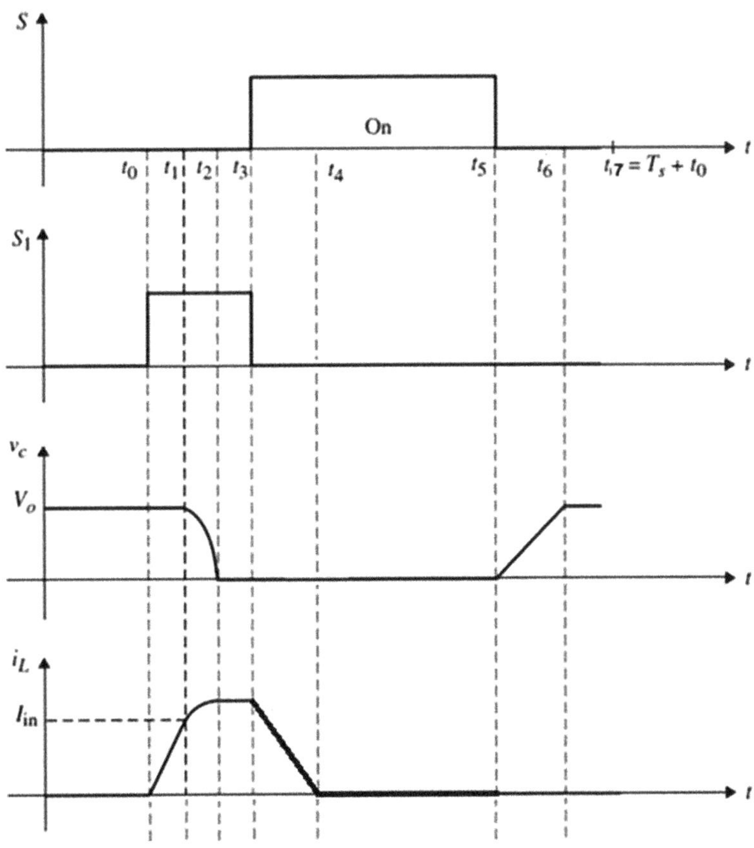

Fig. 6.46 Steady-state waveforms for ideal ZVT boost converter of Fig. 6.43a

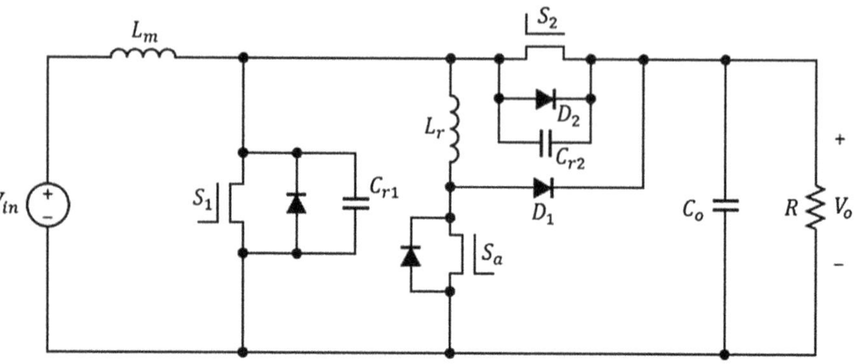

Fig. 6.47 The ZVT boost converter topology including practical device implantation and line inductor

6.6.4 *The Practical ZVT Boost Converter*

As shown earlier, one possible practical implementation for the boost ZVT is to include switching devices with their associated junction capacitors. In this section, for illustration purposes, we will present the detailed steady-state analysis for Fig. 6.47 by including junction capacitor C_r of the main switch, *S1*, the junction capacitor C_{r2} pf the output switch *S2*, and the line inductor L_m. It will be shown below that the steady-state analysis also consist of several modes.

6.6.4.1 Steady-State Analysis

Let us begin the discussion before for $t < t_0$, before the steady state-cycle begins. Let us assume that initially the circuit has its dynamic current and voltage is in static state with $i_{Lm}(t_0) = I_0$ $i_{Lr}(t_0) = 0$ and $v_{cr1}(t)$ and $v_{cr2}(t)$ are fixed at V_o and zero, respectively, as shown in the equivalent circuit in Fig. 6.48. Hence, we have the main switch S_1, and the auxiliary switch S_a was turned off. So, the current flowing through the main inductor will go through the parallel diode of switch S_2. At this point, capacitor C_{r1} is charged to output voltage as V_0.

The equivalent steady-state circuits for the seven modes of operations are shown in Fig. 6.48. The following gives the details for the steady-state modes of operation for the ZVT boost converter topology of Fig. 6.47 by including the device junction capacitors and line inductor.

Mode 1 ($t_0 \leq t < t_1$)
Mode 1 starts at $t = t_0$ when the auxiliary switch is turned on with the initial values at $t = t_0$, with $v_{cr1}(t_0) = V_o$ and $i_{Lr}(t_0) = 0$, as shown in the equivalent circuit for Mode 1 in Fig. 6.49. The current in the input inductor, L_m, slowly discharge since $V_o > V_{in}$. The initial current flowing through main inductor L_m is I_0. The current

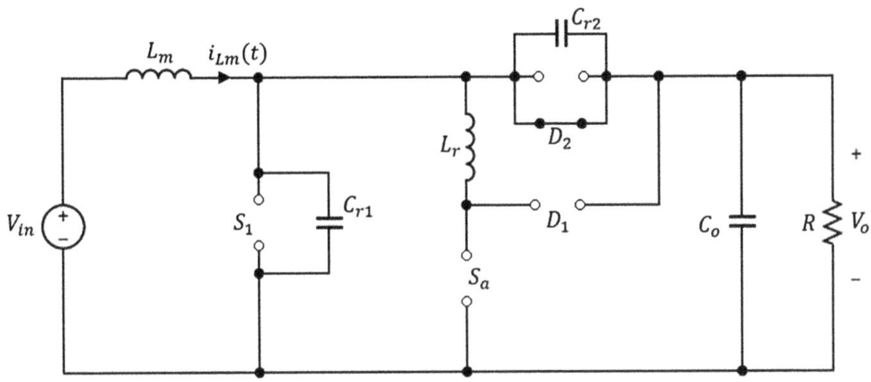

Fig. 6.48 The steady-state equivalent circuit for Fig. 6.47 for $t < t_o$

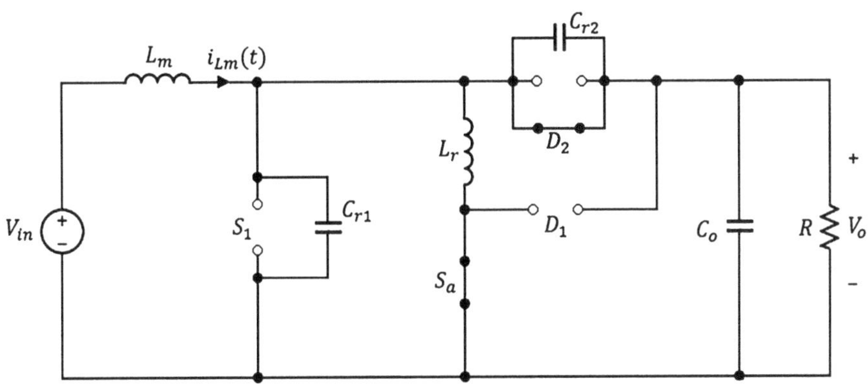

Fig. 6.49 Equivalent circuits for the seven modes of operation: Mode 1

through resonant inductor L_r is linearly increased until it reaches the same value as that in main inductor L_m at t_1. At the same time, the body diode of S_2 is turned off with ZCS. The mathematic equations are shown below:

$$L_m \frac{di_L}{dt} = V_{in} - V_0$$

The current $i_{Lm}(t)$ for $t > t_0$ is given by, with the its initial value of I_0:

$$i_{Lm}(t) = \frac{V_{in} - V_0}{L}(t - t_0) + I_0$$

where as the resonant current, $i_{Lr}(t)$, is obtained from the following equation:

$$L_r \frac{di_{Lr}}{dt} = V_0$$

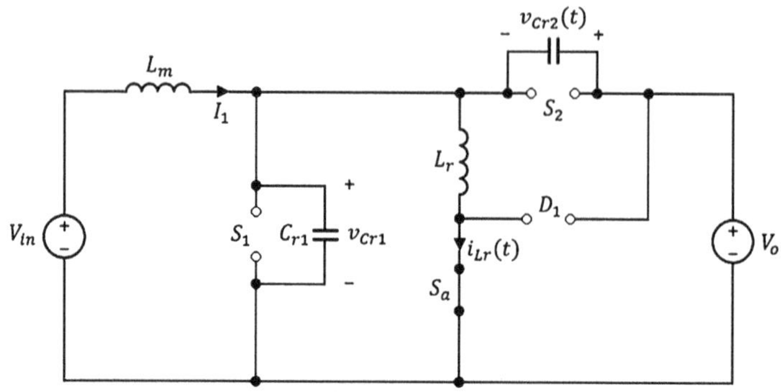

Fig. 6.50 Equivalent circuits for the seven modes of operation: Mode 2

Since its initial value is zero, the expression for $i_{\mathrm{Lr}}(t)$ is given by:

$$i_{\mathrm{Lr}}(t) = \frac{V_0}{L_{\mathrm{r}}}(t - t_0)$$

We assume Mode 1 remains until $t = t_1$ when $i_{\mathrm{Lr}}(t_1) = I_0$ to enter Mode 2. The final values at $t = t_1$ are given by, $i_{\mathrm{Lm}}(t_1) = I_1$, $i_{\mathrm{Lr}}(t_1) = I_{\mathrm{r1}}$, $v_{\mathrm{cr1}}(t_1) = V_0$, and $v_{\mathrm{cr2}}(t_1) = 0$.

Mode 2 ($t_1 \le t < t_2$)

Mode 2 starts at $t = t_1$, when $i_{\mathrm{Lr}}(t_1) = i_{\mathrm{Lm}}(t_1)$, $i_{\mathrm{D2}}(t_1) = 0$, turning $D2$ off at zero current. In this time interval, resonant current continues to increase due to the resonance between L_{r} and C_{r1}. C_{r1} is discharged until the resonance brings its voltage to zero at t_2 at which time the parallel diode of S_1 will conduct (Fig. 6.50).

Assume $i_{\mathrm{Lm}}(t_1) = I_1$ remains relatively constant. Hence, we have:

$$I_1 = C_{\mathrm{r1}}\frac{dv_{\mathrm{cr1}}}{dt} + C_{\mathrm{r2}}\frac{dv_{\mathrm{cr2}}}{dt} + i_{\mathrm{Lr}}(t)$$

From KVL around capacitors C_{r1} and C_{r2} and V_0, we have:

$$v_{\mathrm{cr1}}(t) - v_{\mathrm{cr2}}(t) = V_0$$

Hence, the above equation:

$$I_1 = C_{\mathrm{r1}}\frac{dv_{\mathrm{cr1}}}{dt} + C_{\mathrm{r2}}\frac{d(v_{\mathrm{cr1}}(t) - V_0)}{dt} + i_{\mathrm{Lr}}(t)$$

$$I_1 = (C_{\mathrm{r1}} + C_{\mathrm{r2}})\frac{dv_{\mathrm{cr1}}}{dt} + i_{\mathrm{Lr}}(t) = C_{\mathrm{r}}\frac{dv_{\mathrm{cr1}}}{dt} + i_{\mathrm{Lr}}(t)$$

where $C_{\mathrm{r}} = C_{\mathrm{r1}} + C_{\mathrm{r2}}$.

Solving for $i_{Lr}(t)$ and $v_{cr1}(t)$, we obtain the following expression:

$$i_{Lr}(t) = \frac{V_0}{Z_0}\sin\omega_o(t - t_1) + I_1$$
$$v_{cr1}(t) = V_o \cos\omega_o(t - t_1)$$

And the voltage across C_{r2} is given by:

$$v_{cr2}(t) = V_o - v_{cr1}(t) = V_o - V_o\cos\omega_o(t - t_1)$$

where $Z_0 = \sqrt{\frac{L_r}{C_r}}$ is the resonant impedance and $\omega_0 = \sqrt{\frac{1}{L_rC_r}}$ is the resonant frequency.

We assume Mode 2 remains until $t = t_2$ when $i_{Lr}(t_1) = \frac{V_0}{Z_0} + I_1 = I_{r2}$ to enter Mode 3. The final values at $t = t_2$ are given by $i_{Lm}(t_2) = I_1$, $i_{Lr}(t_2) = I_{r2}$, $v_{cr1}(t_2) = 0$, and $v_{cr2}(t_2) = V_0$.

Mode 3 ($t_2 \le t < t_3$)

In this mode, with $v_{cr1}(t_2) = 0$, the body diode of *S1* and *D1* turns on, resulting in the equivalent circuit for Mode 3 shown in Fig. 6.57.

The accumulated energy in resonant tank is transferred through auxiliary switch and the parallel diode (Fig. 6.51).

In this mode, $v_{cr1} = 0$, so any $t > t_2$, S_1 can be turned off at zero voltage. This time delay before S_1 is turned on at $t = t_3$ and is labeled as T_{delay} that is given by:

$$t_2 + T_{delay} = t_3 - t_0 = \Delta t_{30}$$

For the above equation:

$$t_2 + T_{delay} \ge \frac{L_rI_1}{V_o} + \frac{\pi}{2}\sqrt{L_rC_r} = (t_1 - t_0) + (t_2 - t_1)$$

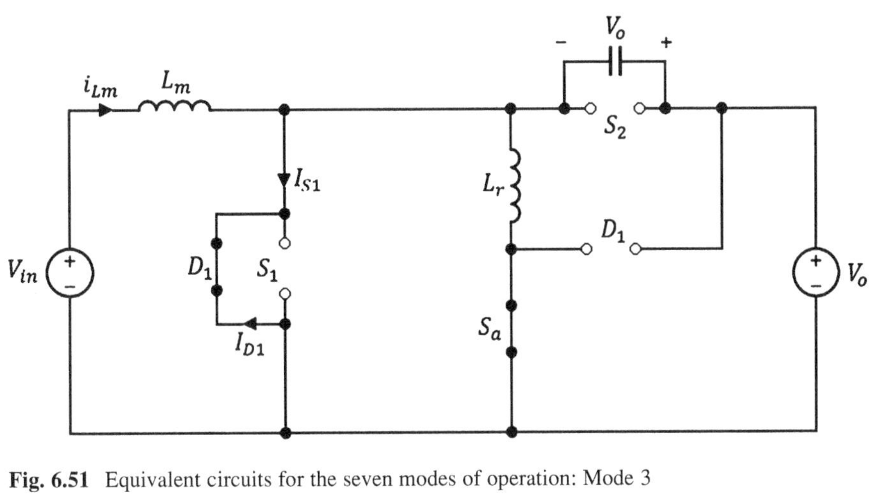

Fig. 6.51 Equivalent circuits for the seven modes of operation: Mode 3

The inductor current between $t_2 \leq t < t_3$ is constant and clamped at $i_{Lr}(t_2) = I_{r2}$. Mode 3 ends when the auxiliary switch Sa is turned off at $t = t_3$.

The currents and voltages at the end of Mode 2 are given by:

$$i_{Lm}(t_3) = I_2, i_{Lr}(t_3) = I_{r2}, v_{cr1}(t_3) = 0, v_{cr2}(t_3) = V_0$$

Let us set the time interval in this mode as 0.1% of the one-switching cycle. The key point is that in order to achieve ZVS, the turn on signal of S_1 should be applied while its diode is conducting.

Mode 4 ($t_3 \leq t < t_4$)

Mode 4 starts at $t = t_3$ when the auxiliary switch S_a is turned off under hard-switching condition, and at the same time, S_1 is turned on at zero voltage condition. Turning off the auxiliary switch S_a will force the diode D_1 to be on in order to transfer the energy in resonant inductor L_r. Notice that in previous mode, the main switch S_1 was turned on through its body diode Ds and the current is flowing opposite to the drain-source current $I_{s1} = i_{Ds}$. However, in this mode, the drain-source current $I_{s1} > 0$ with the body diode current $i_{Ds} = 0$ (Fig. 6.52).

In Mode 4, since Lr is clamped by $+V_o$, it starts linearly to discharge until it reaches zero at $t = t_4$. I_{Lm} will increase slowly as it is clamped by $+V_{in}$ when $i_{Lm}(t_3) = i_{Lm}(t_2) = I_2$, and at $t = t_4$, $i_{Lm}(t_4)$ becomes $i_{Lm}(t_4) = I_3$ and $i_{Lr}(t_4) = 0$.

The inductor current decreases linearly until it reaches zero at t_4:

$$L_r \frac{di_{Lr}}{dt} = -V_o$$

$$i_{Lr}(t) = \frac{-V_o}{L_r}(t - t_3) + I_{r2}$$

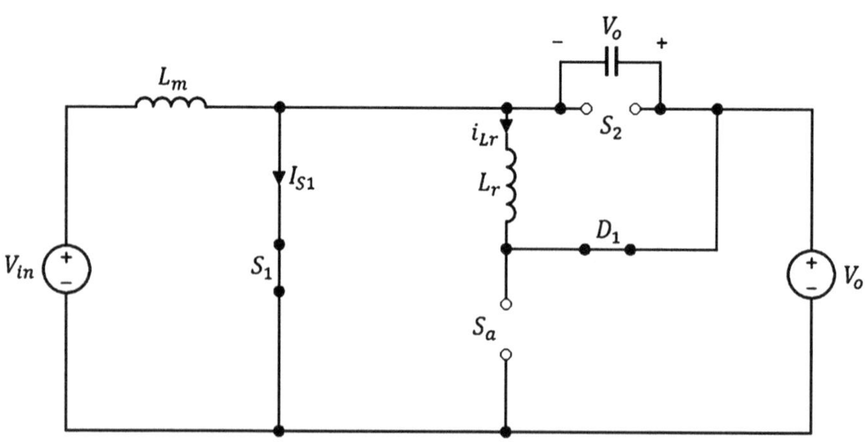

Fig. 6.52 Equivalent circuits for the seven modes of operation: Mode 4

Mode 4 ends at $t = t_4$ when $i_{\mathrm{Lr}} = 0$, and $i_{\mathrm{s1}}(t)$ increases linearly since it is given by:

$$i_{\mathrm{s1}}(t) = i_{\mathrm{Lm}}(t) - i_{\mathrm{Lr}}(t)$$

And the capacitor voltages at the end of this mode are given by:

$$v_{\mathrm{cr1}}(t_4) = 0, v_{\mathrm{cr2}}(t_4) = V_0$$

As expected where both remain at the same values from previous mode.

We assume Mode 4 remains until $t = t_4$ when $i_{\mathrm{Lr}}(t_4)=0$ to enter Mode 5. The final values at $t = t_4$ are given by $i_{\mathrm{Lm}}(t_4) = I_3$, $i_{\mathrm{Lr}}(t_4) = 0$, $v_{\mathrm{cr1}}(t_4) = 0$, and $v_{\mathrm{cr2}}(t_4) = V_0$.

Mode 5 ($t_4 \leq t < t_5$)

This mode starts at $t = t_4$ and ends at $t = t_5$ when L_{r} current drops to zero at which point the diode D_1 will turn off as shown in Figs. 6.53 and 6.59.

i_{Lm} continues to slowly increase by a slope V_{in}/Lm.

Therefore, $i_{\mathrm{Lm}}(t) = \frac{V_{\mathrm{in}}}{L_{\mathrm{m}}}(t - t4) + i_{\mathrm{Lm}}(t_4)$. At $t = t_5$, also S_1 is turned off, and the converter enters Mode 6. The final values at $t = t_5$ are given by:

$$i_{\mathrm{Lm}}(t_5) = I_4, i_{\mathrm{Lr}}(t_5) = 0, v_{\mathrm{cr1}}(t_5) = 0, v_{\mathrm{cr2}}(t_5) = V_0$$

Mode 6 ($t_5 \leq t < t_6$)

At t_5, the main switch S_1 is turned off under ZVS, and its parallel capacitor begins to be charged by main inductor L_{m}. At the end of this mode, the resonant capacitor is charged completely, and the voltage across it is equal to output voltage. Capacitor C_{r2} is discharged to zero. At this time, the body parallel diode of switch S_2 is turned on smoothly under ZVS (Fig. 6.54).

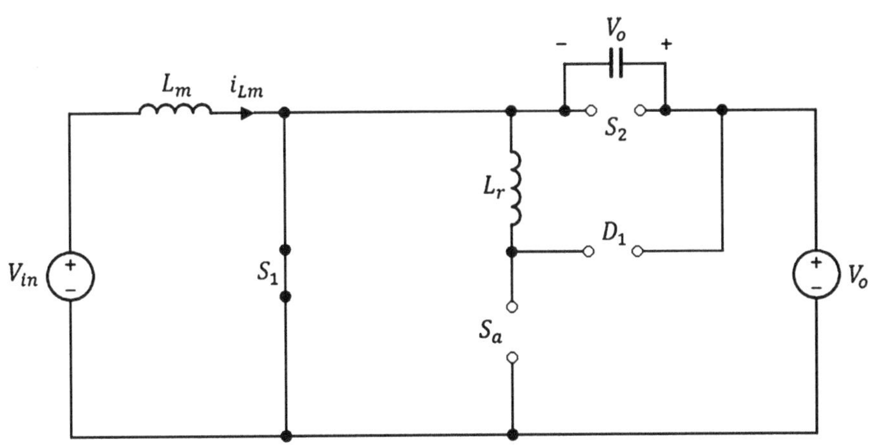

Fig. 6.53 Equivalent circuits for the seven modes of operation: Mode 5

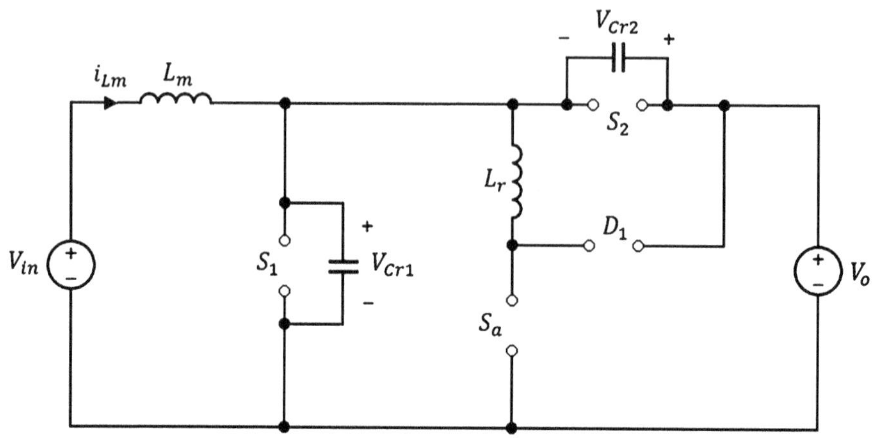

Fig. 6.54 Equivalent circuits for the seven modes of operation: Mode 6

Since $i_{cr2}=0$, so $i_{Lm}(t)$ is given by:

$$i_{Lm}(t) = i_{cr1} = C_{r1}\frac{dv_{cr1}}{dt}$$

Integrating both sides from $t > t_5$, we obtain:

$$\frac{1}{C_{r1}}\int_{t_5}^{t} i_{Lm}(t)dt = v_{cr1}(t) - v_{cr1}(t_5)$$

where $v_{cr1}(t_5) = 0$. Therefore, the voltage:

$$v_{cr1}(t) = \frac{1}{C_{r1}}\int_{t_5}^{t} i_{Lm}(t)dt$$

And from KVL, we have:

$$v_{cr1} + v_{cr2} = V_o$$

This gives us the following relation for $v_{cr2}(t)$:

$$v_{cr2}(t) = V_o - \frac{1}{C_{r1}}\int_{t_5}^{t} i_{Lm}(t)dt$$

At $t = t_6$, the voltage $v_{cr1}(t_6) = V_o$ and $v_{cr2}(t_6) = 0$ with D_2 is turned on at zero voltage and the circuits enter Mode 7. The current in Lm remains almost constant at $i_{Lm}(t_6) = I_5$, and $i_{Lr}(t_6) = 0$.

Therefore, the final values at $t = t_6$ are given by $i_{Lm}(t_6) = I_5$, $i_{Lr}(t_6) = 0$, $v_{cr1}(t_6) = V_o$, and $v_{cr2}(t_6) = 0$.

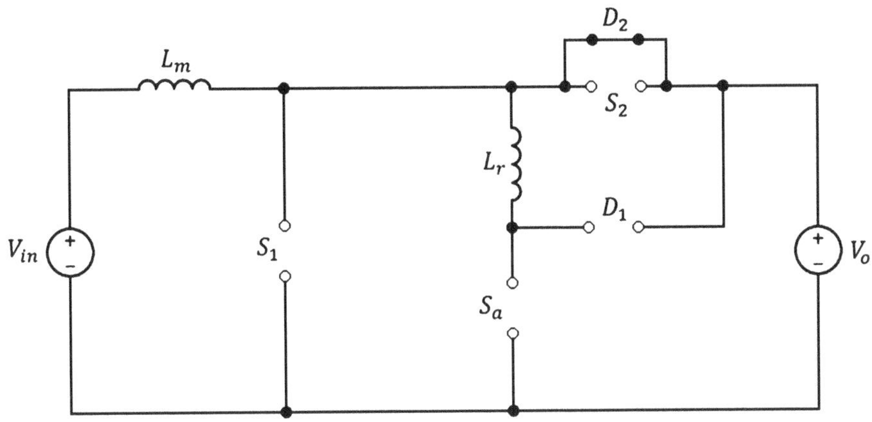

Fig. 6.55 Equivalent circuits for the seven modes of operation: Mode 7

Mode 7 ($t_6 \leq t < t_7$)
In this mode, all switches are turned off. Therefore, current in main inductor L_m is flowing through the body parallel diode of switch S_2. So this mode is identical to the freewheeling stage of the PWM boost converter. At t_7, the auxiliary switch S_{a1} is turned on again, starting another switching cycle (Fig. 6.55).

The voltage across L_m is $V_{in} - V_o$ which is negative causing the current i_{Lm} to decrease slowly while S_1 is off. The mathematical equation in this mode is:

$$L_m \frac{di_L}{dt} = V_{in} - V_0$$

The input current is then given by, $t = t_7 = T_s + t_0$,

$$i_{Lm}(t) = \frac{V_{in} - V_0}{L}(t - t_6) + I_5$$

where the values at the end of Mode 7:

$$i_{Lm}(t_7) = I_6 = I_0, i_{Lr}(t_7) = 0, v_{cr1}(t_7) = V_0, v_{cr2}(t_7) = 0$$

The theoretical waveform for boost mode is provided in Fig. 6.56:

6.6.4.2 Derivation of the Voltage Gain

To simply obtain the gain equation and simplify design steps, it is clear that the input current $i_{Lm}(t)$ remains almost constant since Lm is assumed much higher than L_r. Fig. 6.57 shown the simplified ZVT topology for the boost converter with the input current represented by a dc source I_{in}.

Now the time intervals for each mode is given as:

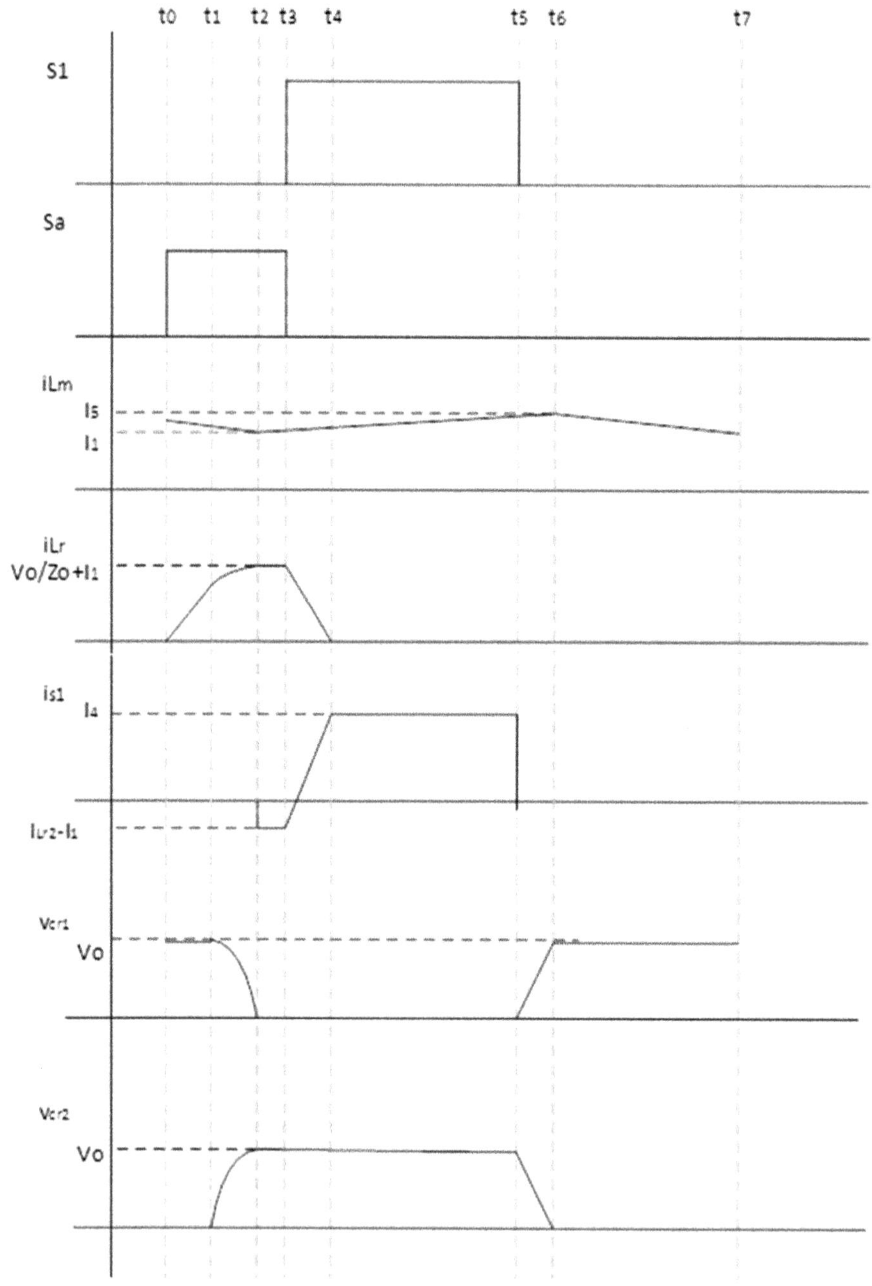

Fig. 6.56 Steady-state waveforms for ZVT boost converter of Fig. 6.47

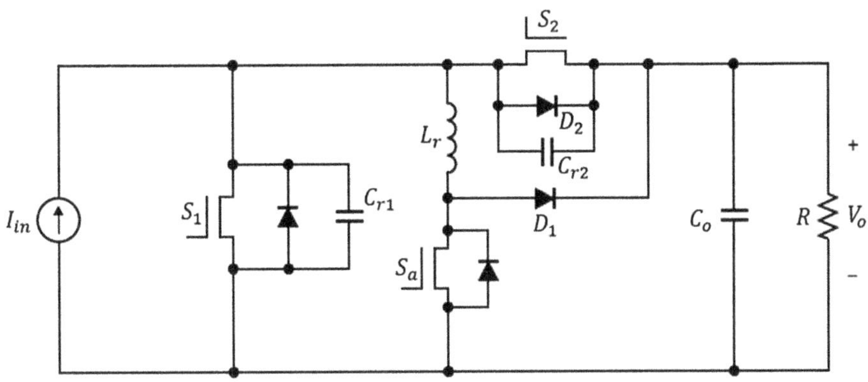

Fig. 6.57 Simplified equivalent circuit for the ZVT boost converter topology

$$\text{Mode 1 time interval}: t_1 - 0 = \frac{L_r I_{in}}{V_o}$$
$$\text{Mode 2 time interval}: t_2 - t_1 = \frac{\pi}{2\omega_o}$$
$$\text{Mode 3 time interval}: t_3 - t_2 = k_1 T_s$$
$$\text{Mode 4 time interval}: t_4 - t_3 = \frac{1}{\omega_o} + \frac{L_r I_{in}}{V_o}$$
$$\text{Mode 5 time interval}: t_5 - t_4 = k_2 T_s$$
$$\text{Mode 6 time interval}: t_6 - t_5 = \frac{V_o C_r}{I_{in}}$$
$$\text{Mode 7 time interval}: t_7 - t_6 = T_s - (t_1 - 0) - (t_2 - t_1) - (t_3 - t_2) - (t_4 - t_3) - (t_5 - t_4) - (t_6 - t_5)$$

In the above time intervals, set $k_1 = 0.1\%$, $k_2 = \left[0.9DT_s - \frac{1}{\omega_o} + \frac{L_r I_{in}}{V_o}\right]/T_s$.

By applying energy balance equation which is $E_{in} = E_{out}$, we can finally obtain the relationship below:

$$\frac{f_{ns}}{2\pi}\left(\frac{\pi}{2} - 1 + \frac{Q}{2M}\right) = 0.099 - \frac{0.1}{M} \tag{6.94}$$

where $f_{ns} = \frac{f_s}{f_o}, Q = \frac{R}{Z_o}, M = \frac{V_o}{V_{in}}$

The design characteristic curve for M vs. f_{ns} for several Q and for $T_{delay} = 0.1\%$ of T_s is shown in Fig. 6.58.

Example 6.7

Let us consider the following design specifications for the ZVT boost converter of Fig. 6.57 that given as below:

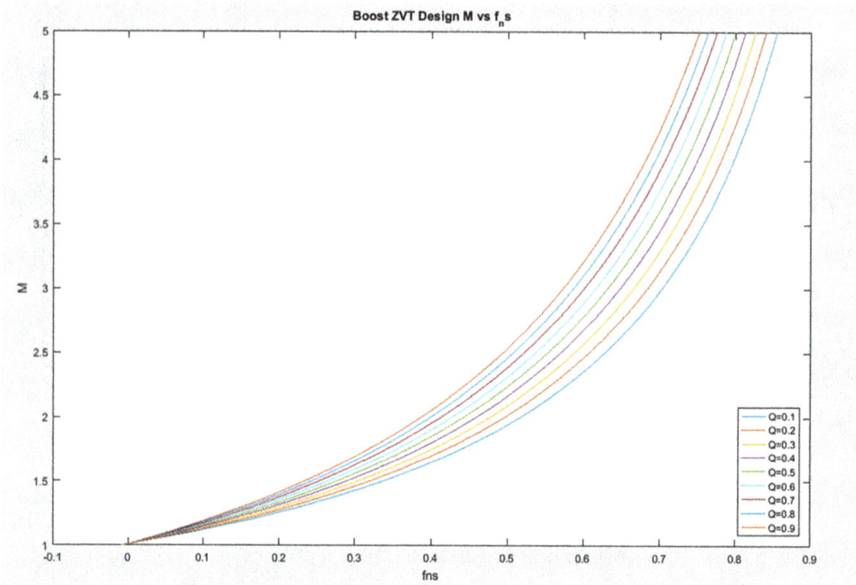

Fig. 6.58 Output characteristics curve for the ZVT boost of Fig. 6.43b

$$V_{\text{in}} = 13\text{ V}$$
$$V_{\text{o}} = 30\text{ V}$$
$$Lm = 20\ \mu H$$
$$Co = 33\ \mu F$$
$$P_{\text{o}} = 100\text{ W}$$
$$f_{\text{sw}} = 133\text{ kHz}$$

Solution

The voltage gain is:

$$M = \frac{V_{\text{o}}}{V_{\text{in}}} = \frac{30}{13} = 2.3$$

And normalized switching frequency is given by:

$$f_{\text{ns}} = \frac{f_{\text{s}}}{f_{\text{o}}}$$

Based on Fig. 6.58, let us choose $Q = 0.7$, which gives us $f_{\text{ns}} = 0.48$.

Based on the equations for f_{ns}, Q and Z_{o} are given above.

It can be shown that the resonant components are given by, $L_{\text{r}} = 7.4$ μH and $C_{\text{r}} = 44.7$ nF. Assume $C_{\text{r1}} = C_{\text{r2}}$, then we obtain $C_{\text{r1}} = 22.35$ nF and $C_{\text{r2}} = 22.35$ nF.

In the ZVT circuit design, the resonant inductor is designed to provide soft turnoff of the upper MOSFET body diode, and the resonant capacitor is selected to provide soft switching of the bottom MOSFET. The resonant inductor controls the *di*/*dt* of the diode by providing an alternate current path for the boost inductor current. When the auxiliary switch turns on, the input current is diverted from the boost diode to the resonant inductor. The resonant value can be calculated by determining how fast the diode can be turned off. The diode's turnoff time is given by its reverse recovery time. A good initial estimate is to allow the inductor current to ramp up to the diode current within three times the diode's specified reverse recovery time. The reverse recovery of the diode is partially a function of its turnoff *di*/*dt*. If a controlled *di*/*dt* is assumed, the reverse recovery time of this diode can be estimated to be approximately 60 ns.

The duty cycle is:

$$D = 1 - \frac{V_{\text{in}}}{V_{\text{o}}} = 1 - \frac{13}{30} = 0.567$$

The output current is:

$$I_{\text{o}} = \frac{P_{\text{o}}}{V_{\text{o}}} = \frac{100}{30} = 3.33\text{A}$$

The input current is:

$$I_{\text{in}} = I_{\text{o}}\left(\frac{1}{1-D}\right) = 3.33^{*}\frac{1}{1-0.567} = 7.7\text{A}$$

The current ripple in the main inductor is given by:

$$\Delta i_{\text{Lm}} = \frac{V_{\text{in}}}{Lm}DT_{\text{s}} = \frac{13}{20\mu}(0.567)\frac{1}{133\text{k}} = 2.77\text{A}$$

So, the peak input current is:

$$I_{\text{in\_peak}} = I_{\text{in}} + \frac{1}{2}\Delta i_{\text{Lm}} = 9.085\text{A}$$

Then:

$$\frac{di}{dt} = \frac{I_{\text{in\_peak}}}{3t_{\text{rr}}} = 5.05^{*}10^{7}\text{A/s}$$

Therefore, the resonant inductance is:

$$L_{\text{r}} \geq \frac{V_{\text{o}}}{\frac{di}{dt}} = \frac{30}{5.05^{*}10^{7}} = 0.59\ \mu H$$

Since the partial energy stored in resonant inductor comes from resonant capacitor and the other comes from the main inductor, the total peak energy stored in resonant inductor should be larger than that in the resonant capacitor.

Peak energy stored in resonant inductor is given as:

$$E_{\mathrm{Lr}} = \frac{1}{2} L_{\mathrm{r}} i_{\mathrm{Lr\_peak}}{}^2$$

Energy stored in resonant capacitor is:

$$E_{\mathrm{Cr}} = \frac{1}{2} C_{\mathrm{r}} V_{\mathrm{o}}{}^2$$

Therefore:

$$\frac{1}{2} C_{\mathrm{r}} V_{\mathrm{o}}{}^2 < \frac{1}{2} L_{\mathrm{r}} i_{\mathrm{Lr\_peak}}{}^2$$

Because the peak current flowing through resonant inductor is larger than the peak input current, we can simply select $I_{\mathrm{in\_peak}}$ to do the calculation. According to mathematic knowledge, if the inequalities:

$$\frac{1}{2} C_{\mathrm{r}} V_{\mathrm{o}}{}^2 < \frac{1}{2} L_{\mathrm{r}} i_{\mathrm{in\_peak}}{}^2$$

and

$$\frac{1}{2} L_{\mathrm{r}} i_{\mathrm{in\_peak}}{}^2 < \frac{1}{2} L_{\mathrm{r}} i_{\mathrm{Lr\_peak}}{}^2$$

are true, then we can say:

$$\frac{1}{2} C_{\mathrm{r}} V_{\mathrm{o}}{}^2 < \frac{1}{2} L_{\mathrm{r}} i_{\mathrm{Lr\_peak}}{}^2$$

is also correct.

Hence, we can have:

$$C_{\mathrm{r}} < \frac{L_r i_{\mathrm{in\_peak}}{}^2}{V_{\mathrm{o}}{}^2} = 54\ \mathrm{nF}$$

Thus, we can see our choice for $L_{\mathrm{r}} = 7.4\ \mu\mathrm{H}$ and $C_{\mathrm{r}} = 44.7\ \mathrm{nF}$ meet the above requirement.

Exercise 6.6

Figure 6.59a, b shows the ZVT buck-boost converter and its simplified equivalent circuit. Derive the steady-state analysis and voltage gain, respectively.

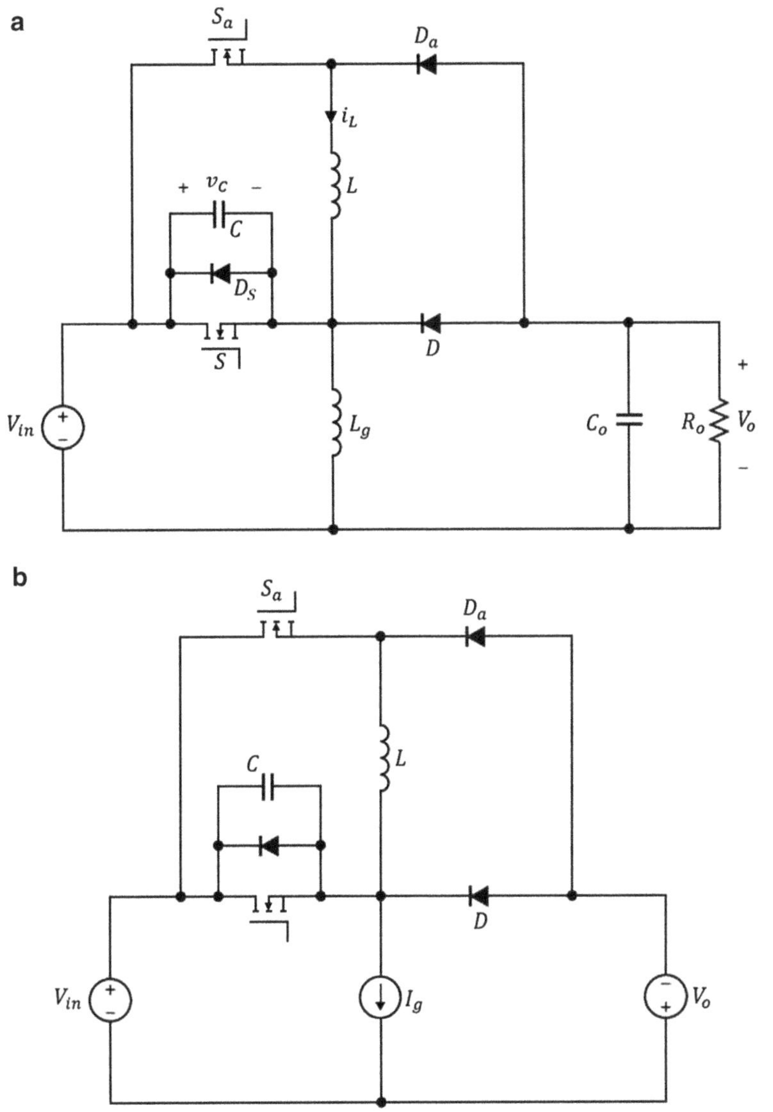

Fig. 6.59 (**a**) The ZVT buck-boost converter and (**b**) its equivalent simplified circuit

6.7 Generalized Analysis for ZCS

It can be shown from the preceding analysis of the quasi-resonant ZCS and ZVS-PWM converters that all dc-dc converter families share the same switching network, with the orientation depending on the topology type (buck, boost, buck-boost, Cuk, Zeta, SEPIC, etc.). As a result, the switching network representation and analysis, including the switching waveforms, can be generalized for each converter family.

The generalized analysis means that in order to derive the analysis and the design curves for a family of converters, only the generalized switching cell with the generalized and normalized parameters for that family need to be analyzed. The equations for the generalized cell will be the generalized equations that describe any converter that uses this specific cell. By using generalized parameters, it is possible to generate a single transformation table from which the voltage ratios and other important design parameters for each converter can be obtained directly.

6.7.1 The Generalized Switching Cell

Figure 6.60a, b shows the generalized switching cells of the quasi-resonant PWM ZCS and ZVS converters, respectively. Note that these cells are all of the common ground three-terminal two-port network type. The following parameters, which will be used throughout this discussion, are defined as follows:

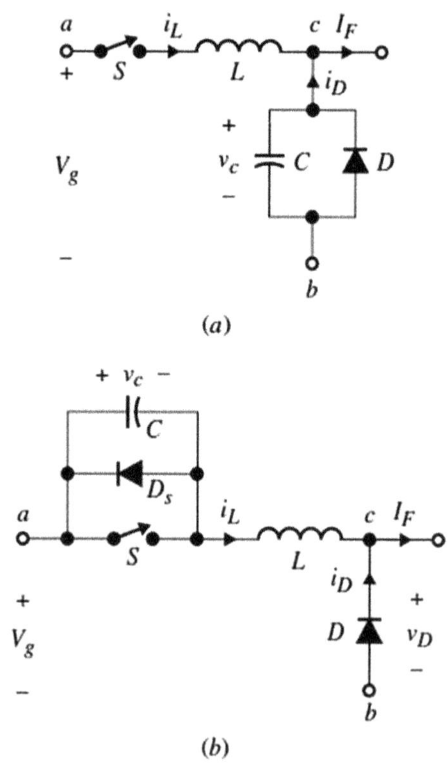

Fig. 6.60 Switching cells: (**a**) ZCS-QRC cell and (**b**) ZVS-QRC cell

- The normalized cell input voltage (V_{ng}):

$$V_{\mathrm{ng}} = \frac{V_{\mathrm{g}}}{V_{\mathrm{in}}}$$

where V_{g} is the switching-cell average input voltage as shown in Fig. 6.38.

- The normalized cell output current (I_{nF}):

$$I_{\mathrm{nF}} = \frac{I_{\mathrm{F}}}{I_{\mathrm{o}}}$$

where I_{F} is the switching-cell average output current.

- The normalized filter capacitor voltage (V_{nF}):

$$V_{\mathrm{nF}} = \frac{V_{\mathrm{F}}}{V_{\mathrm{in}}}$$

where V_{F} is the filter capacitor average voltage.

- The normalized filter inductor current (I_{nT}):

$$I_{\mathrm{nT}} = \frac{I_{\mathrm{T}}}{I_{\mathrm{o}}}$$

where I_{T} is the filter inductor average current (in the ZCS-QSW CC family).

- The normalized cell output average voltage (V_{nbc}):

$$V_{\mathrm{nbc}} = \frac{V_{\mathrm{bc}}}{V_{\mathrm{in}}}$$

where V_{bc} is the switching-cell average output voltage.

- The normalized current entering node *b* in the switching cell(I_{nb}):

$$I_{\mathrm{nb}} = \frac{I_{\mathrm{b}}}{I_{\mathrm{o}}}$$

where I_{b} is the average current entering node *b*.

Note that the generalized transformation table that will be presented next includes the generalized parameters V_{ng}, I_{nF}, V_{nbc}, I_{nb}, V_{nF}, and I_{nT}, which are the normalized versions of the parameters V_g, I_F, V_{bc}, I_b, V_F, and I_T. It will be noted that I_{nb}, V_{nF}, and I_{nT} have the same normalized quantity, as do V_{ng} and I_{nF}.

6.7.2 The Generalized Transformation Table

The derivation of the generalized transformation table for the converter families is beyond the scope of this book. The generalized transformation table is shown in Table 6.1.

Table 6.1 The generalized transformation table

	V_{ng}, I_{nF}	V_{nF}, I_{nT}, I_{nb}	V_{nbc}
Buck	1	$1-M$	$-M$
Boost	M	1	$1-M$
Buck-boost, Cuk, Zeta, and Sepic	$1+M$	1	$-M$

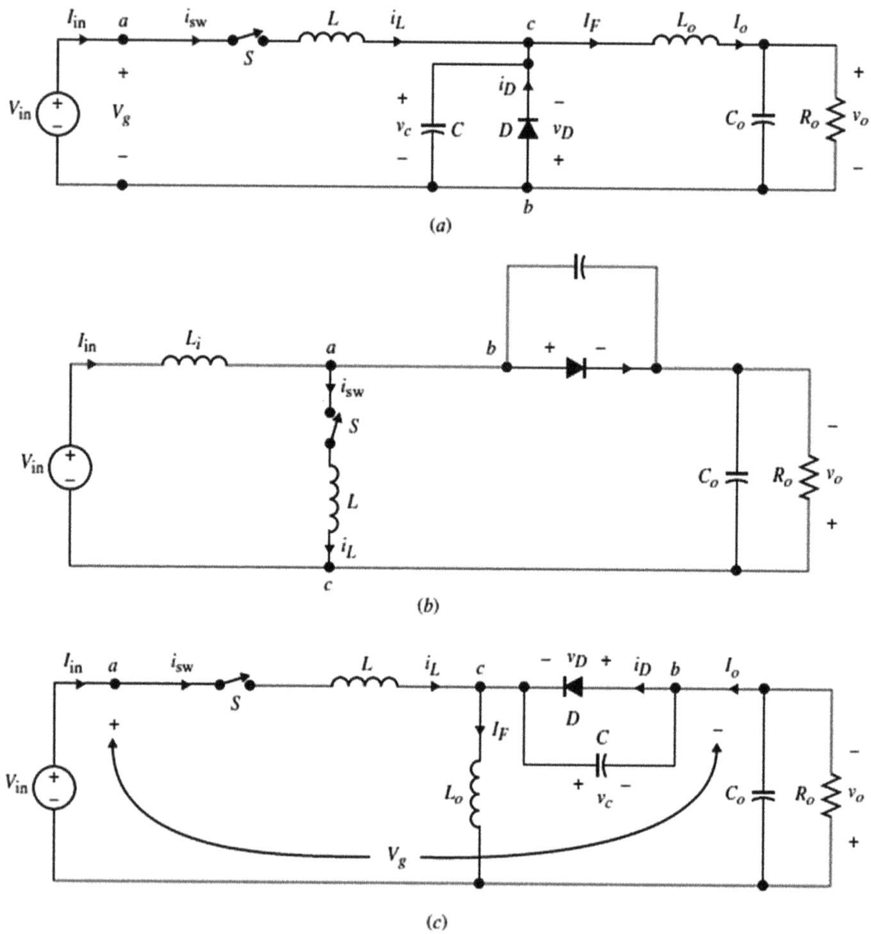

Fig. 6.61 The dc-dc ZCS-QRC family: (**a**) Buck (**b**) Boost (**c**) Buck-boost

By applying the appropriate cell to the conventional dc-dc converters, the ZCT quasi-resonant converter (QRC) family can be formed as shown in Fig. 6.61.

In the next section, the modes of operation for the ZVS-QRC switching network will be discussed briefly, and the main switching waveforms will be drawn in terms of the generalized parameters.

6.7.3 The Basic Operation of the ZCS-QRC Cell

The typical switching waveforms for the cell in Fig. 6.60a are shown in Fig. 6.62. Table 6.2 shows the condition of the switches and diodes in each mode. It can be shown that there are four modes of operation and their analysis is summarized as follows:

Mode 1 ($t_0 \leq t < t_1$)
It is assumed that before $t = t_0$, S was off and D was on in order to carry I_F. The resonant inductor L was carrying no current, and the resonant capacitor C voltage was zero. Mode 1 starts when S is turned on while D is on, which causes L to charge up linearly until the current through it becomes equal to I_F at $t = t_1$, causing D to turn off.

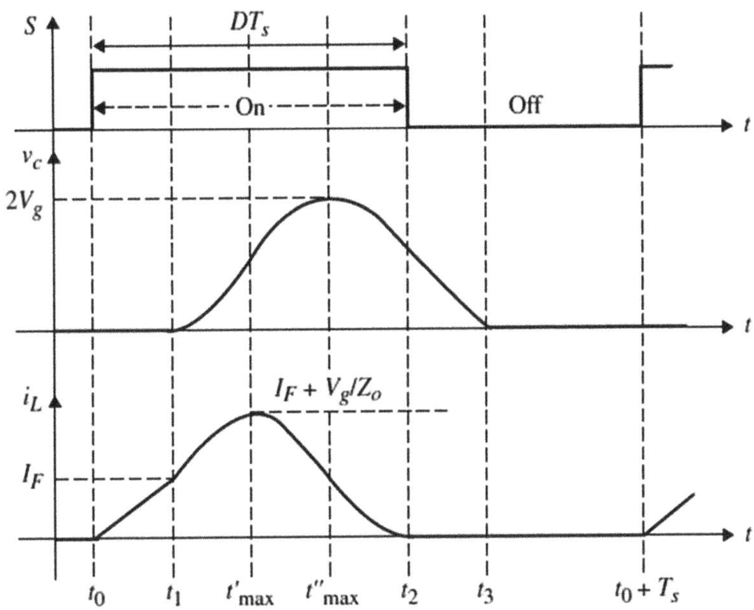

Fig. 6.62 Main ZCS-QRC switching cell waveforms

Table 6.2 Switches conditions

	S	D
Mode 1 ($t_0 \leq t < t_1$)	On	On
Mode 2 ($t_1 \leq t < t_2$)	On	Off
Mode 3 ($t_2 \leq t < t_3$)	Off	Off
Mode 5 ($t_3 \leq t < t_0 + T_s$)	Off	On

Mode 2 ($t_1 \le t < t_2$)
Mode 2 starts when D turns off while S is on, initiating a resonant stage between C and L until the current through L drops to zero at $t = t_2$, causing S to turn off at zero current (soft switching).

Mode 3 ($t_3 \le t < t_0 + T_s$)
Mode 3 starts when S turns off at zero current. The resonant capacitor starts discharging linearly, causing the voltage across it to drop to zero again, in turn causing D to turn on at zero voltage at $t = t_3$.

Mode 4 ($t_3 \le t < t_0 + T_s$)
Mode 4 is a steady-state mode, and nothing happens in it until S is turned on again to start the next switching cycle.

6.7.3.1 Generalized Steady-State Analysis

From the description of the modes of operation, the equivalent circuit for each mode can be drawn as shown in Fig. 6.63. These equivalent circuits can be used

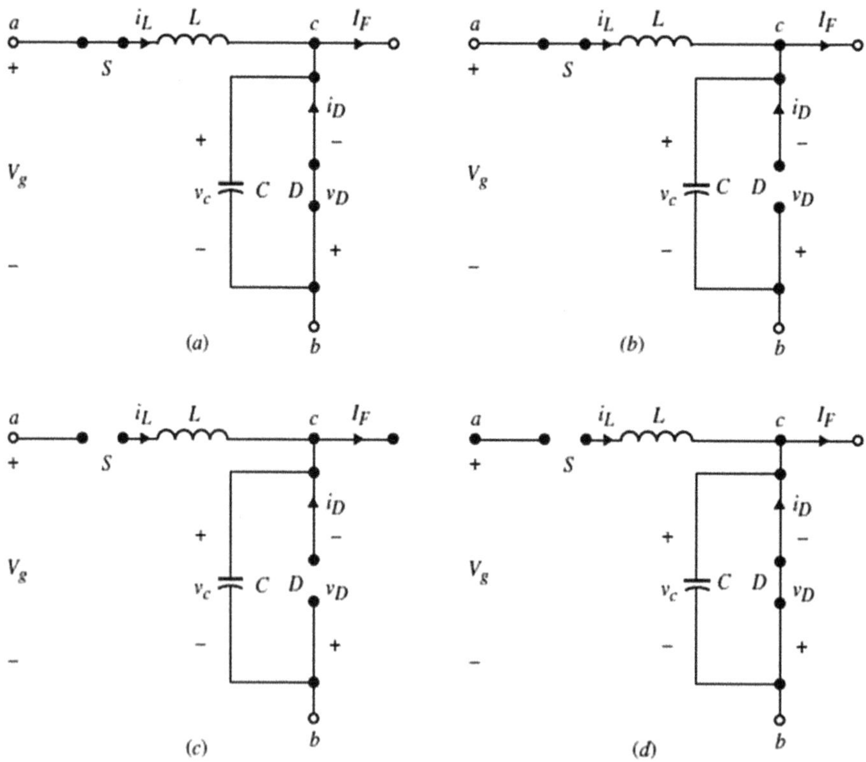

Fig. 6.63 The equivalent circuits for (**a**) Mode 1, (**b**) Mode 2, (**c**) Mode 3, and (**d**) Mode 4

along with the description of the modes to write the mathematical equations for each mode as follows (knowing that $v_c(t_0) = 0$ and $i_L(t_0) = 0$):

Mode 1 ($t_0 \le t < t_1$)

$$v_c(t) = 0$$

$$i_L(t) = \frac{V_g}{L}(t - t_0) \tag{6.95}$$

with the initial conditions at $t = t_1$ given by:

$$v_c(t_1) = 0$$
$$i_L(t_1) = I_F$$

Mode 2 ($t_1 \le t < t_2$)

$$v_c(t) = V_g[1 - \cos \omega_o(t - t_1)] \tag{6.96}$$

$$i_L(t) = I_F + \frac{V_g}{Z_o} \sin \omega_o(t - t_1) \tag{6.97}$$

At $t = t_2$, we have $i_L(t_2) = 0$

Mode 3 ($t_2 \le t < t_3$)

$$v_c(t) = -\frac{I_F}{C}(t - t_2) + V_g(1 - \cos \beta) \tag{6.98}$$

$$i_L(t) = 0 \tag{6.99}$$

At $t = t_3$, we have $v_c(t_3) = 0$

Mode 4 ($t_3 \le t < t_0 + T_s$)

$$v_c(t) = 0 \tag{6.100}$$

$$i_L(\mathrm{t}) = 0 \tag{6.101}$$

6.7.3.2 Generalized Interval Equations

To simplify the analysis, the following time intervals are defined:

$$\alpha = \omega_o(t_1 - t_0)$$

$$\beta = \omega_o(t_2 - t_1)$$

$$\gamma = \omega_o(t_3 - t_2)$$

$$\delta = \omega_o((t_0 + T_s) - t_3)$$

These intervals can be derived as follows:

$$\alpha = \omega_o(t_1 - t_o) = \frac{Z_o I_F}{V_g}$$

- From Eqs. (6.95) and (6.96), α is given by:
  By using the normalized parameters, we have:

$$\alpha = \omega_o(t_1 - t_o) = \frac{MI_{nF}}{QV_{ng}} \tag{6.102}$$

- From Eqs. (6.98) and (6.99), β is given by:

$$\beta = \omega_o(t_2 - t_1) = \sin^{-1}\left(-\frac{MI_{nF}}{QV_{ng}}\right) \tag{6.103}$$

- From Eqs. (6.100) and (6.101), γ is given by:

$$\gamma = \omega_o(t_3 - t_2) = \frac{QV_{ng}}{MI_{nF}}(1 - \cos\beta)$$

- From Fig. 6.62 and the intervals α, β, and γ, we have δ given by:

$$\delta = \omega_o((t_o + T_s) - t_3) = \frac{2\pi}{f_{ns}} - \alpha - \beta - \gamma \tag{6.104}$$

6.7.3.3 Generalized Gain Equation

The cell output to input generalized gain can be found using the average output diode D voltage as follows:

$$\begin{aligned} V_{D,ave} &= -V_{c,ave} \\ &= -\frac{1}{T_s}\int_{t_o}^{t_o+T_s} v_c(t)dt \\ &= -\frac{1}{T_s}\left[V_g\left((t_2 - t_1) - \frac{\sin\beta}{\omega_o}\right) - \frac{I_F}{2C}(t_3 - t_2)^2 + V_g(1 - \cos\beta)(t_3 - t_2)\right] \end{aligned}$$

By using the normalized parameters defined previously, we have:

$$V_{nD} = \frac{f_{ns}}{2\pi}\left[\frac{MI_{nF}}{2Q}\gamma^2 - V_{ng}(\beta + \gamma - \sin\beta - \gamma\cos\beta)\right] \tag{6.105}$$

By substituting for the generalized parameters (V_{nD}, V_{ng}, and I_{nF}) from Table 6.1 in Eq. (6.105), we will have the gain equation for each converter in the family.

6.7.3.4 Generalized ZCS Condition

It can be noted from Fig. 6.62 that S can be turned *off* at any time after $t = t_2$. The generalized condition to achieve zero-current switching can be expresses as follows:

$$\frac{2\pi}{f_{ns}} D \geq \alpha + \beta \tag{6.106}$$

6.7.3.5 Generalized Peak Resonant Inductor Current (Peak Switch Current)

The peak resonant inductor current or peak switch current occurs at $t = t'_{max}$, when $\omega_o\left(t'_{max} - t_1\right) = \pi/2$. By using Eq. (6.98) at $t = t'_{max}$:

$$I_{nL-peak} = I_{nF} + \frac{QV_{ng}}{M} \tag{6.107}$$

The peak resonant capacitor voltage or peak diode voltage occurs at $t = t''_{max}$ when $\omega_o\left(t''_{max} - t_1\right) = \pi$. By using Eq. (6.97) at $t = t''_{max}$:

$$V_{nL-peak} = 2V_{ng}$$

6.7.3.6 Design Control Curves

By substituting for the generalized parameters from Table 6.1 in the generalized equations, design equations for each topology in the family can be found. Using computer software, design curves can be derived.

Fig. 6.64 shows the control characteristic curves of M vs. f_{ns} for the ZCS-QRC family as an example. Fig. 6.65 shows the average and the *rms* switch currents as a function of the voltage gain.

6.7.4 The Basic Operation of the ZVS-QRC Cell

Figure 6.60b shows the generalized ZVS-QRC switching cell, which is formed by adding the resonant capacitor C (which can be considered as the switch's internal capacitor) and the resonant inductor L to the conventional switching cell. As discussed before, this is also known as an M-type resonant switch.

By applying this cell to the conventional dc-dc converters in Fig. 6.7, the ZVS-QRC family can be formed as shown in Fig. 6.66. In the next section, the modes of operation for the ZVS-QRC switching network will be discussed briefly, and the main switching waveforms will be drawn.

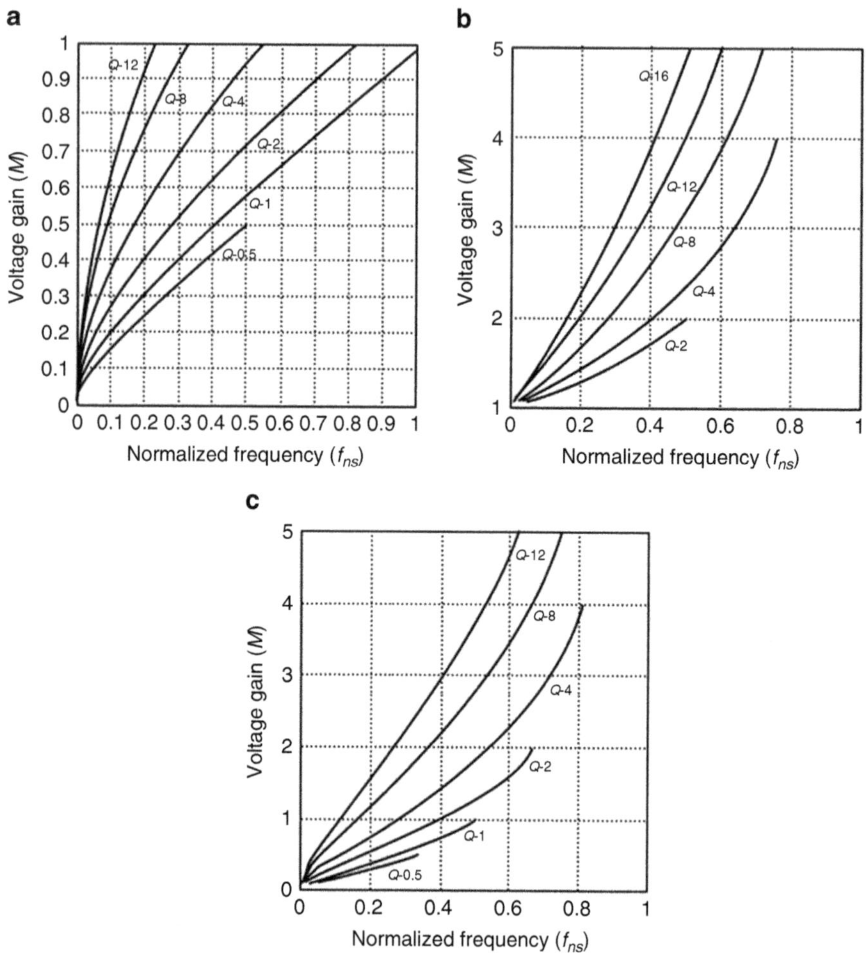

Fig. 6.64 Control characteristic curves of M vs. f_{ns} for: (**a**) ZCS-QRC buck, (**b**) ZCS-QRC boost, (**c**) ZCS-QRC buck-boost, Cuk, Zeta, and SEPIC

The typical switching waveforms for the ZVS-QRC cell are shown in Fig. 6.67. Table 6.3 shows the conduction status of the switches and diodes in each mode. As shown before, there are four modes of operation. It is assumed that before $t = t_0$, S was on and D was *off*. The resonant inductor L was carrying a current equal to I_F, and the resonant capacitor C voltage was zero (Fig. 6.68).

Mode 1 ($t_0 \leq t < t_1$)

Mode 1 starts when S is turned *off* while D is *off*, which causes C to charge up linearly until its voltage reaches a value equal to V_g at $t = t_1$, causing D to start conducting. The resonant inductor current during this mode doesn't change and it is equal to I_F.

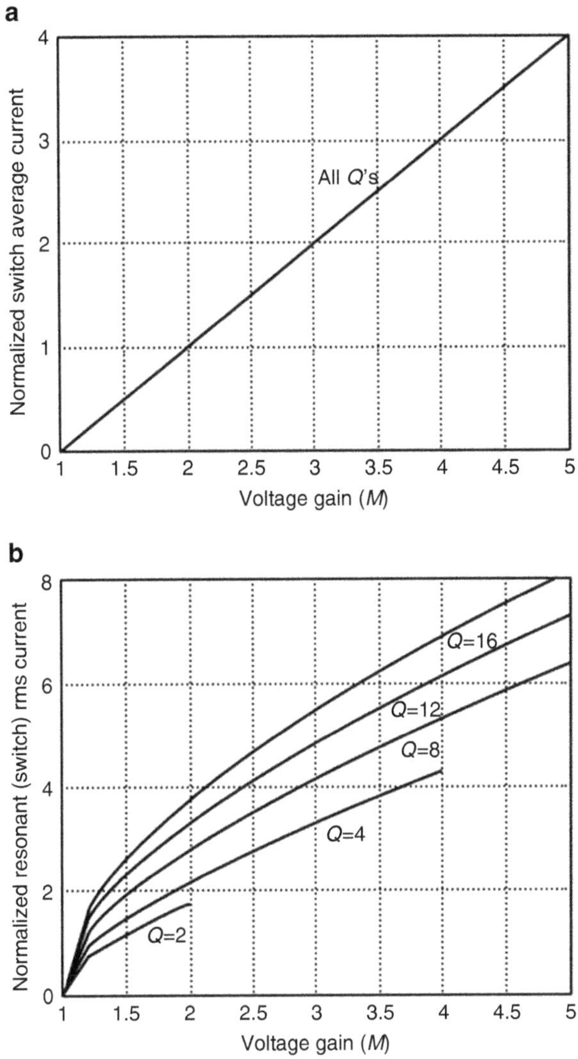

Fig. 6.65 Some of the ZCS-QRC boost main switch (S) normalized stresses: (**a**) normalized average current and (**b**) normalized rms current

Mode 2 ($t_1 \le t < t_2$)
Mode 2 starts when D turns *on* while S is *off* causing a resonant stage between C and L to start until the voltage across C tends to go negative forcing the switch diode D_s to turn *on* at $t = t_2$. After this time, S can be turned *on* at zero voltage.

Mode 3 ($t_2 \le t < t_3$)
Mode 3 starts when S is turned *on* at zero voltage (zero-voltage switching). The resonant inductor current starts charging up linearly until it reaches I_F causing D to turn *off* at zero current at $t = t_3$.

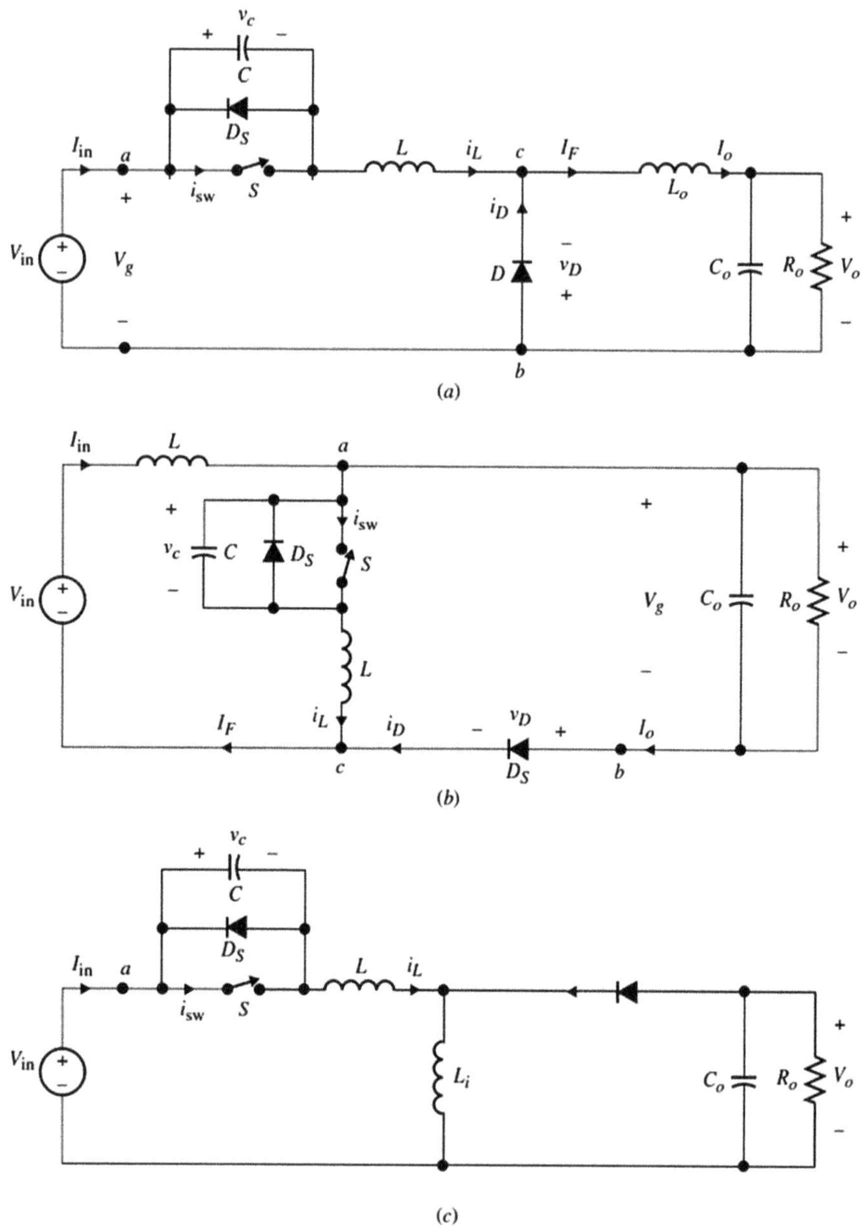

Fig. 6.66 The dc-dc ZVS-QRC family: (**a**) Buck (**b**) Boost (**c**) Buck-boost

Mode 4 ($t_3 \leq t < t_0 + T_s$)

Mode 4 is a steady-state mode, and nothing happens in it until S is turned *off* again to start the next switching cycle .

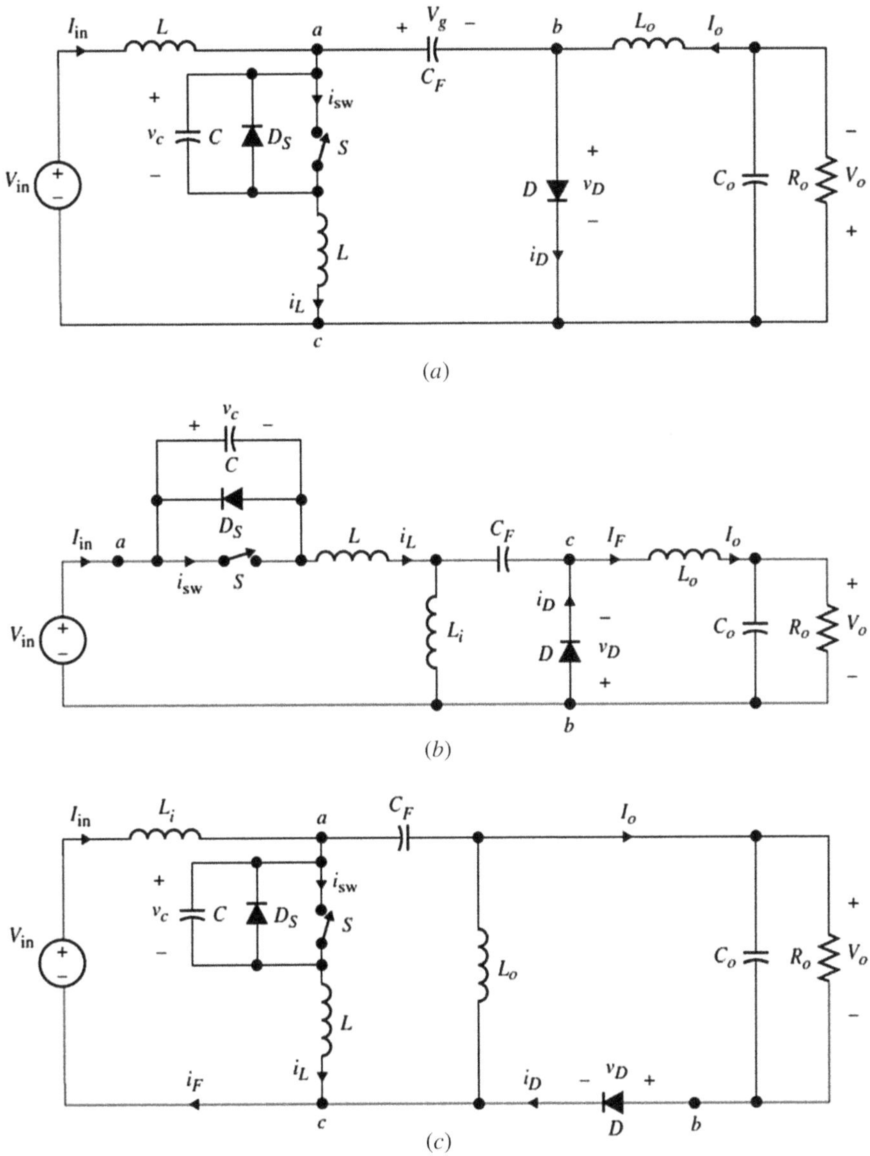

Fig. 6.67 The dc-dc ZVS-QRC family: (**a**) Cuk (**b**) Zeta (**c**) SEPIC

Table 6.3 Switches conditions

	S	D	D_s
Mode 1 ($t_0 \le t < t_1$)	Off	Off	Off
Mode 2 ($t_1 \le t < t_2$)	Off	On	Off
Mode 3 (($t_2 \le t < t_3$)	On	On	On
Mode 5 ($t_3 \le t < t_0 + T_s$)	On	Off	Don't care

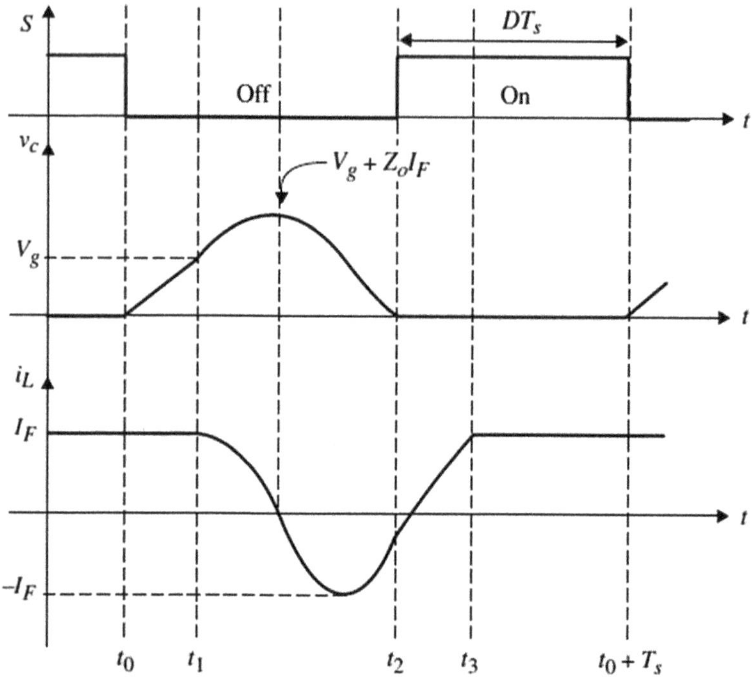

Fig. 6.68 Main ZVS-QRC switching cell waveforms

6.7.4.1 Generalized Steady-State Analysis

From the description of the modes of operation, the equivalent circuit for each mode can be drawn as shown in Fig. 6.69. These equivalent circuits can be used along with the description of the modes to write the mathematical equations for each mode as follows (knowing that $v_c(t_0) = 0$ and $i_L(t_0) = I_F$:

Mode 1 ($t_0 \le t < t_1$)

$$v_c(t) = \frac{I_F}{C}(t - t_9) \qquad (6.108)$$
$$i_L(t) = I_F$$
$$v_c(t_1) = V_g \qquad (6.109)$$
$$i_L(t_1) = I_F$$

Mode 2 ($t_1 \le t < t_2$)

$$v_c = V_g + Z_oI_F \sin \omega_o(t - t_1) \qquad (6.110)$$
$$i_L(t) = I_F \cos \omega_o(t - t_1)$$
$$v_c(t_2) = 0 \qquad (6.111)$$
$$i_L(t_2) = I_F \cos \omega_o(t_2 - t_1) = 0$$

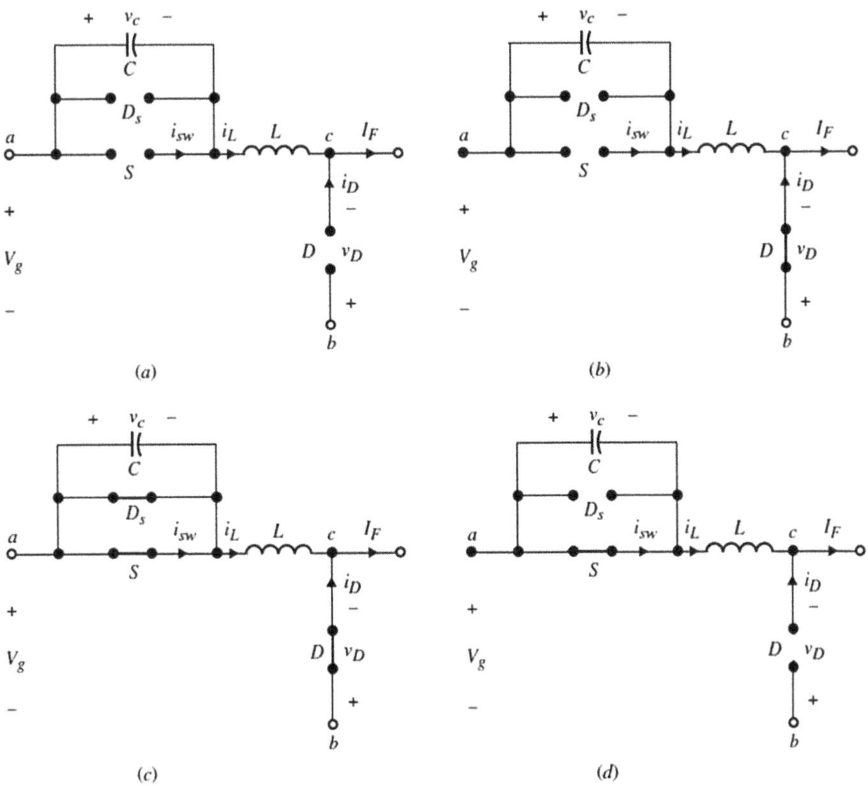

Fig. 6.69 The equivalent circuits for: (**a**) Mode 1, (**b**) Mode 2, (**c**) Mode 3, and (**d**) Mode 4

Mode 3 ($t_2 \leq t < t_3$)

$$v_c(t) = 0$$
$$i_L(t) = I_F \cos \omega_o (t_2 - t_t) - \frac{V_g}{L}(t - t_2) \tag{6.112}$$
$$v_c(t_3) = 0$$
$$i_L(t_3) = I_F \tag{6.113}$$

Mode 4 ($t_3 \leq t < t_0 + T_s$)

$$v_c(t) = 0$$
$$i_L(t) = I_F$$

6.7.4.2 Generalized Interval Equations

To simplify the analysis, the following time intervals are defined:

$$\alpha = \omega_o(t_1 - t_0)$$
$$\beta = \omega_o(t_2 - t_1)$$
$$\gamma = \omega_o(t_3 - t_2)$$
$$\delta = \omega_o((t_0 + T_s) - t_3)$$

These intervals can be derived as follows:

- From Eqs. (6.108) and (6.109):

$$\alpha = \omega_o(t_1 - t_0) = \frac{V_g}{Z_o I_F}$$

By using the normalized parameters defined earlier:

$$\alpha = \omega_o(t_1 - t_0) = \frac{QV_g}{MI_{nF}}$$

- From Eqs. (6.110) and (6.111):

$$\beta = \omega_o(t_2 - t_1) = \sin^{-1}\left(-\frac{QV_g}{MI_{nF}}\right)$$

- From Eqs. (6.112) and (6.113):

$$\gamma = \omega_o(t_3 - t_2) = \frac{MI_{nF}}{QV_g}(1 - \cos\beta)$$

- From Fig. 6.68 and the above intervals:

$$\delta = \omega_o((t_0 + T_s) - t_3) = \frac{2\pi}{f_{ns}} - \alpha - \beta - \gamma$$

6.7.4.3 Generalized Gain Equation

The cell output to input generalized gain can be found using the average output diode D voltage as follows:

$$\begin{aligned} V_{\mathrm{D,ave}} &= V_{\mathrm{s,ave}} - V_g \\ &= \left[\frac{1}{T_{\mathrm{s}}}\int_{t_o}^{t_o+T_{\mathrm{s}}} v_{\mathrm{s}}(t)dt\right] - V_g \\ &= \frac{1}{T_{\mathrm{s}}}\left[\frac{I_{\mathrm{F}}}{2C}(t_1 - t_0)^2 + V_{\mathrm{g}}(t_2 - t_1) - \frac{Z_{\mathrm{o}}I_{\mathrm{F}}}{\omega_{\mathrm{o}}}\cos\omega_{\mathrm{o}}(t_2 - t_1) - 1\right] - V_{\mathrm{g}} \end{aligned}$$

By using the normalized parameters, we will have:

$$V_{\mathrm{nD}} = \frac{f_{\mathrm{ns}}}{2\pi}\left[\frac{MI_{\mathrm{nF}}}{2Q}\alpha^2 + V_{\mathrm{ng}}\beta + \frac{MI_{\mathrm{nF}}}{Q}(1 - \cos\beta)\right] \tag{6.114}$$

By substituting for the generalized parameters (V_{nD}, V_{ng}, and I_{nF}) from Table 6.1 in Eq. (6.114), we will have the gain equation for each converter in the family.

6.7.4.4 Generalized ZVS Condition

It can be noted from Fig. 6.68 that S must be turned on after $t = t_2$ and before $t = t_3$, in other words, before $i_{\mathrm{L}}(t) = I_{\mathrm{F}}$, which causes D to turn off ($i_{\mathrm{D}}(t) = I_{\mathrm{F}} - i_{\mathrm{L}}(t)$) at $t = t_3$) causing $v_{\mathrm{c}}(t)$ to start charging up linearly again. The generalized condition to achieve zero-voltage switching can be expressed as follows:

$$\alpha + \beta \leq \frac{2\pi}{f_{\mathrm{ns}}}(1 - D) \leq \alpha + \beta + \gamma$$

6.7.4.5 Design Control Curves

By substituting for the generalized parameters from Table 6.1 in the generalized equations, design equations for each topology in the family can be found. Using computer software, design curves can be plotted. Fig. 6.70 shows the control characteristic curves for the ZVS-QRC family. Many other curves can also be plotted; as an example, Fig. 6.71 shows the peak and the rms switch currents as a function of the voltage gain, M.

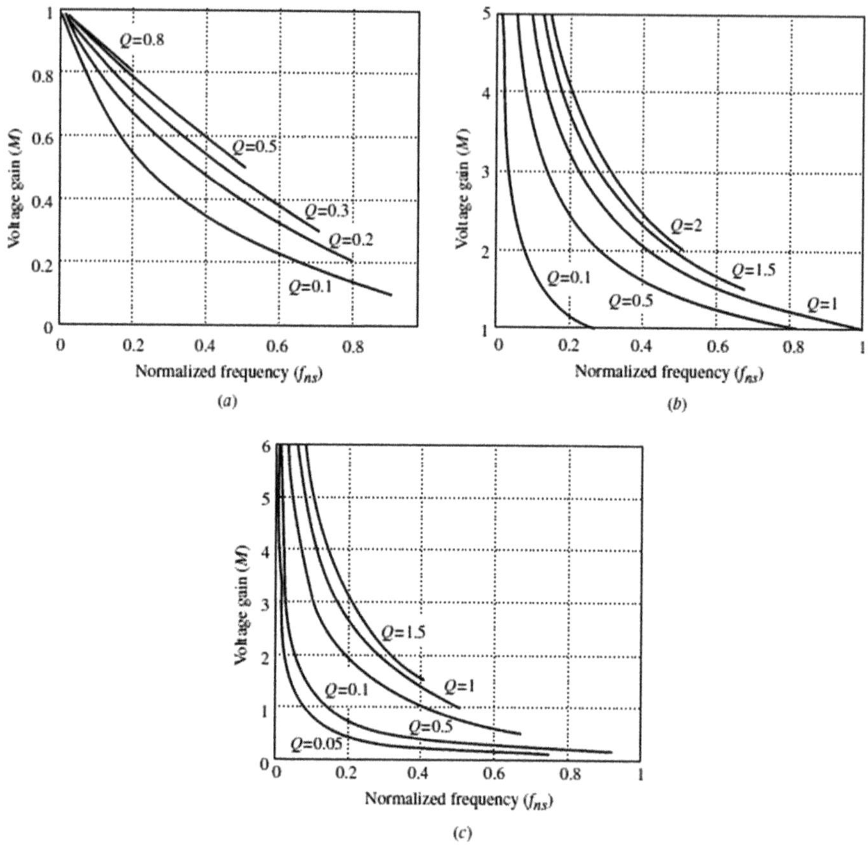

Fig. 6.70 Control characteristic curves of M vs. f_{ns} for: (**a**) ZVS-QRC buck, (**b**) ZVS-QRC boost, (**c**) ZVS-QRC buck-boost, Cuk, Zeta, and SEPIC

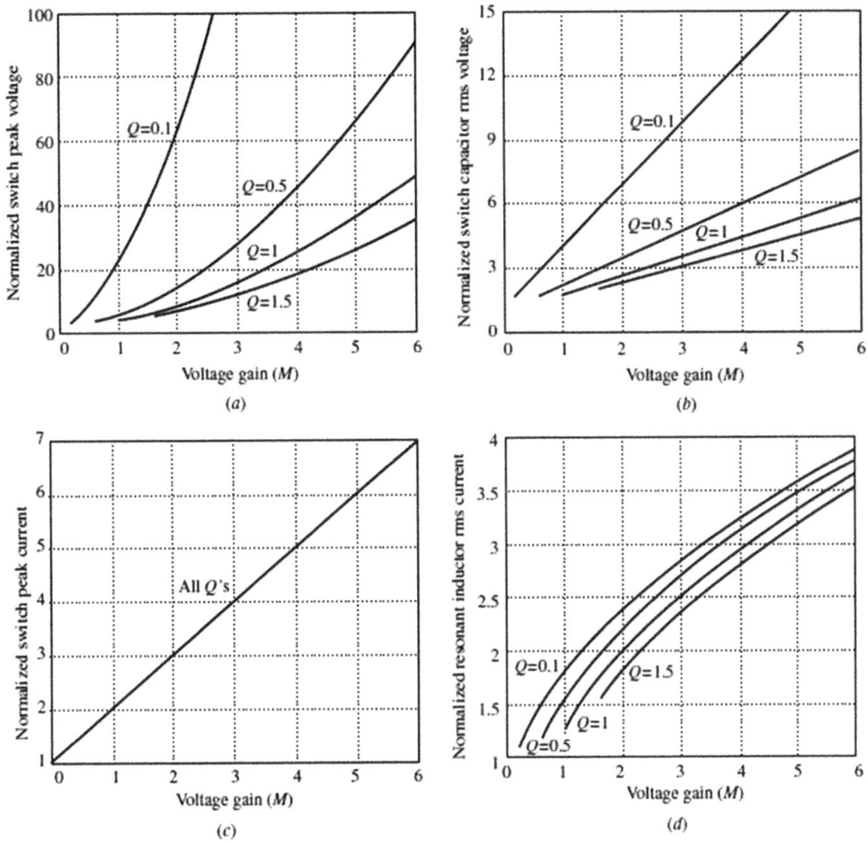

Fig. 6.71 Some of the ZVS-QRC buck-boost stresses: (**a**) normalized switch peak voltage, (**b**) normalized switch rms voltage, (**c**) normalized switch peak current, and (**d**) normalized switch rms current

Problems

ZCS Quasi-Resonant Buck Converters

6.1 Consider a ZCS buck converter whose steady-state waveforms are shown in Fig. 6.10 Assume that the converter has the following parameters: $V_{in} = 25$V, $I_o = 1$A, $L = 3$ μH, and $C = 0.02$ μF. Determine the time intervals at t_1, t_2, and t_3.

6.2 Consider the ZCS buck converter shown in Fig. P6.2. Determine the output power.

Fig. P6.2

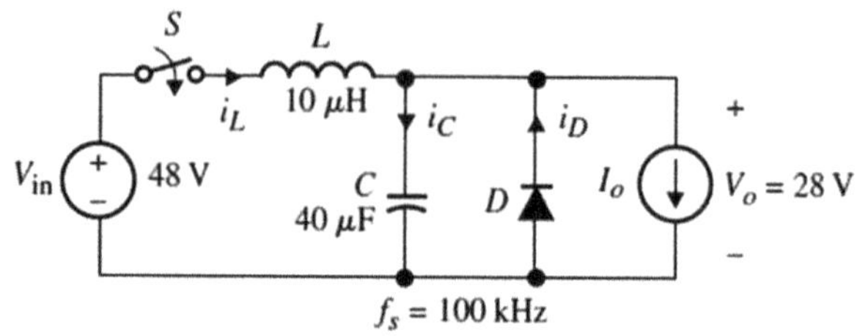

6.3 Determine M and Q for a ZCS buck converter with an L-type switch that has the following converter components: $L = 15$ μH, $C = 60$ nF, $f_s = 100$ kHz, $V_{in} = 40$ V, and $I_o = 0.74$ A.

6.4 Consider the ZCS buck converter shown in Fig. P6.4a with $L = 20$ μH, $C = 5$ μF, $V_{in} = 50$ V, $V_o = 30$ V, and $I_o = 20$ A. Assume that $L_o/R_o \gg 1/T$. The switching waveform of the transistor is shown in Fig. P6.4b.

(a) Determine the minimum t_{ON} in μs to achieve ZCS and determine the switching period T.

(b) Repeat Problem 6.73 by adding a diode across the transistor to allow a bidirectional current flow to produce a full-wave ZCS buck converter. Assume all values are the same as in Problem 6.73.

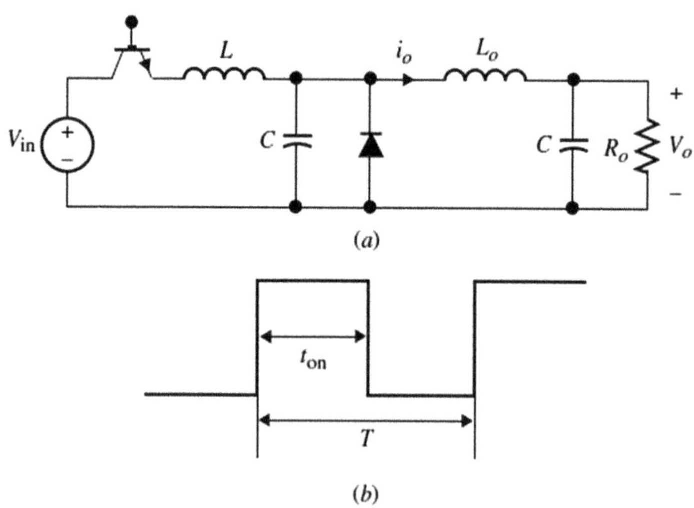

Fig. P6.4

6.5 Determine the average output voltage for the ZCS buck converter shown in Fig. P6.5.

Fig. P6.5

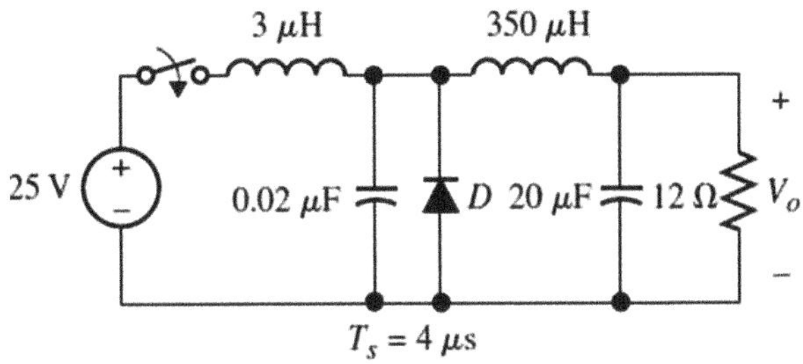

6.6 Consider a ZCS buck converter with an L-type switch with $M = 0.35$ and $Q = 1$. Design for the resonant tank L and C for $V_{\text{in}} = 40$ V, $f_s = 250$ kHz, and $I_o = 0.7$ A.

6.7 Consider a ZCS buck converter shown in Fig. P6.7 that has the following design parameters: $V_{\text{in}} = 25$ V, $V_o = 12$ V, $f_s = 250$ kHz, $I_o = 1$ A, and $f_o = 625$ kHz. Design for L and C and draw the steady-state waveforms for i_L, i, v_c, v_{sw}, v_D, i_D, and i_o.

Fig. P6.7

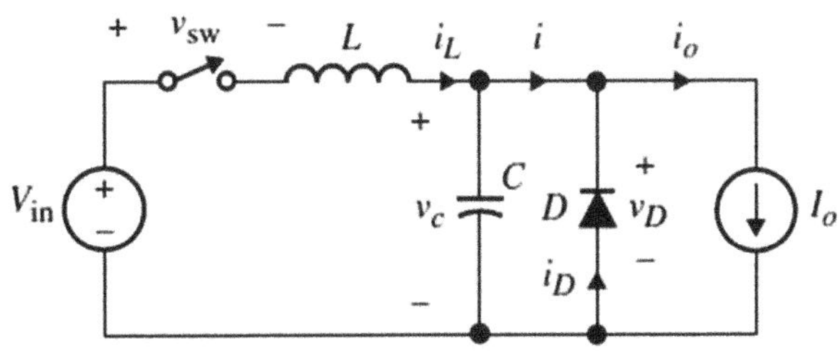

6.8 Consider a ZCS buck converter with a unidirectional switch and the following specifications:

- Maximum load power 750 W
- Nominal output voltage 5 V
- Nominal input voltage 12 V
- Switching frequency 85 kHz
- Maximum peak resonant inductor current is twice the average load current

(a) Design for the resonant tank L and C.

(b) If the load current changes by ±20% what is the new minimum and maximum switching frequencies required to maintain the output voltage at 5 V?

(c) Repeat part (b) for the variation of ±20% in the average input voltage.

6.9 Design a ZCS buck converter with an L-type switch that has the following parameters: $V_{\text{in}} = 50$ V, $f_s = 100$ kHz, $I_o = 0.2$ A, $V_o = 49$ V, and $f_{ns} = 0.5$. Design for L, C, L_o, C_o, and R. Design for an output ripple current within 15% of its dc value and an output ripple voltage not to exceed 2%.

6.10 Consider the ZCS buck converter shown in Fig. 6.8(b) with the following parameters: $L = 25\ \mu H$, $C = 4.7\ \mu F$, $V_{in} = 58$ V, and $I_o = 18$ A. Determine the output voltage and f_s to deliver an average output power equal to 580 W in

ZCS Quasi-Resonant Boost Converters

6.11 Design a ZCS boost converter for the following parameters: $V_{in} = 20$ V, $V_o = 36$ V, $I_o = 0.7$ A, and $f_s = 250$ kHz. What is the range of f_s needed to keep V_o constant when I_o changes between 0.2 and 1 A?

6.12 Design a ZCS boost converter with an M-type switch with the following design parameter: $M = 2.73$ and $f_{ns} = 0.6$. It is desired to deliver 0.5 A load current to $V_o = 68$ V.

6.13 Design an M-type switch boost quasi-resonant converter with the following parameters: $M = 1.8$, $Q = 2.5$, and $f_{ns} = 0.5$. The input voltage varies between 18 V and 26 V, and I_o varies between 0.5 A and 1 A. What is the range of f_s to keep the converter output voltage constant at $V_o = 38$?

ZCS Quasi-Resonant Buck-Boost Converters

6.14 Consider a ZCS buck-boost converter with an L-type switch shown in Fig. P6.14. Find M, Q, and f_s. What is the new f_s needed to keep V_o constant if the average load changes to 1.8 A?

Fig. P6.14

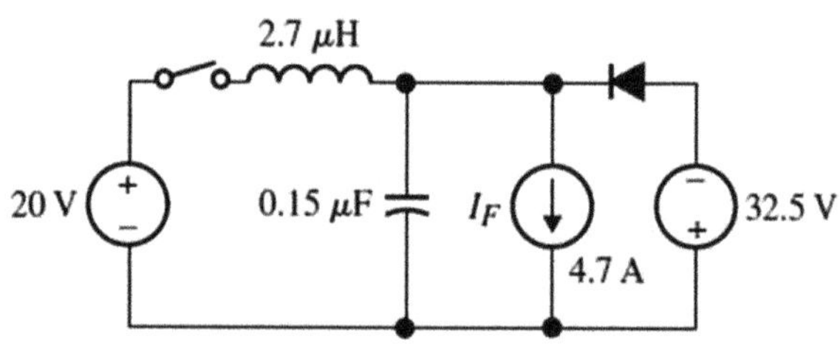

6.15 Determine the value of I_F in the ZCS buck-boost converter given in Fig. P6.15.

Fig. P6.15

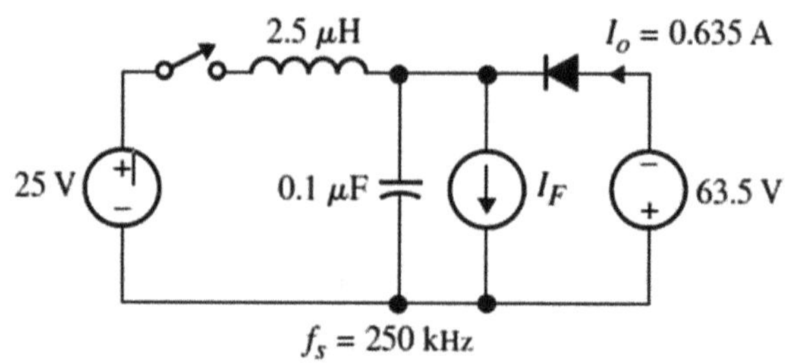

6.16 Consider a ZCS buck-boost converter with an L-type switch to be designed for the following specifications: $V_{in} = 40$ V, $V_o = 20$ V, $I_o = 4$ A, and $f_s = 250$ kHz.

6.17 Design a ZCS buck-boost converter with L-type converter for the following specifications $V_{in} = 30$ V, $V_o = 20$ V@ $I_o = 2.8$ A, and $f_s = 50$ kHz.

6.18 Design a ZCS buck-boost converter with an L-type switch that operates at 150 kHz and delivers 48 W to $V_o = 47$ V from a 10 V dc source.

6.19 Figure P6.19 shows a ZCS buck-boost with S being unidirectional. (a) Sketch i_L and v_c, (b) discuss the four modes of operation, and (c) derive the expression for i_L and v_c in terms of the circuit parameters.

Fig. P6.19

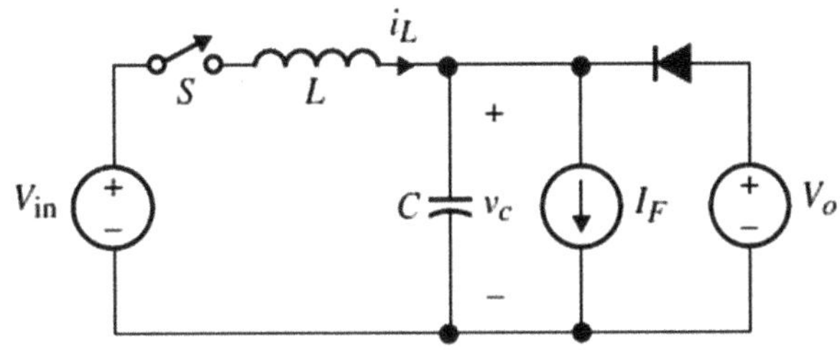

6.20 Repeat Problem 6.79 by using a ZCS buck-boost L-type converter. Select any set of M, Q, and f_{ns} you see fit for your design.

ZVS Quasi-Resonant Buck Converters

6.21 Derive the voltage gain expression for the M-type ZVS buck converter.

6.22 Consider the buck ZVS shown in Fig. P6.22 with an alternative way of implementing the L-type resonant switch. Assume L_o/L is very large.

(a) Discuss the modes of operation over one-switching cycle.
(b) Draw the typical waveform for i_L and v_c.
(c) Derive the expression for V_o/V_{in}.

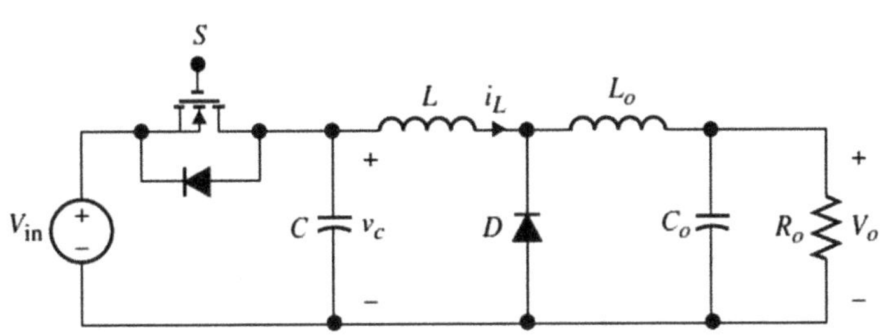

Fig. P6.22

6.23 Consider the resonant capacitor voltage and inductor current waveforms shown in Fig. P6.23 for a ZVS buck. Determine:

(a) L, C, t_1, t_3, V_o
(b) Q, f_{ns}, M
(c) The Maximum voltage across the switch and the diode.

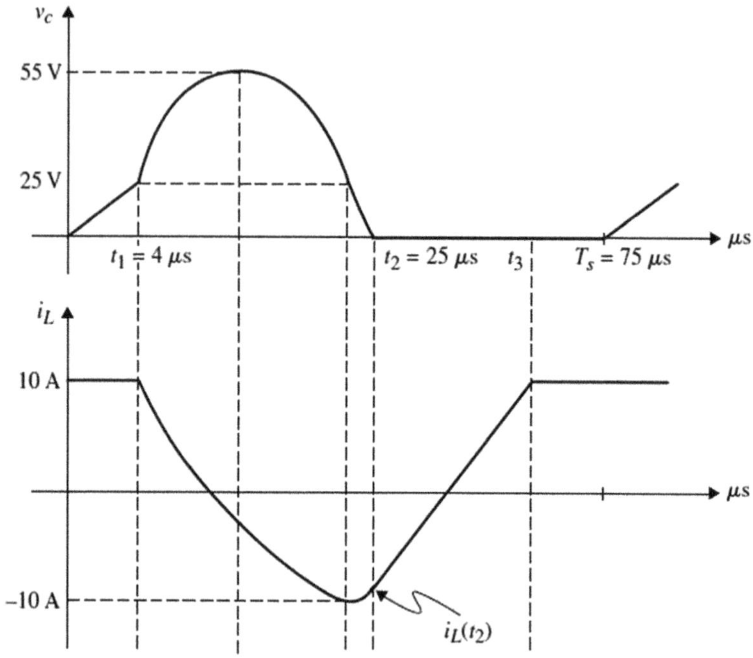

Fig. P6.23

ZVS Quasi-Resonant Boost Converters

6.24 Consider the ZVS boost quasi-resonant converter shown in Fig. P6.24. Assume S is bidirectional.

(a) Sketch the waveforms for v_C and i_L.
(b) Discuss the four modes of operation.
(c) Derive the following expression for the voltage gain:

$$M = \frac{V_o}{V_{in}} = \frac{1}{\frac{f_{ns}}{2\pi}\left(\alpha - \frac{Q}{2M} + \frac{M}{Q}(1 - \cos\alpha)\right)}$$

where:

$$\sin\alpha = \frac{-Q}{M},\ f_{\mathrm{ns}} = \frac{f_s}{f_o},\ Q = \frac{R_{\mathrm{o}}}{Z_{\mathrm{o}}}$$

$$Z_{\mathrm{o}} = \sqrt{L/C},\ f_{\mathrm{o}} = \frac{1}{2\pi\sqrt{LC}}$$

Fig. P6.24

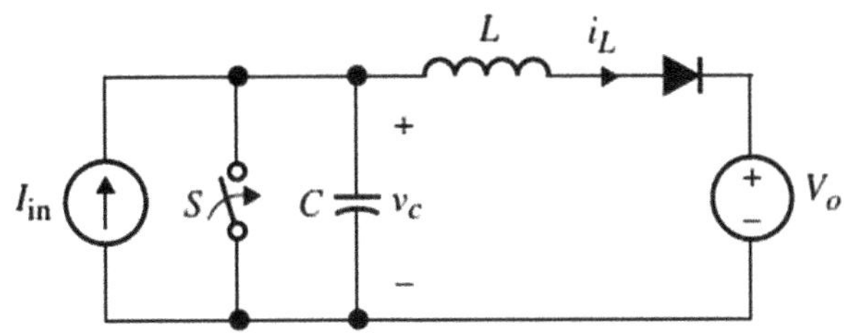

6.25 Design a ZVS boost converter with the following specifications:

- Maximum load power 750 W
- Nominal input voltage 12 V
- Nominal input voltage 5 V
- Switching frequency 100 kHz
- Maximum peak capacitor voltage is 1.5 times the average input voltage

6.26 Figure P6.26a shows an isolated boost converter with switches that operates at 50% duty cycle. If it is assumed that L_{in} and C_{o} are large, then the input current, I_{in}, and the output voltage, V_{o}, may be assumed constant as shown in Fig. P6.26b. The switching waveforms for S_1 and S_2 are shown in Fig. P6.26c. It can be shown that there are four modes of operation in the steady state with S_2 operating at ZVS.

(a) Discuss the modes of operation.

(b) Show that $I_{\mathrm{o}}/I_{\mathrm{in}} = n(1-D)$ where I_{in} and I_{o} are the average input and output currents.

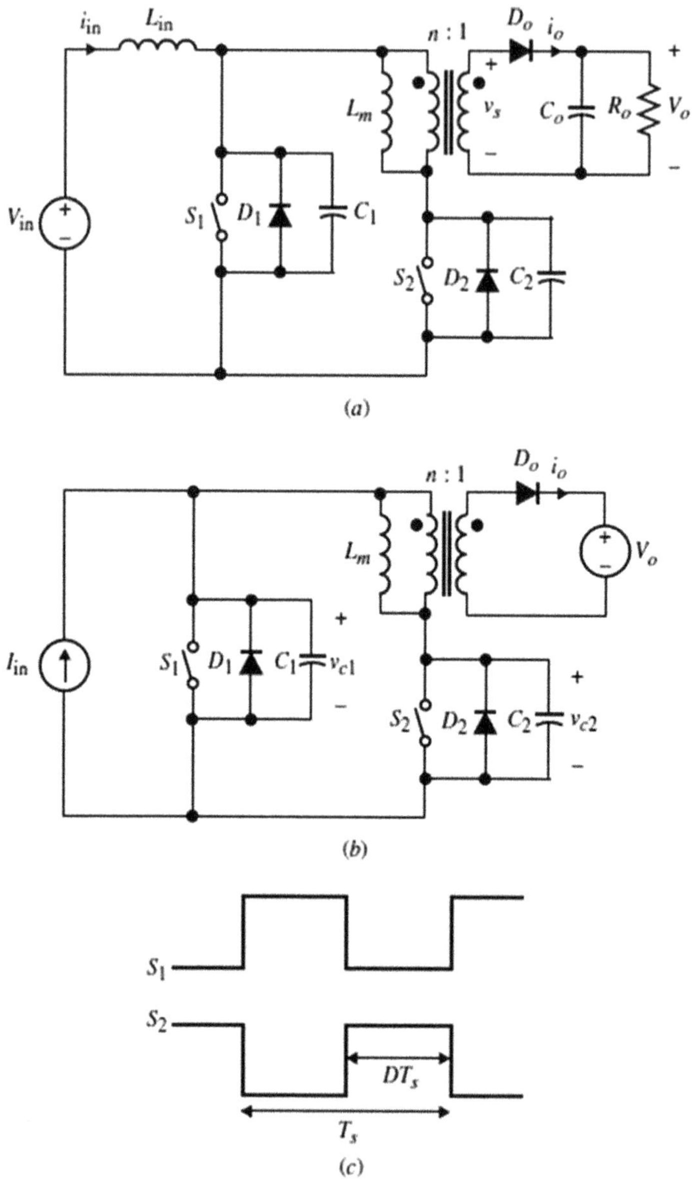

Fig. P6.26

ZVS Quasi-Resonant Buck-Boost Converters

6.27 Determine the output voltage for the ZVS converter shown in Fig. P6.27 and sketch the waveforms for i_L, v_c, and i_D.

Fig. P6.27

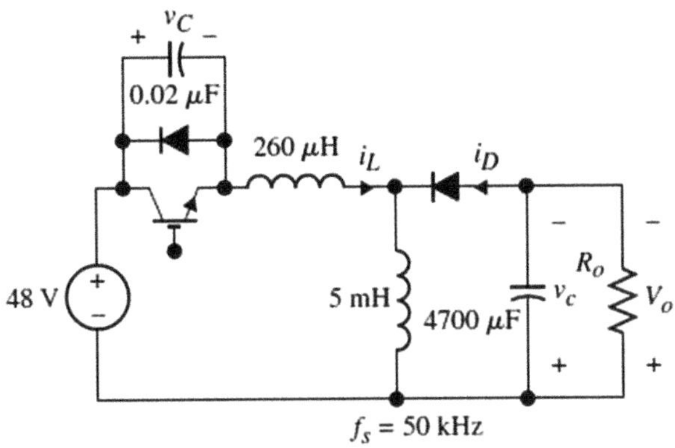

6.28 Design a ZVS buck-boost converter with the following steady-state operating point $M = 0.55, f_{ns} = 0.3$, and $Q = 0.1$. Assume $V_{in} = 30$ V and $T_s = 10$ μs and the output current is at least 1 A.

General Soft-Switching Converters

6.29 Figure P6.29 shows a zero-current transition (ZCT) buck converter.

(a) Discuss the modes of operation and show the typical waveforms for all the currents and voltages. Assume I_o is constant.

(b) Give the expression for the circulating energy in the resonant tank.

(c) What are the major features of this converter?

Fig. P6.29

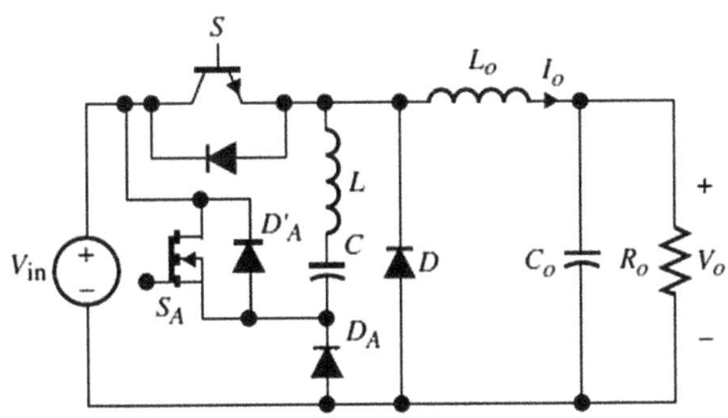

6.30 Consider the circuit given in Fig. P6.30 that operates in the ZVS mode with S_1 and S_2 operating alternatively. Discuss the operation of the circuit and sketch the waveforms for i_{sw}, i_{Lk}, v_{c1}, v_{c2}, i_{Lm}, i_o, and v_s.

Fig. P6.30

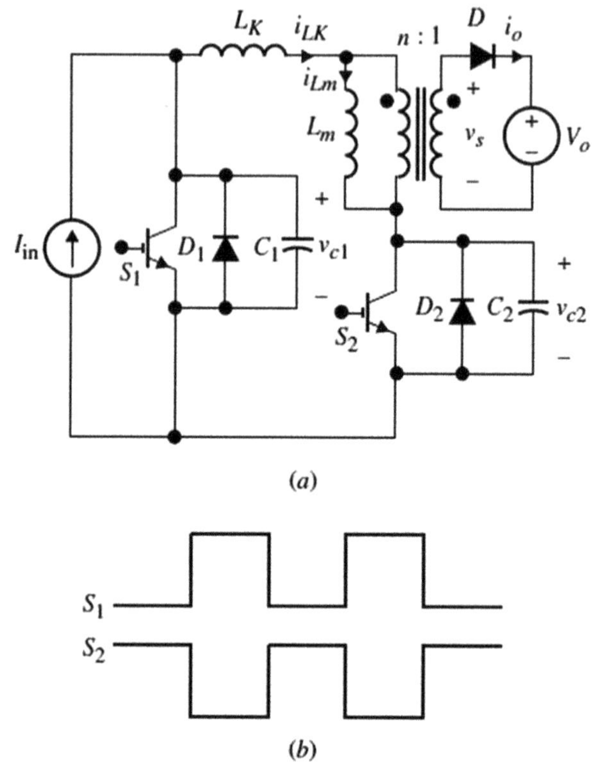

6.31 Figure P6.31a shows a ZVS soft-switching capacitor voltage clamped converter. Assume $C_o \gg C_1$ and $C_o \gg C_2$. It can be shown that in the steady state, there are four modes of operation. Assume S_1 and S_2 are switched as shown in Fig. 6.31b.

(a) Sketch the waveforms for i_L, v_{c1}, v_{C2}, and i_{D2}.
(b) Discuss the four modes of operation.
(c) Derive the output voltage gain in terms of the circuit parameters.

Fig. P6.31

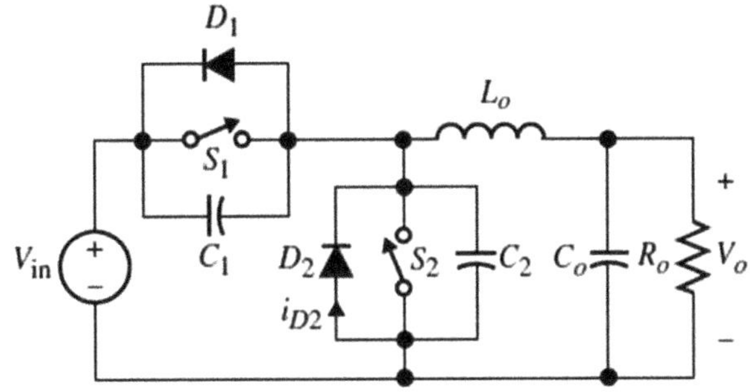

(a)

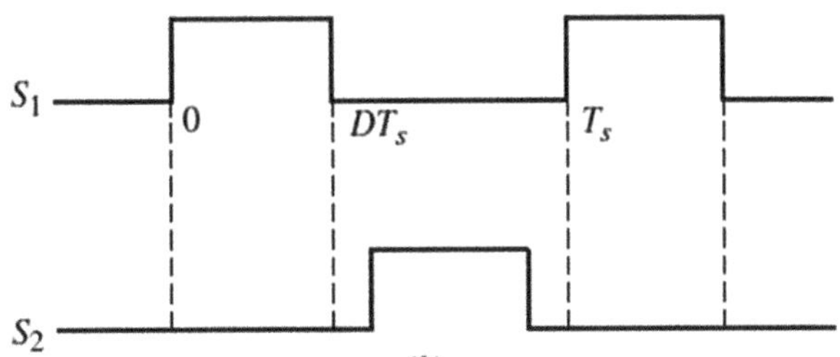

(b)

6.32 Figure P6.32 shows a soft-switching technique that is based on the concept of clamped current (CC) soft switching. Such converter family is known as the ZCS quasi-square wave resonant converter. It is assumed that L_f is large enough so its current may be represented as a current source. S and D_s form a unidirectional switch. This topology offers several advantages. Discuss these advantages.

(a) Sketch the steady-state waveforms for i_L, i_{sw}, i_D, and v_c. Assume $L_F < I_o$ and the first mode starts at $t = t_0$ when S is *on* and D is *on* initially. It can be shown that there are four modes of operation.

(b) Show that during the resonant mode (Mode 3), the resonant inductor current and resonant capacitor voltage equations are given by:

$$i_L(t) = I_F(\cos \omega_o(t - t_2) + 1)$$
$$v_c(t) = V_{in} + Z_o I_F \sin \omega_o(t - t_2)$$

where $Z_o = \sqrt{L/C}$ and $t = t_2$ is the starting of Mode 3.

Fig. P6.32

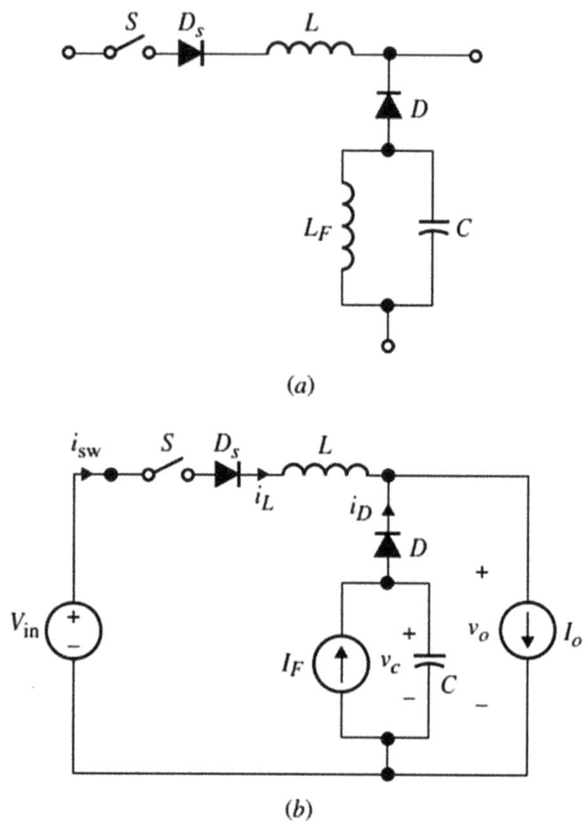

6.33 Figure P6.33 shows a soft-switching converter cell with the diode replaced by a switch S_D. By allowing the rectifier to become a controllable switch, the converter voltage gain can be controlled by the pulse width of S. This soft-switching family is known as PWM quasi-square wave converters (PWM-QSW). Unlike the quasi-resonant converters, which are frequency-controlled converters, the QSW converters are PWM controlled at constant frequency. Assume S and S_D are bidirectional. Discuss the four modes of operation.

Fig. P6.33

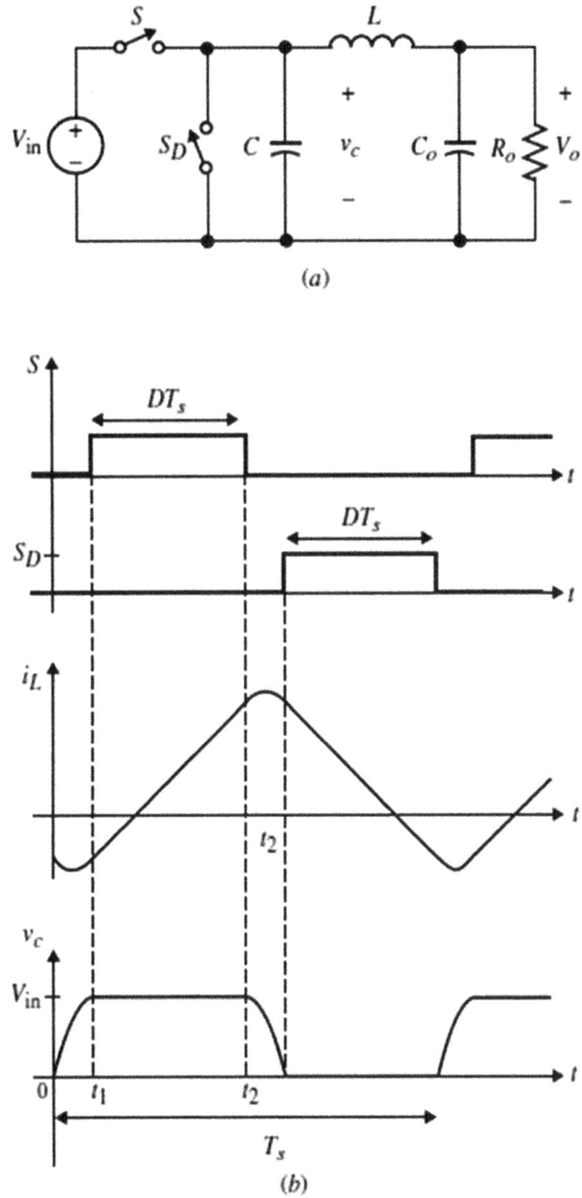

6.34 The topology shown in Fig. P6.34a is known as the isolated full-bridge LLC resonant converter. It can be shown that under certain driving conditions, this topology operates in zero-voltage switching for wide range of load. Assume ideal components with the resonant tank consist of two resonant inductors L_r and Lm and one capacitor C_r. The inductor Lm can be considered as the parasitic magnetizing inductance. Assume the output voltage is constant; derive the six equivalent modes of operation and their corresponding waveforms. Also, derive the expressions for the two resonant currents and the voltage of the resonant capacitor. Assume the switching waveforms for $\varphi_1 - \varphi_4$ are shown in Fig. P6.34b.

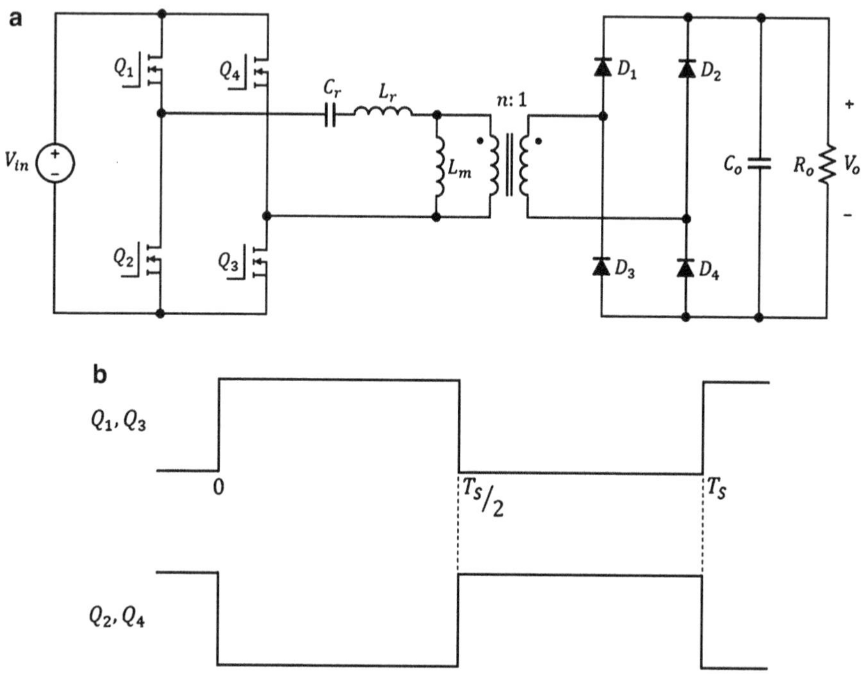

Fig. P6.34a (**a**) Full-bridge LLC ZVS converter (**b**) Driving waveforms for the full-bridge switches

6.35 Derive Eq. (6.94) for the ZVT boost of Fig. 6.50.

Chapter 7
Uncontrolled Diode Rectifier Circuits

7.1 Introduction

For nearly a century, rectifier circuits have been the most common power electronic circuits used to convert ac to dc. The word rectification is used not because these circuits produce dc but rather because the current flows in one direction; only the average output signal (voltage or current) has a dc component. Moreover, since these circuits allow power to flow only from the source to the load, they are often termed unidirectional converters. As will be shown shortly, when rectifier circuits are used solely, their outputs consist of dc along with high-ripple ac components. To significantly reduce or eliminate the output ripple, additional filtering circuitry is added at the output. In the majority of applications, diode rectifier circuits are placed at the front end of the power electronic 60 Hz systems, interfaced with the sine-wave voltage produced by the electric utility. In dc-dc application, at the rectified side or the dc side, a large filter capacitor is added to reduce the rectified voltage ripple. This dc voltage maintained across the output capacitor is known as raw dc or *uncontrolled dc*.

The principal circuit operations of the various configurations of rectifier circuits are similar, whether half or full wave, single phase, or three phase. Such circuits are said to be *uncontrolled* since the rectified output voltage and current are a function only of the applied excitation, with no mechanism for varying the output level. In such circuits, regardless of the configuration, diodes are almost always used to achieve rectification. This is because diodes are inexpensive, are readily available with low- and high-power capabilities, and have terminal characteristics that are simple and well understood. However, since diode circuits are nonlinear, i.e., their circuit topological modes vary with time, their analysis can be challenging, as illustrated in Chap. 3. To simplify the analysis, it will be assumed throughout this chapter that the diodes are ideal as discussed in Chap. 2. In this chapter, we will discuss the single- and three-phase uncontrolled rectifier circuits of both the half-

I. Batarseh, A. Harb, *Power Electronics*,
https://doi.org/10.1007/978-3-319-68366-9_7

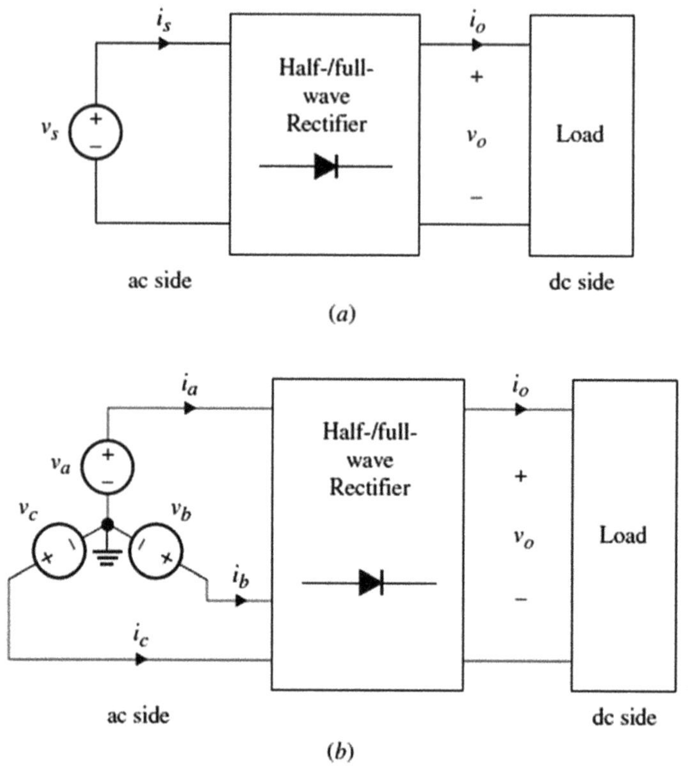

Fig. 7.1 Block diagram representation for (**a**) single-phase and (**b**) three-phase rectifier circuits

and full-bridge configurations as shown in the general block diagram representation in Fig. 7.1.

The analyses of rectifier circuits will include resistive, inductive, and capacitive loads, as well as loads with dc sources. We should point out that rectification is a term generally used for converting ac to dc with power flowing unidirectionally to the load, whereas inversion is a term used when dc power is converted to ac power. In diode and SCR circuits (SCR will be discussed in Chap. 8), inversion refers to the process in which power flow is from the dc side to the ac side. Another important point is that the circuits in this chapter are also known as naturally commutating or line-commutating converters since the diodes are naturally turned off and on by being connected directly to a single- or three-phase ac line supply.

7.2 Single-Phase Rectifier Circuits

7.2.1 Resistive Load

Figure 7.2a shows the simplest single-phase, half-wave rectifier circuit under a resistive load. The diode conducts only when the source voltage v_s is positive, as shown in Fig. 7.2b. Throughout this chapter and the next, the source voltage v_s is given by $v_s(t) = V_s \sin \omega t$, where V_s is the peak voltage of the source and ω is its frequency. At $t = T/2$, when v_s becomes negative, the diode seizes to conduct since it does not allow a negative current to flow through it.

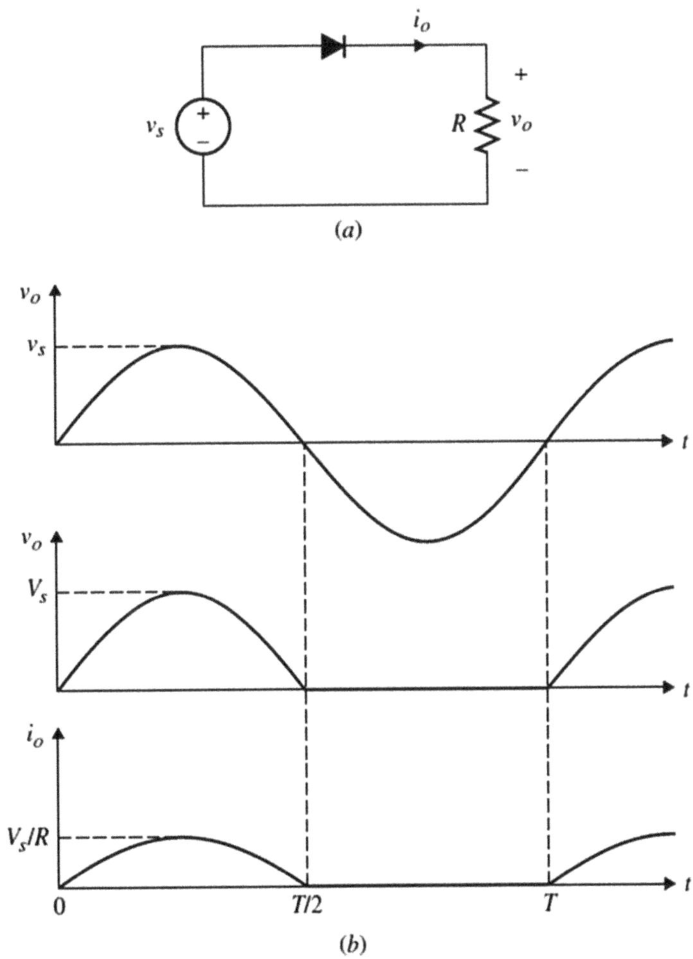

Fig. 7.2 (**a**) Half-wave rectifier circuit (**b**) voltage and current waveforms

The average value of the output voltage, V_o, which represents the dc component of the output signal, v_o, is calculated by evaluating the following integral over one period, T,

$$\begin{aligned} V_o &= \frac{1}{T}\int_0^T v_o(t)dt \\ &= \frac{1}{T}\int_0^{T/2} V_s \sin \omega t dt \\ &= \frac{V_s}{\pi} \end{aligned} \tag{7.1}$$

The average load current is given by:

$$\begin{aligned} I_o &= \frac{V_o}{R} \\ &= \frac{V_s}{\pi R} \end{aligned} \tag{7.2}$$

We observe that the output voltage waveform contains an ac component (ripple) at the same frequency as the ac source; hence, the output dc component is only 32% of the peak of the applied voltage ($V_o = 0.32V_s$). Because of its low dc output and high-ripple voltages, half-wave rectifier circuits are not practical and limited to low-voltage applications.

Example 7.1

Consider a half-wave rectifier circuit of Fig. 7.2a with a resistive load of 25 Ω and a 60 Hz ac source of 110 V rms.

(a) Calculate the average values of v_o and i_o.
(b) Calculate the rms values of v_o and i_o.
(c) Calculate the average power delivered to the load.
(d) Repeat part (c) by assuming that the source has a resistance of 60 Ω.

Solution

(a) The average values of v_o and i_o are given by:

$$\begin{aligned} V_o &= \frac{V_s}{\pi} \\ &= \frac{\sqrt{2}(110\text{ V})}{\pi} \\ &= 49.52\text{ V} \end{aligned}$$

and

$$I_o = \frac{V_o}{R}$$
$$= 1.98\ \text{A}$$

(b) The rms value of the output voltage is expressed as:

$$\begin{aligned} V_{o,\text{rms}} &= \sqrt{\frac{1}{T}\int_0^{T/2} v_o^2 dt} \\ &= \sqrt{\frac{1}{T}\int_0^{T/2} V_s^2 \sin^2\omega t\, dt} \end{aligned} \tag{7.3}$$

Substituting for $\sin^2\omega t = (1 - \cos t)/2$, Eq. (7.3) gives $V_{o,\text{rms}} = 77.78$ V. The rms current is given by:

$$I_{o,\text{rms}} = \frac{V_{o,\text{rms}}}{R}$$
$$= 3.11\ \text{A}$$

(c) The average power delivered to the load over one cycle is obtained from the following relation:

$$\begin{aligned} P_o &= \frac{1}{T}\int_0^T i_o v_o\, dt \\ &= \frac{1}{\text{TR}}\int_0^{T/2} v_o^2\, dt \\ &= \frac{V_{o,\text{rms}}^2}{R} \\ &= 242\ \text{W} \end{aligned}$$

(d) With a 60 Ω source resistance, R_s, the average power is given by:

$$P_o = \frac{RV_{o,\text{rms}}^2}{(R + R_s)^2}$$

$$=20.93\ \text{W}$$

Exercise 7.1
Calculate the efficiency of the circuit of Example 7.1 for parts (c) and (d).

Answer: 100 %, 54.5 %.

A more practical circuit is the full-wave rectifier circuit shown in Fig. 7.3a. Its typical waveforms are shown in Fig. 7.3b.

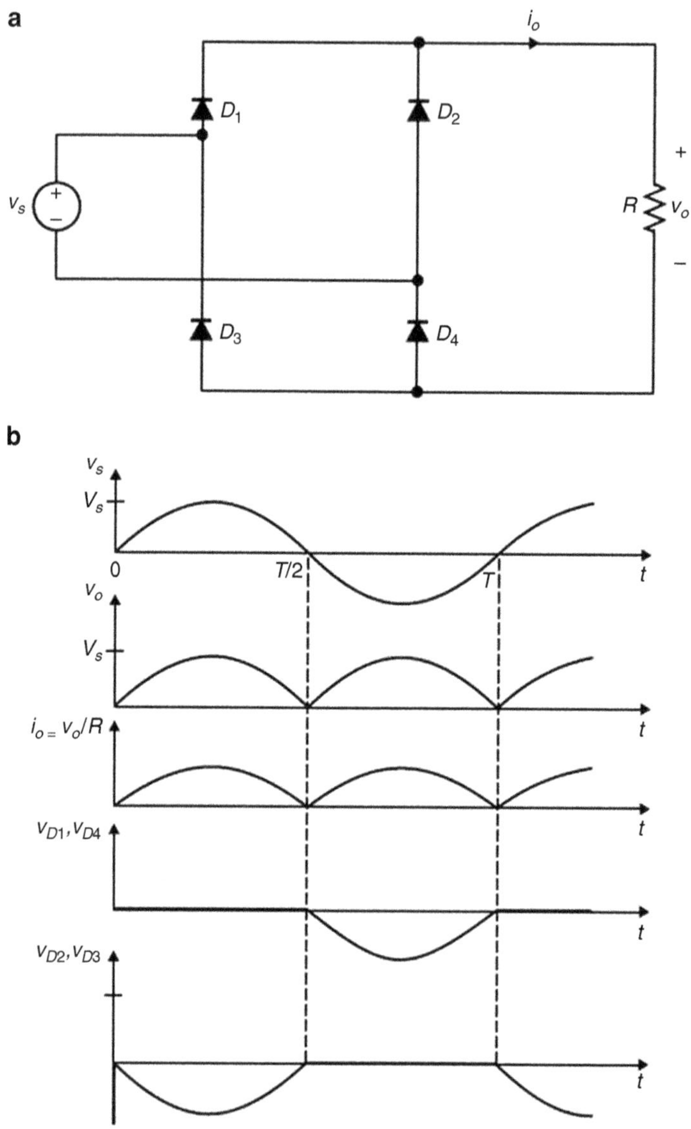

Fig. 7.3 (**a**) Full-wave rectifier with resistive load, (**b**) current and voltage waveforms

The circuit operates as follows: When $v_s > 0$, D_1 and D_4 turn on and D_2 and D_3 turn off. The voltage across D_2 and D_3 is $-v_s$, which keeps them off. At $t = T/2$, the negative source voltage stops D_1 and D_4 from conducting and causes D_2 and D_3 to begin conducting, until $t = T$ where the cycle repeats. The average output voltage and average output current are twice that of the half-wave under a resistive load.

Exercise 7.2
Calculate the rms and average diode current for $v_s = 120\sin(2\pi 60t)$ V and $R = 25\Omega$ in Fig. 7.2a.

Answer: 3.4 A rms, 2.16 A.

Exercise 7.3
Repeat Example 7.1 for the full-wave rectifier circuit given in Fig. 7.3a

Answer: 99 V, 3.96 A, 110 V rms, 4.4 A rms, 484 W, 41.87 W.

7.2.2 Inductive Load

An increase in the conduction period of the load current of Fig. 7.2 can be achieved by adding an inductor in series with the load resistance as shown in Fig. 7.4a, resulting in a half-wave rectifier circuit under an inductive load. This means the load current flows not only during $v_s > 0$ but also for a portion of $v_s < 0$. The diode is kept in the *on-state* by the inductor's voltage, which offsets the negative voltage of $v_s(t)$. The load current exists between $T/2$ and T, but never for the entire period, regardless of the inductor size. This can be easily explained by assuming that the diode conducts for the entire period. Consequently, the output voltage v_o must equal v_s, since the diode voltage is zero. This can only occur when the load current is alternating. This is clearly a contradiction, and there must be a time in which the diode stops conducting. Fig. 7.4b shows the steady-state waveforms for the circuit of Fig. 7.4a. The equivalent circuit mode for $t > t_2$ and $t < t_2$ are shown in Fig. 7.4c, d, respectively, where t_2 is the time when D stops conducting.

In steady state, the average value of the inductor voltage is zero. This is indicated by the equal positive and negative areas of v_L in Fig. 7.4b. At $t = t_1$, the load current reaches the peak at which $v_L = 0$. The average value of v_o is calculated from Fig. 7.4b as follows:

$$\begin{aligned} V_o &= \frac{1}{T}\int_o^{t_2} V_s \sin \omega t \, dt \\ &= \frac{V_s}{2\pi}(1 - \cos \omega t_2) \end{aligned} \tag{7.4}$$

The times t_1 and t_2 can be determined from the output current equation to be derived next. A mathematical expression for the output current, i_o, can be obtained by considering the equivalent circuit for $0 < t < t_2$ as shown in Fig. 7.4c. The first-order differential equation that mathematically represents the above circuit is given by:

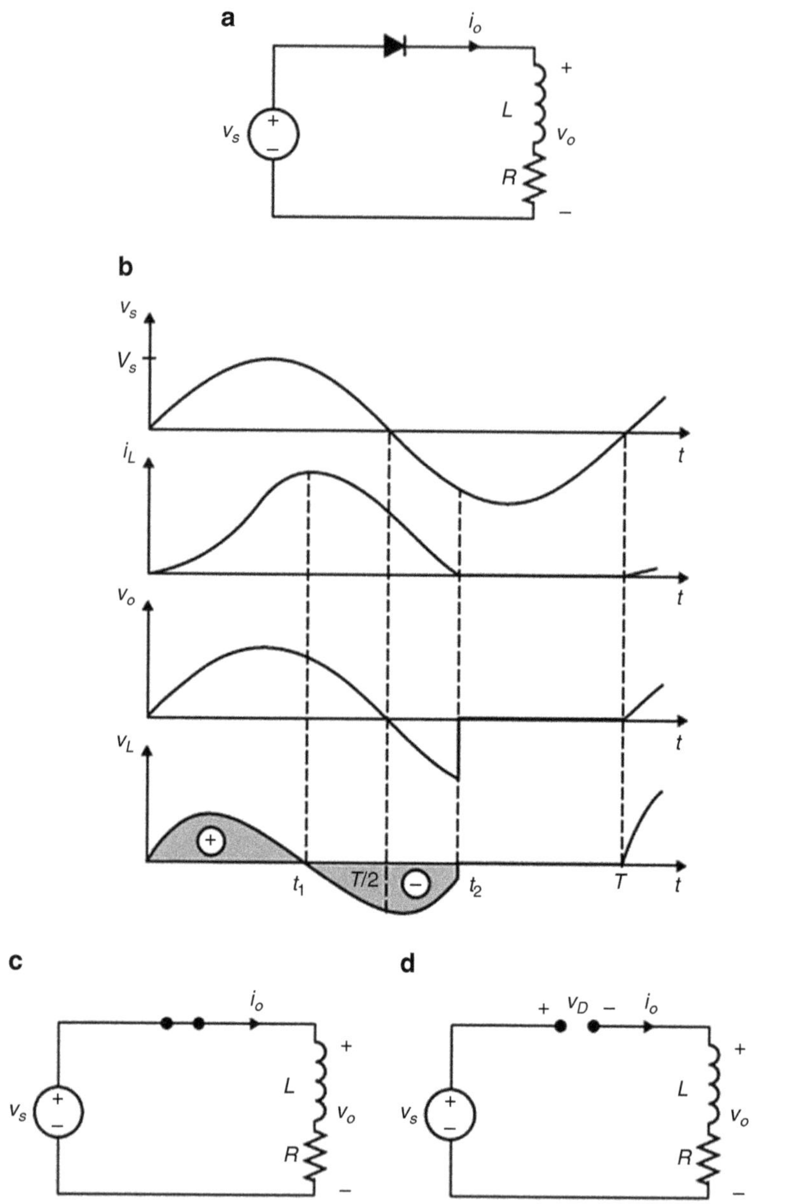

Fig. 7.4 (**a**) Half-wave rectifier circuit with inductive load (**b**) current and voltage waveforms (**c**) equivalent circuit for $t > t_2$ (**d**) equivalent circuit for $t < t_2$

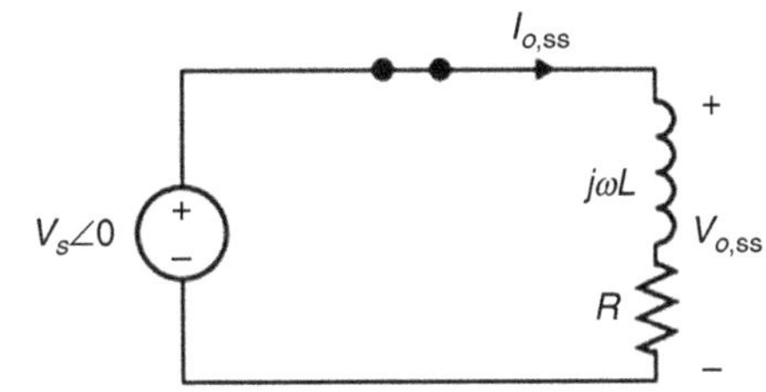

Fig. 7.5 Equivalent phasor circuit

$$\frac{di_o}{dt} + \frac{R}{L} i_o = \frac{1}{L} V_s \sin \omega t \tag{7.5}$$

Solving Eq. (7.5) for i_o requires obtaining both the transient,$i_{o,\mathrm{tr}}(t)$, and the steady-state, $i_{o,\mathrm{ss}}(t)$, solutions, as given by:

$$i_o(t) = i_{o,\mathrm{tr}}(t) + i_{o,\mathrm{ss}}(t)$$

The transient current response, also known as the *natural response*, is obtained by setting the forced excitation,v_s, to zero, resulting in the following current equation:

$$i_{o,\mathrm{tr}}(t) = I_{o,i} e^{-t/\tau} \tag{7.6}$$

where

$I_{o,i}$ Initial value of $i_{o,\mathrm{tr}}(t)$ at $t = 0$.
τ Circuit time constant equals L/R.

The steady-state or *forced* response is obtained by using the equivalent phasor circuit shown in Fig. 7.5. The phasor current is given by:

$$I_{o,\mathrm{ss}} = \frac{V_s \angle 0^\circ}{|Z| \angle \theta} = \frac{V_s}{|Z|} \angle -\theta \tag{7.7}$$

where

$$|Z| = \sqrt{(\omega L)^2 + R^2}$$
$$\theta = \tan^{-1} \frac{\omega L}{R}$$

In the time domain, we obtain the steady-state output current response as follows:

$$i_{o,\mathrm{ss}}(t) = \frac{V_s}{|Z|} \sin(\omega t - \theta) \tag{7.8}$$

The total solution for i_o is obtained by adding Eqs. (7.6) and (7.8) to yield:

$$i_o(t) = I_{o,i}e^{-t/\tau} + \frac{V_s}{|z|}\sin(\omega t - \theta) \tag{7.9}$$

The initial value, $I_{o,i}$, is obtained from evaluating Eq. (7.9) at $t = 0^+$ with $i_o(0^+) = 0$ to give:

$$I_{o,i} = \frac{V_s \omega L}{|Z|^2} \tag{7.10}$$

The time at which the diode stops conducting, $t = t_2$, must be determined in order to evaluate the average load voltage given in Eq. (7.4). Evaluating Eq. (7.9) at $t = t_2$, with $i_o(t_2) = 0$, we obtain the following expression in terms of t_2:

$$t_2 = -\tau \ln\left[-\sqrt{1 + \left(\frac{R}{\omega L}\right)^2}\sin(\omega t_2 - \theta)\right] \tag{7.11}$$

Since t_2 is in both sides of Eq. (7.11), a numerical analysis method is required to solve for t_2. By normalizing the average output voltage as $V_{no} = V_o/V_s$, we can obtain a plot of this normalized value verses a normalized load resistance (R_n), which is defined by:

$$R_n = \frac{\omega L}{R} \tag{7.12}$$

In terms of R_n, the average output voltage is expressed by:

$$V_{no} = \frac{1 - \cos\omega t_2}{2\pi} \tag{7.13a}$$

where

$$\omega t_2 = -R_n \ln\left[-\sqrt{1 + \frac{1}{R_n^{\ 2}}}\sin(\omega t_2 - \theta)\right] \tag{7.13b}$$

$$\theta = \tan^{-1} R_n \tag{7.13c}$$

Note that, unlike the resistive half-wave rectifier circuit, this circuit experiences load regulation since the average output voltage varies with the change of the load.

One important design criterion in power converters is what is known as *load regulation*. This parameter gives a measure of how the output voltage deviates from its desired value when the load changes. If we assume the output voltage takes on a minimum value at low-load, V_{ol}, and a maximum value at a full-load, V_{of}, then the load regulation (LR_{load}) is defined by:

$$LR_{\mathrm{load}} = \left| \frac{V_{\mathrm{of}} - V_{\mathrm{ol}}}{V_{\mathrm{of}}} \right| \tag{7.14}$$

Another important parameter is known as *line regulation*, LR_{line}, which gives a measure of the output variation when the line input voltage changes. The line regulation is defined as:

$$LR_{\mathrm{line}} = \frac{V_{o,\max} - V_{o,\min}}{V_{o,\min}} \tag{7.15}$$

Where $V_{\mathrm{o,\,max}}$ and $V_{\mathrm{o,\,min}}$ are the average output voltages at maximum and minimum line voltages, respectively.

Example 7.2
Consider the inductive-load half-wave rectifier circuit of Fig. 7.4a with the $\mathrm{v_s} = 120$ sin377t, L = 60 mH, and R = 12 Ω. Determine (a) V_{o}, (b) I_{o}, (c) $\mathrm{LR_{line}}$ if V_{s} is changed by ±20% and (d) LR_{load} if R decreases by 50%.

Solution

(a) The normalized average output voltage for $\omega L/R \approx 1.9$ is obtained from equations (7.11) and (7.12b).

$$V_{\mathrm{no}} = 0.213$$

Then we obtain the average output voltage as:

$$\begin{aligned} V_{\mathrm{o}} &= V_{\mathrm{s}} V_{\mathrm{no,\,ave}} \\ &= 25.56 \end{aligned}$$

(b) The average output current is:

$$\begin{aligned} I_{\mathrm{o}} &= \frac{V_{\mathrm{o}}}{R} \\ &= 2.13\ \mathrm{A} \end{aligned}$$

(c) The minimum and maximum peak input voltages are given as:

$$\begin{aligned} V_{\mathrm{s,\,min}} &= 96\ \mathrm{V} \\ V_{\mathrm{s,\,max}} &= 144\ \mathrm{V} \end{aligned}$$

The corresponding average outputs are:

$$\begin{aligned} V_{\mathrm{o,\,min}} &= 20.44\ \mathrm{V} \\ V_{\mathrm{o,\,min}} &= 30.67\ \mathrm{V} \end{aligned}$$

The line regulation is obtained from Eq. (7.14) as $LR_{\mathrm{line}} = 50\,\%$.

(d) With the load resistance decreasing by 50%, the new value is $R = 6\ \Omega$, resulting in $\omega L/R \approx 3.77$ and $V_{no} = 0.156$. This gives a new value of the average output voltage equal to 18.72 V. The resultant load regulation is $LR_{load} = 26.73\%$. This is considered a very poor load regulation.

A more useful half-wave rectifier circuit is one in which an additional diode is added across the output as shown in Fig. 7.6a. Unlike Fig. 7.4a, the load current in Fig. 7.6a is permitted to be continuous and the output voltage cannot go negative.

A careful investigation of Fig. 7.6a shows that D_1 and D_2 cannot conduct simultaneously. For example, if we assume both diodes are *on*, then the average output voltage will be zero and V_s simultaneously. This is not possible. In the steady state condition, this circuit has two topological modes of operation:

Mode 1: D_1 is on and D_2 is off:

This mode occurs when v_s is positive, causing D_1 to become forward biased and D_2 reveres biased. The equivalent circuit model is shown in Fig. 7.6b. The time interval for this mode is $0 < t < T/2$.

The equation that governs this mode is given by

$$\frac{di_o}{dt} + \frac{R}{L} i_o = \frac{1}{L} v_s(t) \tag{7.16}$$

The solution of Eq. (7.15) is similar to Eq. (7.5), which has the following complete solution:

$$i_{o1}(t) = \frac{V_s}{|Z|} \sin(\omega t - \theta) + I_1 e^{-t/\tau} \tag{7.16}$$

Where

$$|Z| = \sqrt{(\omega L)^2 + R^2}$$
$$\tau = \frac{L}{R}$$
$$\theta = \tan^{-1}\frac{\omega L}{R}$$

and I_1 is constant and to be determined from the initial conditions. The subscript of the load current i_{o1} denotes the current i_s of Mode 1.

Mode 2: D_1 is off and D_2 is on:

This topological mode occurs for $v_s < 0$ at which D_1 becomes reverse biased and D_2 turns on to pick up the load current. This way the load current commutates between D_1 and D_2 to maintain its continuity. The equivalent circuit is shown in Fig. 7.6c, and the governing equation for this topology is given by:

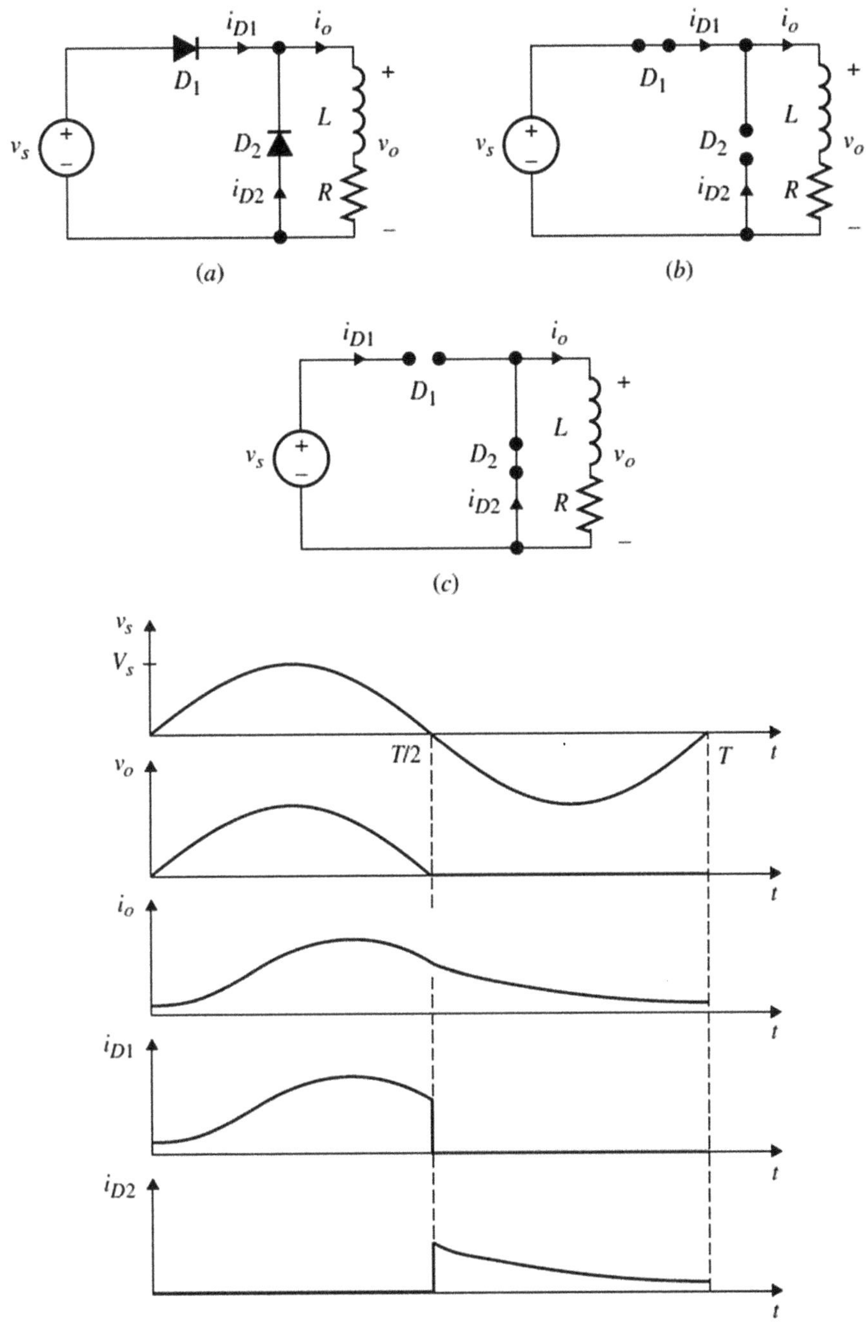

Fig. 7.6 (**a**) Half-wave rectifier with a flyback diode (**b**) Mode 1: D_1 is *on* and D_2 is *off* (**c**) Mode 2: D_1 is *off* and D_2 is *on* (**d**) voltage and current waveforms

$$\frac{di_o}{dt} + \frac{R}{L} i_o = 0 \tag{7.17}$$

The solution of Eq. (7.17) consists of only transient response as shown in Eq. (7.18):

$$i_{o2}(0) = I_2 e^{-(t-T/2)/\tau} \tag{7.18}$$

where I_2 is constant and can be solved for, along with I_1, from the following relations that must hold in the steady state:

$$i_{o1}(0) = i_{o2}(T) \tag{7.19a}$$
$$i_{o1}(T/2) = i_{o,2}(T/2) \tag{7.19b}$$

The left side currents of Eqs. (7.19a) and (7.19b) are obtained from Eq. (7.16), and the right side currents are obtained from Eq. (7.18). Consequently, we obtain the solution for I_1 and I_2 as follows:

$$I_1 = \frac{V_s}{|Z|} \sin\theta \frac{1 + e^{-T/2\tau}}{1 - e^{-T/\tau}} \tag{7.20}$$

$$I_2 = \frac{V_s}{|Z|} \sin\theta \frac{1 + e^{-T/2\tau}}{1 - e^{-T/\tau}} \tag{7.21}$$

The waveforms for i_o, i_{D1}, i_{D2},and v_o are shown in Fig. 7.6d.

If the load inductance is assumed infinitely large, then there will exist no ripple in the load current, $i_o(t)$, and both Eqs. (7.16) and (7.18) become equal, and a constant load current is obtained as shown in the waveforms of Fig. 7.7.

The load resistance according to the following relation limits the value of the dc output current,

$$\begin{aligned} I_o &= \frac{V_o}{R} \\ &= \frac{V_s}{\pi R} \end{aligned} \tag{7.22}$$

One important observation from Fig. 7.7 is that the source current is a pulsating square wave, an unattractive feature in power electronic circuits. The above circuit is also known as a *current-doubler* since the average output current is twice the average input current.

Exercise 7.4

Consider the circuit shown in Fig. 7.6a. Determine V_o, I_o, and the inductor currents at $t = 0$ and at $t = T/2$ in the steady state by using $v_s = 220 \sin 377t$, $R = 15\Omega$ use (a) $L = 20$ mH, (b) $L = 200$ mH, and (c) L infinitely large.

Answer: 70 V, 4.67 A, 16 mA, 5.9 A, 3.2 A, 6 A, 4.67 A, 4.67 A.

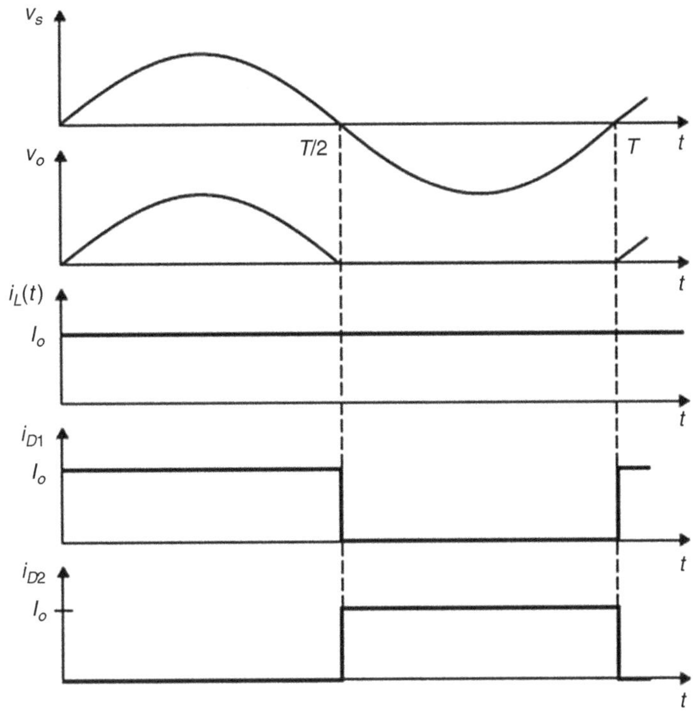

Fig. 7.7 Waveform for Fig. 7.6a with an infinite load inductance

Exercise 7.5
Determine the rms currents of diodes D_1 and D_2 in Fig. 7.3a.

Answer: 3.4 A rms.

A final observation to be made about the half-wave inductive-load rectifier circuit is that unlike in Fig. 7.4a, the inductor current exists for the entire period. As a result, rectifier circuits of Figs. 7.4a and 7.6a are said to operate in the discontinuous and continuous conduction modes, respectively.

7.2.3 Capacitive Load Rectifiers

In some applications in which a constant output voltage is more desired, a series inductor is replaced by a parallel capacitor as shown in the half-wave rectifier circuit of Fig. 7.8a.

Like the single-diode inductive-load half-wave rectifier, the circuit of Fig. 7.8a can be analyzed by considering the two topological modes resulting from the conducting states of the diode. Let Mode 1 corresponds to the time interval in

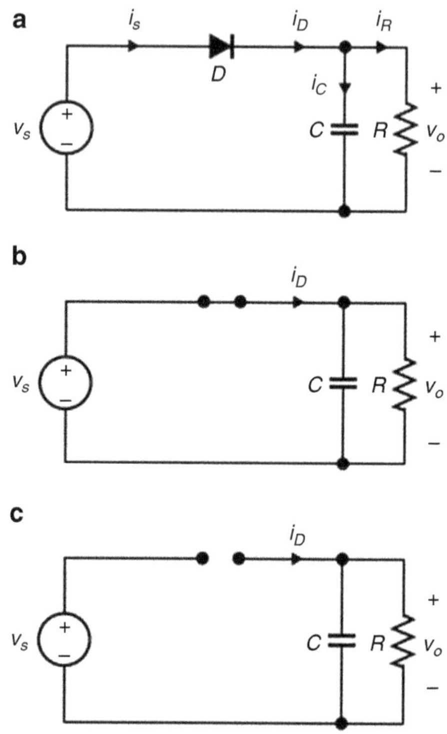

Fig. 7.8 (**a**) Half-wave rectifier with capacitive load (**b**) equivalent circuit mode when the diode is *ON* (**c**) equivalent circuit mode when the diode is *OFF*

which the diode conducts, resulting in the equivalent circuit shown in Fig. 7.8b. Figure 7.8c shows the equivalent circuit when the diode is off, representing Mode 2.

The circuit will be in Mode 1 as long as the capacitor voltage is lower than the input voltage, v_s. If we assume the diode starts conducting at $t = t_1$, then the next time it will turn off to enter Mode 2 of operation is when $t = \frac{T}{4}$, at which the input voltage reaches its peak. The diode will stay off until the capacitor and the input voltages become equal again at $t = T + t_1$. Figure 7.9 shows the steady-state waveforms for the output voltage and the source current.

The analytical expressions for v_o and i_s are given as follows. For $t_1 < t < T/4$:

$$v_o = V_s \sin \omega t \tag{7.23}$$

$$i_s(t) = \frac{V_s}{R} \sin \omega t + \omega C V_s \cos \omega t \tag{7.24}$$

For $T/4 < t < T + t_1$:

$$v_o(t) = V_s e^{-(t-T/4)/\tau} \tag{7.25}$$

$$i_s(t) = 0 \tag{7.26}$$

where the time constant is given by $\tau = RC$.

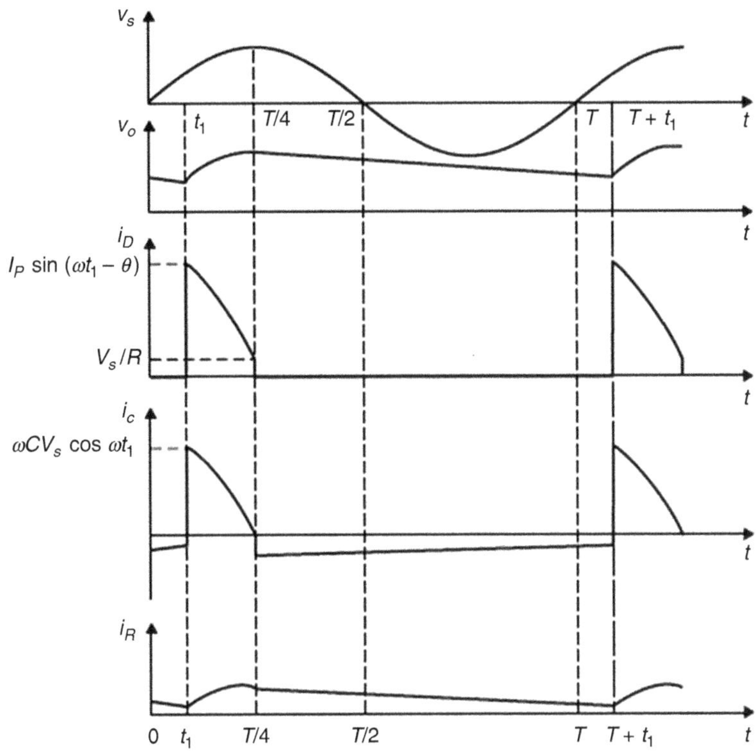

Fig. 7.9 Steady-state waveforms for Fig. 7.8a

The value of $t = t_1$ can be determined by equating Eqs. (7.23) at $t = t_1$ and (7.25) at $t = t_1 + T$. This yields the following relation,

$$\sin \omega t_1 - e^{(t_1 + 3T/4)\tau} \tag{7.27}$$

Equation (7.27) may be evaluated iteratively to obtain t_1. The average value of the output voltage is determined from the following:

$$\begin{aligned} V_o &= \frac{1}{T}\int_{t_1}^{T+t_1} v_o(t)\,dt \\ &= \frac{1}{T}\left[\int_{t_1}^{T/4} V_s \sin \omega t\,dt + \int_{T/4}^{T+t_1} V_s e^{-(t-T/4)/\tau}\,dt\right] \\ &= \frac{V_s}{2\pi}\left[\cos \omega t_1 + \tau\omega\left(1 - e^{(t_1+3T/4)/\tau}\right)\right] \\ &= \frac{V_s}{2\pi}\left[\cos \omega t_1 + \tau\omega\left(1 - e^{(t_1+3T/4)/\tau}\right)\right] \end{aligned}$$

Example 7.3
Consider the Half-Wave Rectifier Circuit of Fig. 7.8a with $v_s(t) = 20 \sin 2\pi 60t$, $R = 10\ \text{k}\Omega$, and $C = 47\ \mu\text{F}$. (a) Sketch the Waveforms for v_o, i_R, i_D, and i_C and (b) Determine the Output Voltage Ripple.

Solution

(a) First we need to find the time t_1 at which the diode turns *ON*. Using $\tau = \text{RC} = 470$ ms and $T = 16.67$ ms, through several iteration, Eq. (7.27) gives $t_1 \approx 3.48$ ms. At this time, the output voltage is 19.3 V and the resistor current is 1.93 mA.

At $t = T/4 = 4.17$ ms, $v_o = 20$ V and $i_R = 2$ mA. When the diode turns *OFF*, $i_C = -i_R = -2$ mA. However, while the diode is conducting i_C is given by:

$$\begin{aligned} i_C(t) &= \omega C V_s \cos \omega t \\ &= 354 \ \cos \omega t \text{ mA} \end{aligned}$$

At $t_1 \approx 3.48$ ms, $i_C = 354 \cos 377(3.48\text{ms}) = 90.6$ mA; hence, $i_D = i_R + i_C = 1.93 + 90.6 = 92.53$ mA. Finally, the output voltage at T or at $t = 0$ is obtained from Eq. (7.25) as follows:

$$\begin{aligned} v_o &= V_s e^{-(t-T/4)/\tau} \\ &= 20e^{-3T/\tau} \\ &= 20e^{-3(16.67)/(4)(470)} \\ &= 19.5 \text{ V} \end{aligned}$$

The waveforms are shown in Fig. 7.10.

(b) The output voltage ripple, V_r, is given by:

$$\begin{aligned} V_r &= v_{o,\max} - v_{o,\min} \\ &= v_o(T/4) - v_o(t_1) \\ &= 20 - 19.3 \\ &= 0.7 \end{aligned}$$

Exercise 7.6
Calculate the average output voltage for Example 7.3.

Answer: 19.68 V.

7.2.4 *Voltage Source in the dc Side*

Figure 7.11a shows a half-wave rectifier circuit with a voltage source in series with the load resistance. This constant voltage may be considered as an *emf* source located in the load side as in the case of dc machine circuit modeling.

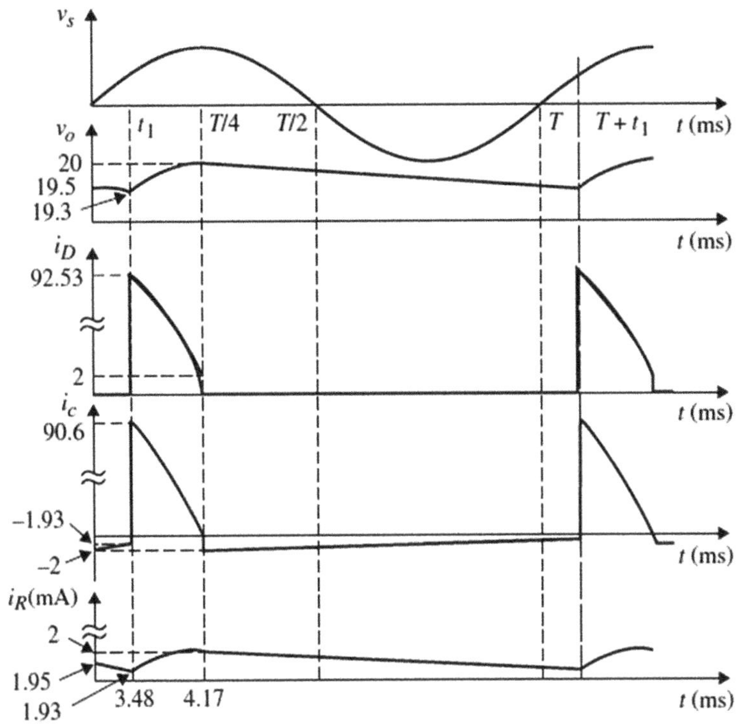

Fig. 7.10 Steady-state waveforms for Example 7.3

Let us assume that the diode starts conducting at $t = t_1$. This time occurs when $v_s = V_{DC}$. Furthermore, the diode stops conducting at $t = t_2$ when the current i_o becomes zero. Typical waveforms are shown in Fig. 7.11b. Please see problem 7.19 to derive the equation for the load current during the conduction period $t_1 < t < t_2$.

7.3 The Effect of the ac-Side Inductance

7.3.1 *Half-Wave Rectifier with Inductive Load*

Let us consider the half-wave diode rectifier circuit by including an ac-side inductance, which is inevitable in all ac systems. Studying the effect of the line inductance on the behavior of diode circuits is important since in practical rectifier systems connected directly to the line voltage, the transformer leakage inductance and the line parasitic inductance might cause line voltage and current disturbances and could affect the average power drawn from the mains. Moreover, an external inductance might also be added in some applications in order to limit di/dt or dv/dt.

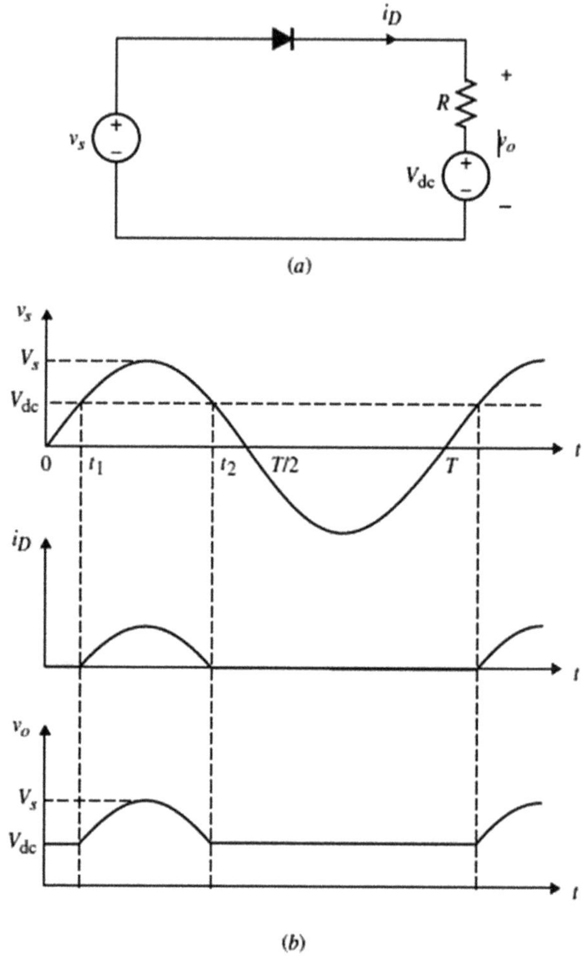

Fig. 7.11 (**a**) Half-wave rectifier with voltage source in the dc side (**b**) voltage and current waveforms

For simplicity, we normally assume that the ac-side inductance is finite and may be represented by a discrete value, L_s, in series with the source voltage given in the block diagram representation in Fig. 7.1. Figure 7.12a shows the circuit for the half-wave rectifier including L_s and under an inductive load. In the following analysis, we will assume that $L/R \gg T/2$ so that the load current, i_o, is constant. This assumption is valid since in many applications, the load inductance is very much larger than the ac-side inductance.

The behavior of the circuit can be easily analyzed by assuming that one of the diodes, D_1, is conducting for some time during the positive cycle of $v_s(t)$, while D_2 is off. Since the current in D_1 is constant, then the voltage across L_s is zero, and the voltage across D_2 is positive. Hence, our assumption is valid. The equivalent circuit under this mode is shown in Fig. 7.12b. During this mode, we have the following current and voltage values:

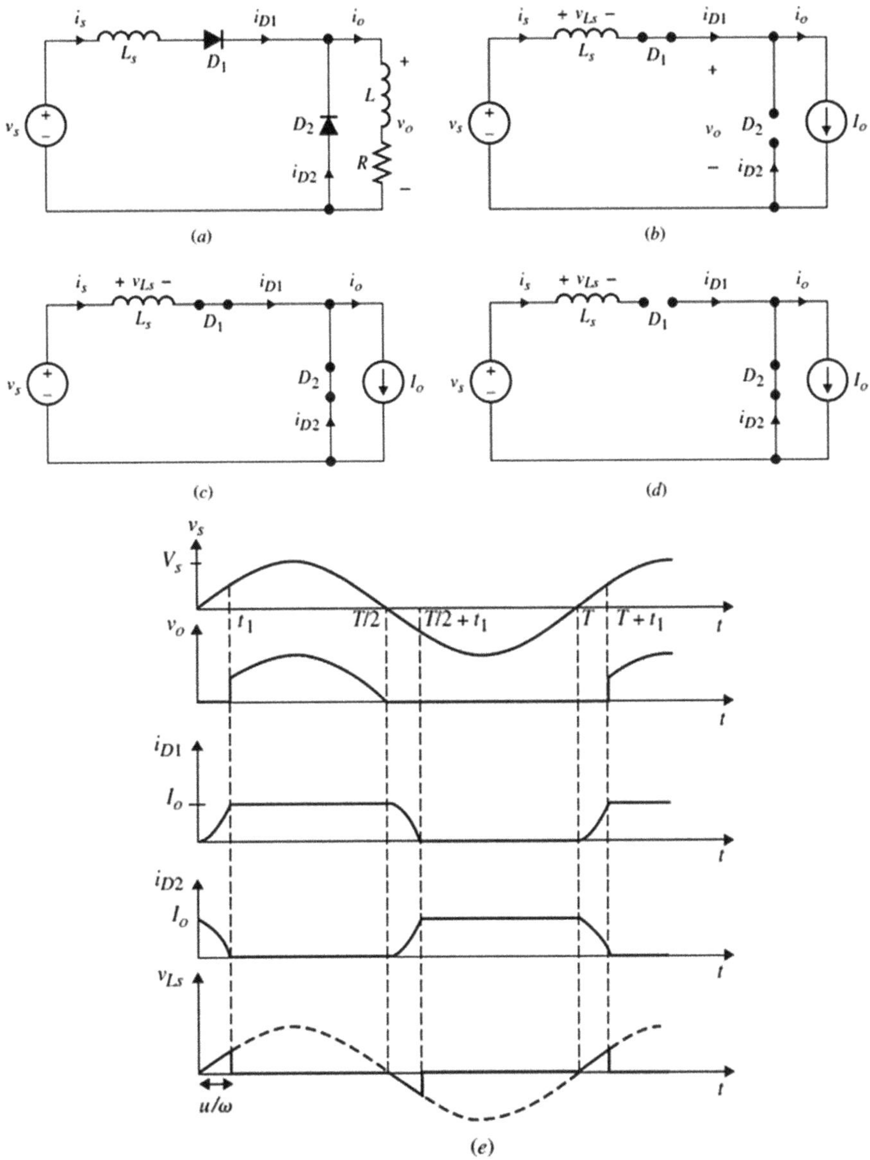

Fig. 7.12 (**a**) Half-wave rectifier with ac-side inductance. (**b**) Mode 1: D_1 is *on* and D_2 is *off* during $v_s(t) > 0$. (**c**) Mode 2: commutation mode—D_1 and D_2 are *on* while $v_s(t) < 0$. (**d**) Mode 3: D_1 is *off* and D_2 is *ON* while $v_s(t) < 0$. (**e**)The waveforms for v_s, v_o, i_{D1}, i_{D2}, and v_{Ls}

$$i_{D1} = i_s(t) = I_o \quad i_{D2} = 0 \;\; v_{Ls} = 0 \quad v_o = v_s$$

At $t = T/2$, $v_s(t)$ starts to become negative, causing D_1 to stop conducting. However, since the current in D_1 is the same as the inductance current, which is not allowed to change instantaneously, D_2 turns on in order to maintain the inductor current's continuity. During this overlapping time, when both diodes are conducting, $i_s(t)$ changes from $+I_o$ to zero, while $i_{D2}(t)$ changes from zero to $+I_o$. This is known as a *commutation process* during which currents are moved from one branch of the circuit to another branch. This is why the ac-side inductance, L_s, is known as the *commutation inductance*. The circuit mode of operation is referred to as commutation mode as shown in Fig. 7.12c. The waveforms for v_s, v_o, i_{D1}, i_{D2},and v_{Ls} are shown in Fig. 7.12e.

During Mode 2, the following equations hold:

$$\begin{aligned} v_{Ls} &= v_s(t) = L_s \frac{di_s}{dt} \\ i_{D1} &= i_s(t) \\ i_{D2} &= I_o - i_{D1} \\ v_o &= 0 \end{aligned}$$

The initial condition for $i_s(t)$ at $t = T/2$ is I_o. Using the above v_{Ls} equation with the given initial condition, we obtain the following input current integral relation:

$$i_s(t) = \frac{1}{L_s} \int_{T/2}^{t} v_{Ls}(t)\, dt + I_o \tag{7.28}$$

Substituting for $v_{Ls}(t) = v_s(t) = V_s \sin \omega t$ in the integral and solving for $i_s(t)$, we obtain:

$$i_s(t) = I_o - \frac{V_s}{L_s \omega}(1 + \cos \omega t) \qquad T/2 \le t \le t_1 + T/2 \tag{7.29}$$

At the end of commutation period, $t = t_1 + T/2$, $i_s(t)$ becomes zero, forcing D_1 to turn *OFF* at zero current and D_2 remains forward biased, carrying the load current as shown in the circuit of Mode 3 given in Fig. 7.12d. In this mode we have the following current and voltage equations:

$$i_{D1} = i_s(t) = 0 \quad i_{D2} = I_o \quad v_{Ls} = 0 \quad v_o = 0$$

To solve for the commutation time, t_1, or the commutation angle, $u = \omega t_1$, we evaluate $i_s(t)$ at $t = T/2 + t_1$ and set it to zero. The load current and t_1 are given by:

$$I_o = \frac{V_s}{L_s \omega}(1 - \cos \omega t_1) \tag{7.30}$$

$$t_1 = \frac{1}{\omega} \cos^{-1}\left(1 - \frac{L_s \omega I_o}{V_s}\right) \tag{7.31}$$

The average output voltage is expressed by:

$$V_o = \frac{1}{T}\int_{t_1}^{T/2} v_s(t)\,dt \\ = \frac{V_s}{\pi}\left(1 - \frac{\omega L_s I_o}{2V_s}\right) \tag{7.32}$$

If we normalize the output voltage by V_s and the load current by Z_o, where $Z_o = \omega L_s$ is the characteristic impedance of the ac-line, we have the following normalized output quantities:

$$V_{no} = \frac{V_o}{V_s} \\ = \frac{1}{\pi}\left(1 - \frac{I_{no}}{2}\right) \tag{7.33}$$

$$I_{no} = \frac{Z_o I_o}{V_s} \\ = 1 - \cos u \tag{7.34}$$

or u is given by:

$$u = \cos^{-1}(1 - I_{no}) \tag{7.35}$$

The maximum voltage across L_s, $V_{Ls,\max}$, occurs at $t = t_1$ which is given by:

$$V_{Ls,\max} = v_s(t_1) \\ = V_s \sin \omega t_1$$

Normalizing $V_{Ls,\max}$ by V_s and substituting the normalized current I_{no} into the above relation, we obtain:

$$V_{nLs,\max} = \sqrt{(2 - I_{no})I_{no}} \tag{7.36}$$

Figure 7.13 shows the curve for $V_{nLs,\max}$ vs. I_{no}, and Fig. 7.14a, b show I_{no} vs. u and I_{no} vs. V_{no} respectively.

Exercise 7.7
Consider the circuit of Fig. 7.12a with $L_s = 0.05$ mH, $=20\ \Omega$, $I_o = 1$ A, and $v_s(t) = 100 \sin 2\pi 60t$. Find (a) the average output voltage, (b) the commutation intervals in

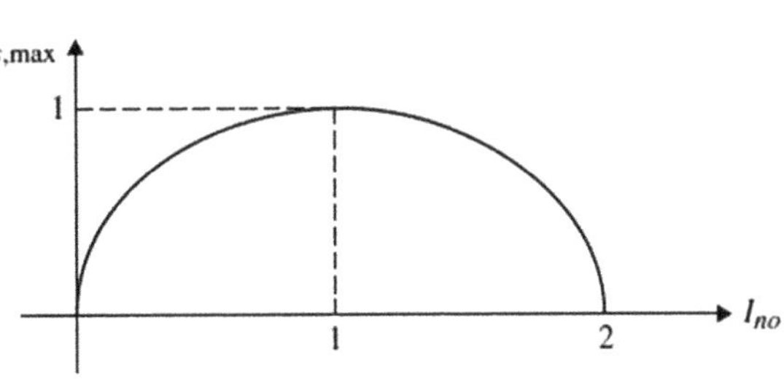

Fig. 7.13 Normalized maximum inductor voltage versus normalized output current

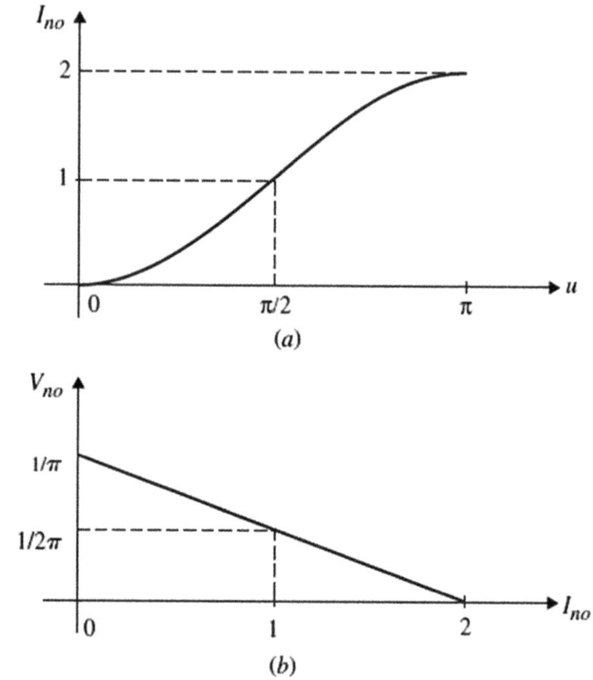

Fig. 7.14 (**a**) Normalized output current versus commutation angel (**b**) normalized output voltage versus normalized output current

seconds and degrees, (c) the rms diode currents, (d) peak inverse diode voltages, and (e) the peak voltage across the inductor L_s.

Answer: 31.83 V, 51.5 μs, 1.1°, 585.26 A rms, 100 V, 1.94 V.

7.3.2 *Half-Wave Rectifier with Capacitive Load*

Consider the half-wave rectifier circuit with a capacitive load including an ac-side inductance as shown in Fig. 7.15a. Again, to simplify the analysis, we assume $RC \gg T/2$, so the output voltage is constant. Its equivalent circuit is given in Fig. 7.15b.

Let us assume first that the diode is off and the input voltage is negative. As shown in Fig. 7.15a, for the diode to conduct, its voltage must be positive. At $t = t_1$, $v_s = V_o$ and the diode turns *ON*. The waveforms are shown in Fig. 7.15c, and the voltage across L_s is given by:

$$v_{Ls} = L_s \frac{di_s}{dt}$$
$$= v_s(t) - V_o$$

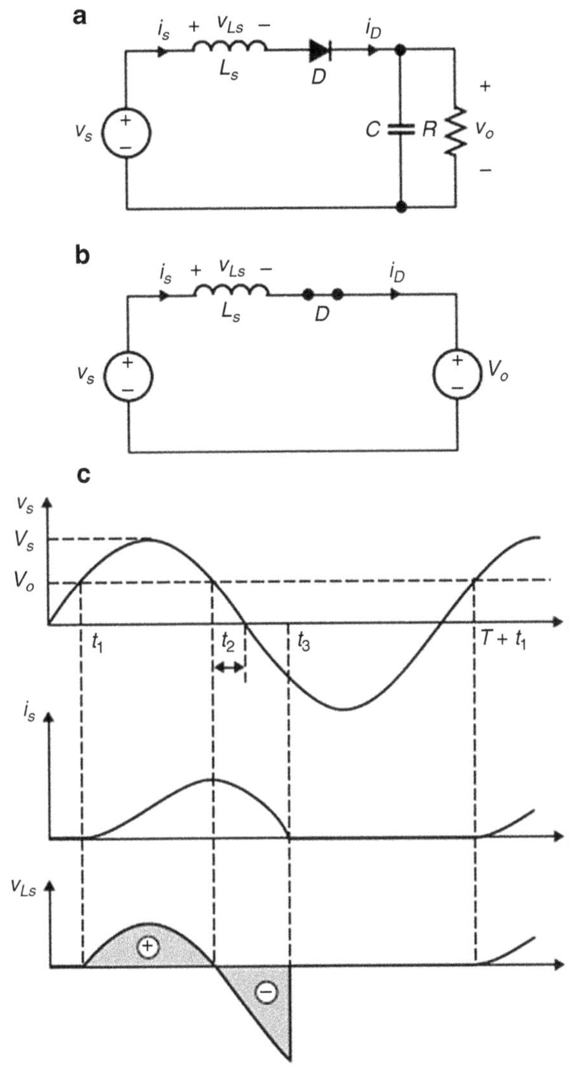

Fig. 7.15 (**a**) Half-wave capacitive load circuit with commutative inductance (**b**) equivalent circuit with RC ≫ $T/2$ (c) current and voltage waveforms

Since the initial inductor current value at $t = t_1$ is zero, the solution for $i_s(t)$ is given by:

$$i_s(t) = \frac{V_s}{\omega L_s}(\cos \omega t_1 - \cos \omega t) - \frac{V_o}{L_s}(t - t_1) \tag{7.37}$$

The peak inductor current occurs when $v_{Ls} = 0$, i.e., when $v_s = V_o$. Let us assume this occurs at $t = t_2$,

$$I_{s,peak} = i_s(t_2) = \frac{V_s}{\omega L_s}\left((\cos\omega t_1 - \cos\omega t) - \frac{V_o}{L_s}(t - t_1)\right) \tag{7.38}$$

where t_2 is obtained from:

$$V_o = V_s \sin\omega t_2 \tag{7.39}$$

Also at $t=t_1$, when the diode first turns on, we have:

$$V_o = V_s \sin\omega t_1 \tag{7.40}$$

where $t_1 + t_2 = T/2$.

As long as $i_s(t)$ is positive, D remains *ON*, and at $t=t_3$, i_s reaches zero, causing D to turn *OFF*. The current at $t=t_3$ is given by:

$$i_s(t_3) = \frac{V_s}{\omega L_s}\left((\cos\omega t_1 - \cos\omega t_3) - \frac{V_o}{L_s}(t_3 - t_1)\right) = 0 \tag{7.41}$$

Equations (7.39), (7.40), and (7.41) can be used to solve for t_1, t_2, and t_3 numerically. The average output current is given by:

$$I_o = \frac{1}{T}\int_{t_1}^{t_3} i_s(t)\,dt = \frac{V_s}{2\pi L_s}\left((t_3 - t_1)\cos\omega t_1 - \frac{\sin\omega t_3}{\omega} + \frac{\sin\omega t_1}{\omega}\right) - \frac{V_o}{2L_oT}(t_3 - t_1)^2$$

Normalizing I_o and V_o, and using the relation (7.39) for $t=t_2$, we obtain

$$I_{no} = \frac{\omega L_s I_o}{V_s} = \frac{1}{2\pi}\left(\alpha\cos\omega t_1 - \frac{\sin\omega t_3}{\omega} + \frac{\sin\omega t_1}{\omega}\right) - \frac{V_o}{2L_sT}(t_3 - t_1)^2 \tag{7.42}$$

where $\alpha = \omega(t_c - t_1)$ which is the diode conducting time.

Figure 7.16 shows I_{no} vs. α and Fig. 7.17 shows V_{no} vs. α.

The effect of L_s on the variation of the average output voltage is shown in Fig. 7.18.

The normalized inductor peak current is given by:

$$I_{nLs,peak} = \frac{\omega L_s I_{Ls,peak}}{V_s} = (\cos\omega t_1 - \cos\omega t_2) - \alpha V_{no}$$

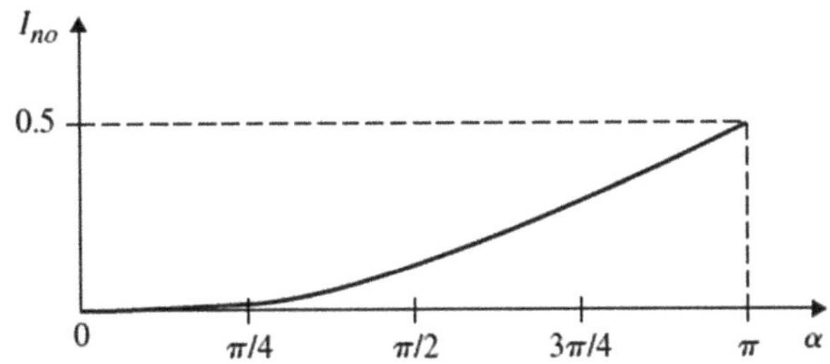

Fig. 7.16 Normalized load current versus the conduction angle

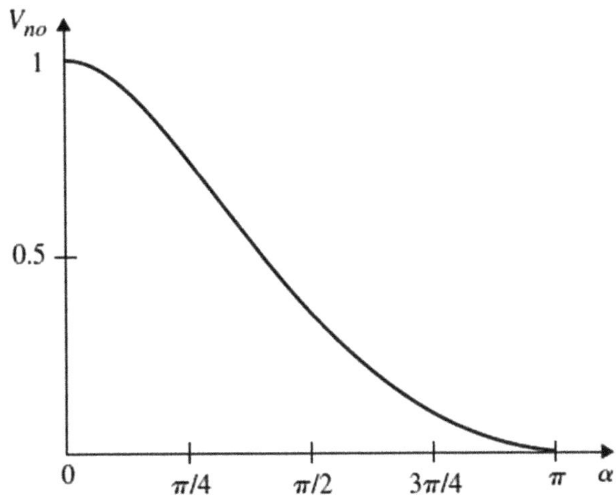

Fig. 7.17 Normalized load voltage versus the conduction angle

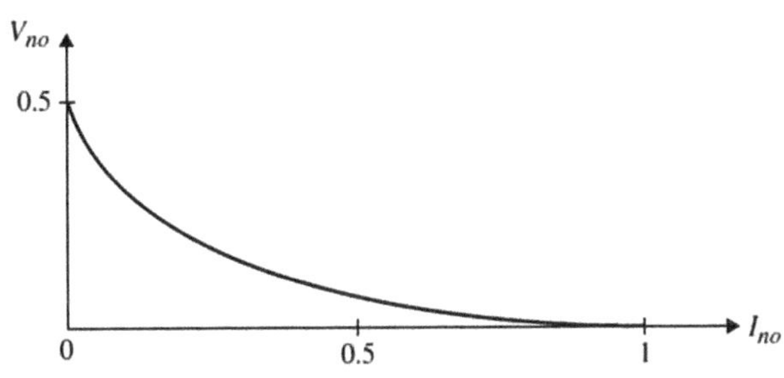

Fig. 7.18 Normalized load voltage versus normalized load current

A plot of the normalized peak inductor current vs. α under different load conditions is shown in Fig. 7.19.

Example 7.4

Consider the Half-Wave Rectifier of Fig. 7.15b with $v_s(t) = 25 \sin 377t, V_o = 20$ V and $L_s = 2$ mH. Determine: V_{no}, I_{no}, α and $I_{nLs,\,peak}$.

Solution

First find the value of t_1

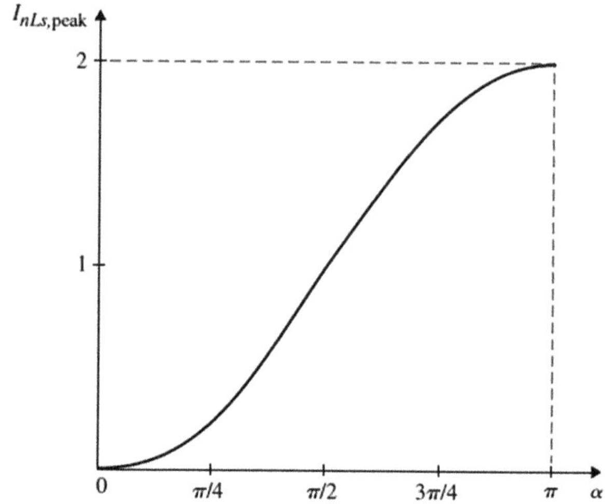

Fig. 7.19 Normalized peak inductor current versus the conduction angle

$$t_1 = \frac{1}{\omega} \sin^{-1}\left(\frac{V_o}{V_s}\right) = 2.46 \text{ ms}$$

Through some numerical iterations, we can find that $t_3 = 7.66$ ms. The normalized average output current can be obtained theoretically as follows:

$$I_o = \frac{V_s}{2\pi L_s}\left((t_3 - t_1)\cos\omega t_1 - \frac{\sin\omega t_3}{\omega} + \frac{\sin\omega t_1}{\omega}\right) - \frac{V_o}{2L_o T}(t_3 - t_1)^2$$
$$= 0.99 \text{ A}$$

Hence,

$$I_{no} = \frac{\omega L_s I_o}{V_s} = 0.03$$

The other parameters are given as

$$\alpha = \omega(t_3 - t_1) = 1.96 \text{ rad/s}$$
$$V_{no} = \frac{V_o}{V_s} = 0.8 \text{ V}$$

$$I_{nLs,\text{peak}} = = (\cos\omega t_1 - \cos\omega t_2) - \alpha V_{no} = 0.17$$

Exercise 7.8

Determine the diode rms current for Example 7.4.

Answer: 2.078 A.

7.3.3 Full-Wave Rectifiers with an Inductive Load

The effect of the commutation inductance on the waveforms and parameters of the full-wave rectifier shown in Fig. 7.20a is investigated next. As before, we will assume $L/R \gg T/2$, resulting in a constant load current, I_o. The typical waveforms are shown in Fig. 7.20b. The behavior of the circuit can be discussed similarly to the case for the half-wave rectifier circuit of Fig. 7.6a. We first assume that D_1 and D_4 are conducting during the positive cycle of $v_s(t)$. Since I_o is constant, the voltage across L_s is zero, and $v_o(t) = v_s(t)$ as shown in Fig. 7.21a, which is represented as Mode 1. Notice that the voltages across D_2 and D_3 are negative and equal to $-v_s(t)$. As long as $v_s(t)$ is positive, D_2 and D_3 remain *off*. The second topological mode

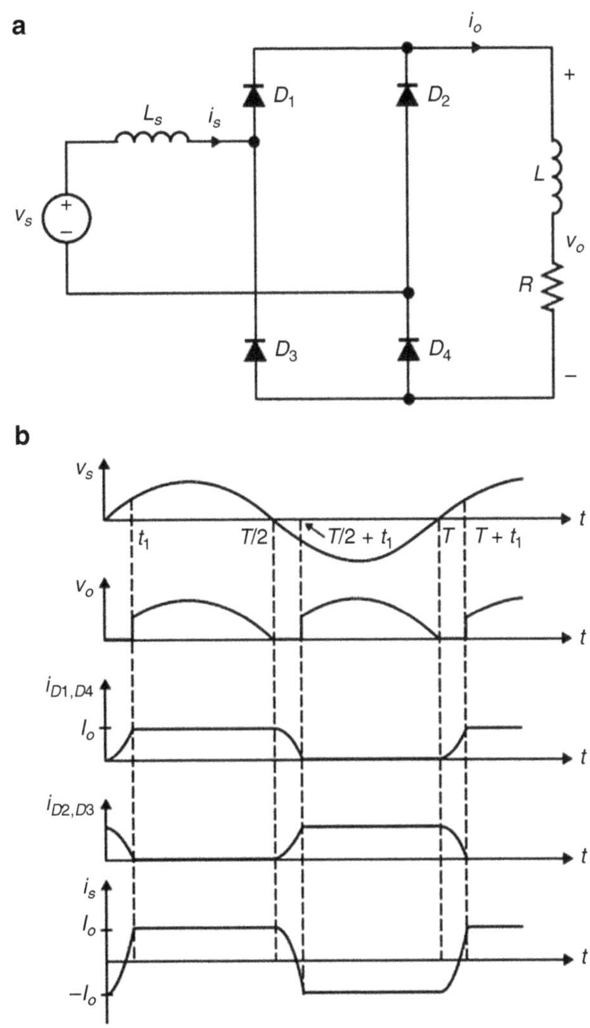

Fig. 7.20 (**a**) Full-wave rectifier with ac-side inductance (**b**) current and voltage waveforms

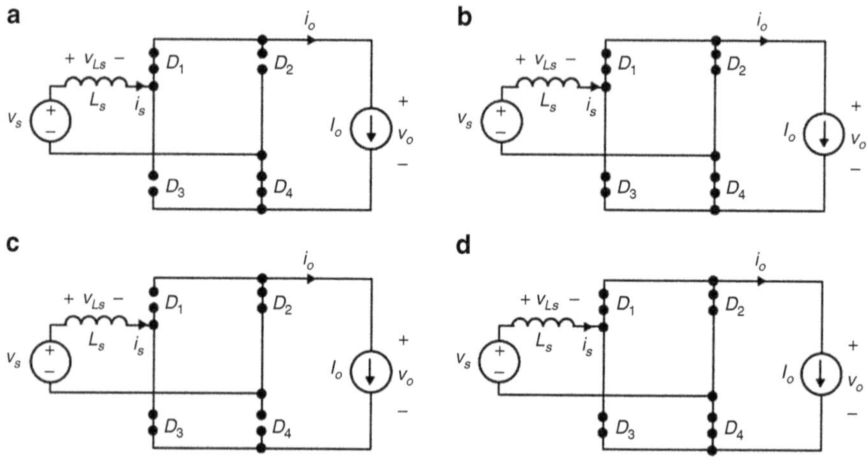

Fig. 7.21 Circuit modes of operation for Fig. 7.20 (**a**) Mode 1: equivalent circuit for $v_s(t) > 0$ (**b**) Mode 2: equivalent circuit for $v_s(t) < 0$ and during diode commutation period (**c**) Mode 3: equivalent circuit for $v_s(t) < 0$ (**d**) Mode 4: equivalent circuit for $v_s(t) > 0$ and during diode commutation period

starts at $t = T/2$ when $v_s(t)$ becomes negative. At this time, D_2 and D_3 start conducting, while D_1 and D_4 remain *on* in order to maintain the continuity of the inductor current. Figure 7.21b shows this mode when all diodes are conducting. The following current and voltage relations are given in this mode:

$$\begin{aligned} v_{Ls} &= L_s \frac{di_s}{dt} \\ v_o &= 0 \\ i_{D1} &= i_{D4} = \frac{I_o + i_s}{2} \\ i_{D2} &= i_{D3} = \frac{I_o - i_s}{2} \end{aligned} \tag{7.43}$$

Using the initial inductor value of $i_s(T/2) = +I_o$ in the above equation for $i_s(t)$, we obtain:

$$i_s(t) = I_o - \frac{V_s}{L_s \omega}(1 + \cos \omega t) \tag{7.44}$$

At $t = t_1 + T/2$, we have $i_s(t\_1 + T/2) = -I_o$. This gives the following relation for the commutation period:

$$\cos \omega t_1 = 1 - \frac{2L_s \omega I_o}{V_s} \tag{7.45}$$

In terms of normalized values, $u = \omega t_1$ is given by:

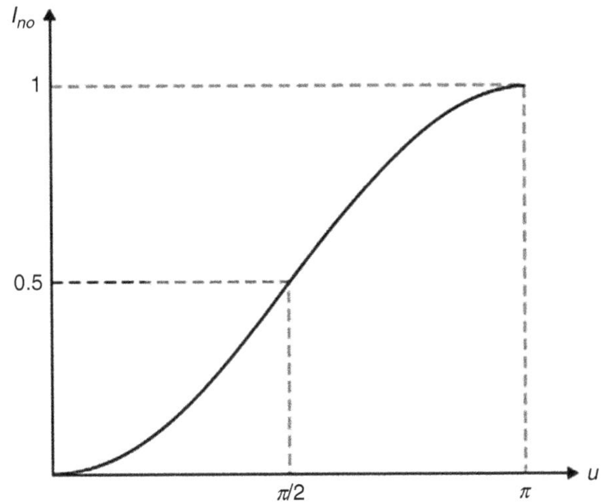

Fig. 7.22 Normalized output current vs. commutation angle

$$u = \cos^{-1}(1 - 2I_{\text{no}})$$

Figure 7.22 shows the curves for the normalized output current as a function of the commutation angle. The larger the commutation inductance, the less the average output current. At $t = T/2 + t_1$, the currents in D_1 and D_4 become zero and the currents in D_2 and D_3 reach I_o, as shown in the topology of Mode 3 given in Fig. 7.21c. The output voltage is equal to $v_s(t)$. This mode remains until $t = T$, when the voltage becomes positive again. Since $v_s(t) > 0$, D_1 and D_4 turn on and D_2 and D_3 remain on until the current commutates from D_2 and D_3 to D_1 and D_4 at the end of the commutation period u, and the circuit enters Mode 4 as shown in Fig. 7.21d. The current and voltage equations are the same as those for Mode 2. The initial condition for $i_s(t)$ is $i_s(T) = -I_o$, and the solution for $i_s(t)$ is given by

$$i_s(t) = -I_o + \frac{V_s}{L_s\omega}(\cos\omega t - 1) \tag{7.46}$$

We notice from the above relations that the average output voltage, inductor peak current, and rms value of $i_s(t)$ are directly related to the normalized load current.

To illustrate the source of L_s in the full-wave rectifier, we refer to Fig. 7.23. This is a center-tap transformer including a leakage inductance L_K reflected to the primary side. The equivalent circuit for Fig. 7.23a can be redrawn as shown in Fig. 7.23b where the inductance is reflected to the primary side. The voltage relations are given by:

$$v_1 = \frac{N_2}{N_1} v_s$$

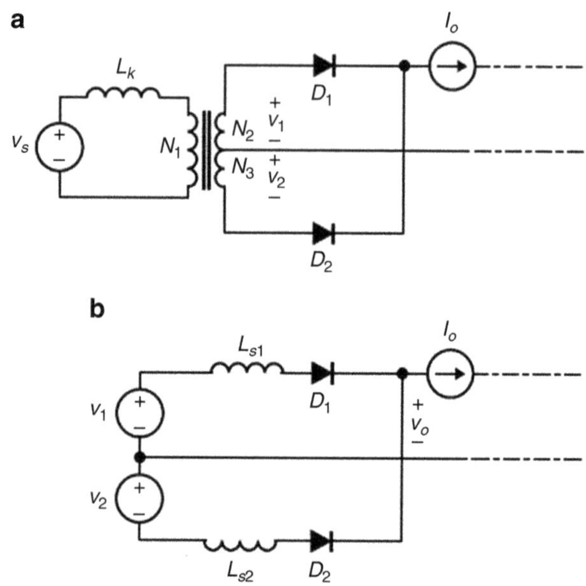

Fig. 7.23 (**a**) Full-wave center-tap transformer with leakage inductance and (**b**) simplified equivalent circuit

$$v_2 = \frac{N_3}{N_1} v_s$$

where $L_{s1} = (N_1/N_2)^2 L_k$ and $L_{s2} = (N_1/N_3)^2 L_k$.

Exercise 7.9

Consider the center-tap full-wave rectifier of Fig. 7.23a with $N_1 = 10$, $N_2 = 5$, $N_3 = 15$, $L_k = 10\ \mu H$, $I_o = 20\ A$, $v_s = 100 \sin 2\pi(60)t$. Assume the diodes have no forward voltage drops. (a) Determine the average output voltage and (b) repeat part (a) by assuming each diode has 1 V forward drop.

Answer: 63.61 V, 62.63 V.

7.4 Three-Phase Rectifier Circuits

Many industrial applications require high power that a single-phase system is unable to provide. Three-phase diode rectifier circuits are used widely in high-power applications with low output ripple. It is important that we understand the basic concepts of a three-phase rectifier circuit. In this section, we will cover both half- and full-wave rectifier circuits under resistive and high inductive loads.

7.4.1 Three-Phase Half-Wave Rectifier

Figure 7.24 shows the general configuration for an m-phase half-wave rectifier connected to a single load. The explanation of the circuit is quite simple since all diode cathodes are connected to the same point, creating a diode-OR arrangement. At any given time, the highest anode voltage will cause its corresponding diode to conduct, with all other diodes in the reverse-bias state. In other words, the output voltage will ride on the peak voltage at all times. Figure 7.25 shows four random sine functions and the output voltage.

The half-wave three-phase resistive load rectifier circuit is shown in Fig. 7.26a. We assume that the three-phase voltage source is a Δ-configuration with the three balance voltages are given by

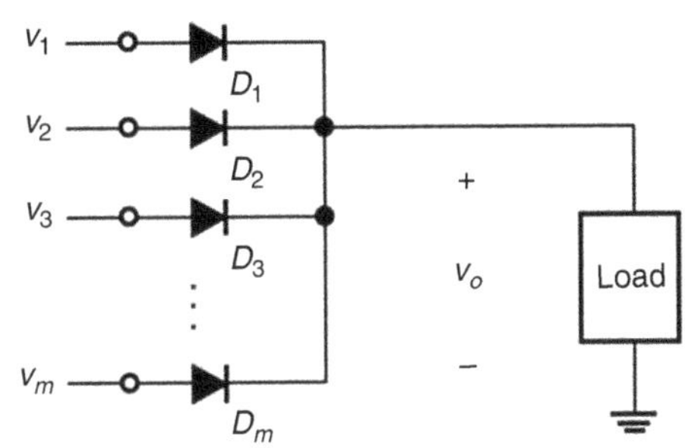

Fig. 7.24 m-phase half-wave rectifier connected to a single load

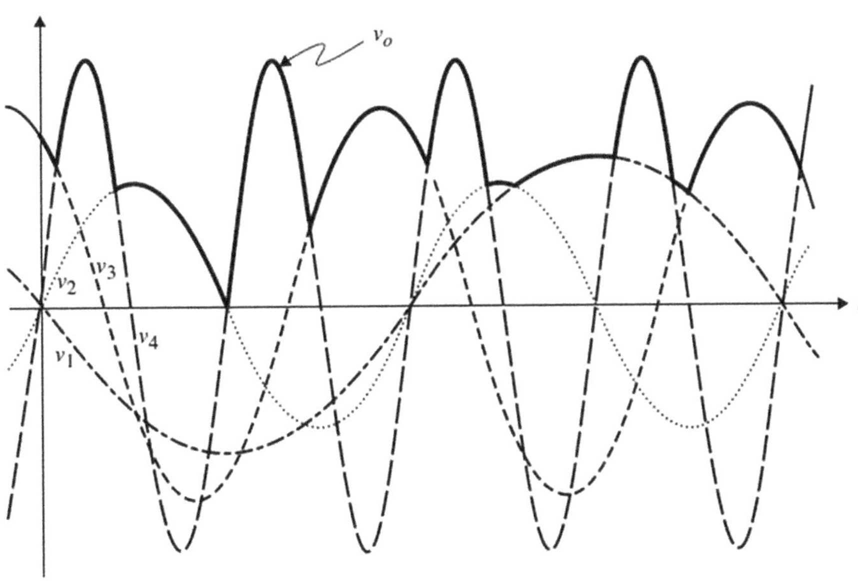

Fig. 7.25 Random four-phase sinusoidal voltages and the output voltage

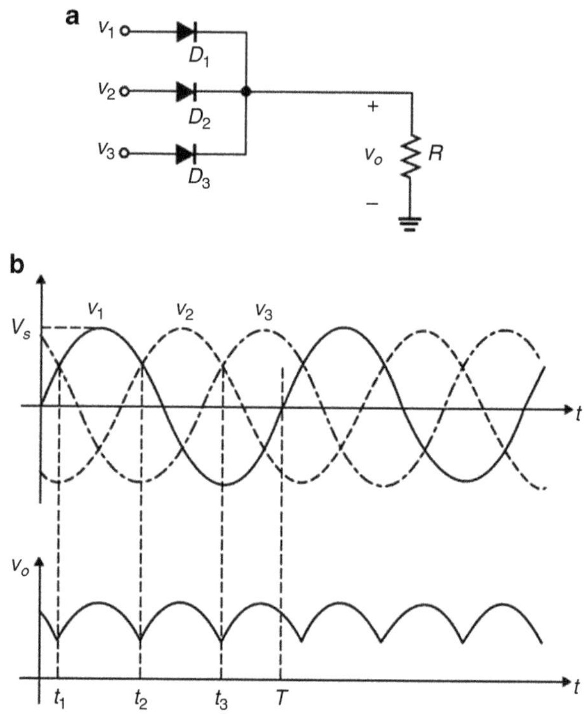

Fig. 7.26 (**a**) Half-wave three-phase resistive load rectifier (**b**) waveform for v_1, v_2, v_3, and the output voltage

$$v_1 = V_s \sin \omega t \qquad v_2 = V_s \sin\left(\omega t - 120^\circ\right) \qquad v_3 = V_s \sin\left(\omega t - 240^\circ\right)$$

Figure 7.26b shows the output voltage waveform. This circuit is also known as a *three-pulse rectifier* circuit. Here, the number of pulses refers to the number of voltage peaks in a given cycle.

The circuit works as follows: At $t = t_1$, we have $v_1 = v_3$. This occurs when $t_1 = T/12$, and at $t_2 = 5T/12$ we have $v_2 = v_1$. The average output voltage is given by:

$$\begin{aligned} V_o &= \frac{1}{T}\int_{t_1}^{t_2} v_o(t)\ dt \qquad (7.47) \\ &= \frac{3}{T}\int_{t_1}^{t_2} v_1(t)\ dt \\ &= \frac{3}{T}\int_{T/12}^{5T/12} v_1(t)\ dt \\ &= \frac{3\sqrt{3}}{2\pi} V_s \end{aligned}$$

The source voltage arrangement of v_1, v_2 and v_3 can also be of the *delta* (Δ) or the *wye* (Y) connection as shown in Fig. 7.27a, b, respectively. The voltages v_a, v_b, and v_c are the balanced, positive sequence phase voltages given by:

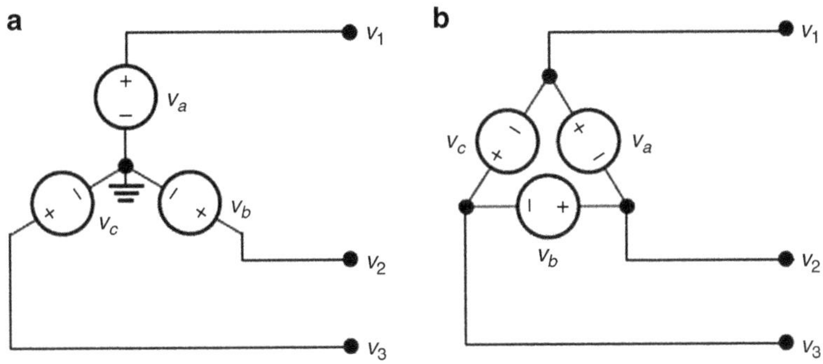

Fig. 7.27 Three-phase connections: (**a**) Y- and (**b**) Δ-connections

$$v_a = V_s \sin \omega t \qquad v_b = V_s \sin\left(\omega t - 120^\circ\right) \qquad v_c = V_s \sin\left(\omega t - 240^\circ\right)$$

The voltage set $v_1, v_2,$ and v_3 are line voltages. In the Δ-configuration, the line-to-line voltages are the same as the phase voltages, whereas the line-to-line voltages in Y-configuration are given by:

$$\begin{aligned} v_{12} &= v_1 - v_2 \\ &= v_a - v_b \\ &= v_{ab} \\ &= \sqrt{3}V_s \sin\left(\omega t + 30^\circ\right) \\ v_{23} &= v_2 - v_3 \\ &= v_b - v_c \\ &= v_{bc} \\ &= \sqrt{3}V_s \sin\left(\omega t - 90^\circ\right) \\ v_{31} &= v_3 - v_1 \\ &= v_c - v_a \\ &= v_{ca} \\ &= \sqrt{3}V_s \sin\left(\omega t + 150^\circ\right) \end{aligned}$$

The phase diagram representation for the phase and the line-to-line voltages is given in Figure 7.28.

Figure 7.29a, b show the equivalent circuit for a half-wave circuit under a highly inductive load and its waveforms, respectively.

The half-wave circuit with a capacitive load is shown in Fig. 7.30a, and its output waveform is shown in Fig. 7.30b. To obtain an expression for the output voltage ripple of Fig. 7.30b, we will make the assumption that the load time constant, RC, is much larger than $T/3$. Let $t = t_1$ be the time at which the diode D_1 turns *ON* when v_a is the most positive. At $t = T/4$, $v_a = V_s$ and the diode D_1 stop conducting; since the capacitor voltage at this instant is equal to the peak voltage V_s, all diodes remain *OFF* and the capacitor starts discharging through R. At $t = t_1 + T/3$, v_o becomes

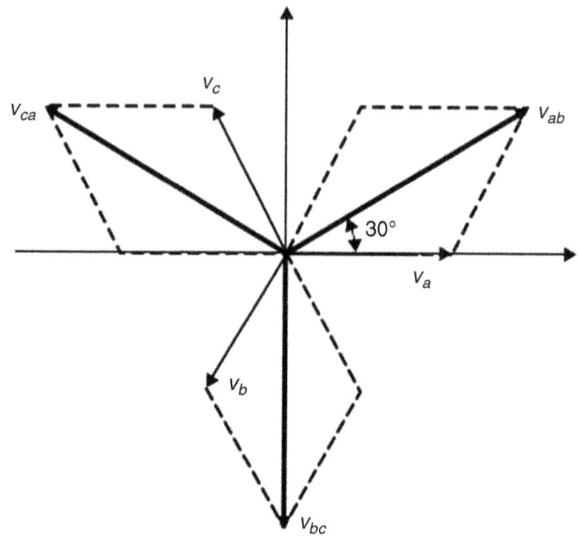

Fig. 7.28 Phasor diagram representation for phase and line-to-line three-phase voltages

equal to v_b and D_2 turns *ON*. Once again, the capacitor voltage increases with v_b until it equals V_s at which the cycle repeats. For $T/4 < t < T/3 + t_1$, the output voltage is given by:

$$v_o = V_s e^{-(t-T/4)/RC}$$

At $t = t_1 + T/3$, the output voltage equals v_b, i.e.

$$\begin{aligned} v_b(t_1 + T/3) &= V_s \sin\left(\omega(t_1 + T/3) - 120^\circ\right) \\ &= V_s e^{-(t_1 - T/12)/RC} \end{aligned}$$

The above equation can be rewritten as follows:

$$\sin \omega t_1 = e^{-(t_1 + T/12)/RC}$$

As in the case of the single-phase rectifier circuit, we can solve for t_1 numerically. The ripple voltage, V_r, is obtained from the following relation:

$$\begin{aligned} V_r &= V_s - v_b(t_1 + T/3) \\ &= V_s(1 - \sin \omega t_1) \end{aligned} \tag{7.48}$$

Using the approximation $t_1 \approx T/4$, the above equation becomes:

$$\begin{aligned} v_o(t_1 + T/3) &= V_s e^{-(T/4 + T/12)/RC} \\ &= V_s e^{-T/3RC} \end{aligned}$$

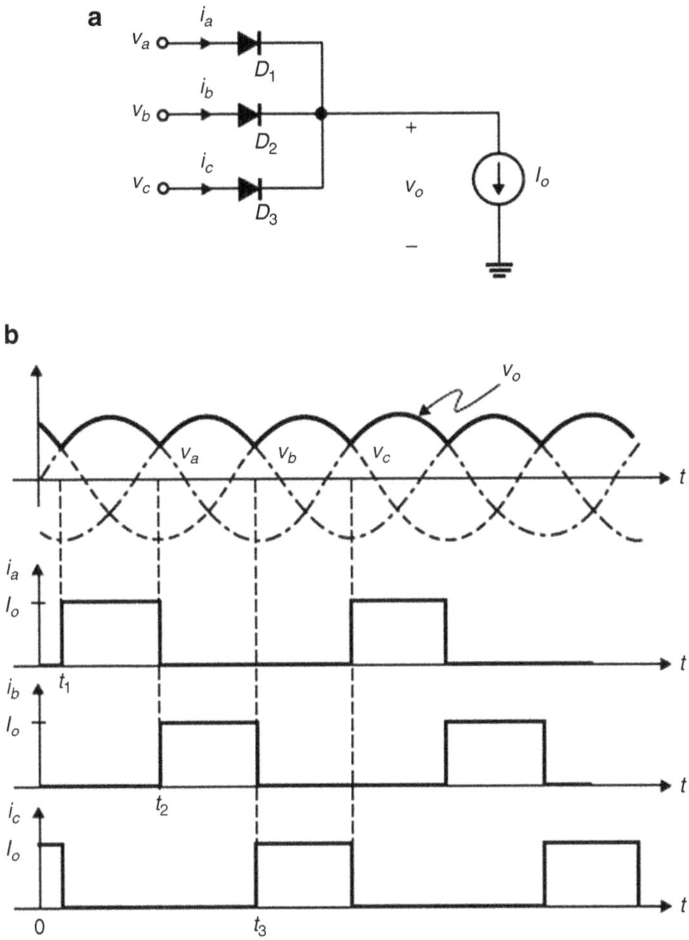

Fig. 7.29 (**a**) Half-wave three-phase rectifier with infinite load inductance (**b**) current and voltage waveforms

$$\cong V_s\left(1-\frac{T}{3RC}\right)$$

so the ripple is given by:

$$\begin{aligned} v_r &= V_s - v_o(t_1 + T/3) \\ &= \frac{T}{3RC}V_s \\ &= \frac{1}{3fRC}V_s \end{aligned} \tag{7.49}$$

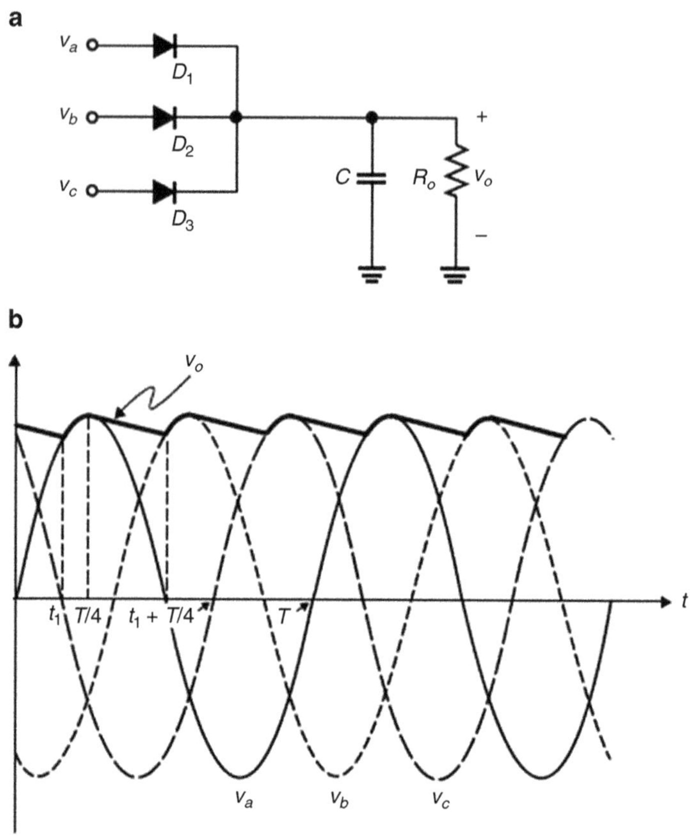

Fig. 7.30 (**a**) Three-phase, half-wave capacitive load rectifier (**b**) output voltage waveform

It can be shown that this ripple is one-third of the voltage ripple for the single phase circuit.

Exercise 7.10

Consider the three-phase rectifier shown in Fig. 7.30a with $f = 60$Hz and $R_o = 10$ kΩ. Design for C so that the output voltage ripple does not exceed 1% of the peak voltage.

Answer: $C \geq 55.6\ \mu\text{F}$.

It can be shown that the diode conduction time can be approximated by:

$$\frac{T}{2} - t_1 - \sqrt{\frac{2V_r}{V_s}} \tag{7.50}$$

Using the above expression, it is possible to obtain approximated average and rms current values for the diodes.

7.4.2 *Three-Phase Full-Wave Rectifiers*

Let us now consider the full bridge rectifier circuit including the commutation inductance. The full bridge is more common since it provides a large output voltage and less ripple. First let us consider the full-bridge circuit under a resistive load as shown in Fig. 7.31. Let us assume that v_a, v_b, and v_c are the three phase voltages in Fig. 7.32. The easiest way to approach the bridge rectifier circuit of Fig. 7.31 is to consider it as a combination of a positive commutating diode group D_1, D_2, and D_3 and a negative commutating diode group D_4, D_5, and D_6. Since no commutating inductance is included, at any given time only two diodes are conducting simultaneously, one from the positive group and the other from the negative group. To show that this is the case, let us redraw Fig. 7.31 as that shown in Fig. 7.33. It is clear to notice from Fig. 7.32 that only one diode from the upper group can be on during the positive cycle and only one diode among the lower group can be on during the negative cycle. The output voltage, v_o, is given by:

$$v_o = v_{o1} - v_{o2}$$

where v_{o1} and v_{o2} are the output voltages of the positive and negative commuting diode groups to ground, respectively.

The average output voltage is given by:

$$\begin{aligned} V_o &= \frac{1}{T/6} \int_{T/12}^{T/4} V_a \, dt \\ &= \frac{3\sqrt{3}}{\pi} V_s \end{aligned} \tag{7.51}$$

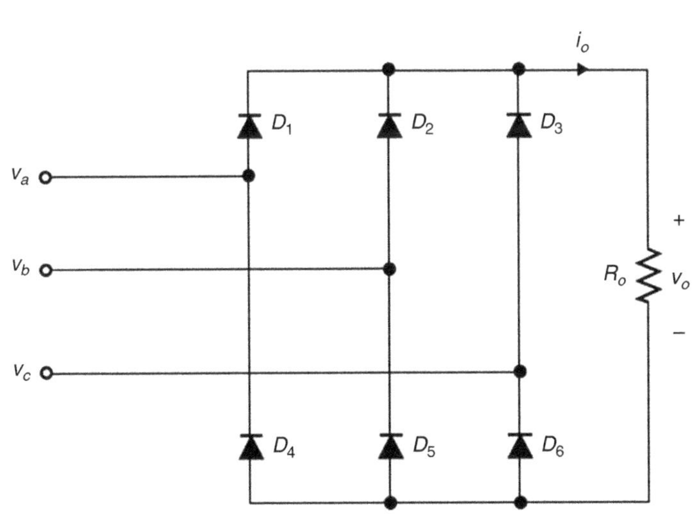

Fig. 7.31 Full-bridge rectifier circuit under resistive load

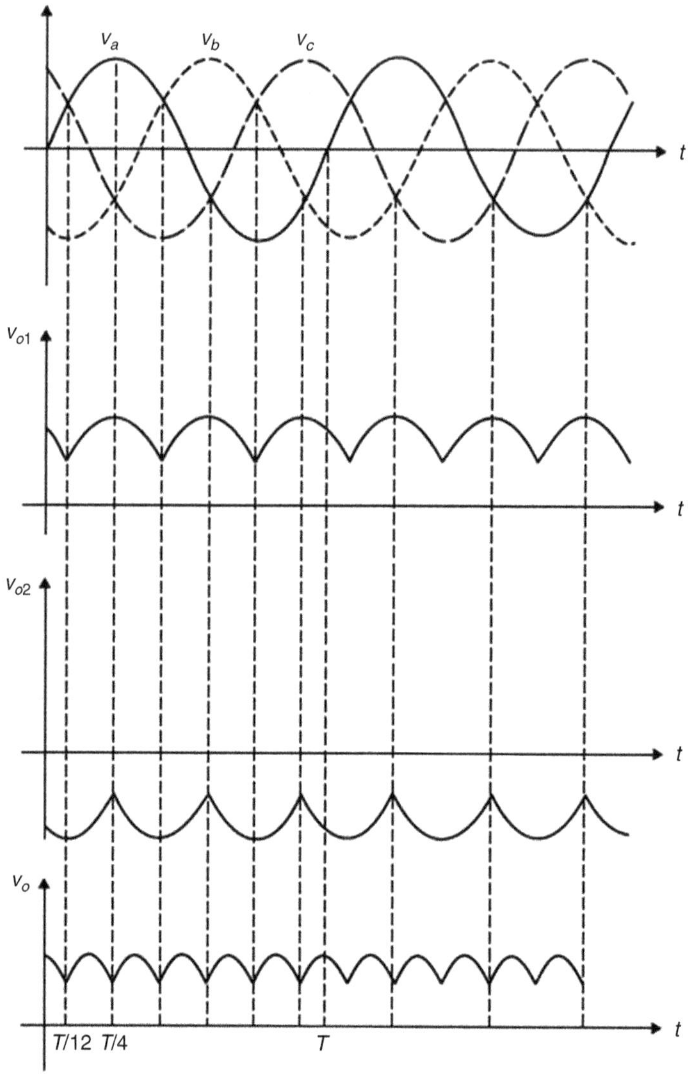

Fig. 7.32 Output voltage waveform

Let us consider the bridge rectifier under a highly inductive load using the Y-connected voltage source with $L/R \gg T/6$ as shown in Fig. 7.34a. Table 7.1 shows all six modes of operation with the corresponding diode conduction angles and currents, where $\theta = \omega t$.

The waveforms for the output voltage and the diode and line currents are shown in Fig. 7.34b. The output voltage is the same as that in the resistive case given in Fig. 7.31.

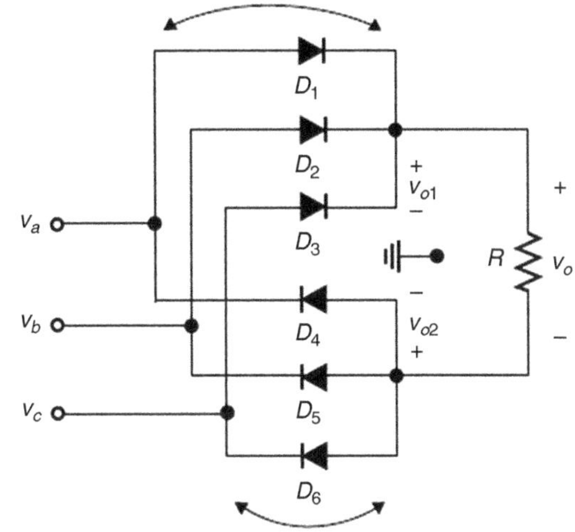

Fig. 7.33 Equivalent circuit for Fig. 7.31

7.5 Ac-Side Inductance in Three-Phase Rectifier Circuits

7.5.1 Half-Wave Rectifiers

Figure 7.35a shows a three-phase, half-wave rectifier circuit including an ac-side commutating inductance under a highly inductive load. The phase voltages v_a, v_b, and v_c are shown in Fig. 7.36a. The operation of the circuit can be discussed as follows: Prior to $\pi/6$, v_c is the most positive and D_3 is conducting, with D_1 and D_2 *off*, as shown in the equivalent circuit of Fig. 7.35b. During this interval, $i_c = +I_o$ and $v_o = v_c$.

At $t = T/12$, v_a and v_c become equal and D_1 starts conducing as in the single phase case. Since i_c cannot change to zero instantaneously, D_3 remains on while the load current is being moved from D_3 to D_1 as shown in Fig. 7.35c. At this instant, $i_a = 0$, $i_c = +I_o$ and v_o is the average value between v_a and v_c, to be shown next.

Assuming equal ac-side inductances, the following relations are obtained:

$$v_a = L_s \frac{di_a}{dt} + v_o \tag{7.52a}$$

$$v_c = L_s \frac{di_c}{dt} + v_o \tag{7.52b}$$

Since $i_a + i_c = I_o$, then $di_c/dt = -di_a/dt$, resulting in the following relation for v_o:

$$v_o = \frac{v_c + v_a}{2} \tag{7.53}$$

Substituting for v_o in Eq. (7.53), we obtain the following differential equation for i_a:

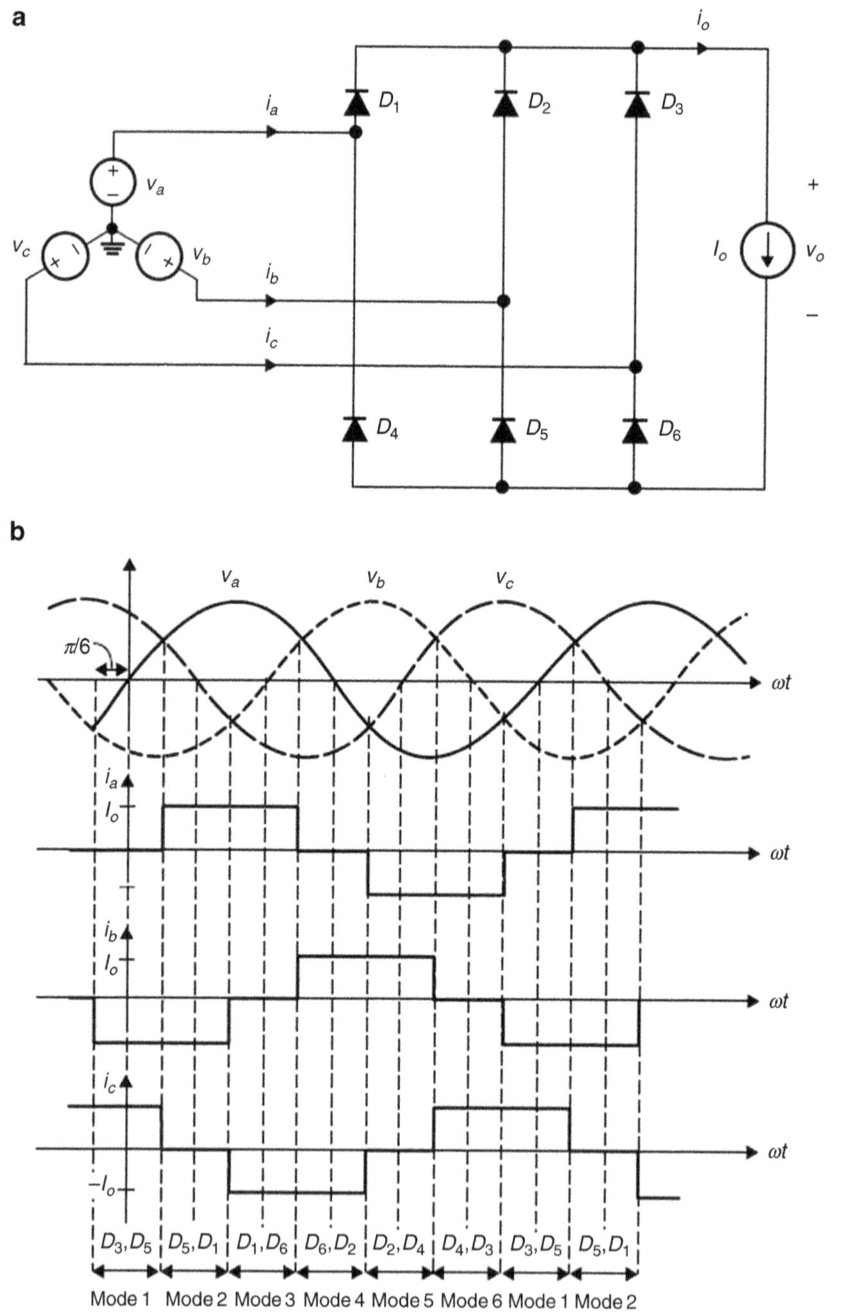

Fig. 7.34 (**a**) Three-phase bridge rectifier with highly inductive load (**b**) current waveforms

Table 7.1 Conduction modes for Fig. 7.34

Mode	Conduction angle	Diodes conducting	i_a	i_b	i_c	Most positive absolute voltage
I	$-\pi/6 < \theta < \pi/6$	D_3, D_5	0	$-I_o$	I_o	$\lvert v_b \rvert, v_c$
II	$\pi/6 < \theta < \pi/2$	D_5, D_1	I_o	$-I_o$	0	$v_a, \lvert v_b \rvert$
III	$\pi/2 < \theta < 5\pi/6$	D_1, D_6	I_o	0	$-I_o$	$v_a, \lvert v_c \rvert$
IV	$5\pi/6 < \theta < 7\pi/6$	D_6, D_2	0	I_o	$-I_o$	$v_b, \lvert v_c \rvert$
V	$7\pi/6 < \theta < 9\pi/6$	D_2, D_4	$-I_o$	I_o	0	$\lvert v_a \rvert, v_b$
VI	$9\pi/6 < \theta < 11\pi/6$	D_4, D_3	$-I_o$	0	I_o	$\lvert v_a \rvert, v_c$

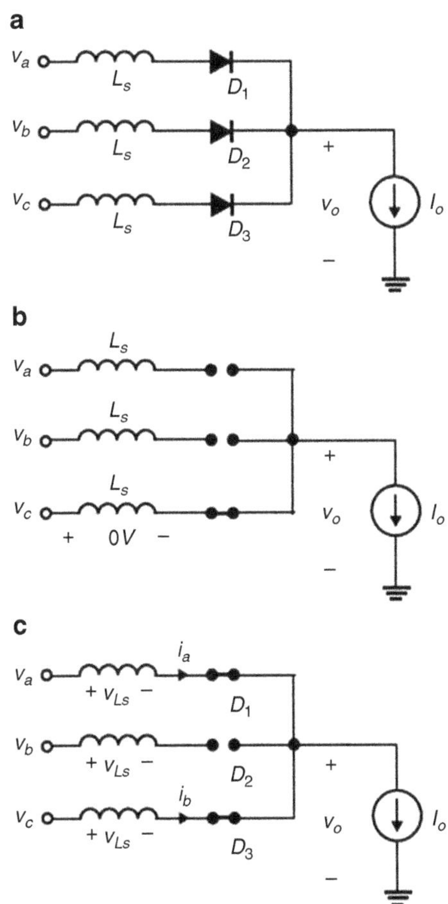

Fig. 7.35 (**a**) Three-phase, half-wave rectifier with constant load current (**b**) equivalent circuit with v_c most positive (**c**) equivalent circuit during current commutation from D_3 to D_1

$$\frac{di_a}{dt} = \frac{1}{2L_s}(v_a - v_c)$$

Substitute for $v_a = V_s \sin(\omega t)$ and $v_c = V_s \sin(\omega t - 240^\circ)$ and use the initial values of $i_c(T/12) = I_o$, $i_a(T/12) = 0$. The solution for i_c is then given by

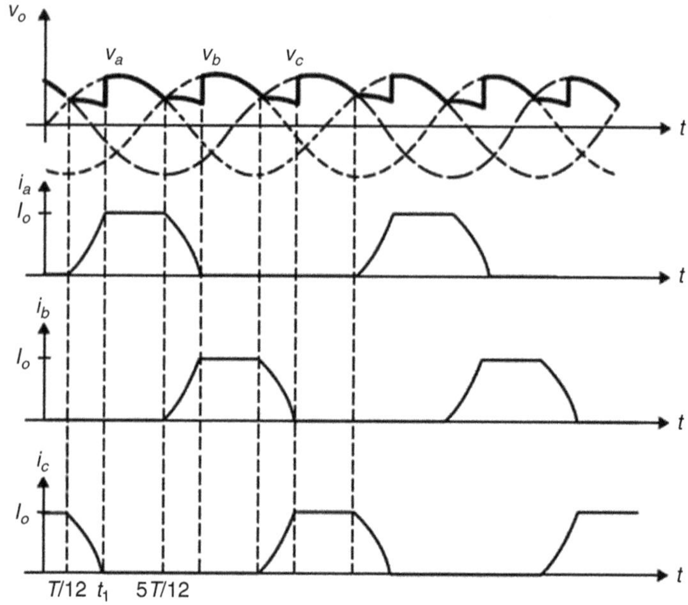

Fig. 7.36 Waveforms for v_o, i_a, i_b, i_c

$$i_c(t) = \frac{V_s\sqrt{3}}{2L_s\omega}\left[1 - \cos\left(\omega t - 30^\circ\right)\right]$$

The waveforms for v_o, i_a, i_b, and i_c are given in Fig. 7.36.

The commutation angle is obtained by evaluating $i_a(t)$ at t_1 when the current reaches I_o, as shown in the commutating waveform from phase c to phase a, in Fig. 7.37.

$$i_a(t) == \frac{V_s\sqrt{3}}{2L_s\omega}[1 - \cos(\omega t_1 - T/12)] = I_o \tag{7.54}$$

$$i_c(t) = I_o - i_a = I_o - \frac{V_s\sqrt{3}}{2L_s\omega}\left[1 - \cos\left(\omega t - 30^\circ\right)\right] \tag{7.55}$$

Hence, u can be found from the following expression:

$$u = \cos^{-1}\left(1 - \frac{2L_s\omega I_o}{V_s\sqrt{3}}\right) \tag{7.56}$$

Since the pulse width of the phase current is smaller than that of the case for a single-phase case, the power factor for a three phase is lower. Finally, the output voltage is given by the following integral:

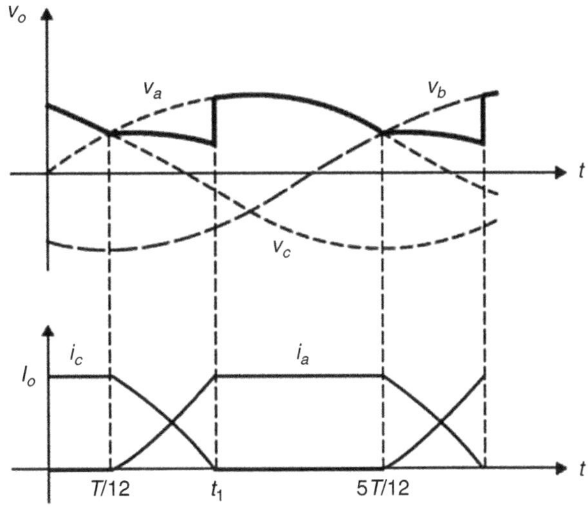

Fig. 7.37 Commutation waveforms from phase "c" to phase "a" in Fig. 7.35

$$V_o = \frac{3}{T}\left[\int_{T/12}^{t_1} v_o\,dt + \int_{t_1}^{5T/12} v_a dt\right] \tag{7.57}$$

Substituting for V_o from the above equation, Eq. (7.57) can be rewritten as:

$$\begin{aligned} V_o &= \frac{3}{T}\left[\int_{T/12}^{t_1} \frac{v_a + v_c}{2}\,dt + \int_{t_1}^{5T/12} v_a dt\right] \\ &= \frac{3}{T}\left[\int_{T/12}^{t_1} v_a\,dt + \int_{T/12}^{t_1} \frac{v_a - v_c}{2}\,dt + \int_{t_1}^{5T/12} v_a dt\right] \\ &= \frac{3}{T}\left[\int_{t_1}^{5T/12} v_a dt - \int_{T/12}^{t_1} \frac{v_a - v_c}{2}\,dt \int_{t_1}^{5T/12} v_a dt\right] \end{aligned} \tag{7.58}$$

We recognize that the first integral is the average output voltage for the half-wave three-phase converter with $L_s = 0$. The second integral is the average voltage loss due to the commutation. Substituting for $(v_a - v_c)/2$ in Eq. (7.58) by $L_s di_a/dt$, we have

$$\begin{aligned} V_o &= V_o\,|_{L_s=0} - \frac{3}{T}\int_{T/12}^{t_1} L_s \frac{di_a}{dt} dt \\ &= V_o\,|_{L_s=0} - \frac{3}{T}\int_{\pi/6}^{\pi/6+u} L_s \frac{di_a}{dt} d\theta \\ &= V_o\,|_{L_s=0} - \frac{3}{T} L_s I_o \end{aligned} \tag{7.59}$$

Recall that the average value of the voltage when $L_s = 0$ is given by:

$$V_o \left.\right|_{L_s=0} = \frac{3\sqrt{3}V_s}{2\pi}$$

Then V_o of Eq. (7.59) may be written as

$$V_o = \frac{3\sqrt{3}}{2\pi} V_s \left(1 - \frac{\omega L_s I_o}{\sqrt{3}V_s}\right)$$

Normalizing the output voltage by V_s and I_o by$V_s/(\omega L_s)$ gives:

$$V_{no} = \frac{3\sqrt{3}}{2\pi}\left(1 - \frac{I_{no}}{\sqrt{3}}\right) \tag{7.60}$$

7.5.2 Full-Wave Bridge Rectifiers

Let us consider the full-wave three-phase rectifier circuit by including an ac-side inductance. The unidirectional source current in the half-wave circuit has a significant effect on the input power factor. This is why the bridge connection is widely used. Figure 7.38 shows the full bridge with an ac-side commutation inductance from a three-phase Y-connected voltage source. Again, we assume a large inductive load.

The three source line-to-line voltages are

$$v_{ab} = \sqrt{3}V_s \sin\left(\omega t + 30^{\circ}\right) \quad v_{bc} = \sqrt{3}V_s \sin\left(\omega t - 90^{\circ}\right) \quad v_{ca} = \sqrt{3}V_s \sin\left(\omega t + 150^{\circ}\right)$$

The operation of this circuit can be discussed in a similar way to that of the case with $L_s = 0$, except in this case, the commutation inductance will present an additional commutation angle, u, during which three diodes are on simultaneously. From Mode 1 in Table 7.1, for $\omega t < \pi/6$, $|v_{bc}|$ is the maximum voltage with D_3 and D_5 conducting as shown in Figure 7.39a. In this mode, we have:

$i_a = 0 \quad i_b = -I_o \quad i_c = I_o \quad v_o = -v_{bc} \quad v_{Ls} = 0.$

From Table 7.1, the next mode is when D_5 and D_1 turn on, since the next most positive voltage in this mode is V_{ab}. However, the current will commute from D_3 to D_1 during a commutation period, u, during which D_3 remains on while D_1 and D_5 are conducting. Figure 5.39b shows the equivalent circuit during this commutation period. From Mode 1 to Mode 2, the currents commutate from zero to $+I_o$ in i_a; hence, D_3 remains on for a short time until its current goes to zero at $t = t_1$. In this time interval, we have:

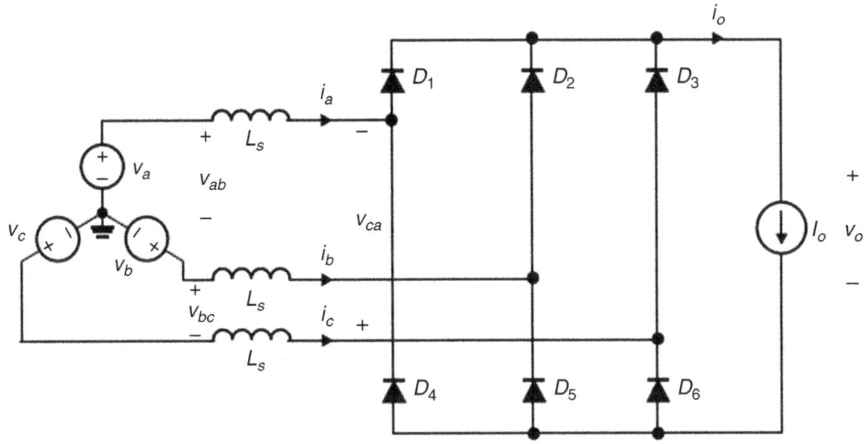

Fig. 7.38 Three-phase bridge rectifier with ac-side inductance

$$-v_{\mathrm{b}}-v_{\mathrm{o}}-L_{\mathrm{s}}\frac{di_{\mathrm{c}}}{dt}+v_{\mathrm{c}}=0 \tag{7.61a}$$

$$v_{\mathrm{b}}-v_{\mathrm{o}}-L_{\mathrm{s}}\frac{di_{\mathrm{c}}}{dt}+v_{\mathrm{a}}=0 \tag{7.61b}$$

Since $i_{\mathrm{a}}+i_{\mathrm{c}}=I_{\mathrm{o}}$, then $di_{\mathrm{c}}/dt=-di_{\mathrm{a}}/dt$. Hence, the above relation becomes:

$$v_{\mathrm{o}}=\frac{v_{\mathrm{ab}}+v_{\mathrm{cb}}}{2}=\frac{v_{\mathrm{ab}}+v_{\mathrm{bc}}}{2} \tag{7.62}$$

From Eqs. (7.61) and (7.62), we have the following first-order differential equations in terms of the line-to-line voltages and currents:

$$-v_{\mathrm{cb}}-v_{\mathrm{o}}-L_{\mathrm{s}}\frac{di_{\mathrm{c}}}{dt}=0 \tag{7.63a}$$

$$v_{\mathrm{ab}}-v_{\mathrm{o}}-L_{\mathrm{s}}\frac{di_{\mathrm{a}}}{dt}=0 \tag{7.63b}$$

Substituting for v_{o} in the above equations, we obtain:

$$\frac{v_{\mathrm{cb}}-v_{\mathrm{ab}}}{2L_{\mathrm{s}}}=\frac{di_{\mathrm{c}}}{dt}=-\frac{di_{\mathrm{a}}}{dt}$$

$$\frac{v_{\mathrm{cb}}-v_{\mathrm{ab}}}{2L_{\mathrm{s}}}=\frac{di_{\mathrm{a}}}{dt}$$

Substituting for $v_{\mathrm{cb}}-v_{\mathrm{ab}}=v_{\mathrm{ca}}=\sqrt{3}V_{\mathrm{s}}\sin\left(\omega t+150^{\circ}\right)$, and since at $t=0$, $i_{\mathrm{c}}(\pi/6)=I_{\mathrm{o}}$, we have:

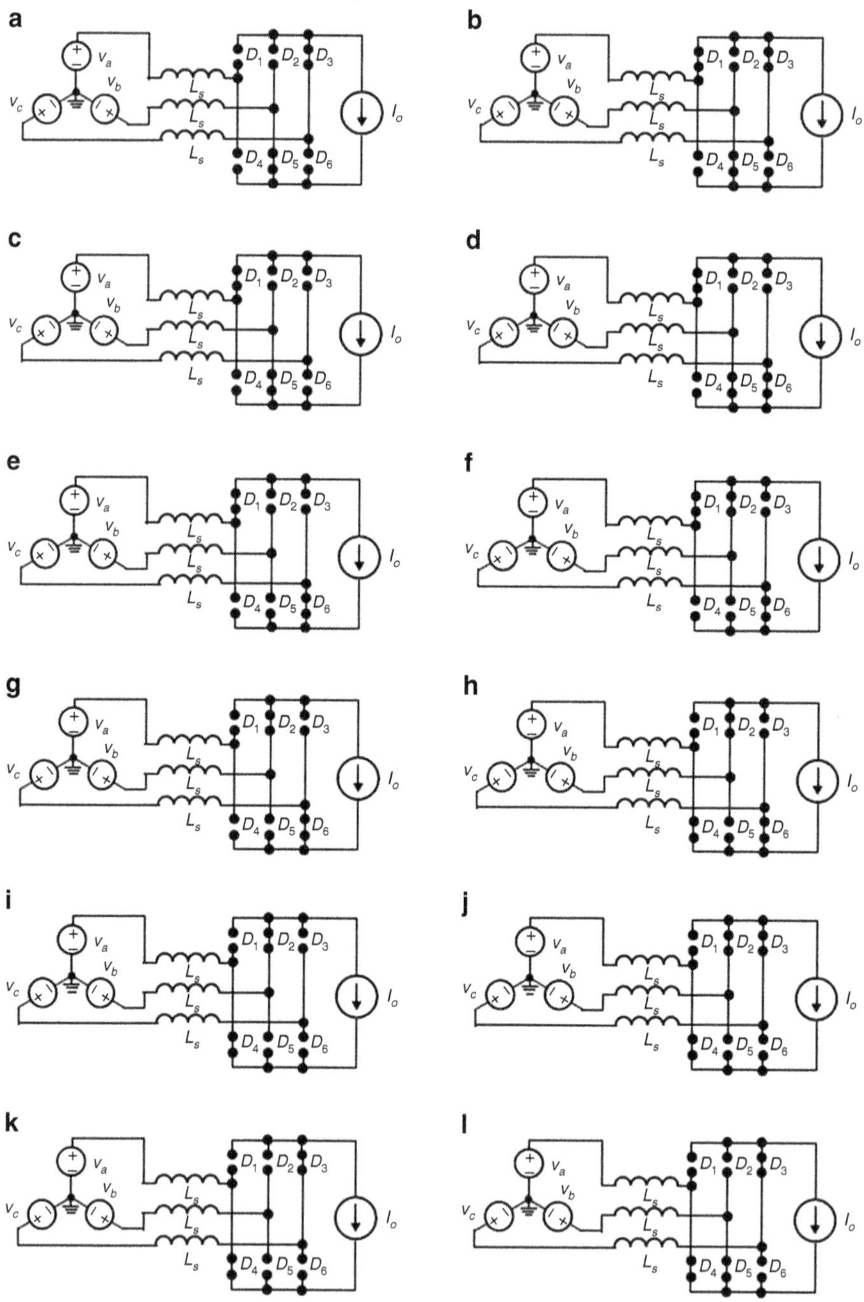

Fig. 7.39 Equivalent circuit modes for Mode 1 to Mode 6 (**a**) Mode I D_3 and D_5 are conducting (**b**) Mode I and Mode II commutation (**c**) Mode II D_1 and D_5 are conducting (**d**) Mode II and Mode III commutation (**e**) Mode III D_1 and D_6 are conducting (**f**) Mode III and Mode IV commutation (**g**) Mode IV D_2 and D_6 are conducting (**h**) Mode IV and Mode V commutation (**i**) Mode V D_2 and D_4 are conducting (**j**) Mode V and Mode VI commutation (**k**) Mode VI D_3 and D_4 are conducting (**l**) Mode VI and I are commutation

$$i_c(t) = \frac{1}{2L_s}\int_{\pi/6}^{\omega t} \sqrt{3}V_s \sin\left(\omega t + 150^\circ\right) d\omega t + I_o$$

At $\omega t = u + \pi/6$, we have $i_c(u + \pi/6) = 0$, hence, u is given by

$$u = \cos^{-1}\left(1 - \frac{2L_s\omega I_o}{V_s\sqrt{3}}\right) \tag{7.64}$$

At the end of the commutation period u, i_c becomes zero and D_3 stops conducting. Diodes D_1 and D_5 remain *on* until Mode 3 starts at $\pi/2$ when $|v_{ca}|$ becomes the most positive, turning diodes D_1 and D_6 *on* and in turn turning D_5 *off*. In this mode, the current i_b will commute from I_o to 0 through D_5, and i_c goes from 0 to$-I_o$ through D_6. Diodes D_5 and D_6 will overlap during the commutation period u. The remaining modes and their corresponding mode-to-mode transition are shown in Fig. 7.39a–l. The current waveforms of i_a, i_b, and i_c and the output voltage are shown in Fig. 7.40.

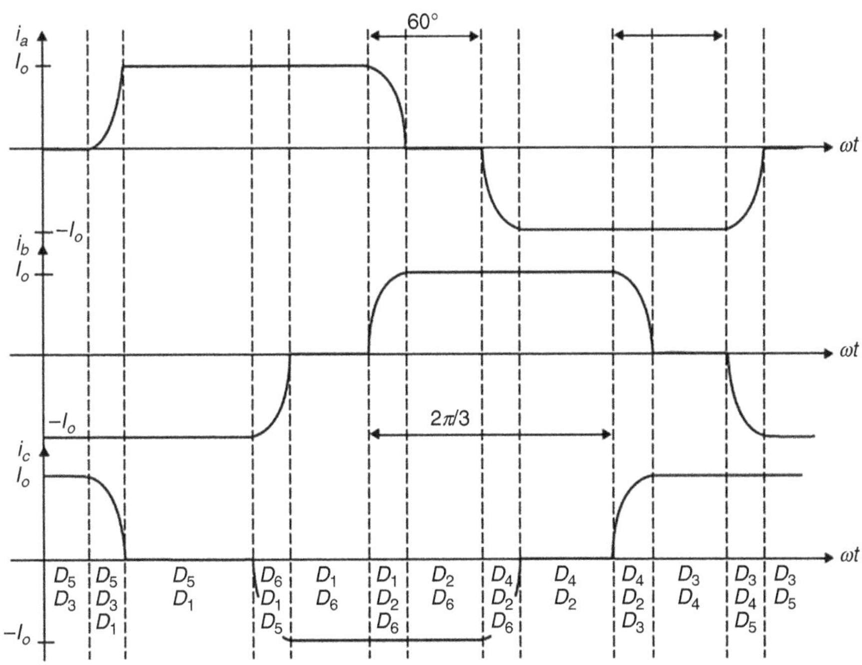

Fig. 7.40 Current waveforms fori_a, i_b, and i_c for Fig. 7.38

Problems

In all the problems assume ideal diodes unless stated otherwise.

Single-Phase Rectifiers

Half-Wave Rectifiers

7.1 Determine the power factor and the THD for the waveforms of Example 7.1.

7.2 Consider the half-wave rectifier in Fig. P7.2 with $v_s = 110 \sin \omega t$ V and $R = 1$ kΩ. Assume ideal diode except the forward resistance (r_d) is 10 Ω. Sketch v_o and determine (a) the average output voltage, (b) the rms current in the diode, (c) the peak current in the diode, and (d) the input power factor.

Fig. P7.2

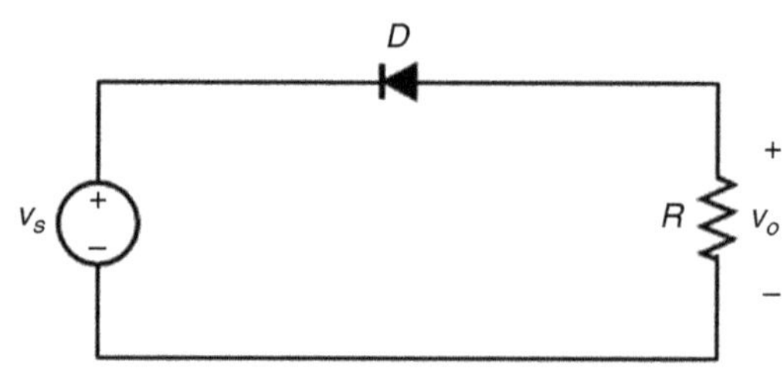

7.3 Derive Eqs. (7.9) and (7.11).

7.4 Consider the half-wave inductive-load rectifier circuit of Fig. 7.4a with $v_s = 110\sqrt{2} \sin \omega t$ V, $\omega = 60$ Hz, $L = 800$ mH, and $R = 100$ Ω. Find (a) the expression for $i_L(t)$, (b) the time t_1 when i_L becomes zero, (c) the average load current, (d) the peak inductor current, (e) the rms current in the inductor, and (f) the average power delivered to the load.

7.5 Repeat Problem 7.4 with the diode reversed.

7.6 Repeat Problem 7.4 by placing a free-wheeling diode in parallel with the load as shown in Fig. 7.6a.

7.7 Repeat Example 7.2 for the circuit of Fig. 7.6a.

D7.8 Design for L and R for the half-wave rectifier of Fig. 7.6a to provide a ripple current not to exceed 5% of its average value at $V_o = 20$ V. Assume $v_s = 40 \sin 120\omega t$ V.

D7.9 The single-phase, half-wave rectifier of Fig. 7.6a is required to supply the load with an output ripple current not to exceed 5% of its dc value. (a) Design for L to meet this requirement. Use $R = 20$ Ω, $v_s = 150 \sin 120\pi t$ V, (b) what is the average load voltage.

7.10 Derive Eqs. (7.20) and (7.21) for the half-wave rectifier with an $L-R$ load.

D7.11 Design a half-wave rectifier of Fig. 7.4a with an $L-R$ load to supply 25 W to the load with a minimum output ripple current of 10% and an average output of 85 V. Use $v_s = 120\sin(120\pi t)$.

D7.12 Design a dc power supply with an average dc output of 20 V with a 2% ripple. The power supply should be capable of providing 500 mA to the load. Use the rectifier of Fig. P7.12 for your design with $v_s = 110\sin(2\pi 60t)$ V and assume the diode's forward voltage drop is 1V. Specify the diode's ratings including its peak reverse voltage and rms values.

Fig. P7.12

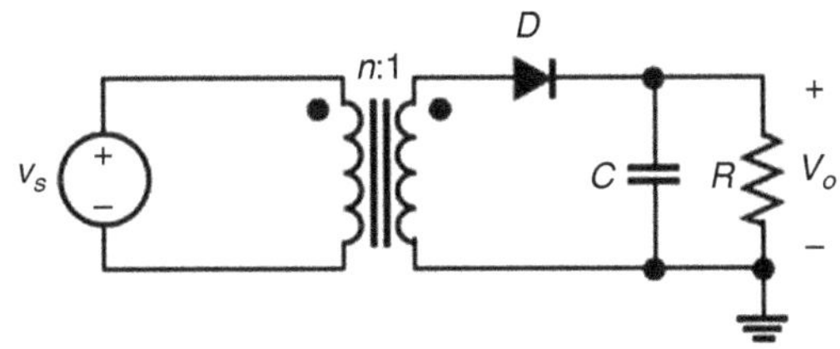

D7.13 Design the half-wave rectifier with an R-C load with a 10% output ripple voltage and which provides 10 W to the load. Assume $v_s = 20\sin(120\pi t)$V.

7.14 Sketch the waveforms for i_s, i_c, i_R, and v_o for Fig. 7.8a with $R = 10$ kΩ, $C = 100\mu$F, and $v_s = 10\sin(377\pi t)$V.

Full-Wave Rectifiers

7.15 Determine the power factor and the THD for the waveforms of Fig. 7.3b.

7.16 Consider the full-wave rectifier with a resistive load of Fig. 7.3a with $v_s = 120\sin\omega t$ and $R = 20\ \Omega$. Find: (a) the average output voltage, (b) the average output current, (c) the rms output current, (d) the input power factor, and (e) the load power.

7.17 Consider a full-wave rectifier with an infinite inductive load.

(a) Show that the dc and harmonic components of the output voltage are given by:

$$v_o(t) = \frac{2V_s}{\pi} - \frac{4V_s}{\pi}\sum_{n=2}^{\infty}\frac{1}{(n-1)(n+1)}\cos\omega t$$

For $n = 2, 4, 6, \ldots$

(b) Show that the nth harmonics of the source current is given below:

$$i_s(t) = \frac{4I_o}{\pi}\sum_{n=1}^{\infty}\frac{\sin n\omega t}{n}$$

D7.18 It is required that the circuit shown in Fig. P7.18 provides an output power of 2 kW at an average load current of 20 A. Design for the transformer turns ratio and R to meet this requirement. Assume $L/R \gg T$. Use $v_s = 100 \sin(2\pi 60t)$ V.

Fig. P7.18

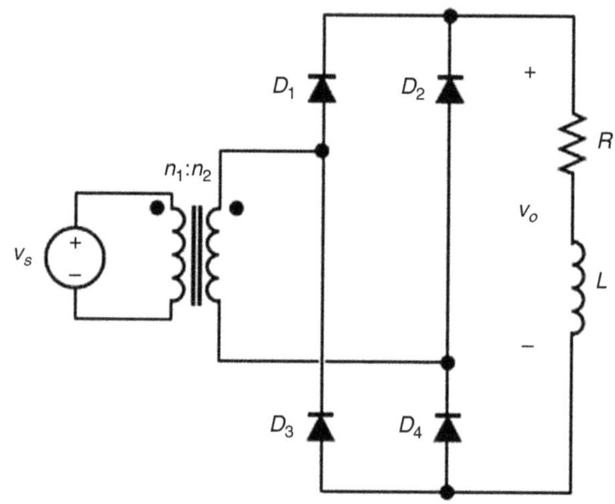

7.19 Consider the generalized-type load with a dc source representing a dc motor load full-wave rectifier as shown in Fig. P7.19. Assume $v_s = V_s \sin \omega t$ V and $V_{DC} < V_s$. Derive (a) the expression for $i_o(t)$, (b) the rms value of $i_o(t)$, (c) the average load current, and (d) the ripple load current and (e) calculate the above values for $V_s = 25$ V, $\omega = 377$ rad/s, $R = 2\ \Omega$, and $L = 5$ mH.

Fig. P7.19

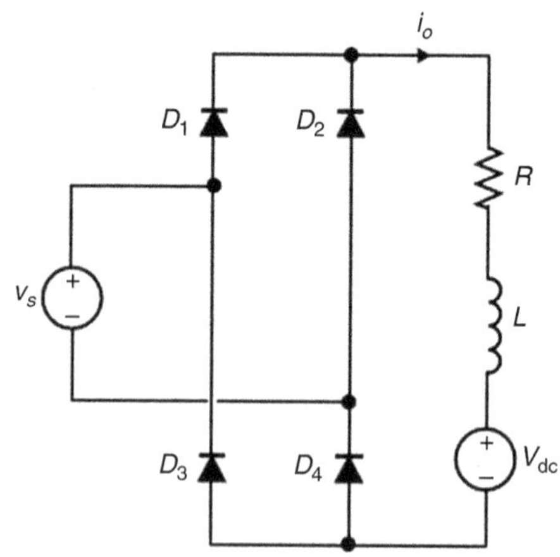

7.20 Consider the center-tap full-wave rectifier of Fig. P7.20. (a) Sketch v_o, (b) determine V_o, (c) repeat parts (a) and (b) by including each diode drop V_D, and (d) determine the peak inverse voltage (PIV) for the two diodes.

Fig. P7.20

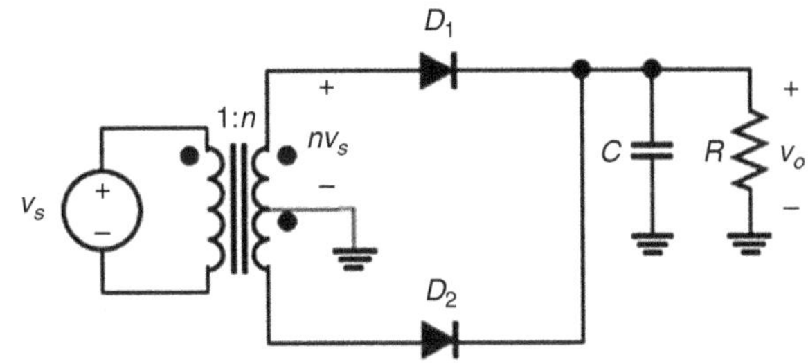

D7.21 If it is required to design the full-wave rectifier of Problem 7.20 to provide an average output voltage of 48 V with an output power of 25 W. Design for the required transformer turns ratio n. Assume $v_s = 110\sqrt{2}\sin 377t$ V.

D7.22 Redesign Problem 7.21 by assuming $V_s(t)$ fluctuates by±20%. Design for the worst case.

7.23 Consider the full-wave center-tap diode rectifier circuit shown in Fig. P7.23. (a) Sketch v_{o1} and v_{o2} and (b) determine the peak inverse voltage for each diode.

Fig. P7.23

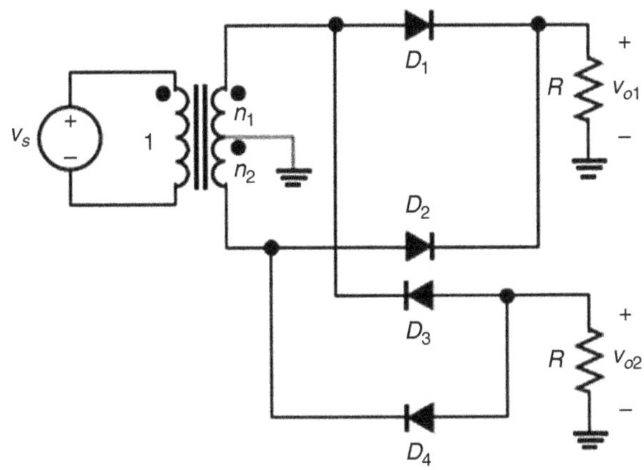

D7.24 Design the dual center-tap full-wave rectifier of Fig. P7.24 to provide ±15 V dc from a line voltage $v_s = 155\sin\omega t$ V, where $\omega = 377$ rad/s.

Fig. P7.24

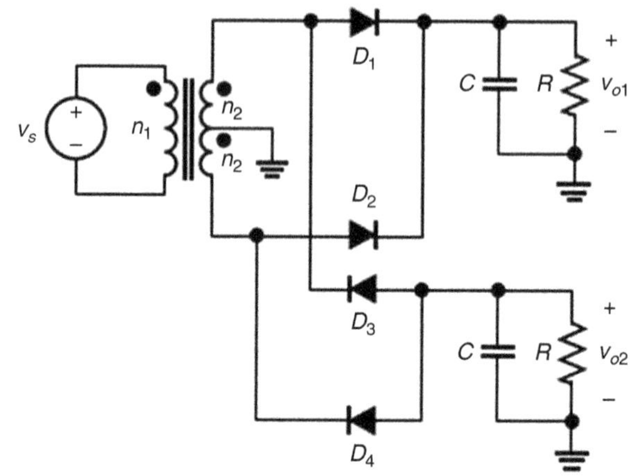

7.25 Repeat Problem 7.9 by using the full-bridge rectifier given in Fig. 7.20a with $L_s = 0$.

7.26 (a) Calculate the output ripple voltage for the full-wave rectifier of Fig. P7.26. Use $R = 10\ \text{k}\Omega$ and $C = 47\ \mu\text{F}$, with $v_s = 20 \sin 120\omega t$ V.

(b) Show that the expression for the average output voltage is given by:

$$V_o = \frac{V_s}{2\pi}\left[\cos\omega t_1 + \tau\omega\left(1 - e^{-(t_1+3T/4)/\tau}\right)\right]$$

where $\tau = RC$ and t_1 $(0 < t_1 < T/4)$ is the time at which the diode starts conducting.

(c) Show that if it is assumed that the ripple voltage is much smaller than the peak input voltage, i.e., $t_1 = T/4$, and $RC \gg T$, then show that $V_r = V_s/2RCf$. Apply this simplified ripple voltage to part (a).

(d) Use the approximation in part (c) to show that the average diode current is given by:

$$I_{D,ave} = \frac{V_s}{2R}\left[1 + \frac{1}{\pi}\sqrt{1 - \left(1 - \frac{V_r}{V_s}\right)^2}\right]$$

Fig. P7.26

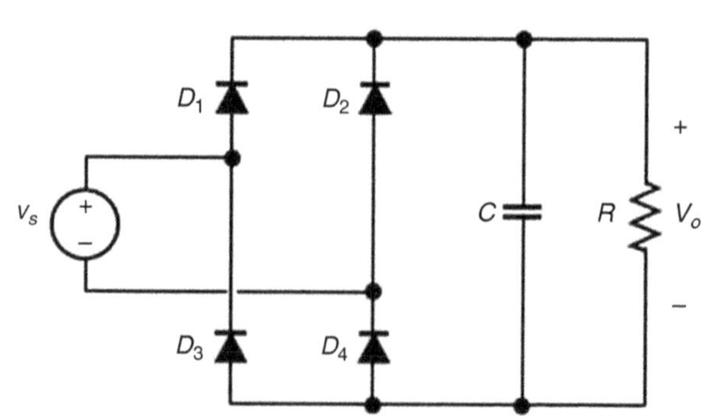

7.27 The circuit shown in Fig. P7.27 is called a voltage doubler. It is used when the line voltage of a given system changes from 110 *VAC* to 220 *VAC* or vice versa. By changing the switch position, the output voltage can be held constant under both a 110 *VAC* and a 220 *VAC* line inputs. Consider the following voltage doubler. (a) Determine the voltages v_{o1} and v_{o2} for both cases when the switch is *on* and *off*. Assume $v_s = V_s \sin \omega t$ V with $R_1C_1 \gg T$ and $R_2C_2 \gg T$.

Fig. P7.27

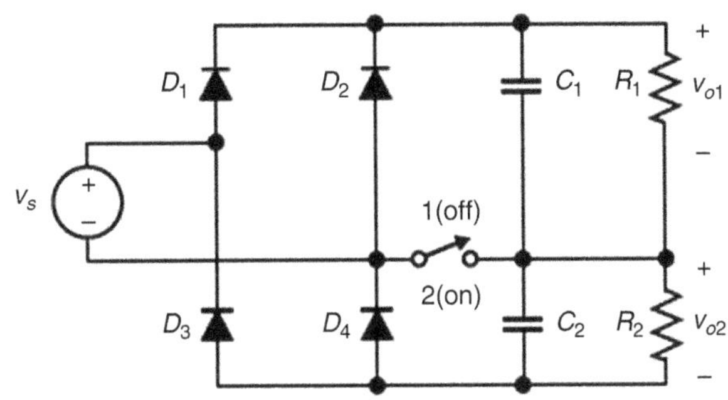

7.28 Consider the full-wave rectifier with an LC output filter shown in Fig. P7.28. (a) Draw the equivalent circuit seen between a and b by assuming $L = 0$ an $C = \infty$d, (b) repeat part (a) by assuming $L = \infty$ and $C = 0$, and (c) calculate the power dissipated in the load in parts (a) and (b). Assume $v_s = V_s \sin \omega t$.

Fig. P7.28

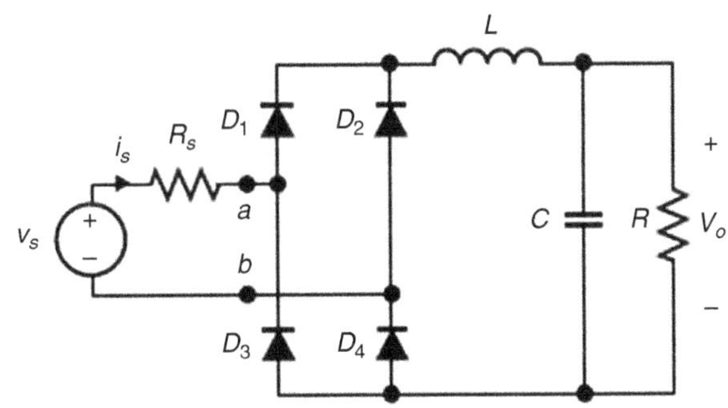

7.29 The single-phase, full-wave rectifier with an R-C load shown in Fig. P7.26. (a) is required to supply the load with an output ripple voltage not to exceed 5% of its dc value. (a) Design for C to meet this requirement. Use $R = 20\ \Omega$, $v_s = 150 \sin 120\omega t$ V and (b) calculate the average load current.

The Effect of ac-Side Inductance

Half-Wave Circuits

7.30 Consider the half-wave rectifier shown in Fig. P7.30 with a flyback diode including L_s and the diode resistances. Assume $v_s = V_s \sin \omega t$ V. (a) Derive the expression for the power dissipated in the rectifier. (b) Sketch i_s, i_{D2}, and v_o.

Fig. P7.30

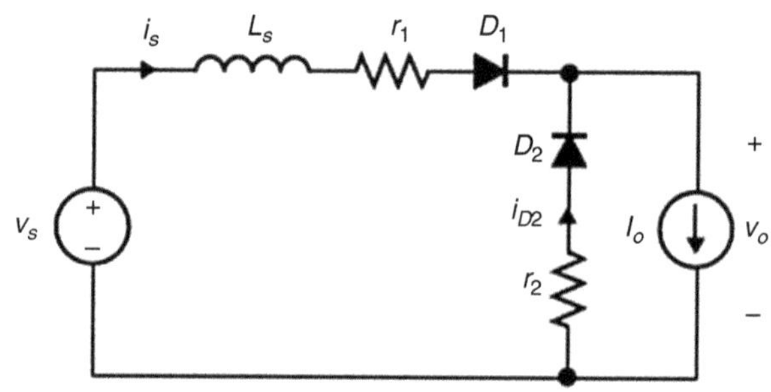

7.31 It is required to design the circuit of Fig. 7.12a to deliver a 1kW output power at $I_o = 20$ A. Assume L_s fluctuates in the range of $\pm 20\%$. Use $L_s = 0.1$ mH and $v_s(t) = 120 \sin 2\pi 60t$ V. Design for the load resistance and inductance.

Full-Wave Circuits

7.32 Derive the rms and fundamental component expressions for the input current $i_s(t)$ of a full-wave rectifier circuit with an ac-side inductance which is redrawn in Fig. P7.32.

Fig. P7.32

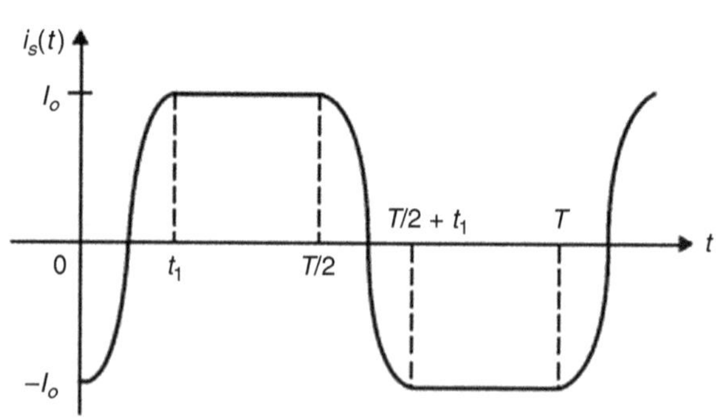

7.33 Repeat Problem 7.32 by the approximation for $i_s(t)$ as shown in Fig. P7.33.

Fig. P7.33

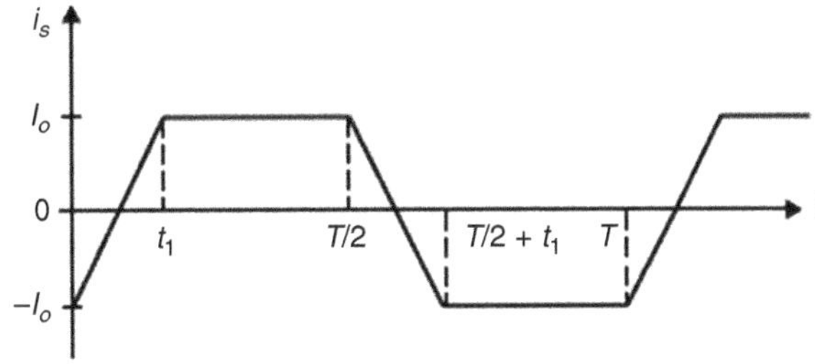

7.34 (a) Discuss the operations of the circuit in Fig. P7.34 and sketch the waveforms for $v_o, i_{s1}, i_{s2}, v_{Ls1}, v_{Ls2}$. Assume $v_{s1} = V_{s1} \sin \omega t$ V, $v_{s2} = V_{s2} \sin \omega t$ V. (b) Derive the expression for $v_o(t)$. (c) What is the difference between this circuit and the full-wave bridge circuit of Fig. 7.21?

Fig. P7.34

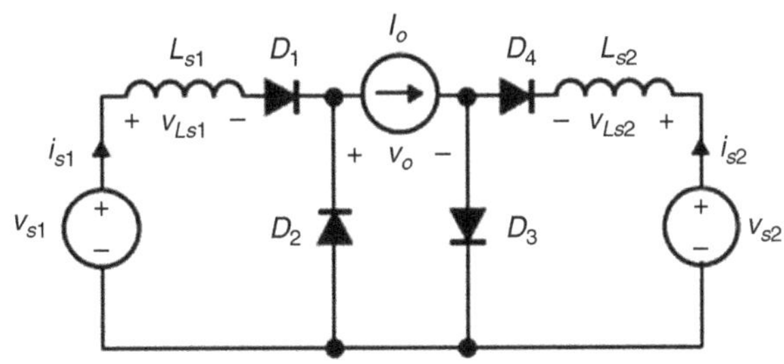

7.35 (a) Find the ratio between the rms currents for the waveforms shown in Fig. P7.35a, b representing the line current for half-wave rectifiers with and without an ac-side inductance, respectively. (b) Sketch the ratios as a function of $\omega t_1 = u$.

Fig. P7.35

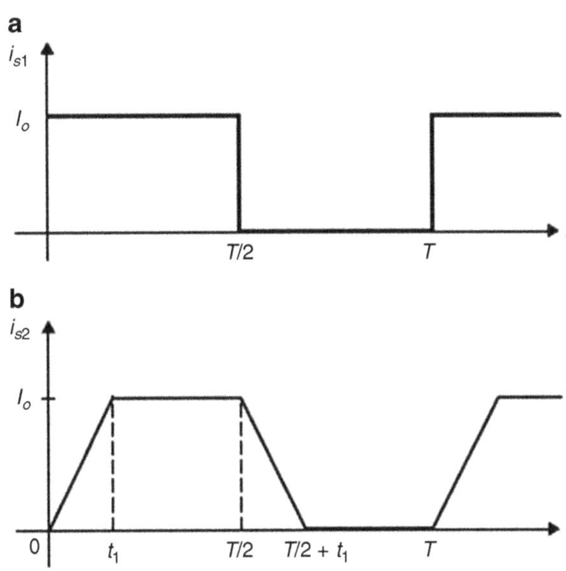

Input Power Factor

7.36 Consider the line current for the half-wave rectifier with L_s as shown in Fig. 7.12e, which is redrawn in Fig. P7.36a. (a) Derive the expression for the power factor for $i_s(t)$, (b) repeat part (a) by using the trapezoidal approximation as shown in Fig. P7.36b, c compare the two solutions.

Fig. P7.36

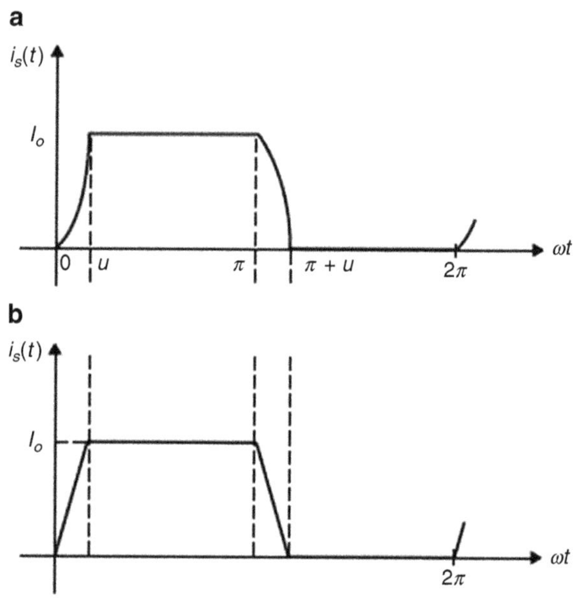

7.37 Consider the one-port converter network shown in Fig. P7.37 with $i_s = 5 \sin(377t + 30^{\circ})$ A, and $v_s = 10 \sin(377t - 45^{\circ})$ V. Determine (a) the instantaneous input power, (b) the input power factor, and (c) the average input power.

Fig. P7.37

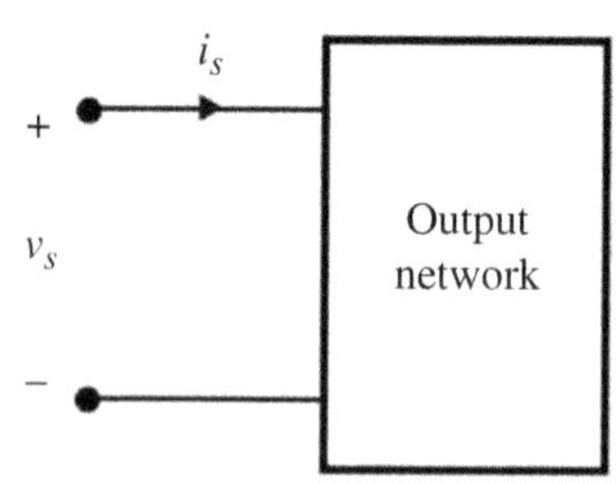

7.38 Derive the input power factor and the THD for the input current waveforms shown in Fig. P7.38. Assume the line voltage is pure sinusoidal.

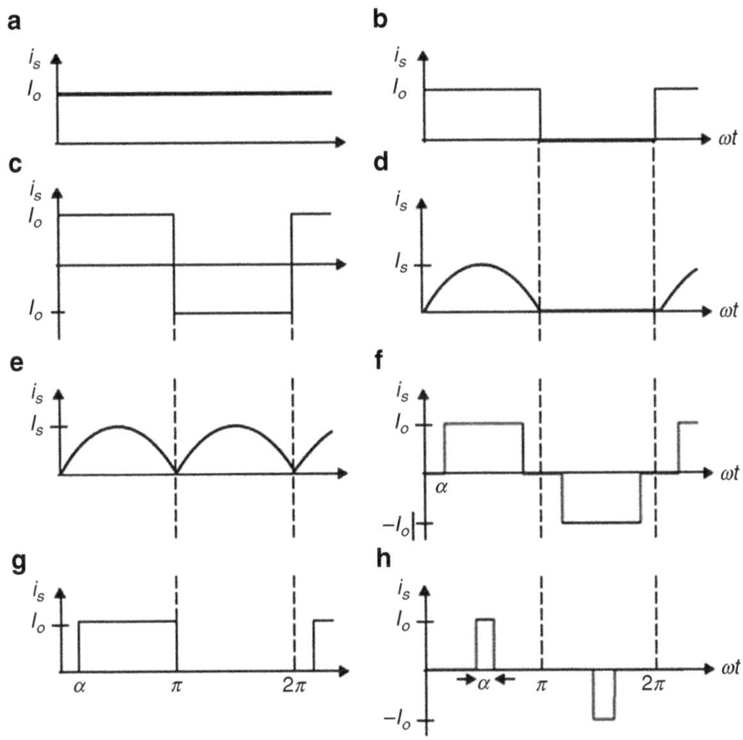

Fig. P7.38

Three-Phase Rectifier Circuit

7.39 (a) Show that the average output voltage for the n-phase, half-wave rectifier of Fig. P7.39 is given by:

$$V_{\mathrm{o}} = V_{\mathrm{s}} \frac{\sin(\pi/2)}{(\pi/n)}$$

where V_{s} is the phase peak voltage. What is the value of V_{o} as $n \to \infty$?

(b) Show that the rms and average current in each diode is given by:

$$I_{\mathrm{Di,rms}} = \frac{I_{\mathrm{o}}}{\sqrt{2}}$$

$$I_{\mathrm{D,ave}} = \frac{I_{\mathrm{o}}}{n}$$

Fig. P7.39

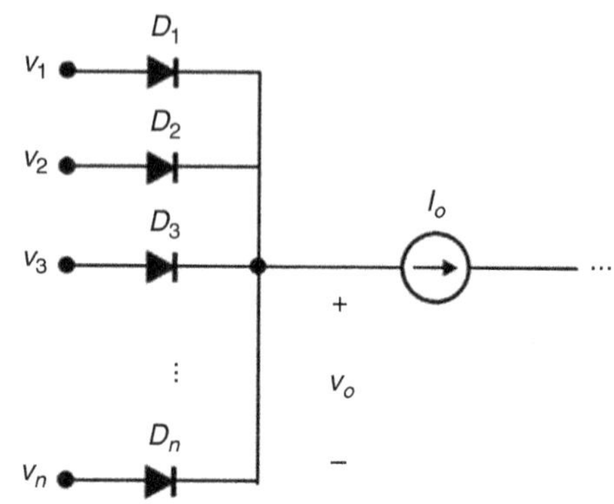

7.40 Consider the balanced three-phase rectifier circuit shown in Fig. P7.40 with the delta-connected voltage sources given below:

$$v_a = V_s \sin \omega t \qquad v_b = V_s \sin\left(\omega t - 120^\circ\right) \qquad v_c = V_s \sin\left(\omega t - 240^\circ\right)$$

(a) Sketch the waveform for i_1, i_2, i_3, v_o, (b) derive the expression for the average V_o, and (c) determine the first fundamental component for the line current.

Fig. P7.40

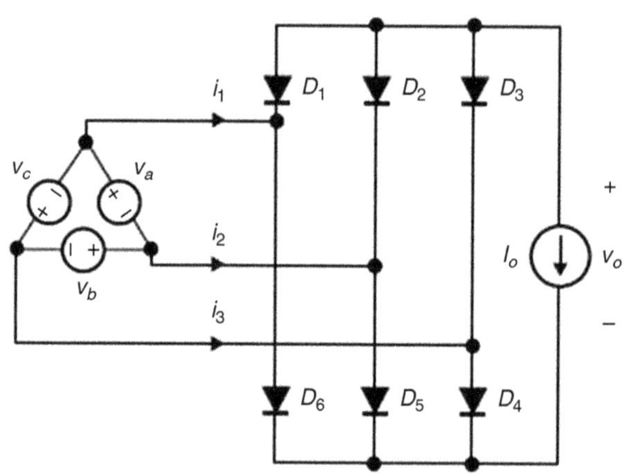

7.41 Consider the half-wave three-phase capacitive load circuit of Fig. P7.41. (a) Determine the rms value for v_o, (b) calculate the input power factor for each phase, and (c) sketch v_o for $RC = T/3$.

Fig. P7.41

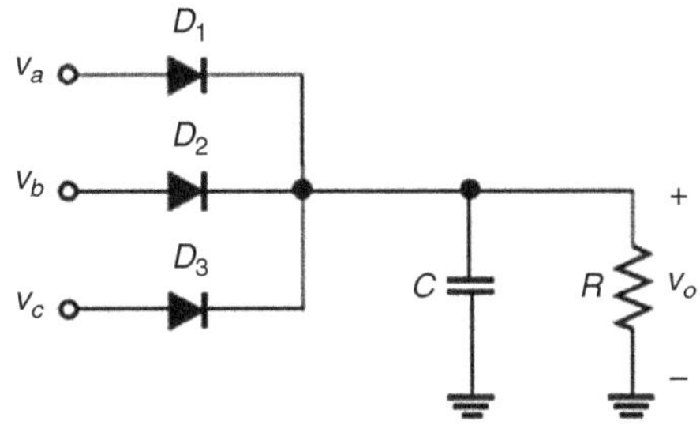

Three-Phase Rectifiers with ac-Side Inductance

7.42 Derive Eq. (7.60) for the average output voltage for the half-wave, three-phase converter.

7.43 Design the transformer turns ratio of the circuit shown in Fig. P7.43 to deliver a 10 A load current with a 120 V output voltage. Assume the line-to-line voltages are 580 VAC, with $\omega = 2\pi 60$ and $L_s = 1$ mH.

Fig. P7.43

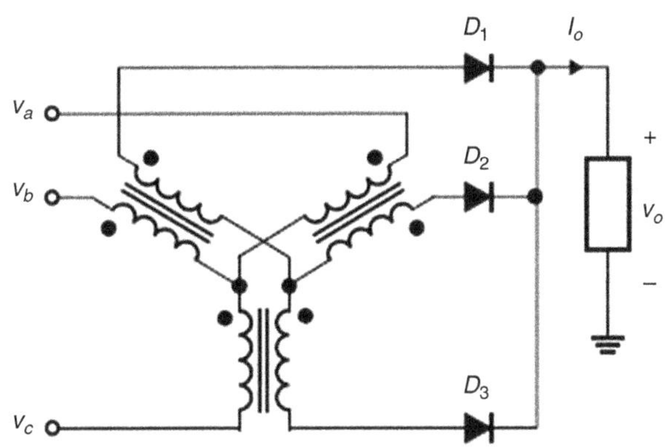

7.44 Derive Eq. (7.64) for the commutation angle u for the full-wave converter.

General Problem

7.45 Sketch the input current, i_s, and output voltage, v_o, for the circuit shown in Fig. P7.45.
Assume $v_s(t) = V_s \sin \omega t$ V.

Fig. P7.45

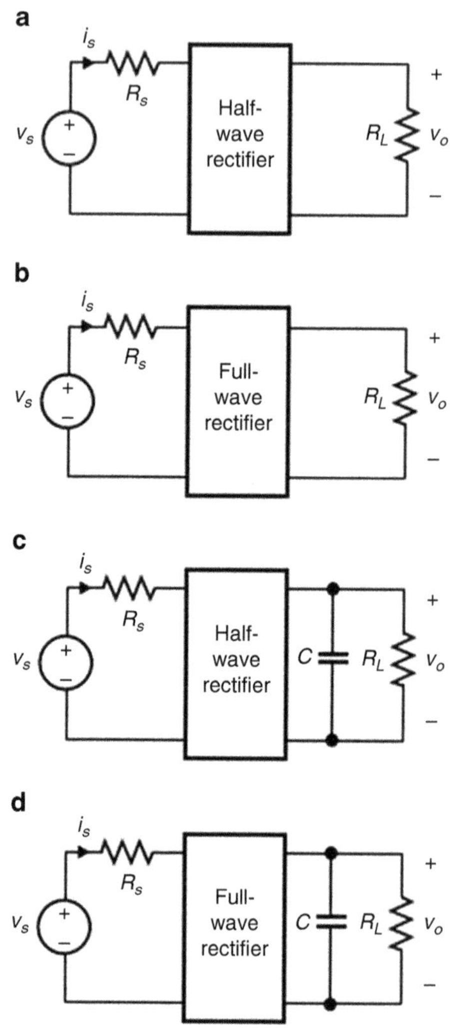

7.46 Derive the expression for the average output power and the commutation angle u for the circuit shown in Fig. P7.46. Find the average and rms values of each diode current. Assume $v_s(t) = V_s \sin \omega t$ V.

Fig. P7.46

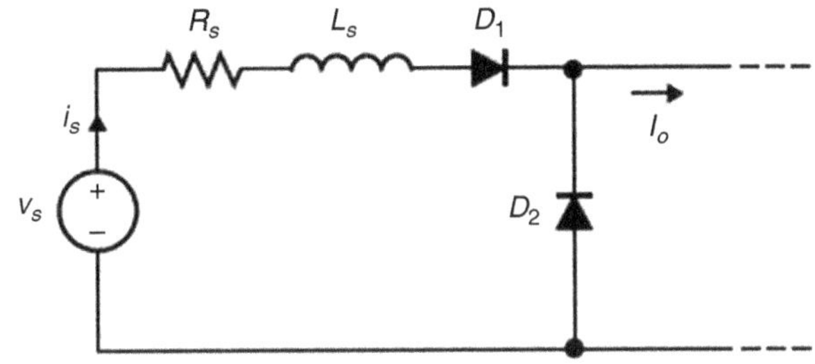

7.47 In problem 7.46, if we assume that D_1 and D_2 have forward resistances r_{D1} and r_{D2}, respectively, show that the commutation angle is given by

$$u = \frac{\omega L_s}{r_{D1} + r_{D2}} \ln \frac{V_s + r_{D2} I_o}{(r_{D1} + r_{D2}) I_o}$$

7.48 Find the average output voltage for the circuit of Fig. P7.48. Determine the percentage change in the average output voltage if the ac-side line inductance doubles. Assume $v_s = 110\sqrt{2} \sin 377t$ V.

Fig. P7.48

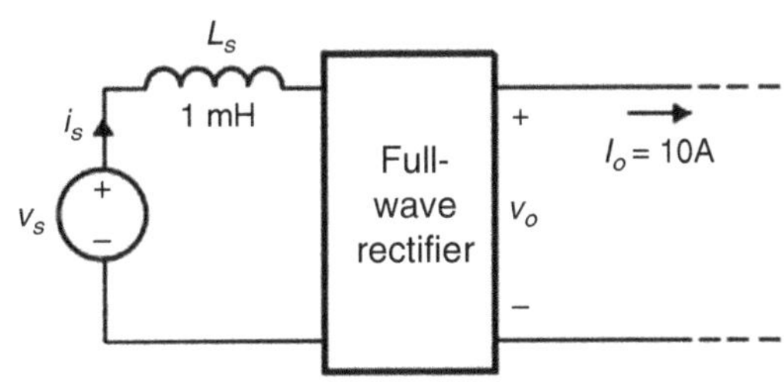

7.49 Consider the full-wave rectifier with an ac-side line inductance and a dc-source in the load side as shown in Fig. P7.49. (a) Sketch $i_o, i_s,$ and v_{Ls}, (b) calculate the average output power, and (c) calculate the percentage change in the average output power if the ac-side line inductance doubles in value. Assume $v_s = 50 \sin 2\pi 60t$ V.

Fig. P7.49

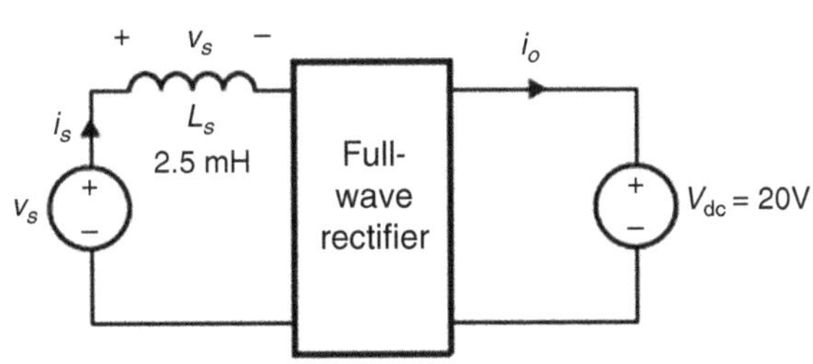

7.50 Repeat Problem 7.49 for v_s being a square wave with ± 50 V peaks and $T = 20$ ms.

7.51 Sketch the waveforms for i_{s1}, i_{s2}, and v_s' for the circuit in Fig. P7.51. Find the THD for v_s'.

Fig. P7.51

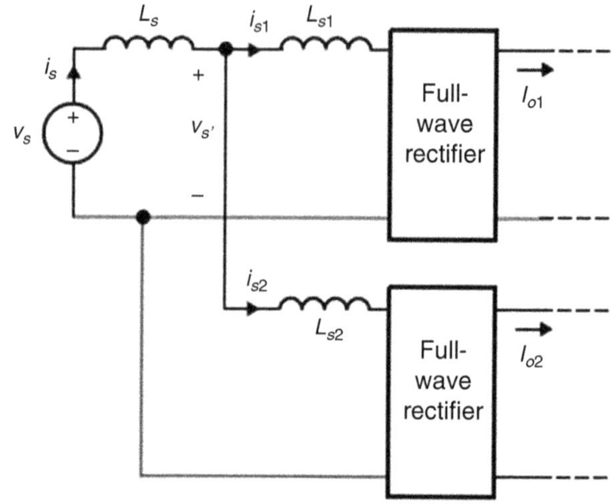

7.52 Derive the expression for the average output voltage in Fig. P7.52 in terms of the circuit parameters, assuming $v_s(t) = V_s \sin \omega t$.

Fig. P7.52

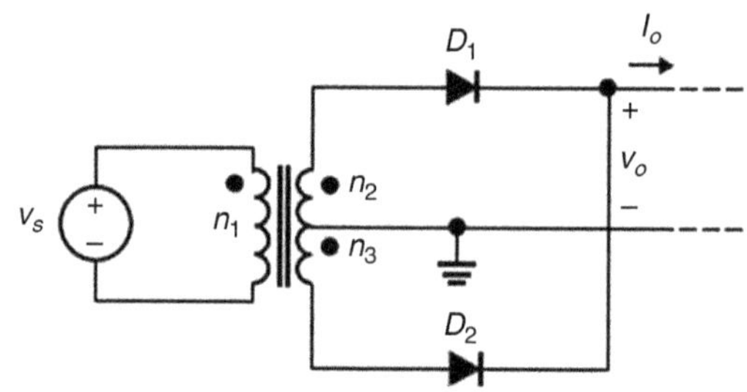

7.53 Repeat problem 7.52 by assuming that there exist an ac-side line inductance and diode resistors r_{D1} and r_{D2} as shown in Fig. P7.53.

Fig. P7.53

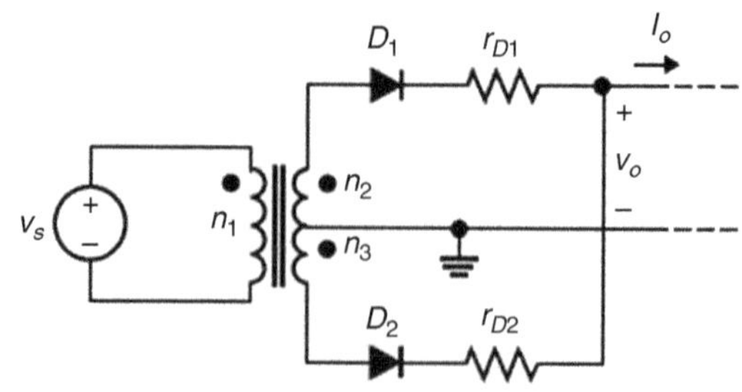

Chapter 8
Phase-Controlled Converters

8.1 Introduction

We saw in the preceding chapter how diodes can be used to rectify an ac input voltage to produce an uncontrolled dc output. These circuits—whether half-wave or full-wave configurations under resistive, inductive, or capacitive loads—have one common feature: The level of the output voltage is a function only of the circuit parameters and the peak voltage and frequency of the applied voltage source. For this reason such circuits are known as uncontrolled rectifier circuits. In this chapter, controlled rectifier circuits, using the silicon-controlled rectifier (SCR) instead of the diode, will be discussed. Unlike in diode rectifier circuits, in controlled circuits, the power may flow from the load side (dc side) to the source side (ac side) under some control condition. This negative direction of power flow is known as inversion, and the circuits are known as controlled inverter circuits.

In controlled rectifier circuits, diodes are replaced by SCRs to allow control of the conduction period or the phase of the conducting waveform. Since the voltage or current at the output is controlled by varying or delaying the phase of the conducting waveforms, such circuits are also known as phase-controlled converters. As shown in Chap. 2, unlike a diode, an SCR does not turn on when only the anode-cathode voltage becomes positive; rather, an additional signal must be applied to a third terminal (gate). The source of control stems from the fact that the gate signal can be applied at any time in the period during which the anode-cathode voltage is positive.

Controlled SCR rectifiers have a wide range of industrial and residential applications, especially applications in which power flows in both directions. For example, in the electrochemical industry, phase-controlled rectifiers are used to control the power in electroplating to the dc side, and in rotary machines, they are used to control the speed of dc and ac motors in both directions. In residential applications, medium-power phase-controlled rectifiers are used in light dimmers and variable-speed appliances.

I. Batarseh, A. Harb, *Power Electronics*,
https://doi.org/10.1007/978-3-319-68366-9_8

As in Chap. 7, half- and full-wave circuit configurations under resistive, inductive, and capacitive loads will be discussed. Also, because of their wide use in high-power applications, three-phase controlled circuits will be investigated. The SCR is used in such circuits because of its high current and voltage capabilities, its simple gating circuits, and its ability to control large anode currents with a small gate signal. Throughout this chapter, it will be assumed that the gate signal used to turn on the SCR is very short in duration and does not interfere with the circuit operation, that the turn-on and turn-off times are negligible, and that the anode-cathode voltage is zero in the conducting state (on-state) and the anode current is zero in the nonconducting state (off-state).

8.2 Basic Phase-Controlled Concepts

Figure 8.1 shows block diagram representation for single- and three-phase SCR-controlled circuits.

The ac side consists of the line-frequency input voltage supply, which is either a single-phase or a three-phase (delta or wye connections), with an ac-side inductance, L_s. Like the case in Chap. 7, the input power factor will be defined.

The SCR circuit block consists of one or more SCRs and possibly diodes, resulting in half- and full-wave and other configurations. The control signals are externally applied gating signals to turn on the SCRs in order to control the average output voltage. There are various integrated circuits commercially available to generate gating signals for such applications. Such ICs and other gating techniques are outside the scope of this textbook. The final block is labeled load, which represents the dc side of the converter.

Depending on the type of the load used, the load current can be either continuous or discontinuous. Both cases will be analyzed in this chapter. Under highly inductive loads, i_o will be assumed constant. This will simplify the analysis significantly.

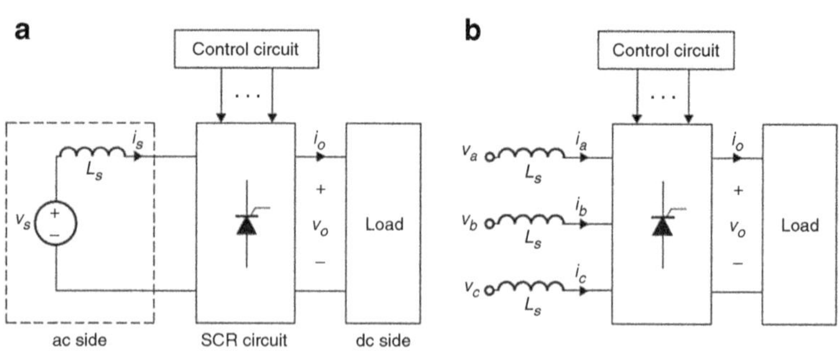

Fig. 8.1 Block diagram representation for phase-controlled converters: (**a**) single-phase and (**b**) three-phase

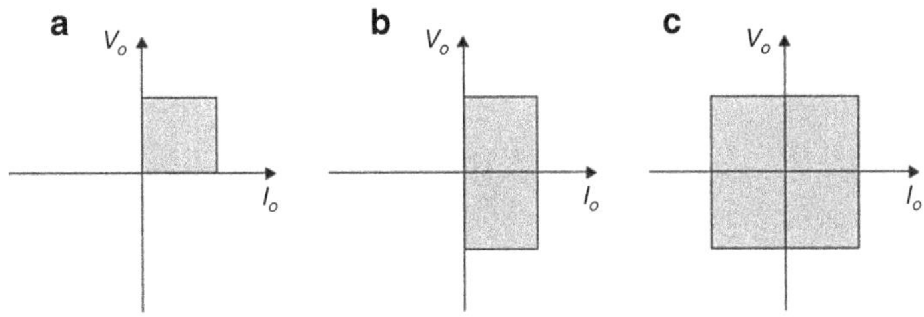

Fig. 8.2 Quadrant diagram representation for three different modes of operation for Fig. 8.1 (**a**) one-quadrant mode, (**b**) two-quadrant mode, and (**c**) four-quadrant mode

Fig. 8.3 (**a**) SCR representation with gate-trigger signals: i_{g_1} and i_{g_2} (**b**) Example with firing angles at α and π/ω

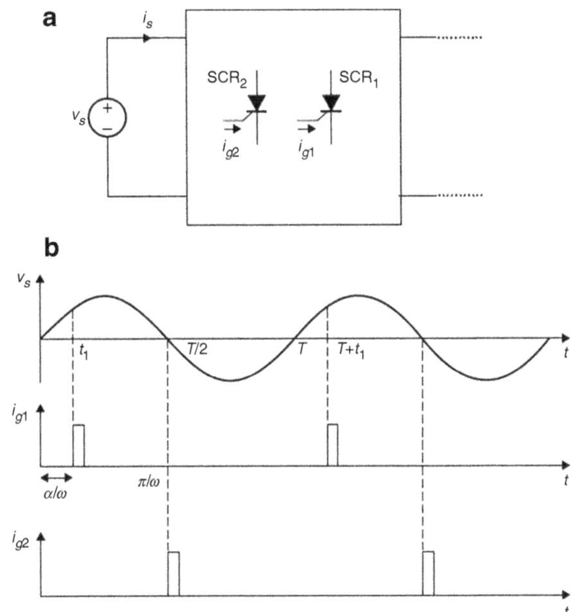

Depending on the SCR circuit configurations, the control signals, and the type of the load, it is possible to produce a dc load current, I_o, and a dc load voltage, V_o, with two polarities, resulting in a four-quadrant modes of operation as shown in Fig. 8.2.

In the two- and four-quadrant modes of operation, power can flow from the dc side to the ac side, resulting in what is known as a *phase-controlled inverter*. It will be shown later that for the inversion process to take place, the dc side must contain a type of energy source capable of delivering energy from the dc circuit to the ac supply side. The four-quadrant mode of operation is attainable if two circuits of the two-quadrant type are connected in a series or a back-to-back configuration.

Gating the SCR will be presented by a gate-trigger signal shown in Fig. 8.3b. The time, t_1, is known as the firing time or the delay time. It is common to represent the instant in terms of α, which is known as the *firing angle* or the *delay angle*,

$$\alpha = \omega t_1$$

where ω is the frequency (rad/s) of the supplied voltage. For example, the delay angle for SCR_1 is α and for SCR_2 is π as shown in Fig. 8.3b.

Since we assume ideal SCR, the height and width of the gate-trigger signal are not of importance. We should point out that both time and angle notations will be used interchangeably throughout this textbook.

8.3 Half-Wave Phase-Controlled Rectifiers

In this section, we will discuss half-wave phase-controlled rectifiers under resistive and inductive loads. These circuits are mainly used in medium-power application involving several kilowatts.

8.3.1 *Resistive Load*

When the diode is replaced by an SCR in the single-phase half-wave rectifier circuit, the resultant circuit known as a half-wave phase-controlled rectifier is shown in Fig. 8.4.When the source voltage is positive, the SCR will not conduct until the gate signal is applied at $t = t_1$ as shown in Fig. 8.5. At this time, the output voltage becomes equal to the input voltage. The output voltage, v_o, is sketched in Fig. 8.5.

We must restate that once the SCR starts conducting, the gate signal can be removed, and the *delay angle* (α) range is between 0° and 180°. At $t = T/2$ the input voltage becomes negative causing the SCR to turn off. The average output voltage is evaluated from the following equation:

$$\mathrm{V_o} = \frac{1}{T}\int_0^T v_o dt = \frac{1}{T}\int_{t_1}^{T/2} v_s\, dt = \frac{V_s}{2\pi}(1 + \cos\alpha) \tag{8.1}$$

where the normalized average voltage is given by:

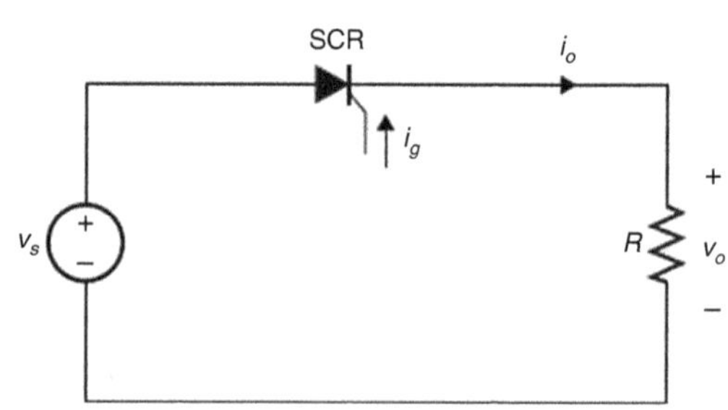

Fig. 8.4 Half-wave phase-controlled rectifier

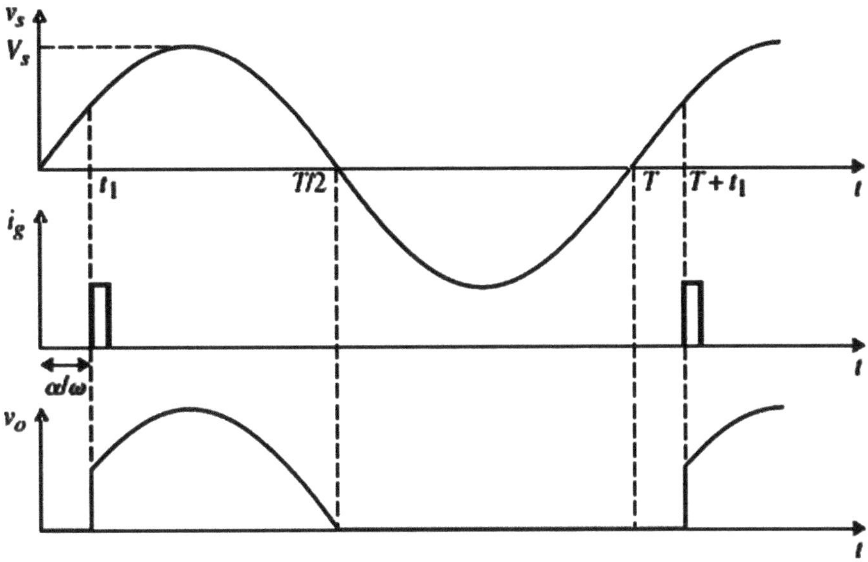

Fig. 8.5 Waveforms of Fig. 8.4 when the SCR is triggered at $t = t_1$

$$V_{\text{no}} = \frac{V_{\text{o}}}{V_{\text{s}}/\pi} = \frac{1 + \cos\alpha}{2} \tag{8.2}$$

where $\alpha = \omega t_1$ is the firing angle, the varying of which the controllability of the output voltage. Figure 8.6 shows the normalized average output voltage V_{no} as a function of the delay angle α. This control characteristic curve will allow us to design the rectifier firing angle for a desired normalized output.

The rms value can be evaluated from Eq. (8.3) or Eq. (8.4) as follows:

$$V_{\text{o,rms}} = \sqrt{\frac{1}{T}\int_{t_1}^{T/2} (V_{\text{s}} \sin\omega t)^2 dt} \tag{8.3}$$

$$= \frac{V_{\text{s}}}{2}\sqrt{\left(1 - \frac{\alpha}{\pi}\right) + \frac{\sin 2\alpha}{\pi}} \tag{8.4}$$

The input power factor (pf) and the total harmonic distortion (THD) for Fig. 8.4 are given by Eqs. (8.5) and (8.6), respectively:

$$\text{pf} = \frac{\sqrt{2}}{2}\sqrt{\left(1 - \frac{\alpha}{\pi}\right) + \frac{\sin 2\alpha}{\pi}} \tag{8.5}$$

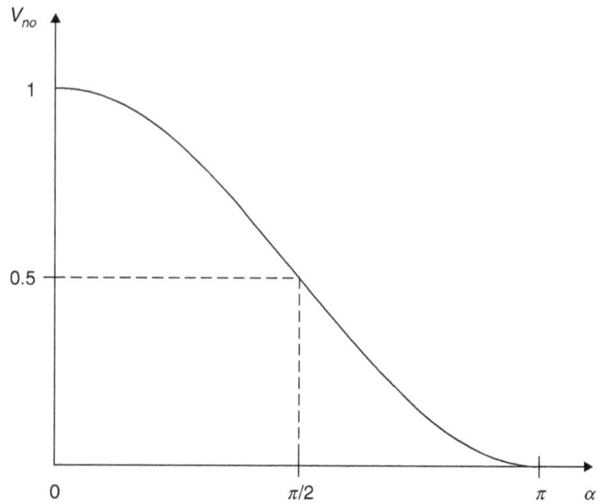

Fig. 8.6 Normalized average output voltage as a function of the delay angle

$$\text{THD} = \sqrt{\frac{2}{\left(1 - \frac{\alpha}{\pi}\right) + \frac{\sin 2\alpha}{\pi}} - 1} \tag{8.6}$$

The reader is invited to verify these equations (see Problem 8.1). Since the average output voltage is positioned for $0 \le \alpha \le \pi$ and the average output current is V_o/R , this inverter operates in the first quadrant. Comparing this inverter with the uncontrolled half-wave diode rectifier circuit, we note that the average output voltage can be controlled by varying α. The phase-controlled current and voltage waveforms are related by the delay angle, α. This is why α is also known as the *angle of retard*. Both the THD and the input power factor are reduced in the phase-controlled rectifier due to the presence of the displacement angle α.

8.3.2 *Inductive Load*

Figure 8.7a shows a half-wave phase-controlled converter with an inductive-resistive load. The relevant waveforms are depicted in Fig. 8.7b.

When the SCR is fired at $t = t_1$, the load current starts flowing until it reverses direction at $t = t_2$, at which the SCR is naturally turned off. It can be shown that the output current equation in the interval $t_1 < t < t_2$ is given by:

$$i_o(t) = \frac{V_s}{|Z|}\left[\sin(\omega t - \theta) - \sin(\alpha - \theta)e^{-(t - t_1)/\tau}\right] \tag{8.7}$$

where $\alpha = \omega t_1$, $\tau = L/R$, $\theta = \tan^{-1}(L/R)$, and $|Z| = \sqrt{R^2 + (\omega L)^2}$.

The average value of $v_o(t)$ can be obtained by evaluating the following integral:

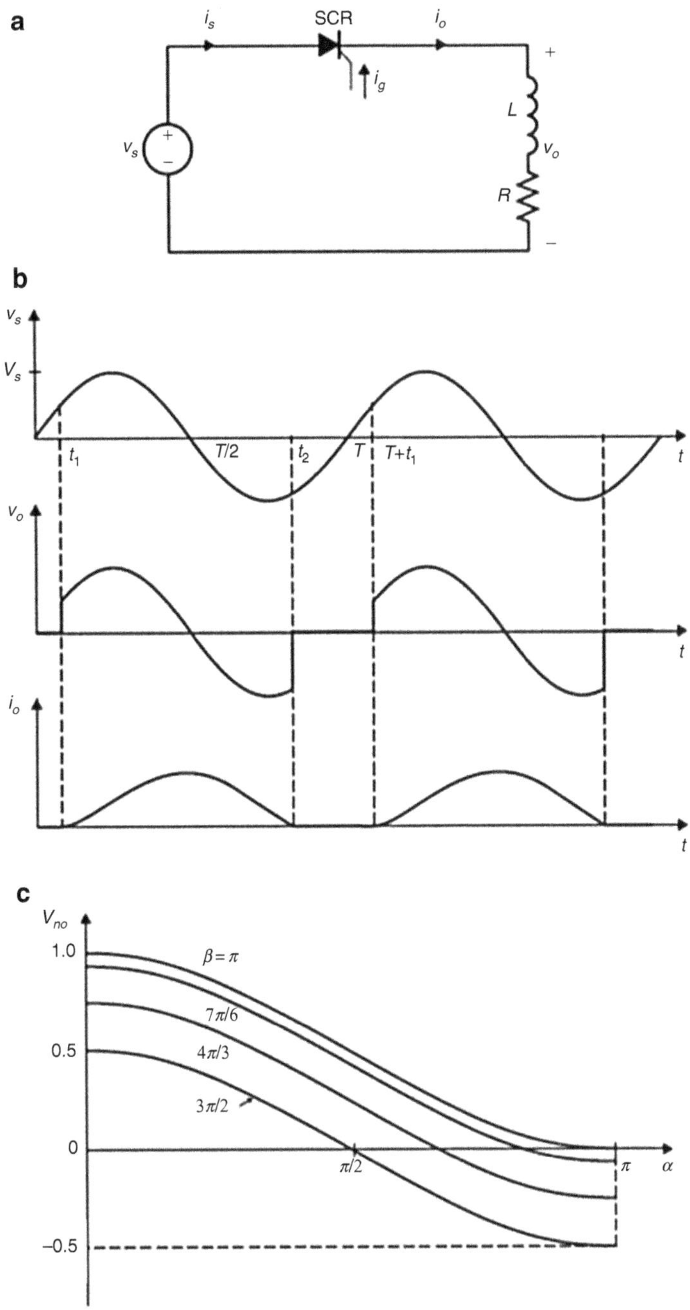

Fig. 8.7 (**a**) Half-wave phase-controlled circuit with inductive-load (**b**) relevant waveforms and (**c**) V_{no} verses α under different values of β

$$V_o = \frac{1}{T}\int_{t_1}^{t_2} V_s \sin \omega t\, dt \tag{8.8}$$

$$= \frac{V_s}{2\pi} \tag{8.9}$$

The numerical value of $\beta = \omega t_2$ can be obtained by evaluating Eq. (8.7) at $t = t_2$, with $i_o(t_2) = 0$.

Since the average value of the inductor voltage is zero in the steady state, the output average value is the same as the value across the load resistance. As a result, the average value of the load current, I_o, is given by:

$$I_o = \frac{V_s}{2\pi R}(\cos\alpha - \cos\beta) \tag{8.10}$$

Similarly, we normalize the output average voltage by:

$$V_{no} = \frac{V_o}{V_s/\pi}$$

Figure 8.7c shows the plot of V_{no} verses α under different values of β. Notice that for $\beta = \pi$, the curve gives the resistive-load case given in Fig. 8.6.

Exercise 8.1

For the half-wave phase-controlled rectifier of Fig. 8.4, assume the on-state voltage drop of the SCR is V_{SCR} and firing angle α:

(a) Sketch the waveforms for v_o and i_o.
(b) Derive the expression for the average output voltage V_o.
(c) Calculate V_o, *pf*, THD, and $V_{o,rms}$ using the following values: $V_s = 25$ V, $R = 10\ \Omega$, $V_{SCR} = 1.5$ V, and $\alpha = \pi/5$.

Answer: 6.59 *V*, 0.743, 104.9 %, 11.312 *V*.

Exercise 8.2

For the half-wave phase-controlled rectifier of Fig. 8.7a:

(a) Use $v_s = 78 \sin 377t$, L = 10 mH, $R = 5\ \Omega$, $V_o = 16$ V, to design for α.
(b) Obtain t_1 and t_2 in Fig. 8.7b.

Answer: 61°, 2.83 ms, 10 ms.

Consider the case when a free-wheeling diode is added to Fig. 8.7a as shown in Fig. 8.8a. The presence of the diode results in a continuous load current, and the output voltage is always positive. The corresponding waveforms are shown in Fig. 8.8b.

Notice that the principle operation of this circuit is similar to that of the half-wave diode rectifier with free-wheeling diode discussed in Chap. 3, with the exception that the intervals, in which the two circuit equations are applied, are

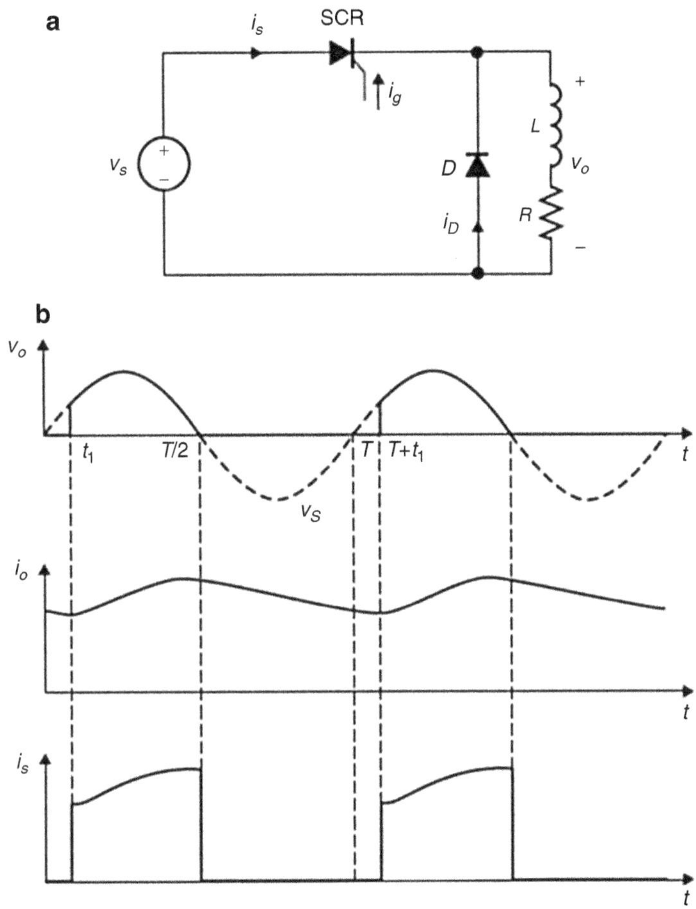

Fig. 8.8 (**a**) Half-wave phase-controlled rectifier with free-wheeling diode (**b**) typical waveforms for (a)

shifted by α. For the time interval in which the free-wheeling diode D is conducting, the circuit equation is given by:

$$L\frac{di_o}{dt} + Ri_o = 0 \qquad \pi \leq \omega t \leq 2\pi + \alpha \tag{8.11}$$

and for the time interval in which the SCR is conducting and D is reverse biased is:

$$L\frac{di_o}{dt} + Ri_o = V_s \sin \omega t \qquad \pi \leq \omega t \leq 2\pi + \alpha \tag{8.12}$$

By applying the proper boundary conditions at $\omega t = \alpha$ and $\omega t = \pi$, an expression for i_o can be derived (see Problem 8.3). It is interesting to notice that since v_o does

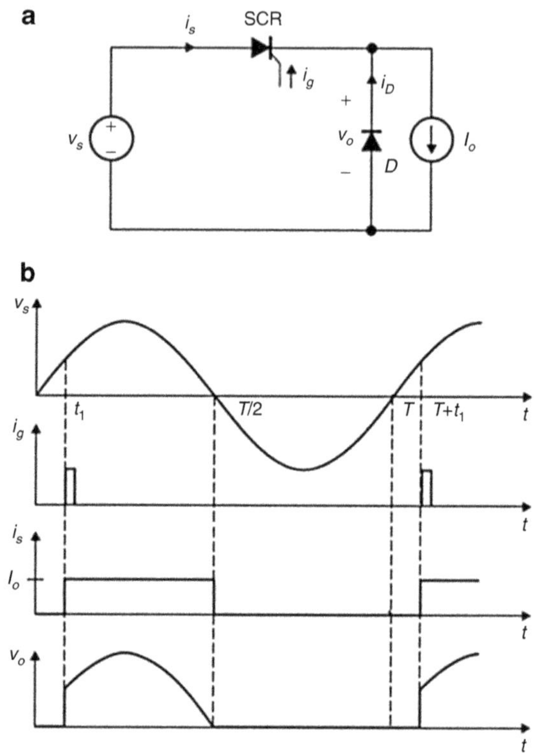

Fig. 8.9 (**a**) Equivalent circuit for Fig. 8.8a with infinitely large L, (**b**) relevant waveforms

not go negative, the average output is no longer a function of the load inductance. It is given by:

$$V_o = \frac{1}{T}\int_{t_1}^{T/2} V_s \sin \omega t \, dt = \frac{V_s}{2\pi}(1 + \cos \alpha) \tag{8.13}$$

Note that this circuit does not support a negative output voltage (no inversion).

Example 8.1

Draw the waveforms for i_s and v_o in Fig. 8.8a. Assume $L/R \gg T/2$.

Solution

The equivalent circuit of Fig. 8.8a is shown in Fig. 8.9a, and the relevant waveforms are shown in Fig. 8.9b. The control characteristic curve is the same as the one given in Fig. 8.6.

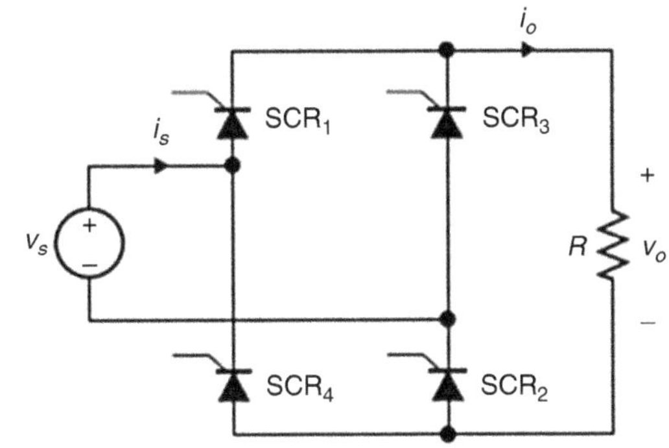

Fig. 8.10 Full-bridge phase-controlled rectifier

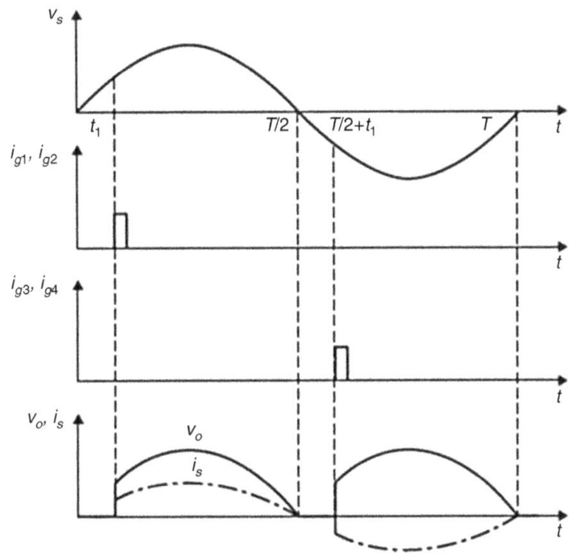

Fig. 8.11 Waveforms for Fig. 8.10

8.4 Full-Wave Phase-Controlled Rectifiers

8.4.1 Resistive Load

A phase-controlled full-wave bridge rectifier is shown in Fig. 8.10. This circuit is more practical than the phase-controlled half-wave rectifier shown in Fig. 8.4.

Like the uncontrolled full-wave bridge rectifier, the diagonal pairs SCR$_1$–SCR$_2$ and SCR$_3$–SCR$_4$ conduct during the positive and negative half-cycles, respectively. Assume the gating signal pairs are both delayed by α in their perspective half-cycles as shown in Fig. 8.11.

The output average value is calculated from the following integral:

$$V_o = \frac{2}{T}\int_{t_1}^{T/2} V_s \sin\omega t dt = \frac{V_s}{\pi}(1+\cos\alpha) \tag{8.14}$$

From Eq. (8.14), it is clear that the control characteristic curve of the average output voltage verses α for this full-wave topology is doubled when compared to the half-wave topology given in Fig. 8.6.

Example 8.2
Determine the input power factor and the THD for the full-bridge topology of Fig. 8.10.

Solution
The rms value of v_s is $V_s/(\sqrt{2})$, and the rms value for i_s, calculated from the waveform of Fig. 8.11, is:

$$I_{s,rms} = \sqrt{\frac{1}{T/2}\int_{t_1}^{T/2}\left(\frac{V_s}{R}\sin\omega t\right)^2 dt} = \frac{V_s}{\sqrt{2}R}\sqrt{1-\frac{\alpha}{\pi}+\frac{\sin 2\alpha}{2\pi}} \tag{8.15}$$

The average input power can be calculated from the following relation:

$$P_{in} = \frac{1}{T/2}\frac{1}{R}\int_{t_1}^{T/2}(V_s\sin\omega t)^2 dt = \frac{V_s^2}{2R}\left[\left(1-\frac{\alpha}{\pi}\right)+\frac{\sin 2\alpha}{2\pi}\right] \tag{8.16}$$

From the definition of the power factor, and using Eqs. (8.15) and (8.16), we can obtain the following relation for the power factor:

$$pf = \frac{P_{ave}}{I_{s,rms}V_{s,rms}} = \sqrt{\left(1-\frac{\alpha}{\pi}\right)+\frac{\sin 2\alpha}{2\pi}} \tag{8.17}$$

A plot of the pf verses α is given in Fig. 8.12.
The THD can be found to be:

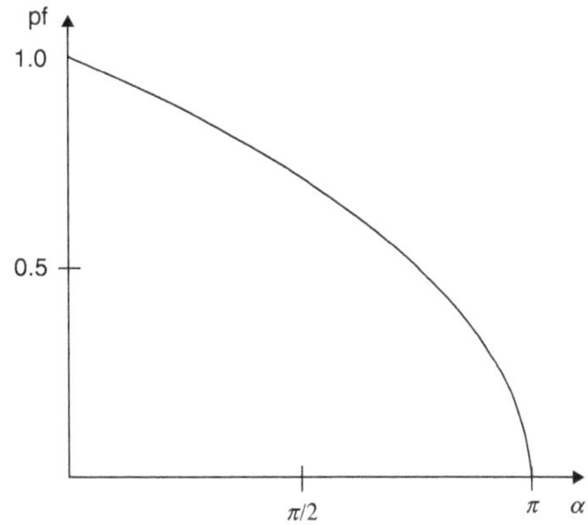

Fig. 8.12 Power factor verses the delay angle for circuit of Fig. 8.10

$$\mathrm{THD} = \sqrt{\frac{2\pi\alpha - \pi\sin(2\alpha) - 2\alpha^2 - 1 + 2\alpha\sin(2\alpha) + \cos(2\alpha)}{1 + 2(\pi-\alpha)^2 + 2(\pi-\alpha)\sin(2\alpha) - \cos(2\alpha)}}$$

Exercise 8.3
Calculate the input power factor for $\alpha = 30^\circ$ and 150° using Fig. 8.12.

Answer: $0.985, 0.17$.

Exercise 8.4
Consider the waveform for $i_s(t)$ given in Fig. 8.11:

(a) Derive the expression of the fundamental component, $i_{s_1}(t)$, for the input current i_s of Fig. 8.11.
(b) Calculate the THD for $\alpha = 30^\circ$ and $\alpha = 150^\circ$.

Answer: $10.2\,\%$, $18.9\,\%$.

8.4.2 Inductive Load

The full-wave phase-controlled rectifier under an inductive load is shown in Fig. 8.13. Let us first consider the waveforms for i_o and v_o and assume that the time constant L/R, which is not large compared to $T/2$. Again, the thyristor pairs $SCR_1 - SCR_2$ and $SCR_3 - SCR_4$ are triggered diagonally, each delayed by α in the respective half-cycles. It can be shown that, as far as the local current is concerned, there are two modes of operation: discontinuous and continuous conduction modes. If we assume that when in steady-state operation, the inductor current value reaches zero before $t = T/2 + t_1$, and then the rectifier is known to operate in the discontinuous conduction mode (dcm). For dcm to occur, the load current must reach zero before the opposite pair of SCRs is fired. However, if the inductor current (load) is not allowed to reach zero, i.e., the second diagonal pair of SCRs is triggered while

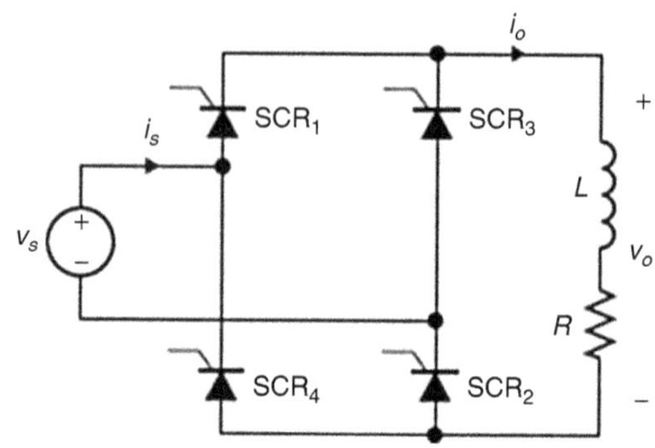

Fig. 8.13 Full-wave phase-controlled rectifier with inductive load

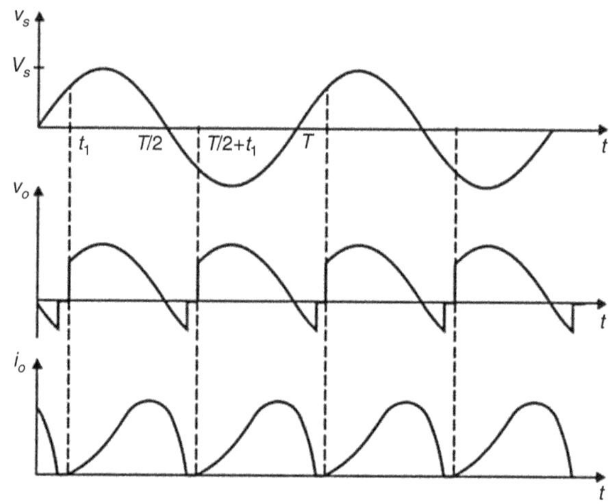

Fig. 8.14 Waveforms for Fig. 8.13 under discontinuous conduction mode

the current is flowing, then since the current in the inductor does not change instantaneously, the current commutates instantaneously in the other thyristors. Under this case, the rectifier is known to operate in the continuous conduction mode (ccm). Figures 8.14 and 8.15 show the waveforms for v_o and i_o under both dcm and ccm, respectively.

Example 8.3

Derive the expression for $i_o(t)$ under both the continuous and discontinuous conduction modes, and determine the condition at which the boundary between ccm and dcm occurs.

Solution

Since the dcm is a special case of the ccm, let us first consider the continuous conduction mode case. As shown in the previous chapter, the total solution for $i_o(t)$ is given by:

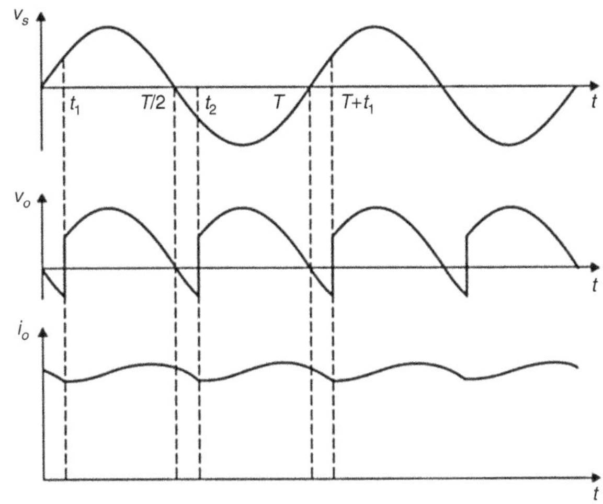

Fig. 8.15 Waveforms for Fig. 8.13 under continuous conduction mode

$$i_o(t) = Ae^{-(t-t_1)/\tau} + \frac{V_s}{|Z|}\sin(\omega t - \theta) \tag{8.18}$$

where A is a constant and must be determined from the initial condition, and the other parameters are given by:

$$\tau = L/R$$
$$|Z| = \sqrt{(\omega L)^2 + R^2}$$
$$\theta = \tan^{-1}\frac{\omega L}{R}$$

If we let the initial condition of $i_o(t)$ at $t = t_1$ be I_{o1}, from Eq. (8.18), we have:

$$i_o(t_1) = I_{o1} = A + \frac{V_s}{Z}\sin(\alpha - \theta)$$

solving for A, we obtain:

$$A = I_{o1} - \frac{V_s}{|z|}\sin(\alpha - \theta)$$

Therefore, the expression for the load current is given by:

$$i_o(t) = \left[I_{o1} - \frac{V_s}{|z|}\sin(\alpha - \theta)\right]e^{-(t-t_1)/\tau} + \frac{V_s}{|Z|}\sin(\omega t - \theta) \tag{8.19}$$

In the steady-state operation, $i_o(t_1)$ must equal $i_o(T/2 + t_1)$; hence, evaluating Eq. (8.19) at $t = T/2 + t_1$ and solving for I_{o1}, we obtain:

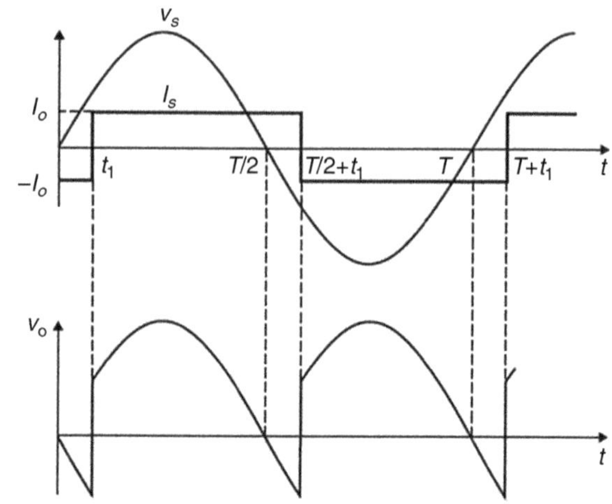

Fig. 8.16 Waveforms of Fig. 8.13 under infinite load inductance

$$I_{o1} = \frac{-\frac{V_s}{|z|}\left(e^{-\pi/\omega t} - 1\right)\sin\left(\alpha - \theta\right)}{1 - e^{-\pi/\omega t}} \tag{8.20}$$

This initial condition will determine the mode of operation as follows:

$$I_{o1} : \begin{cases} < 0 & \text{for } \alpha > \theta \quad \text{dcm} \\ = 0 & \text{for } \alpha = \theta \text{ boundary condion} \\ > 0 & \text{for } \alpha > \theta \quad \text{ccm} \end{cases}$$

If we assume that the load time constant L/R is much larger than the half-period of the applied voltage, then we can model the load current by a constant current source, I_o. In this case, the waveform of the current i_s is a square wave of magnitude $\pm I_o$, shifted by the triggering angle $\alpha = \omega t_1$ as shown in Fig. 8.16.

The average output voltage of the full-bridge phase-controlled circuits with the waveforms of Figs. 8.14, 8.15, and 8.16 are the same and given by:

$$\begin{aligned} V_o &= \frac{1}{T/2}\int_{t_1}^{T/2+t_1} V_s \sin\omega t \, dt \\ &= \frac{2V_s}{\pi}\cos\alpha \end{aligned} \tag{8.21}$$

The control characteristic curve for the normalized average output voltage $V_{no} = \pi V_o/V_s$ is shown in Fig. 8.17. Notice that when $\pi/2 < \alpha < \pi$, this circuit produces a negative average output voltage. Since the average load current is always positive, the direction of the power flow in this range of α is from the dc output side to the ac input side. Consequently, as stated before, the circuit is known to operate in the inversion mode, whereas for $0 < \alpha < \pi/2$, the circuit operates in the rectification mode.

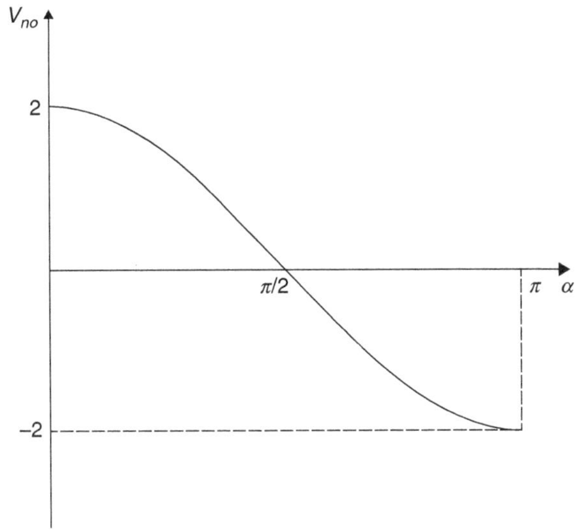

Fig. 8.17 Control characteristic curve for Fig. 8.16

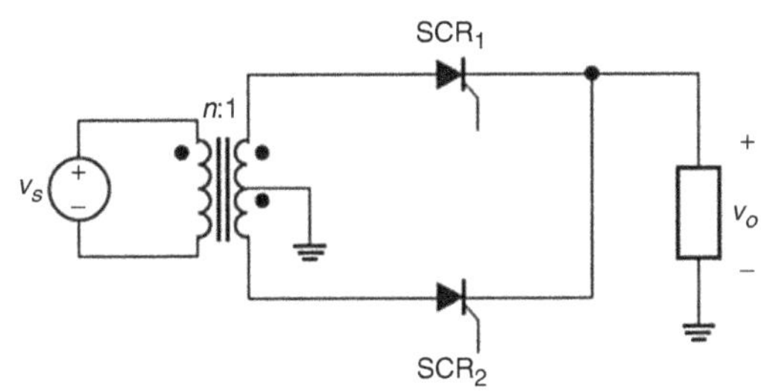

Fig. 8.18 Full-wave controlled rectifier with center-tap transformer

If the application requires the use of an isolated transformer, a center-tap full-wave controlled rectifier can be obtained using two SCRs as shown in Fig. 8.18. All waveforms are similar to the waveforms under resistive and inductive loads.

Example 8.4

Derive the power factor for the phase-controlled full-wave bridge rectifier under a constant load current as shown in Fig. 8.19a.

Solution

The waveform for $i_s(t)$ is redrawn in Fig. 8.19b. From the Fourier series, we have:

$$i_s(t) = I_{dc} + \sum_{n=1}^{\infty} a_n \cos n\omega t + b_n \sin (n\omega t)$$

From the symmetry of $i_s(t)$, the dc average value is zero and only its odd harmonics exist.

The fundamental component is given by:

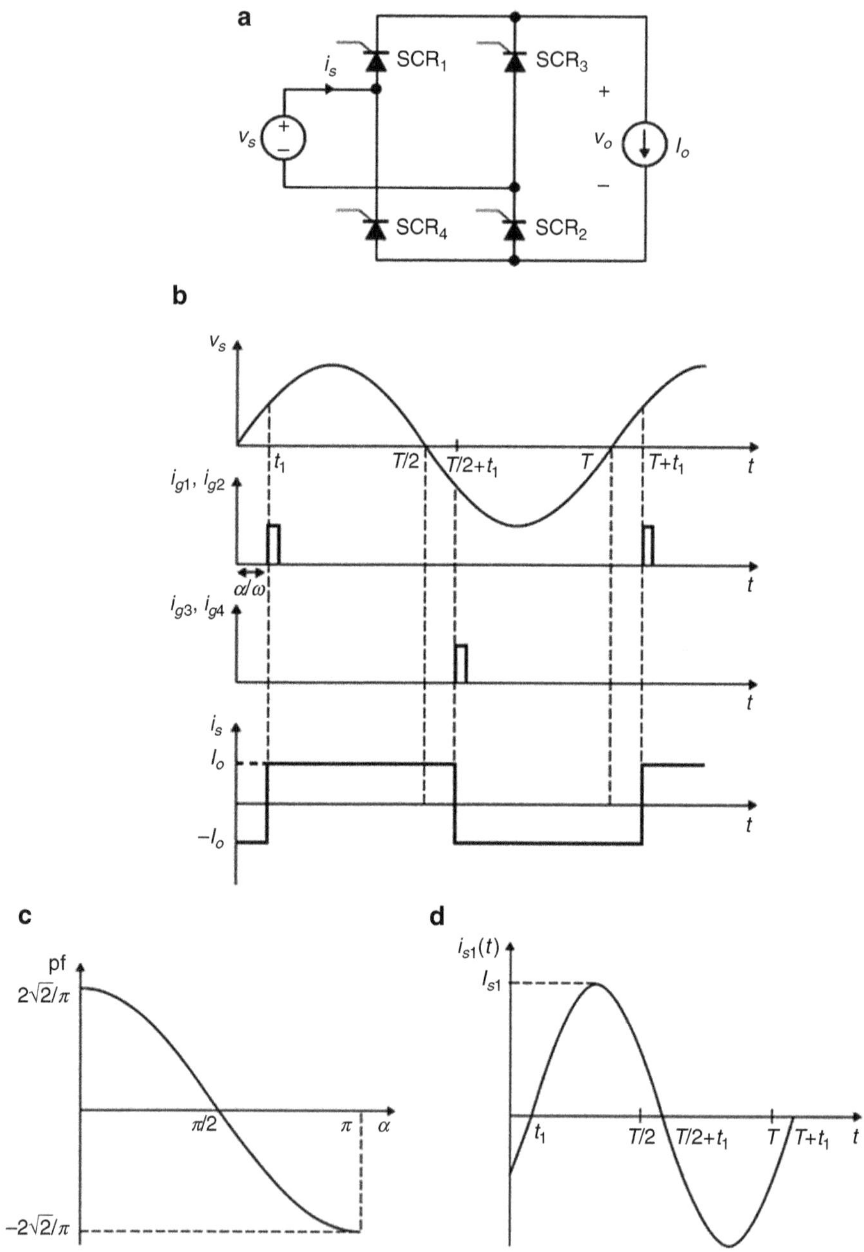

Fig. 8.19 (**a**) Full-bridge phase-controlled circuit under constant load current, (**b**) the source current waveforms, (**c**) power factor, and (**d**) fundamental component of $i_s(t)$

$$i_{s_1}(t) = I_{s_1} \sin(\omega t + \theta)$$

where

$$I_{s_1} = \sqrt{a_1^2 + b_1^2}$$

$$\theta = \tan^{-1}\frac{a}{b}$$

The coefficients a_1 and b_1 are evaluated from the following relations:

$$a_1 = \frac{2}{T}\left[\int_{t_1}^{T/2+t_1} I_o \cos(\omega t)dt + \int_{T/2+t_1}^{T+t_1} -I_o \cos(\omega t)dt\right]$$

$$= -\frac{4I_o}{\pi}\sin\alpha$$

$$b_1 = \frac{2}{T}\left[\int_{t_1}^{T/2+t_1} I_o \sin(\omega t)dt + \int_{T/2+t_1}^{T+t_1} -I_o \sin(\omega t)dt\right]$$

$$= -\frac{4I_o}{\pi}\cos\alpha$$

Hence, the peak value of the fundamental component is given by:

$$I_{s_1} = \frac{4I_o}{\pi}$$

and its rms value is:

$$I_{s_1,\mathrm{rms}} = \frac{4I_o}{\sqrt{2}\pi}$$

The rms value of $i_s(t)$ is I_o.

From the above expression, we obtain the power factor:

$$pf = \frac{4}{\sqrt{2}\pi}\cos\theta$$

where the displacement angle is given by:

$$\theta = -\tan^{-1}\left(\frac{\sin\alpha}{\cos\alpha}\right) = -\tan^{-1}(\tan\alpha) = -\alpha$$

The power factor is expressed in terms of the delay angle α as:

$$pf = \frac{4}{\sqrt{2}\pi} \cos \alpha$$

which is sketched in Fig. 8.19c. The fundamental component is shown in Fig. 8.19d.

Example 8.5
Consider the full-wave phase-controlled rectifier given in Fig. 8.20a with two SCRs, two diodes D_1, D_2, and a flyback diode D_3.

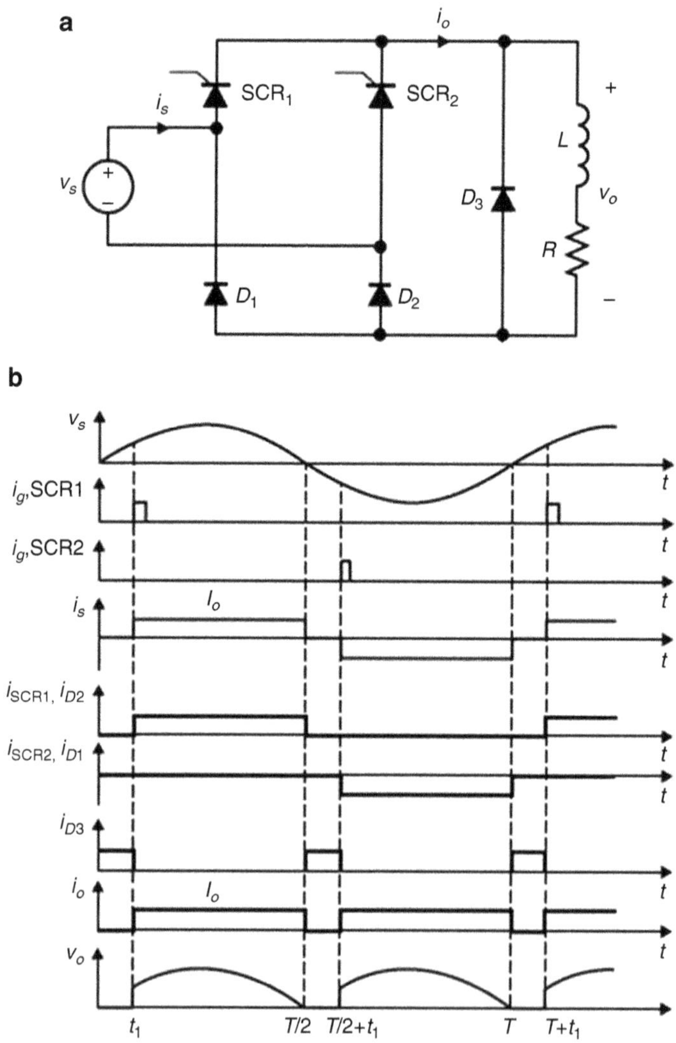

Fig. 8.20 (**a**) Full-wave rectifier with two SCRs and flyback diode and (**b**) voltage and current waveforms

(a) Assume $L/R \gg T/2$, draw the waveforms for i_s, i_{SCR1}, i_{SCR2}, $i_{D_1}, i_{D_2}, i_{D_3}, i_o$, and v_o for $\alpha = 30^\circ$.
(b) Show that the fundamental component of $i_s(t)$ is given by:

$$i_{s_1}(t) = I_{s_1} \sin\left(\omega t - \frac{\alpha}{2}\right)$$

where

$$I_{s_1} = \frac{2\sqrt{2}I_o}{\pi}\sqrt{1 + \cos 2\alpha}$$

I_o is the load current, assumed constant, and $\alpha = \omega t_1$.

(c) Derive the expression for the power factor and THD.

Solution

(a) To sketch the waveforms, we must understand the basic circuit operation. Since we assume $L/R \gg T/2$, then i_o is considered constant as I_o. Let the firing times for SCR_1 and SCR_2 occur at t_1 and $T/2 + t_1$, respectively, as shown in Fig. 8.20b. Before SCR_1 is gated at $t = t_1$, the load current is flowing in the flyback diode D_3, since SCR_2 is off because of the negative source voltage. At $t = t_1$, SCR_1 and D_2 turn on. At $t = T/2$, the negative source voltage turns D_3 on, and SCR_1 and D_2 seize to conduct. At $t = T/2 + t_1$, SCR_2 is gated, turning it on and causing D_3 to turn off and D_1. This mode of operation continues until $t = T$ when SCR_2 and D_1 turn off, causing D_3 to turn on again. The cycle repeats at $t = T = t_1$ when $SCR_1$is gated again.
(b) The fundamental component of $i_s(t)$ is given by:

$$i_{s_1}(t) = I_{s_1} \sin(\omega t + \theta)$$

The parameters I_{s_1} and θ are obtained from:

$$I_{s_1} = \sqrt{a_1^2 + b_1^2}$$
$$\theta = \tan^{-1}\frac{a_1}{b_1}$$

where

$$\begin{aligned} a_1 &= \frac{2}{T}\int_0^T i_s(t)\cos\omega t dt \\ &= \frac{2}{T}\left[\int_{t_1}^{T/2} I_o \cos(\omega t)dt - \int_{T/2+t_1}^{T} -I_o \cos(\omega t)dt\right] \\ &= \frac{-2I_o}{\pi}\sin\alpha \end{aligned}$$

and

$$\begin{aligned} b_1 &= \frac{2}{T}\int_0^T i_s(t)\sin\omega t dt \\ &= \frac{2}{T}\left[\int_{t_1}^{T/2+t_1} I_o \sin(\omega t)dt + \int_{T/2+t_1}^{T+t_1} -I_o \sin(\omega t)dt\right] \\ &= \frac{2I_o}{\pi}(1+\cos\alpha) \end{aligned}$$

Hence, I_{s_1} is given by:

$$I_{s_1} = \frac{2\sqrt{2}I_o}{\pi}\sqrt{1+\sin 2\alpha}$$

(c) The rms expression is given by:

$$\begin{aligned} I^2_{s,\text{rms}} &= \frac{1}{2\pi}\left[\int_\alpha^\pi I_o^2 d\omega t + \int_{\pi+\alpha}^{2\pi} I_o^2 d\omega t\right] \\ &= \frac{1}{2\pi}\left[I_o^2(\pi-\alpha) + I_o^2(2\pi-(\pi+\alpha))\right] \\ &= \frac{I_o^2(-2\alpha+2\pi)}{2\pi} \\ &= I_o^2\left(1-\frac{\alpha}{\pi}\right) \end{aligned}$$

Hence, the power factor is expressed as:

$$\text{pf} = \frac{2}{\pi}\frac{\sqrt{1+\cos\alpha}}{\sqrt{1-\frac{\alpha}{\pi}}}\cos\frac{\alpha}{2}$$

and the total harmonic distortion is given by:

$$\text{THD} = \sqrt{\frac{\pi(\pi-\alpha)}{4(1+\cos\alpha)} - 1}$$

The pf and THD plots are given in Fig. 8.21.

It is possible to design a full-wave phase-controlled converter that can provide bidirectional flow of the output voltage and the output current. The system of Fig. 8.22a can operate in the four quadrants as shown in Fig. 8.22b.

The preceding general characteristics can be achieved by using dual full-wave phase-controlled converters connected in parallel, with their SCR connections opposite to each other. Two other configurations for full-wave bridge converters

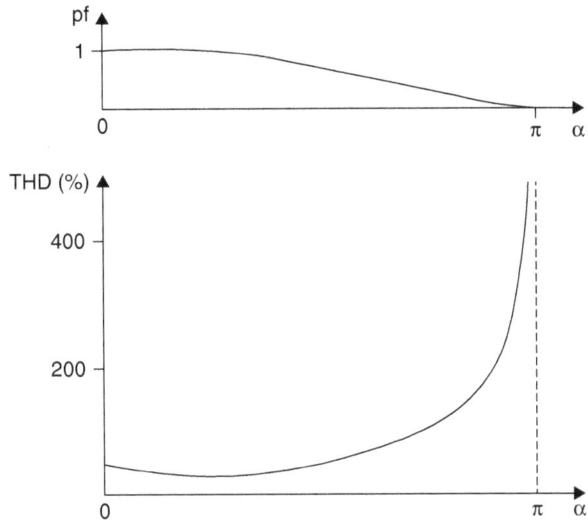

Fig. 8.21 pf and THD as a function of α

are known as series and parallel converters, shown in Figs. 8.23 and 8.24, respectively.

Figure 8.25 shows a phase-controlled full-wave circuit with a dc voltage source in the load side. Again, in the analysis, we assume that the time constant *L*/*R* is very large compared to the period of the applied voltage. Hence, we can model the inductor load current as a constant I_o. The waveforms of the output voltage, v_o, and the source current, i_s, are the same as those of Fig. 8.19a. In the steady state, the average output current is given by:

$$I_o = \frac{V_o}{R} - \frac{V_{dc}}{R} = 2\frac{V_s}{\pi R}\cos\alpha - \frac{V_{dc}}{R} \tag{8.22}$$

Exercise 8.5

Consider the phase-controlled full-wave rectifier shown in Fig. 8.25 with $v_s = 110\sin 2\pi(60)t$, $L = 38$ mH, $R = 25\ \Omega$, and $V_{dc} = 12\ V$. Assume the firing angle is at 45°, and the load current is constant. Find (a) the average output voltage, (b) the average output current, (c) the input and output average powers, and (d) the power factor.

Answer: $49.5\ V, 1.5\ A, 74.26\ W, 0.64$.

Example 8.6

Determine the range of the delay angle α in Fig. 8.25 so that the input power factor is at least 0.75 and the minimum output power delivered to a 5 Ω load resistor is 1.2 kW. Assume $v_s = 240\sin 377t$, $L = 125$ mH, and $V_{dc} = 20$ V.

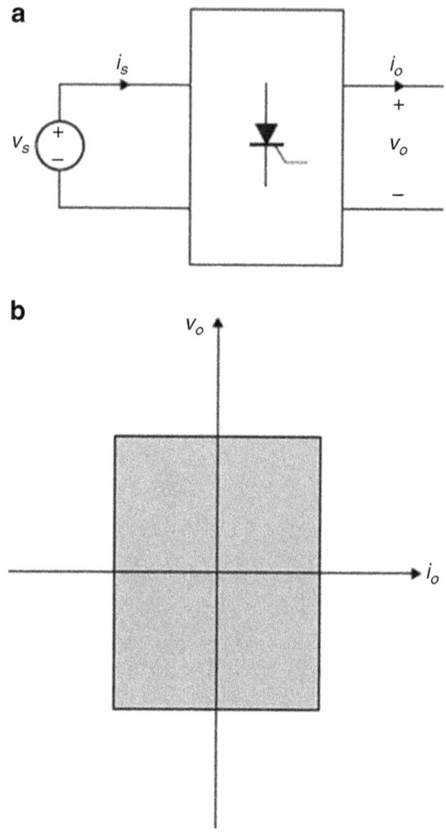

Fig. 8.22 (**a**) Block diagram representation for a single-phase phase-controlled rectifier and (**b**) four-quadrant operation

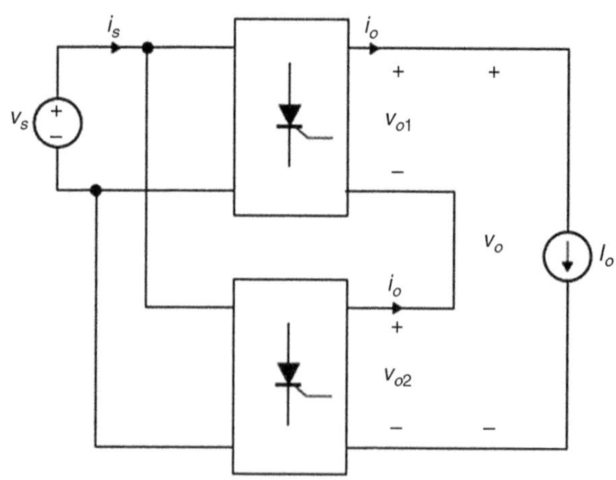

Fig. 8.23 Series-connected full-wave SCR circuits

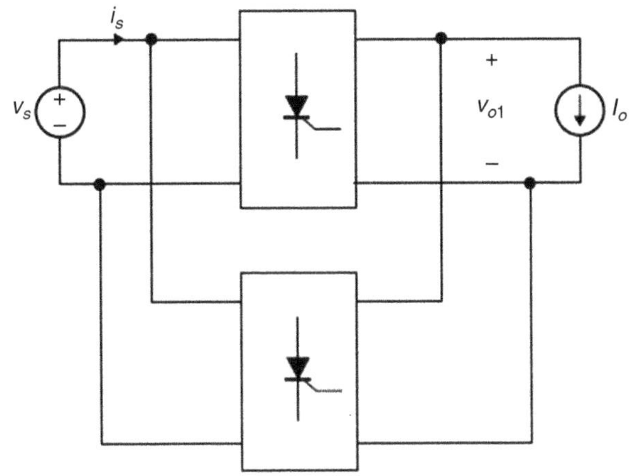

Fig. 8.24 Parallel-connected full-wave SCR circuits

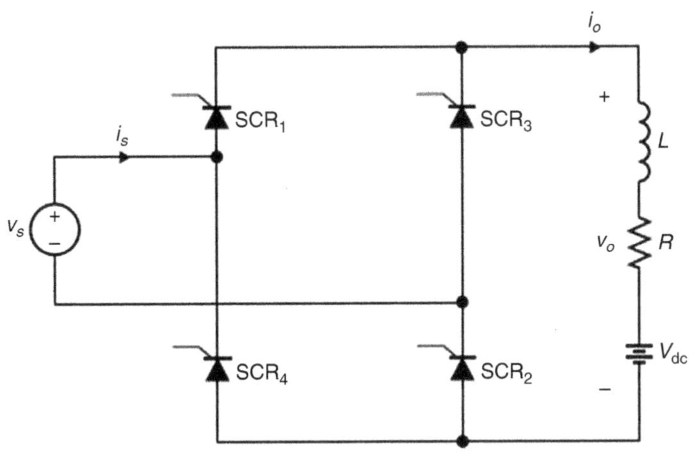

Fig. 8.25 Phase-controlled full-wave circuit with load voltage source

Solution

The L-R time constant is $L/R = 25\text{ms}$ which is larger than $T/2$. Thus the ripple current is considered very small. For an output power equal to 1.2 kW, the load current I_o equals 2.83 A from Eq. (8.22), and $\alpha = 50.37^{\circ}$. For a power factor of 0.75, the conduction angle must be smaller than 33.6°. As a result, the resultant range of α is $0^{\circ} \leq \alpha \leq 33.6^{\circ}$. The power delivered to the load at $\alpha = 0^{\circ}$ and $\alpha = 33.6^{\circ}$ is 3.53 kW and 2.304 kW, respectively.

8.5 Effect of AC-Side Inductance

In this section, we turn our attention to the analysis of the SCR rectifier and inverter circuits by including the ac-side inductance L_s. As with the diode rectifier circuits, here, both the half-wave and the full-wave configurations will be considered.

8.5.1 *Half-Wave Circuits*

Figure 8.26a shows the half-wave phase-controlled circuit including L_s with an inductive load. Without a flyback SCR across $R-L$, the circuit behaves like the half-wave rectifier under an inductive load discussed in Section. 8.3. Figure 8.26b shows the above circuit by including a flyback SCR_2.

The analysis of this circuit is easily carried out by assuming $L/R \gg T$. Figure 8.26c shows the waveforms for i_s, i_{SCR_1}, i_{SCR_2}, and v_o under a constant load current.

Notice that at $t = T/2$, SCR_2 cannot be turned on because its gate signal is not available until time $t = t_1 + T/2$. The operation of this circuit is similar to that of the diode circuit with an ac-side inductance discussed in the previous chapter. At $t = t_1 + T/2$, when the SCR_2 is fired, the current in SCR_1 cannot change to zero because of L_s; hence, both SCR_1 *and* SCR_2 remain on for a period given by $u/w = t_2 - t_1$ in Fig. 8.26c:

$$V_o = \frac{1}{T}\int_{t_1}^{T/2+t_1} V_s \sin \omega t \, dt \tag{8.23}$$

During this interval, the current in SCR_1 decreases to zero, and the current in SCR_2 increases to I_o at the same rate. The average output voltage is given by evaluating Eq. (8.23) and using $\alpha = \omega t_1$ and $u = \omega(t_2 - t_1)$, to obtain the following equation for V_o:

$$V_o = \frac{V_s}{2\pi}(\cos \alpha + \cos(\alpha + u)) \tag{8.24}$$

The load current in terms of the commutation angle u and the firing angle α is obtained by integrating Eq. (8.25) from $t = t_1$ to t:

$$L_s \frac{di_s}{dt} = v_s(t) \tag{8.25}$$

Equation (8.25) is obtained from Fig. 8.26a when both SCR_1 and SCR_2 are conducting. From Eq. (8.25), we obtain $i_s(t)$ using the initial condition $i_s(t_1) = 0$, to yield:

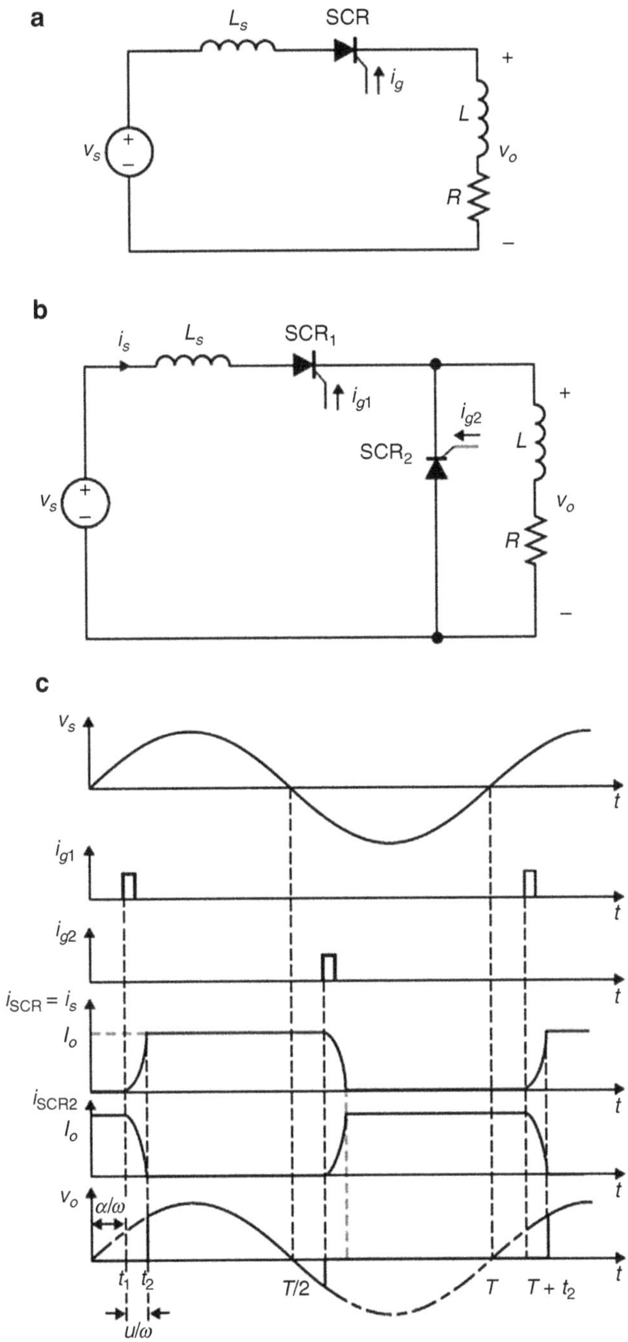

Fig. 8.26 Half-wave SCR rectifier with ac-side reactance: (**a**) without flyback SCR, (**b**) with flyback SCR_2, (**c**)voltage, and current waveform

$$\begin{aligned} i_s(t) &= -\frac{V_s}{L_s\omega}(\cos\omega t - \cos\omega t_1) \\ &= \frac{V_s}{L_s\omega}(\cos\alpha - \cos\omega t) \end{aligned} \tag{8.26}$$

Evaluating Eq. (8.26) at $t = t_2$, where $i_s(t_2) = I_o$, we obtain:

$$I_o = \frac{V_s}{\omega L_s}[\cos\alpha - \cos(u+\alpha)] \tag{8.27}$$

Substitute the above equation into Eq. (8.24) to give:

$$V_o = \frac{V_s}{\pi}\left(\cos\alpha - \frac{\omega L_s I_o}{2V_s}\right) \tag{8.28}$$

In terms of the normalized output voltage and the normalized output current, Eq. (8.28) may be written as follows:

$$V_{no} = \frac{1}{\pi}\left(\cos\alpha - \frac{I_{no}}{2}\right) \tag{8.29}$$

The characteristic curve for I_{no} as a function of V_{no} under different firing angles is shown in Fig. 8.27.

8.5.2 *Full-Wave Rectifier Circuits*

Figure 8.28 shows the controlled full-wave bridge rectifier with an ac-side inductance. Assuming constant I_o, the waveforms for v_s, i_s, and v_o are shown in Fig. 8.29.

The basic circuit operation is similar to that of the full-wave diode rectifier with an ac-side inductance. We first assume that SCR_1 and SCR_2 are off during the negative cycle of $v_s(t)$ for $t \le t_1$. The voltage across L_s is zero; hence, the entire source voltage appears across the output as shown in Fig. 8.30a, which is represented as Mode 1 (M_1).

In this mode, we have $v_{Ls} = 0$, $i_s = -I_o$, and $v_o = -V_s$. This mode continues, while SCR_1 and SCR_2 remain off, until $t = t_1$ when they are triggered on. At this time, SCR_3 and SCR_4 remain on to maintain continuity of the inductor current. The resultant circuit is shown in Fig. 8.30b and labeled as Mode 2.

The inductor current, i_s, is obtained from the following relation:

$$\begin{aligned} v_{Ls} &= L_s\frac{di_s}{dt} \\ &= v_s(t) \end{aligned} \tag{8.30}$$

Using the initial condition $i_s(t_1) = -I_o$ in Eq. (8.30), we obtain:

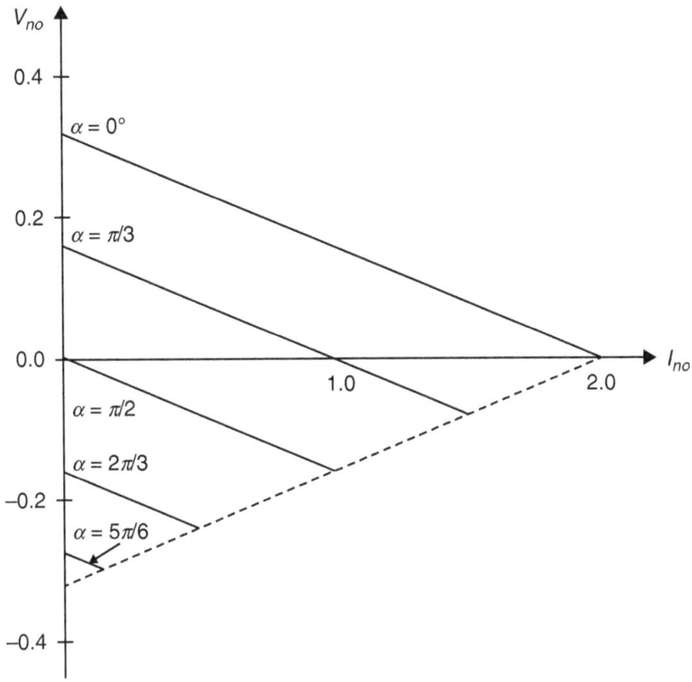

Fig. 8.27 Characteristic curves for I_{no} versus V_{no} different values of α

Fig. 8.28 Controlled full-bridge rectifier with ac-side reactance

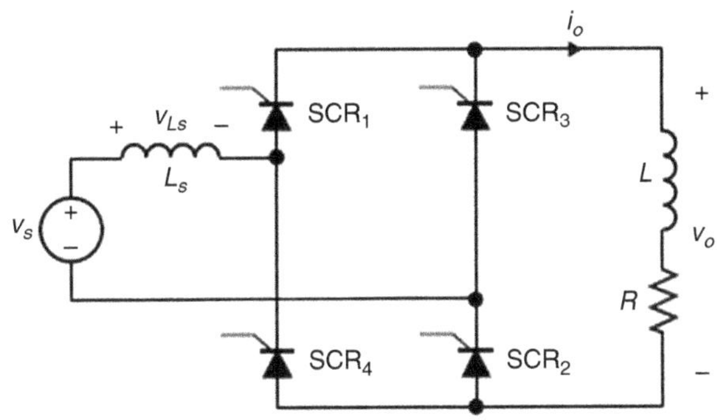

$$
\begin{aligned}
i_s(t) &= \int_{t_1}^{t} \frac{V_s}{L_s} \sin \omega t \, dt - I_o \\
&= \frac{V_s}{L_s \omega} (\cos \alpha - \cos \omega t) - I_o
\end{aligned}
\tag{8.31}
$$

At $t = t_2$, the currents in SCR_3 and SCR_4 reach zero, and the circuit enters Mode 3 as shown in Fig. 8.30c. In this mode we have $i_s = +I_o$, $v_{Ls} = 0$, and $v_o = v_s$.

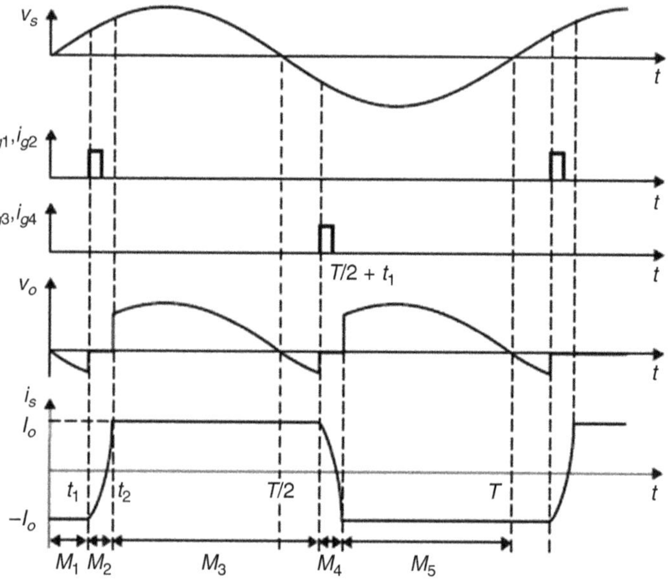

Fig. 8.29 Output voltage and source current waveforms

Mode 4 starts at $t = T/2 + t_1$ when SCR_3 and SCR_4 are turned on, resulting in the same equivalent circuit of Mode 2 with $i_s(t)$ given by:

$$i_s(t) = \frac{-V_s}{L_s\omega}(\cos\alpha + \cos\omega t) + I_o \tag{8.32}$$

At $t = T/2 + t_2$, the current $i_s(t)$ reaches $-I_o$, resulting in Mode 5, where SCR_3 and SCR_4 are on and SCR_1 and SCR_2 are off, which is similar to Mode 1. The cycle repeats at $t = T + t_1$ when SCR_1 and SCR_2 are triggered. To obtain an expression for the commutation angle, u, we evaluate Eq. (8.32) at $t = T/2 + t_2$ with $i_s(T/2 + t_2) = -I_o$; hence, we have:

$$-I_o = I_o - \frac{V_s}{L_s\omega}[\cos\omega(T/2 + t_2) + \cos\omega t_1]$$

Using $u = \omega\,(t_2 - t_1)$, the commutation angle may be given by:

$$u = \cos^{-1}\left(\cos\alpha - \frac{2I_o\omega L_s}{V_s}\right) - \alpha \tag{8.33}$$

In terms of the normalized load current, Eq. (8.33) gives:

$$u = \cos^{-1}(\cos\alpha - 2I_{no}) - \alpha \tag{8.34}$$

and the average output voltage is:

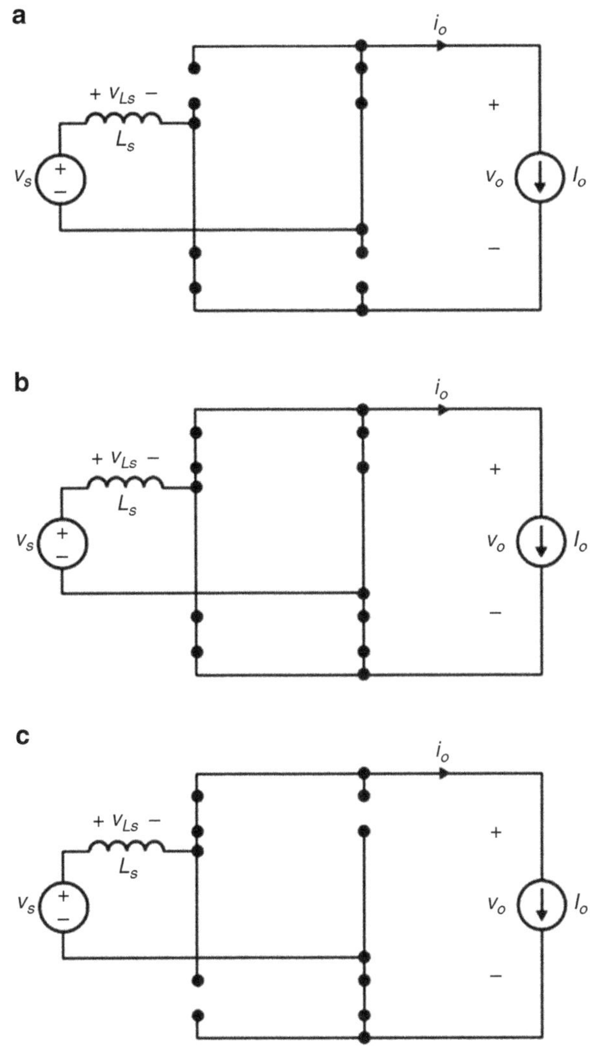

Fig. 8.30 (**a**) Mode 1: $0 \leq t \leq t_1$, (**b**) Mode 2: $t_1 \leq t \leq t_2$, (**c**) Mode 3: $t_2 \leq t < T/2 + t_1$

$$\begin{aligned} V_o &= \frac{2}{T}\int_{t_1}^{T/2+t_1} v_s(t)dt \\ &= \frac{2V_s}{2\pi}(\cos\omega t_1 + \cos\omega t_2) \\ &= \frac{2V_s}{2\pi}(\cos\omega t_1 + \cos(\alpha+u)) \end{aligned} \tag{8.35}$$

Substituting for $\cos(\alpha + u) = \cos\alpha - 2L_sI_o/V_s$, we get:

$$V_{\text{o}} = \frac{2V_{\text{s}}}{2\pi}\left(\cos\alpha - \frac{\omega L_{\text{s}} I_{\text{o}}}{V_{\text{s}}}\right) \tag{8.36}$$

In terms of the normalized values, Eq. (8.36) becomes:

$$V_{\text{no}} = \frac{2}{\pi}(\cos\alpha - I_{\text{no}}) \tag{8.37}$$

Figure 8.31 shows the regulation curve for I_{no} vs. V_{no} under different firing angles.

Exercise 8.6
Consider the circuit given in Fig. 8.28. Assume $v_{\text{s}} = 110\sin 2\pi(60)t$, $\alpha = 60^{\circ}$, $L_{\text{s}} = 20\ mH$, and $I_{\text{o}} = 10\ A$. Determine the average output voltage V_{o} and the approximate rms input current. (Hint: Assume current commutation is linear.)

Answer: $-13\ V, 8.15\ A$.

Exercise 8.7
Consider the full-wave circuit of Fig. 8.28 with $L_{\text{s}} = 5$ mH, $V_{\text{s}} = 110$ V, and $\omega = 377$ rad/s. Use Fig. 8.31 to determine (a) the average output voltage for $I_{\text{o}} = 10\ A$. (b) Range of α if the load current changes from $10\ A$ to $20\ A$ while V_{o} remains the same as that in part (a). (c) Range of α if we assume the input voltage V_{s}

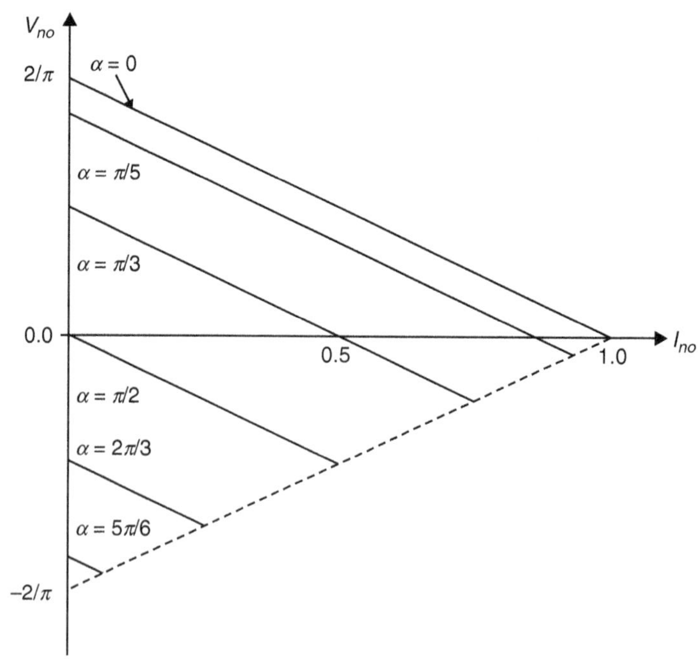

Fig. 8.31 Regulation curve for the phase-controlled full-wave bridge converter with ac-side inductance

changes by ±20% of its nominal value of 110 V while the output voltage remains constant as given in part (a).

Answer: 67 V, $47.8^\circ < \alpha < 60^\circ$, $51.3^\circ < \alpha < 37^\circ$.

8.6 Three-Phase Phase-Controlled Converters

8.6.1 *Half-Wave Converters*

Figure 8.32a shows a phase-controlled half-wave rectifier circuit under a resistive load. Its current and voltage waveforms are shown in Fig. 8.32b. Figure 8.32c shows the waveforms under a highly inductive load.

If we assume all SCRs are triggered simultaneously at $\alpha = \omega t_1$, then it is apparent from the circuit that only one positive source current can flow at any given time. This type of arrangement resembles the operation of an OR gate in digital systems. In practice, each SCR is triggered at α within its cycle, i.e., SCR_1, SCR_2, and SCR_3 are triggered at α, $2\pi/3 + \alpha$, and $4\pi/3 + \alpha$, respectively, as shown in Fig. 8.32b. Figure 8.32b, c shows relevant current and voltage waveforms under a purely resistive load and a highly inductive load, respectively. Here we assumed the three-phase voltages are position sequence with this line to natural voltages which are given as $v_a = V_s \sin \omega t$, $v_b = V_s \sin(\omega t - 120^\circ)$, and $v_c = V_s \sin(\omega t - 240^\circ)$.

The average output voltage is given by:

$$\begin{aligned} V_o &= \frac{3}{T}\int_{t_1}^{T/3+t_1} V_s \sin \omega t dt \\ &= \frac{-3V_s}{T\omega}\left[\cos \omega(T/3 + t_1) - \cos \omega t_1\right] \\ &= \frac{3V_s}{2\pi}\left[\cos \alpha - \cos \omega(T/3 + t_1)\right] \end{aligned} \quad (8.38)$$

where $\alpha = \omega t_1$.

8.6.2 *Full-Wave Converters*

Figure 8.36a shows the phase-controlled full-wave three-phase converter under an inductive-resistive load with a Y-connected three-phase voltage source.

The waveforms for $L = 0$ and $L = \infty$are shown in Fig. 8.33b and c, respectively, using $\alpha = 15^\circ$. To illustrate how the full-wave bridge circuit works, we will give a detailed mode-by-mode operation for $\alpha = 15^\circ$. If we select a positive set of voltages as in the half-wave case, then we must obtain a line-to-line voltage sequence for the converter operation.

The line-to-line voltage set is v_{ab}, v_{bc}, and v_{ca}, given in Chap. 7 and whose waveforms are redrawn in Fig. 8.34a, showing Mode 1 to Mode 6 for $\alpha = 0^{\circ}$ (SRCs act like diodes), with the corresponding conducting SCRs. Figure 8.34b shows the firing sequence for $\alpha = 0^{\circ}, 30^{\circ}, 60^{\circ}, 90^{\circ}, 120^{\circ}, 150^{\circ}$, and 180°. For $\alpha > 90^{\circ}$,

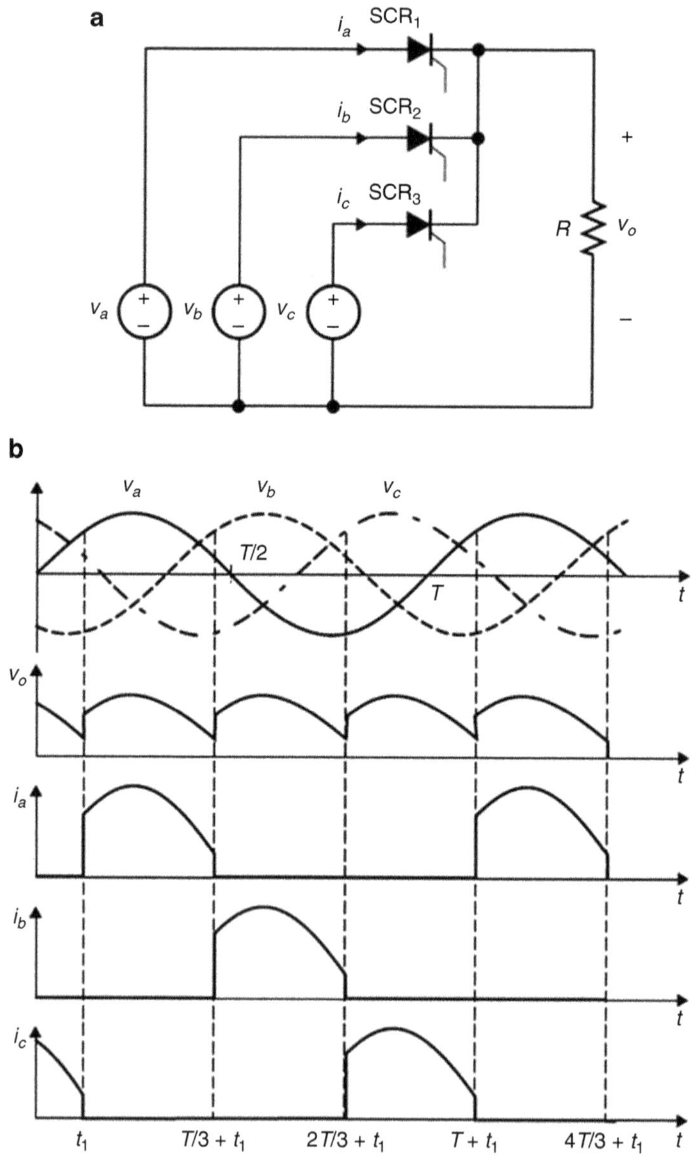

Fig. 8.32 (**a**) Three-phase phase-controlled half-wave rectifier (**b**) output voltage and line current waveforms under resistive load (**c**) waveforms under high inductive load

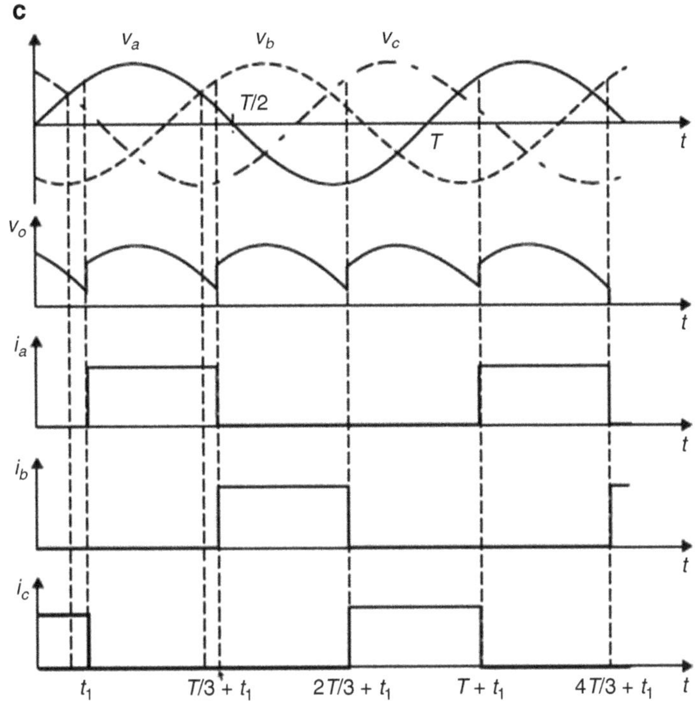

Fig. 8.32 (continued)

the voltage starts to become negative, and the converter operates in the *inversion mode*. For simplicity, these waveforms are plotted against angle ωt. From Fig. 8.34b, the most rectified positive voltage in Mode 1 (M_1) is $|v_{bc}|$. Hence, the conducting SCRs are SCR_3 and SCR_5, and the equivalent circuit is shown in Fig. 8.35a. In this mode, $v_o = -v_{bc}$, $i_a = 0, i_b = -i_c = -I_o$ and the SCR voltages are $v_{SCR1} = -v_{ca}, v_{SCR2} = v_{bc}, v_{SCR4} = v_{bc}, v_{SCR6} = -v_{ab}$. It is clear from these voltages that all the voltages across the other SCRs are negative.

At $\omega t = 30^\circ + \alpha$, v_{ab} becomes the most positive, and SCR_1 and SCR_5 are triggered to start Mode 2. At 60° later, as shown in Fig. 8.35, once again it can be shown that all the SCR voltages are negative in this mode also. Note from Fig. 8.35 that during each voltage phase, two gating signals are applied during the positive and negative cycles. For example, during the positive cycle of $v_{bc}(t)$, SCR_2 and SCR_6 are triggered with $\alpha = 45^\circ$ at $\omega t = \pi/2 + \pi/4$, and SCR_3 and SCR_5 are triggered in the negative cycle at $\omega t = 3\pi/2 + \pi/4$.

The output voltages for $\alpha = 0^\circ$, $\alpha = 30^\circ$, $\alpha = 60^\circ$, and $\alpha = 90^\circ$ are shown in Fig. 8.36a. For clarity, Fig. 8.36 shows the absolute voltage values for v_{ab}, v_{bc}, and v_{ca}.

The average output voltage can be evaluated over any of the given six pulse intervals. Let us select the second pulse between t_1 and t_2 indicated in Fig. 8.36. The average output voltage is given by:

$$V_o = \frac{6}{T}\int_{t_1}^{t_2} v_{ab}dt$$

Substituting for $v_{ab} = \sqrt{3}V_s \sin\left(\omega t + 30^\circ\right)$ and $\omega t_1 = 30^\circ + \alpha$ and $\omega t_2 = 90^\circ + \alpha$, we have:

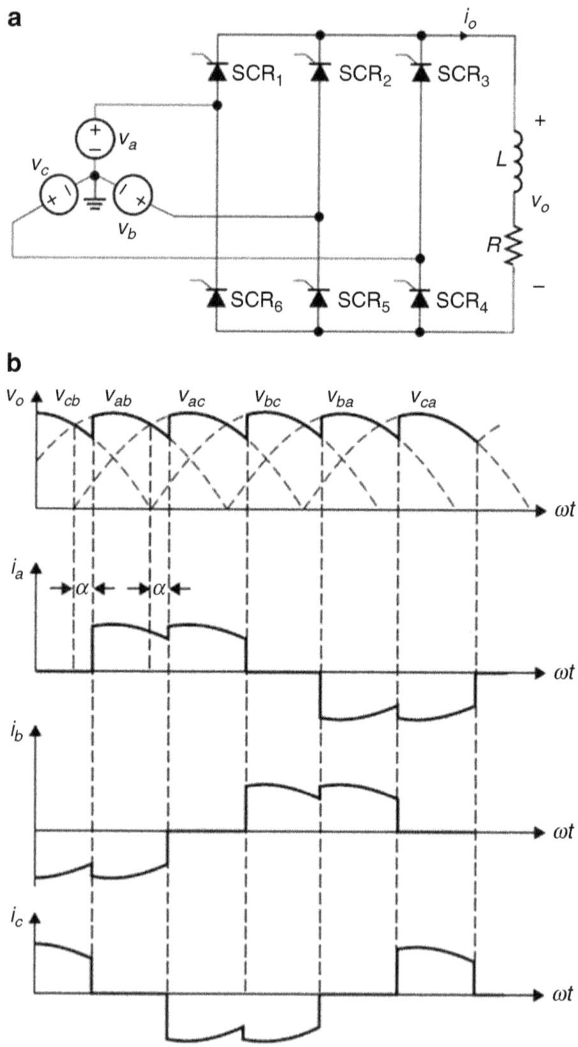

Fig. 8.33 (**a**) Full-bridge three-phase controlled rectifier (**b**) output voltage and current waveforms under purely resistive load (**c**) Output voltage and current waveforms under highly inductive load

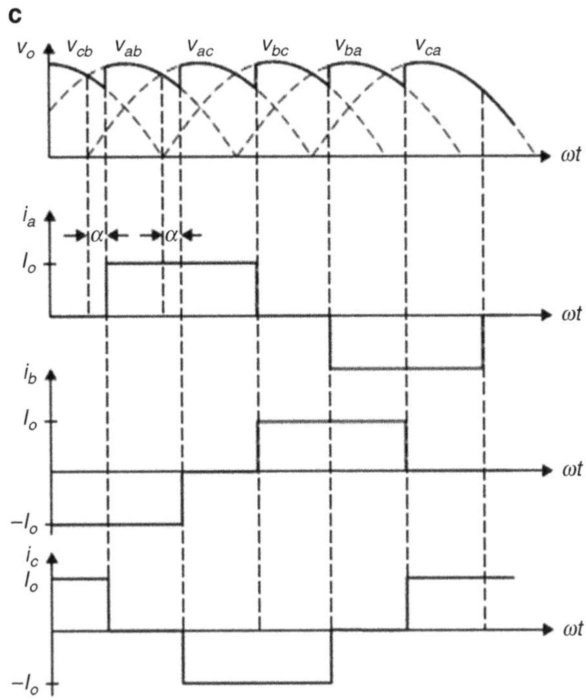

Fig. 8.33 (continued)

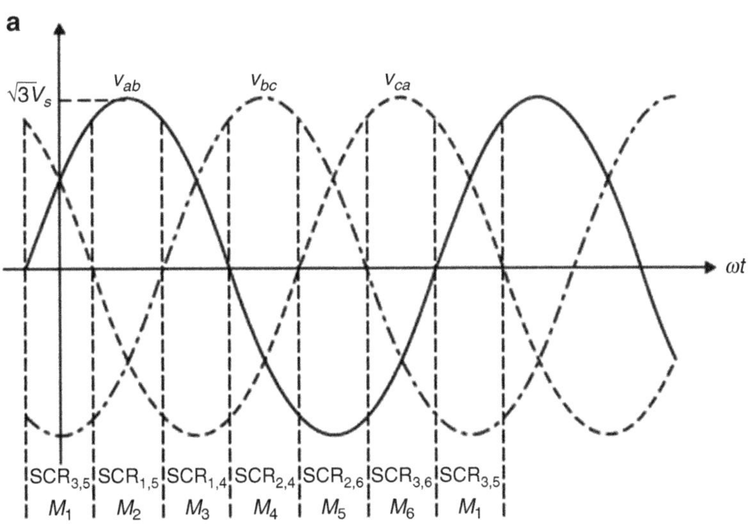

Fig. 8.34 Three-phase line-to-line voltage: (**a**) corresponding conduction modes Three-phase line-to-line voltage (**b**) corresponding firing angles

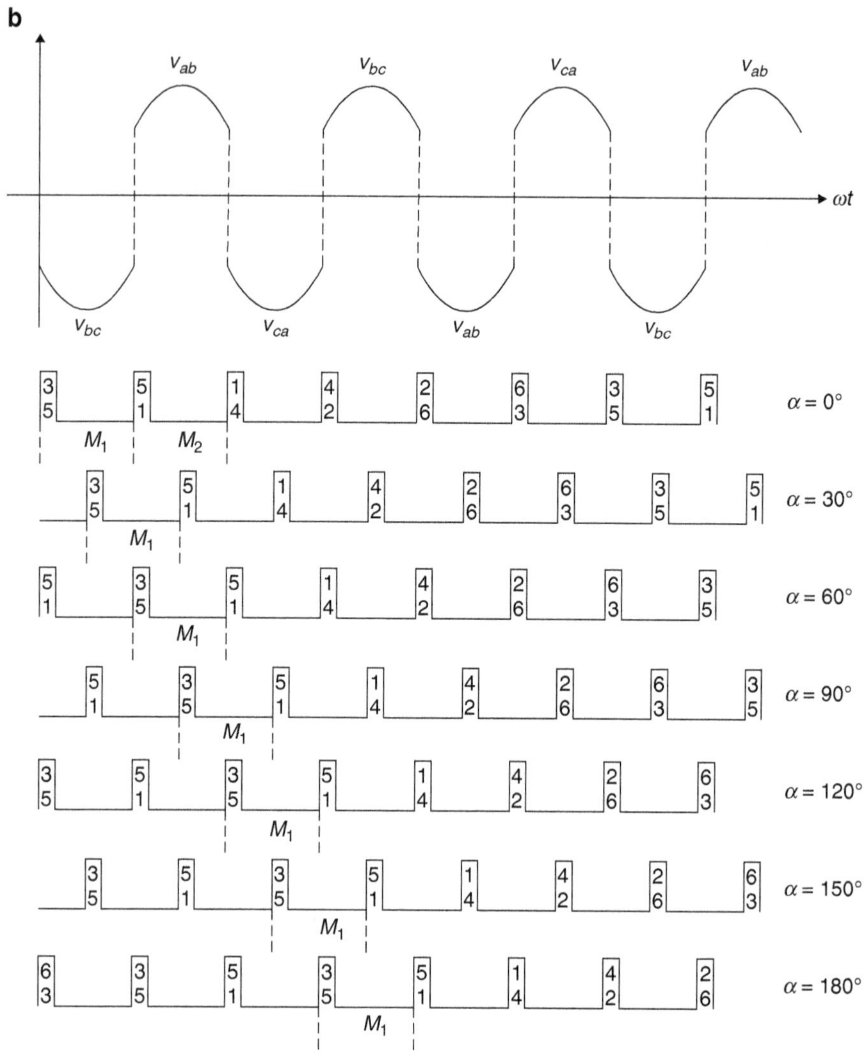

Fig. 8.34 (continued)

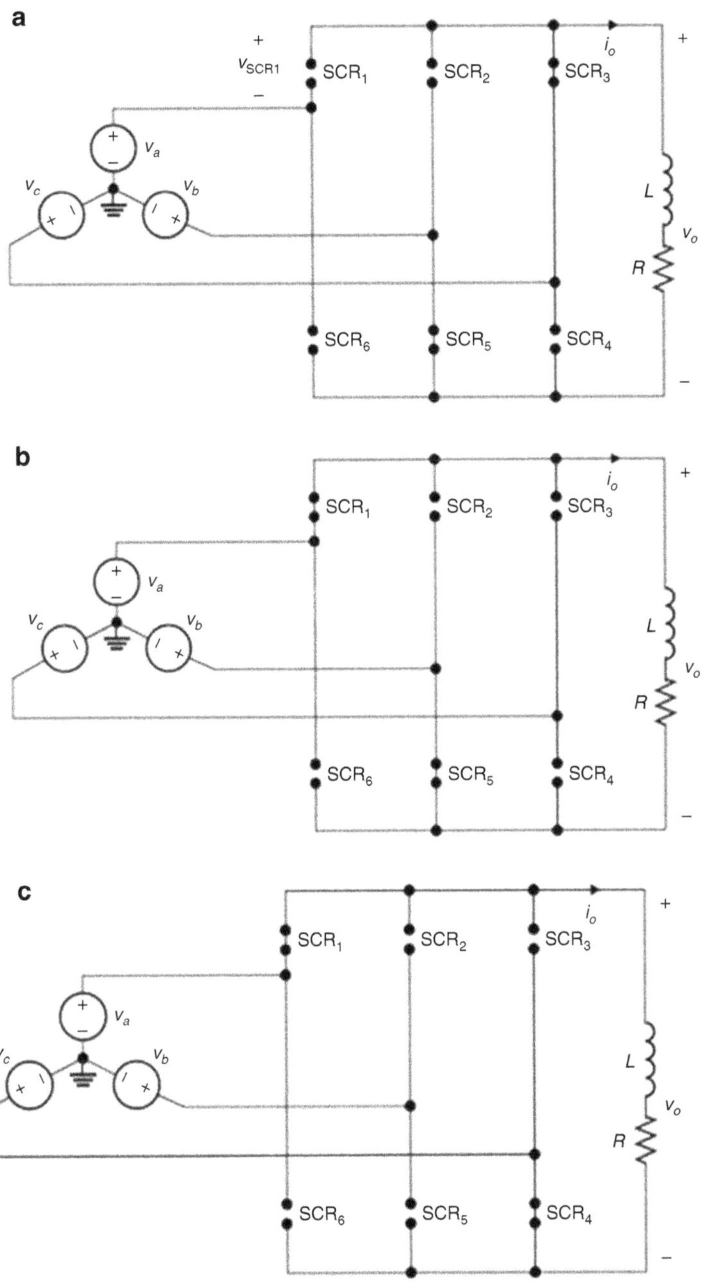

Fig. 8.35 Equivalent circuits for the six modes of operation (**a**) Mode 1. (**b**) Mode 2. (**c**) Mode 3 Equivalent circuits for the six modes of operation. (**d**) Mode 4. (**e**) Mode 5. (**f**) Mode 6

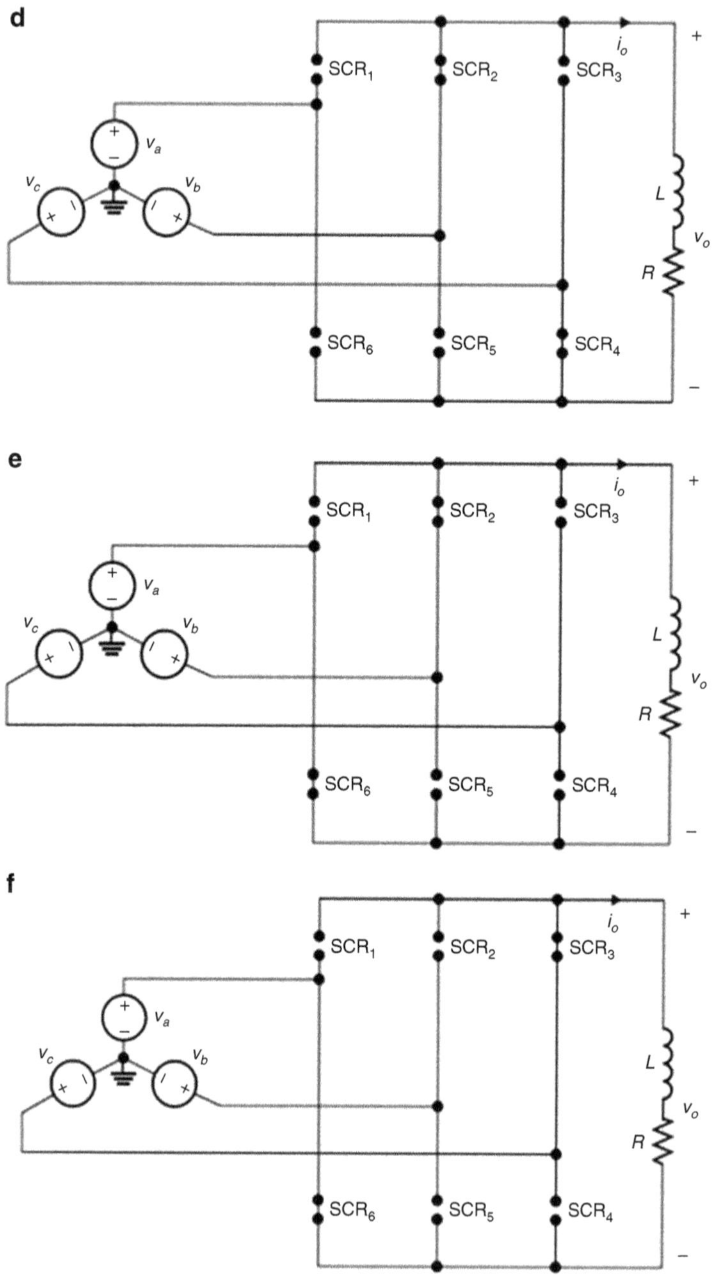

Fig. 8.35 (continued)

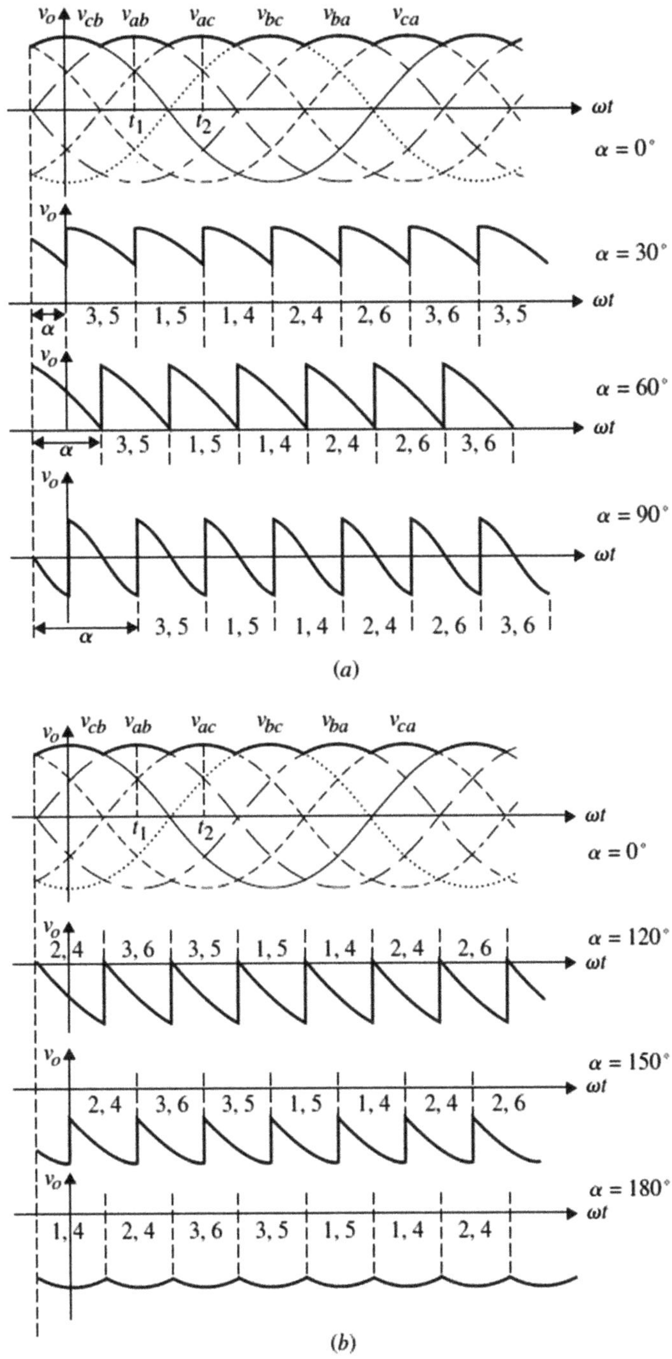

Fig. 8.36 Output voltages for firing angles: $\alpha = 0^{\circ}$, $\alpha = 30^{\circ}$, $\alpha = 60^{\circ}$, and $\alpha = 90^{\circ}$

$$
\begin{aligned}
V_o &= \frac{3}{\pi}\int_{\pi/6+\alpha}^{\pi/2+\alpha} \sqrt{3}V_s \sin\left(\omega t + \frac{\pi}{6}\right) d\omega t \\
&= \frac{-3\sqrt{3}}{\pi} V_s \left[\cos\left(\alpha + \frac{2\pi}{3}\right) \cos\left(\alpha + \frac{\pi}{6}\right)\right] \\
&= \frac{-3\sqrt{3}V_s}{\pi} \cos\alpha
\end{aligned}
\tag{8.39}
$$

The natural conduction is at $\omega t = \pi/6$ if all SCRs are triggered continuously. In this case modes 1 to 6 are the same as those in the previous chapter. In this section we are investigating the effect of delaying the trigger by α with respect to the angle of natural conduction. In Fig. 8.36, the natural conduction angle in $M1$ is $\pi/6$ when SCR_3 and SCR_5 are acting like diodes. Now let us consider $M1$ by delaying the triggering of SCR_3 and SCR_5 by α. Prior to $\omega t = \alpha t$, v_{ca} is the most positive; hence, SCR_3 and SCR_6 are conducting.

For illustration purposes, consider the case for $\alpha = 60^\circ$. We notice that for $\alpha > 60^\circ$, the output voltage starts becoming negative. For $\alpha = 90^\circ$, the positive and negative areas are equal resulting in a zero average output voltage.

Exercise 8.8
Show that the average output power for Fig. 8.33a with $L = \infty$ is given by:

$$P = \frac{3\sqrt{3}}{\pi} V_s I_o \cos\alpha$$

Problems

Single-Phase, Half-Wave Rectifiers

Resistive Load

8.1 Derive the power factor and the THD expressions for the half-wave phase-controlled rectifier given in Eqs. (8.5) and (8.6).

8.2 Calculate the average power delivered to a resistive load in a half-wave phase-controlled rectifier with $v_s = 200 \sin 377t$, $\alpha = 30^\circ$, and $R_L = 25\ \Omega$.

Inductive Load

8.3 Derive the load current equation, i_o, for the inductive-load phase-controlled rectifier with a free-wheeling diode in Fig. 8.8a.

8.4 Determine the delay angle α so that the total power to be delivered to a 20 Ω load resistance with $\omega L = 20\ \Omega$ in Fig. 8.8a equals to 40 W. Assume $v_s = 100 \sin \omega t$.

Single-Phase, Full-Wave Rectifiers

8.5 An alternative way to make a full-wave rectifier is to use a center-tap transformer as shown in Fig. P8.5. (a) Draw the waveforms for i_s, i_1, i_2, and v_o and determine the power factor, (b) determine α for a power factor equal to 0.95. Assume $v_s = 100 \sin 377t$ and $\alpha = 40^\circ$.

Fig. P8.5

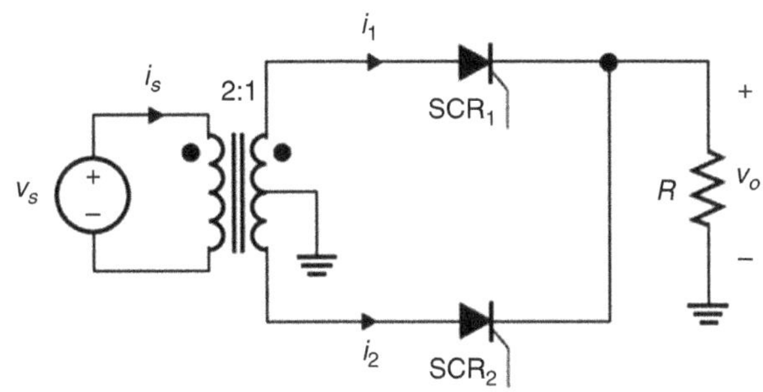

8.6 (a) Determine the power factor for the inductive-load phase-controlled full-wave rectifier shown in Fig. P8.6. Assume $v_s = 100 \sin 377t$, $\alpha = 60^\circ$, and i_0 is constant. (b) Determine α for a power factor equal to 0.75.

Fig. P8.6

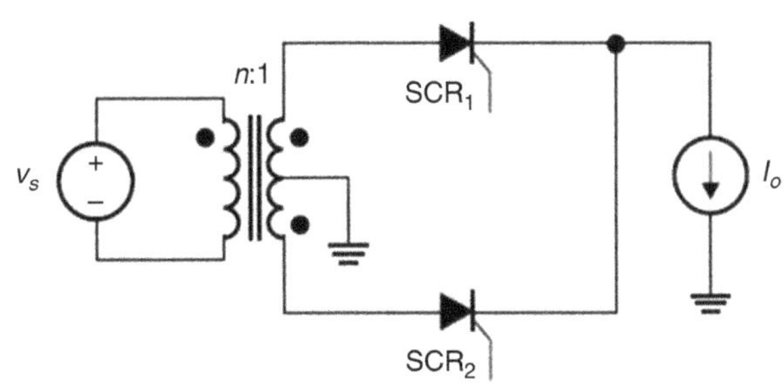

8.7 Figure P8.7 shows a full-wave phase-controlled rectifier with an inductive load and a free-wheeling diode. (a) Sketch the waveforms for v_o and give the expression for the average output voltage. (b) Show that the load current is given by the following two equations:

$$i_o(t) = I_{o1}e^{-(t-t_1)/\tau} + \frac{V_s}{|Z|}\sin\left[\omega(t-t_1)\right] \qquad t_1 < t < \frac{T}{2}$$

$$i_o(t) = I_{o2}e^{-(t-T/2)/\tau} \qquad \frac{T}{2} < t < T - t_1$$

where

$$|Z| = \sqrt{(\omega L)^2 + R^2},\ \alpha = \omega t_1,\ \theta = \tan^{-1}\tfrac{\omega L}{R},\ \tau = L/R$$

$$I_{o1} = e^{-(\pi+\alpha/\tau\omega)} + e^{\alpha/\tau\omega}\sin(\alpha+\theta)$$
$$I_{o2} = e^{-(\pi/\tau\omega)} + e^{\alpha/\tau\omega}\sin(\alpha+\theta)$$

Fig. P8.7

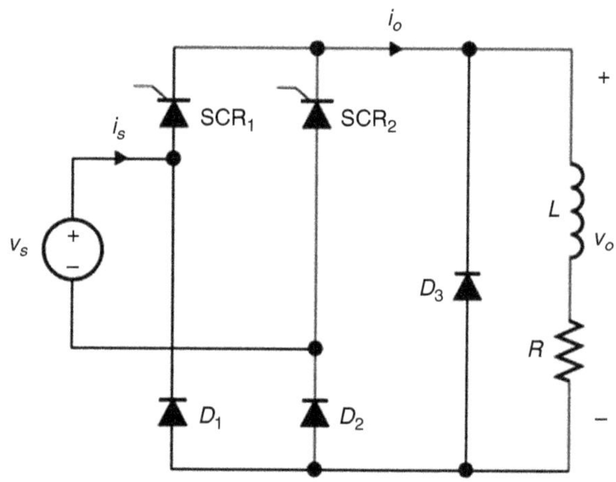

8.8 Derive the THD for the full-bridge phase-controlled rectifier of Fig. 8.13.
8.9 Show that the average power delivered to the load in Fig. 8.25 is zero when:

$$\alpha = \cos^{-1}\frac{V_{dc}\pi}{2V_s}$$

8.10 Sketch the waveforms for i_s, i_o, and v_o, and determine the average output voltage for the circuit of Fig. P8.10, under the conditions:

(a) $C = 0$ and
(b) $C = \infty$. Assume the devices are triggered in a similar way to those of Fig. P8.10.

Fig. P8.10

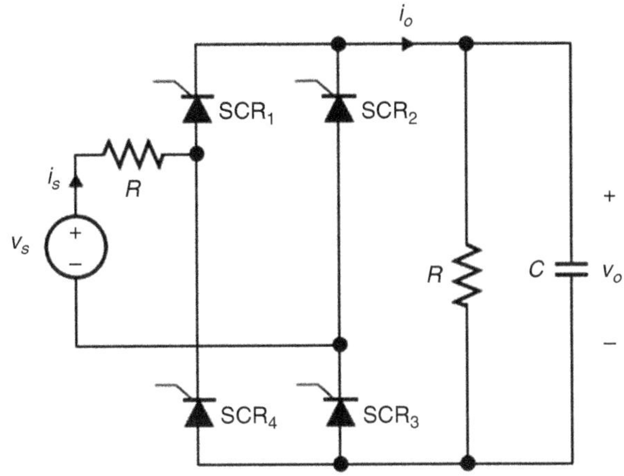

8.11 Consider a power electronic circuit that generates the waveforms i_s and v_s shown in Fig. P8.11. Determine the THD.

Fig. P8.11

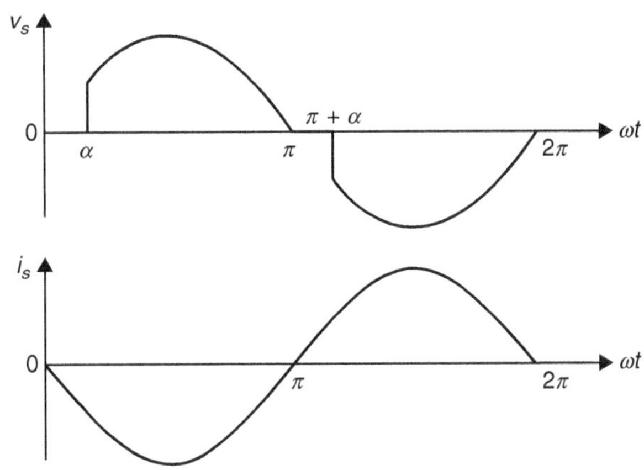

8.12 Sketch the waveforms for i_s and v_o in Fig. P8.12. Assume $L/R \gg T$ and $v_s = V_s \sin \omega t$ and $V_{dc} < V_s$.

Fig. P8.12

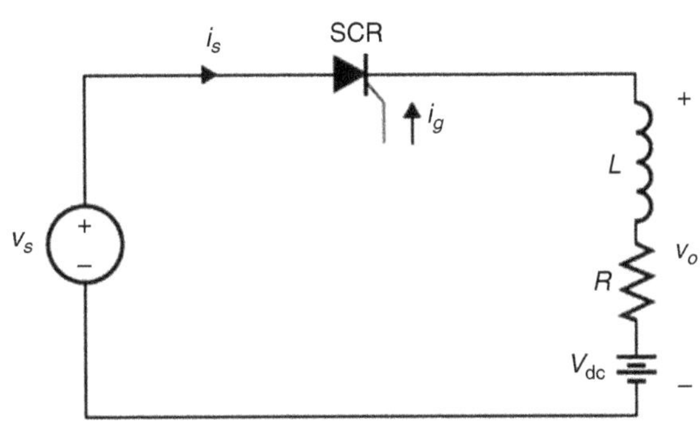

8.13 Repeat Exercise 8.5 for $\alpha = 150^{\circ}$.

8.14 Sketch the waveforms for v_o, i_{SCR1}, and i_{SCR2} using the circuit shown in Fig. P8.15, and obtain the expression for the fundamental component of $i_s(t)$. Assume v_s is a square wave of amplitude $\pm V_s$.

8.15 Calculate the input *pf* for $I_o = 5\ A$ and $\alpha = 60^{\circ}$ and the current THD using the above values.

Fig. P8.15

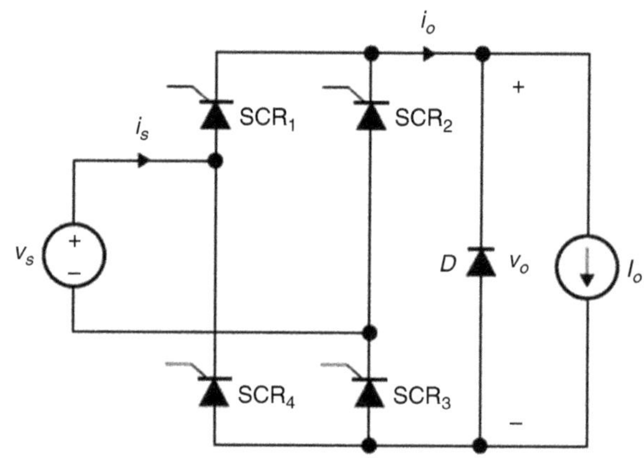

8.16 Consider the full-wave phase-controlled circuit shown in Fig. P8.16. (a) Calculate the average output voltage V_o, (b) the average output current, I_o, (c) the rms values of $i_s(t)$, and (d) the input power factor. Assume $L/R \gg T/2$ and $v_s = 110 \sin 377t$ and $\alpha = 45^{\circ}$.

Fig. P8.16

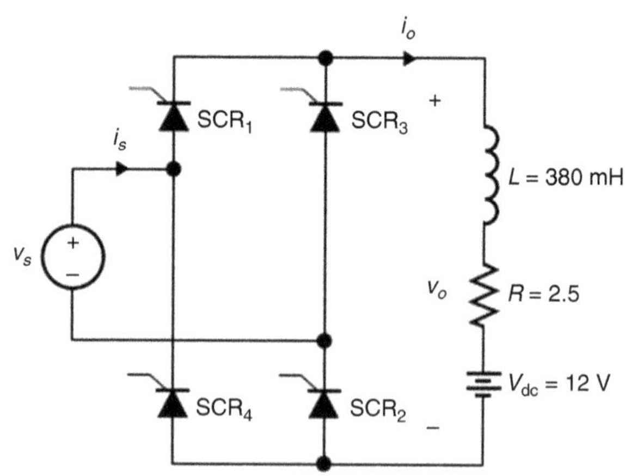

8.17 Repeat Problem 8.16 for $\alpha = 150^{\circ}$.

8.18 Figure P8.18 shows a half-wave phase-controlled converter with a flyback SCR_2. Assume $v_s = 110 \sin 377t$. (a) Calculate V_o, u, and the power delivered to the load for $\alpha = 60^\circ$; (b) design for R and L under the above load conditions; (c) if the load resistance is reduced by 50%, find the new value of α for maintaining the same load power; and (d) repeat part (c) by assuming L_s has increased by 20%.

Fig. P8.18

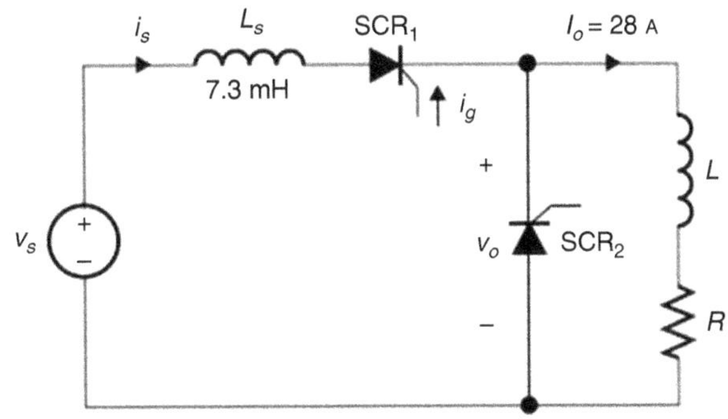

8.19 Calculate the rms value for $i_s(t)$ given in Problem 8.18 by assuming that the commutation periods are linear.

8.20 Calculate the average load current in the circuit shown in Fig. P8.20. Assume $\alpha = 60^\circ$ and $v_s = 110 \sin 377t$.

Fig. P8.20

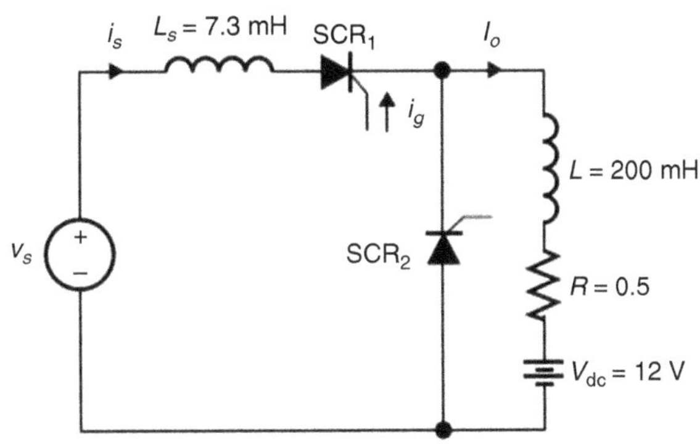

8.21 Draw the waveforms for i_s, i_{D_1}, i_{S_1}, and v_o in Fig. P8.21 under the following load conditions: (a) $L = 0$ and (b) $L = \infty$. Label your axes and show which device(s) conduct for each time interval. Assume $v_s = V_s \sin \omega t$ and the firing angle is 30°.

Fig. P8.21

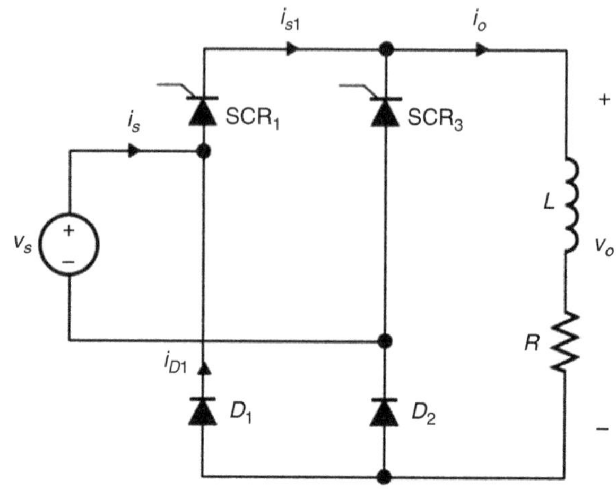

8.22 Consider the SCR circuit in Fig. P8.22 with $\alpha = 30^\circ$. Assume v_{S_1} and v_{S_2} are 180° out of phase:

(a) Sketch the waveforms for i_{S_1}, i_{S_2}, and v_O for two periods.
(b) Derive the expressions for i_{S_1}, i_{S_2}, and v_O.
(c) Find the commutation interval during which both SCRs are conducting.

Fig. P8.22

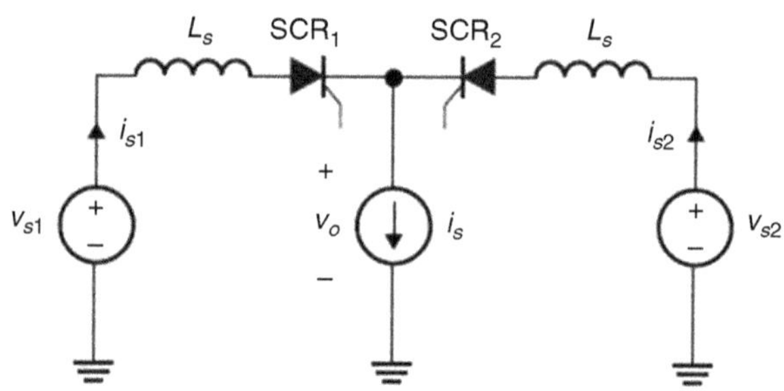

8.23 Consider the half-controlled diode/SCR full-bridge circuit shown in Fig. P8.23. Assume the input is given by $v_s(t) = V_s \sin \omega t$. Assume the firing angle is $\alpha = 30^\circ$. (a) Sketch v_O, $i_{D_1}, i_{D_2}, i_{S_1}, i_{S_2}$, and i_s, (b) derive the expression for v_O, and (c) design for α so that the average output voltage is 48 V. Assume $v_s(t) = 100 \sin 377t$ V.

Fig. P8.23

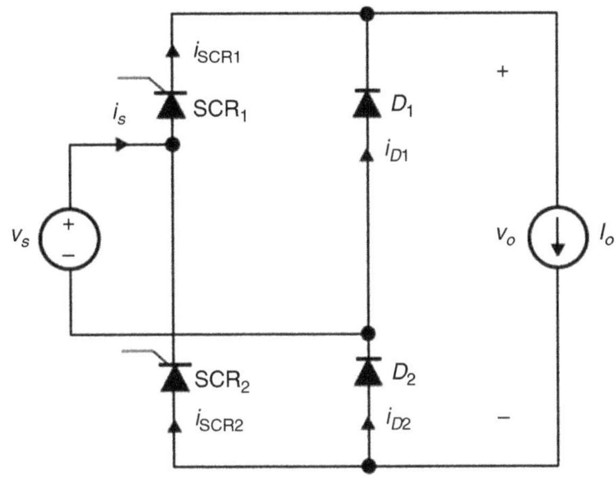

8.24 Consider the controlled half-wave rectifier of Fig. 8.26a with $L_s = 7.3$ mH, $\alpha = 60^\circ$,which supplies a load current of 20 A. Assume $v_s = 110\sin(2\pi 60t)$.

(a) Use Fig. 8.27 to determine the average output voltage.
(b) Determine the average power delivered to the load.
(c) Find the commutation angle.
(d) Design for the inductor L in the load side, $V_o = 8.75$ V, and $R = 0.44\ \Omega$.
(e) If the load resistance is decreased by 50%, what is the new α needed to maintain the same average output voltage.
(f) If the source voltage is increased by 20 %, what is the new α needed to maintain the same average load.

Three-Phase Converters

8.25 The three-phase half-wave controlled rectifier with a commutating diode across the load is shown in Fig. P8.25. Assume that the voltage drops across the thyristor are negligible and the supply voltage is 100 V rms. For a firing delay angle of 45°, sketch the waveform of $v_o$and determine the DC voltage across the load.

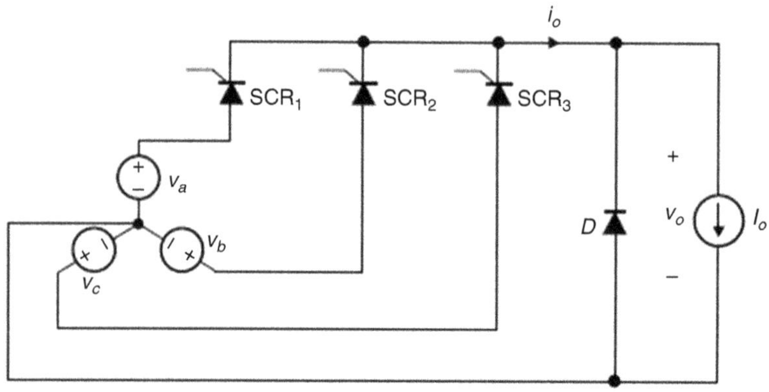

Fig. P8.25

8.26 (a) Sketch the waveforms for v_o, i_{SCR3}, and $i_o$for the three-phase controlled rectifier shown in Fig. P8.26 for $\alpha = 30^\circ$ and $\alpha = 60°$. (b) Derive the expression for the average output voltage. (c) Sketch the waveform of the voltage across SCR_6 and SCR_3. Assume $v_a = V_s \sin \omega t$, $v_b = V_s \sin(\omega t - 2\pi/3)$, and $v_c = V_s \sin(\omega t - 4\pi/3)$.

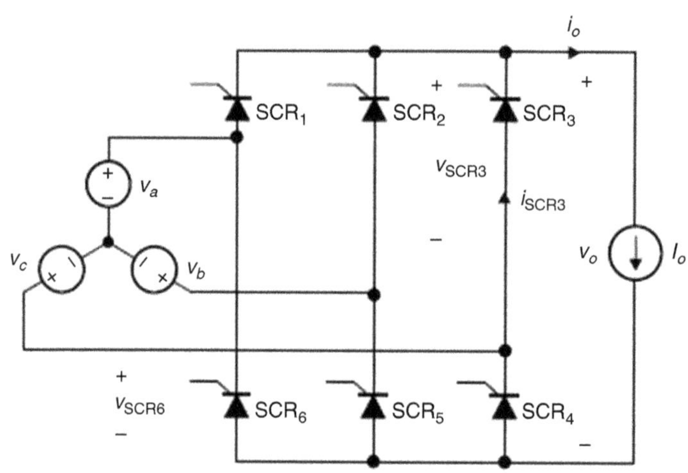

Fig. P8.26

8.27 Determine the average output voltage for Problem 8.26 using $V_s = 110\,V$.

Chapter 9
dc-ac Inverters

In this chapter, we will consider power electronic circuits that produce variable-frequency ac output voltages from dc sources. This functionality in power electronics is becoming very important for the ever-increasing penetration of photovoltaic (PV) systems. All kinds of inverters at various power ratings, and output voltage levels, have been designed and commercialized over the last two decades. Moreover, one class of converters known as microinverters is designed at the module level. The first introduction of fully integrated microinverter and a PV panel (known as *ac module*) was first introduced and commercialized by Dr. Batarseh's team at the University of Central Florida (UCF). More recently the team at UCF is developing one module that takes the input from four PV panels, in the process introducing a new class of inverters may be best to be called mini-inverters since their power range is between the microinverter and string inverter.

As discussed in Chap. 3, depending on whether the source is dc or ac, power electronic circuits with ac output voltages are referred to as *dc-ac inverters* or *ac-ac cycloconverters*. In converting ac-ac, if the output voltage frequency is different from the source frequency, the converter is called an *ac voltage controller*. Traditionally, dc-ac inverters (also known as *static inverters*) use fixed dc sources to produce symmetrical ac output voltages at fixed or variable frequency or magnitude. The output ac voltage system can be of the single-phase or three-phase type at frequencies of $50, 60$, and 400 Hz with a voltage magnitude range of $110 - 380$ VAC. Inverter circuits are used to deliver power from a dc source to a passive or active ac load employing conventional SCRs or gate-driven semiconductor devices such as GTOs, IGBTs, and MOSFETs. Due to their increased switching speed and power capabilities coupled with complex control techniques, today's inverters can operate in wide ranges of regulated output voltage and frequency with reduced harmonics. Medium- and high-power half-bridge and full-bridge switching devices using MOSFETs, IGBTs, and SiC- and GaN-based-devices are available as packages.

Dc-ac inverters are used in applications where the only source available is a fixed dc source and the system requires an ac load such as in uninterruptible power supply

I. Batarseh, A. Harb, *Power Electronics*,
https://doi.org/10.1007/978-3-319-68366-9_9

(UPS). Applications where dc-ac inverters are used include aircraft power supplies, variable-speed ac motor drives, and lagging or leading VAR generation. For example, an inverter used to provide necessary changes in the frequency of the ac output is used to regulate the speed of an induction motor and is also used in a UPS system to produce a fixed ac frequency output when the main power grid system is out.

9.1 Basic Block Diagram of dc-ac Inverters

Figure 9.1 shows a typical block diagram of a power electronic circuit utilizing a dc-ac inverter with input and output filters used to smooth the output ac signal.

The feedback circuit is used to sense the output voltage and compare it with a sinusoidal reference signal as shown in Fig. 9.1. Depending on the arrangement of the power switches and their types, various control techniques are normally adopted in industry. The control objective is to produce a controllable ac output from an uncontrollable dc voltage source. Even though the desired output voltage waveform is purely sinusoidal, practical inverters are not purely sinusoidal but include significant high-frequency harmonics. This is why inverters normally employ a high-frequency switching technique to reduce such harmonics. Two alternative control methods of inversions will be discussed in this chapter: uniform pulse width modulation (PWM) and sinusoidal PWM. Depending on the output power level, both techniques find their way into practice. Sinusoidal PWM is widely used in motor drive applications with high-frequency operation.

The front end of the power electronic circuit in Fig. 9.1 is the line ac-dc converter discussed in Chaps. 7 and 8. Unlike the line ac-dc converter circuits in

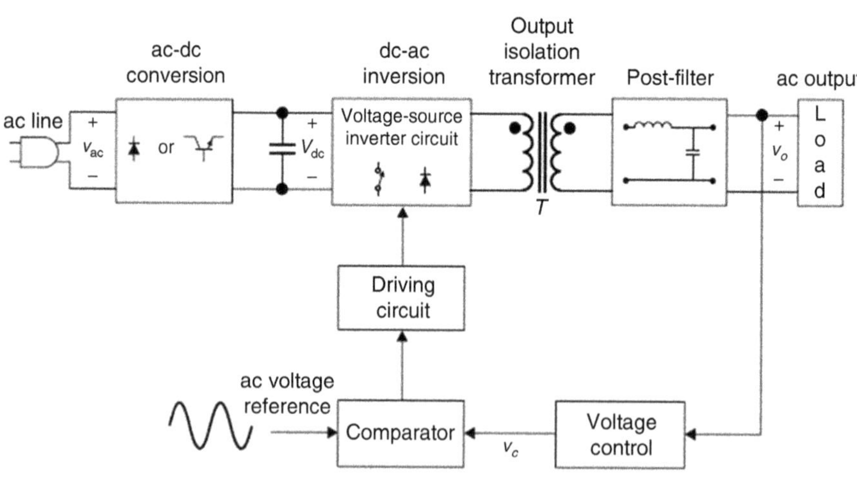

Fig. 9.1 Block diagram of a typical power electronic circuit with dc-ac inverter

which diodes and SCRs are commutated naturally, in dc-ac inverters forced commutation is used to turn on and off the power switches. The negative portion of the source voltage is used naturally to turn off the diode and/or the SRC in line ac-dc converters. As a result, the ac-dc converter circuit has one ac frequency—the line frequency—and the device relies on line commutation for switching. In dc-ac inverters, the ac frequency is not necessarily the line frequency. Hence, if an SCR is used, we must devise a means to force it to turn off in order to produce the desired ac output. This feature, the variable-frequency output voltage, causes these circuits to play a very important role in all kinds of industrial applications, such as controlling the speed of motors. Variable-speed drives are used to control the speed of electric vehicles, pumps, rollers, and conveyors. Therefore, inverter circuits require more elaborate control signals to shape the ac voltage.

The load in Fig. 9.1 is broadly classified as either passive or active. If the load consists of impedance only (i.e., is passive), its time-domain response is determined by the nature of the load and cannot be controlled externally. If the load consists of a source (i.e., is active), its time-domain information can be controlled externally, as in the case of machine loads.

One final note should be made about power switches. The simplest inverter is one that employs a switching device that can be gate-controlled to interrupt current flow, resulting in a naturally commutating or self-commutated inverter. These switching devices are the gate-driven types such as GTOs, IGBTs, MOSFETs, and power BJTs. When SCRs are used, the converter requires an additional external circuit to force the SCR to turn off. The reason for the mandatory forced commutation in such inverters is that the dc input voltage across the SCR devices causes them to be in the forward conduction state. The higher the power rating of the gate-controlled devices, the less SCRs are used in the design of inverters.

9.1.1 Voltage- and Current-Source Inverters

Since a practical source can provide either a constant voltage or a constant current, broadly speaking, inverters are divided into either Voltage-Source Inverters (VSI) or Current-Source Inverters (CSI). The dc source in VSI is a fixed voltage such as battery, fuel cells, solar cells, dc generators or rectified dc sources. In CSI, the dc source is nearly a constant current source. The nature of the source, whether it is a dc current source or a dc voltage source, makes the power inverter clearly distinguishable and its practical application more defined. A block diagram for a voltage- and a current-source inverter is shown in Fig. 9.2a, b, respectively.

In the voltage-source inverter (VSI), the output voltage, v_o, is a function of the inverter operation; the load current, i_o, is a function of the nature of the load; and the dc input, V_{dc}, is a constant input voltage. In the current-source inverter, the output voltage is a function of the inverter operation; the load current, i_o, is a function of the nature of the load; and the source, I_{dc}, is a constant input current.

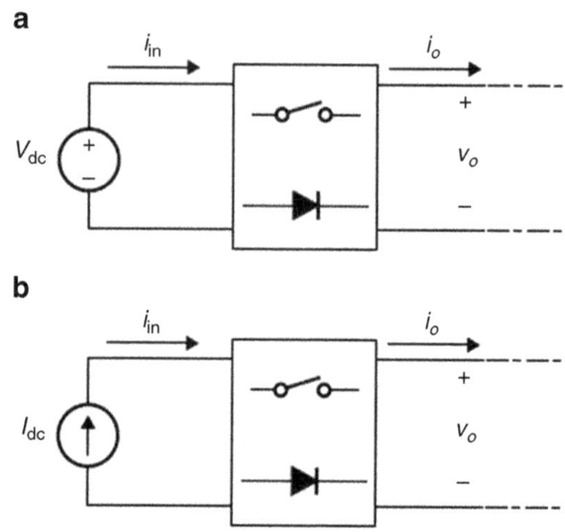

Fig. 9.2 Block diagram representations: (**a**) voltage-source inverter and (**b**) current-source inverter

9.1.2 Inverter Configurations

Figure 9.3a–c shows, respectively, three possible single-phase inverter arrangements: biphase, half-bridge, and full-bridge. The biphase inverter, also known as a push-pull inverter, is drawn in two different ways in Fig. 9.3a.

9.1.3 Output Voltage Control

If the output voltage is controlled by varying the dc source voltage, then this can be accomplished by either controlling the dc input by using a dc-dc converter as shown in Fig. 9.4a or by using an ac-dc phase control converter as shown in Fig. 9.4b. In many applications, varying the input dc voltage is not possible and costly.

9.2 Basic Half-Bridge Inverter Circuit

9.2.1 Resistive Load

To illustrate the basic concept of a dc-ac inverter circuit, we consider a half-bridge voltage-source inverter circuit under a resistive load as shown in Fig. 9.5a. The switching waveforms for S_1, S_2 and the resultant output voltage are shown in Fig. 9.5b.

The circuit operation is very simple since S_1 and S_2 are switched on and off alternatively at a 50% duty cycle as shown in the switching waveform in Fig. 9.5b. This shows that the circuit generates a square ac voltage waveform across the load

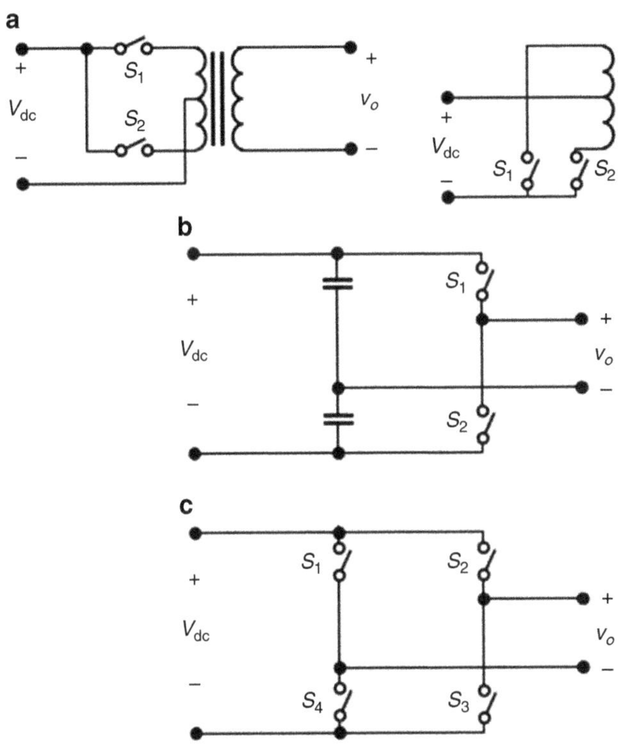

Fig. 9.3 Single-phase inverter arrangements (**a**) Biphase inverter (**b**) half-bridge inverter (**c**) full-bridge inverter

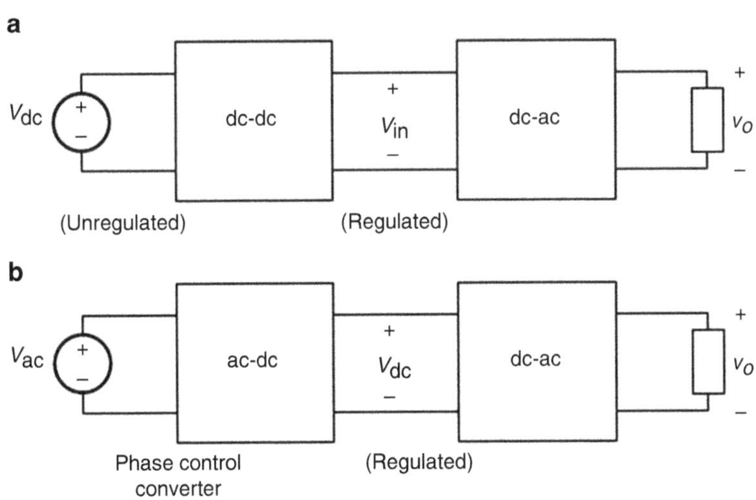

Fig. 9.4 Controlling the dc input using (**a**) dc-dc converter or (**b**) ac-dc phase-controlled rectifier

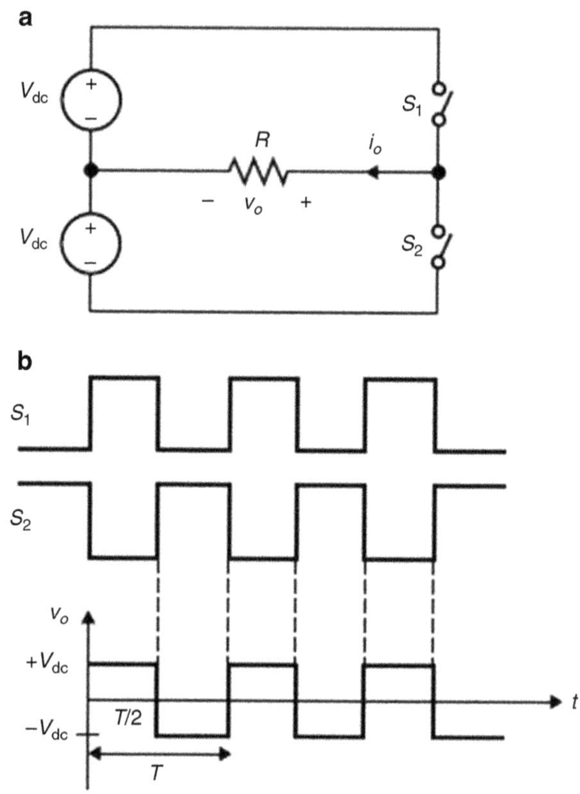

Fig. 9.5 (**a**) Half-bridge inverter under resistive load (**b**) Switching and output voltage waveform

from a constant dc source. The voltages, V_{dc} and $-V_{dc}$, are applied across R when S_1 is ON and S_2 is OFF and when S_2 is ON and S_1 is OFF, respectively. One observation to be made here is that the frequency of the output voltage is equal to $f = 1/T$ and is determined by the switching frequency. This is true as long as S_1 and S_2 are switched complementarily. Moreover, the *rms* value of the output voltage is simply V_{dc}. Hence, to control the *rms* value of the output voltage, we must control the rectified V_{dc} voltage source. Another observation is that the load power factor is unity since we have a purely resistive load. That is rarely encountered in practical application.

Finally, we should note that in practice, the circuit does not require two equal dc voltage sources as shown in Fig. 9.5a. Instead, large splitting capacitors are used to produce two equal dc voltage sources as shown in Fig. 9.6. The two capacitors, C_∞, are equal and very large, so that RC_∞ is much larger than the half switching period. This will guarantee that the midpoint, a, between the capacitors has a fixed potential at one-half of the supply voltage V_{dc}. The current from the source, V_{dc}, equals one-half of the load current, i_o. In steady state, the average capacitor currents are zero; hence, capacitors C_∞ are used to block the dc component of i_o.

The limitation of the above half-bridge configuration is that varying the switching sequence cannot control the output voltage.

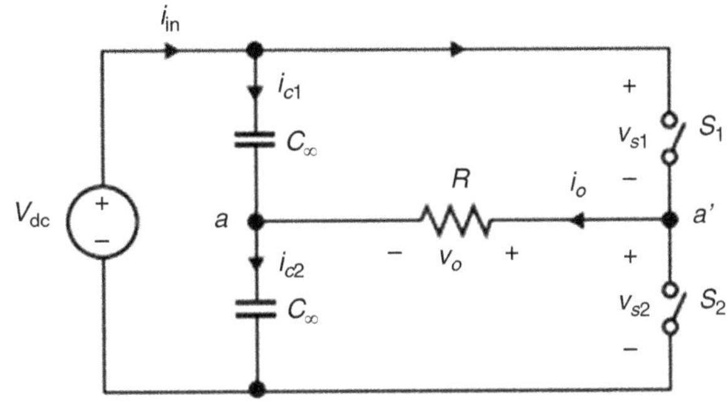

Fig. 9.6 Half-bridge inverter circuits with large splitting capacitors

Example 9.1
Sketch the current and voltage waveforms for $i_{in}, i_o, v_{s1}, v_{s2}$, and v_o in the circuit shown in Fig. 9.6 for $\theta = 0$ and $\theta \neq 0$ by using that the switching waveform for S_1 and S_2 is shown in Fig. 9.7. Determine the average output voltage in terms of V_{dc} and θ when the inverter operates in the steady state.

Solution
Let us consider Mode 1 when S_1 is ON and S_2 is OFF, and then the output voltage and current equations are given by:

$$v_o = \frac{V_{dc}}{2}$$
$$i_o = \frac{v_o}{R} = \frac{V_{dc}}{2R}$$

Since we assumed that the capacitors are equal, then the load current, i_o, splits equally, i.e.:

$$i_{c1} = -\frac{1}{2} i_o = -\frac{V_{dc}}{4R}$$
$$i_{c2} = i_{in} = -i_{c1} = \frac{V_{dc}}{4R}$$

The voltages across the switches are:

$$v_{s1} = 0$$
$$v_{s2} = v_o + \frac{V_{dc}}{2} = V_{dc}$$

Mode 2 starts when S_1 and S_2 are OFF during the short interval θ:

$$v_o = 0$$
$$i_o = 0$$
$$i_{c1} = i_{c2} = i_{in} = 0$$
$$v_{s1} = v_{s2} = \frac{V_{dc}}{2}$$

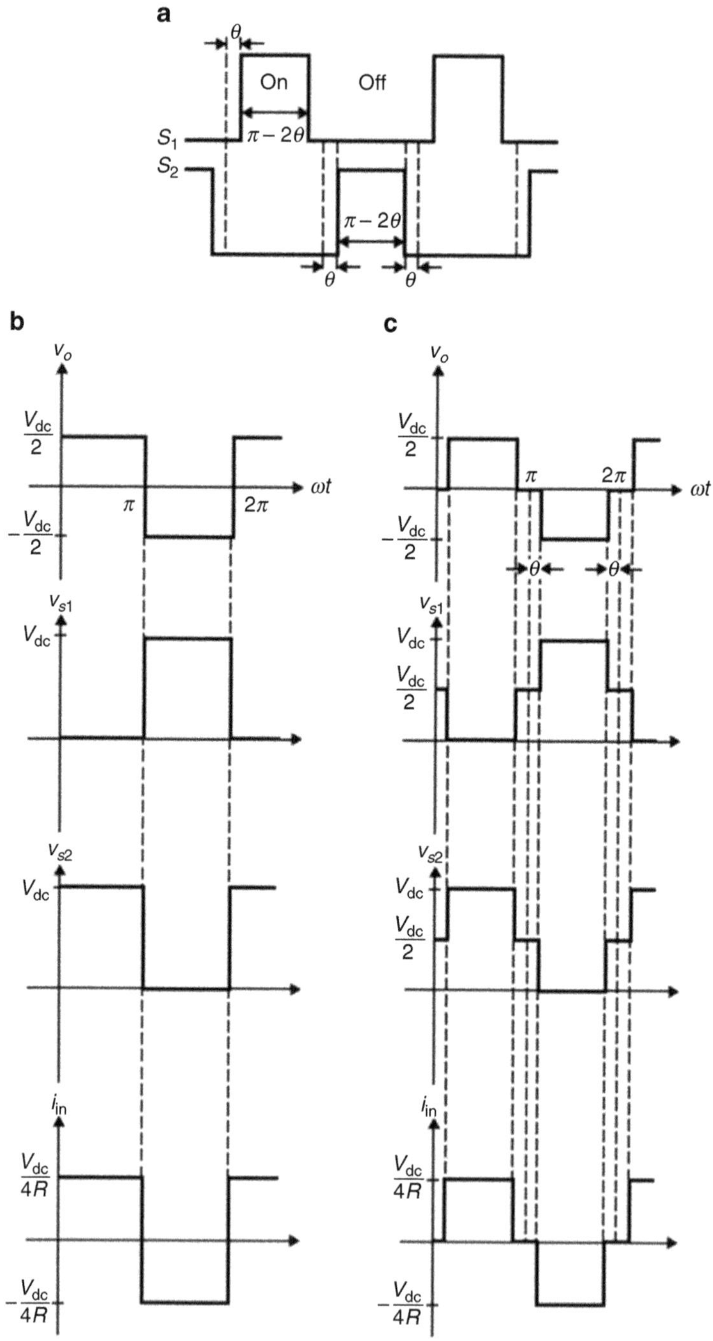

Fig. 9.7 (**a**) Switching waveform for Example 9.1 (**b**) Current and voltage waveform for $\theta = 0$ (**c**) Current and voltage waveform for $\theta \neq 0$

Mode 3 starts when S_2 is ON and S_1 is OFF, which yields the following equations:

$$v_o = -\frac{V_{dc}}{2}$$
$$i_o = \frac{v_o}{R} = -\frac{V_{dc}}{2R}$$
$$i_{c1} = \frac{i_o}{2} = -\frac{V_{dc}}{4R}$$
$$i_{c2} = -i_{c1} = \frac{V_{dc}}{4R}$$
$$i_{in} = i_{c1} = -\frac{V_{dc}}{4R}$$
$$v_{s1} = \frac{V_{dc}}{2} - v_o = V_{dc}$$
$$v_{s2} = 0$$

Mode 4 is similar to Mode 2 since both switches are open.

The average output voltage is given by:

$$v_{o,\mathrm{rms}} = \sqrt{\frac{1}{T}\int_o^T v_o^2(t)dt}$$

$$v_{o,\mathrm{rms}} = \sqrt{\frac{1}{2\pi}\left[\int_\theta^{\pi-\theta}\left(\frac{V_{dc}}{2}\right)^2 d\theta + \int_{\pi+\theta}^{2\pi-\theta}\left(\frac{-V_{dc}}{2}\right)^2 d\theta\right]} \tag{9.1}$$

$$v_{o,\mathrm{rms}} = \frac{V_{dc}}{2}\sqrt{\left(1-\frac{2\theta}{\pi}\right)}$$

Notice that when $\theta = 0$, the rms value of the output is $V_{dc}/2$, as expected. Notice that in a practical situation under an inductive load, the two switches are not allowed to switch off simultaneously. Figure 9.7b, c shows the currents and voltages for $\theta = 0$ and $\theta \neq 0$, respectively. Since the waveform is symmetric, the average output voltage will always be zero.

9.2.2 Inductive-Resistive Load

Figure 9.8a shows a half-bridge inverter under an inductive-resistive load, with its equivalent circuit, and the output waveforms are shown in Fig. 9.8b, c, respectively.

With S_1 and S_2 switched complementarily, each at a 50% duty cycle at a switching frequency f, then the load between terminal a and a' is excited by a square voltage waveform $v_{in}(t)$ of amplitudes $+V_{dc}$ and $-V_{dc}$ as shown in Fig. 9.8b, i.e., $v_{in}(t)$ is defined as follows:

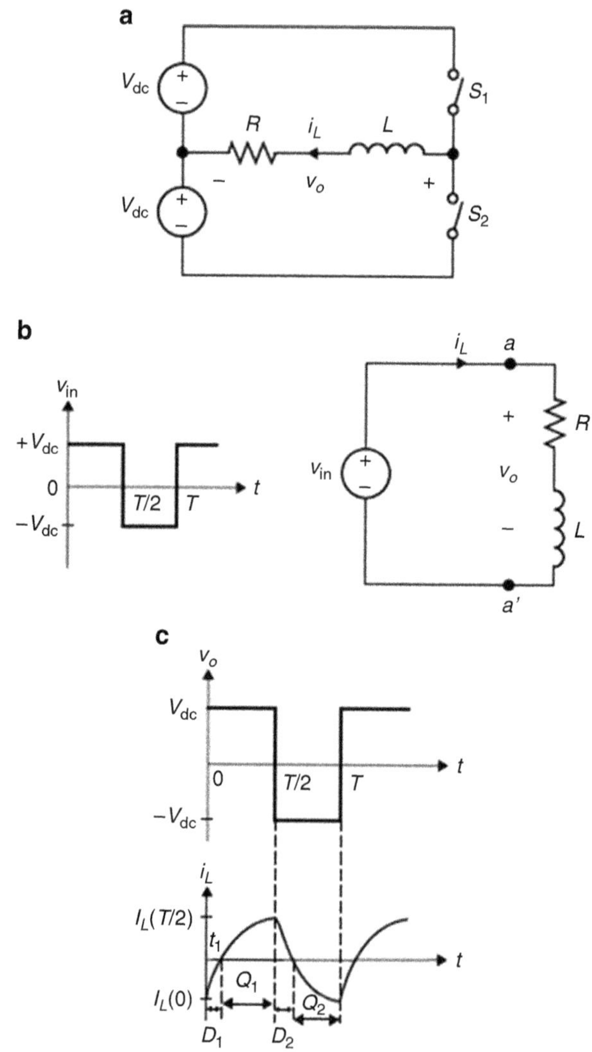

Fig. 9.8 (**a**) Half-bridge with inductive-resistive load (**b**) Equivalent circuit and (**c**) steady-state waveforms

$$v_{\text{in}}(t) = \begin{cases} +V_{dc} & 0 \le t < T/2 \\ -V_{dc} & T/2 \le t < T \end{cases} \tag{9.2}$$

The switches are implemented using a conventional SCR (requiring an external forced commutation circuit) or fully controlled power switching devices such as IGBTs, GTOs, BTJs, or MOSFETs. Notice that from the direction of the load current i_L, these switches must be bidirectional. An example is the half-bridge inverter circuit shown in Fig. 9.9 with S_1 and S_2 implemented by MOSFETs.

Assuming the inverter operates in the steady state and its inductor current waveform is shown in Fig. 9.8c for $0 \le t < t_1$, the inductor current is negative

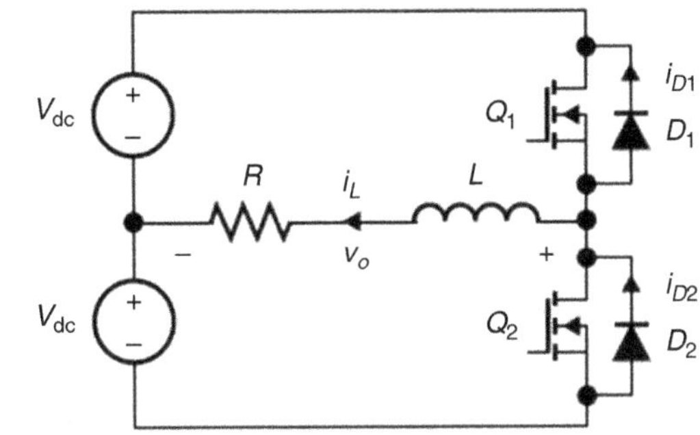

Fig. 9.9 MOSFET implementation for S_1 and S_2 in the half-bridge inverter of Fig. 9.8a

which means that while S_1 is ON the current actually flows in the reverse direction, i.e., in the body diode (flyback) of the bidirectional switch S_1. At $t = t_1$, the current flows through the transistor Q_1, as shown in Fig. 9.8c. At $t = T/2$, when S_2 is turned ON, since the current direction is positive, the flyback diode, D_2, turns ON until $t = T/2 + t_1$ when Q_2 starts conducting.

In steady state, the following conditions must hold:

$$i_L(0) = -i_L(T/2)$$
$$i_L(0) = i_L(T)$$

During the first interval ($0 \leq t < T/2$) when S_1 is *on* and S_2 is *off*, $v_{in}(t) = +V_{dc}$, resulting in the following equation for $i_L(t)$:

$$L\frac{di_L}{dt} + Ri_L = V_{dc} \tag{9.3}$$

If the inductor initial value equals $I_L(o)$, the solution for $i_L(t)$is given by:

$$i_L(t) = -\left(I_L(0) + \frac{V_{dc}}{R}\right)e^{-t/\tau} + \frac{V_{dc}}{R} \tag{9.4}$$

where $\tau = L/R$. Since $I_L(T/2) = -I_L(0)$, then the initial condition at $t = 0$ is constant and given by:

$$i_L(t) = -\frac{V_{dc}}{R}\frac{1 - e^{-T/2\tau}}{1 \mp e^{-T/2\tau}} \tag{9.5}$$

The second half-cycle $t > T/2$ produces the following expression for $i_L(t)$ with the initial condition at equaling$-I_L(0)$:

$$i_L(t) = \left(I_L(0) + \frac{V_{dc}}{R}\right)e^{-(t-T/2)/\tau} - \frac{V_{dc}}{R} \tag{9.6}$$

This expression equals $-i_L(t)$ of Eq. (9.4), which is valid for the interval ($T/2 \leq t < T$).

The exact expression for the average power delivered to the load can be obtained from the following relation:

$$\begin{aligned} P_{o,ave} &= \frac{1}{T}\int_0^T i_L(t)v_o(t)dt \\ &= \frac{2V_{dc}}{T}\int_0^{T/2}\left[-\left(I_L(0)+\frac{V_{dc}}{R}\right)e^{-t/\tau}+\frac{V_{dc}}{R}\right] \end{aligned} \tag{9.7}$$

where $I_L(0)$ is given by Eq. (9.5).

To determine the device's ratings, we must find the average and *rms* current values through the switching devices and flyback diodes. It is clear that when i_L changes polarity at $t = t_1$ during the first interval, the load current changes polarity, hence, commutes from D_1 to Q_1. Similarly, at $t = t_1 + T/2$, the current commutation goes from D_2 to Q_2 as shown in Fig. 9.8c. At the time at which $i_L(t)$ becomes zero, $t = t_1$ is obtained by setting $i_L(t)$ in Eq. (9.4) to zero at $t = t_1$, to yield:

$$t_1 = \tau \ln \frac{2}{1 + e^{T/2\tau}} \tag{9.8}$$

Obtaining the value of t_1 allows us to find the average and *rms* values for the switching devices and flyback diodes.

9.2.2.1 Average Transistor and Diode Currents

To help us obtain quantitatively the expressions for the diode and transistor currents, we represent the load voltage and current by their fundamental components as shown in Fig. 9.10a.

Let the fundamental component of $v_o(t)$ and $i_L(t)$ be given by:

$$v_{o1}(t) = V_{o1} \sin \omega t \tag{9.9}$$

$$i_{L1}(t) = I_{o1} \sin(\omega t + \theta) \tag{9.10}$$

where $V_{o1} = 4V_{dc}/\pi$ and I_{o1} and θ are the peak current and the phase angle as shown in Fig. 9.10a, which are given by:

$$I_{o1} = \frac{2V_{dc}}{\pi|Z|}, \quad \theta = \tan^{-1}\left(\frac{\omega L}{R}\right) \tag{9.11}$$

where $|Z| = \sqrt{(\omega L)^2 + R^2}$. Using the fundamental component expression, the *rms* values of the diode and transistor currents are given by:

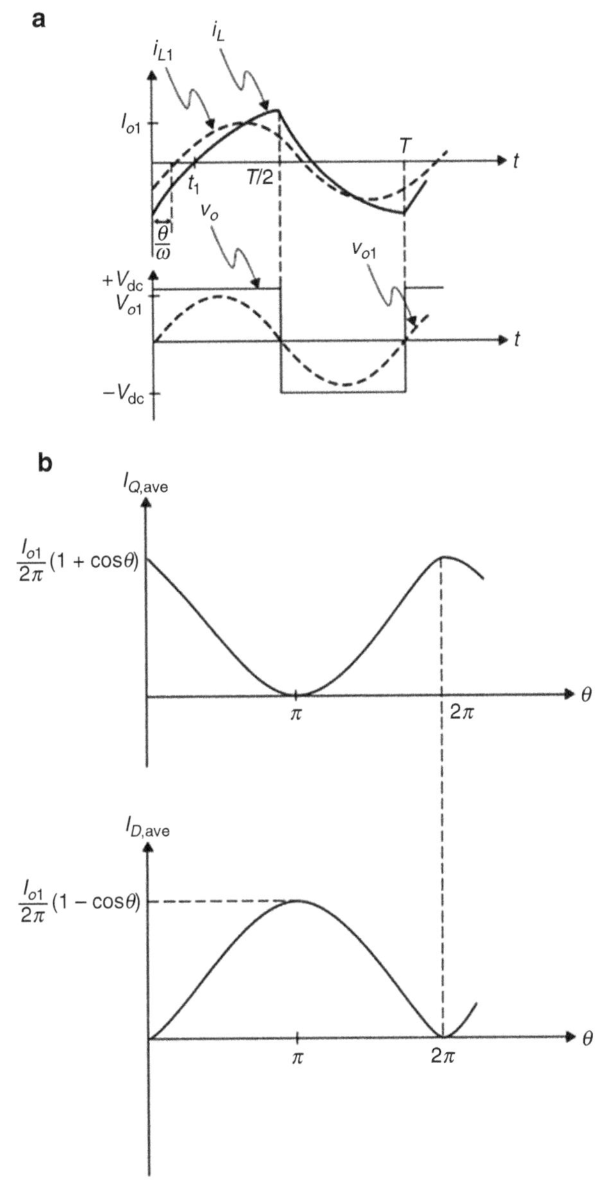

Fig. 9.10 (**a**) Output voltage and current waveforms (**b**) Average transistor and diode current waveforms

$$I_{\mathrm{D,rms}} = \sqrt{\frac{1}{2\pi}\int_{o}^{\theta} i_{\mathrm{L1}}^{2}(t)d\omega t}$$
$$= \sqrt{\frac{1}{2\pi}\int_{o}^{\theta} I_{o1}^{2}\sin^{2}\omega t d\omega t} \tag{9.12}$$
$$= \sqrt{\frac{I_{o1}^{2}}{2\pi}\left(-\frac{\sin 2\theta}{2}+\theta\right)} = \frac{I_{o1}}{2}\sqrt{\frac{\theta}{\pi}-\frac{\sin\theta\cos\theta}{2\pi}}$$

$$I_{\mathrm{Q,rms}} = \sqrt{\frac{1}{2\pi}\int_{o}^{\pi} I_{o1}^{2}\sin^{2}\omega t d\omega t}$$
$$= \sqrt{\frac{I_{o1}^{2}}{4\pi}(\pi + \cos\theta\sin\theta - \theta)} = \frac{I_{o1}}{2}\sqrt{1-\frac{\theta}{\pi}+\frac{\sin\theta\cos\theta}{\pi}} \tag{9.13}$$

and the average values of the diode and transistor currents are given by:

$$I_{\mathrm{D,ave}} = \frac{1}{2\pi}\int_{o}^{\theta} I_{o1}\sin\omega t d\omega t$$
$$= \frac{I_{o1}}{2\pi}(1-\cos\theta) \tag{9.14}$$

$$I_{\mathrm{Q,ave}} = \frac{I_{o1}}{2\pi}(1+\cos\theta) \tag{9.15}$$

Notice that under a resistive load, the lagging phase angle is zero and the entire load current flows through Q_1 and Q_2 with D_1 and D_2 always reverse biased. This is clear from the zero average diode current obtained from the preceding equation. Under a purely inductive load, the phase angle is 90°, and the load current flows through the diode and transistors for an equal time, resulting in equal average and rms current values for the diode and transistor. Figure 9.10b shows the average transistor and diode current values as functions of the phase angle.

The average output power delivered to the load is given by:

$$P_{\mathrm{o,ave}} = V_{\mathrm{o1,rms}} I_{\mathrm{o1,rms}} \cos\theta \tag{9.16}$$

where:

$$V_{\mathrm{o1,rms}} = \frac{V_{o1}}{\sqrt{2}} = \frac{\sqrt{2}V_{dc}}{\pi}$$
$$I_{\mathrm{o1,rms}} = \frac{I_{o1}}{\sqrt{2}}$$

Equation (9.16) yields the following expression $P_{o,\text{ave}}$:

$$P_{o,\text{ave}} = \frac{2{V_{dc}}^2}{\pi^2|Z|}\cos\theta \tag{9.17}$$

The ripple voltage in the dc-ac inverter is defined by:

$$V_{o,\text{ripple}} = \sqrt{V_{o,\text{rms}}^2 - V_{o,\text{ave}}^2} \tag{9.18}$$

Similarly, the input ripple current expression in the dc-ac inverter is defined by:

$$I_{\text{in,ripple}} = \sqrt{I_{o,\text{rms}}^2 - I_{o,\text{ave}}^2} \tag{9.19}$$

$$I_{\text{in,ripple}} \approx I_{o1,\text{rms}} = \frac{\sqrt{2}V_{dc}}{\pi|Z|} \tag{9.20}$$

Example 9.2

Consider the half-bridge inverter of Fig. 9.9 with the following circuit components: $V_{dc} = 408$ V, $R = 8\ \Omega$, $f = 400$ Hz, and $L = 40$ mH.

(a) Derive the exact expression $i_L(t)$.
(b) Derive the expression the fundamental component of $i_L(t)$.
(c) Determine the average diode and transistor currents.
(d) Determine the average power delivered to the load.
(e) Determine the ripple current.

Solution

(a) The exact solution for $i_L(t)$ is derived before and given again in Eq. (9.21):

$$i_L(t) = \begin{cases} -\left(I_L(0) + \frac{V_{dc}}{R}\right)e^{-t/\tau} + \frac{V_{dc}}{R} & 0 \le t < T/2 \\ +\left(I_L(0) + \frac{V_{dc}}{R}\right)e^{-(t-T/2)/\tau} - \frac{V_{dc}}{R} & T/2 \le t < T \end{cases} \tag{9.21}$$

Since $i_L(T/2) = -I_L(0)$, we have:

$$I_L(0) = -\frac{V_{dc}}{R}\frac{1 - e^{-T/2\tau}}{1 \mp e^{-T/2\tau}}, \text{ where } \tau = \frac{L}{R} \tag{9.22}$$

(b) To calculate the fundamental component of $i_L(t)$, we first determine the a_1 and b_1 coefficients:

$$I'_{L1}(t) = \frac{2}{T}\left[\int_0^{T/2} i_L(t)\cos\omega t\, dt + \int_{T/2}^{T} i_L(t)\cos\omega t\, dt\right] \tag{9.23a}$$

$$I''_{L1}(t) = \frac{2}{T}\left[\int_0^{T/2} i_L(t)\sin\omega t\,dt + \int_{T/2}^{T} i_L(t)\sin\omega t\,dt\right] \tag{9.23b}$$

The fundamental component of $i_L(t)$ is given by:

$$i_{L1}(t) = I'_{L1}(t)\cos\omega t + I''_{L1}(t)\sin\omega t \tag{9.24}$$

$$i_{L1}(t) = -0.75\cos\left(2513.3t + 57.8^\circ\right)\text{A}$$

The *rms* value of $i_L(t)$:

$$I_{L,rms} = \sqrt{\frac{1}{T}\left[\int_0^{T/2} i_L^2(t)\,dt + \int_{T/2}^{T} i_L^2(t)\,dt\right]}$$

$$I_{L,rms} = 64.1\ \text{A}$$

(c) The average diode current is given by Eq. (9.14) as:

$$I_{D,ave} = \frac{I_{o1}}{2\pi}(1 - \cos\theta)$$

with $\theta = \tan^{-1}(\omega L/R) = 85.45^\circ$ and:

$$I_{o1} = \frac{2V_{dc}}{\pi|Z|} = \frac{2\times 408}{\pi\times 100.85} = 2.58\ \text{A}$$

$$|Z| = \sqrt{(\omega L)^2 + R^2} = 100.85\ \Omega$$

If we substitute these values, then $I_{D,ave}$ turns out to be:

$$I_{D,ave} = \frac{2.58}{2\pi}\left(1 - \cos 85.45^\circ\right) = 0.38\ \text{A}$$

Similarly, $I_{Q,ave}$ is equal to:

$$I_{Q,ave} = \frac{I_{o1}}{2\pi}(1 + \cos\theta) = 0.44\ \text{A}$$

(d) The average power is given in Eq. (9.17):

$$P_{o,ave} = V_{o1,rms}I_{o1,rms}\cos\theta = \frac{2{V_{dc}}^2}{\pi^2|Z|}\cos\theta$$

$$P_{o,ave} = \frac{2(408)^2}{\pi^2(100.85)}\cos 85.45 = 26.53\ \text{W}$$

Exercise 9.1

Fig. E9.1 shows a half-bridge inverter under a purely inductive load with $V_{dc} = 24$ V, $L = 10$ mH, and $f = 60$ Hz. Assume Q_1 and Q_2 are operating at a 50% duty cycle, and the circuit has reached the steady-state operation.

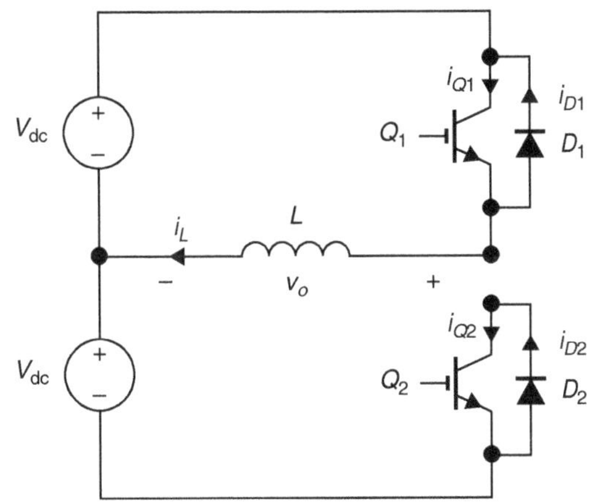

Fig. E9.1 Half-bridge with pure inductive load

(a) Sketch the waveforms for $i_{\mathrm{L}}, i_{\mathrm{Q1}}, i_{\mathrm{D1}}, i_{\mathrm{Q2}}, i_{\mathrm{D2}}$, and v_{o}.
(b) Calculate the *rms* value of $i_{\mathrm{L}}(t)$.
(c) Determine the average output power delivered to the load.

Answer: $I_{\mathrm{L,\,rms}} = 11.54$ A and $P_{\mathrm{ave}} = 0$ W.

Exercise 9.2

Consider the single-phase inverter of Fig. 9.8a with an inductive-resistive load that delivers 400 W to a 60 Hz load from a dc source of 420 V. Assume the output voltage and current are represented by their fundamental components with a lagging power factor of 0.6.

(a) Determine the average and *rms* current values for the transistors and diodes.
(b) Determine the ripple value for i_{in}.

Answer: $I_{\mathrm{D,\,rms}} = 0.81$ A, $I_{\mathrm{D,\,ave}} = 0.45$ A, $I_{\mathrm{Q,\,rms}} = 2.31$ A, and $I_{\mathrm{Q,\,ave}} = 1.80$ A

Example 9.3

Draw the output voltage and v_{s1}waveforms for the center-tap biphase inverter shown in Fig. 9.11. Assume S_1 and S_2 are bidirectional switches and are switched at a 50% duty cycle. It is used in a low input voltage application to reduce losses, since the current only flows half-period in a section of the transformer (the transformer is not fully utilized). The two modes of operations are shown in Fig. 9.11b, c, and the waveforms are shown in Fig. 9.11d.

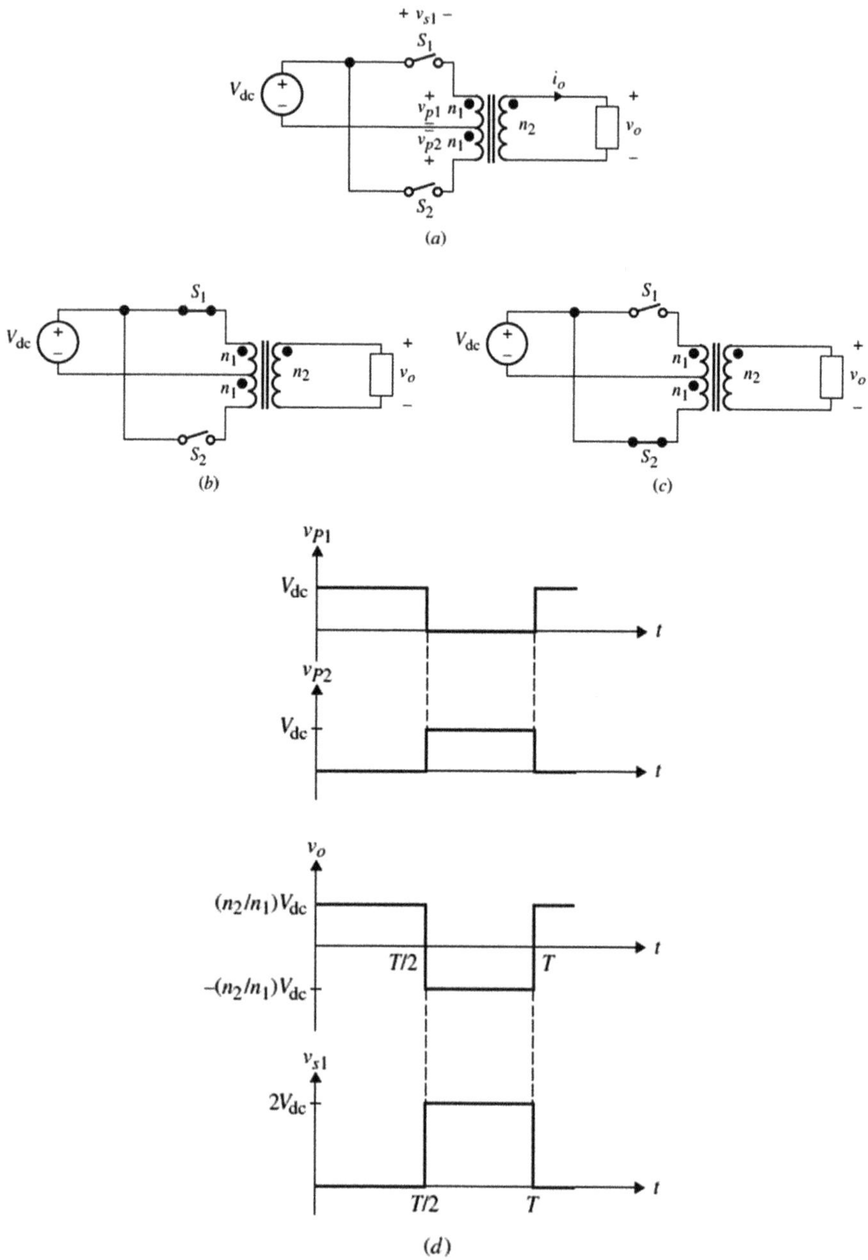

Fig. 9.11 (**a**) Center-tap biphase inverter for Example 9.2, (**b**) Mode 1, (**c**) Mode 2, and (**d**) voltage waveforms

Solution

The equivalent circuit for Mode 1 when switch S_1 is *on* is shown in Fig. 9.11b. The output voltage is given by:

$$\frac{v_o}{V_{dc}} = \frac{n_2}{n_1}$$
$$v_o = \frac{n_2}{n_1} V_{dc}$$

Figure 9.11c shows the equivalent circuit for Mode 2, with v_o given by:

$$\frac{v_o}{V_{dc}} = -\frac{n_2}{n_1}$$
$$v_o = -\frac{n_2}{n_1} V_{dc}$$

The waveforms for v_o are shown in Fig. 9.11d

9.3 Full-Bridge Inverters

Figure 9.12 shows the full-bridge circuit configuration for a voltage-source inverter under resistive load.

Depending on the switching sequence of S_1, S_2, S_3, and S_4, the output voltage can be controlled by either varying V_{dc} only or by controlling the phase shift between the switches. If $S_1 - S_3$ and $S_2 - S_4$ are switched ON and OFF at a 50% duty cycle as shown in Fig. 9.13a, the output voltage, shown in Fig. 9.13b, is a symmetrical square wave whose fundamental *rms* value is controlled only by varying V_{dc}. The fundamental value of $v_o(t)$ is given by:

$$v_{o1}(t) = \frac{V_{dc}}{\pi} \sin \omega t \tag{9.25}$$

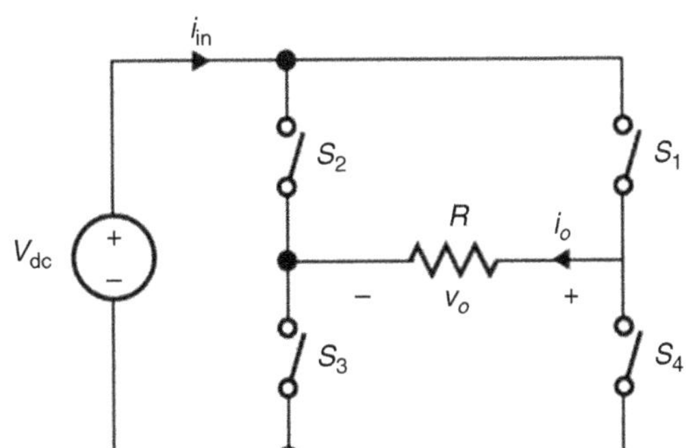

Fig. 9.12 Full-bridge inverter under a purely resistive load

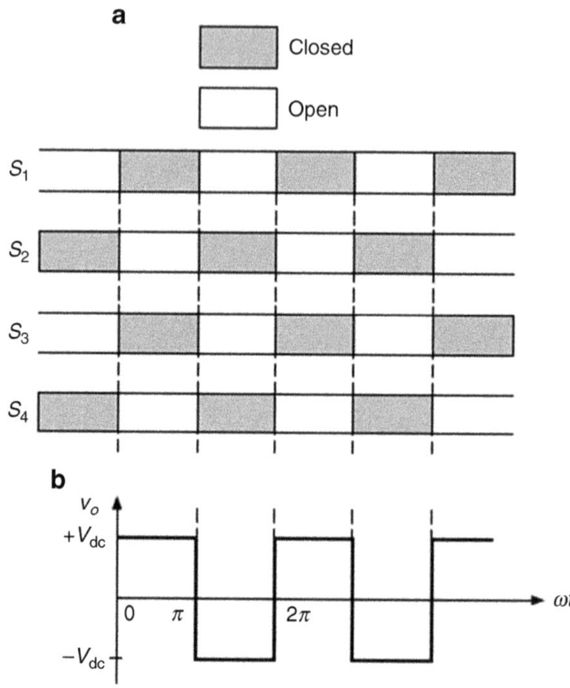

Fig. 9.13 Switching sequence for full-bridge voltage-source inverter (**a**) 50% duty cycle and (**b**) output voltage

The *rms* value of the fundamental component is $V_{dc}/\sqrt{2}\pi$. The use of SCR rectifier circuits or dc-dc switch-mode converters normally accomplishes varying of the dc voltage source.

Another method to control the *rms* value of the output voltage is to use a switching sequence shown in Fig. 9.14a. This switching sequence is obtained by shifting the timing sequence of S_1 and S_4 in Fig. 9.13a to the left by a phase angle α and S_2 and S_3 to the right by the same angle.

The fundamental component of the output voltage, $v_{o1}(t)$, is given by:

$$v_{o1}(t) = V_{dc}\sqrt{\frac{1}{2} - \frac{\alpha}{\pi}\sin(\omega t - \theta)} \tag{9.26}$$

And the *rms* value is given by $V_{dc}\sqrt{\frac{1}{4} - \alpha/(2\pi)}$. It is clear from this relation that the *rms* value or the peak of the fundamental component can be controlled by the amount of the phase shift between the switching signals. We notice that, unlike the switching sequence given in Fig. 9.13b, which has two states for the output voltage, $+V_{dc}$ and $-V_{dc}$, the switching sequence in Fig. 9.14a gives three states for the output voltage, $+V_{dc}$, 0, and $-V_{dc}$; such inverters are known as *tri-state inverters*. Since the load is resistive, the four switches can be implemented by SCRs with a unidirectional current flow. This is because the load current can reverse direction instantaneously as the voltage v_o reverses its direction. However, under an inductive load,

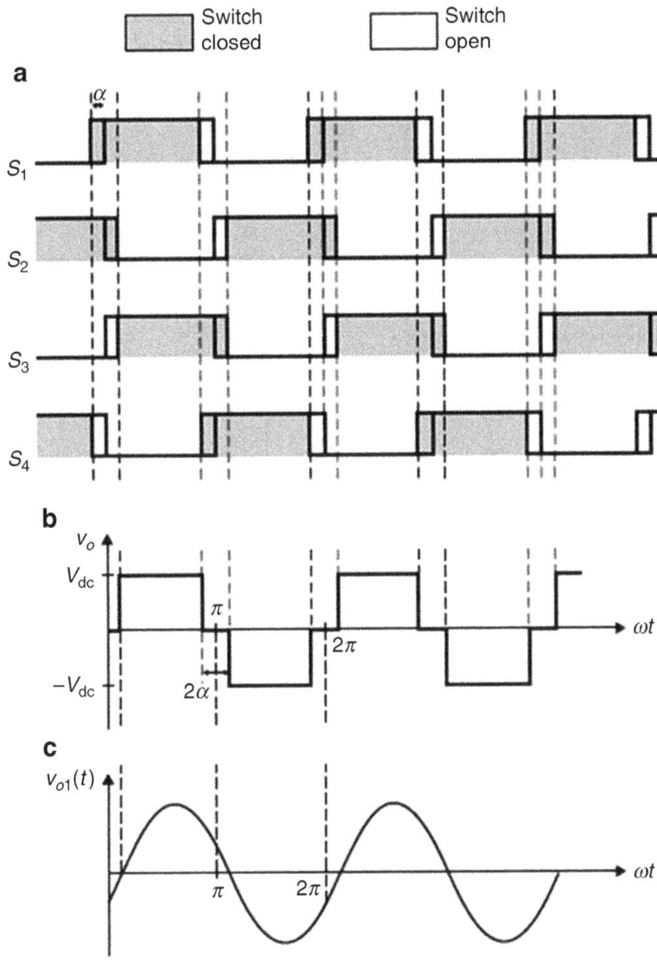

Fig. 9.14 (**a**) Switching sequence with α-phase shift, (**b**) output voltage, (**c**) fundamental component for $v_o(t)$

for the circuit to work using SCRs, a diode must be added in parallel with each SCR as shown in Fig. 9.15.

It can be seen from the switching sequence of Fig. 9.14a that in the steady state, there exist four modes of operation as shown in Fig. 9.16. The switch implementation for S_1, S_2, S_3, and S_4 is quite simple since there is no need for the bidirectional current flow. The average power delivered to the load is $\left(V_{\mathrm{o,rms}}^2/R\right)$ when $V_{\mathrm{o,rms}}$ is equal to V_{dc} or $V_{dc}\sqrt{\frac{1}{4}-\alpha/(2\pi)}$, depending on whether the switching sequence of Fig. 9.13a or Fig. 9.14a is used, respectively.

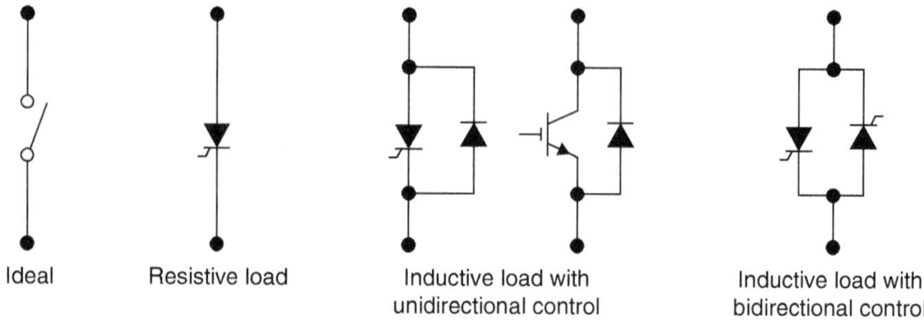

Fig. 9.15 Possible switch implementation

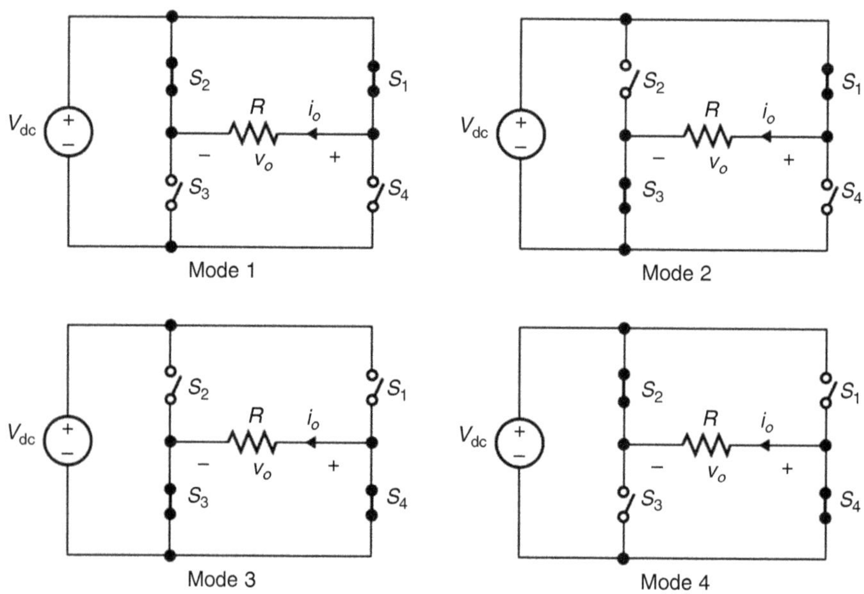

Fig. 9.16 Modes of operation

Example 9.4

Consider the resistive load full-bridge voltage-source inverter shown in Fig. 9.12 with the following circuit parameters: $V_{dc} = 150$ V, $R = 12\ \Omega$, and $f_s = 60$ Hz. Sketch the waveforms for v_o and i_{in} and determine the average power delivered to the load for the two switching sequences shown in Figs. 9.13a and 9.14a, with $\alpha = 10^{\circ}$.

Solution

For the switch sequence shown in Fig. 9.13a, v_o and i_{in} are symmetric and given by:

$$v_o = \begin{cases} +V_{dc} & 0 \le t < T/2 \\ -V_{dc} & T/2 \le t < T \end{cases}$$

$$i_o = \begin{cases} +\dfrac{V_{dc}}{R} & 0 \le t < T/2 \\ -\dfrac{V_{dc}}{R} & T/2 \le t < T \end{cases}$$

The average output power is given by:

$$P_{o,ave} = \frac{1}{T}\int_0^T v_o i_o dt$$

For the switch sequence shown in Fig. 9.14a, the average output power as given by:

$$P_{o,ave} = \frac{V_{o,rms}^2}{R}$$

where the *rms* value is expressed as:

$$V_{o,rms} = V_{dc}\sqrt{\frac{1}{4} - \frac{\alpha}{2\pi}}$$

The resultant average output power for $\alpha = 10^{\circ}$ is given by:

$$P_{o,ave} = \frac{V_{dc}^2\left(1 - \dfrac{2\alpha}{\pi}\right)}{R} = 1666.67 \text{ W}$$

As stated earlier, practical loads do not consist of a simple resistor with a unity power factor but rather have some sort of an inductance. Figure 9.17a shows a full-bridge inverter under an inductive-resistive load. If the switches are operating at a 50% duty cycle with a two-state output, then the current and voltage waveforms are as shown in Fig. 9.17b. The analysis of this inverter is similar to that for the half-bridge voltage-source inverter discussed earlier.

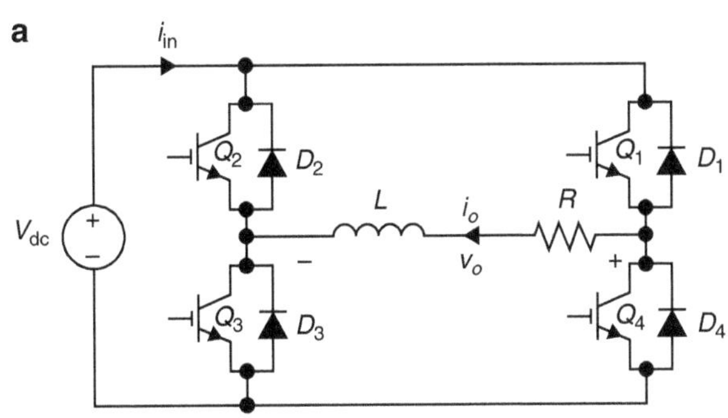

Fig. 9.17 (**a**) Full-bridge inverter under R-L load (**b**) Waveforms under 50% duty cycle (**c**) Fundamental component of the inductor current

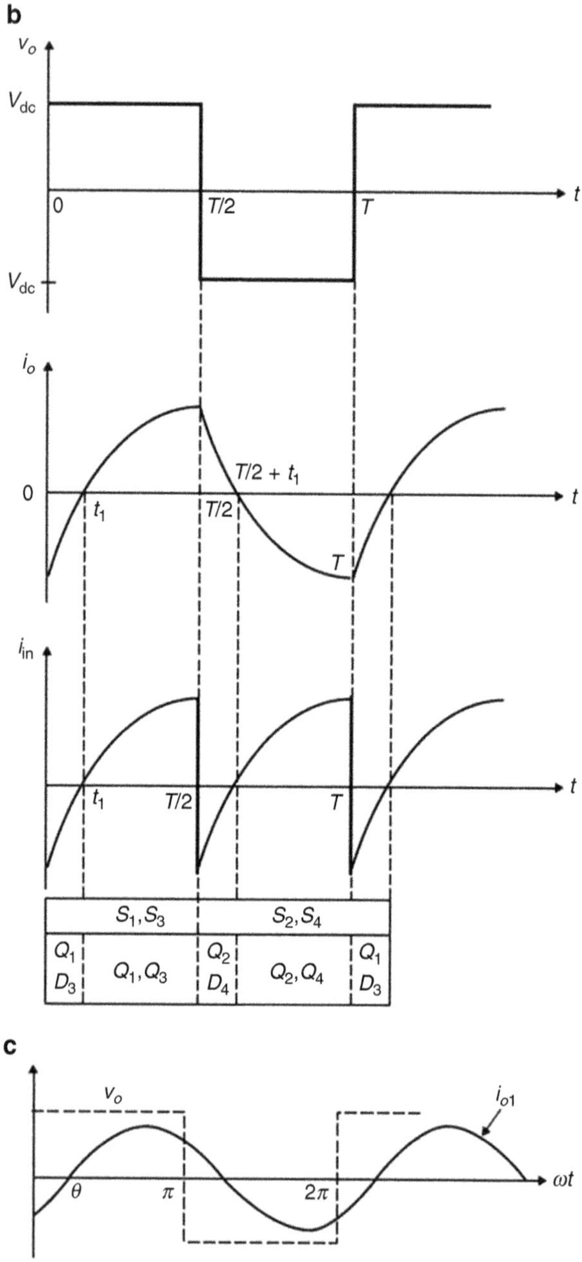

Fig. 9.17 (continued)

To obtain an expression for the average power absorbed by the load, we revert to using the Fourier series as our analysis technique. The *rms* values for v_o and $i_o$of Fig. 9.17b are based on a 50% square-wave output given by:

$$I^2_{\text{o.rms}} = I^2_{\text{1.rms}} + I^2_{\text{2.rms}} + \ldots + I^2_{\text{n.rms}} \tag{9.27a}$$

$$V^2_{\text{o.rms}} = V_{dc} \tag{9.27b}$$

where $I_{\text{n,rms}} = I_{\text{n}}/\sqrt{2}$ and I_{n} is the peak current of the *n*th harmonic of $i_o(t)$.

Since the output voltage is a square wave with a 50% duty cycle, its Fourier series may be expressed as follows:

$$v_o(t) = \frac{4V_{dc}}{\pi}\left(\sin\omega t + \frac{\sin 3\omega t}{3} + \frac{\sin 5\omega t}{5} + \ldots + \frac{\sin n\omega t}{n}\right) \tag{9.28}$$

Therefore, $i_o(t)$ is given by:

$$i_o(t) = \frac{4V_{dc}}{\pi}\left(\frac{\sin\omega t}{\sqrt{R^2 + (\omega L)^2}} + \frac{\sin 3\omega t}{3\sqrt{R^2 + (3\omega L)^2}} + \frac{\sin 5\omega t}{5\sqrt{R^2 + (5\omega L)^2}} + \ldots + \frac{\sin n\omega t}{n\sqrt{R^2 + (n\omega L)^2}}\right) \tag{9.29}$$

And the *rms* value for the *n*th current component is given by:

$$I_{\text{n,rms}} = \frac{2\sqrt{2}V_{dc}}{n\pi|Z_{\text{n}}|}$$

where:

$$|Z_{\text{n}}| = \sqrt{R^2 + (n\omega L)^2}$$

If we assume that the major part of the average output power is delivered at the fundamental frequency, then $P_{\text{o,ave}}$ is given by:

$$P_{\text{o,ave}} = \frac{8V^2_{dc}}{\pi^2\sqrt{R^2 + (\omega L)^2}} \tag{9.30}$$

9.3.1 Approximate Analysis

An approximate solution for the load current can be obtained by assuming that $L/R \gg T/2$. This will allow us to represent the load current by its first harmonic. As illustrated in the previous example, Fig. 9.17c shows the inductor current being represented by its fundamental component; the load current may be approximated by:

$$i_{o1} \approx I_{o1} \sin(\omega t - \theta) \tag{9.31}$$

where:

$$\theta = \tan^{-1}\frac{\omega L}{R}$$

$$I_{o1} = \frac{V_{o1}}{\sqrt{(\omega L)^2 + R^2}}$$

$$V_{o1} = \frac{4V_{dc}}{\pi}$$

The average power delivered to the load is given by:

$$P_{ave} = I_{o1,rms}V_{o1,rms}\cos\theta$$

$$= \frac{I_{o1}V_{o1}}{2}\cos\theta$$

$$P_{ave} = \frac{8V_{dc}^2 R}{\pi^2\sqrt{R^2 + (\omega L)^2}}\cos\theta \tag{9.32}$$

where $I_{o1,rms}$ is the *rms* of the fundamental component of the inductor current.

Including the contribution of higher harmonics to the power delivered to the loads leads to more complex equations for the power factor and total harmonic distortion.

Exercise 9.3

Consider the full-bridge voltage-source inverter under an R-L load of Fig. 9.17a with $V_{dc} = 220$ V, $L = 6$ mH, $R = 16\ \Omega$, and $f_s = 50$Hz. Calculate the average power delivered to the load up to the seventh harmonics.

Answer: 2.73 W

9.3.2 *Generalized Analysis*

As in the half-bridge configuration, the above analysis under a 50% duty cycle control with no α overlap does not allow output control. Figure 9.18a shows a typical output voltage under α control that is produced using the switching sequence of Fig. 9.14a. The equivalent circuit for the single-phase bridge inverter is shown in Fig. 9.18b.

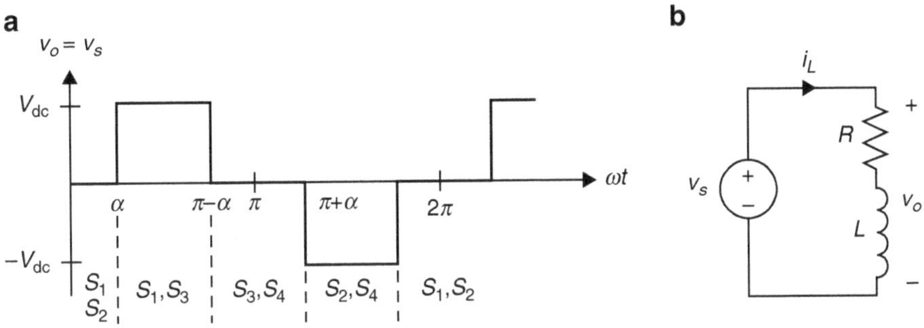

Fig. 9.18 (**a**) Output voltage using switching sequence given in Fig. 9.14a, (**b**) Equivalent circuit for the full-bridge inverter

Fig. 9.19 Generalized load representation

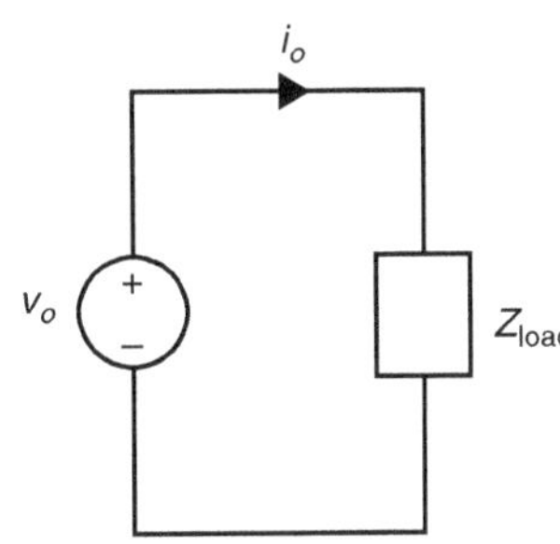

Let us assume we have a generalized impedance load, Z_{load}, as shown in Fig. 9.19. The Fourier analysis representation for $v_o(t)$ is given by the following equation:

$$v_o(t) = \sum_{n=1,3,5,\ldots}^{\infty} V_n \sin n\omega t \tag{9.33}$$

where:

$$V_n = \frac{4V_{dc}}{n\pi} \cos n\alpha \tag{9.34}$$

and $i_o(t)$ is obtained from the following equation:

$$i_o(t) = \sum_{n=1,3,5,\ldots}^{\infty} I_n \sin(n\omega t - \theta_n) \tag{9.35}$$

where:

$$I_n = \frac{V_n}{|Z_n|} = \frac{4V_{dc}}{n\pi|Z_n|} \cos n\alpha \tag{9.36a}$$

$$\theta = \angle Z_n \tag{9.36b}$$

where $|Z_n|$ is the magnitude of the nth harmonic impedance and $\angle Z_n$ is the angle of the nth harmonic impedance. The overall *rms* output voltage is given by:

$$V_{o,\mathrm{rms}} = V_{dc}\sqrt{1 - \frac{2\alpha}{\pi}} \tag{9.37}$$

In terms of the nth harmonics, the *rms* value of the voltage for each harmonic is given by:

$$V_{on,\mathrm{rms}} = \frac{4V_{dc}}{n\sqrt{2}\pi}\sqrt{\sum_{n=1,3,5,\ldots}^{\infty} \cos^2 n\alpha} \tag{9.38}$$

and the *rms* value of the current for each harmonic is given by:

$$I_{on,\mathrm{rms}} = \sqrt{\sum_{n=1,3,5,\ldots}^{\infty} I^2_{\mathrm{n,rms}}} \tag{9.39}$$

where the *rms* for the nth harmonic is given by:

$$I_{on,\mathrm{rms}} = \frac{4V_{dc}}{\sqrt{2}n\pi|Z|}\cos n\alpha \tag{9.40}$$

and the *rms* of the fundamental output current is given by:

$$I_{o1,\mathrm{rms}} = \frac{4V_{dc}}{\sqrt{2}n\pi|Z_1|}\cos\alpha \tag{9.41}$$

where:

$$|Z_1| = \sqrt{R^2 + (\omega_0 L)^2}$$

And ω_0 is the fundamental frequency.

The total average power delivered to the load resistance is given by:

$$P_{o,\mathrm{ave}} = \frac{1}{T}\int_0^T i_o v_o dt \tag{9.42}$$

The voltage and current's THD are given in Eqs. (9.43a, 9.43b), respectively.

$$\begin{aligned} THD_v &= \sqrt{\left(\frac{V_{o,\mathrm{rms}}}{V_{o1,\mathrm{rms}}}\right)^2 - 1} \\ &= \frac{\pi}{2\cos\alpha}\sqrt{\frac{1}{2} - \frac{\alpha}{\pi}} \end{aligned} \tag{9.43a}$$

$$THD_i = \sqrt{\left(\frac{I_{\rm o,rms}}{I_{\rm o1,rms}}\right)^2 - 1}$$
$$= \sqrt{\left(\sum_{n=1,3,5,\ldots}^{\infty} \frac{|Z_1|}{n|Z_n|} \frac{\cos\alpha}{\cos n\alpha}\right)^2 - 1} \tag{9.43b}$$

For the nth harmonic, the average power of Eq. (9.42) is given by:

$$P_{\rm on,ave} = V_{\rm on,rms} I_{\rm on,rms} \cos\theta_n$$

So the total average power is given by:

$$P_{\rm o,ave} = \sum_{n=1,3,5,\ldots}^{\infty} V_{\rm on,rms} I_{\rm on,rms} \cos\theta_n \tag{9.44}$$

Example 9.5
Consider the full-bridge inverter whose equivalent circuit is represented in Fig. 9.19 with the four different loads shown in Fig. 9.20 with $R = 8\ \Omega$, $L = 30$ mH, $C = 147\ \mu$F, $f_{\rm o} = 60$ Hz, and $V_{dc} = 120$ V.

(a) Determine the *rms* for $i_{\rm o}$ and $v_{\rm o}$ for the first, third, fifth, and seventh harmonics.
(b) Determine the total average power delivered to the load for each of the above harmonics.
(c) Determine the output current and voltage total harmonic distortion.

Solution
To determine the *rms* for $i_{\rm o}$ and $v_{\rm o}$ for the first, third, fifth, and seventh harmonics, we use the nth harmonic given by:

$$v_{n,\rm o} = \frac{4V_{dc}}{n\pi} \cos n\alpha \sin n\omega t \tag{9.45}$$
$$I_{\rm n,rms} = \frac{4V_{dc}}{\sqrt{2}n\pi|Z_n|} \cos n\alpha$$

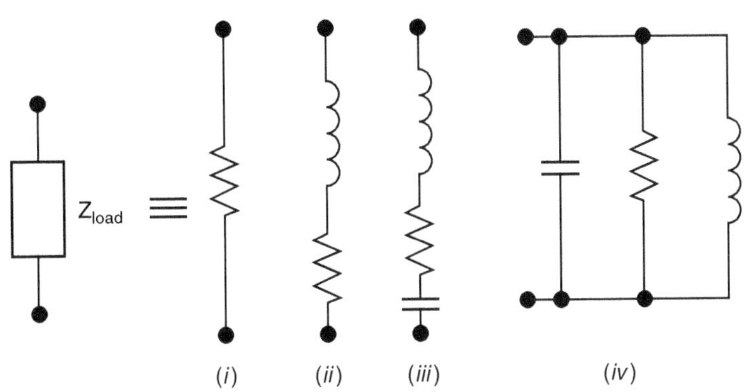

Fig. 9.20 Various types of loads

i. *For the resistive load*:

$$V_{n,\text{rms}} = \frac{2\sqrt{2}V_{dc}}{n\pi}\cos n\alpha \qquad I_{n,\text{rms}} = \frac{V_{n,\text{rms}}}{|Z_n|}, \qquad |Z_n| = R = 8\,\Omega,\ \theta_n = 0,\ \alpha = \pi/6$$

$$V_{1,\text{rms}} = \frac{2\sqrt{2}.\ 120}{\pi}\cos\left(\frac{\pi}{6}\right) = 93.56\,\text{V} \qquad I_{1,\text{rms}} = \frac{V_{1,\text{rms}}}{|Z_1|} = \frac{V_{1,\text{rms}}}{R} = 11.69\,\text{A}$$

$$V_{3,\text{rms}} = \frac{2\sqrt{2}.\ 120}{3\pi}\cos\left(\frac{\pi}{2}\right) = 0\,\text{V} \qquad I_{3,\text{rms}} = \frac{V_{3,\text{rms}}}{|Z_3|} = \frac{V_{3,\text{rms}}}{R} = 0\,\text{A}$$

$$V_{5,\text{rms}} = \frac{2\sqrt{2}.\ 120}{5\pi}\cos\left(\frac{5\pi}{6}\right) = -18.71\,\text{V} \qquad I_{5,\text{rms}} = \frac{V_{5,\text{rms}}}{|Z_5|} = \frac{V_{5,\text{rms}}}{R} = -2.34\,\text{A}$$

$$V_{7,\text{rms}} = \frac{2\sqrt{2}.\ 120}{7\pi}\cos\left(\frac{7\pi}{6}\right) = -13.37\,\text{V} \qquad I_{7,\text{rms}} = \frac{V_{7,\text{rms}}}{|Z_7|} = \frac{V_{7,\text{rms}}}{R} = -1.67\,\text{A}$$

Using the first four harmonics, the approximate *rms* values for v_o and i_o are given by:

$$V_{o,\text{rms}} = \sqrt{V_{1,\text{rms}}^2 + V_{3,\text{rms}}^2 + V_{5,\text{rms}}^2 + V_{7,\text{rms}}^2} = 96.34\ \text{V}$$

$$I_{o,\text{rms}} = \sqrt{I_{1,\text{rms}}^2 + I_{3,\text{rms}}^2 + I_{5,\text{rms}}^2 + I_{7,\text{rms}}^2} = 12.04\ \text{A}$$

The exact *rms* values are given by:

$$V_{o,\text{rms}} = \sqrt{\frac{2}{\pi}\int_{\alpha}^{\pi} V_{dc}^2\, d\omega t} = \sqrt{\frac{5}{6}}V_{dc} = 109.59\ \text{V}$$

$$I_{o,\text{rms}} = \frac{V_{o,\text{rms}}}{|Z_n|} = 13.69\ \text{A}$$

ii. *For the series R-L load*: The *rms* value for the output voltage is the same as the one above, since the output voltage is independent of the load.

$$I_{n,\text{rms}} = \frac{V_{n,\text{rms}}}{|Z_n|}, \qquad |Z_\text{n}| = \sqrt{R^2 + (n\omega L)^2} \quad \theta_n = \tan^{-1}\left(\frac{n\omega L}{R}\right)$$

$$I_{1,\text{rms}} = \frac{V_{1,\text{rms}}}{\sqrt{R^2 + (\omega L)^2}} = 6.75\ \text{A}$$

$$I_{3,\text{rms}} = \frac{V_{3,\text{rms}}}{\sqrt{R^2 + (3\omega L)^2}} = 0\ \text{A}$$

$$I_{5,\text{rms}} = \frac{V_{5,\text{rms}}}{\sqrt{R^2 + (5\omega L)^2}} = -0.33\ \text{A}$$

$$I_{7,\text{rms}} = \frac{V_{7,\text{rms}}}{\sqrt{R^2 + (7\omega L)^2}} = -0.168\ \text{A}$$

iii. *For the series RLC load:* The *rms* value for the output voltage is the same as in (i), since the output voltage is independent of the load.

$$|Z_n| = \sqrt{R^2 + \left(n\omega L - \frac{1}{n\omega C}\right)^2}, \qquad \theta_n = \tan^{-1}\left(\frac{n\omega L - \frac{1}{n\omega C}}{R}\right)$$

$$I_{1,\text{rms}} = \frac{V_{1,\text{rms}}}{\sqrt{R^2 + \left(\omega L - \frac{1}{\omega C}\right)^2}} = 8.95 \text{ A}$$

$$I_{3,\text{rms}} = \frac{V_{3,\text{rms}}}{\sqrt{R^2 + \left(3\omega L - \frac{1}{3\omega C}\right)^2}} = 0 \text{ A}$$

$$I_{5,\text{rms}} = \frac{V_{5,\text{rms}}}{\sqrt{R^2 + \left(5\omega L - \frac{1}{5\omega C}\right)^2}} = -0.35 \text{ A}$$

$$I_{7,\text{rms}} = \frac{V_{7,\text{rms}}}{\sqrt{R^2 + \left(7\omega L - \frac{1}{7\omega C}\right)^2}} = -0.17 \text{ A}$$

iv. *For the parallel RLC load*:

The *rms* value for the output voltage is the same as the one shown above, since the output voltage is independent of the load.

$$|Z_n| = \frac{1}{\sqrt{\left(\frac{1}{R}\right)^2 + \left(\frac{n^2\omega^2 LC - 1}{n\omega L}\right)^2}}, \qquad \theta_n = \tan^{-1}\left(\frac{n^2\omega^2 LC - R}{n\omega L}\right)$$

$$I_{1,\text{rms}} = \frac{V_{1,\text{rms}}}{1/\sqrt{\left(\frac{1}{R}\right)^2 + \left(\frac{1^2\omega^2 LC - 1}{\omega L}\right)^2}} = 12.1 \text{ A}$$

$$I_{3,\text{rms}} = \frac{V_{3,\text{rms}}}{1/\sqrt{\left(\frac{1}{R}\right)^2 + \left(\frac{3^2\omega^2 LC - 1}{3\omega L}\right)^2}} = 0 \text{ A}$$

$$I_{5,\text{rms}} = \frac{V_{5,\text{rms}}}{1/\sqrt{\left(\frac{1}{R}\right)^2 + \left(\frac{5^2\omega^2 LC - 1}{5\omega L}\right)^2}} = -5.39 \text{ A}$$

$$I_{7,\text{rms}} = \frac{V_{7,\text{rms}}}{1/\sqrt{\left(\frac{1}{R}\right)^2 + \left(\frac{7^2\omega^2 LC - 1}{7\omega L}\right)^2}} = -5.29 \text{ A}$$

(b) Power calculations are obtained from the following equation:

$$P_{\text{on, ave}} = V_{\text{on, rms}} I_{\text{on, rms}} \cos\theta_n$$

i. *For the resistive load*:

$$\cos\theta_n = 1 \quad \text{For } n = 1, 3, 5\ldots$$

$$P_{\text{o1, ave}} = (93.56)(11.69) = 1094.65 \text{ W}$$

$$P_{\text{o3, ave}} = 0 \text{ W}$$

$$P_{\text{o5, ave}} = (-18.71)(-2.34) = 43.78 \text{ W}$$

$$P_{\text{o7, ave}} = (-13.37)(-1.67) = 22.33 \text{ W}$$

$$P_{\text{o}} = \sum_{n=1,3,5,\ldots}^{\infty} I_{\text{on, rms}} V_{\text{on, rms}} \cos\theta_n = 1160.76 \text{ W}$$

ii. *For the series R-L load*:

$$P_{\text{o1, ave}} = (93.56)(11.69)\cos\left(54.73^{\circ}\right) = 364.67 \text{ W}$$

$$P_{\text{o3, ave}} = 0 \text{ W}$$

$$P_{\text{o5, ave}} = (-18.71)(-2.34)\cos\left(81.95^{\circ}\right) = 0.865 \text{ W}$$

$$P_{\text{o7, ave}} = (-13.37)(-1.67)\cos\left(84.23^{\circ}\right) = 0.226 \text{ W}$$

$$P_{\text{o, ave}} = \sum_{n=1,3,5,\ldots}^{\infty} I_{\text{on, rms}} V_{\text{on, rms}} \cos\theta_n = 365.75 \text{ W}$$

iii. *For the series RLC load*:

$$P_{\text{o1, ave}} = (93.56)(11.69)\cos\left(40.093^{\circ}\right) = 640.85 \text{ W}$$

$$P_{\text{o3, ave}} = 0 \text{ W}$$

$$P_{\text{o5, ave}} = (-18.71)(-2.34)\cos\left(81.36^{\circ}\right) = 0.98 \text{ W}$$

$$P_{\text{o7, ave}} = (-13.37)(-1.67)\cos\left(84.037^{\circ}\right) = 0.24 \text{ W}$$

$$P_{\text{o, ave}} = \sum_{n=1,3,5,\ldots}^{\infty} I_{\text{on, rms}} V_{\text{on, rms}} \cos\theta_n = 641.8 \text{ W}$$

iv. *For the parallel RLC load*:

$$P_{\text{o1,ave}} = (93.56)(11.69)\cos\left(-14.76^{\circ}\right) = 1094.2\ \text{W}$$

$$P_{\text{o3,ave}} = 0\ \text{W}$$

$$P_{\text{o5,ave}} = (-18.71)(-2.34)\cos\left(-64.274^{\circ}\right) = 43.78\ \text{W}$$

$$P_{\text{o7,ave}} = (-13.37)(-1.67)\cos\left(-71.579^{\circ}\right) = 22.35\ \text{W}$$

$$P_{\text{o,ave}} = \sum_{n=1,3,5,\ldots}^{\infty} I_{\text{on,rms}} V_{\text{on,rms}} \cos\theta_{\text{n}} = 1160.3\ \text{W}$$

(c) The total harmonic distortion can be obtained from Eq. (9.43a) given by:

$$THD_i = \sqrt{\left(\frac{I_{\text{o,rms}}}{I_{\text{o1,rms}}}\right)^2 - 1} = 0.31$$

Exercise 9.4
Consider an inductive-resistive load inverter shown in Fig. 9.17a with the following parameters: $\alpha = \pi/3$, $L = 100$ mH, $R = 16\Omega$, $V_{dc} = 250$ V, and $f_{\text{o}} = 60$ Hz. Determine the fundamental current and voltage components.

Answer: $i_{\text{o}} = \frac{12.2}{\pi}\sin\left(\omega t - 67^{\circ}\right)\text{A}$
The general analysis presented in this section is based on a passive load Z_{L} that produces a fixed load phase shift $\angle Z_{\text{L}} = \theta$; therefore, the output voltage control can be achieved only by varying α. However, when the load contains a voltage source as in motor and utility grid applications, then it is an active load that allows both the phase shift θ and the switching displacement α to be used as control variables.

Example 9.6
Consider the active load in a bridge inverter that consists of an R-L load and an ac sinusoidal voltage source as shown in Fig. 9.21 with the load voltage v_{o} as shown. Obtain the expression for the fundamental load current and the average power delivered to the load for the following circuit parameters: $v_{ac} = 100\sin(2\pi 100t - 30^{\circ})$, $V_{dc} = 180$ V, $L = 42$ mH, $R = 0.5\ \Omega$, $\alpha = 15^{\circ}$, and $f_{\text{o}} = 60$ Hz.

Solution

(a) To obtain the expression for the fundamental load current, we can use the following equations:

$$\overrightarrow{I_{\text{o1}}} = \frac{\overrightarrow{V_{\text{o1}}} - \overrightarrow{V_{ac}}}{R + j\omega L}$$

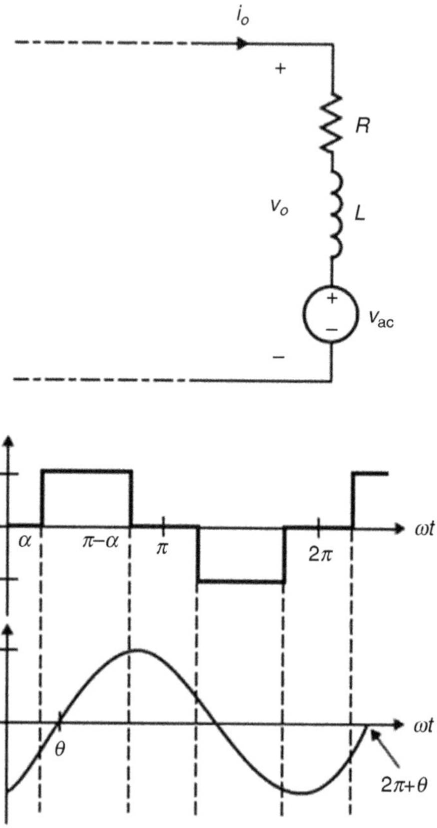

Fig. 9.21 Inverter example with an active load

We have:

$$V_{o1} = \frac{4V_{dc}}{\pi}\cos\alpha = \frac{4\times 180}{\pi}\cos\left(15^{\circ}\right) = 221.37\ \text{V}$$

Hence:

$$\overrightarrow{V_{o1}} = 221.37\angle 0^{\circ}$$

$$\overrightarrow{V_{ac}} = 100\angle 30^{\circ}$$

$$\overrightarrow{I_{o1}} = \frac{221.37\angle 0^{\circ} - 100\angle 30^{\circ}}{(0.5 + j2\pi(60))\left(42\times 10^{-3}\right)} = 9.08\angle - 110.25^{\circ}$$

The total power delivered to the load is given by the following equation:

$$\overrightarrow{P_{T}} = \frac{1}{2}\overrightarrow{I_{o1}}\ \overrightarrow{V_{o1}}$$

$$\overrightarrow{P_{T}} = \frac{1}{2}\left(221.37\angle 0^{\circ}\right)\left(9.08\angle - 110.25^{\circ}\right) = 1005.2\angle - 110.25^{\circ}$$

9.4 Harmonic Reduction

Harmonic reduction includes the elimination and cancelation of certain harmonics of the output voltage. Reducing the harmonic content of the ac output is one of the most difficult challenges of dc-ac inverter design. If the inverter drives an electromechanical load (ac motor), the harmonics can excite a mechanical resonance, causing the load to emit acoustic noise.

Unlike the case in dc-dc converters, harmonics in the output waveforms are very significant. This is why the output filters perform different functions and their design is quite different in terms of complexity and size. In designing output filters in dc-dc converters, the objective is to limit the output voltage ripple to a certain desired percentage of the average output voltage. In other words, this is a passive filter approach that is limited by the physical size of inductors and capacitors. In dc-dc inverters, the reduction or cancelation of the output harmonics is done actively by controlling the switching technique of the inverter. Compared with ac-dc conversion, harmonic filtering in dc-ac is harder since it will affect the attenuation and/or the phase shift of the fundamental component.

The harmonics that are present in the inverter's output voltage are high for many practical applications, especially when the output voltage needs to be near sinusoidal. To help produce a near sinusoidal output, electronic low-pass filters are normally added at the output to remove third and higher harmonics. The design of such filters tends to be challenging when the third and fifth harmonics close to the fundamental are high. In most cases, effective filters require large size and high numbers of filtering capacitive and inductive components, resulting in bulkiness. By controlling the width or the number of pulses, certain harmonic content can be removed without the need for complex harmonic filtering circuits.

It is possible to cancel certain harmonics by simply selecting the duration of the pulse in the half-cycle of the output voltage. Recall that the peak component of the nth harmonic for Fig. 9.18a is given by:

$$v_n = \frac{4V_{dc}}{n\pi} \cos n\alpha \tag{9.46}$$

Considering the third harmonic, we have:

$$v_3 = \frac{4V_{dc}}{3\pi} \cos 3\alpha \tag{9.47}$$

To cancel the third harmonics, we set $\alpha = \pi/6$; this results with the cancelation of all the harmonics of the order of $3n$.

Consider the case for a two-pulse output as shown in Fig. 9.22 with angles of α_1 and α_2. It can be shown that the nth harmonic is given by:

$$V_{\text{on}} = \frac{4V_{dc}}{n\pi}(1 - \cos n\alpha_1 + \cos n\alpha_2) \tag{9.48}$$

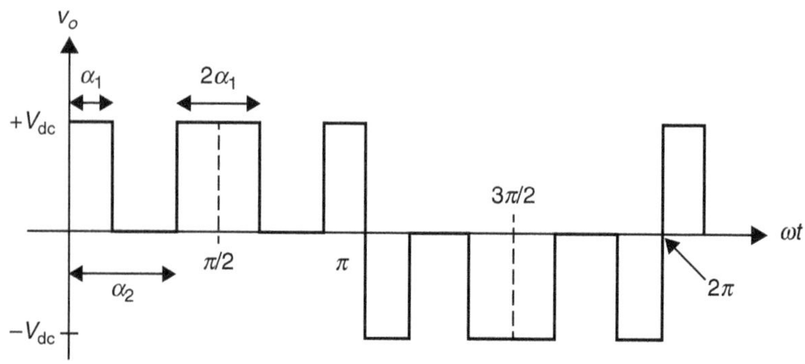

Fig. 9.22 Two-angle shift control of inverter output voltage

For example, to eliminate the third and fifth harmonics, we make $V_{on} = 0$ at two frequencies as follows:

$$1 - \cos 3\alpha_1 + \cos 3\alpha_2 = 0$$
$$1 - \cos 5\alpha_1 + \cos 5\alpha_2 = 0$$

Solve for α_1 and α_2 to obtain $\alpha_1 = 17.8^\circ$ and $\alpha_2 = 38^\circ$.

Example 9.7
Consider two cascaded push-pull inverters as shown in Fig. 9.23a; the switching waveforms for S_1–S_4 are shown in Fig. 9.23b. Sketch the waveforms for the outputs v_{o1}, v_{o2}, and v_o.

Solution
The waveforms for the output voltage waveforms are shown in Fig. 9.23b.
The fundamental component of v_o is given by:

$$v_{o1}(t) = \frac{8V_{dc}}{\pi} \cos \alpha \sin \omega t$$

where $\alpha = \omega t_1$. It is clear that the magnitude of the fundamental component of v_{o1} has been reduced by $\cos\alpha$ when compared to the case when no phase is present ($\alpha = 0$).

9.4.1 Harmonic Analysis

The harmonics contents present in the output of the inverter could be significant. Depending on the application, reducing the effect of these harmonics is very important. Recall that the nth component for a square-wave output ($\alpha = 0$) is $4V_{dc}/n\pi$.

Fig. 9.23 (**a**) Two push-pull inverters for Example 9.7 (**b**) Typical switching waveform

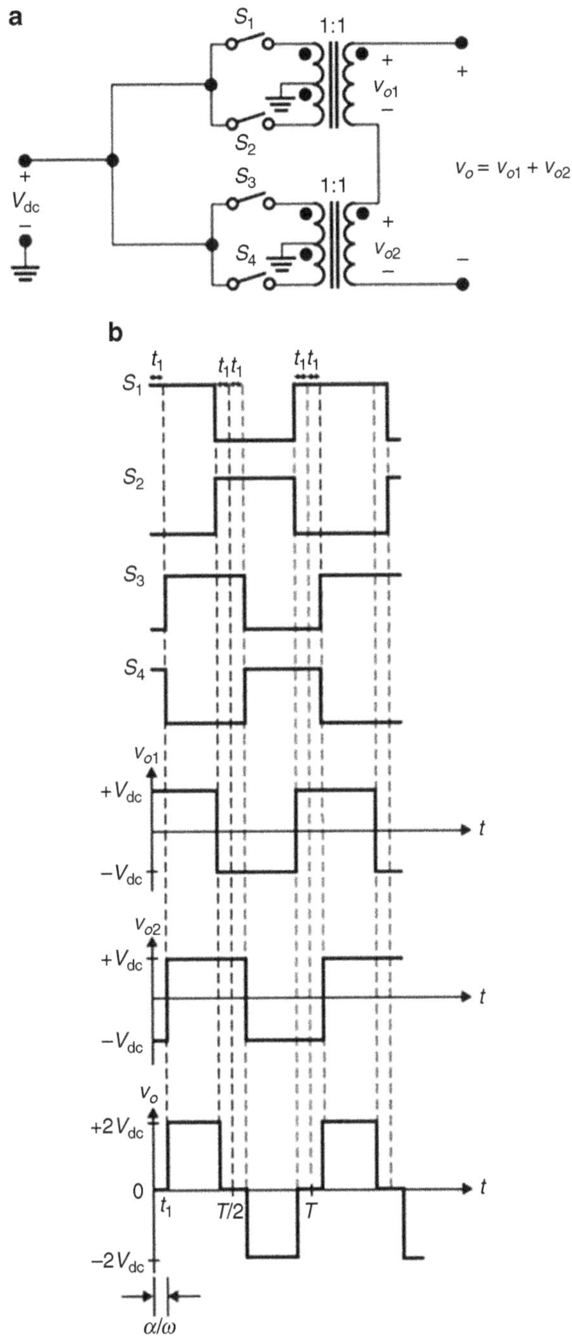

For $\alpha \neq 0$, the expression for $v_o(t)$ is given by:

$$v_o(t) = \sum_{n=1,3,5,\ldots} V_n \sin \omega t \tag{9.49}$$

where:

$$V_n = \frac{4V_{dc}}{n\pi} \cos n\alpha$$

The fundamental output voltage component is given by:

$$v_{o1} = \frac{4V_{dc}}{\pi} \cos \alpha \sin \omega t \tag{9.50}$$

Figure 9.24a–c shows the resultant output voltage where harmonics up to the ninth are included for $\alpha = 0, \alpha = \pi/6$, and $\alpha = \pi/3$, respectively. The sketch for the magnitude of the harmonics of Fig. 9.24, as a function of α, is shown in Fig. 9.25.

It is clear from the above equation that the magnitude of the harmonics is inversely proportional to the magnitude of n and α. Under a wide range of variation of α, the filters of the harmonics might not be an easy task.

The total harmonic distortion for $v_o(t)$ is given by:

$$THD_V = \sqrt{\left(\frac{V_{o,rms}}{V_{o1,rms}}\right)^2 - 1} \tag{9.51}$$

$$\begin{aligned} V_{o,rms}^2 &= V_{o1,rms}^2 + V_{o2,rms}^2 + \ldots + V_{on,rms}^2 \\ &= \sum_{n=1,3,5,\ldots}^{\infty} V_{on,rms}^2 \end{aligned} \tag{9.52}$$

where:

$$V_{on,rms} = \frac{V_{on}}{\sqrt{2}} \tag{9.53a}$$

$$V_{on} = \frac{4V_{dc}}{n\pi} \cos n\alpha \tag{9.53b}$$

Substitute Eqs. (9.52) and (9.53a) into Eq. (9.51) to yield:

$$THD_V = \sqrt{\left(\frac{\sum_{n=1,3,5,\ldots}^{\infty} V_{on,rms}^2}{V_{o1,rms}^2}\right) - 1}$$

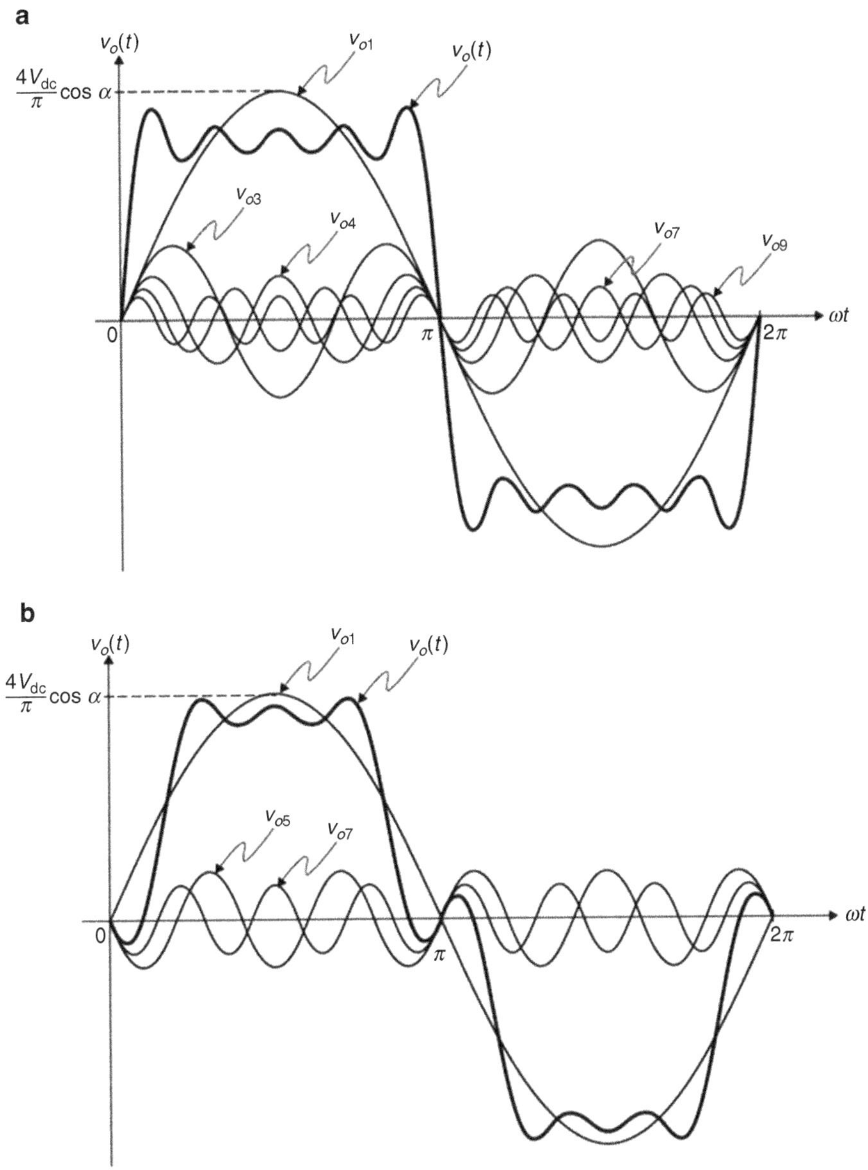

Fig. 9.24 The first nine output harmonics (**a**) $\alpha = 0$ and (**b**) $\alpha = \pi/6$ (**c**) $\alpha = \pi/3$

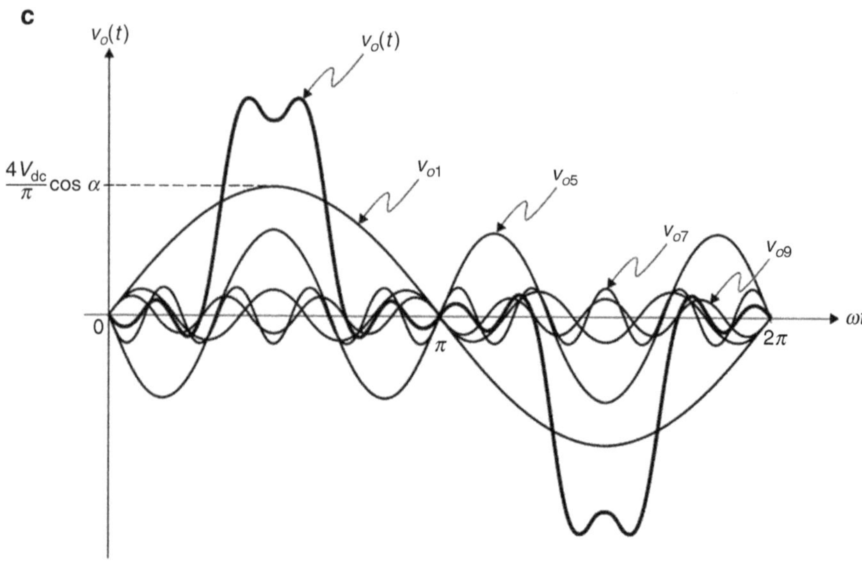

Fig. 9.24 (continued)

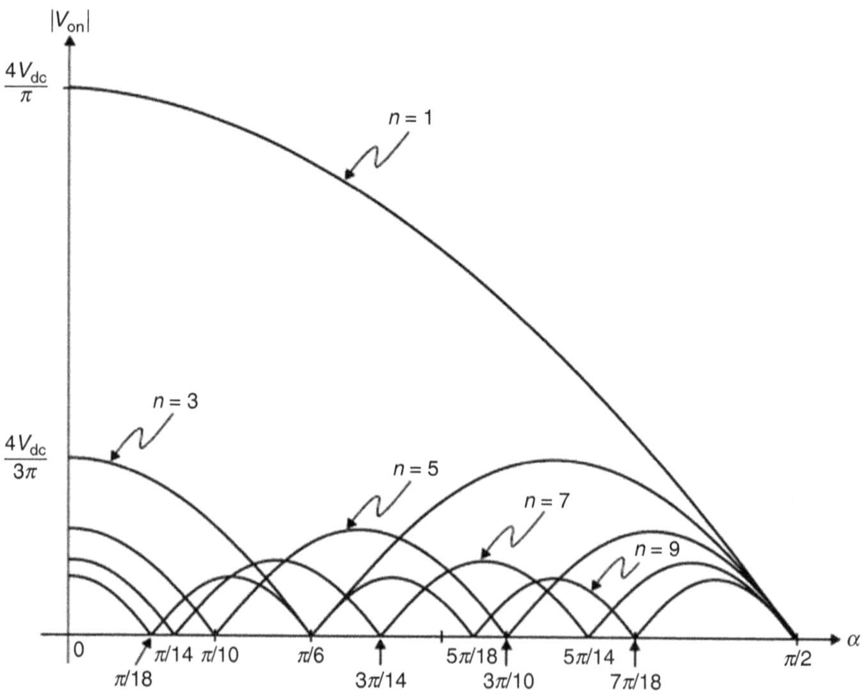

Fig. 9.25 Magnitude of the output harmonics as a function of α

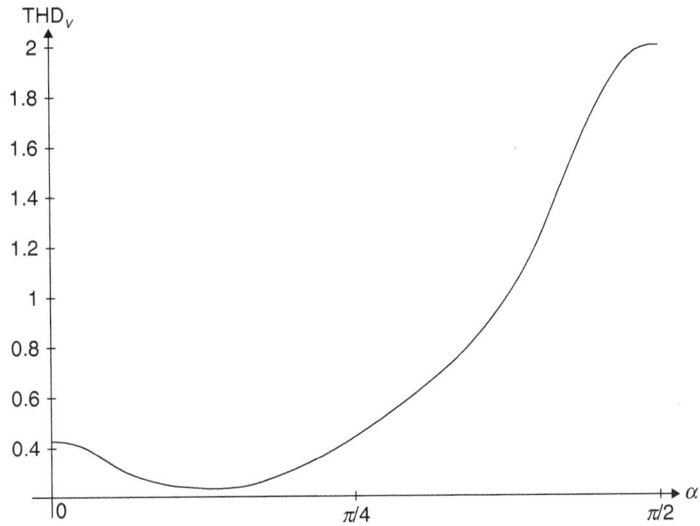

Fig. 9.26 THD plots as a function of α

$$THD_V = \sqrt{\left(\frac{\sum_{n=1,3,5,\ldots}^{\infty} V_{on,rms}^2}{V_{o1,rms}^2}\right)} \tag{9.54}$$

$$THD_V = \frac{1}{\cos\alpha}\sqrt{\sum_{n=3,5,\ldots}\frac{1}{n^2}\cos^2 n\alpha}$$

The plots of the THD for $\alpha = 0$ to $\alpha = \pi/2$ are shown in Fig. 9.26.

Exercise 9.5

Show that the ratio of the harmonic magnitude with respect to its fundamental component for Fig. E9.5 is given by:

$$\frac{V_{on}}{V_{o1}} = \frac{\sin(n\theta/2)}{\sin(\theta/2)}$$

Fig. E9.5 Inverter voltage with α-control

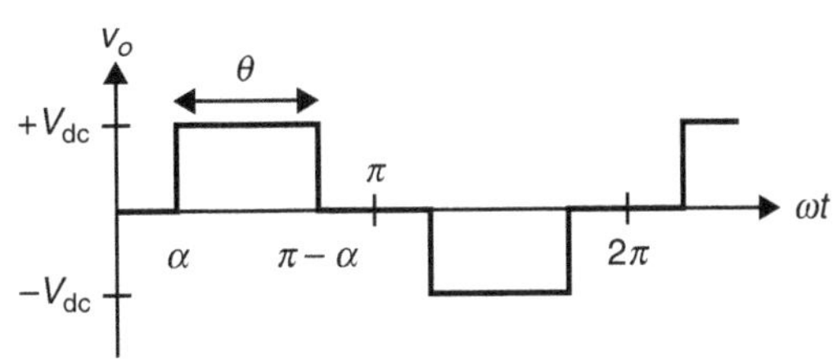

Exercise 9.6

Determine the value of $\alpha > 0$ that produces the largest THD in Eq. (9.54) when only the first and third harmonics are included.

Answer: $\alpha = \pi/2$

9.5 Pulse Width Modulation

Figure 9.27 shows the simplified block diagram representation for a single-phase switching mode inverter. The output $v'_o(t)$ shows different types of possible output waveforms that can be produced depending on the pulse width modulation (PWM) control technique employed.

In the single-phase inverter, one or more pulses in a given half-cycle are used to control the output voltage. Since varying the width of these pulses within the half-cycle carries out the control, the process is appropriately known as pulse width modulation (PWM). By modulating the width of several pulses per half-cycle, a more efficient method of controlling the output voltage of the inverter is obtained. Using the PWM process, we can extract a low-frequency signal from a train of high-frequency square waves.

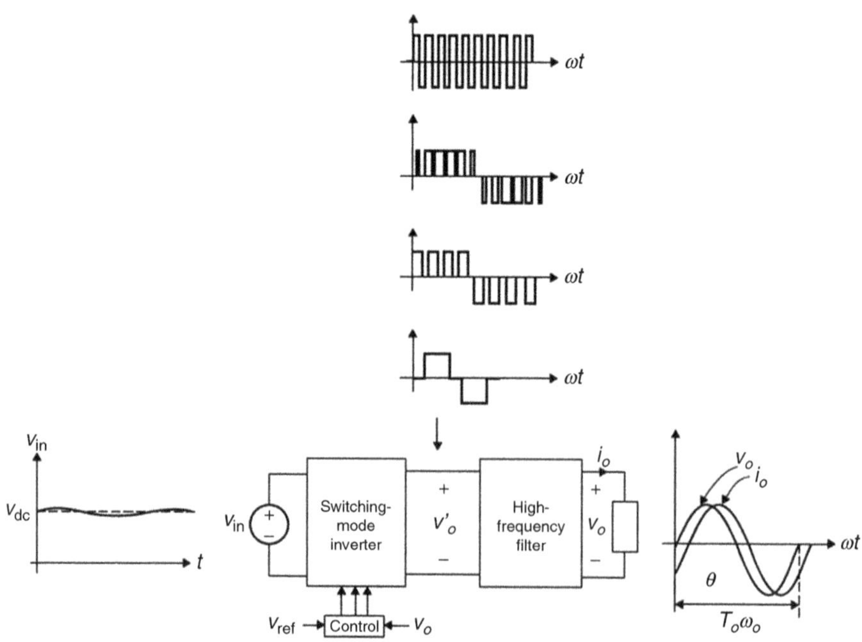

Fig. 9.27 Simplified block diagram of single-phase switching mode inverter

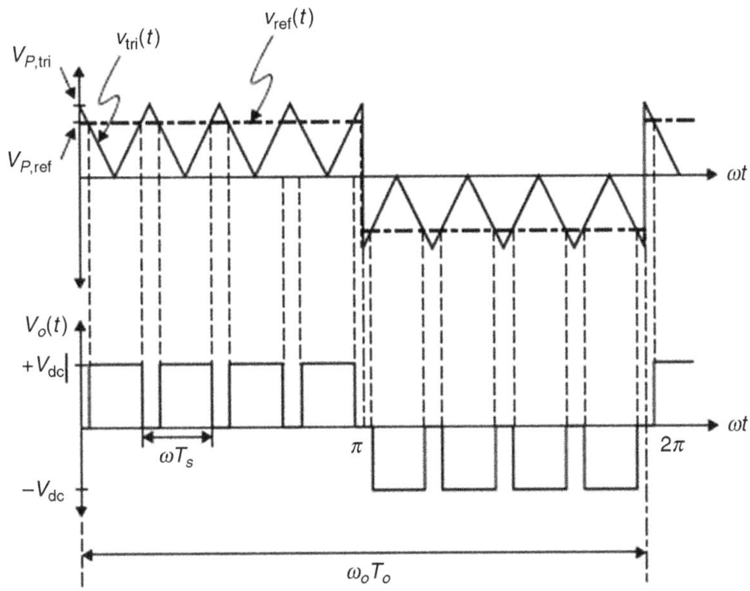

Fig. 9.28 Typical waveform for equal-pulse PWM techniques

Generally speaking, the PWM method may be grouped into classes, depending on the modulation technique:

1. Non-sinusoidal PWM, in which all pulses have the same width and are normally modulated equally to control the output voltage, as shown in the four-pulse-per-half-cycle example in Fig. 9.28. The widths of these pulses are adjusted equally to control the output voltage. Eliminating a selected number of series harmonics as discussed before requires a very complex control technique to generate the required switching sequence.
2. Sinusoidal PWM, which allows the pulse width to be modulated sinusoidally; i.e., the width of each pulse is proportional to the instantaneous value of a reference sinusoid whose frequency equals that of the fundamental components as shown in the six-pulse-per-half-cycle example in Fig. 9.29.

Normally the pulses in the non-sinusoidal PWM method are arranged to produce an odd function output voltage that is symmetrical around $T/2$. This results with the cancelation of all the cosine terms and the even harmonics of the sine terms.

We notice that the reference voltage $v_{ref}(t)$ is square and sinusoidal waveforms for the equal-pulse and sinusoidal PWM, respectively. It is important that first we define some terms that pertain to square and sinusoidal PWM inverters that will be discussed next.

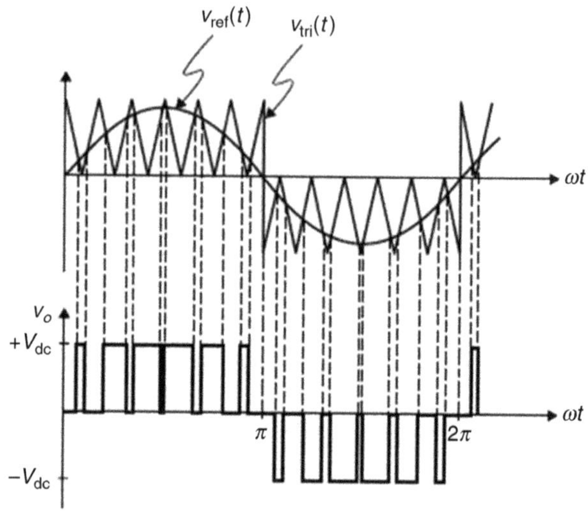

Fig. 9.29 Typical waveforms for sinusoidal PMW technique

$v_{tri}(t)$: Repetitive triangular waveform (also known as a carrier signal)
$V_{P,\,tri}$: The peak value of the triangular waveform
T_s,f_s: The period and the frequency of the triangular waveform (also known as a carrier or switching frequency)
v_{ref}: Reference signal that can be either a square or a sinusoidal waveform (also known as a control signal)
$V_{p,\,ref}$: The peak value of the reference signal
T_o,f_o: The desired inverter output period and output frequency, which are equal to the period and frequency of the reference or control signal
m_a: Inverter amplitude modulation index
m_f: Inverter frequency modulation index
k: Number of pulses per half-cycle

The amplitude and frequency modulation indices are defined as follows:

$$m_a = \frac{V_{P,\,ref}}{V_{P,\,tri}} \tag{9.55}$$

$$m_f = \frac{f_s}{f_o} \tag{9.56}$$

Next we discuss the two well-known methods of PWM techniques.

9.5.1 *Equal-Pulse (Uniform) PWM*

The equal-pulse PWM technique, known also as a single-pulse PWM control, is very old and less popular nowadays. The technique is very simple and requires simple control since all generated pulses have equal-pulse widths. Generating the equal and multiple pulses is achieved by comparing a square-wave reference voltage waveform $v_{ref}(t)$ to a triangular control (carrier) voltage waveform, $v_{cont}(t)$, as shown in Fig. 9.28. The op-amp produces a triggering signal every time the carrier signal goes below or above the reference signal as shown in Fig. 9.28. It is clear that the frequency of the reference voltage waveform determines the frequency of the output voltage and the frequency of the control signal determines the number of equal pulses in each half-cycle.

Figure 9.30a–d shows examples of one-, two-, three-, and seven-pulse outputs, respectively. As the magnitude of the reference signal increases, the pulse width

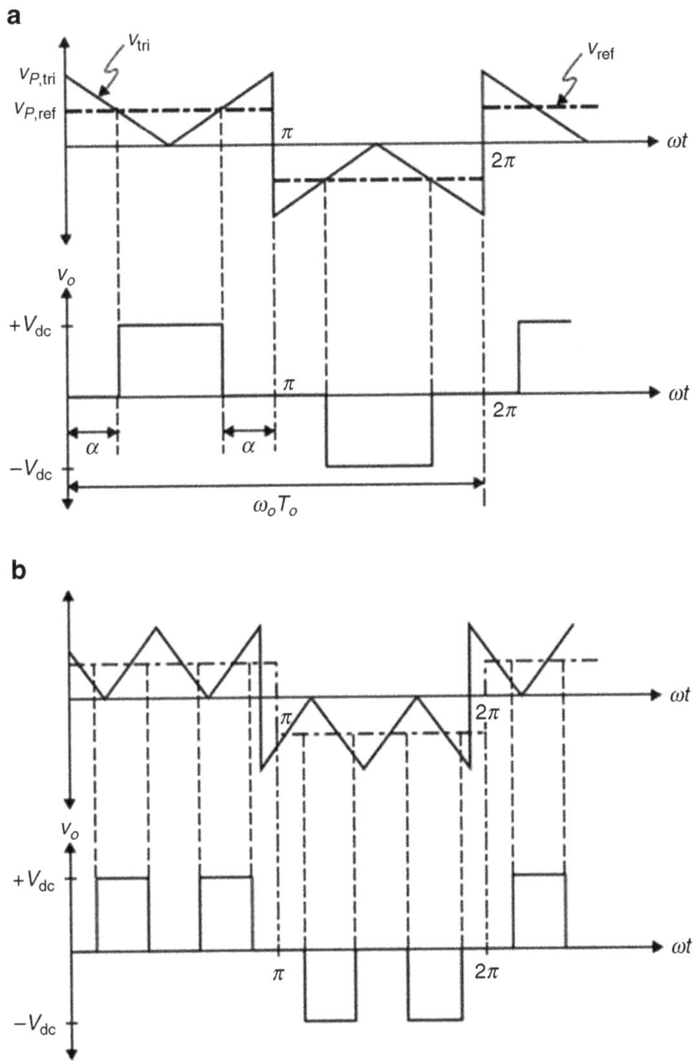

Fig. 9.30 Examples of equal pulses: (**a**) one-pulse output, (**b**) two-pulse output, (**c**) three-pulse output, and (**d**) seven-pulse outputs

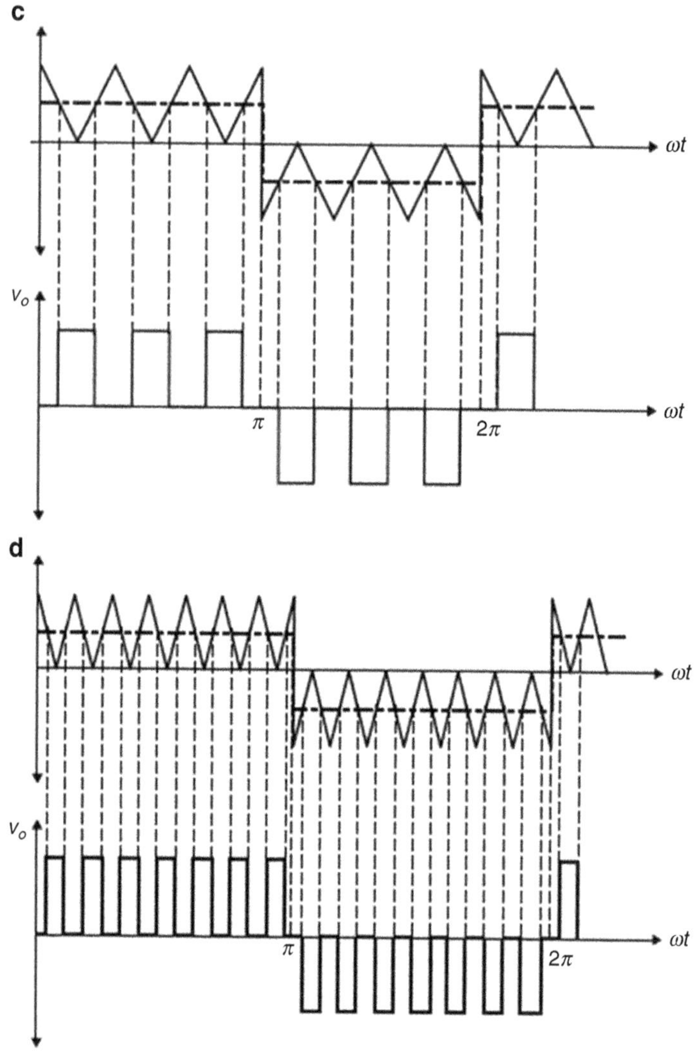

Fig. 9.30 (continued)

increases as well, reaching its maximum value of π when the magnitude of the reference signal becomes equal to the peak of the modulating signal.

It can be shown the α can be expressed in terms of $V_{\mathrm{P,\,tri}}$ and $V_{\mathrm{P,\,ref}}$ as follows:

$$\alpha = \frac{\pi}{2}\left(\frac{V_{\mathrm{P,ref}}}{V_{\mathrm{P,tri}}} - 1\right) = \frac{\pi}{2}(1 - m_a) \tag{9.57}$$

Notice that the frequency of the control signal, f_{tri}, is twice the frequency of the reference signal, f_{ref}. It is clear from these waveforms that the number of pulses, k, is equal to the number of periods of the control signal per half a period of the reference signal, i.e., k is the number of switching periods, T_s, in the $T_o/2$ period that can be expressed as:

$$k = \frac{1}{2}\frac{f_s}{f_o} \tag{9.58}$$

In terms of the frequency modulation index, m_f, k may be expressed by the following relation:

$$k = \frac{1}{2}m_f \tag{9.59}$$

Notice that $k = 1$ is a special case where $f_s/f_o = 2$.

We should point out that the output frequency is equal to the frequency of the reference signal (i.e., $f_o = f_{ref}$) and the switching frequency is equal to the frequency of the carrier or triangle signal ($f_s = f_{tri}$).

For example, in Fig. 9.30d, $k = 7$ and $m_f = 14$. The maximum width of each pulse occurs when $m_a = 1$ and is given by:

$$t_{width,\,max} = \frac{T_o}{2k} \tag{9.60}$$

The maximum conduction angle width of each pulse is given by:

$$\theta_{width,\,max} = \omega_o t_{width,\,max} = \frac{\pi}{k} \tag{9.61}$$

Exercise 9.7

Derive Eq. (9.57)

Next we will derive a general expression the ith pulse width in a given k-pulse output in terms of i, k, and m_a. Referring to Fig. 9.31 that shows a k-pulse inverter output, the start of the ith pulse is given by:

$$t_i = (i-1)T_s + \frac{T_s}{2}\left(1 - \frac{V_{P,\,ref}}{V_{P,\,tri}}\right) \tag{9.62}$$

Substituting for $f_s = 1/T_s$ from Eq. (9.58) into Eq. (9.62), t_i becomes:

$$t_i = \frac{T_o}{2k}(i-1) + \frac{T_o}{4k}(1 - m_a) \tag{9.63}$$

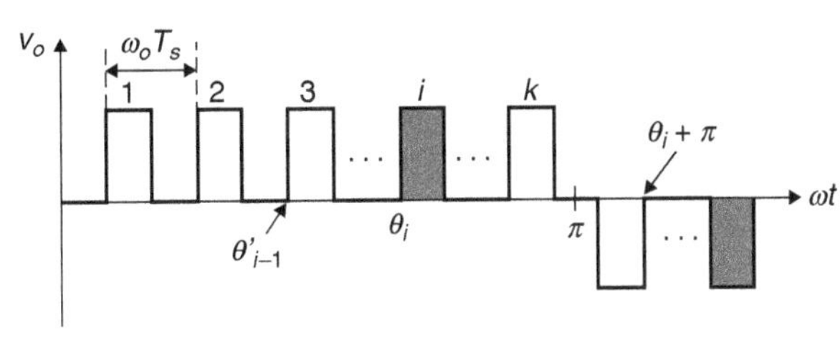

Fig. 9.31 k-pulse inverter output in a half-cycle

In terms of the starting angle, $\theta_i = \omega_o t_1$, of the *i*th pulse, Eq. (9.63) maybe written as follows:

$$\theta_i = \frac{\pi}{k}\left[i - \frac{m_a}{2} - \frac{1}{2}\right] \quad (9.64)$$

For example, for the two-pulse waveform, $k = 2$ and $m = 0.5$, the angles at which the pulses start are given by:

$$\theta_1 = \frac{\pi}{2}\left[1 - \frac{1}{4} - \frac{1}{2}\right] = \frac{\pi}{8} \qquad \theta_3 = \frac{\pi}{2}\left[3 - \frac{1}{4} - \frac{1}{2}\right] = \frac{9\pi}{8}$$

$$\theta_2 = \frac{\pi}{2}\left[2 - \frac{1}{4} - \frac{1}{2}\right] = \frac{5\pi}{8} \qquad \theta_4 = \frac{\pi}{2}\left[4 - \frac{1}{4} - \frac{1}{2}\right] = 13\frac{\pi}{8}$$

Notice that because of the symmetry, $\theta_3 = \theta_1 + \pi$ and $\theta_4 = \theta_2 + \pi$. It can be shown, in general, that the width of each pulse is given by:

$$\theta_{\mathrm{width}} = \pi\frac{m_a}{k} \quad (9.65)$$

For $k = 2$ and $m = 0.5$, $\theta_{\mathrm{width}} = \pi/4$ as expected.

9.5.1.1 The Output Voltage

It can be shown that the average output voltage over a period of T_s is given by:

$$V_{o,\mathrm{ave}} = m_a V_{dc} \qquad 0 < m_a \leq 1 \quad (9.66)$$

The *rms* value for the *i*th pulse is given by:

$$\begin{aligned} V_{o,\mathrm{rms}} &= \sqrt{\frac{1}{2\pi}\int_{\theta_i}^{\theta_i + \theta_{\mathrm{width}}} 2kV_{dc}^2 \, d\omega t} \\ &= V_{dc}\sqrt{\frac{k}{\pi}\theta_{\mathrm{width}}} \end{aligned} \quad (9.67)$$

Recall that from Eq. (9.65), we have $\theta_{\mathrm{width}} = \pi(m_a/k)$; hence, Eq. (9.67) becomes:

$$V_{o,\mathrm{rms}} = V_{dc}\sqrt{\frac{k}{\pi}\frac{\pi m_a}{k}} = V_{dc}\sqrt{m_a} \quad (9.68)$$

The *rms* of the output voltage is a function of the modulation index.

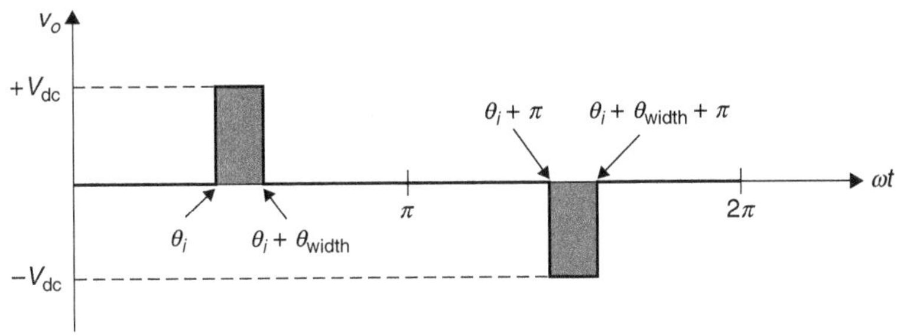

Fig. 9.32 Symmetrical representation of an *i*th pulse for a given inverter output

9.5.1.2 Harmonics of k Equal Pulses

Let us determine the harmonic components for the *i*th pulse shown in Fig. 9.32.

Since it is an odd function only, the odd harmonics exist in the b_n coefficients. The harmonics of the output voltage due to the *i*th pulse acting alone are given by:

$$V_{on,i} = \frac{V_{dc}}{\pi}\left[\int_{\theta_i}^{\theta_i+\theta_{width}} \cos n\omega t \, d\omega t - \int_{\theta_i}^{\theta_i+\theta_{width}+\pi} \cos n\omega t \, d\omega t\right]$$
$$= \frac{2V_{dc}}{n\pi} \sin\left(\frac{n\theta_{width}}{2}\right)\left[\sin\left(n\left(\theta_i + \frac{\theta_{width}}{2}\right)\right) - \sin\left(n\left(\pi + \theta_i + \frac{\theta_{width}}{2}\right)\right)\right] \quad (9.69)$$

For the total k pulses, V_{on} is given by:

$$V_{on} = \sum_{i=1}^{k} \frac{2V_{dc}}{n\pi} \sin\left(\frac{n\theta_{width}}{2}\right)\left[\sin\left(n\left(\theta_i + \frac{\theta_{width}}{2}\right)\right) - \sin\left(n\left(\pi + \theta_i + \frac{\theta_{width}}{2}\right)\right)\right] \quad (9.70)$$

for $n = 1, 3, 5, \ldots$

In terms of m_a and θ_{width}, θ_i can be expressed as:

$$\theta_i = \frac{\pi}{k}\left(i - \frac{1}{2}\right) - \frac{\theta_{width}}{2} \quad \text{where} \quad i = 1, 2, \ldots, k \quad (9.71)$$

$$\theta_{width} = \frac{\pi}{k} m_a \quad (9.72)$$

It can be shown that the *n*th harmonic component for the *k-pulse* output voltage maybe expressed as follows:

$$V_{on} = \frac{4V_{dc}}{n\pi} \sin\left(\frac{m_a\pi}{2k} n\right) \sum_{i=1}^{k} \sin\left(\frac{\pi}{k}(i - 1/2)n\right) \quad (9.73)$$

For illustration purposes, let us consider the example with $m = 0.5$ and the first nine harmonics, and then we evaluate the total harmonic distortion as a function of the number of pulses.

To verify the above equation a single pulse, we calculate the harmonic components for $k = 1$, $i = 1$, and $\theta_1 = \alpha$, to yield:

$$V_{\text{on}} = \frac{4V_{dc}}{n\pi} \sin\left(\frac{n\pi}{2} m_{\text{a}}\right) \sum_{i=1}^{1} \sin\left(\frac{\pi}{k}(i - 1/2)n\right) \tag{9.74}$$

Substituting for m_{a} from Eq. (9.72), Eq. (9.74) becomes:

$$V_n = \left(\frac{4V_{dc}}{n\pi}\right) \sin\left(n\frac{\theta_{\text{width}}}{2}\right), \quad \text{for odd } n \tag{9.75}$$

Equation (9.75) represents the harmonic components of the output voltage for $k = 1$ as a function of the width θ_{width}. This equation is similar to what we have derived previously as a function of α.

Example 9.8
For a uniform PWM with a value of $k = 5$ and a modulation index $m_{\text{a}} = 0.2$, calculate the output harmonic components up to the fifteenth harmonic.

Solution
Using Eq. (9.73), Table 9.1 shows the values of the first 15 harmonics. Figure 9.33 shows the plot for the harmonic contents of Table 9.1. Figure 9.34 shows the harmonic ratios with respect to the fundamentals for $m = 0.2$ and $k = 1$ to $k = 7$ pulses per half-cycle.

From Table 9.1 for $m_{\text{a}} = 0.2$, we observe that as the number of pulses increases per half a cycle, the magnitude of the lower harmonics (third, fifth, seventh) decreases with respect to the fundamental component. Furthermore, there is an increase in magnitude for the higher-order harmonics with respect to the fundamental; however, such higher-order harmonics produce a negligible ripple that can be easily filtered out. However still, the ratio of the harmonic to the fundamental is relatively unchanged as the number of pulses increases within a half-cycle. The THD for this example is 223%! The higher the modulation index, the lower the THD.

Table 9.1 Normalized harmonics for $k = 5$ and $m_{\text{a}} = 0.2$

Magnitude harmonic coefficient	V_n/V_{dc}
V_1	0.258715
V_3	0.098301
V_5	0.078691
V_7	0.095728
V_9	0.245304
V_{11}	−0.238761
V_{13}	−0.088251
V_{15}	−0.068671

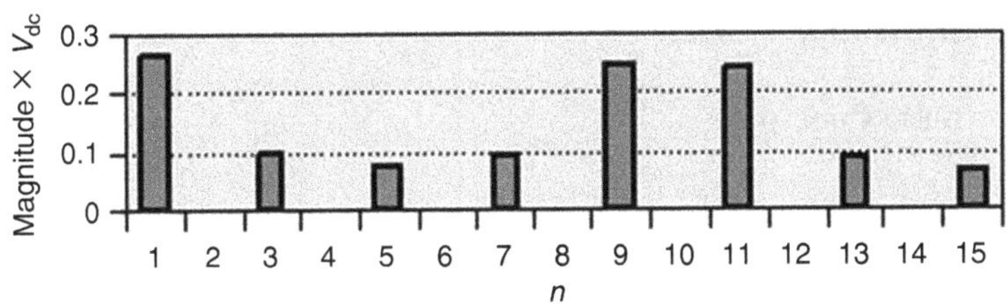

Fig. 9.33 Harmonic contents for $k = 5$ and $m_a = 0.2$

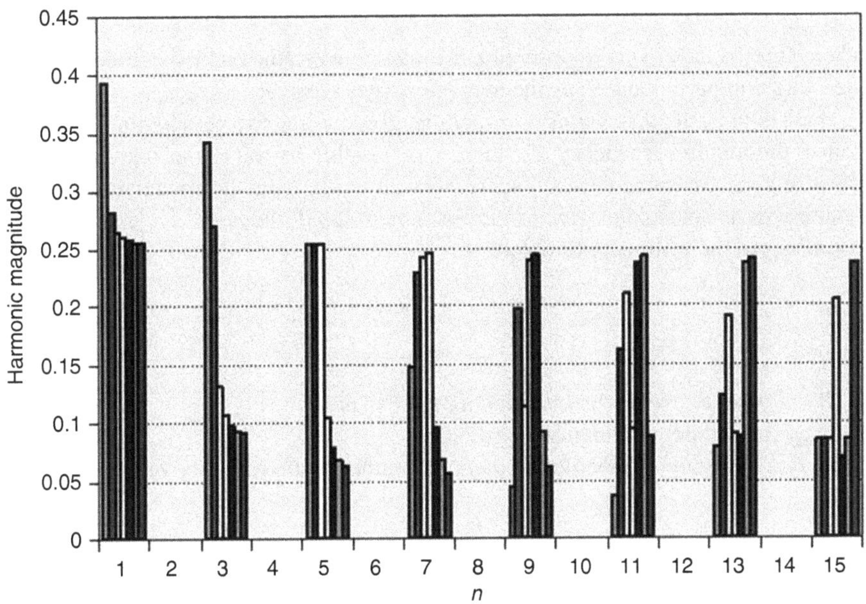

Fig. 9.34 Harmonic contents for $k = 1$ to $k = 7$ and $m_a = 0.2$

Exercise 9.8

Consider a pair of equal pulses placed at $\omega t = \theta_i$ and $\omega t = \theta_i + \pi$, respectively, each with a pulse width of θ_{width} as shown in Fig. 9.32. Show that the nth harmonic component is given by:

$$v_{\text{on}} = V_n \sin n\omega t$$

where:

$$V_n = \frac{2V_{dc}}{n\pi} \sin\left(\frac{n\theta_{\text{width}}}{2}\right)\left[\sin n\left(\theta_i + \frac{\theta_{\text{width}}}{2}\right) - \sin n\left(\pi + \theta_i + \frac{\theta_{\text{width}}}{2}\right)\right]$$

9.5.2 *Sinusoidal PWM*

9.5.2.1 Basic Concept

To illustrate the process of sinusoidal PWM, we refer to a simplified buck converter shown in Fig. 9.35. Recall that in PWM dc-dc converters the duty cycle is modulated between 0 and 1 in order to regulate the dc output voltage. In the steady state, the duty cycle in PWM switch-mode converters is relatively constant and does not vary with time as shown in Eq. (9.76)

$$V_o = DV_{dc} \tag{9.76}$$

where D is the duty cycle representing the ratio between the on time of the switch to the switching period and V_o is the average output voltage.

If the duty cycle, $d(t)$, varies or is modulated according to a certain time function, with a modulating frequency, f_o, then it is possible to shape the output voltage waveform, v_o, in such a way that its average value over the modulating period synthesizes a sinusoidal waveform. For example, if the duty cycle is defined according to the following function:

$$d(t) = D_{dc} + D_{\max} \sin \omega_o t \tag{9.77}$$

where:

D_{dc}: dc duty cycle when no modulation exists.
$D_{\max}$: maximum modulation constant.
ω_o: frequency of modulation.then, the output voltage,v_o, is given by:

$$\begin{aligned} v_o &= d(t)V_{dc} \\ &= V_{dc}D_{dc} + V_{dc}D_{\max} \sin \omega_o t \end{aligned} \tag{9.78}$$

For a buck converter, since the output voltage cannot be negative, then $D_{\max} \leq D_{dc}$ as shown in Fig. 9.36.

As an example, if $D_{dc} = 0.5$, $D_{\max} = 0.8D_{dc}$, and $f_o = f_s/12$, the duty cycle, $d(t)$, is given by:

$$d(t) = 0.5 + 0.4 \sin (2\pi f_o t_i)$$

where $t_i = 0, 1, 2, 3, \ldots, 12$.

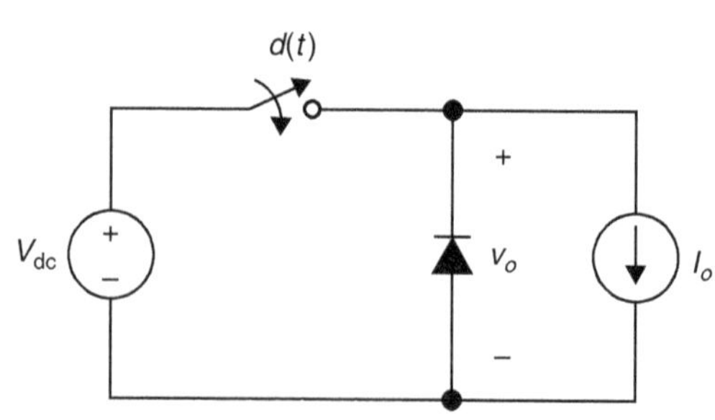

Fig. 9.35 Simplified buck converter

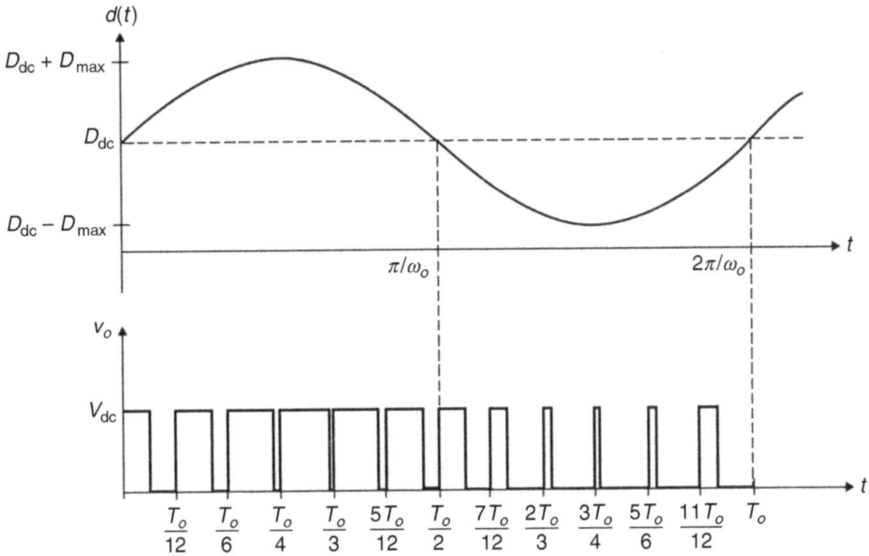

Fig. 9.36 Example of $D_{dc} = 0.5$, $D_{\max} = 0.8D_{dc}$, and $f_o = f_s/12$

Table 9.2 Caption missing

Pulse	Time, t_i	$d(t) = 0.5 + 0.4\ \sin(2\pi f_o t_i)$
1	0	0.50
2	$T_o/12$	0.70
3	$T_o/6$	0.85
4	$T_o/4$	0.90
5	$T_o/3$	0.85
6	$5T_o/12$	0.70
7	$T_o/2$	0.50
8	$7T_o/12$	0.30
9	$2T_o/3$	0.15
10	$9T_o/12$	0.10
11	$5T_o/6$	0.15
12	$11T_o/12$	0.30

Since the switching frequency is 12 times faster than the modulating frequency, f_o, then $d(t)$ is sampled 12 times between $0 \leq t < T_o$. This is illustrated in Table 9.2 for $0 \leq t < T_o/2$ and six pulses in a half-cycle.

The above concept will now be applied to sinusoidal PWM inverters.

9.5.2.2 PWM Modulation Function

Since the waveform is symmetrical around the $T_o/2$ point, we only define $m(t)$ in the first half-cycle. In the single-pulse inverter with a two-state output of Fig. 9.13b, the modulation function is simply unity.

$$v_o = m(t)V_{dc} \quad m(t) = 1 \quad 0 \leq t < T/2 \tag{9.79}$$

In the single-pulse inverter with a tri-state output of Fig. 9.18a, the modulation function in a half-cycle is given by:

$$v_o = m(t)V_{dc} \qquad m(t) = \begin{cases} 0 & 0 \leq t < \alpha/\omega \\ +1 & \alpha/\omega \leq t < T_o/2 - \alpha/\omega \\ 0 & T_o/2 - \alpha/\omega \leq t < T_o/2 \end{cases} \tag{9.80}$$

In a k-pulse inverter with a constant duration, the modulation function is given by:

$$v_o = m(t)V_{dc} \qquad m(t) = \begin{cases} +1 & 0 \leq t < d(t) \\ 0 & d(t) \leq t < T_s \end{cases} \tag{9.81}$$

In the sinusoidal PMW technique, the pulse durations are adjusted to change slowly to follow the sinusoidal function. The modulation function is normally limited between 0 and 1. Since it is desired to have an output voltage with zero dc, the modulation function must be symmetrical around zero.

For the sinusoid PWM waveforms, we define $m(t)$ as follows:

$$m(t) = M_{max} \cos \omega_o t \tag{9.82}$$

where ω_o is the modulation frequency, which must be less than the switching frequency; i.e., $\omega_o \ll \omega_s$. M_{max} is the gain of the modulation function, which varies between 0 and 1. Normally the switching frequency is in the range of a few hertz to a 100 kHz, while the modulation frequency is less than 500 Hz. As stated before, using the PWM process, we can extract a low-frequency signal from a train of high-frequency square waves. The higher the switching frequency of the square-wave output with respect to the desired low-frequency output signal, the more the output waveform approximates a sinusoidal.

A modulation gain of unity represents the largest possible output voltage, whereas a modulation gain of zero means the output waveform frequency equals the switching frequency, and the output voltage is the smallest.

9.5.2.3 Switching Schemes

Depending on the switching sequence, the output voltage in PWM inverters can be either bipolar or unipolar. Figure 9.37 shows a bipolar output voltage in a PWM inverter. When the reference sinusoidal signal is larger or smaller than the triangular wave, the output equals $+V_{dc}$ or $-V_{dc}$, respectively. Both the half-bridge and full-bridge configurations are used in practice to generate a PWM output voltage waveform.

In the bipolar voltage switching, m_f is an odd number that is the same switching frequency, f_s. The output frequency, f_o, in the unipolar voltage switching, is twice that of the frequency in the bipolar voltage switching (m_f is doubled).

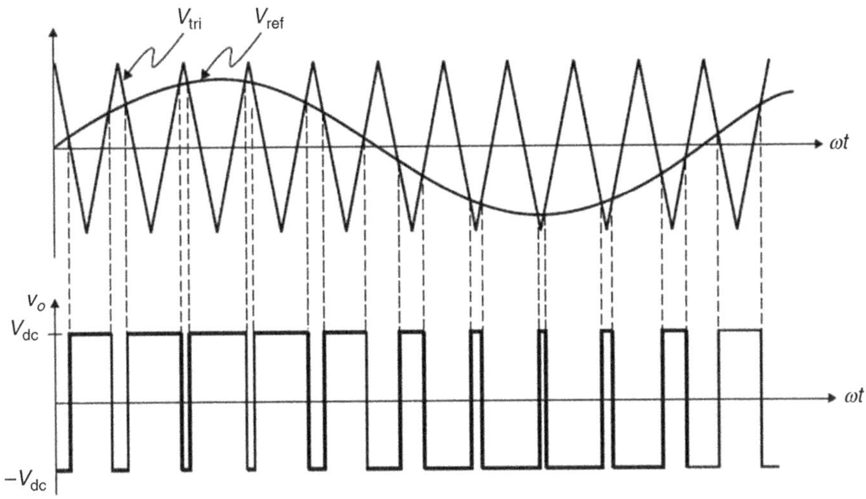

Fig. 9.37 Example of a bipolar PWM output waveform

If both positive and negative sinusoidal control signals are available, then the switching sequence will produce a unipolar output waveform as shown in Fig. 9.38. The output waveforms for v_{o1} and v_{o2} are shown in Fig. 9.38a, b, respectively, and Fig. 9.38c shows $v_o = v_{o1} - v_{o2}$. Here the triangular signal is chosen to be a sawtooth function.

9.5.2.4 Signal Generation

Advanced digital and analog techniques exist in today's inverters to generate the driving signals that produce a sinusoidal PWM. For illustration purposes, Fig. 9.39 shows a comparator that compares a triangular signal to a sinusoidal reference signal.

9.5.2.5 Analysis of Sinusoidal PWM

As stated before, the width of each pulse is varied in proportion to the instantaneous integrated value of the required fundamental component at the time of its event. In other words, the pulse width becomes a sinusoidal function of the angular position. In sinusoidal PWM, the lower-order harmonics of the modulated voltage waveform are highly reduced compared with the use of the uniform pulse width modulation.

The output voltage signal in sinusoidal PWM can be obtained by comparing a control signal, v_{cont}, against a sinusoidal reference signal, v_{ref}, at the desired frequency. At the first half of the output period, the output voltage takes a positive value ($+V_{dc}$) whenever the reference signal is greater than the control signal. In the

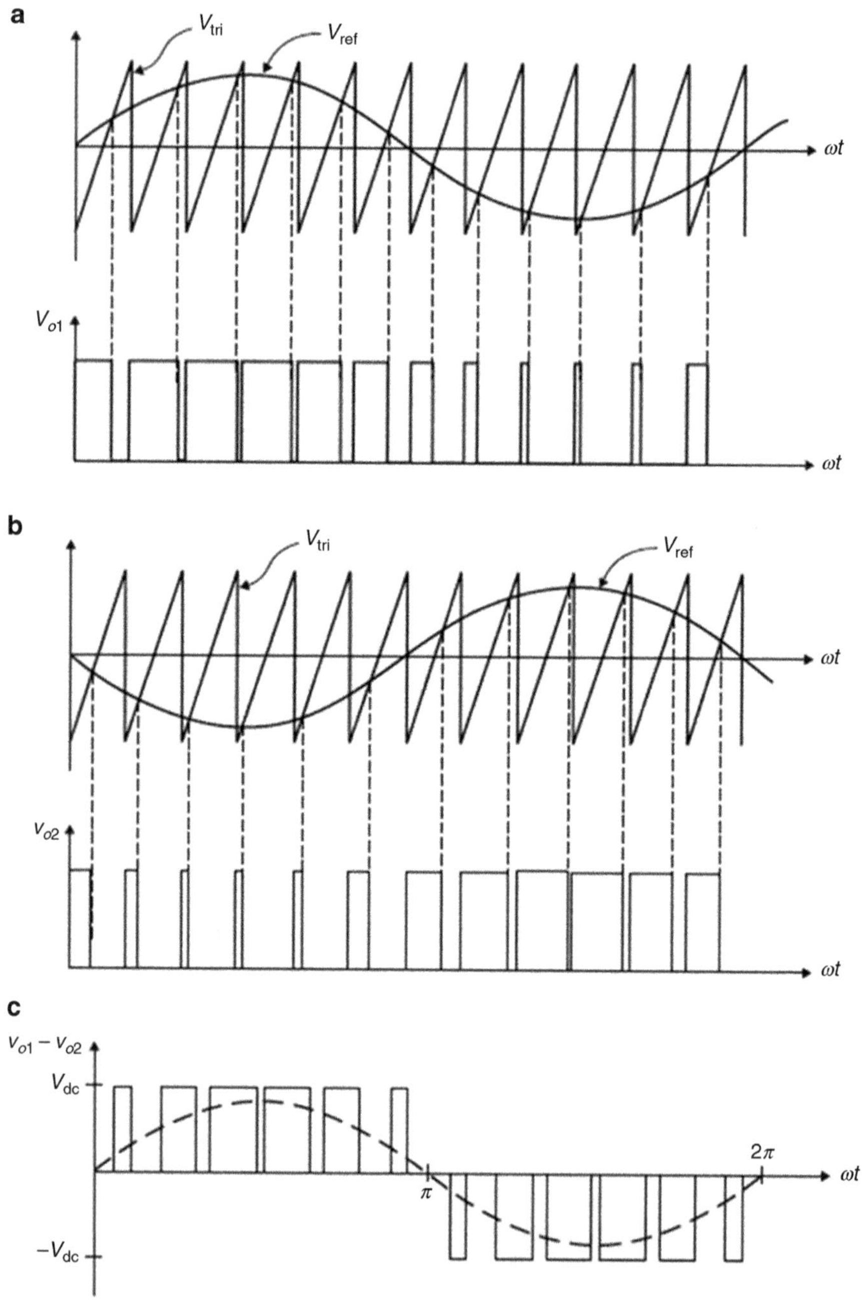

Fig. 9.38 Unipolar PWM output (**a**) A positive sinusoidal reference to produce v_{o1} and (**b**) positive sinusoidal reference to produce v_{o2} (**c**) The differential output $v_o = v_{o1} - v_{o2}$

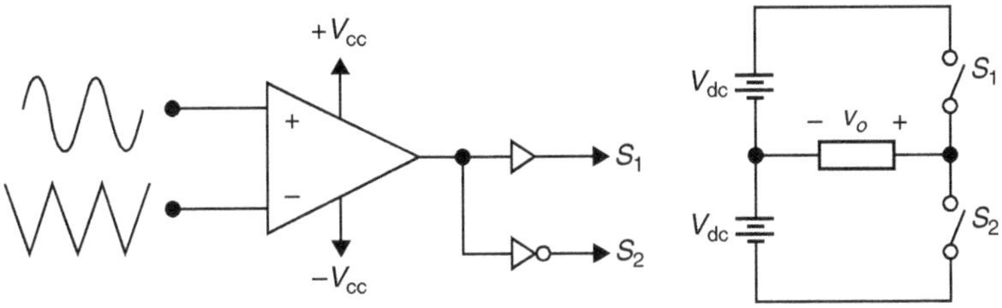

Fig. 9.39 Simplified circuit shows how signals are generated in sinusoidal PWM inverters

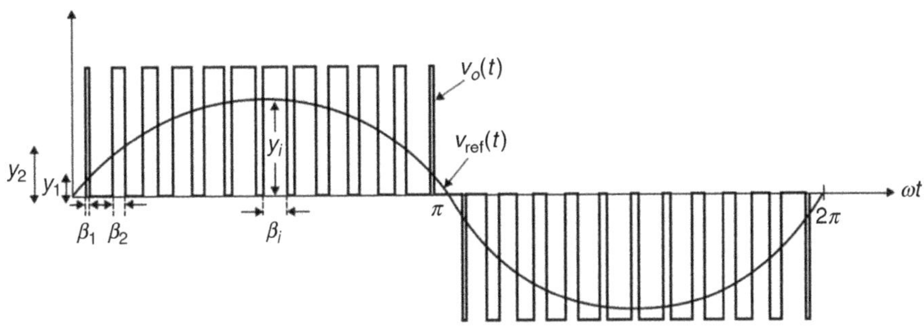

Fig. 9.40 PWM figure to illustrate the constant ratio between the width and height of a given pulse

second half of the output period, the output voltage takes a negative value ($-V_{dc}$) whenever the reference signal is less than the control signal.

Similar to the case of equal-pulse PWM, the control frequency f_{cont} (equal to the switching frequency) determines the number of pulses per half-cycle of the output voltage signal. Also, the output frequency f_{o} is determined by the reference frequency f_{ref}. The amplitude modulation index, m_{a}, is defined as the ratio between the sinusoidal magnitude and the control signal magnitude:

$$m_{\text{a}} = \frac{V_{\text{P,ref}}}{V_{\text{P,cont}}} \tag{9.83}$$

The duration of the pulses is proportional to the corresponding value of the sine wave at that corresponding position. Then, the ratio of any pulse duration to its corresponding time duration is constant as shown in Fig. 9.40:

$$\frac{\beta_1}{y_1} = \frac{\beta_2}{y_2} = \frac{\beta_3}{y_3} \Rightarrow \frac{\beta_i}{y_i} = \text{constant} \tag{9.84}$$

The most important simplifying assumption here is that if the control frequency signal is very high with respect to the reference frequency signal, $m_{\text{f}} \gg 1$, then the value of the reference signal between two consecutive intersections with the control signal is almost constant. This is illustrated in Fig. 9.41.

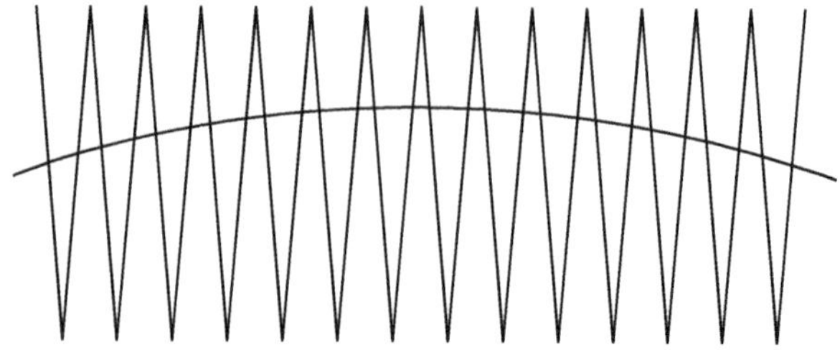

Fig. 9.41 High-frequency sinusoidal PWM

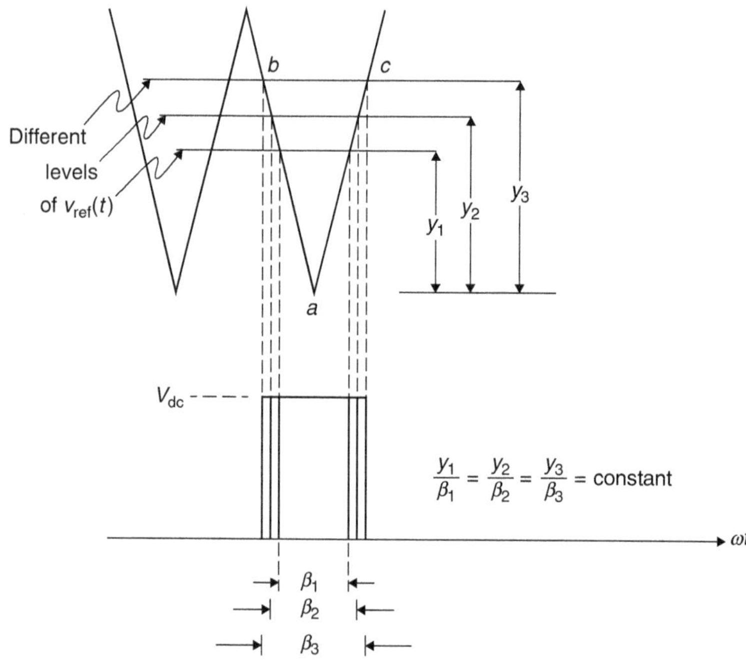

Fig. 9.42 Triangular approximation

The proportional variation of each pulse width with respect to the corresponding sine wave amplitude could be seen by applying a triangular relationship as shown in Fig. 9.42.

9.5.2.6 Output Voltage Harmonics

The calculation of the sinusoidal PWM output voltage is the same as that of the uniform PWM output voltage. However, for a sinusoidal PWM, the width of each pulse varies according to its position. The expression for the output voltage is obtained using a Fourier series transformation for v_o, given by:

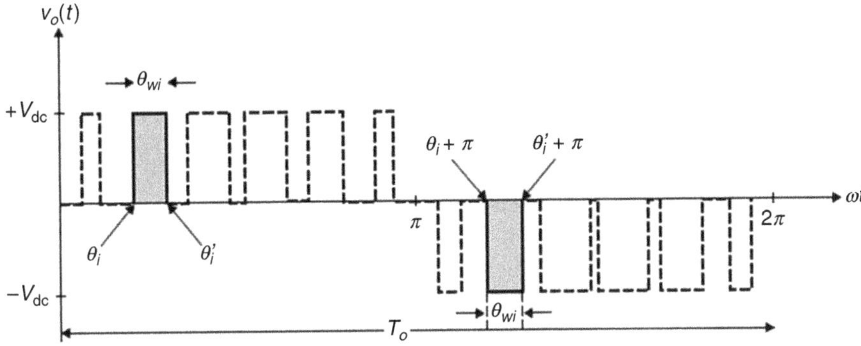

Fig. 9.43 Single sinusoidal PWM pair of pulses

$$v_o(t) = \sum_{n=1,2,\ldots}^{\infty} (a_n \cos n\omega t + b_n \sin n\omega t) \tag{9.85}$$

Since the inverter output voltage is an odd function, only odd harmonics exist.

The calculation of the output voltage harmonic components can be done using one single pair of pulses as shown in Fig. 9.43.

$$\begin{aligned} V_{n,i} &= \frac{1}{\pi}\int_0^{2\pi} v_o(\omega t)\sin(n\omega t)d(\omega t) \\ V_{n,i} &= \frac{1}{\pi}\int_{\theta_i}^{\theta_i+\theta_{wi}} V_{dc}\sin(n\omega t)\, d(\omega t) + \int_{\pi+\theta_i}^{\pi+\theta_i+\theta_{wi}} -V_{dc}\sin(n\omega t)\, d(\omega t) \end{aligned} \tag{9.86}$$

Using this trigonometric relationship:

$$\cos x - \cos y = -\left(2\sin\frac{x+y}{2}\right)\left(\sin\frac{x-y}{2}\right)$$

it can be shown that the harmonic component for a single pair of pulses is given by:

$$V_n = \left(\frac{2V_{dc}}{n\pi}\right)\sin\left(n\frac{\theta_{wi}}{2}\right)\left[\sin n\left(\theta_i + \frac{\theta_{wi}}{2}\right) - \sin n\left(\theta_i + \frac{\theta_{wi}}{2} + \pi\right)\right] \tag{9.87}$$

Adding the contribution from all other pulses, the ith component of v_o is given by:

$$V_n = \sum_{i=1}^{k}\left(\frac{2V_{dc}}{n\pi}\right)\sin\left(n\frac{\theta_{wi}}{2}\right)\left\{\sin n\left(\theta_i + \frac{\theta_{wi}}{2}\right) - \sin n\left(\theta_i + \frac{\theta_{wi}}{2} + \pi\right)\right\} \tag{9.88}$$

where θ_i is the starting angle of the ith pulse and θ_{wi} is the pulse width at the corresponding angular position. Next, we estimate the width of each pulse, θ_{wi}, for each ith pulse.

9.5.2.7 Approximating the Pulse Width θ_{wi}

Assume each pulse is located at the discrete value of θ_i, which represents the first intersection for the generation of the ith pulse. Then, the approximated mathematical relation for the width is found using a geometrical relation as shown in Fig. 9.44.

From the geometry of the triangle ABC in Fig. 9.44, we have:

$$h_y = V_{\mathrm{p,\,tri}},\ h_x = V_{\mathrm{p,\,ref}} \sin\theta,\ y = \frac{T_{\mathrm{o}}}{2k}\omega_{\mathrm{o}} = \frac{\pi}{k}$$

The approximated width of the ith pulse, $\theta_{\mathrm{wi,\,app}}$, is given by:

$$\theta_{\mathrm{wi,app}} = x \tag{9.89}$$

Since the angles $\alpha = \beta$, then:

$$\frac{h_x}{x} = \frac{h_y}{y} \tag{9.90}$$

Substituting for h_y, h_x, y, and x in Eq. (9.90) and using $m_{\mathrm{a}} = V_{\mathrm{P,\,ref}}/V_{\mathrm{P,\,tri}}$, Eq. (9.89) becomes:

$$\theta_{\mathrm{wi,app}} = \left(\frac{\pi}{k}\right) m_{\mathrm{a}} \sin\theta_i \tag{9.91}$$

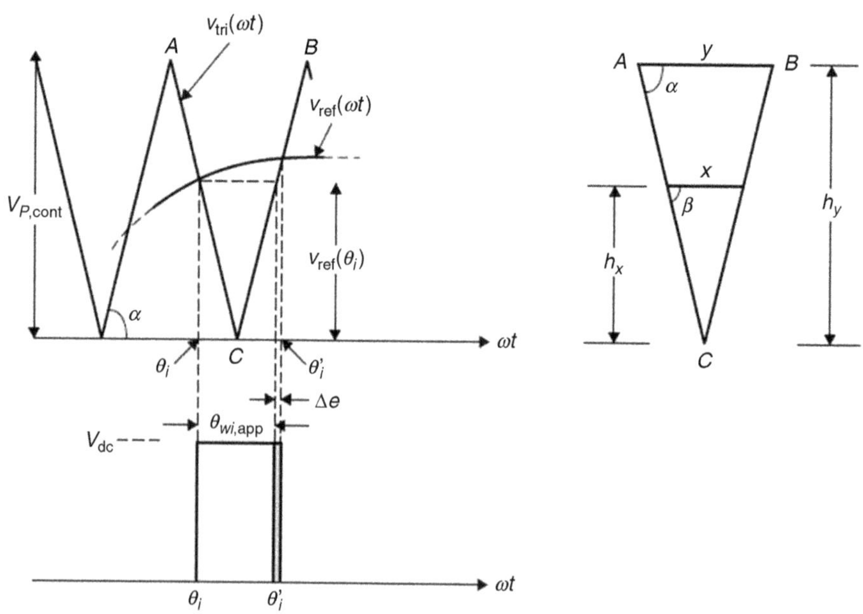

Fig. 9.44 Approximated pulse width, i.e., $\Delta e \approx 0$

It can be seen that the exact width is produced for the consecutive intersections of the control signal and the sinusoidal signal at positions θ_i and θ'_i. Therefore, an error is originated when the signal is considered constant and the value of the width is found using the approximated procedure. To reduce the width approximation error, it is necessary to increase the carrier frequency. Increasing the index frequency m_f produces a considerable reduction in the difference of the reference signal values evaluated at θ_i and θ'_i, respectively.

9.5.2.8 Exact Expression for θ_{wi}

For the ith pulse with an angle θ_i, it can be easily shown that the point of intersection between the control signal and the ωt axis is $((2i-1)\pi/2k, 0)$ as shown in Fig. 9.45.

The general expression for $v_{cont}(\omega t)$ having all the lines of negative slopes is given by:

$$v_{cont}(\omega t) = -\frac{m_f}{\pi} V_{P,tri}\left(\omega t - (2i-1)\frac{\pi}{m_f}\right) \qquad i = 0, 1, 2, \ldots, k. \tag{9.92}$$

The expression for the reference signal is given by:

$$v_{ref}(\omega t) = V_{P,ref} \sin \omega t \tag{9.93}$$

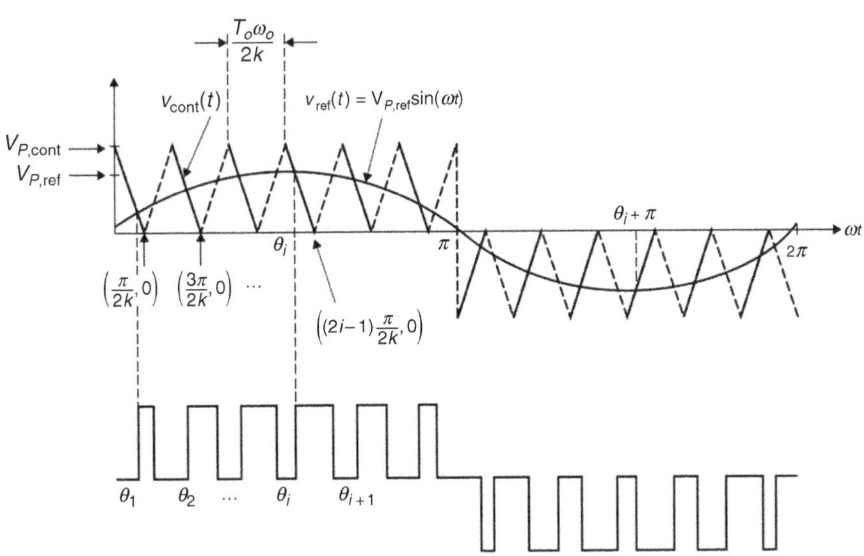

Fig. 9.45 Intersections between v_{ref} and v_{cont}

Evaluating Eqs. (9.92) and (9.93) at $\omega t = \theta_i$, $v_{\text{cont}}(\theta_i) = v_{\text{ref}}(\theta_i)$ yields Eq. (9.94):

$$-\frac{m_{\text{f}}}{\pi} V_{\text{P,tri}} \left(\theta_{\text{i}} - (2i-1)\frac{\pi}{m_{\text{f}}} \right) = V_{\text{P,ref}} \sin \theta_i \tag{9.94}$$

Equation (9.94) can be rewritten as:

$$m_{\text{a}} \sin \theta_i = -\frac{m_{\text{f}}}{\pi} \theta_i + (2i-1) \tag{9.95}$$

The value of θ_i can be found when solving Eq. (9.95) numerically.

Consider the cases for a different number of pulses per half-cycle as shown in Fig. 9.46. The exact width for each pulse is found by adding the approximated width with the corresponding error Δe in each interval. As shown in Fig. 9.46, for the two-pulse case, we have $\Delta e_1 = \Delta e_2$ by symmetry, as well as for the three and four pulses.

Generalizing for any symmetric pair of pulses, the calculation for any width can be done as shown in Fig. 9.47.

From the triangle ABC:

$$\tan \alpha = \frac{v_{\text{ref}}(\theta'_i) - v_{\text{ref}}(\theta_i)}{\Delta e_i} \tag{9.96}$$

From the triangle DEF:

$$\tan \alpha = \frac{V_{\text{P,tri}}}{\left(\frac{T_{\text{o}}\omega_{\text{o}}}{4k}\right)} = \frac{V_{\text{P,tri}}}{\frac{1}{2}\frac{\pi}{k} = \frac{2k}{\pi} V_{\text{P,tri}}} \tag{9.97}$$

Equating Eqs. (9.96) and (9.97), Δe_i maybe expressed by:

$$\Delta e_i = \left(\frac{\pi}{2k}\right) \frac{v_{\text{ref}}(\theta'_i) - v_{\text{ref}}(\theta_i)}{V_{\text{P,tri}}} \tag{9.98}$$

Substituting for $v_{\text{ref}}(\omega t) = V_{\text{P,tri}} \sin(\omega t)$, $k = \frac{1}{2} m_{\text{f}}$ and $m_{\text{a}} = V_{\text{P,ref}}/V_{\text{P,cont}}$ in Eq. (9.98), Δe_i becomes:

$$\Delta e_i = \frac{\pi m_{\text{a}}}{2k} \left(\sin \theta'_i - \sin \theta_i \right) \tag{9.99}$$

From the supplementary angle relation, $\sin x = \sin(\pi - x)$, Eq. (9.99) becomes:

$$\Delta e_i = \left(\frac{\pi m_{\text{a}}}{2k}\right) \sin \theta_{k+1-i} - \sin \theta_i$$

and since $\theta_{\text{wi}} = \theta_{\text{wi, app}} \pm \Delta e_i$, go to the general expression for the exact width which is given by:

$$\theta_{\text{wi}} = \frac{\pi m_{\text{a}}}{2k} \left(\sin \theta_i \pm \frac{1}{2}(\sin \theta_i - \sin \theta_{k+i-1}) \right) \tag{9.100}$$

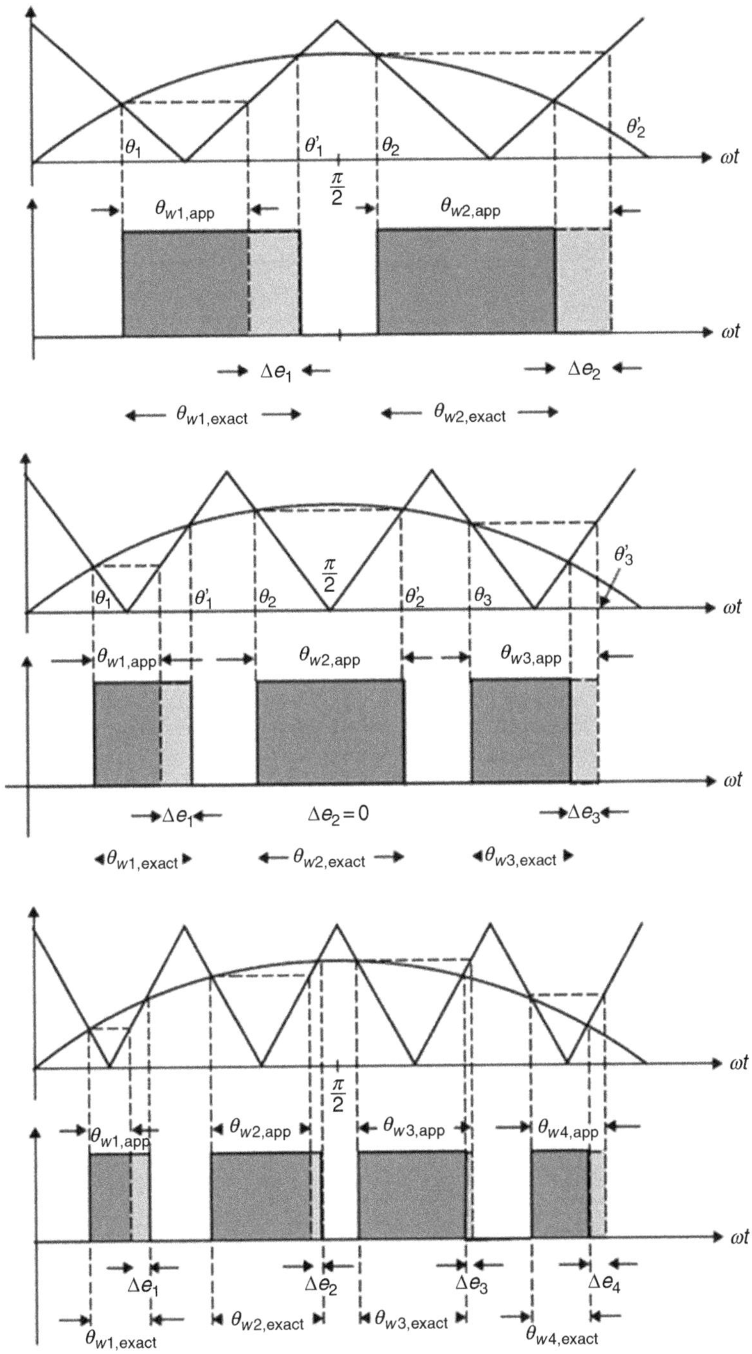

Fig. 9.46 Illustration for two, three, and four PWM pulses per half-cycle

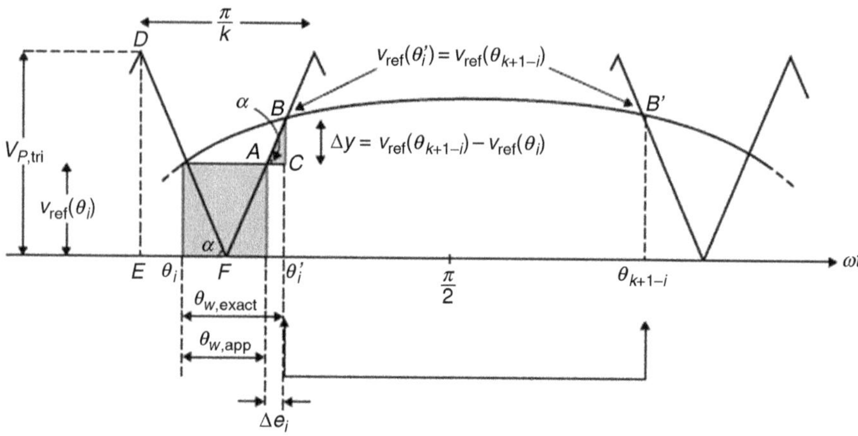

Fig. 9.47 Symmetric pair of pulses about $\pi/2$

An increase in the control frequency causes a decrease in the value of Δe and $\theta_{wi,\,app}$. Therefore, when the control frequency tends to go to infinity, the width of Δe tends to go to zero. It can be shown that Δe decreases at a higher rate than does $\theta_{wi,\,app}$.

Example 9.9
Consider a full-bridge inverter with an R-L load with an output sinusoidal PWM voltage with a reference frequency f_o of 60 Hz, $V_{dc} = 280$ V, $m_a = 0.6$, and $m_f = 24$.

(a) Find the carrier frequency (triangular wave).
(b) Find the number of pulses per half-cycle.
(c) Find the angles of intersection and the pulse widths in a half-cycle.
(d) Find the harmonic components.
(e) Find the total harmonic distortion.

Solution

(a) $f_s = m_f \times f_o = 24 \times 60$ Hz $= 1.44$ kHz
(b) $k = \frac{24}{2} = 12$ pulses
(c) The exact angles were numerically calculated using Mathcad as shown in the table below:

Starting angle θ_i (degrees)	Arriving angle θ_i (degrees)	Pulse width θ_{wi} (degrees)
6.955	8.137	1.182
20.895	24.364	3.469
34.926	40.419	5.493
49.099	56.242	7.143
63.474	71.773	8.299
78.093	86.994	8.901
93.006	101.903	8.897
108.226	116.526	8.300

(continued)

Starting angle θ_i (degrees)	Arriving angle θ_i (degrees)	Pulse width θ_{wi} (degrees)
123.759	130.901	7.106
139.582	145.076	5.494
155.644	159.104	3.460
171.863	173.045	1.182

(d) Harmonic components in volts (Fig. 9.48):

$$
\begin{array}{lll}
V_1 = 167.931 & V_{11} = -0.019 & V_{21} = 19.909 \\
V_3 = -0.048 & V_{13} = 0.085 & V_{23} = 103.541 \\
V_5 = 0.127 & V_{15} = -0.021 & V_{25} = -103.74 \\
V_7 = 0.03 & V_{17} = -0.071 & V_{27} = -19.736 \\
V_9 = -0.118 & V_{19} = 1.058 & V_{29} = -0.898
\end{array}
$$

Using the approximated width we obtain the following, using Matlab:

$$
\begin{array}{lll}
V_1 = 167.742 & V_{11} = 0.02 & V_{21} = 16.071 \\
V_3 = 0.769 & V_{13} = -0.021 & V_{23} = 108.129 \\
V_5 = 6.917 \times 10^{-3} & V_{15} = 8.579 \times 10^{-3} & V_{25} = -99.036 \\
V_7 = 7.869 \times 10^{-3} & V_{17} = -4.769 \times 10^{-3} & V_{27} = -23.464 \\
V_9 = -6.94 \times 10^{-3} & V_{19} = 0.564 & V_{29} = -2.089
\end{array}
$$

These results are plotted in Fig. 9.49.

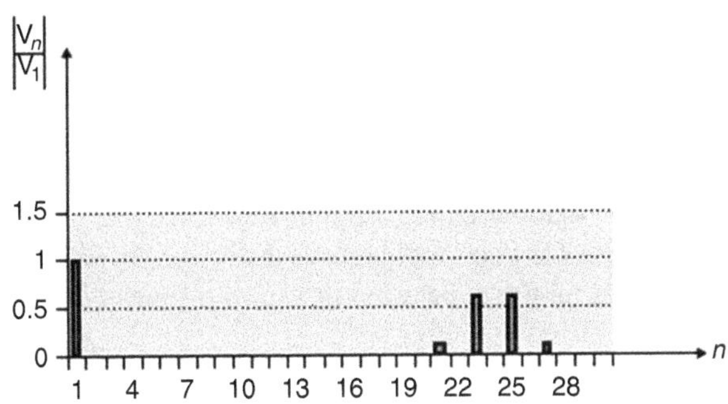

Fig. 9.48 First 29 harmonics using exact analysis for Example 9.9

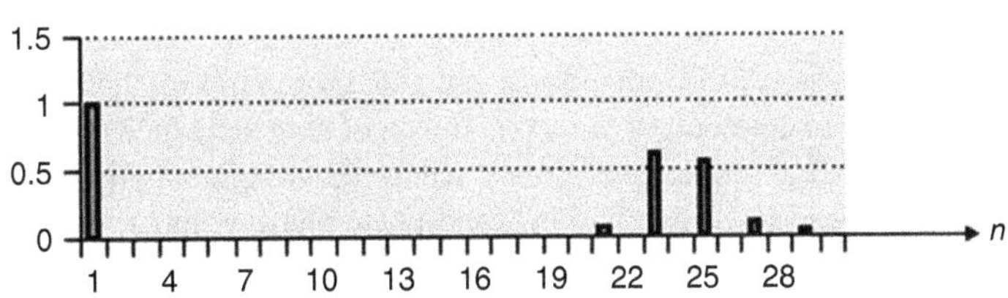

Fig. 9.49 A spectrum plot for the first 29 harmonics using the approximate analysis

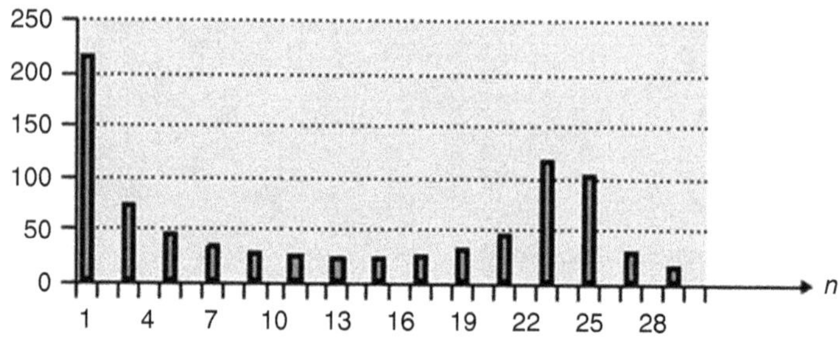

Fig. 9.50 Uniform PWM for Example 9.9

(e) The total harmonic distortion is approximated by:

$$THD = \sqrt{\frac{1}{V_1^2}\left(V_{21}^2 + V_{23}^2 + V_{25}^2 + V_{27}^2 + V_{29}^2\right)} = 0.89$$

If we compare the uniform PWM approach with the sinusoidal PWM approach, we can notice a considerable reduction in the harmonic components. The next graph, Fig. 9.50, shows a plot of the normalized magnitude of the harmonic components using uniform pulse width modulation with a modulation index of 0.6 for 12 pulses per half-period.

As we know, the inverter will produce a rectangular pulse waveform. To transform the rectangular waveform generated by the switching process to an output waveform that resembles a sinusoid wave, we need to use a harmonic filter. A low-pass filter can be used due to the high attenuation property without affecting the fundamental harmonic.

However, a perfect low-pass filter is impractical to realize, and the lower harmonics that are close to the fundamental will affect our desired output. Therefore, we need to maximally reduce the lower-order harmonics so the filter will allow the first harmonic to pass in our variable-frequency output.

9.6 Three-Phase Inverters

Consider the three-phase full-bridge dc-ac inverter shown in Fig. 9.51. To obtain a set of balanced line-to-line output voltages, the switching sequence of the switches $S_1 - S_6$ should produce a sequence of pulses whose summation at any given time is zero. As a result, it can be shown that in a one-pulse phase voltage, the conduction angle is $\pi/3$. The switch numbering follows the sequence of switching. The bidirectional switch implementation of $S_1 - S_6$ allows an inductive-load current flow.

Figure 9.52a, b shows two switching sequences for $S_1 - S_6$ with each switch conducting for π or $\pi/3$, respectively. Both sequences produce a similar output

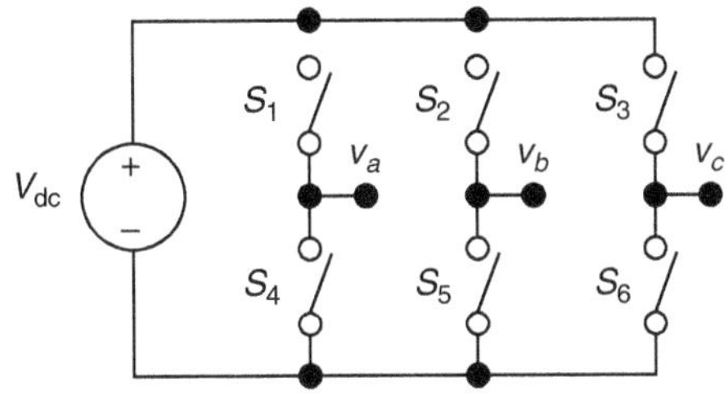

Fig. 9.51 Three-phase inverter

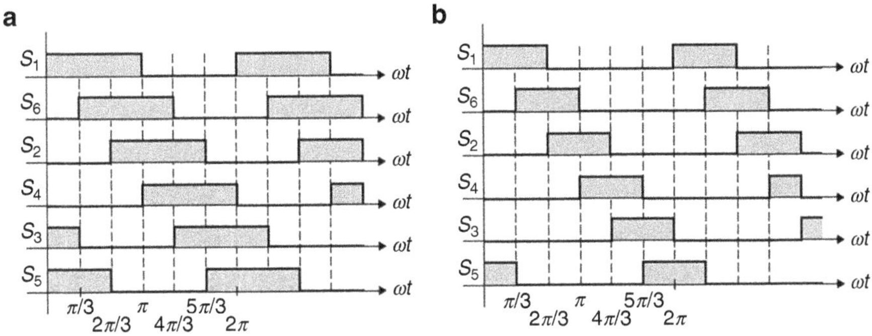

Fig. 9.52 Switching sequence (**a**) conduction equals π and (**b**) conduction equal π/3

voltage. To avoid shorting the voltage source V_{dc}, the switching sequence of $S_1 - S_6$ must make sum; $S_1 - S_4$, $S_3 - S_6$, and $S_5 - S_2$ pairs are not switched ON at the same time.

Figure 9.53a shows the circuit configuration for a three-phase inverter with a wye load type and splitting input capacitors. The inverters generate three-phase output voltages. The switches are switched in such a way that the voltages, v_b and v_c, are shifted by 2π/3. Figure 9.53b shows the three-phase output voltages v_a, v_b, and v_c. Figure 9.53c shows the three-phase line-to-line voltages v_{ab}, v_{bc}, and v_{ca}.

Figure 9.54 shows another line-to-line voltage using the switching sequences of Fig. 9.52a.

Three-phase inverters are implemented by employing three single-phase inverters as shown in Fig. 9.55. The transformer's primary windings must be isolated from one another, and the transformer's secondary windings may be connected in delta or wye configuration.

From Fourier analysis, v_{ab} is given by:

$$\begin{aligned} v_{ab}(t) &= \sum_{n=1,3,5,\ldots}^{\infty} \frac{4V_{dc}}{n\pi} \cos\frac{n\pi}{6} \sin n\left(\omega t + \frac{\pi}{6}\right) \\ v_{bc}(t) &= \sum_{n=1,3,5,\ldots}^{\infty} \frac{4V_{dc}}{n\pi} \cos\frac{n\pi}{6} \sin n\left(\omega t - \frac{\pi}{2}\right) \\ v_{ca}(t) &= \sum_{n=1,3,5,\ldots}^{\infty} \frac{4V_{dc}}{n\pi} \cos\frac{n\pi}{6} \sin n\left(\omega t - \frac{7\pi}{6}\right) \end{aligned} \tag{9.101}$$

Notice all triple harmonics $n = 3, 6, 9, 12\ldots$ are zero.

Similarly, the output voltage control can be implemented using three-phase inverters. Figure 9.56 shows one possible three-phase outputs under α control.

Example 9.10

Consider the three-phase inverter circuit shown in Fig. 9.57 under an inductive-resistive load. Sketch the voltage waveform, for the phase voltages v_{an}, v_{bn}, and v_{cn} and line-to-line voltages v_{ab}, v_{bc}, and v_{ca}, using the switching sequence shown in Fig. 9.35.

Solution

Figure 9.58 shows the three-phase voltages v_{an}, v_{bn}, and v_{cn}. Fig. 9.59 shows the line-to-line voltage v_{ab}, v_{bc}, and v_{ca}.

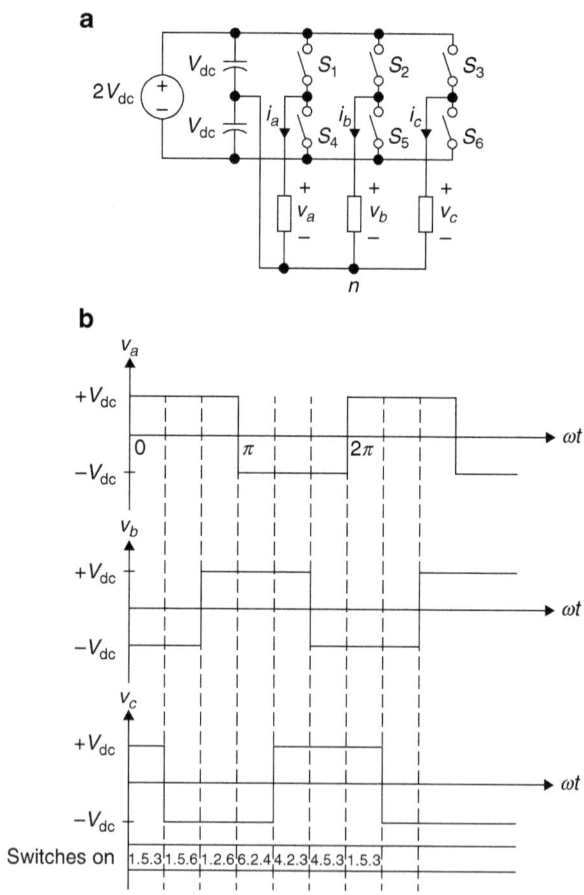

Fig. 9.53 (**a**) Three-phase inverter with wye load (**b**) Line phase-voltage waveforms using switching sequence Fig. 9.52a (**c**) Line-to-line voltages v_{ab}, v_{bc}, *and* v_{ca}

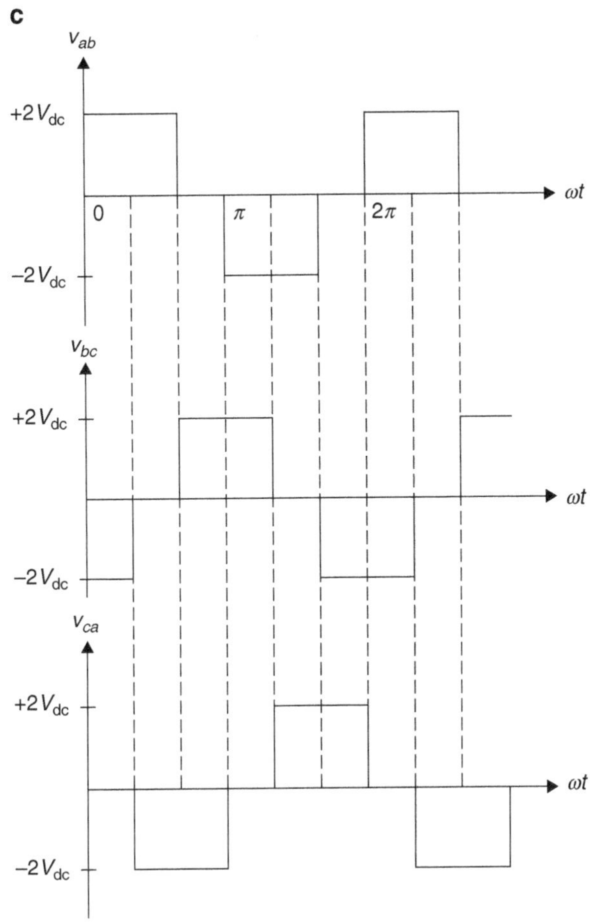

Fig. 9.53 (continued)

Mode 1: $0 \leq \omega t < \pi/3$. S_5, S_6, and S_1 are on. S_2, S_3, and S_4 are off.

It can be shown that since the loads are the same, the pulse voltages are given by:

$$v_{\text{an}} = \frac{1}{3}V_{dc}, \qquad v_{\text{bn}} = \frac{2}{3}V_{dc}, \qquad v_{\text{cn}} = \frac{1}{3}V_{dc}$$

The equivalent circuit mode is shown in Fig. 9.60a.

The line-to-line voltages are given by:

$$v_{\text{ab}} = V_{dc}, \qquad v_{\text{bc}} = 0, \qquad v_{\text{ca}} = 0$$

Mode 2: $\pi/3 \leq \omega t < 2\pi/3$ S_1, S_2, and S_6 are on. S_3, S_4, and S_5 are off. The equivalent circuit for this mode is shown in Fig. 9.60b

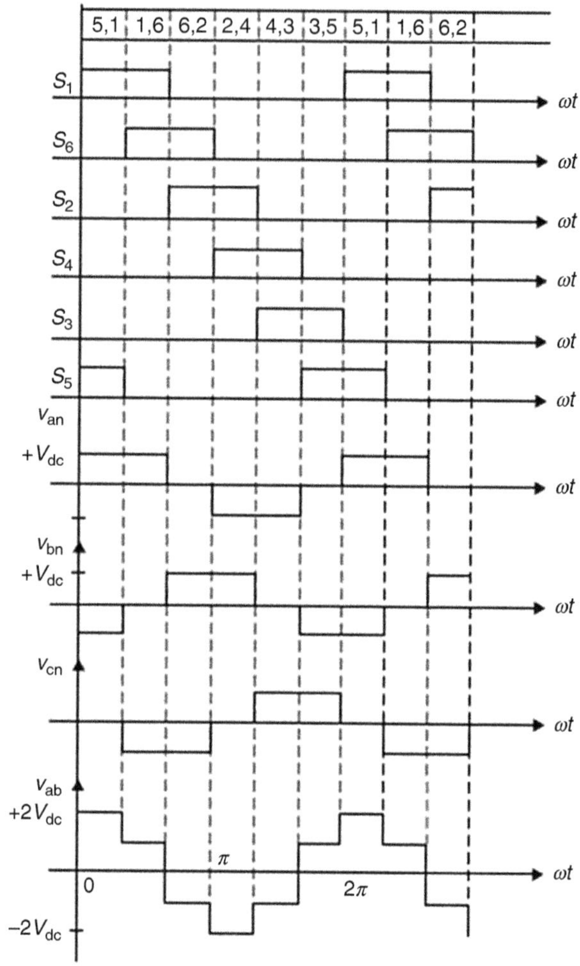

Fig. 9.54 Switching sequence for three-phase inverter to produce v_{ab}

The line-to-line voltages are given by:

$$v_{ab} = V_{dc}, \quad v_{bc} = 0, \quad v_{ca} = -V_{dc}$$

And the phase voltages are given by:

$$v_{an} = \frac{2}{3} V_{dc}, \quad v_{bn} = -\frac{V_{dc}}{3}, \quad v_{cn} = \frac{-V_{dc}}{3}$$

It can be shown that the remaining modes will produce the waveform in Fig. 9.60b.

Exercise 9.9

Sketch the load currents i_{ab}, i_{bc}, and i_{ca} in the three-phase inverter shown in Fig. E9.9, and assume the device conduction sequence is given in Fig. 9.58.

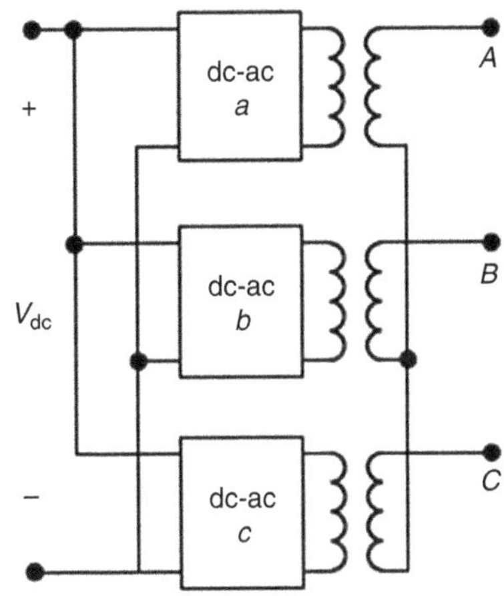

Fig. 9.55 Three-phase inverter by using three single-phase dc-ac inverters

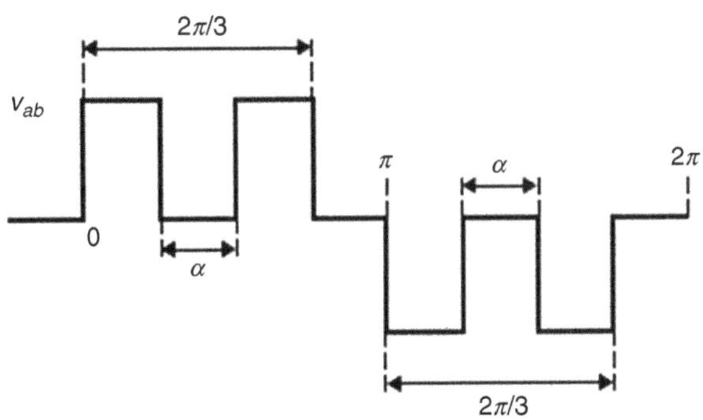

Fig. 9.56 α-control in line-to-line phase voltage

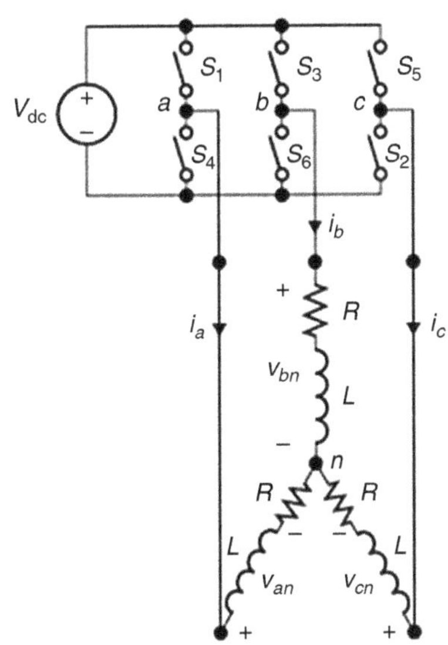

Fig. 9.57 Three-phase inverter under an inductive-resistive load

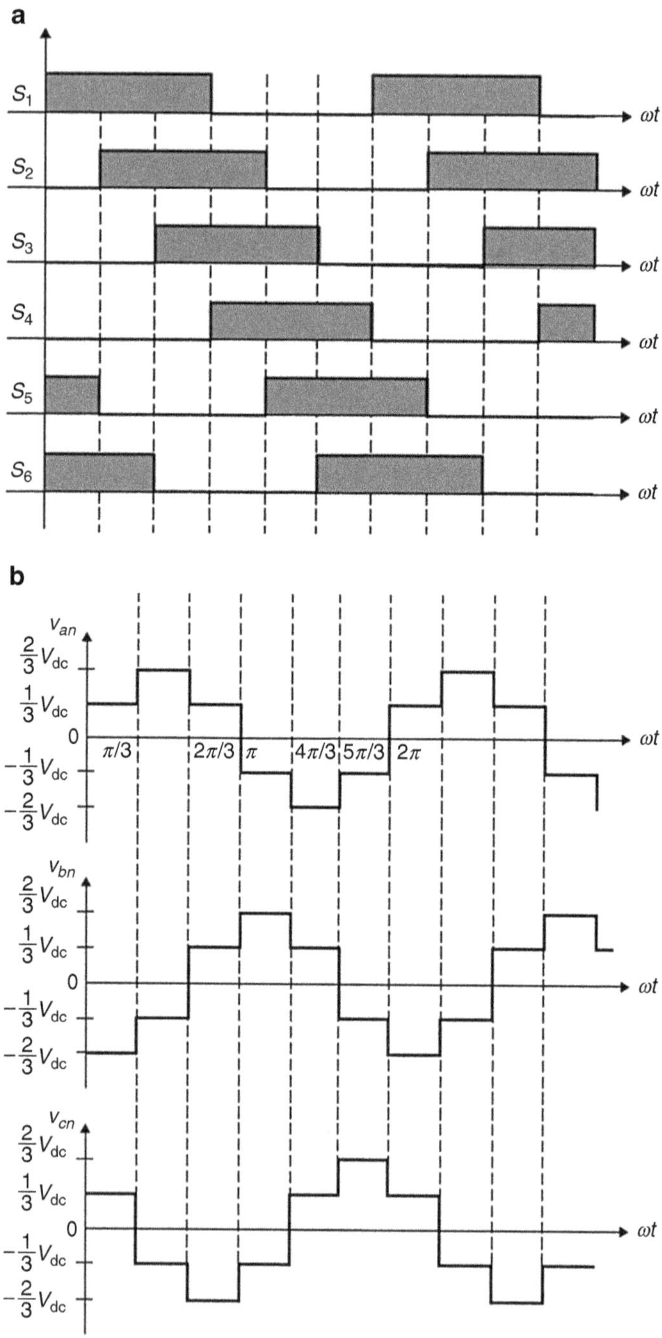

Fig. 9.58 (**a**) Switching sequence for Fig. 9.57 (**b**) Phase voltages v_{an}, v_{bn}, and v_{cn}

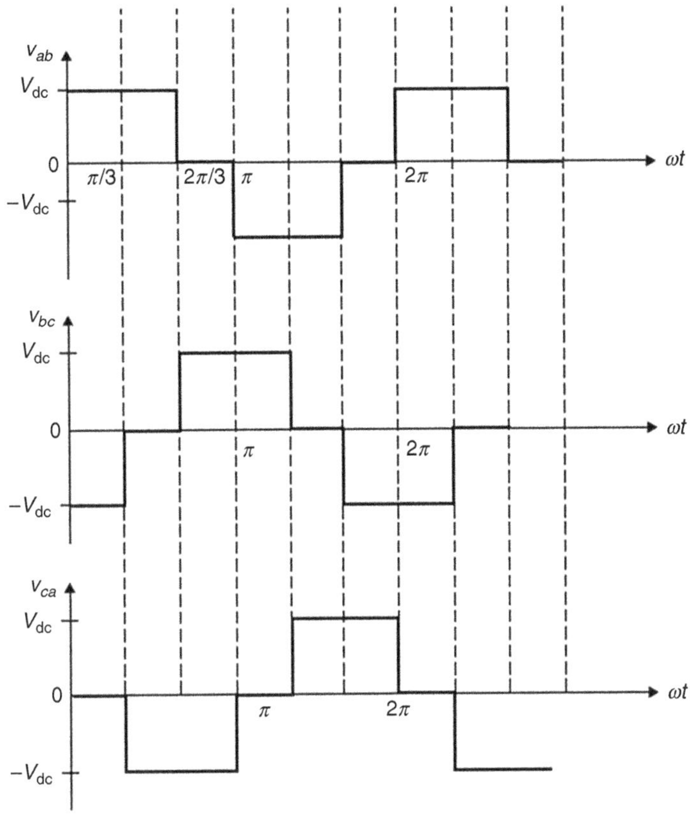

Fig. 9.59 Three-phase line-to-line voltages v_{ab}, v_{bc}, and v_{ca}

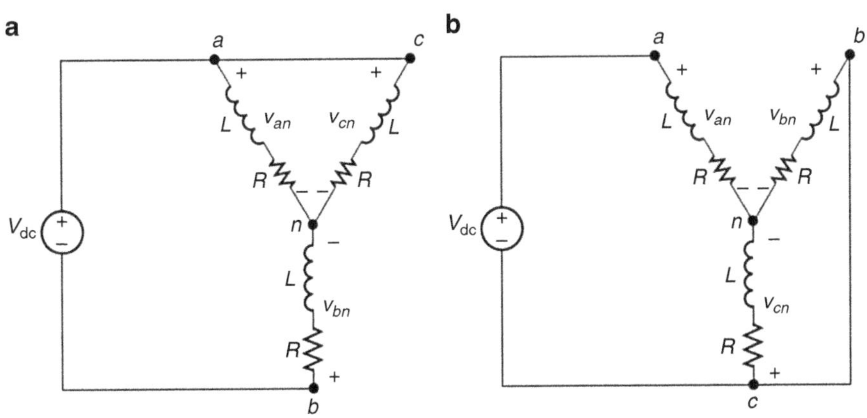

Fig. 9.60 (**a**) Equivalent circuit for Mode 1 (**b**) Equivalent circuit for Mode 2

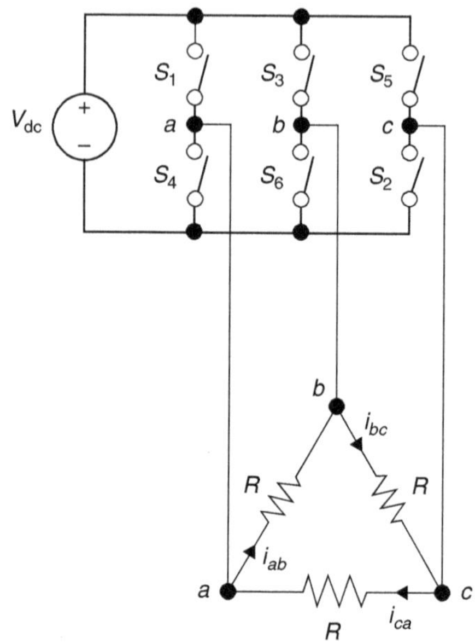

Fig. E9.9 Three-phase inverter for Exercise 9.8

Exercise 9.10

Show that the fundamental component for the load current in E9.9 is given by:

$$i_{ab,1} = \frac{4V_{dc}}{R\pi}\cos\frac{\pi}{6}\sin(\omega t + \pi/6)$$
$$i_{bc,1} = \frac{4V_{dc}}{R\pi}\cos\frac{\pi}{6}\sin(\omega t - \pi/2)$$
$$i_{ca,1} = \frac{4V_{dc}}{R\pi}\cos\frac{\pi}{6}\sin(\omega t - 7\pi/6)$$

9.7 Current-Source Inverters

In the current-source inverter, the dc input is a current source regardless of the input voltage variation. Practically, the dc current source is implemented by using a large dc inductor in series with the dc voltage source as shown in Fig. 9.61. The dc supply is of a high impedance (because of the high input inductor). Since the dc inductor (L_{dc}) is large, I_{dc} is nearly constant.

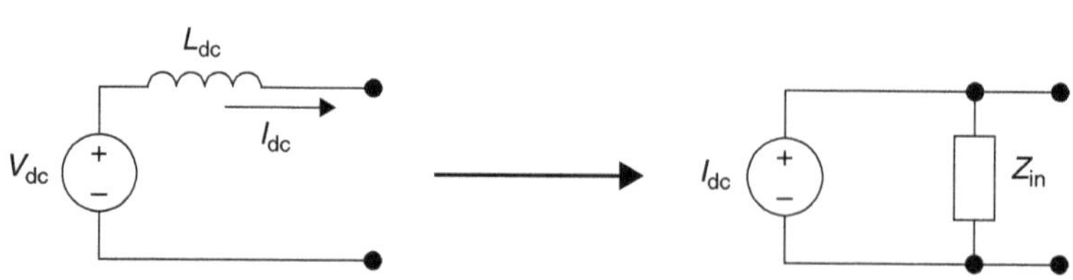

Fig. 9.61 dc current implementation

The output current waveform is determined by the circuit's topology and switching sequence, while the output voltage waveform is determined by the nature of the load. Loads with low impedance to the harmonics are normally used. The switches $S_1 - S_4$ are implemented using GTO and diodes, whereas the SCR implementation is difficult in CSI since the turnoff time is hard to set (Fig. 9.62).

As with the voltage-source inverter, depending on the switching sequences of S_1, S_2, S_3, and S_4, two possible output currents may be obtained as shown in Fig. 9.63a, b. Figure 9.63a uses a 50% duty cycle, and the output current and its rms value are controlled by varying the value of I_{dc}. Figure 9.63b shows a duty cycle less than 50%. There are three output states I_{dc}, 0, and $-I_{dc}$. In both cases, the output voltage waveforms will depend on the nature of the load.

From the Fourier analysis, the load current is given by:

$$i_o(t) = \frac{4I_{dc}}{\pi}\left[\sin \omega t + \frac{\sin 3\omega t}{3} + \ldots + \frac{\sin n\omega t}{n}\right] \quad n = 1, 3, 5, 7, \ldots \tag{9.102}$$

The fundamental component of $i_o(t)$is given by:

$$i_{o1}(t) = \frac{4I_{dc}}{\pi} \sin i_o(t) \tag{9.103}$$

The peak and *rms* values of the fundamental component are:

$$i_{o1,\text{peak}} = \frac{4I_{dc}}{\pi} \tag{9.104a}$$

$$i_{o1,\text{rms}} = \frac{2I_{dc}\sqrt{2}}{\pi} \tag{9.104b}$$

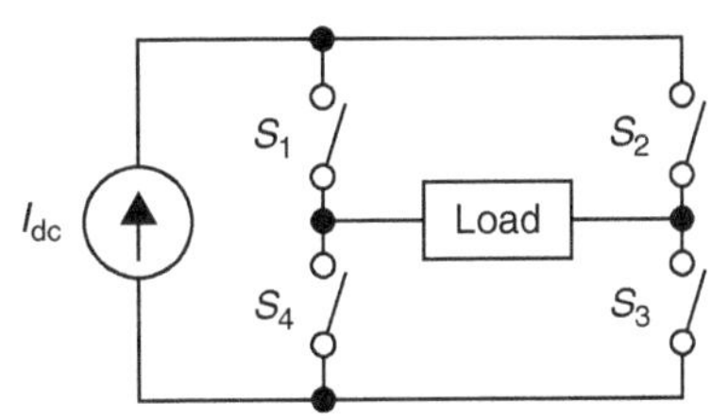

Fig. 9.62 Full-bridge current-source inverter

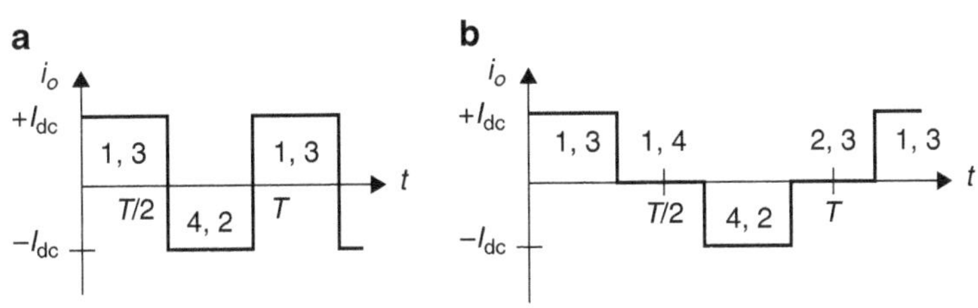

Fig. 9.63 Two possible output current waveforms

Let the normalized *rms* of the *i*th harmonic be given by:

$$I_{\text{no,rms}} = \frac{I_{\text{no,rms}}}{I_{\text{no,rms}}} \quad i = 1, 3, 5, \ldots \tag{9.105}$$

Harmonics	1	3	5	7	9
$I_{\text{no,rms}}$	1	1/3	1/5	1/7	1/9

The frequency spectrums for the normalized rms current harmonics are shown in Fig. 9.64.

Consider the ideal current-source inverter shown in Fig. 9.65a. Assume $v_o = V_o \sin(\omega t + \theta)$ and $S_1 - S_4$ are switched according to the sequence shown in Fig. 9.65b which has 50% switching duty cycle with no α-control. By shifting S_1

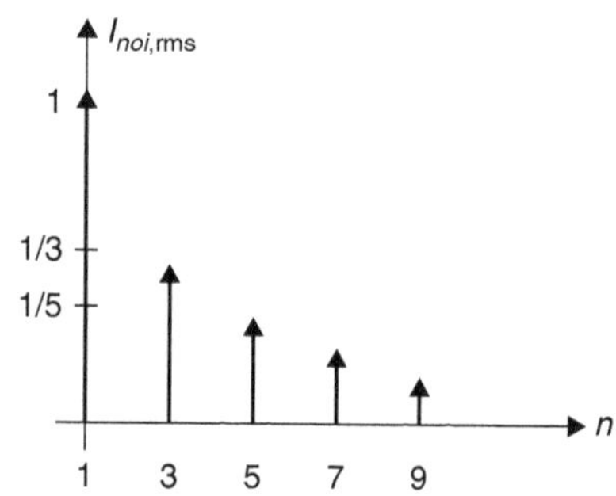

Fig. 9.64 Frequency spectrum for the normalized *rms* current harmonics

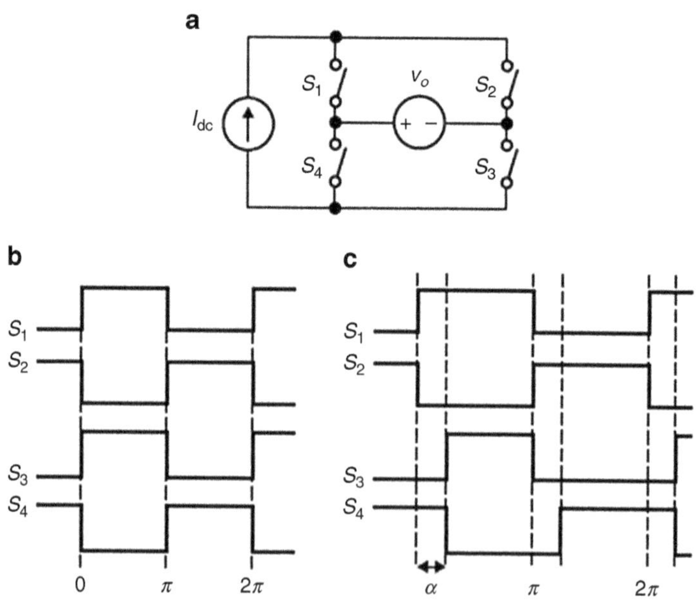

Fig. 9.65 (**a**) Ideal current-source inverter ($\alpha = 0$), (**b**) 50% switch ($\alpha = 0$), (**c**) less than 50% switch ($\alpha \neq 0$)

to the left by α and S_3 to the right by α, we allow an interval during which the input current source is shorted as shown in Fig. 9.65c.

Problems

Half-Bridge and Full-Bride Inverters

9.1. Derive Eq. (9.5), which gives the initial inductor current in a half-bridge inverter, under a resistive-inductive load.

9.2. Repeat Exercise 9.2 by including the first, third, and fifth harmonics.

9.3. Consider the half-bridge inverter and its driven waveforms shown in Fig. P9.3 with the following inverter parameters, $V_{dc} = 185$ V, $R = 10\ \Omega$, and $f = 60$ Hz.

(a) Determine the average diode current for $L = 1, 5, 10,$ and 20 mH.
(b) As the load's time constant increases with respect to $T/2$, what is the effect on the load's harmonics?

Fig. P9.3

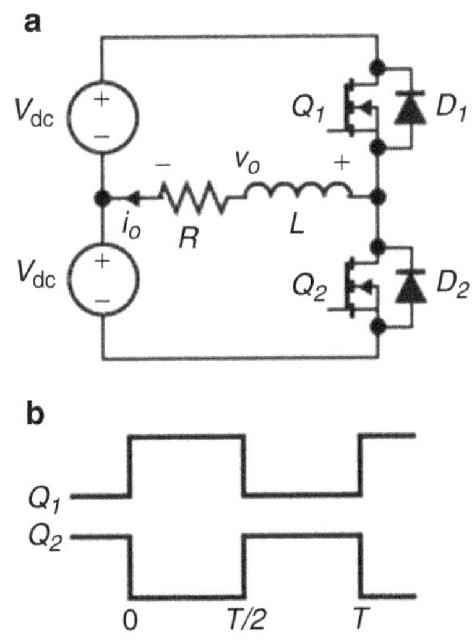

9.4. Consider the half-bridge inverter of Fig. 9.9 with the following circuit components: $V_{dc} = 408$ V, $R = 6\ \Omega, f = 60$ Hz, and $L = 20$ mH.

(a) Derive the exact expression for i_L.
(b) Derive the expression for the fundamental component of $i_L(t)$.
(c) Determine the average diode and transistor currents.
(d) Determine the average power delivered to the load.
(e) Determine the ripple current.

9.5. Consider the full-bridge inverter shown in Fig. P9.5a with the given parameters. Assume the switching sequence of $S_1 - S_4$ produces the output voltage waveform shown in Fig. P9.5b.

(a) Determine the expression for $i_o(t)$and sketch it.
(b) With $S_1 - S_4$ replaced by a parallel combination of switch and diode, find the peak and average switch and diode currents.
(c) Find the average power delivered to the load.
(d) Determine the fundamental and third harmonic peak voltage. What is the percentage of the third harmonic with respect to the fundamental?
(e) Repeat (d) for $L = 50$, 100, and 200 mH.

Fig. P9.5

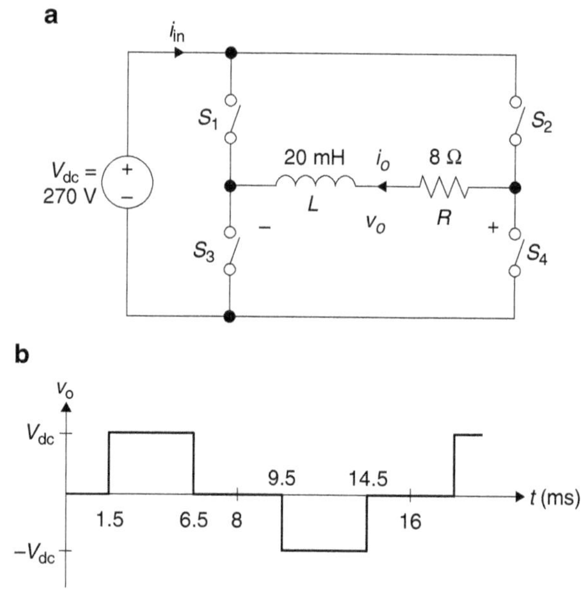

9.6. Consider the ideal dc-ac inverter shown in Fig. P9.6. Assume $S_1 - S_4$ and $S_2 - S_3$ are switched alternatively at a 50% duty cycle with a switch period of T. Let the output current be given by $i_o(t) = I_P \sin \omega t$, where $\omega = 2\pi/T$.

(a) Derive the expression for the instantaneous power delivered to the load.
(b) Determine the average power delivered to the load.
(c) Discuss the requirement for a practical switch implementation for $S_1 - S_4$.
(d) Repeat part (c) for $i_o(t) = I_P \sin(\omega t - \theta)$.

Fig. P9.6

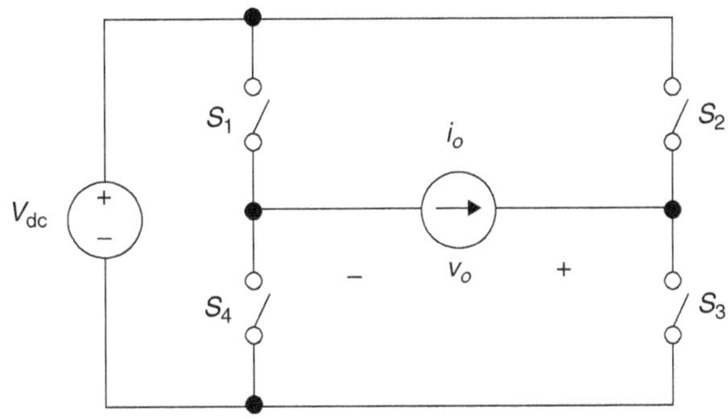

9.7. Consider the full-bridge inverter circuit with an RLC load as shown in Fig. P9.7a. Use $L = 42$ mH, $R = 18\ \Omega$, $C = 900\ \mu$F, and $V_{dc} = 270$ V. Assume the switch sequence of $S_1 - S_4$ produces the output voltage waveform shown in Fig. P9.7b. Determine the peak fundamental component for the load current $i_o(t)$.

Fig. P9.7

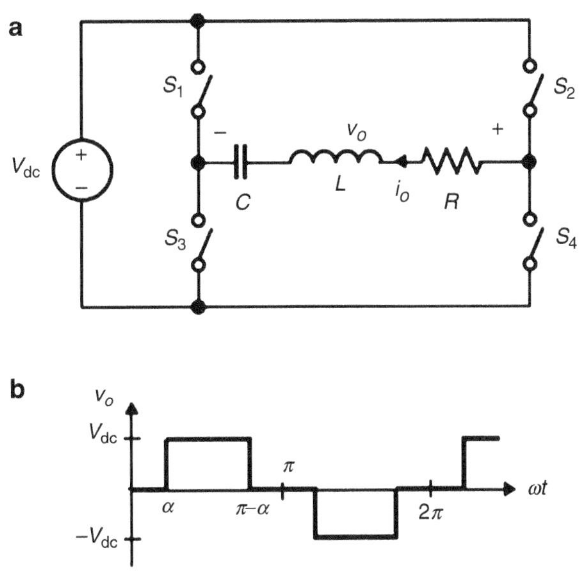

9.8. *(Approximation analysis for an R-L load)*

Consider the bridge inverter of Fig. 9.8b with an R-L load, and assume the load's time constant, L/R, is longer than half of the switching period, i.e., $L/R \gg T/2$, so that the third and the higher harmonics are neglected. Show that the average power delivered to the load is given by:

$$P_{\text{ave}} = |S| \cos \theta$$

where $|S|$ is the magnitude of the reactive power, which is given by:

$$|S| = \frac{8V_{dc}^2}{\pi^2 |Z|} \cos^2 \alpha$$

$$\theta = \tan^{-1}(\omega L/R)$$

and

$$|Z| = \sqrt{R^2 + (\omega L)^2}$$

9.9. Consider the switching R-L circuit shown in Fig. P9.9a with v_s given in Fig. P9.9b.

(a) Derive the expression for $i_L(t)$ in terms of V_{dc} and τ_n, where $\tau_n = \tau/T = L/RT$.
(b) Give the expression for $i_L(0)$, $i_L(t_1)$, and $i_L(T/2)$ in terms of V_{dc} and τ_n.

Fig. P9.9

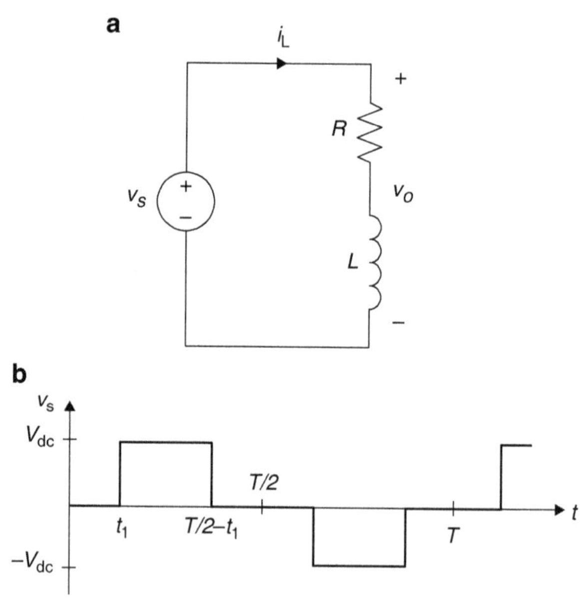

Harmonics

9.10. Show that the harmonics for the generalized square wave with a pulse width equal to $\pi - 2\alpha$ and a delay angle α as shown in Fig. P9.10 is given by:

$$v_o(t) = V_1 \sin(\omega t) + V_3 \sin(3\omega t) + V_5 \sin(5\omega t) + \ldots + V_n \sin(n\omega t)$$

where:

$$V_n = \frac{-4}{n\pi} \cos n\alpha$$

Fig. P9.10

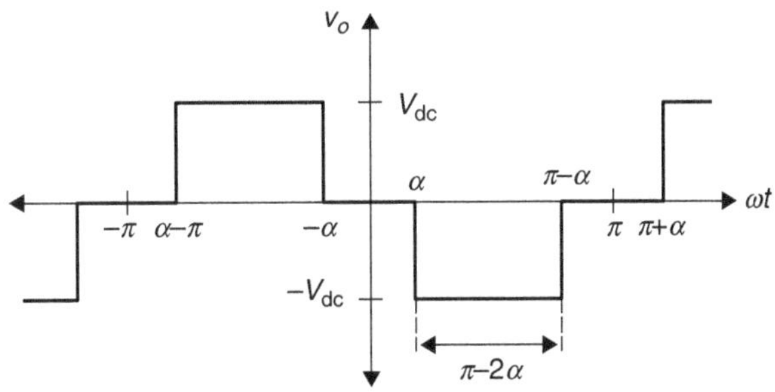

9.11. Consider the output voltage of an inverter shown in Fig. P9.11 that uses the α-control to eliminate two harmonics from the output.

(a) For $\beta_1 = \pi/4$, $\beta_2 = \pi/3$, and $\beta_3 = 3\pi/4$, find the fundamental, third, and fifth harmonics.
(b) Redesign the problem for α_1 and α_2 to accomplish this.

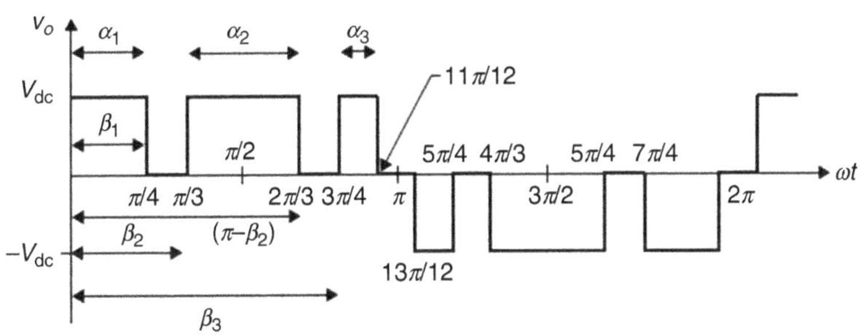

Fig. P9.11

9.12. Consider an inverter circuit that produces the stepped output waveform shown in Fig. P9.12. It can be shown by properly selecting α_1 and α_2 that the third and fifth harmonics can be eliminated from the output voltage. Determine α_1 and α_2 to accomplish this.

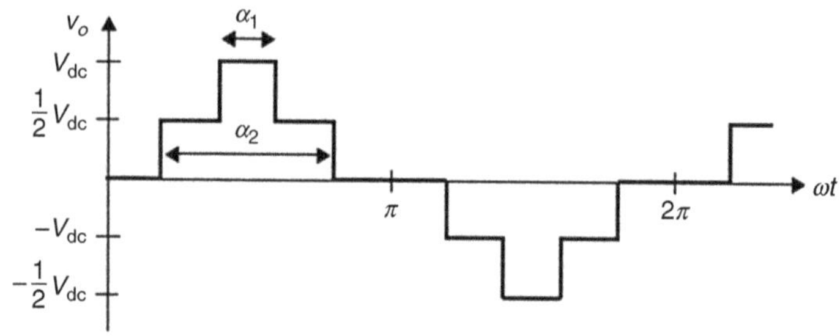

Fig. P9.12

9.13. Determine the THD for the stepped waveform shown in Problem 9.12 by using only up to the ninth harmonic.

9.14. In a certain output voltage in an inverter, it is desired to eliminate the third, fifth, and seventh harmonics by controlling the switching sequence of the power devices. It was shown that the following equation was obtained for the *n*th output voltage harmonic of Fig. P9.14.

$$v_{on} = \frac{4V_{dc}}{n\pi}(1 - \cos n\alpha_1 + \cos n\alpha_2 - \cos n\alpha_3)$$

Determine α_1, α_2, and α_3 for a zero value of third, fifth, and seventh harmonic. (Hint: use a numerical solution method or an iterative technique).

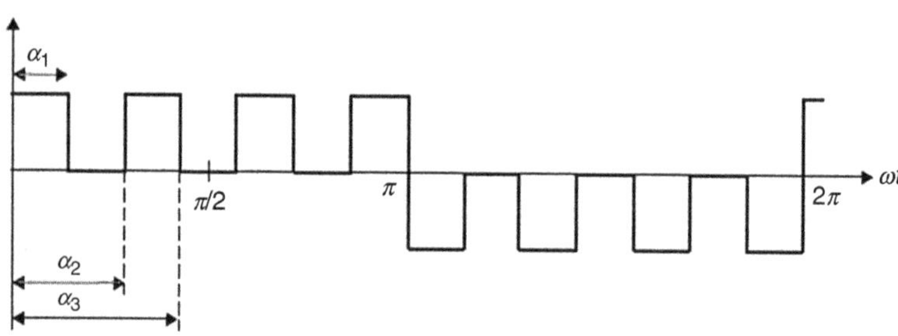

Fig. P9.14

Uniform Pulse Width Modulation

9.15. Derive the relation for θ_i given by Eq. (9.64), given again below:

$$\theta_i = \frac{\pi}{k}\left[i - \frac{m_a}{2} - \frac{1}{2}\right]$$

for the *i*th pulse in a *k*-pulse output.

9.16. Derive the expression given in Eq. (9.65), which shows that the width of the pulse in a uniform PWM output is constant and is given by $\theta_{\text{width}} = \pi m_a/k$.
9.17. Determine the rms value for a uniform PWM inverter output with $V_{dc} = 260$ V and $m_a = 0.8$. Determine the width of each pulse if the output has 24 pulses per cycle.
9.18. Derive Eq. (9.75).
9.19. Repeat Example 9.8 for $k = 10$ and $m_a = 0.4$. How does the THD compare to the THD obtained in that example?

Sinusoidal Pulse Width Modulation

9.20. Sketch the output voltage for the buck converter whose output is modulated according to Eq. (9.78). Assume $D_{dc} = D_{\max} = 0.5$, $V_{dc} = 1$ V, and $f_o = f_s/10$.
9.21. Derive Eq. (9.88).
9.22. Derive the expression for θ_i given in Eq. 9.95.
9.23. Determine the approximate pulse width of the 14th pulse in a sinusoidal PWM inverter output of Fig. 9.44 with $k = 18$ and $m_a = 0.5$.
9.24. Repeat Example 9.9 for $m_a = 1.0$.

Three-Phase Inverters

9.25. The waveforms shown in Fig. P9.25 represent one possible way to implement a line-to-line output voltage in a three-phase inverter control. By varying the angle α, the *rms* value of the output can be controlled.

Show that the nth harmonic component for the line-to-line voltage is given by:

$$V_{\text{ab},n} = \frac{4}{n\pi} V_{\text{in}} \left(\sin n\frac{\pi}{3} - \sin\frac{n\alpha}{2} \right) \left(\cos \left(\omega t - \frac{\pi}{3} - \frac{\alpha}{2} \right) \right)$$

and show that the total harmonic distortion is given by:

$$THD_{\text{V}}^2 = \frac{\pi(2\pi/3 - \alpha)}{8(\sin(\pi/3) - \sin(\alpha/2))^2} - 1$$

Fig. P9.25

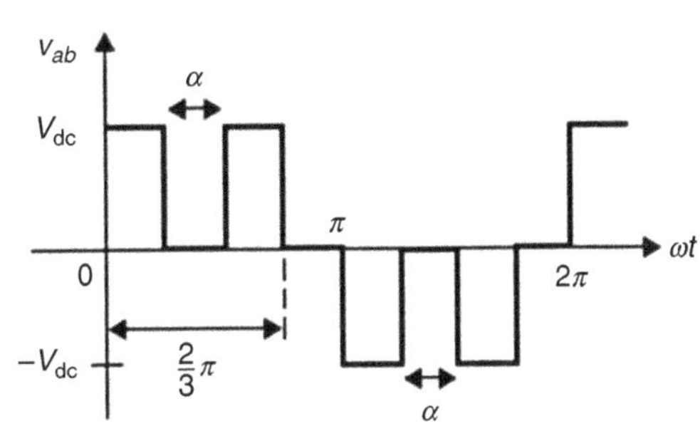

9.26. Derive the harmonic component for v_{ab}, v_{bc}, and v_{ca} shown in Fig. 9.53c and for v_{ab} of Fig. 9.53c.

9.27. One way to reduce the harmonics of a line-to-line voltage in a three-phase inverter is to produce an output waveform like the one shown in Fig. 9.54. Determine the nth harmonic component of v_{ab}.

9.28. Drive the harmonic component for v_{ab} of Fig. P9.28. (Hint: See Appendix B).

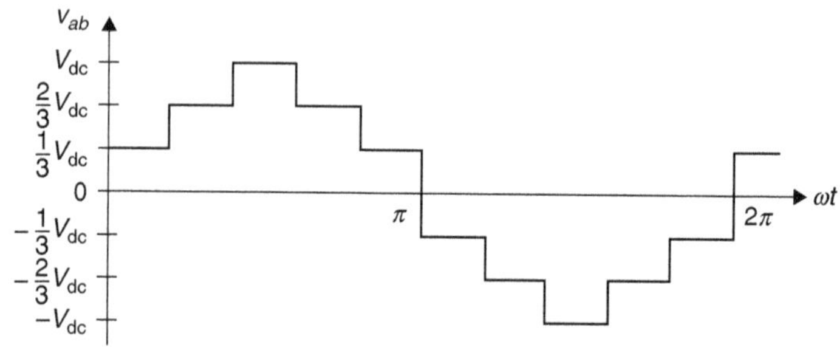

Fig. P9.28

9.29. The equivalent circuit for a line-to-line voltage in a three-phase inverter is given in Fig. P9.29

(a) Show that the steady-state inductor current at $\omega t = 0$ is given by:

$$I_L(0) = -\frac{V_{dc}}{3R}\frac{\left(1 + e^{-1/6\tau_n} - e^{-1/3\tau_n} - e^{-1/2\tau_n}\right)}{1 + e^{-1/2\tau_n}}$$

where:

$$\tau_n = \frac{\tau}{T} = \frac{L}{RT}, \qquad T = \frac{2\pi}{\omega}$$

(b) Show that the steady-state inductor current value at $\omega t = \pi/3$ is given as a function of $I_L(0)$ as follows:

$$I_L(\pi/3) = \left(I_L(0) - \frac{V_{dc}}{3R}\right)e^{-1/6\tau_n} + \frac{V_{dc}}{3R}$$

(c) Determine the time at which the flyback diode of S_1 stops conducting, assuming $L = 12$ mH, $R = 18\ \Omega$, $f = 400$ Hz, and $V_{dc} = 260$ V.

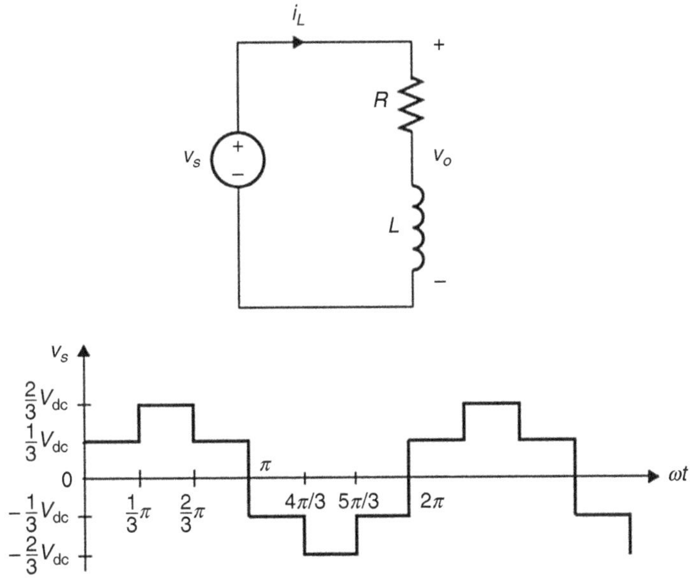

Fig. P9.29

General Problems

9.30. Consider the single-phase ac controller, shown in Fig. P9.30, with the induction motor load modeled by a resistor, an inductor, and an *emf* ac source.

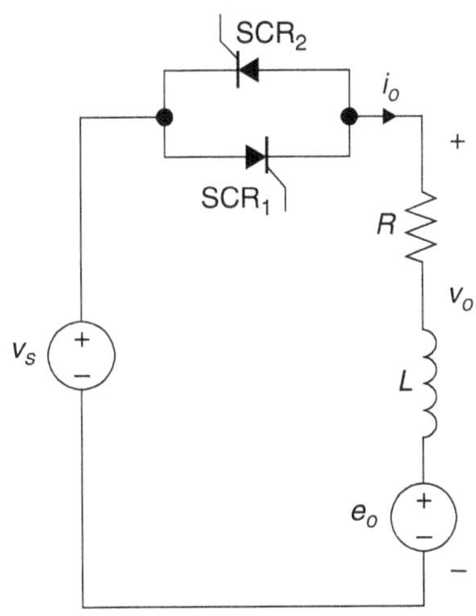

Fig. P9.30 Single-phase controller

Assume the *emf* voltage, $e_o(t)$, is given by:

$$e_o(t) = V_o \sin(\omega t + \theta_o)$$

and

$$v_s(t) = V_s \sin(\omega t)$$

Here we assume that the source frequency and the *emf* voltage frequency are equal. Also assume SCR_1 and SCR_2 are triggered continuously at $\alpha_1 = 30^\circ$ and $\alpha_2 = 330^\circ$, respectively. Use $R = 0.85\ \Omega$, $L = 10.2$ mH, $V_s = 120$ V, and $V_o = 85$ V.

(a) Sketch the waveform of $i_o(t)$ for the full cycle.
(b) Drive the expression for $i_o(t)$ for the first half-cycle.

9.31. Figure P9.31a shows a single-phase power electronic circuit with an $R-L$ load known as an *ac controller*. Ac controllers are widely used in single- or three-phase arrangements for various residential and commercial applications such as heating, motor speed control, and power factor correction. Possible switch implementation is shown in Fig. P9.31b with a back-to-back SCR to produce a bidirectional current flow and a bidirectional voltage blocking.

Discuss the operation of the circuit that produces the output waveform shown in Fig. P9.31c.

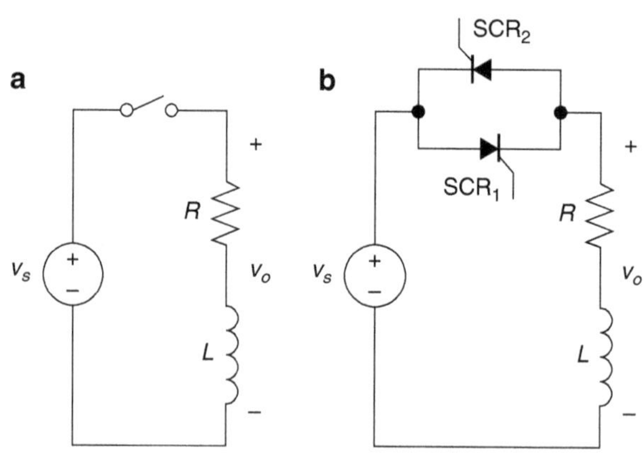

Fig. P9.31 (**a**) Ac controller with inductive-resistive load (**b**) SCR implementation (**c**) output waveform

9.32. (a) Show that if the exact expression for $i_L(t)$ is used in Fig. 9.9, then the fundamental component is given by:

$$i_L(t) = I_{L1} \sin(\omega t + \theta)$$

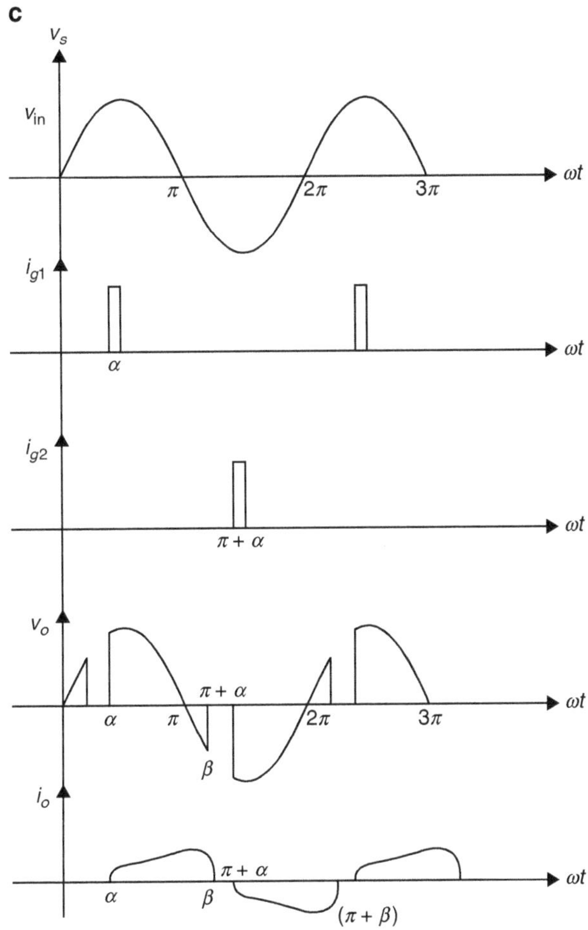

Fig. P9.31 (continued)

where:

$$I_{L1} = \frac{4V_{dc}}{R\pi}\sqrt{\frac{4\pi^2\tau_n^2(e^{-1/\tau_n} - e^{-1/2\tau_n}) + 4\pi^2\tau_n^2 + 1}{4\pi^2\tau_n^2 + 1}}$$

$$\theta = \tan\left[2\pi\tau_n e^{-1/2\tau_n} - \frac{1}{2\pi\tau_n} - 2\pi\tau_n\right]e^{1/(2\tau_n)}$$

(b) Show that the simplified expression when $L/R \gg T$ is given by:

$$I_{L1} \approx \frac{4V_{dc}}{\pi}\frac{1}{\sqrt{R^2 + (\omega L)^2}}$$

Bibliography

Introduction to Power Electronics

Branko L. Dokić and Branko Blanuša. *Power Electronics: Converters and Regulators*, Springer; 3 edition, November 26, 2014
Rim, Chun Taek. *Phasor Power Electronics*, Springer Singapore 1st edition, 2016
Dorin O. Neacsu. *Switching Power Converters: Medium and High Power*, Second Edition March 29, 2017 by CRC Press, Taylor and Francis
Bacha, Seddik, Munteanu, Iulian, and Bratcu, Antoneta Iuliana. *Power Electronic Converters Modeling and Control*, Springer-Verlag London 1st edition, 2014
Euzeli dos Santos, and Edison R. da Silva. *Advanced Power Electronics Converters: PWM Converters Processing AC Voltages*, January 2015, Wiley-IEEE Press
Gonzalo Abad. *Power Electronics and Electric Drives for Traction Applications*, November 2016, Wiley
Sergio Alberto Gonzalez, Santiago Andres Verne, and Maria Ines Valla. *Multilevel Converters for Industrial Applications*, March 29, 2017 by CRC Press, Taylor and Francis
Chengshan Wang, Jianzhong Wu, and Janaka Ekanayake, Nick Jenkins. *Smart Electricity Distribution Networks*, February 24, 2017 by CRC Press, Taylor and Francis
Ang, S. *Power Switching Converters*. Marcel Dekker, 1995.
Bird, B., K. King, and D. Pedder. *An Introduction to Power Electronics*, 2nd ed. John Wiley, 1993.
Bradly, D. A. *Power Electronics*, 2nd ed. Chapman & Hall, 1995.
Dewan, S. B., and A. Straughen. *Power Semiconductor Circuits*. John Wiley, 1975.
Erickson, R. *Fundamentals of Power Electronics*. Chapman & Hall, 1997.
Fisher, M. *Power Electronics*. PWS-KENT, 1991.
Hart, D. *Introduction to Power Electronics*. Prentice-Hall, 1997.
Heumann, Klemens. *Basic Principles of Power Electronics*. Springer-Verlag, 1986.
Hoft, Richard. *Semiconductor Power Electronics*. Van Nostrand Reinhold, 1986.
Kassakian, J., M. Schlecht, and G. Vergese. *Principles of Power Electronics*. Addison-Wesley, 1991.
Krein, Philip. *Elements of Power Electronics*. Oxford University Press, 1998.
Lander, Gyril. *Power Electronics*, 2nd ed. McGraw-Hill, 1987.
Mohan, N., T. Undeland, and W. Robbins. *Power Electronics: Converters, Applications, and Design*, 3rd ed. John Wiley, 2002.
Motto, J. W. *Introduction to Solid State Power Electronics*. Westinghouse, 1977.
Ohno, Eiichi. *Introduction to Power Electronics*. Clarendon Press, 1988.

I. Batarseh, A. Harb, *Power Electronics*,
https://doi.org/10.1007/978-3-319-68366-9

Ramshaw, R. S. *Power Electronics Semiconductor Switches*, 2nd ed. Chapman & Hall, 1993.
Rashid, M. H. *Power Electronics: Circuits, Devices, Applications*, 2nd ed. Prentice Hall, 1993a.
Seyuier, Guy. *Power Electronic Converters: AC/ DC Conversion*. McGraw-Hill, 1986.
Tarter, R. E. *Principles of Solid-State Power Conversion*. Howard Sams & Co., 1985.
Thorborg, K. *Power Electronics*. Prentice Hall, 1988.
Trzynadlowski, A. *An Introduction to Modern Power Electronics*. John Wiley, 1998.
Vithayathil, J. *Power Electronics: Principles and Applications*. McGraw-Hill, 1995.
Williams, B. W. *Power Electronics: Devices, Drivers, and Applications*. John Wiley, 1987.
Wood, Peter. *Switching Power Converters*. Krueger, 1981.

Switching-Mode Power Supplies

Billings, K. *Switch-Mode Power Supply Hand-book*. McGraw-Hill, 1989.
Brown, M. *Power Supply Cookbook*, 2nd ed. Newnes, 2001.
Chryssis, George. *High-Frequency Switching Power Supplies: Theory and Design*, 2nd ed. McGraw-Hill, 1989.
Hnatek, E. *Design of Solid State Power Supplies*, 3rd ed. Van Nostrand Reinhold, 1989.
Kilgenstein, O. *Switched-Mode Power Supplies in Practice*. John Wiley & Sons, 1989.
Middlebrook, R. D., and S. Cuk. *Advances in Switch-Mode Power Conversion*. Vols. I and II. TESLAco, 1981.
Mitchell, Daniel. *DC-DC Switching Regulator Analysis*. McGraw-Hill, 1988.
Pressman, A. I. *Switching Power Supply Design*. McGraw-Hill, 1991.
Severns, R., and G. Bloom. *Modern DC-to-DC Switchmode Power Converter Circuits*. Van Nostrand Reinhold, 1985.
Sum, Kit. *Switch Mode Power Conversion: Basic Theory and Design*. Marcel Dekker, 1984.

dc-ac Inverters and ac-ac Converters

Bedforf, B., and R. Hoft. *Principles of Inverter Circuits*. John Wiley, 1964.
Griffith, D. *Uninterruptible Power Supplies*. Marcel Dekker, 1993.
Gyugyi, L., and B. R. Pelly. *Static Power Frequency Changers*. John Wiley, 1976.
McMurray, W. *The Theory and Design of Cycloconverters*. MIT Press, 1972.
Pelly, B. R. *Thyristor Phase-Controlled Converters and Cycloconverters*. John Wiley, 1971.
Rombaut, C., Guy Seguier, and R. Bausiere. *Power Electronics Converters: AC/AC Conversion*. McGraw-Hill, 1987.

Machines and Drives

Anderson, Leonard. *Electric Machines and Transformers*. Reston, 1981.
Barton, T. H. *Rectifiers, Cycloconverters, and AC Controllers*. Clarendon Press, 1994.
Bose, B. K. *Microcomputer Control of Powerful Electronics and Drives*. IEEE Press, 1987.
Bose, B. K. *Power Electronics and AC Drives*. Prentice Hall, 1986.
Chapman, Stephen J. *Electric Machinery Fundamentals*, 3rd ed. McGraw-Hill, 1999.
Dewan, S. B., G. R. Slemon, and A. Straughen. *Power Semiconductor Drives*. John Wiley, 1984.
Dubey, G. *Power Semiconductor Controlled Drives*. Prentice Hall, 1989.
Fitzgerald, A. E., C. Kingsley, and S. Umans. *Electric Machinery*, 5th ed. McGraw-Hill, 1990.
Leonard, W. *Control of Electric Drives*. Springer-Verlag, 1985.
Mohan, Ned. *Electric Drives: An Integrative Approach*. MNPERE, 2000.

Sarma, M. *Electric Machines*, 2nd ed. West Publishing, 1994.
Sen, P. C. *Principles of Electric Machines and Power Electronics*, 2nd ed. John Wiley, 1997.
Slemon. G. *Electric Machines and Drives*. Addison-Wesley, 1992.
Vas, P. *Vector Controlled AC Machines*. Clarendon Press, 1990.
Wildi, T. *Electric Machines, Drives, and Power Systems*. John Wiley, 1989.

Magnetics

Cheng, David. *Field and Wave Electromagnetics*, 2nd ed. Addison-Wesley, 1989.
Chi-Sha, Liang, and Jin Au Kong. *Applied Electromagnetism*. PWS Engineering, 1987.
Kraus, John. *Electromagnetics*. McGraw-Hill, 1992.
Lowdon, Eric. *Practical Transformer Design Handbook*. TAB Professional and Reference Books, 1989.
McLyman, W. T. *Transformer and Inductor Design Handbook*, 2nd ed. Marcel Dekker, 1993.
MIT Staff. *Magnetic Circuits and Transformers*. MIT Press, 1965.
Ozenbaugh, R. *EMI Filter Design*. Dekker, 1996.
Schwarz, Steven E. *Electromagnetics for Engi-neers*. Saunders College Publishing, HRW, 1990.
Tihanyi, L. *Electromagnetic Compatibility in Power Electronics*. IEEE Press, 1995.

Power Devices

Baliga, B. J. *Modern Power Devices*. John Wiley & Sons, 1997.
Baliga, B. J., and D. Chen. *Power Transistors: Devices, Design, and Applications*. IEEE Press, 1984.
Blicher, B. *Field Effect and Bipolar Power Transistor Physics*. Academic Press, 1981.
Cobbold, Richard. *Theory and Applications of Field Effect Transistor*. John Wiley, 1970.
Dubey, G. K., S. R. Doradla, A. Joshi, and R. M.K. Sinha. *Thyristorised Power Controllers*. John Wiley, 1986.
Ghandhi, S. *Semiconductor Power Devices*. John Wiley, 1977.
Grant, D. A., and J. Gowar. *Power MOSFETS: Theory and Applications*. John Wiley, 1989.
Jaecklin, A. *Power Semiconductors Devices and Circuits*. Plenum Press, 1992.
Oxner, E. *Power FETs and Their Applications*. Prentice-Hall, 1982.
Streetman, B. *Solid-State Electrical Devices*, 3rd ed. Prentice Hall, 1990.
Tarter, R. E. *Solid-State Conversion Handbook*. John Wiley, 1993a.
Taylor, B. E. *Power MOSFET Design*. John Wiley, 1993.
Warner, R. M., and B. L. Grung. *MOSFET: Theory and Design*. Oxford, 1999.

PSPICE Modeling

Basso, C. *Switch-Mode Power Supply Spice Cook book*. McGraw-Hill, 2001.
Connelly, J., and P. Choi. *Macromodeling with SPICE*. Prentice Hall, 1992.
Gottling, James G. *Hands on PSPICE*. Houghton Mifflin, 1995.
Massobrio, G., and P. Antognetti. *Semiconductor Device Modeling with PSPICE*. McGraw-Hill, 1993.
Ramshaw, R., and D. Schuurman. *PSPICE Simulation of Power Electronics Circuits*. Chapman & Hall, 1997.
Rashid, M. *SPICE for Power Electronics and Electric Power*. Prentice Hall, 1993b.

Roborts, G., and A. Sedra. *SPICE*, 2nd ed. Oxford University Press, 1997.
Tuinenga, P. *SPICE: A Guide to Circuit Simulation and Analysis Using PSPICE*, 3rd ed. Prentice Hall, 1995.

Other Related Textbooks

Bose, B. K. *Modern Power Electronics: Evaluation, Technology, and Applications*. IEEE Press, 1992.
Datta, S. *Power Electronics and Controls*. Reston, 1985.
Heydt, G. *Electric Power Quality*. Stars in a Circle Publications, 1991.
Kazimierczuk, M., and D. Czarkowski. *Resonant Power Converters*. John Wiley, 1995.
Kislovski, R. Redl, and N. Sokal, *Dynamic Analysis of Switching-Mode DC/DC Converters*. Van Nostrand Reinhold, 1991.
Tarter, R. E. *Solid State Power Conversion Handbook*. John Wiley & Sons, 1993b.

Index

I. Batarseh, A. Harb, *Power Electronics*,
https://doi.org/10.1007/978-3-319-68366-9

N

O

P

T

GPSR Compliance
The European Union's (EU) General Product Safety Regulation (GPSR) is a set of rules that requires consumer products to be safe and our obligations to ensure this.

If you have any concerns about our products, you can contact us on

ProductSafety@springernature.com

In case Publisher is established outside the EU, the EU authorized representative is:

Springer Nature Customer Service Center GmbH
Europaplatz 3
69115 Heidelberg, Germany

www.ingramcontent.com/pod-product-compliance
Lightning Source LLC
Chambersburg PA
CBHW060744150926
51501CB00003B/86
9783319683652